CATIONIC POLYMERIZATIONS

PLASTICS ENGINEERING

Founding Editor

Donald E. Hudgin

Professor
Clemson University
Clemson, South Carolina

20. Rubber Compounding: Principles, Materials, and Techniques, *Fred W. Barlow*
21. Thermoplastic Polymer Additives: Theory and Practice, *edited by John T. Lutz, Jr.*
22. Emulsion Polymer Technology, *Robert D. Athey, Jr.*
23. Mixing in Polymer Processing, *edited by Chris Rauwendaal*
24. Handbook of Polymer Synthesis, Parts A and B, *edited by Hans R. Kricheldorf*
25. Computational Modeling of Polymers, *edited by Jozef Bicerano*
26. Plastics Technology Handbook: Second Edition, Revised and Expanded, *Manas Chanda and Salil K. Roy*
27. Prediction of Polymer Properties, *Jozef Bicerano*
28. Ferroelectric Polymers: Chemistry, Physics, and Applications, *edited by Hari Singh Nalwa*
29. Degradable Polymers, Recycling, and Plastics Waste Management, *edited by Ann-Christine Albertsson and Samuel J. Huang*
30. Polymer Toughening, *edited by Charles B. Arends*
31. Handbook of Applied Polymer Processing Technology, *edited by Nicholas P. Cheremisinoff and Paul N. Cheremisinoff*
32. Diffusion in Polymers, *edited by P. Neogi*
33. Polymer Devolatilization, *edited by Ramon J. Albalak*
34. Anionic Polymerization: Principles and Practical Applications, *Henry L. Hsieh and Roderic P. Quirk*
35. Cationic Polymerizations: Mechanisms, Synthesis, and Applications, *edited by Krzysztof Matyjaszewski*

Additional Volumes in Preparation

Polyimides: Fundamentals and Applications, *edited by Malay K. Ghosh and K. L. Mittal*

Thermoplastics Melt Rheology and Processing, *A. V. Shenoy and D. R. Saini*

Practical Thermoforming: Second Edition, Revised and Expanded, *John Florian*

Macromolecular Design of Polymeric Materials, *edited by Koichi Hatada, Tatsuki Kitayama, and Otto Vogl*

Prediction of Polymer Properties: Second Edition, *Jozef Bicerano*

CATIONIC POLYMERIZATIONS

MECHANISMS, SYNTHESIS, AND APPLICATIONS

EDITED BY

KRZYSZTOF MATYJASZEWSKI

Carnegie Mellon University
Pittsburgh, Pennsylvania

Marcel Dekker, Inc. New York • Basel • Hong Kong

Library of Congress Cataloging-in-Publication Data

Cationic polymerizations: mechanisms, synthesis, and applications / edited by Krzysztof Matyjaszewski.
p. cm. — (Plastics engineering ; 35)
Includes index.
ISBN 0-8247-9463-X (alk. paper)
1. Addition polymerization. I. Matyjaszewski, K. (Krzysztof) II. Series: Plastics engineering (Marcel Dekker, Inc.) ; 35.
QD281.P6C384 1996
660'.28448—dc20 96-5421
CIP

The publisher offers discounts on this book when ordered in bulk quantities. For more information, write to Special Sales/Professional Marketing at the address below.

This book is printed on acid-free paper.

MARCEL DEKKER, INC.
270 Madison Avenue, New York, New York 10016

Current printing (last digit):
10 9 8 7 6 5 4 3 2 1

PRINTED IN THE UNITED STATES OF AMERICA

Preface

Cationic polymerizations are among the most important synthetic methods in polymer chemistry. They are used to prepare a variety of commodity and specialty polymers. More recently, controlled cationic polymerizations have been used to synthesize novel functional polymers, block copolymers, and macromolecules with new topologies. The importance of reactions involving cationic active species is continuously increasing, and was recently recognized by the awarding of the 1994 Nobel Prize in Chemistry to George Olah.

This book covers both chain-growth and step-growth cationic polymerizations. It provides the first comprehensive overview of cationic polymerizations of vinyl and heterocyclic monomers, as well as a unified view on conventional and controlled/"living" carbocationic polymerizations. The focus is on understanding the mechanisms of the elementary reactions and explaining the controversial aspects. All of the reactions are analyzed from the perspective of modern organic and physical organic chemistry of carbenium and onium ions. This useful reference reviews synthetic and commercial applications of cationic polymerizations and contains synthetic "recipes" for preparing several homo- and block copolymers.

The book is divided into eight chapters. The Introduction is a primer for both synthetic polymer chemistry in general, and cationic polymerizations in particular. More advanced readers may go directly to the following chapters. The second chapter covers the reactions of carbenium ions with various nucleophiles and focuses on the ionization of covalent species and the addition of carbenium ions to alkenes, arenes, and other π-nucleophiles. Chapter 3 discusses the four elementary reactions of initiation, propagation, transfer, and termination in cationic polymerizations of alkenes. It also describes the active species involved in carbocationic polymerizations. Chapter 4 deals with the newest developments in "living" carbocationic polymerizations, which is a hot, albeit controversial, topic. The fifth chapter details the synthetic aspects of cationic polymerizations of alkenes. Chapter 6 discusses the fundamentals of cationic ring-opening polymerizations and the recent advances in the field. The similarities between heterocyclic and alkene polymerizations should become clear at this point. The seventh chapter primarily describes step polymerizations based on aromatic compounds, as well as other cationic processes. Finally, Chapter 8 summarizes the commercial applications of cationic polymerizations and offers some perspective on future industrial developments.

Written by internationally recognized experts, this book is the most recent and comprehensive source for academic and industrial researchers, as well as for graduate and upper level undergraduate students interested in cationic polymerizations.

Krzysztof Matyjaszewski

Contents

Contributors

Dale H. Hill Department of Macromolecular Science, The W. M. Keck Laboratories for Organic Synthesis, Case Western Reserve University, Cleveland, Ohio

Przemysław Kubisa Center for Molecular and Macromolecular Studies, Polish Academy of Sciences, Lodz, Poland

Krzysztof Matyjaszewski Department of Chemistry, Carnegie Mellon University, Pittsburgh, Pennsylvania

Herbert Mayr Institute for Organic Chemistry, Technical University of Darmstadt, Darmstadt, Germany

Virgil Percec Department of Macromolecular Science, The W. M. Keck Laboratories for Organic Synthesis, Case Western Reserve University, Cleveland, Ohio

Coleen Pugh Department of Chemistry, The University of Michigan, Ann Arbor, Michigan

Mitsuo Sawamoto Department of Polymer Chemistry, Kyoto University, Kyoto, Japan

Nicolas Spassky Laboratoire de Chimie Macromoléculaire, Université Pierre et Marie Curie, Paris, France

Jean-Pierre Vairon Laboratoire de Chimie Macromoléculaire, Université Pierre et Marie Curie, Paris, France

1

Introduction

KRZYSZTOF MATYJASZEWSKI Carnegie Mellon University, Pittsburgh, Pennsylvania

COLEEN PUGH The University of Michigan, Ann Arbor, Michigan

I. INTRODUCTION

Most cationic polymerizations are chain polymerizations involving positively charged or electrophilic active centers at the growing chain end. As discussed in the next section, polymerization mechanisms are classified as either chain growth or step growth, with cationic polymerizations being one of the three major types of chain polymerizations. Both alkenes and heterocyclic monomers can be polymerized cationically. The two systems differ in that the active species in the polymerization of alkenes are carbenium-ions, whereas onium ions are the active species in the polymerization of heterocycles. Nevertheless, there are more similarities between the two systems than differences. Therefore, this book develops a unified description of cationic polymerizations for both alkenes and heterocycles. In addition to cationic chain polymerizations, there are also cationic step polymerizations involving oxidative coupling and Friedel-Crafts reactions.

Cationic polymerizations are not only important commercial processes, but, in some cases, are attractive laboratory techniques for preparing well-defined polymers and copolymers. Polyacetal, poly(tetramethylene glycol), poly(ϵ-caprolactam), polyaziridine, polysiloxanes, as well as butyl rubber, poly(*N*-vinyl carbazol), polyindenes, and poly(vinyl ether)s are synthesized commercially by cationic polymerizations. Some of these important polymers can only be prepared cationically. Living cationic polymerizations recently have been developed in which polymers with controlled molecular weights and narrow polydispersity can be prepared.

Some of these living systems, especially those which generate block copolymers and polymers with reactive end groups, should be commercialized in the near future.

II. FUNDAMENTALS OF POLYMER SYNTHESIS

Although many chemists are familiar with polymeric products only as the insoluble and intractable material formed as an undesirable side product in their reactions, the synthesis of well-defined high molecular weight polymer requires state-of-the-art chemistry. For example, the chemoselectivity of propagation relative to chain breaking reactions (transfer and termination) must be 99.9% to prepare a polymer with a degree of polymerization of 1000 (molecular weight $\approx$ 100,000). Although few organic reactions can be run with such high chemoselectivity, many polymers are prepared with molecular weights more than 10 times this value.

The synthesis of well-defined polymers is also called macromolecular engineering. Indeed, the placement of all fragments of the chain must be controlled. The initiator is placed at the macromolecule's head group. The terminator is incorporated at the other end group. Transfer agents will affect the structures of both polymer end groups. These polymers or oligomers may subsequently be used to prepare block or multiblock copolymers if they were prepared using functionalized initiators and transfer/terminating agents. The regioselectivity and stereoselectivity of propagation affects the microstructure of the chain and determines its tacticity. Chains of different lengths will form if there are multiple propagating species with varying activities which do not exchange rapidly. In addition to the rate of exchange of active sites, the reversibility of propagation and the relative rates of initiation and propagation determine the molecular weight distribution. Thus, the major events in the "life of the macromolecule" can be discerned by careful analyses of the polymer's structure, microstructure, and polydispersity. Such events include the polymer's birth, growth, generation of offspring (transfer), and death (termination). This structural information is very helpful in designing new macromolecules and more controlled polymerizations.

Many new developments in cationic polymerizations and in polymer synthesis in general are due to the synthesis of new monomers and therefore new polymers with novel properties. However, the most significant developments are in more controlled polymerizations that enable the synthesis of materials with well-defined properties. After a brief introduction to polymer synthesis, we will focus on syntheses involving only cationic intermediates. Much of the basis of this chapter is covered in general polymer chemistry textbooks and in several seminal review articles on cationic polymerizations. For example, readers interested in a more in-depth coverage of basic polymer chemistry may consult references [1–5]; those interested in the developments of carbocationic and cationic heterocyclic polymerizations may consult references [6–10] and [5,11–13], respectively. A few excellent textbooks and reviews deal with the chemistry of carbocations and oxonium ions [14–19]. We also recommend textbooks related to ionic polymerization in the most general sense [20–22].

A. Chain Growth and Step Growth Polymerizations

Polymers and polymerizations can be classified either by the polymer composition or by the polymerization mechanism from which the polymers are obtained. The classification based on composition separates polymers into *condensation* and *addition* polymers or polymerizations. The original definition of condensation polymers by Carothers included polymers formed by typical organic condensation reactions in which a small molecule by-product is generated by the chain-forming reaction. This includes esterifications, amidations, imidations, etherifications, etc. However, this classification based on a by-product forming during the polymerization is problematic when the same polymer can be formed by either a condensation or an addition process. For example, polyethylene can be prepared by radical polymerization of ethylene (polyaddition) or by a polycondensation reaction from diazomethane. Poly(tetramethylene glycol) can also be prepared by either condensation of the corresponding glycol or by cationic ring-opening polymerization of tetrahydrofuran.

The less ambiguous classification is based on the polymerization mechanism, which can be either *chain growth* or *step* (nonchain) *growth*. In the latter case, a given functional group has similar if not identical reactivity, whether it is in the monomer or at the polymer chain end. In a chain growth mechanism, only monomer adds to the active species at the growing chain end; i.e., two monomer molecules will not react with each other.

Propagation is the only elementary reaction in a step polymerization [Eq. (1)].

$$\mathrm{A{\sim\sim}B} + \mathrm{A{\sim\sim}B} \xrightarrow{k} \mathrm{A{\sim\sim\sim\sim}B} + \mathrm{AB} \quad (1)$$

Although a low molecular weight by-product such as water or alcohol is usually generated by each propagation reaction, polyadditions may also proceed by a step mechanism. For example, no by-product is generated by the reaction of diols and diisocyanates to form polyurethane. Because the product always contains reactive end groups, unless one of the reactants is a monofunctional end-capping agent, a polymer resulting from a step polymerization can undergo further chain extension. However, there are also reactive sites present in the condensation polymer's backbone which can react and randomize under appropriate conditions; for example, by transesterification, transamidation, etc.

A step polymerization yields high molecular weight polymer only after very high conversion is reached (Fig. 1) as described by Carother's equation [Eq. (2)] relating the number average degree of polymerization ($\overline{DP}_n$) to conversion (p).

$$\overline{DP}_n = \frac{1}{1 - p} \tag{2}$$

Polymer is built up slowly because species of all sizes, including monomer, oligomer, and polymer, have reactive end groups of approximately equal reactivity. In general, two functional groups of the same type have equal reactivity if neither are conjugated with the functional group at the other chain end, and if the reactive sites are separated by at least three methylenic units. Therefore, monomer first reacts with monomer to give dimer,

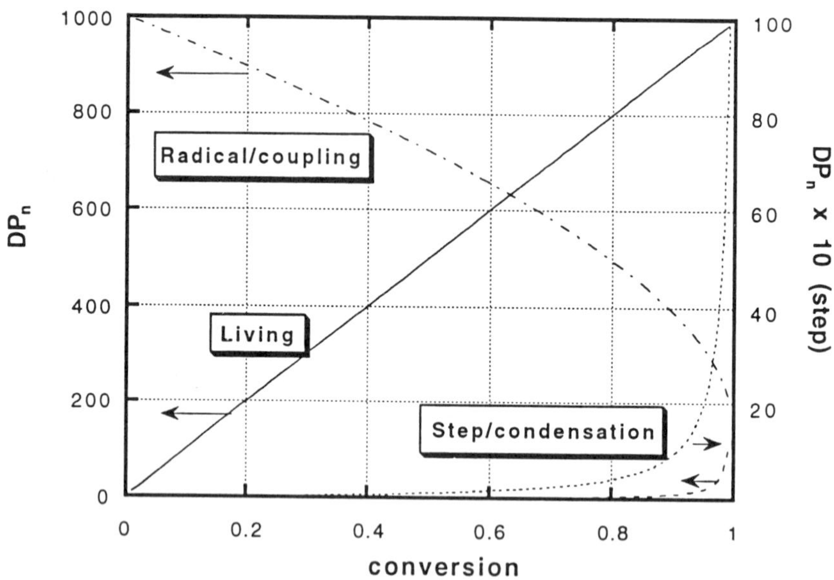

Figure 1 Polymer molecular weight as a function of monomer conversion in chain growth, step growth, and living polymerizations.

dimer can then react with monomer to give trimer or it can react with dimer to give tetramer, etc, and the molecular weight doesn't increase significantly until the two reactants are themselves of significantly high molecular weight. This results in rates of polymerization that are usually second order (catalyzed) or third order (self-catalyzed) in monomer.

In contrast to step polymerizations, chain polymerizations require an initiator (I) to produce reactive centers. Because monomer (M) reacts exclusively with the active center (M*) and not with another monomer molecule, the polymerization rate is usually first order in monomer. In addition to initiation and propagation, chain polymerizations may also undergo transfer and termination reactions (e.g., with reagent A) in which inactive chains (P) are formed [Eq. (3)].

$$\begin{array}{llll} 1. & \text{Initiation} & \mathrm{I + M} \xrightarrow{k_i} \mathrm{IM^*} & \\ 2. & \text{Propagation} & \mathrm{IM^* + }n\mathrm{M} \xrightarrow{k_p} \mathrm{I(M)_{\mathit{n}}M^*} & \\ 3. & \text{Transfer} & \mathrm{I(M)_{\mathit{n}}M^* + A} \xrightarrow{k_{tr}} \mathrm{P + A^*} & (3) \\ 4. & \text{Termination} & \mathrm{I(M)_{\mathit{n}}M^* + A} \xrightarrow{k_t} \mathrm{P} & \end{array}$$

There is much effort to control or ideally eliminate the two latter reactions to yield *living* polymerizations. As shown in Fig. 1, high molecular weight polymer forms rapidly if initiation is slow and propagation is relatively fast; the molecular weight is limited finally by one of the chain-breaking reactions (transfer and termination). In this case, the polymerization system consists of monomer and high molecular weight polymer at all stages of the polymerization, with only the yield of polymer increasing with increasing monomer conversion. In the particular case of a radical polymerization with a constant rate of initiation and with termination by coupling, the molecular weight of the first polymer formed (<1% conversion) is high, but then the average molecular weight decreases with conversion. As explained in the next section, the number average degree of polymerization is determined by the ratio of the rate of propagation to that of initiation. Because the rate of initiation does not vary significantly in most radical reactions and only a small fraction of the initiator is consumed at complete monomer conversion, the rate of propagation and therefore the degree of polymerization decrease with declining monomer concentration as shown in Figure 1. If initiation is faster than, or at least comparable to, propagation as in most living polymerizations, all chains begin growing simultaneously and the molecular weight of the polymer increases linearly with conversion, provided that transfer is negligible.

Most carbocationic and cationic ring-opening polymerizations are chain processes proceeding with carbocations and/or onium ions as the active species. Nonchain processes which occur via cationic and electrophilic intermediates will be discussed in Chapter 7.

B. Molecular Weights and Molecular Weight Distributions

In contrast to the synthesis of small molecules, it is impossible to obtain a completely pure product from a step or chain polymerization that has a unique chain length and therefore a unique molecular weight. Instead, a distribution of chain lengths and molecular weights are obtained. These are described by the moments of the statistical distribution of molecular weights [Eq. (4)]. That is, the first moment of the distribution is the number average molecular weight ($\overline{M}_n$), which is the total weight of the sample (Σw_i) divided by the number of chains in that sample (Σn_i). The weight ($\overline{M}_w$) and z-average ($\overline{M}_z$) molecular weights are the second and third moments of the distribution, respectively. This demonstrates that $\overline{M}_z$ describes a higher molecular weight average than does $\overline{M}_w$, which in turn describes a higher molecular weight average than does $\overline{M}_n$ ($\overline{M}_z > \overline{M}_w > \overline{M}_n$).

$$\begin{aligned} \overline{M}_n &= \frac{\sum n_i M_i}{\sum n_i} = \frac{\sum w_i}{\sum n_i} \\ \overline{M}_w &= \frac{\sum n_i M_i^2}{\sum n_i M_i} = \frac{\sum w_i M_i}{\sum w_i} \\ \overline{M}_z &= \frac{\sum n_i M_i^3}{\sum n_i M_i^2} = \frac{\sum w_i M_i^2}{\sum w_i M_i} \end{aligned} \tag{4}$$

where n_i = number of chains of length i; M_i = molecular weight of chains of length i.

The number of repeating units in the chains corresponding to the number, weight, and z-average molecular weights are called the number ($\overline{\mathrm{DP}}_n$), weight ($\overline{\mathrm{DP}}_w$), and z-average ($\overline{\mathrm{DP}}_z$) degrees of polymerization, respectively [Eq. (5)].

$$\overline{M}_n = \overline{\mathrm{DP}}_n M_0 \qquad \overline{M}_w = \overline{\mathrm{DP}}_w M_0 \qquad \overline{M}_z = \overline{\mathrm{DP}}_z M_0 \tag{5}$$

where M_0 = molecular weight of the repeating unit and/or monomer.

The number average degree of polymerization of a polymer sample is most easily understood as the number of monomer repeating units in the sample, divided by the number of polymer chains. As mentioned in the preceding section, the degree of polymerization and polymer molecular weight are determined in a step polymerization by the extent of conversion of functional groups. Thus, the degree of polymerization can be limited by adding a small quantity of a monofunctional reagent, which limits the conversion of the complementary functional group. Because the number of polymer chains resulting from a chain polymerization is proportional to the amount of initiator used, the degree of polymerization is inversely proportional to initiator concentration. The ratio of the concentration of reacted monomer and polymer chains is determined by the ratio of the rates of propagation (R_p) and all chain-forming reactions, including transfer (R_{tr}) [Eq. (6)].

$$\overline{\mathrm{DP}}_n = \frac{\Delta[\mathrm{M}]}{[\mathrm{Polymer}]} = \frac{R_p}{R_i + R_{tr} + \cdots} \tag{6}$$

For example, the number average degree of polymerization resulting from a chain polymerization which involves both chain transfer to monomer and chain transfer to a chain transfer agent, A, is described by Eq. (7).

$$\overline{\mathrm{DP}}_n = \frac{\Delta[\mathrm{M}]}{[\mathrm{I}]_0 + \dfrac{k_{trM}}{k_p}(\Delta[\mathrm{M}]) + \dfrac{k_{tr}}{k_p}[\mathrm{A}]\ln([\mathrm{M}]_0/[\mathrm{M}])} \tag{7}$$

The first term in the denominator accounts for the concentration of polymer chains formed by initiator ($[\mathrm{I}]_0$), the second term accounts for the concentration of polymer chains formed by chain transfer to monomer, and the final term accounts for the concentration of chains formed by transfer to molecules other than monomer (e.g., a transfer agent, A, by a bimolecular reaction, or counterion by a unimolecular reaction). Therefore, molecular weight can be limited in chain polymerizations by intentionally adding a chain transfer agent. If functionalized chain transfer or terminating agents are used, the polymer chain ends will be functionalized. Similarly, the initiator may also be functionalized to generate polymers with functional end groups.

The breadth of the molecular weight distribution is described by the ratio of the weight and number average molecular weights or degrees of polymerization, and is referred to as the polydispersity index (PDI) or molecular weight distribution (MWD) [Eq. (8)].

$$\mathrm{PDI} = \frac{\overline{M}_w}{\overline{M}_n} = \frac{\overline{\mathrm{DP}}_w}{\overline{\mathrm{DP}}_n} \tag{8}$$

Step polymerizations as well as chain polymerizations that do not terminate by bimolecular chain coupling result in polymers with the most probable molecular weight distribution of 2.0; free radical polymerizations that terminate by bimolecular chain coupling result in initial PDI = 1.5. However, the polydispersities resulting from cationic polymerizations are often much broader than 2.0 due to several factors that are discussed in Chapter 4, including transfer, slow initiation, and especially slow exchange between active species of different reactivities.

C. Radical and Ionic Polymerizations

Radical and ionic polymerizations involve the four elementary reactions of initiation, propagation, transfer, and termination [cf, Eq. (3)]. For a vinyl monomer to polymerize, it must contain a substituent capable of stabilizing the resulting active species. Therefore, most 1-substituted alkenes undergo facile radical polymerizations (although α-olefins do not), whereas an electron-donating substituent is required to activate olefins to cationic intiation and propagation, and an electron-withdrawing substituent is required to activate olefins to anionic reactions. Typical vinyl monomers for cationic polymerization are vinyl ethers, *N*-vinylcarbazole, styrenes (especially those substituted with electron-donating groups), and 1,1-disubstituted alkenes such as isobutylene and α-methylstyrene.

In ionic polymerizations, monomer is mixed typically with a solution of the ionic initiator at ambient or subambient temperature; initiation is thus first order in monomer and first (or fractional) order in initiator. In contrast, a radical initiator is generated in situ by homolytic cleavage of weak chemical bonds using either heat, light, or high-energy radiation; radical initiation is thus zero order in monomer. In all cases, propagation is usually regioselective and occurs by addition of the growing chain to the less hindered position of the monomer's double bond, resulting in the more stable active species [Eq. (9)].

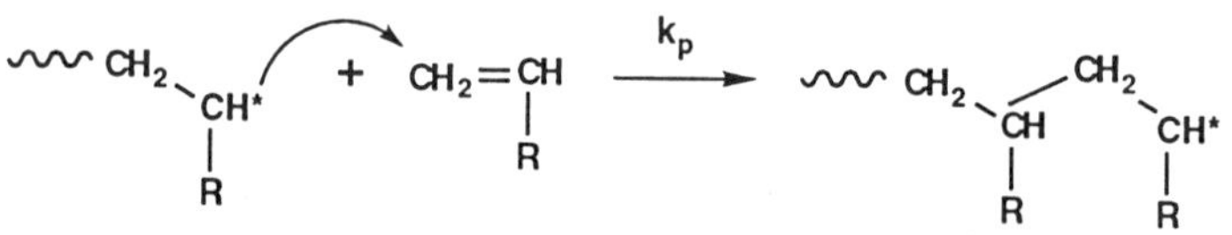

(9)

The stereoselectivity in radical and most ionic polymerizations is poor, with the tacticity successfully controlled in only a few systems in which the monomer forms a complex with the active site. So far, the tacticity has not been controlled satisfactorily in most cationic polymerizations.

Typical values of the rate constants of propagation and the concentrations of active sites differ substantially in the three types of chain polymerizations as shown in Table 1. (These values are not all encompassing.) Table 1 also lists representative polymerization times as determined by

Table 1 Characteristics of Chain Polymerizations

	Radical	Anionic	Cationic	Ring opening
k_p (mol^{-1}·L sec^{-1})	$\approx 10^2$–10^4	≈ 1–10^4	$\approx 10^4$–10^6	$\approx 10^{-3}$–10^{-1}
[Active Center] (mol/L)	$\approx 10^{-7}$	$\approx 10^{-4}$	$\approx 10^{-5}$	$\approx 10^{-3}$–10^{-2}
Reaction time	≈ 1 hr	≈ 1 min	≈ 1 sec	>1 hr

the inverse product of the rate constants of propagation and the concentrations of active sites. However, in systems involving equilibria between active and dormant species, the concentration of the true active centers are often much smaller than those listed in Table 1 resulting in much longer reaction times. Values from Table 1 can be compared to those of step polymerizations, with typical rate constants of polymerization $k_p \sim 10^{-3}$ mol^{-2} L^2 sec^{-1} and polymerization times greater than 1 hr.

In all cases, chain transfer may occur to monomer, solvent, polymer, or a chain transfer agent. For this to be a transfer reaction rather than termination, the molecule generated (A*) must be reactive enough to reinitiate polymerization [Eq. (10)].

$$\mathrm{I(M)}_n\mathrm{M}^* + \mathrm{A} \xrightarrow{k_{tr}} \mathrm{P} + \mathrm{A}^* \tag{10}$$

Solvent is not required in radical polymerizations, although it retards autoacceleration and helps to diffuse heat; solvent is required in ionic polymerizations to control the heat generated. Because solvent is the component present in the highest concentration in most ionic polymerizations, solvents that either terminate the active sites or that act as chain transfer agents must be avoided. Although protonic solvents can be used for radical polymerizations, they result in either transfer or termination in most ionic polymerizations. For the same reason, halogenated solvents are not used in anionic polymerizations, and amines are not used in cationic polymerizations. Common solvents for anionic polymerizations are hydrocarbons and ethers. Hydrocarbons are also used for cationic polymerizations, as are halogenated and nitro-containing solvents.

In contrast to chain transfer to solvent which would be prevalent from the initial stages of a polymerization due to the solvent's high concentration, chain transfer to polymer often does not compete noticeably with propagation until the end of the polymerization when monomer is depleted. In addition, chain transfer and termination reactions generally have higher activation energies than propagation, and therefore can be suppressed relative to propagation by performing the polymerization at lower temperatures. Chain transfer by β-proton elimination from the propagating carbocations is the most common and detrimental side reaction in cationic polymerizations of alkenes. On the other hand, cationic ring opening polymerizations involve both intra- (macrocycle formation) and intermolecular (scrambling of chain lengths) chain transfer to polymer.

Termination generally occurs in cationic polymerizations by unimolecular collapse of the ion pair, or by reaction with impurities and/or terminating agents. Termination sometimes occurs by formation of highly delocalized and therefore unreactive carbocations. However, termination can be avoided in cationic polymerizations under appropriate conditions. In contrast to ionic polymerizations in which the growing species can not react with each other due to their like charge, termination does occur in radical polymerizations by combination of two growing polymeric radicals, or by disproportionation of the two radicals. Therefore, termination is unavoidable in radical polymerizations. If both transfer and termination are absent as in some ionic polymerizations, the polymerization is *living*.

D. Living Polymerizations

Since the 1960s [20], much of the research in polymer synthesis has been directed at establishing living conditions for chain polymerizations. The only requirements for a polymerization to be considered living are that no chain-breaking reactions occur during the polymerization. That is, the rate constants of both chain transfer and termination should be equal to zero ($k_{tr} = 0$, $k_t = 0$).

It is much easier to analyze living systems and prepare controlled polymers if initiation is fast compared to propagation. In this case, the degree of polymerization is determined by the ratio of the concentrations of reacted monomer and initiator [Eq. (11)].

$$\overline{DP}_n = \frac{\Delta[M]}{[I]_0} = \frac{[M]_0 - [M]_t}{[I]_0} \tag{11}$$

Although this formally requires that $k_i = \infty$, controllable molecular weight is obtained if the rate constant of initiation is comparable to that of propagation ($k_i \geq k_p$) (see Chapter 4). As demonstrated subsequently, if initiation is slow, the number average molecular weight is initially too high but becomes equal to the theoretical value once initiator is consumed.

The number of active species continuously increases with conversion if initiation is slow, until initiation is complete, at which time the concentration of active sites becomes equal to that of the initiator introduced. As shown in Fig. 2, this does not happen until all monomer is consumed

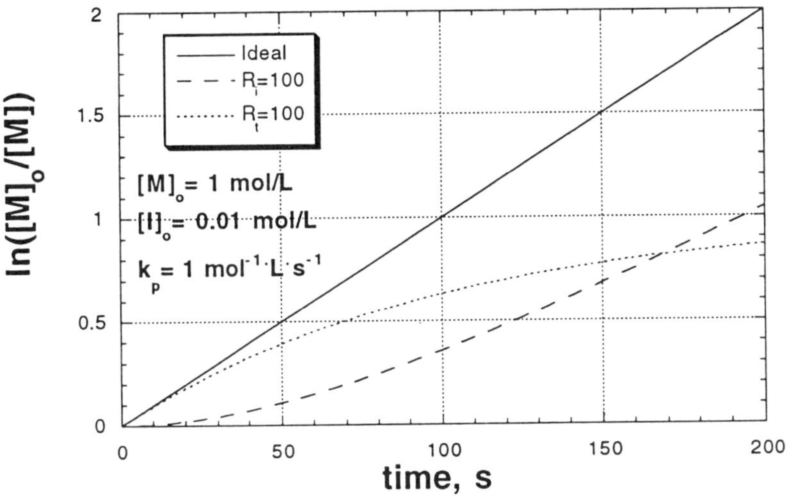

Figure 2 The effect of slow initiation ($R_i = k_p/k_i$) and termination ($R_t = k_p/k_t$) on kinetics.

if the rate constant of initiation is 100 times smaller than that of propagation ($R_i = k_p/k_i = 100$), resulting in continuous acceleration. The tangent to the semilogarithmic anamorphoses shown in Fig. 2 is the product of the rate constant of propagation and the concentration of growing species [Eq. (12)].

$$\frac{d \ln [M]}{dt} = k_p[P^*]_t \tag{12}$$

It continuously increases if initiation is slow, until it reaches the same value as that for the ideal case of instantaneous initiation. If initiation is fast but termination occurs, the tangent decreases continuously following the declining concentration of growing species as shown in Fig. 2, which often leads to only partial monomer conversion. For example, if $[M]_0 = 1$ mol/L, $[I]_0 = 0.01$ mol/L and $R_t = k_p/k_t = 100$ mol/L, then 37% monomer will remain unreacted at infinite time.

In contrast, transfer reactions may not be detected by following the monomer conversion if the rate of reinitiation is comparable to that of propagation. In this case, transfer is detected by a nonlinear dependence of the polymer molecular weight as a function of monomer conversion or polymer yield (Fig. 3); termination does not affect the number of chains

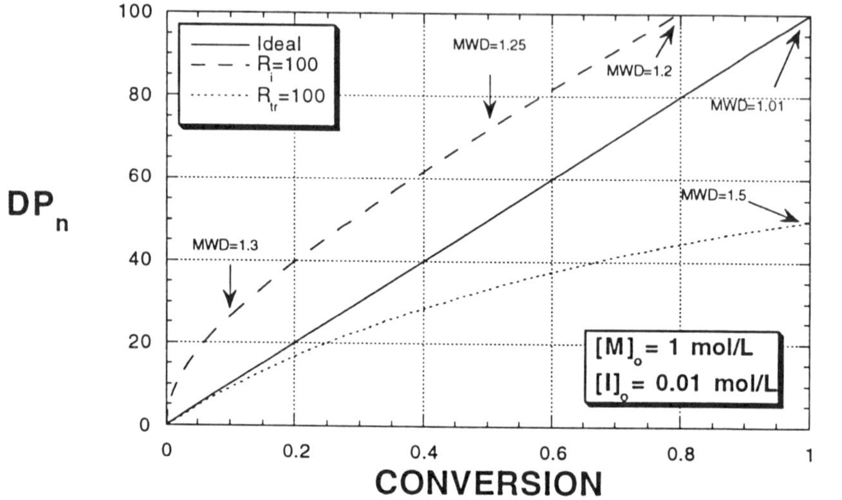

Figure 3 The effect of slow initiation ($R_i = k_p/k_i$) and transfer reactions ($R_{tr} = k_p/k_{tr}$) on molecular weights and polydispersities.

in the system and is therefore not detected by following the average molecular weight.

Both slow initiation and chain-breaking reactions (transfer and termination) lead to increased polydispersity. However, the highest polydispersity caused by slow initiation is only $\overline{M}_w/\overline{M}_n \approx 1.3$. Whereas transfer to monomer has a limiting $\overline{M}_w/\overline{M}_n \approx 2$, transfer to counterion or transfer agent may result in even broader polydispersities. The highest polydispersity due to termination is $\overline{M}_w/\overline{M}_n \approx 2$.

In practice, linear semilogarithmic kinetic plots and linear dependencies of molecular weight on monomer conversion require only that the rate constants of chain transfer and termination are much less than that of propagation ($k_{tr} \ll k_p$, $k_t \ll k_p$). This is therefore the practical requirement for the synthesis of well-defined polymers, such that complete monomer conversion can be reached and the chain ends can be functionalized quantitatively. However, because chain-breaking reactions are actually present, we prefer to call such systems controlled polymerizations rather than living polymerizations.

In addition to the absence of chain-breaking reactions and fast initiation, two additional constraints must be met to obtain polymers with narrow molecular weight distributions: all chains must have equal reactivity ($k_p{}^x = k_p{}^y = k_p{}^z$ etc.) or the active sites must exchange rapidly, and the growth must be irreversible ($k_p \gg k_d$). In this case, the molecular weight distribution depends on only the degree of polymerization as defined by a Poisson distribution [Eq. (13), cf., also Chapter 4].

$$\frac{\overline{\mathrm{DP}}_w}{\overline{\mathrm{DP}}_n} = 1 + \frac{\overline{\mathrm{DP}}_n}{(1 + \overline{\mathrm{DP}}_n)^2} \cong 1 + \frac{1}{\overline{\mathrm{DP}}_n} \cong 1 \tag{13}$$

Living polymerizations in which initiation is fast and quantitative and which have irreversible growth offer several advantages over conventional polymerizations. In addition to the ability to obtain polymers with controlled molecular weights and narrow molecular weight distributions, it is also possible to control the polymer architecture and chain end functionality (F = 1.0, 2.0, etc.). For example, diblock, triblock, and multiblock copolymers are prepared routinely by living polymerizations, as are telechelic, star, and comb polymers. Cyclic polymers and cyclic block copolymers also have been prepared recently by terminating a living polymer growing in two directions with a difunctional terminating agent [23–28].

E. Ring Opening Polymerizations

Ring opening polymerization provides a synthetic method for introducing functional groups typical of condensation polymers into a polymer backbone, separated by varying lengths of methylenic units [Eq. (14)]. Typical functional groups include ethers, sulfides, esters, amides, double bonds, etc.

(14)

Therefore, polymers prepared by ring opening polymerization always contain heteroatoms or other reactive functional groups along the polymer backbone. By providing sites capable of "trans" reactions (such as transesterification, transetherification, transamidation, etc.) subsequent to polymerization, these sites participate in depolymerization, transfer, and termination reactions in competition with propagation. For example, intra- and intermolecular chain transfer of the growing chain end to polymer results in less strained macrocyclic and branched onium ions, respectively [Eq. (15)]. If the resulting onium ion is not strained enough to react with monomer, or if its formation is irreversible, the reaction is termination rather than transfer.

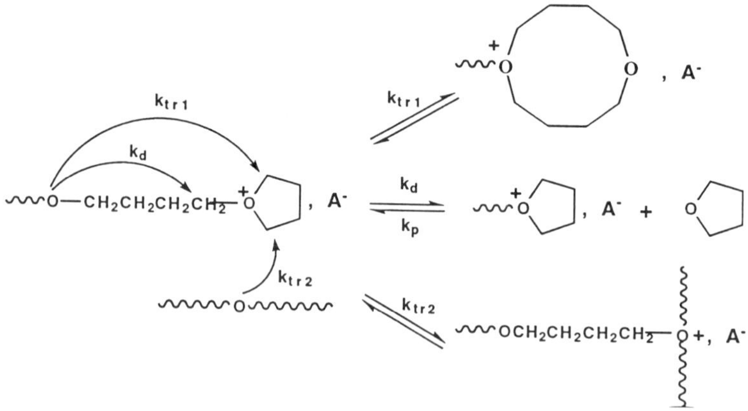

(15)

F. Thermodynamics of Polymerization

For a polymerization to occur, its change in free energy must be negative ($\Delta G_p < 0$). Because the change in free energy depends only on the initial and final states of

$$\Delta G_p{}^o = \Delta H_p{}^o - T\Delta S_p{}^o \quad (16)$$

the system [Eq. (16)], it is independent of the polymerization mechanism (° = standard conditions = 1 M). Most polymerizations are exoentropic due to the loss of translational entropy (~130 J/mol K) upon going from several molecules of monomer to one polymer chain. Therefore, the loss in entropy must be compensated for by a loss in enthalpy. This can be provided by either relief of ring strain in a ring opening polymerization, or by isomerization of the double bond in an alkene to single bonds in the polymer.

In exoenthalpic and exoentropic polymerizations, $\Delta G_p{}^o$ eventually becomes positive above a temperature known as the *ceiling* temperature ($\Delta G_p{}^o = \Delta H_p{}^o / T_c \Delta S_p{}^o > 0$). For example, the ceiling temperature of bulk styrene is ≈400° C, whereas that of methyl methacrylate is ≈200° C. In contrast, $\Delta G_p{}^o$ becomes positive below a *floor* temperature when the polymerization is endoenthalpic and endoentropic ($\Delta G_p{}^o = \Delta H_p{}^o / T_f \Delta S_p{}^o > 0$), such as in the polymerization of S_8 ($T_f \approx 160°$ C).

When the system is at equilibrium, the rate of polymerization equals the rate of depolymerization [Eq. (17)].

$$M_n^* + \mathrm{M} \underset{k_d}{\overset{k_p}{\rightleftharpoons}} M_{n+1}^* \quad (17a)$$

$$k_p[\mathrm{M}^*][\mathrm{M}]_e = k_d[\mathrm{M}^*] \quad (17b)$$

$$[\mathrm{M}]_e = \frac{k_d}{k_p} = \frac{1}{K_e} \tag{17c}$$

Therefore, $K_e = 1/[\mathrm{M}]_e$. The equilibrium constant is assumed to be equal for all chain lengths. For example, the equilibrium constant of the dimer is assumed to be equal to that of both tetramer and high molecular weight polymer ($K_e = K_2 = K_4 = K_\infty$, etc.). However, as demonstrated for α-methylstyrene ($K_2 = 2.4 \times 10^2$ L/mol, $K_4 = 2.9$ L/mol, $K_\infty = 0.41$ L/mol at 25° C) [20], the equilibrium constants for at least the shortest oligomers may differ from that of high molecular weight polymer. The equilibrium monomer concentrations calculated from this example demonstrates that although it may not be possible to form high molecular weight polymer at a given temperature ($[\mathrm{M}]_{e,\infty} = 2.4$ mol/L at 25° C), the monomer may still oligomerize ($[\mathrm{M}]_{e,4} = 0.34$ mol/L) or at least dimerize ($[\mathrm{M}]_{e,2} = 4.2 \times 10^{-3}$ mol/L).

The ceiling (or floor) temperature varies with the equilibrium monomer concentration, with the highest (or lowest) temperature corresponding to that of a bulk polymerization [Eq. (18)].

$$\Delta G_p = \Delta G_p{}^0 - \mathrm{R}T_c \ln [\mathrm{M}]_e = 0 \tag{18a}$$

$$T_c = \frac{\Delta H_p{}^0}{\Delta S_p{}^0 + R \ln [\mathrm{M}]_e} \tag{18b}$$

The terms *ceiling* and *floor* temperatures usually refer to bulk conditions unless stated otherwise. Because polymerization does not occur when the initial monomer concentration is at or below the equilibrium monomer concentration, $[\mathrm{M}]_e$ in the above equations can be replaced with $[\mathrm{M}]_0$. This demonstrates that ΔG_p will be negative even if $\Delta G_p{}^0$ is positive if $[\mathrm{M}]_0$ is high enough, and that polymerization will occur. For example, although tetrahydrofuran (THF) does not polymerize at 25° C when $[\mathrm{M}]_0 = 1$ mol/L ($\Delta G_p{}^0 > 0$), it does polymerize at concentrations $[\mathrm{M}]_0 > 5$ mol/L ($\Delta G_p < 0$) because at 25° C $[\mathrm{M}]_e \approx 5$ mol/L [29].

Other variables can also be used to influence the thermodynamics of a polymerization. For example, most step polymerizations involve equilibrium reactions, which may be driven to completion by removing the small molecule by-product in an open system. Addition polymerizations are influenced by the solvent used. That is, $[\mathrm{M}]_e$ depends on both the nature of the solvent and on $[\mathrm{M}]_0$. For example, the equilibrium monomer concentration of THF increases as the acidity of the solvent increases due to complex formation. In other cases, solvation of a polymer segment may be more exothermic than monomer solvation, resulting in a more exothermic ΔH_p compared to the bulk polymerization.

The polymerizability of a monomer is also influenced by the physical state of the polymerization. For example, crystallization of poly(oxymethylene) provides the driving force for trioxane polymerization. In this case, propagation occurs at active sites on the crystal lattice rather than in solution, and ΔG_p includes the change in free energy of the phase transition as well as that of the solution polymerization [Eq. (19)].

$$\Delta G_p = \Delta G_{p,\mathrm{ss}} + \Delta G_{\mathrm{cryst}} \tag{19}$$

Thus, it is important to specify the polymerization conditions. Here, the subscript ss stands for solution-solution in which both monomer and polymer are in solution. Other common conditions are liquid monomer-condensed (amorphous) polymer (subscript lc) and the hypothetical gas-gas system (subscript gg).

The thermodynamic parameters $\Delta S_p{}^0$ and $\Delta H_p{}^0$ may also be affected by the microstructure of the resulting polymer or copolymer. In particular, low values of $\Delta H_p{}^0$ lead to reversible propagation, which in turn results in significant deviation of the copolymer composition as described by the terminal copolymerization model discussed below. On the other hand, the microstructure of the polymer affects $\Delta S_p{}^0$, with atactic polymers and more random copolymers having higher entropies than tactic polymers and more regular copolymers, respectively.

G. Chain Copolymerization

Polymers that contain more than one type of monomeric repeat unit are called copolymers. By copolymerizing two or more monomers in varying ratios and arrangements, polymeric products with an almost limitless variety of properties can be obtained. As shown in Eq. (20), there are four basic types of copolymers as defined by the distribution of comonomers A and B, for example. Although the properties of a random copolymer are intermediate between those of the two homopolymers, block and graft copolymers exhibit the properties of both homopolymers. The properties of an alternating copolymer are usually unique.

RANDOM	poly(A-*ran*-B)	AABAABBBABBAAABABAABB
BLOCK	poly(A-*block*-B)	AAAAAAAAAAABBBBBBBBBB
ALTERNATING	poly(A-*alt*-B)	ABABABABABABABABABABA
GRAFT	poly(A-*graft*-B)	AAAAAAAAAAAAAAAAAAAA

(20)

The first three types of copolymers can be prepared by polymerizing the two monomers simultaneously. In this case, the distribution of comonomers is determined by their relative concentrations and reactivity ratios. The reactivity ratios (r_1 and r_2) are the ratios of the rate constants of homopropagation and cross-propagation [Eq. (21)].

$$r_1 = \frac{k_{11}}{k_{12}} \qquad r_2 = \frac{k_{22}}{k_{21}} \tag{21}$$

Using the terminal model of copolymerization, there are four possibilities for propagation [Eq. (22)].

$$\begin{aligned}
&\sim\!\sim M_1^* + M_1 \xrightarrow{k_{11}} \sim\!\sim M_1^* \\
&\sim\!\sim M_1^* + M_2 \xrightarrow{k_{12}} \sim\!\sim M_2^* \\
&\sim\!\sim M_2^* + M_2 \xrightarrow{k_{22}} \sim\!\sim M_2^* \\
&\sim\!\sim M_2^* + M_1 \xrightarrow{k_{21}} \sim\!\sim M_1^*
\end{aligned} \tag{22}$$

To solve the kinetics of this four-equation scheme in order to determine the copolymer composition, two assumptions must be made: (1) there are only two active sites (M_1^* and M_2^*) whose concentrations are at steady state; and (2) high polymer is formed which requires that monomer is consumed entirely by propagation. In this case, the instantaneous molar ratio of the two monomer units in the polymer ($d[M_1]/d[M_2]$) is defined by Eq. (23), in which $[M_1]$ and $[M_2]$ are the concentrations of the two monomers in the polymer feed.

$$\frac{d[M_1]}{d[M_2]} = \frac{[M_1]}{[M_2]} \cdot \frac{r_1[M_1] + [M_2]}{r_2[M_2] + [M_1]} \tag{23}$$

Random copolymers in which the comonomer distribution follows Bernoullian or zero-order statistics are formed by ideal copolymerizations in which $r_1 r_2 = 1$. However, a truely random distribution of the two units results only if the Bernoullian distribution is symmetric; i.e., when $r_1 = r_2 = 1$. In this case, the two monomers have equal probability of reacting with a given active center, regardless of the monomer from which it is derived, and the copolymer composition equals the comonomer feed composition. Random copolymers are generally formed by radical copolymerizations, whereas ionic copolymerizations tend to favor propagation of one of the comonomers much more than the other, yielding blocky sequences of that comonomer. That is, it is usually difficult to copolymerize monomers by an ionic mechanism because of large differences in their reactivity ratios: $r_1 \gg 1$, $r_2 \ll 1$. Alternating copolymers result when both reactivity ratios are close to zero due to little tendency of either monomer to homopolymerize ($r_1 = r_2 = 0$).

True block copolymers containing long blocks of each homopolymer in a diblock, triblock, or multiblock sequence are formed by simultaneous polymerization of the two monomers when $r_1 \gg 1$ and $r_2 \gg 1$. However, block copolymers are prepared more effectively by either sequential monomer addition in living polymerizations, or by coupling two or more telechelic homopolymers subsequent to their homopolymerization. Alternatively, if the two monomers do not polymerize by the same mechanism, a block copolymer can still be formed by sequential monomer addition if the active site of the first block is transformed to a reactive center capable of initiating polymerization of the second monomer.

Graft copolymers are prepared by initiating polymerization of monomer B from several active sites along the backbone of homopolymer A (grafting from), or by terminating the polymerization of homopolymer B with homopolymer A containing many terminating sites (grafting onto). Graft copolymers are also prepared by copolymerizing monomer A with macromonomers of B (grafting through) [Eq. (24)].

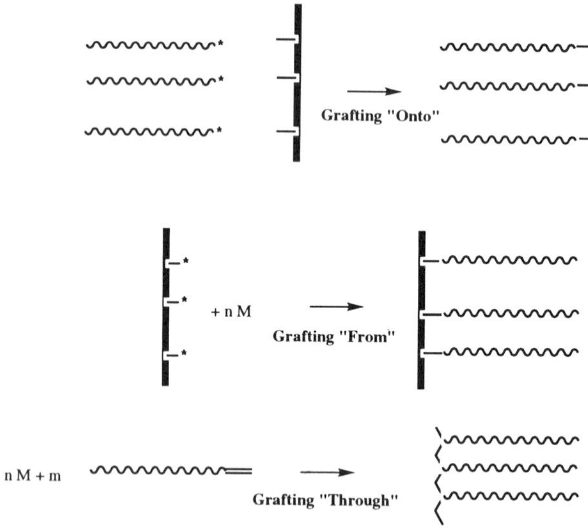

(24)

In addition to their synthetic utility, copolymerization studies provide a method for determining relative monomer reactivities and relative reactivities of the resulting active centers in chain polymerizations. That is, relative monomer reactivity ($1/r_1 = k_{12}/k_{11}$) is determined by copolymerizing different monomers in the presence of the same reference polymeric active center 1. The relative reactivity of various growing active centers are determined from the same experiments using a single reference monomer. However, this can only be calculated if the true homopropagation rate constant k_{11} is known.

H. Cationic Condensation Processes

Polymers such as polyetherketones and polyethersulfones can be prepared by electrophilic aromatic substitution using aromatic acid chlorides and aromatic sulfonyl chlorides, respectively [Eq. (25)]. However, due to ortho-substitution in addition to the desired para-substitution, it is difficult for these Friedel-Crafts acylations to compete with nucleophilic aromatic substitution of activated aromatic halides which are usually used for their synthesis.

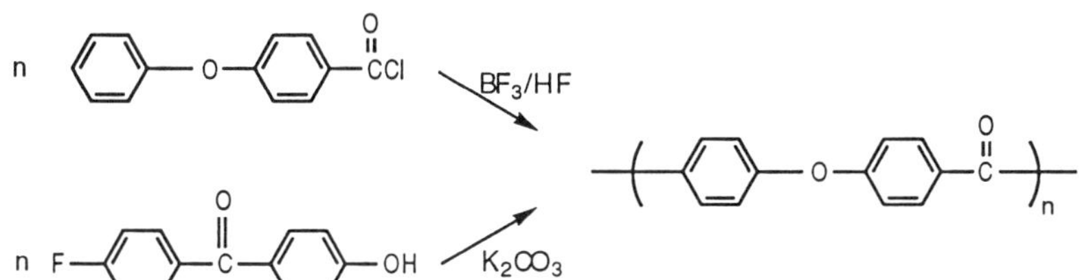

(25)

In other systems, such as in some Friedel-Crafts alkylations, ortho-substitution is desirable. For example, extensive alkylation at both the ortho- and para-positions of phenol with formaldehyde in the presence of an acid catalyst yields highly branched *novolac* phenolic resin prepolymers [Eq. (26)].

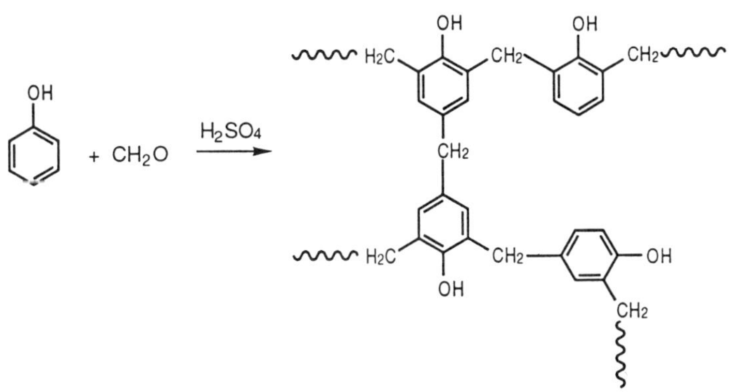

(26)

Ortho-substitution is also used to prepare branched liquid crystalline polyethers of 3,4-(di-*n*-alkoxy)benzyl derivatives in which the cyclotetraveratrylene disc-like mesogen is formed in situ during the polymer-forming Friedel-Crafts alkylation, [30–32].

Rearrangements of the alkyl groups do not occur in the systems above because the alkylating agents are benzyl dervivatives. If highly regular and therefore crystalline polymers are desired, ortho-substitution and branching due to polyalkylation must also be prevented. Although polyalkylation is inherent to Friedel-Crafts alkylations due to the increased reactivity of the products from the first alkylation versus the starting material, linear poly(α-methylbenzyl) can be obtained by performing the polymerization of 1-phenylethyl chloride (α-chloroethylbenzene) at low temperature ($\leq -65°$ C) using $AlCl_3$ complexed with excess nitroethane (6:1) as the catalyst [Eq. (27)] [33].

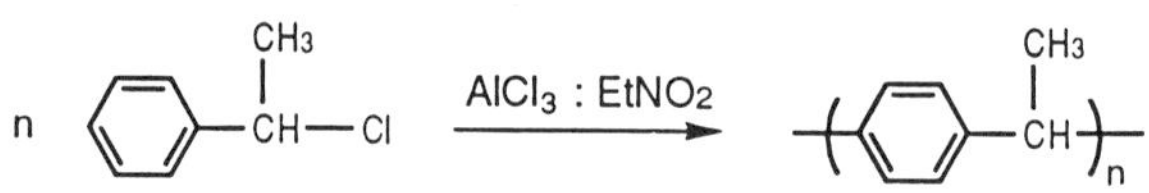

(27)

In this case, the ortho-position is somewhat hindered due to the methyl substituent at the benzyl carbon, and the selectivity and solubility of $AlCl_3$ at low temperature in ethyl chloride is increased by complexation with nitroethane.

Cationic cyclopolymerization of difunctional 1,4-diisopropenylbenzene and other 1,4-bis(1-alkylvinyl)benzenes to form polyindanes is also a step polymerization because protonation and deprotonation occur in every step of monomer addition [Eq. (28)] [34].

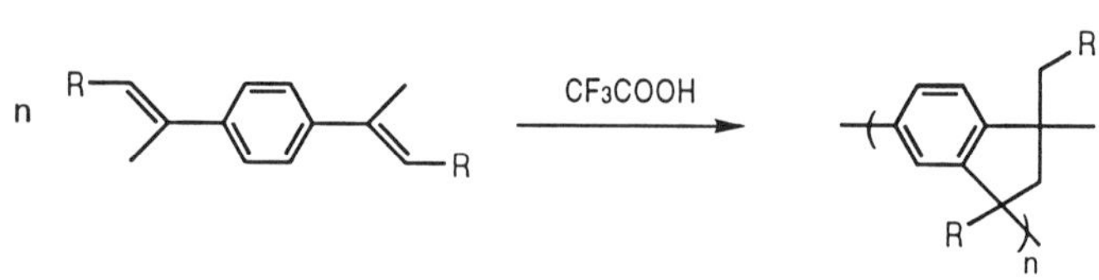

(28)

The indane structure is formed by an intramolecular Friedel-Crafts alkylation. To prevent intermolecular alkylation and standard vinyl polymerization, respectively, the polymerization is performed at high dilution and above the monomer's ceiling temperature.

Oxidative addition polymerizations involving radical cations will be discussed in Chapter 7.

I. Cationic Ring Opening Polymerizations

Sufficiently strained heterocycles such as ethers, acetals, orthoesters, iminoethers, sulfides, amines, siloxanes, and phosphazenes are susceptible to ring opening polymerization under cationic conditions. Typical initiators include strong protonic and Lewis acids, onium and carbenium salts, and some covalent esters and halides. Polymerizations are generally performed in nonnucleophilic solvents such as hydrocarbons and chlorinated solvents, typically at ambient temperatures, although higher temperatures are sometimes necessary. However, more nucleophilic solvents can be used if the monomer itself is very nucleophilic. The elementary reactions of a cationic ring opening polymerization are shown in Eq. (29) using a tetrahydrofuran polymerization as an example.

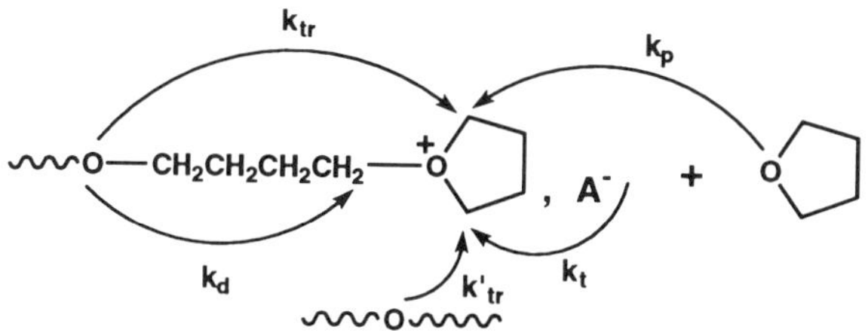

(29)

Propagation involves nucleophilic attack of monomer on the growing onium ion. The rates of propagation of ring opening polymerizations are generally slow, with k_p typically equal to 10^{-3}–10^{1} L/mol·sec^{-1}. Some propagation may also occur by stable ring-opened carbenium ions. For example, small amounts of stabilized carbenium ions are formed in the polymerizations of acetals, orthoesters (stabilization by α-alkoxy group), and 1,1-dimethyloxirane (stabilization by two α-methyl groups).

Alternatively, a heteroatom along the polymer backbone may attack the growing onium ion in a backbiting or an intermolecular chain transfer reaction (k_{tr}). Intramolecular backbiting endocyclic to the ring results in cyclic oligomers. For example, primarily 1,4-dioxane is formed when ethylene oxide is ring opened under cationic polymerization conditions. Depropagation (k_d) occurs by intramolecular backbiting at the last unit exocyclic to the ring. Termination (k_t) occurs by collapse of the ion pair or by attack of an impurity or added nucleophile.

J. Cationic Vinyl Polymerizations

Sufficiently nucleophilic alkenes such as vinyl ethers, styrenes, and isobutylene polymerize cationically to generate somewhat stabilized carbenium ions as the propagating species. Initiators include strong protonic acids and the electrophilic species generated by reaction of a Lewis acid with water, an alcohol, ester, or alkyl halide. The polymerization is usually performed in hydrocarbon and chlorinated solvents at low temperatures to suppress transfer by β-H elimination.

As demonstrated by styrene polymerization [Eq. (30)], propagation occurs in carbocationic polymerizations by electrophilic addition of the vinyl monomer to a growing carbenium ion. The resulting carbenium ions are very reactive and therefore difficult to control, with rate constants of propagation $k_p = 10^4$–10^6 L/mol·sec^{-1}. In addition, a significant amount of the positive charge is distributed at the β-H atoms, making them prone to abstraction by either monomer ($k_{tr,M}$) or counteranion (k_{tr}). β-Proton abstraction results in polymers with unsaturated end groups. However, due to their higher activation energies compared with propagation, these chain transfer reactions can be suppressed by polymerizing at lower temperatures [35]. In the case of styrene polymerizations, transfer also occurs by intramolecular Friedel-Crafts cyclization (k_c) to form polymers with indanyl end groups, especially at high conversion.

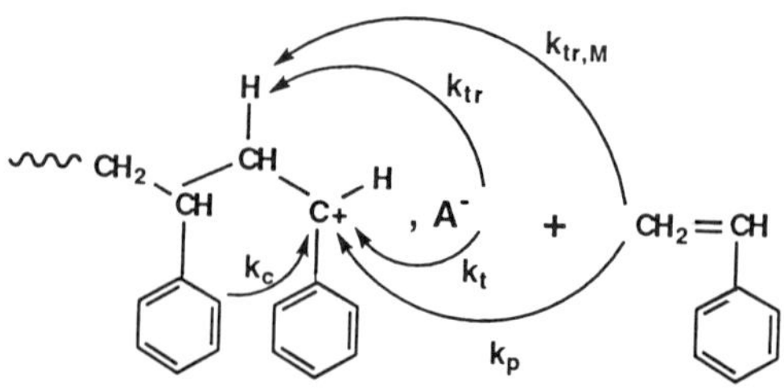

(30)

Termination (k_t) occurs by either formation of stable carbenium ions subsequent to β-proton elimination, or by reaction of the counteranion (X^- or MtX_n^-) with the growing carbenium ion. Depending on the nature of the ligand X, this process may be reversible [Eq. (31)].

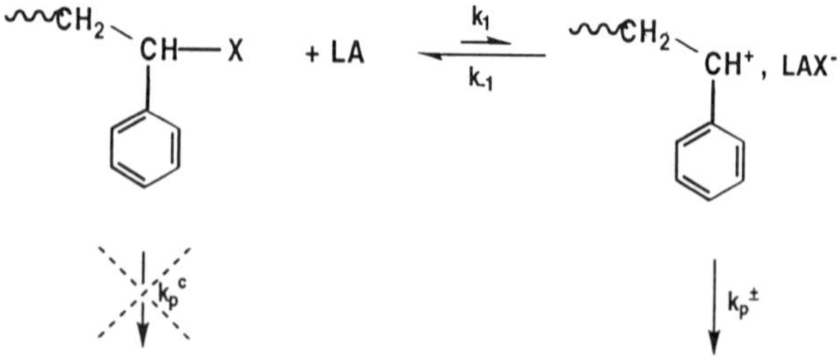

(31)

For example, anions containing chloride and bromide ligands such as $SnCl_5^-$, $SbCl_6^-$, $SnBr_5^-$, BCl_4^- and BBr_4^- usually decompose reversibly. The momentary concentration of propagating carbenium ions is much lower than that of the dormant covalent species, leading to lower polymerization rates and more controlled polymerizations. Reversible systems that exchange quickly in comparison with propagation provide "living" systems with narrow molecular weight distributions.

Because the elementary reactions of cationic alkene polymerizations are directly related to the organic chemistry of carbocations, Chapter 2 will investigate electrophilic additions to double bonds, nucleophilic substitution, electrophilic aromatic substitution, and elimination reactions.

III. MONOMERS FOR CATIONIC POLYMERIZATIONS

Cationic polymerizations are possible when the thermodynamic requirement of a negative change in free energy is met. Most polymerizations are exoentropic (approximately −120J/mol K) due to the loss of three degrees of translational freedom caused by connecting monomeric units together. Therefore, the thermodynamic feasibility generally requires that the polymerization is sufficiently exoenthalpic to more than compensate for the loss in entropy. This is supplied in ring opening polymerizations by the relief of ring strain, and in alkene polymerizations by isomerization of the double bond in the monomer to C—C single bonds in the polymer. Similarly, the driving force for the polymerization of five- and six-membered cyclic imino ethers which have little or no ring strain is provided by isomerization to repeating units containing the more stable carbonyl

(32)

double bond of an alkyl aminoacyl group [Eq. (32)]. However, because intramolecular nonbonded interactions are lower in both cyclic and alkene monomers than in the corresponding polymers, the change in enthalpy becomes less negative as the number and size of substituents increase. If the strain due to nonbonded interactions in the polymer is too high, ΔH may be endothermic or not sufficiently exothermic for the polymerization to occur.

In addition to the thermodynamic feasibility of a polymerization, there must also be a kinetic pathway for the polymerization to occur. A heteroatom in a cyclic monomer provides a site for coordination with an anionic, cationic, or coordination type of initiator [Eq. (33)]

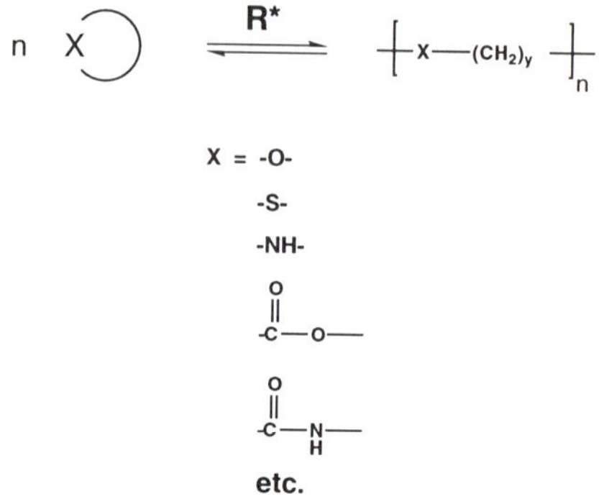

(33)

The double bond in alkenes provides a site for attack by reactive centers generated by an initiator. In addition, the electronic effects within the monomer must be such that the monomer is reactive and that the active sites generated are stabilized. For example, although the thermodynamic feasibility of a polymerization is independent of the mechanism, most monomers do not polymerize by all possible mechanisms. Cyclic monomers which are basic can be initiated by electrophilic species to generate stabilized onium ions. Similarly, alkenes containing electron-donating groups are nucleophilic and therefore able to react with electrophilic initiators to generate stabilized carbenium ions.

A. Alkenes

Alkenes polymerize cationically by electrophilic addition of the monomer to a growing carbenium ion. Therefore, the monomer must be nucleophilic and capable of stabilizing the resulting positive charge. In addition, the double bond must be the most nucleophilic functionality in the monomer. Some vinyl monomers which can polymerize cationically are listed in Eq. (34) in their order of reactivity, which corresponds to the electron-donating ability of their substituents.

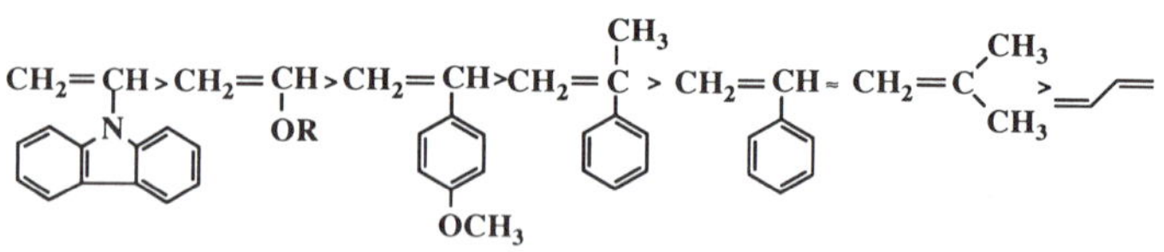

(34)

Due to resonance stabilization and their higher nucleophilicity, heteroatoms stabilize the growing carbenium ions better than alkyl and aryl groups do; *N*-vinyl carbazole is more reactive than vinyl ethers because of nitrogen's higher nucleophilicity. However, the reactivity of the growing carbenium ions follows the opposite order shown above, with the most stable carbenium ions being the least reactive. Although nitrogen is highly stabilizing and could potentially generate unreactive carbenium ions, its high stabilizing ability is apparently balanced in *N*-vinyl carbazole by the inductively electron-withdrawing aromatic rings.

The reactivity sequence shown above corresponds well to Mayr's [18] model reactions of the electrophilic addition of benzhydryl carbenium ions to substituted alkenes. Table 2 lists the second-order rate constants for the addition of a diarylcarbenium ion to various alkenes and dienes [36]. One alkyl group offers little activation of the double bond; α-olefins therefore form only oligomers with isomerized repeat units in low conversions under cationic polymerization conditions. One vinyl group activates the double bond slightly more than alkyl groups do. Table 2 also demon-

Table 2 Rate Constants (−70° C) for the Electrophilic Addition of *p*-Methoxydiphenylcarbenium Tetrachloroborate to Alkenes and Dienes

	k (L mol^{-1} sec^{-1})	k_{rel}
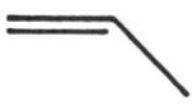	9.39×10^{-4}	1.0
	1.93×10^{-2}	21
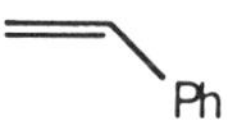	1.09×10^{1}	1.2×10^{4}
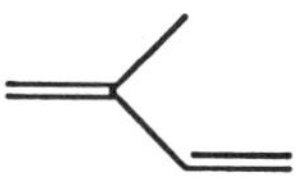	1.56×10^{1}	1.7×10^{4}
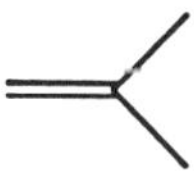	2.33×10^{1}	2.5×10^{4}
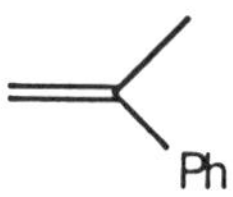	1.45×10^{3}	1.5×10^{6}

Source: Ref. 36.

strates that a phenyl ring stabilizes carbenium ions to about the same extent as two methyl groups or one methyl and one vinyl group. The stability scales of carbenium ions and the reactions they undergo will be covered extensively in Chapter 2.

The increase in the change in enthalpy of polymerization to less negative with increasing size; number of substituents is revealed in Table 3. A single small substituent on the olefin results in only a minor increase in ΔH_p, and propylene polymerizations are nearly as exothermic as that of ethylene. The change in enthalpy of polymerization is slightly lower for styrene due to stabilization of the monomer by conjugation of the phenyl ring with the alkene double bond. All three monomers have negligi-

Table 3 Enthalpies, Ceiling Temperatures, and Equilibrium Monomer Concentration for Bulk Polymerizations of Selected Monomers

Monomer	$-\Delta H$ (kJ/mol)	T_c (° C)	$[M]_e$ @ 25° C (mol/L)
	94	610	10^{-10}[a]
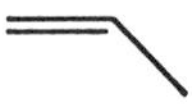	84	492	10^{-8}[a]
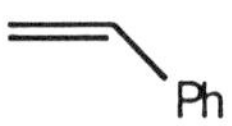	73	400	10^{-6}
	48	175	10^{-2}[a]
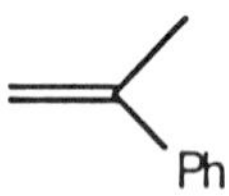	35	63	2.2
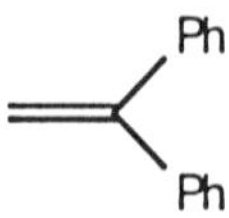	?		>15

[a] Calculated from ΔH and T_c.
Source: Ref. 37.

ble equilibrium monomer concentrations and ceiling temperatures which are higher than the corresponding polymer's thermal decomposition temperatures. It must be noted, however, that the values of the ceiling temperatures depend strongly on the method used for the estimation and vary for styrene from 400° C [37] to 310° C [38] and to 235° C [39] and for isobutene from 175° C [37] to 50° C [39]. Exoentropic polymerizations become reversible when $-\Delta H_p$ is small. The change in enthalpy of polymerization is noticeably smaller in α-methylstyrene and isobutylene polymerizations, which therefore have lower ceiling temperatures and significant equilibrium monomer concentrations. This is because both 1,1- and 1,3-intramolecular interactions are always higher in sp^3-hybridized polymers than in the corresponding sp^2-hybridized 1,1-disubstituted alkenes. For example, sp^3 hybridization results in angles of only approximately 109.5° between 1,1-substituents, rather than approximately 120° as in the sp^2-hybridized monomers. 1,1-Diphenylethylene does not homopolymerize although it does copolymerize and is often used to control initiation and/or termination by decreasing the reactivity of an active species.

The ceiling temperatures listed in Table 3 refer to bulk conditions, whereas the equilibrium monomer concentrations refer to polymerizations performed at 25° C. As discussed in Section F, the equilibrium monomer concentration varies with the polymerization temperature, just as the ceiling temperature depends on the monomer concentration. Table 4 demonstrates that the equilibrium monomer concentration of α-methylstyrene

Table 4 Equilibrium Monomer Concentration for α-Methylstyrene as a Function of the Polymerization Temperature

Temp. (° C)	$[M]_e$ (mol/L)
−80	<0.005
−40	0.060
−30	0.114
−20	0.235
−10	0.431
0	0.747
25	2.17
36	3.50
47	4.53
55	6.65
60	7.52

Source: Ref. 40.

increases with increasing polymerization temperature [40]. That is, less of the monomer is converted to polymer as the temperature increases.

B. Heterocycles

Heterocycles polymerize cationically by nucleophilic attack of the monomer on a cationic initiator or growing onium ion. Therefore, the monomer must be nucleophilic and capable of stabilizing the resulting positive charge. Most of the cyclic monomers that polymerize only cationically are shown in Eq. (35), along with heterocyclic monomers that polymerize using both cationic and anionic initiators. The general reactivity of a homologous series of rings of different sizes is $3 > 4 > 8\text{–}11 > 7 > 5 > 6$. That is, the rate constant of propagation increases with increasing ring strain for a homologous series of heterocycles. In general, six-membered rings are not strained and therefore do not polymerize. Some exceptions include the six-membered cyclic imino ethers, cyclic esters, and cyclic amides. As stated previously, the driving force for the polymerization of six-membered cyclic imino ethers or oxazines is isomerization to repeating units containing the more stable carbonyl double bonds. The six-membered lactones and lactams polymerize apparently because of strain induced by deviation of the —CO—*X*— unit from planarity, cis/trans equilibria, and the resulting decreased resonance stabilization [41]. For example, the amide group in up to the eight-membered lactams is forced to adopt the cis conformation which is 6 kJ/mol less stable than the trans form [42]. The reactivity of the homologous series of cyclic siloxanes and phosphazenes also does not follow the order shown above, with hexamethylcyclotrisiloxane (D_3) being more strained than octamethylcyclotetrasiloxane (D_4). The driving force for the polymerization of strainless rings such as D_4, S_8, hexachlorocyclotriphosphazene, and some cyclic esters of phosphoric acid is an increase in entropy caused by higher rotational and vibrational freedom in linear chains.

Cationic

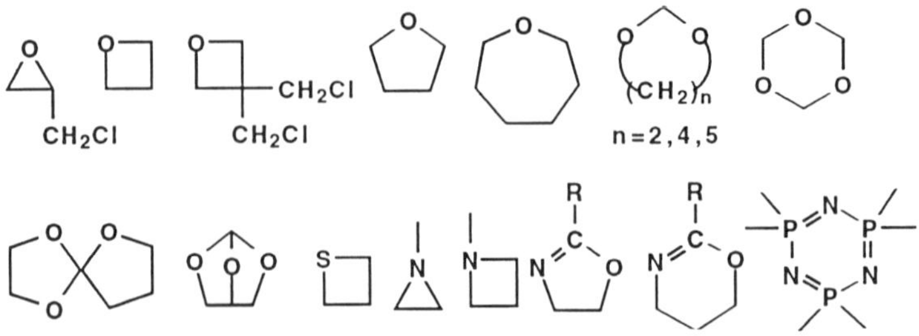

Cationic/Anionic

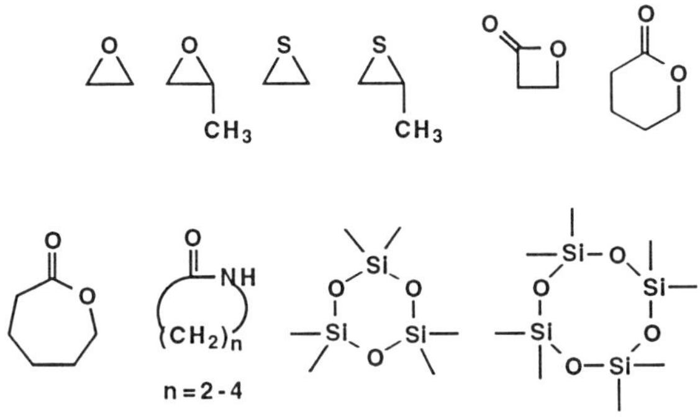

(35)

The kinetic polymerizability of cyclic monomers is also affected by the nature of the heteroatom and its electronegativity. Because the monomer acts as a nucleophile, the monomer reactivity of a series of rings with approximately the same ring strain increases as the nucleophilicity of the heteroatom increases. However, this stabilizes the resulting onium ions which therefore show the opposite order of reactivity. The overall kinetic polymerizability of a given monomer by cationic ring-opening polymerization follows the reactivity order of the onium ions and is therefore inverse to the monomer nucleophilicity. That is, the rate constants of propagation decrease in the following order for rings having approximately the same strain: orthoesters > acetals > ethers > sulfides > imino ethers > amines.

Table 5 demonstrates that substitution in the ring decreases monomer polymerizability because the intramolecular interactions between substituents are always higher in the polymer than in the cyclic monomer. Thus,

Table 5 Enthalpies, Entropies, Ceiling Temperatures, and Equilibrium Monomer Concentration for Bulk Polymerizations of Selected Heterocyclic Monomers

Monomer	$-\Delta H_p^0$ (kJ/mol)	$-\Delta S_p^0$ (J/K mol)	$[M]_e$ @ 25° C (mol/L)	T_c (° C)
Tetrahydrofuran	23	67	4.4	80
3-Methyltetrahydrofuran	23	101		4
1,3-Dioxolane	15	37	1.0	98
4-Methyl-1,3-dioxolane	13	53		−21
4-Isopropyl-1,3-dioxolane	12	61		−74
4,4-Dimethyl-1,3-dioxolane	10	55		−98

Source: Ref. 37.

although unsubstituted five-membered cyclic ethers (THF) and acetals (1,3-DXL) can be easily polymerized at room temperature, the corresponding methyl-substituted rings have much lower ceiling temperatures and most of the dimethyl-substituted rings do not polymerize at all under the same conditions. In the cases shown, neither a small methyl substituent nor a larger isopropyl group have much effect on the change in enthalpy of polymerization due to the great distance between substituents in the polymer; however, the change in entropy of polymerization decreases more significantly. It is worth noting that the thermodynamic parameters and the monomer equilibrium concentration depend also on solvent. In polymerization of THF at 25° C, $[M]_e$ = 3.1 mol/L in bulk ($[M]_0$ = 12.3 mol/L) but it increases to 3.7 mol/L in CCl_4, 3.9 mol/L in benzene, 4.8 mol/L in CH_2Cl_2, and 5.1 mol/L in CH_3NO_2, all at $[THF]_0$ = 8.0 mol/L [29]. Values of $[THF]_e$ increase with the proportion of the solvent added; for example, there is no polymerization of THF in nitromethane at $[THF]_0$ = 6.0 mol/L. The effect of solvent on $[M]_e$ can be explained by the differences in solvent-monomer and solvent-polymer interactions [43]. Substituents also decrease the polymerizability of highly strained rings. For example, the three-membered cyclic amine *N*-benzyl-2,3-dimethylaziridine does not homopolymerize, although it copolymerizes readily with less substituted aziridines [44], demonstrating that the lack of homopolymerization is solely a thermodynamic phenomenon.

The success of a cationic ring-opening polymerization and its ability to compete with other polymerization mechanisms depend on the functional groups in the ring. For example, higher molecular weight polymers are obtained by anionic polymerizations than by cationic polymerizations of rings with two reactive sites such as cyclic lactones, lactams, carbonates, anhydrides, and esters of phosphoric acid. Cationic polymerizations of oxiranes and cyclosiloxanes are also much more difficult to control than the corresponding anionic polymerizations because nucleophilic sites along the polymeric backbone lead to side reactions such as macrocyclization with highly reactive cationic intermediates.

IV. CATIONIC INTERMEDIATES IN ORGANIC REACTIONS AND IN POLYMER SYNTHESIS

Cationic intermediates are considered the active species in many organic reactions, as well as in cationic polymerizations. For example, cationic intermediates are postulated in both electrophilic addition and elimination reactions. They are also involved in electrophilic aromatic substitutions and in some nucleophilic aliphatic substitutions. The latter reactions may involve either onium or carbenium ions. The current understanding of cationic intermediates in organic reactions is the basis for understanding cationic polymerizations and therefore will be analyzed in detail in Chapter 2. It will only be covered briefly here.

Winstein et al. [45] first presented evidence for the concept that different types of electrophilic species, each with distinct reactivities, may participate in reactions involving cationic intermediates. As shown in Eq. (36), Winstein et al. proposed that four species are in equilibrium, including covalent electrophiles, contact ion pairs, solvent-separated ion pairs, and free ions. In addition, ion pairs may aggregate in more concentrated solutions. According to this concept, electrophilic species do not react with a continuous spectrum of charge separation, but rather in well-quantified minima in the potential energy diagram.

$$\begin{array}{ccccccc} & & \text{aggregates} \quad 1/n\left(\sim\sim R^{+}, X^{-}\right)_{n} & & & & \\ & & \Updownarrow & & & & \\ \sim\sim R\text{-}X & \rightleftharpoons & \sim\sim R^{+}, X^{-} & \rightleftharpoons & \sim\sim R^{+} \| X^{-} & \rightleftharpoons & \sim\sim R^{+} + X^{-} \\ \text{covalent} & & \text{contact ion pairs} & & \text{solvent separated ion pairs} & & \text{free ions} \end{array} \qquad (36)$$

Evidence for Winstein's concept is mostly indirect and is based on the overall kinetics of solvolysis reactions, the extent of racemization of optically active compounds, and the extent of scrambling in radiolabeled electrophiles. These studies were performed with a systematic variation in solvents, temperatures, concentrations, substituents, leaving groups, and additives. However, the proposed intermediates in this proposal have since been observed directly [21].

Contact and solvent-separated ion pairs can be distinguished in anionic systems; the interionic distance of the former is usually 1–3 Å, which increases to 4 or even 7 Å in solvent-separated ion pairs [21]. There is apparently no further minimum in the potential energy diagram. The reactivity of solvent-separated ion pairs and free ions in anionic systems are similar, being a few orders of magnitude more reactive than contact ion pairs. In contrast, contact ion pairs in cationic systems are separated by 4–6 Å, and therefore resemble the solvent-separated species of anionic systems in terms of structure, as well as their relative reactivity and ability to dissociate. The existence of solvent-separated ion pairs in cationic polymerization is questionable and has not yet been proven spectroscopically.

The overall rate of polymerization is determined by both the position of the equilibria between the covalent species and various cations, and by their respective reactivities. The dynamics of the exchange reactions among all of these species relative to the rate of propagation determines the molecular weight distribution of the resulting polymer.

A. Carbenium Ions

IUPAC recommends that cationic species with higher than their usual valency be denoted by the suffix ONIUM, and species with lower valencies be denoted by the suffix ENIUM [46,47]. Thus, R_3S^+ is a sulf*onium* ion, whereas RS^+ is a sulf*enium* ion. Although this recommendation was initially ignored by the organic community who continued to call R_3C^+ carbonium ions, the identification of R_5C^+ species has necessitated distinguishing between carbenium and carbonium ions. We will focus primarily on carb*enium* ions (R_3C^+) because carb*onium* ions (R_5C^+) are generally not involved in reactions relevant to polymer synthesis. It is recommended to use the term *carbocation* if the precise structure of C-based cation is not known, for example in the case of bridging or complexation.

Carbenium ions have also been labeled with the suffix YLIUM added to the structure of the parent radical. Thus, Ph_3C^+ is called either a triphenylmethylium or a triphenyl carbenium ion; trityl is also commonly used. Similarly, diphenylmethylium, diphenyl carbenium and benzhydryl all denote Ph_2CH^+. Phenylmethylium, phenyl carbenium, and benzyl denote $PhCH_2^+$. Trimethylmethylium, trimethyl carbenium and *t*-butyl denote $(CH_3)_3C^+$. Analogously, $CH_3CH(Ph)^+$ can be called a 1-phenylethylium cation, a methylphenyl carbenium ion or a styryl cation. The cation $CH_3OCH_2^+$ is methoxymethylium, or in general, an alkoxy carbenium ion.

As shown in Eq. (37), there are several methods of generating carbenium ions (general formula R^+, X^-) from a variety of substrates. One method of generating carbenium ions is ionization of covalent species. This may be a unimolecular reaction which occurs spontaneously if the resulting carbenium ion is sufficiently stable, if the leaving group is sufficiently nucleofugic and/or if the solvent has sufficient ionizing power. Alternatively, ionization of the covalent species may require activation by a protonic or Lewis acid. If carbenium ions are generated by protonating or alkylating an alcohol, ether, ester or sulfide, ionization proceeds through onium ion intermediates. Protonation of alkenes also yields carbenium ions. Finally, carbenium ions can be generated by hydride or alkyl anion abstraction, and new carbenium ions can be formed by rearrangements and addition of other carbenium ions to alkenes.

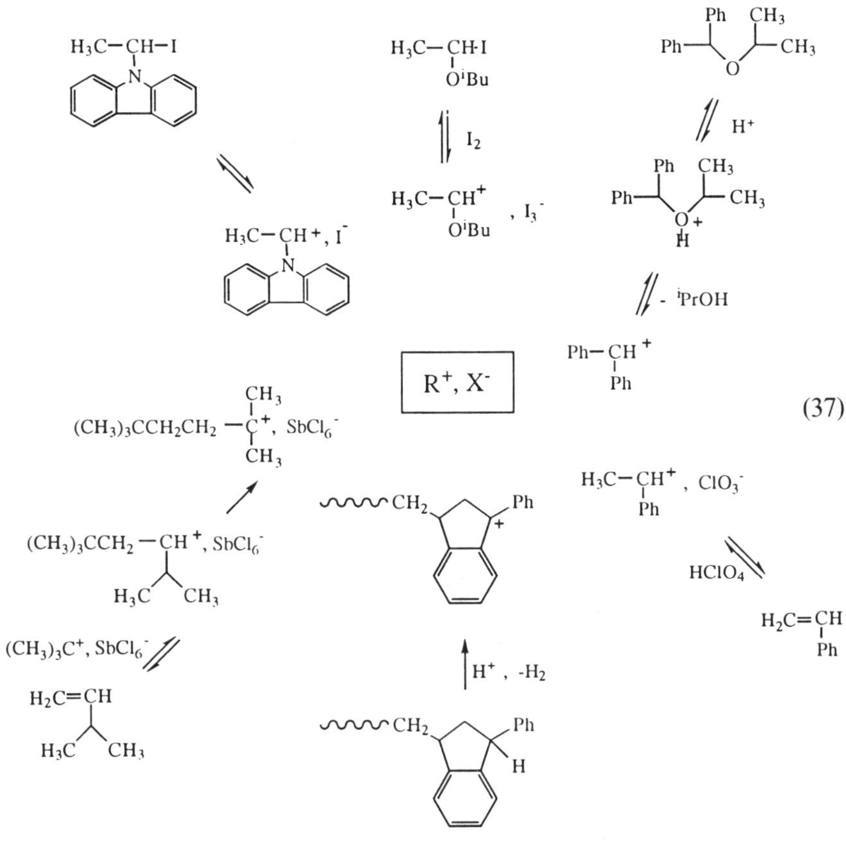

(37)

Carbenium ions are usually unstable, very reactive species which readily rearrange or decompose by β-proton elimination. The propensity for such reactions are decreased by stabilization through inductive effects, through resonance and/or by hyperconjugation. Alkyl substituents are electron donating and stabilize carbenium ions inductively. The carbenium ion stability increases with increasing substitution, and tertiary carbenium ions are much more stable than secondary, which are more stable than primary ions; all are more stable than the unsubstituted methylium ion.

Substituents with α-heteroatoms such as α-alkoxy, thio, halo, and amino groups stabilize through resonance to form, for example, oxonium ($R—O^+{=}CH_2$) or halonium ($^+Br{=}CR_2$) ions. Aryl groups, especially those with electron-donating substituents, also delocalize the positive charge through resonance, with the stability of carbenium ions increasing with the number of aryl substituents: $Ar_3C^+ \gg Ar_2CH^+ \gg ArCH_2^+$. Compounds such as crystal violet [*tris*(*p*-dimethylaminophenyl)methylium] or malachite green [bis(*p*-dimethylaminophenyl)phenylmethylium] are stable even in aqueous media.

Hyperconjugation (38) stabilizes by overlap of the σ-orbital of the C—H_β bond and the carbenium ion's vacant orbital. It is responsible for the unusually high stability of

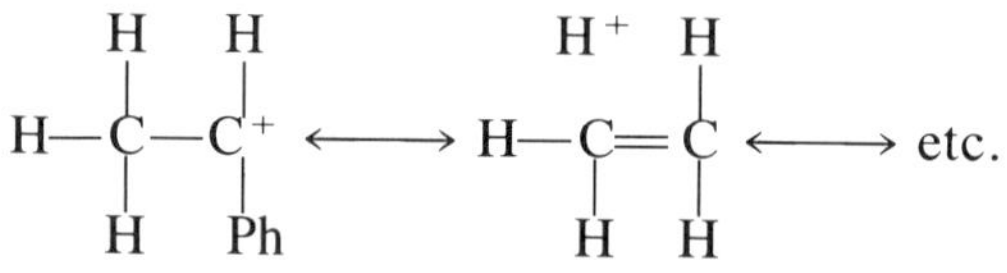

(38)

t-butyl cations [14], which have nine β-hydrogens, and for the fact that two methyl groups stabilize a carbenium ion to about the same extent as one phenyl group (Chapter 2). Unfortunately, hyperconjugation also increases the positive charge on β-H atoms. For example, only 20–25% of the positive charge is concentrated on the sp^2-hybridized carbenium carbon in both cumyl ($PhMe_2C^+$) and styryl ($PhMeHC^+$) ions, whereas as much as 10% of the positive charge is on the β-hydrogens [48]. This increase in the positive charge on β-hydrogens due to hyperconjugation facilitates β-proton elimination.

The stability of a carbenium ion can be described by either its thermodynamic or its kinetic stability. The thermodynamic stability can be estimated using MNDO, AM1 and ab initio calculations and experimentally using thermochemistry. The thermodynamic stability of carbenium can be correlated with rates of solvolysis, proton affinities, electrochemistry, and from exchange reactions such as that shown in Eq. (39) and discussed in more detail in Chapter 2.

$$R_3C{-}X + R'_3C^+ \rightleftharpoons R_3C^+ + R'_3C{-}X. \qquad (39)$$

The kinetic stability describes a carbenium ion's lifetime in terms of its rate of decomposition. That is, although the methylium cation is thermodynamically very unstable, it can not decompose and is therefore kinetically stable. However, it reacts extremely quickly with even the weakest of nucleophiles due to its thermodynamic instability.

In addition to rearranging and undergoing β-proton elimination, a carbenium ion will interact with any available nucleophile to fill its open valency. If the Π-electrons of an alkene or arene are involved in the nucleophilic attack, a Π-complex forms which eventually leads to electrophilic addition or aromatic electrophilic substitution, respectively. Onium ions are formed by reaction with the p-electrons of nucleophiles. Covalent species are formed by reaction of carbenium ions with an anion. All such reactions are demonstrated in Eq. (40).

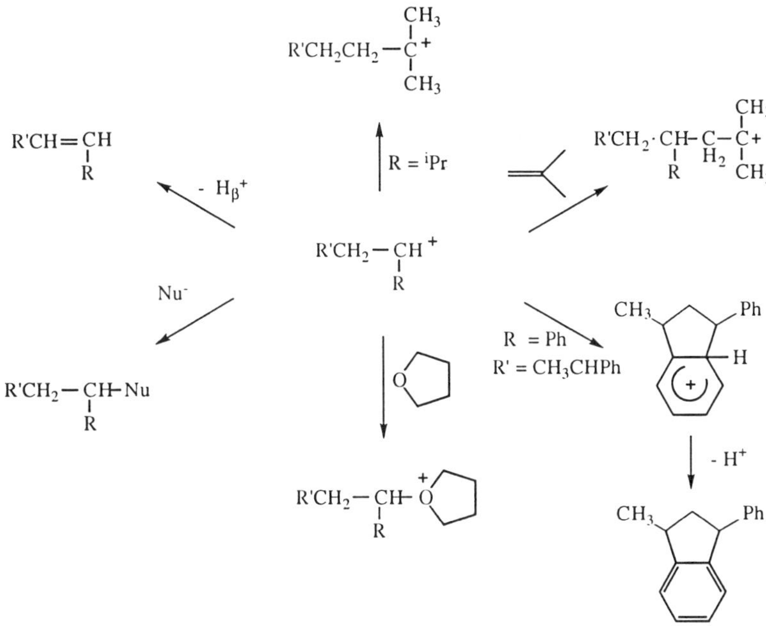

(40)

The equilibrium between carbenium ions and either onium ions or covalent species depends on the stability of the carbenium ion and the nucleophilicity of the nucleophile. Although anions are stronger nucleophiles than their corresponding neutral species, phosphines, amines, and sulfides are much more reactive than anions such as triflate, perchlorate or tosylate, which are relatively nonnucleophilic. In fact, the nucleophilicity of triflate is comparable to that of ethers [49]. Complex MtX_{n+1}^- anions based on strong Lewis acids are even less nucleophilic. Although anions with fluoride ligands are the most stable, their decomposition to unreactive alkyl fluorides is irreversible, leading to a true termination in polymerization reactions. In contrast, complex anions based on chlorides, bromides, and iodides are less stable, but their alkyl halide decomposition products can be reactivated with Lewis acids, leading to reversible termi-

$$\underset{\displaystyle |\atop \displaystyle O^{i}Bu}{CH_3-CH^+}, ZnCl_3^- \rightleftharpoons \underset{\displaystyle |\atop \displaystyle O^{i}Bu}{CH_3-CH-Cl} + ZnCl_2 \qquad (41)$$

nation [Eq. (41)]. Polymerizations involving reversible termination are slow, but steady and controllable.

Carbenium ions are usually sp^2 hybridized, and should therefore be planar with angles of approximately 120° between substituents. Although interaction with a solvent should not change this hybridization, it may affect the rate constants of various reactions by changing the charge distribution or by altering the accessibility of the carbocationic center. Nevertheless, the rate constants of addition of benzhydryl cation to an alkene is increased only less than five times by changing solvent from chloroform to nitromethane [18]. In contrast, formation of onium ions changes the hybridization on carbon from sp^2 to sp^3, and moves the charge away from the carbon atom to a heteroatom and its substituents. This also happens when covalent species are formed. However, there are significant energy barriers to changes in hybridization. Reactions of carbanionic species do not involve such energy barriers because the sp^3-hybridized carbanions react with electrophiles to form sp^3-hybridized covalent species.

Dissociation of ion pairs to free ions is a physical process governed by electrostatic forces. The dissociation constant is regulated by the size of the counterions, temperature, and dielectric constant of the media. Because the counterions in cationic systems are usually much larger than those in anionic systems, the dissociation constants are correspondingly larger. However, these have only been determined in a few model systems, all of which involve very stable and bulky carbenium ions such as trityl, [50–52], benzhydryl [53], and tropylium [51] ions. More reactive carbenium ions such as cumyl or styryl are not stable enough to generate high enough concentrations to be detected by conductivity measurements. They readily undergo side reactions of β-proton elimination and intramolecular electrophilic aromatic substitution. Although alkoxycarbenium ions are more stable and serve as models of the growing species in vinyl ether polymerizations, reliable dissociation constants have not been determined yet [54]. However, dissociation constants of the growing species can be extrapolated from the kinetics of polymerizations performed in the presence of salts with common counteranions. Dissociation constants for ion pairs in cationic polymerization of styrene in CH_2Cl_2 solution are in the range of $K_D \approx 10^{-6}$ mol/L, which is 10–100 times smaller than those of trityl cations. This may be due to the smaller interionic distance within the growing ion pair, its nonsymmetrical structure, or simply due to experimental inaccuracy.

B. Onium Ions

Onium ions are cationic species with expanded valencies. The most important onium ions relevant to cationic polymerizations are oxonium, sulfonium, ammonium, and phosphonium ions. Ammonium, phosphonium, and sulfonium ions are tetrahedral with angles of approximately 109.5° between substituents. These species may be chiral if all

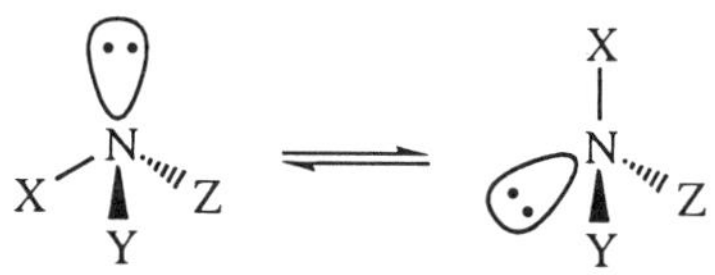

(42)

substituents are different and their pyrimidal inversion is slow. Whereas amines preserve their chirality for less than seconds [Eq. (42)], phosphines can preserve their chirality for several days [55]. Sulfonium ions with three different substituents are also chiral due to slow pyramidal inversion [56]. Oxonium ions are tetrahedral; their pyramidal structure and inversion have been detected by low temperature nuclear magnetic resonance studies [57].

Onium ions can be generated by alkylating or protonating the corresponding ethers, sulfides, amines, and phosphines. Alkylation produces tertiary onium ions. Secondary oxonium ions are produced by protonating an ether or by alkylating an alcohol. Protonation of the alcohol and alkylation of water will generate primary oxonium ions. Onium ions can also be generated by oxidation reactions. New onium ions are generated by rearrangements and by reaction of onium ions with some nucleophiles. For example, cationic propagation occurs in ring-opening polymerizations by reaction of cyclic onium ions with the heterocyclic monomer to regenerate the same type of onium ion. If a noncyclic species reacts with the cyclic onium ion, transfer or termination results, depending on the reactivity of the new onium ion. All such routes are shown in Eq. (43) in which, for oxonium and sulfonium ions, one of the R substituents is an electron pair.

Aromatic onium ions are a unique class of onium ions which are photoactive and decompose in the presence of light to generate radical-cations [58]. These radical-cations react with solvent or residual moisture to release protons that initiate cationic polymerization [Eq. (43)]. Thus, aromatic onium ions are used as cationic photoinitiators in both carbocationic olefin and ring-opening polymerizations, especially in photocuring and photolithographic processes.

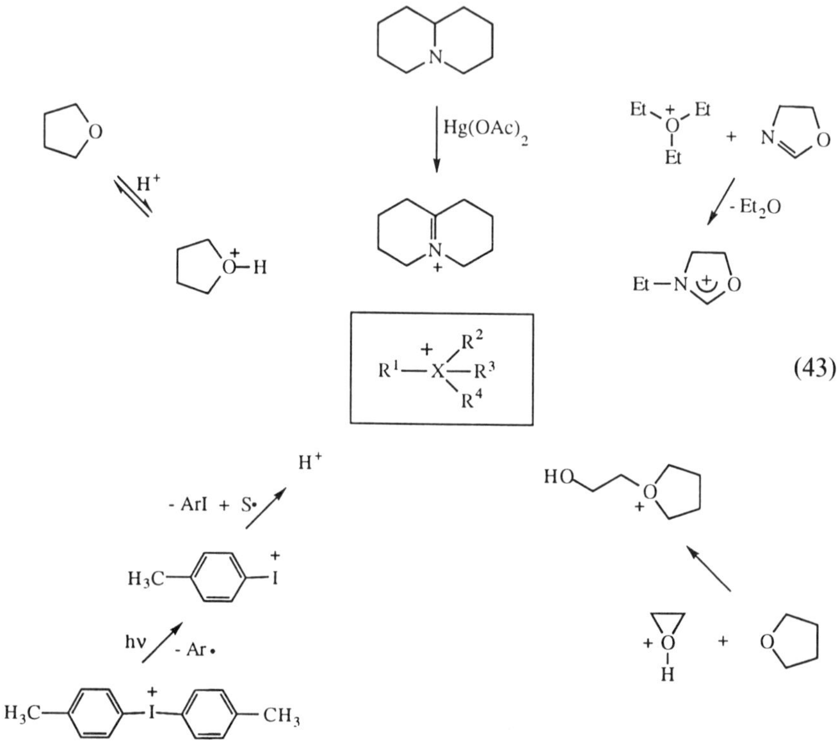

(43)

The stability of onium ions correlates directly with the nucleophilicity of the parent species; the nucleophilicity order of heterocyclic monomers as correlated with their basicity is shown is shown in Eq. (44) [59]. Thus, O-based rings which are the least nucleophilic of the four types of pre-

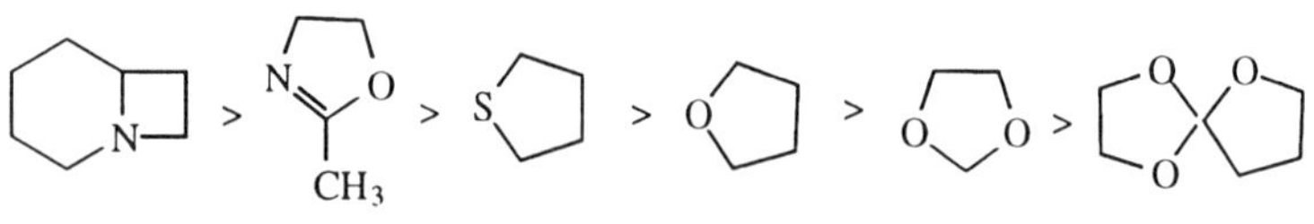

(44)

cursors, form unstable and reactive oxonium ions. The more nucleophilic sulfides generate more stable and less reactive sulfonium ions. Amines and iminoethers are the most nucleophilic precursors, forming very stable and unreactive onium ions. Ammonium ions are stable in water and even in the presence of strong nucleophiles such as chloride or bromide anions. Stronger nucleophiles are obviously required to react with less reactive onium ions. Thus, sulfonium ions do not react with THF and other ethers although oxonium ions react rapidly with sulfides.

Although the formal charge is usually placed on the O, S, N, or P heteroatom of the onium ion, most of the positive charge apparently resides on the α-C and β-H atoms. Nevertheless, onium ion formation greatly reduces the heteroatom's electronegativity. For example, the negative charge on oxygen in THF is reduced from −25% to −2% by conversion to a 1-methyltetrahydrofuranium ion [60]. Compared to carbenium ions, less of the positive charge resides on the β-H atoms, thereby reducing the propensity for β-proton elimination in ring opening polymerizations. Onium ions undergo other side reactions such as Hoffman elimination at higher temperatures in the presence of strong bases.

As stated in the previous section, the counterions in cationic systems are usually much larger than those in anionic systems. For example, the ionic radius of F^-, which is the smallest possible counteranion, is 1.4 Å. This is larger than that of most countercations in anionic polymerizations, including K^+, and is nearly as large as that of Rb^+. The ionic radii of these and other common counteranions [11] in cationic polymerizations are listed in Table 6, along with those of common countercations [61] in anionic polymerizations. The large size of both the counteranions and the onium ions combined with the onium ions' broad charge distribution lead to loose ion pairs with large dissociation constants. Dissociation constants are typically $K_D \approx 10^{-5}$ to 10^{-4} mol/L at room temperature in methylene chloride [11]. Because the ion pairs are loose, their reactivities are very similar to those of free ions and usually do not depend on the structure of the counteranions.

Table 6 Interionic Radii (Å) of Counteranions and Countercations

F^-	1.4	Li^+	0.68
Cl^-	1.81	Na^+	0.97
Br^-	1.96	K^+	1.33
SbF_6^-	2.10	Rb^+	1.47
BF_4^-	2.30	Cs^+	1.67
ClO_4^-	2.40		
$CF_3SO_3^-$	2.96		
$SbCl_6^-$	3.00		

Sources: Refs. 11 and 61.

Most ammonium and phosphonium ions are very stable and do not dissociate unimolecularly to the corresponding carbenium ion. However, unimolecular decomposition is possible if the resulting carbenium ion is unusually stable, as is true with trityl and alkoxycarbenium ions [52,62–66]. Because primary carbenium ions are too unstable to be generated by onium ions, most ring-opening polymerizations propagate only through onium ions. Again, the only exceptions are those cases that generate stable carbenium ions as in ring-opening polymerizations of orthoesters, acetals, and substituted oxiranes. In fact, the resulting carbenium/onium ion equilibrium is dominated by carbenium ions in orthoester polymerizations [67].

In addition, there may be an equilibrium in ring-opening polymerizations between onium ions and covalent species. This equilibrium has been observed in the polymerizations of cyclic ethers [29], oxazolines [68], and phosphorus-containing monomers [69]. The position of the equilibrium and its dynamics depend on the relative nucleophilicities of the monomer and the counteranion, as well as on the nucleofugacity ot the leaving group.

C. Covalent Electrophiles

Covalent species such as alkyl esters and halides may be in equilibrium with either onium or carbenium ions [Eq. (45)]. However, only heterocyclic monomers are

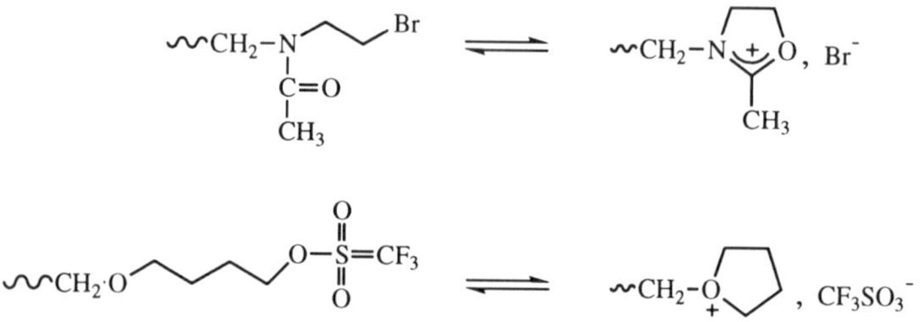

(45)

nucleophilic enough to react directly with covalent electrophiles in covalent propagation. This covalent propagation is similar to the Menshutkin reaction, in which tertiary amines react with alkyl halides to form quaternary ammonium salts. The less nucleophilic alkenes will react with covalent species only after they ionize to carbenium ions [Eq. (46)].

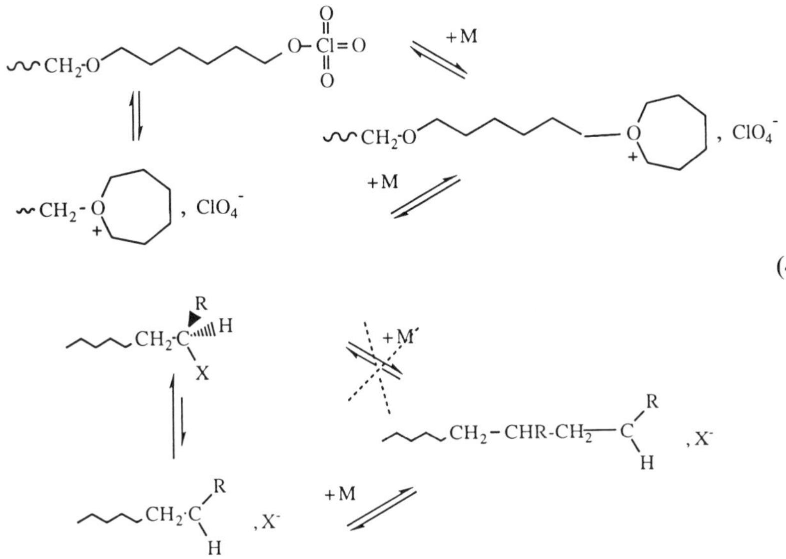

(46)

V. BASIC ORGANIC REACTIONS WITH ELECTROPHILIC ACTIVE CENTERS

All four of the elementary reactions in a cationic polymerization involve electrophilic or cationic intermediates. Thus, initiation, propagation, transfer, and termination may be classified as either nucleophilic substitution, electrophilic addition, elimination, rearrangement, or possibly as a pericyclic reaction. *Initiation* occurs in alkene polymerizations by either addition of acid to the alkene, or by ionization of a covalent initiator followed by addition of the resulting carbocationic intermediate to an olefin's double bond. Although initiation is an electrophilic addition (Ad_E) reaction in

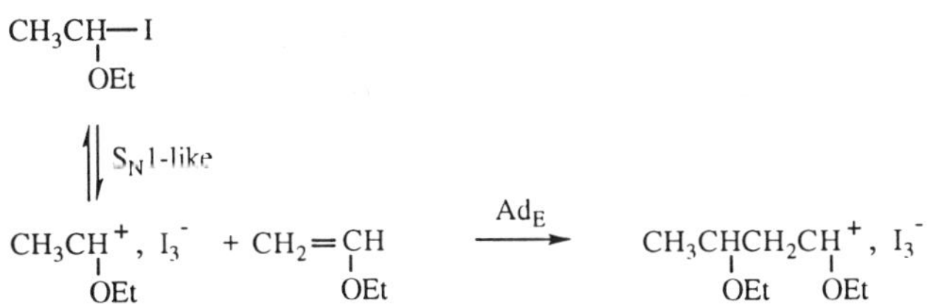

(47)

both cases [Eq. (47)], the latter resembles S_N1 reactions in which the covalent initiator ionizes in the rate-determining step. In contrast, initiation in cationic ring-opening polymerizations is usually by an S_N2 reaction of a covalent species with a heterocyclic monomer [Eq. (48)].

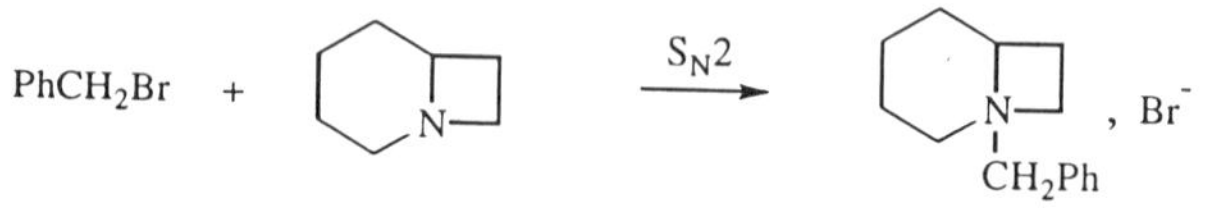

(48)

The polymer chain grows by *propagation* with regeneration of the same type of active species. Thus, propagation is an Ad_E reaction in carbocationic alkene polymerizations [Eq. (49)], and an S_N2 reaction in most ring-opening polymerizations,

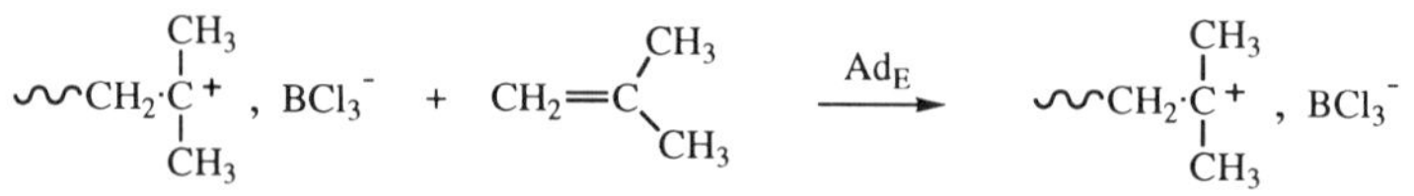

(49)

although propagation after isomerization of the onium ion to a carbenium ion might have partial S_N1 character.

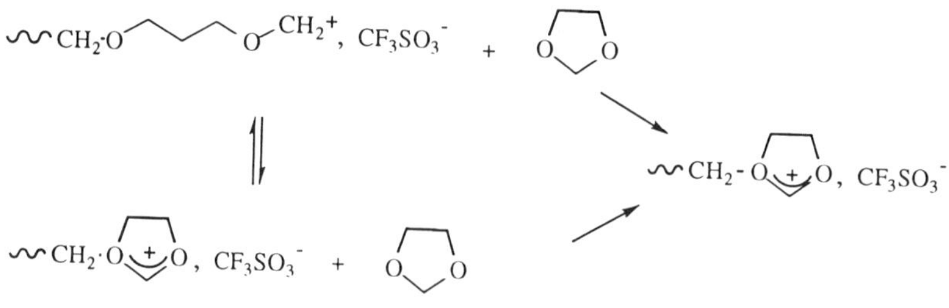

(50)

Alternatively, covalent propagation may proceed by a pericyclic reaction involving a multicenter rearrangement such as a group transfer polymerization; no examples of these type of reactions have been reported so far.

The active species are destroyed in *termination*. As discussed in Section A, this may occur by rearrangement of less stable carbenium ions to more stable and unreactive carbenium ions, by nucleophilic attack on carbenium or onium ions to form nonstrained onium ions, or by reaction of a nucleophile such as the counteranion with a carbenium or onium ion to form inactive covalent species. Carbenium ions also terminate by hydride abstraction from some other molecule in the system.

Although the active site of a growing chain is also destroyed by *transfer* reactions, a new species is generated that is reactive enough to reinitiate polymerization. This process may involve one or more steps. The most frequent transfer reaction in carbocationic alkene polymerizations is β-proton abstraction from the carbenium ion by either the counteranion, solvent, monomer, polymer, or some additive or impurity. Polymerizations of styrenes also transfer by intramolecular electrophilic aromatic to form

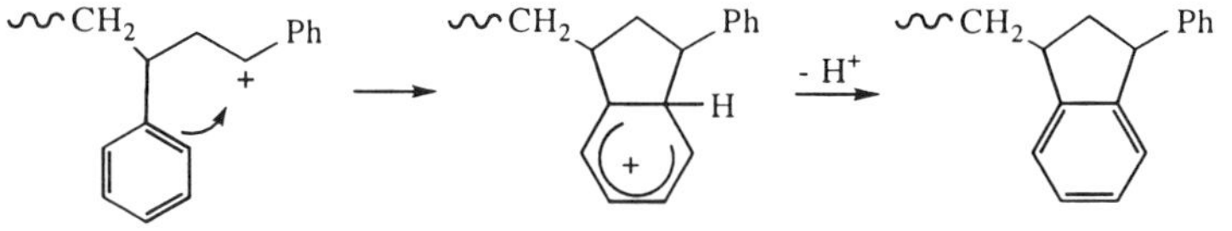

(51)

polymers with indanyl end groups [Eq. (51)]. Heteroatoms along the polymer backbone participate in transfer reactions in ring-opening polymerizations most often by either an intramolecular or intermolecular reaction.

High chemoselectivity is required to successfully control a polymerization in terms of the resulting polymer's molecular weight, chemical structure, and molecular architecture. The chemoselectivity can be defined by the ratio of the rates of the desired propagation and the undesired chain-breaking reactions of transfer and/or termination. That is, this ratio determines the limiting molecular weight of a given polymerization system. The regioselectivity of a polymerization controls the distribution of head-to-tail and head-to-head structures [Eq. (52)]. Most cationic vinyl polymerizations are highly

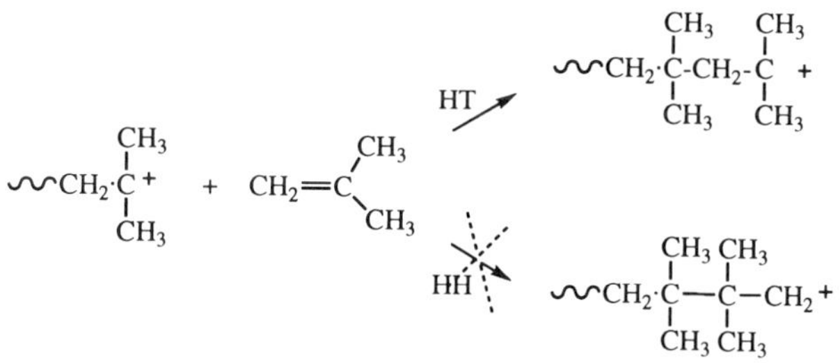

(52)

regioselective due to predominant Markovnikov addition and much higher stability of tertiary and secondary carbenium ions compared to primary carbenium ions. The regioselectivity of ring-opening polymerizations is determined by the extent of S_N1 vs. S_N2 propagation. Most onium ions react by S_N2 reactions with preferential attack at the least hindered and therefore the least substituted carbon. If propagation occurs by an S_N1 reaction in which the oxonium ion first isomerizes to a carbenium ion, it would generate the most stable species resulting in attack at the most substituted position. This occurs with some substituted oxiranes such as 2,2-dimethyloxirane capable of generating very stable carbenium ions. If both reactions occur then the H-H and H-T units can be formed.

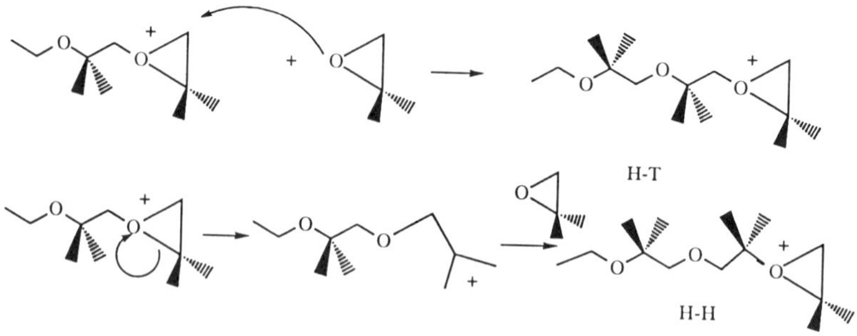

(53)

Finally, the stereoselectivity of most cationic vinyl polymerizations is poor due to the sp^2 hybridization of carbenium ions at carbon. In this case, attack from either side of the plane has approximately equal probability leading to similar proportions of meso and racemic diads. For the sufficiently bulky substituent(s) tacticity control improves; for example, highly syndiotactic (>90%) poly(α-methylstyrene) can be prepared by cationic polymerization [70].

VI. SYNTHETIC ASPECTS OF CATIONIC POLYMERIZATIONS

Many polymers such as poly(vinyl ethers), polyisobutylene, poly(tetrahydrofuran), linear poly(ethylene imine), and poly(1,3-dioxolane) can only be prepared cationically. Nevertheless, other polymers which can be prepared cationically may be synthesized more easily by an alternative mechanism. For example, solution, suspension, and emulsion radical polymerizations of styrene yield high molecular polymers; well-defined polystyrene is best prepared by an anionic polymerization. Cationic polymerization of styrene is limited to the preparation of low molecular weight polymers. However, this is the preferred route to low molecular weight polystyrene functionalized with electrophilic end groups which are sensitive to nucleophilic attack, and to block copolymers with polyisobutylene or poly(vinyl ether) segments. Although well-defined poly(dimethylsiloxane) can be better prepared anionically due to competing macrocyclization in the cationic route, cationic polymerization of cyclic siloxanes with Si—H bonds is the only possible route due to the cleavage of Si—H bonds by nucleophiles. High molecular weight poly(ethylene oxide) and other polyoxiranes can also be prepared relatively easily by anionic and coordinative mechanisms. In contrast, cationic polymerization of oxiranes is accompanied by dimerization and/or cyclooligomerization. Cyclooligomerization is minimized in cationic polymerizations via an "activated monomer" mechanism, which occurs in the presence of an alcohol as initiator and relatively high concentrations of protonic acid [Eq. (54)].

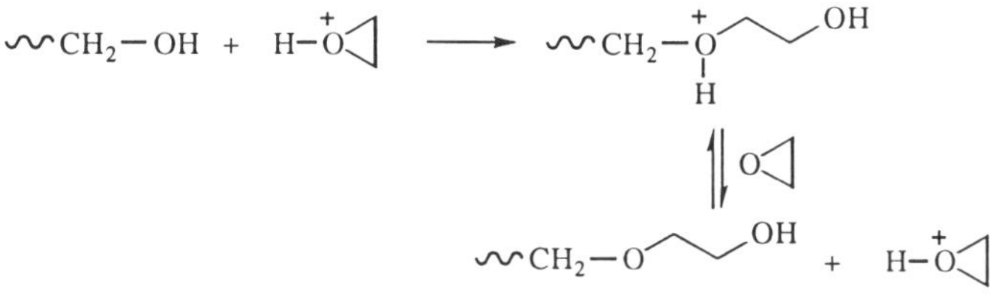

(54)

Because propagation via onium ions competes at lower acid concentrations, the activated monomer mechanism is limited to the synthesis of lower molecular weight polyoxiranes.

Nevertheless, well-defined polymers can sometimes be prepared by a cationic mechanism. These new systems are usually described as "living," even if chain-breaking reactions are detected and determined quantitatively. Polymers with narrow polydispersities and degrees of polymerization determined by the ratio of the concentrations of reacted monomer to initiator can be prepared if the attempted molecular weight is sufficiently low. In principle, degrees of polymerization less than 10% of the ratio of the rate constants of propagation and chain-breaking reactions ($DP_n < 0.1\, k_p/k_{t(tr)}$) can be achieved. In addition to limiting the polymerization to low molecular weights, suppression of chain-breaking reactions may require low temperatures and low levels of impurities such as moisture. The requirements for living polymerizations will be discussed in detail in Chapter 4.

The synthesis of polymers with predetermined molecular weights and narrow polydispersities is the most fundamental advantage of controlled and/or living polymerizations. Controlled polymerizations also enable efficient end functionalization and block copolymerization. Chapter 5 describes three general synthetic methods for incorporating functional groups into polymers. Functional groups may be present not only in the monomer, but may also be introduced via initiation with a functional initiator or by termination with a functional end-capping agent. Functional groups which react with electrophilic propagating species may be introduced if they are protected, although this is less necessary if the group is only present in an end-capping agent. For example, vinyl ethers functionalized with primary amino groups can be polymerized cationically if the amine is protected as inside. Hydroxy groups can be protected by conversion to silyl ethers, etc. Deprotection during or after termination then regenerates the desired terminal and/or pendant functionality.

Diblock, triblock, and multiblock copolymers are typically prepared by sequential monomer addition to polymerization systems in which the chain-breaking reactions are sufficiently suppressed. Polymer properties can thereby be varied by manipulating the constituent blocks' compatibilities, hydrophilicities/hydrophobicities, and hardness/softness. New and/or novel topologies can also be prepared by controlled processes, including cyclic polymers and/or copolymers, comb-like macromolecules, and star polymers. The synthetic range of cationic vinyl polymerizations will be discussed in detail in Chapter 5.

VII. INDUSTRIAL IMPORTANCE OF CATIONIC POLYMERIZATIONS

A number of industrially important polymers are produced by cationic polymerizations. Polymers produced by cationic ring-opening polymerizations include poly(tetramethylene glycol), poly(epichlorohydrin), poly(dimethylsiloxane), poly(ethylene imine), and stabilized poly(oxymethylene). For example, oligomeric poly(tetramethylene glycol) is produced by DuPont, BASF, and others. It is used extensively in the production of thermoplastic elastomers, especially polyesters, and polyurethanes. Epichlorohydrin and the larger cyclic ethers can only be polymerized cationically and in some cases by coordinative polymerizations. Higher molecular weight poly(dimethylsiloxane) is sometimes prepared by cationic polymerization of the cyclic tetramer, D_4. This polymer has a very low Tg, and is thus useful as oils or rubbers, as well as for biomedical applications. Polysiloxanes with Si—H bonds can only be prepared cationically. Poly[oxy(2,2-dichloromethyltrimethylene)] (pBCMO) is a crystalline thermoplastic which was produced previously by Hercules under the trade name Penton. Cationic polymerization of ethylene imine results in highly branched polymers which are water-soluble and useful as flocculants (BASF), whereas linear pEI is produced by hydrolysis of cationically produced poly(oxazoline). Poly(ϵ-caprolactam) or Nylon-6 is usually produced by a hydrolytic polymerization that involves cationic intermediates.

Polymers produced by cationic vinyl polymerizations include poly(*N*-vinylcarbazole) and poly(vinyl ether). However, polyisobutylene and its copolymer with isoprene (butyl rubber) is probably the most important commercial polymer produced by a cationic polymerization. Other industrial polymers such as poly(styrene) can be prepared by cationic polymerization, although they are usually produced radically or anionically. Many low molecular weight polymers produced by cationic polymerizations of indene, vinyl ethers, and α-methylstyrene are used as adhesives. Future applications are expected to include coatings and specialty applications requiring functionalized polymers or block and graft copolymers. The industrial aspects of cationic polymerizations will be covered extensively in Chapter 8.

REFERENCES

1. H. R. Allcock and F. W. Lampe, *Contemporary Polymer Chemistry*, Prentice-Hall., Englewood Cliffs, New Jersey, 1990.
2. G. Odian, *Principles of Polymerization*, Wiley, New York, 1991.
3. P. Rempp and E. W. Merrill, *Polymer Synthesis*, Huthig & Wepf Verlag, Basel, 1991.
4. P. M. Stevens, *Polymer Chemistry. An Introduction*, Oxford University Press, New York, 1990.
5. G. Allen and J. C. Bevington, *Comprehensive Polymer Science*, Pergamon Press, Oxford, 1989.
6. P. H. Plesch, *The Chemistry of Cationic Polymerisation*, Pergamon Press, Oxford, 1963.
7. A. Gandini and H. Cheradame, *Adv. Polym. Sci. 34/35*:1 (1980).
8. J. P. Kennedy, *Cationic Polymerization of Olefins: A Critical Inventory*, Wiley, New York, 1975.
9. J. P. Kennedy and E. Marechal, *Carbocationic Polymerization*, Wiley, New York, 1982.
10. J. P. Kennedy and B. Ivan, *Designed Polymers by Carboctionic Macromolecular Engineering, Theory and Practice*, Hanser, Munich, 1992.
11. S. Penczek, P. Kubisa, and K. Matyjaszewski, *Adv. Polym. Sci.*, *37*:1 (1980).
12. S. Penczek, P. Kubisa, and K. Matyjaszewski, *Adv. Polym. Sci.*, *68/69*:1 (1985).
13. K. J. Ivin and T. Saegusa, *Ring-Opening Polymerization*, Elsevier, Essex, 1984.
14. P. Vogel, *Carbocation Chemistry*, Elsevier, Amsterdam, 1985.
15. H. Perst, *Oxonium Ions in Organic Chemistry*, Verlag Chemie, Weinheim, 1971.
16. G. A. Olah, *Friedel-Crafts and Related Reactions*, Wiley, New York, 1963.
17. G. A. Olah and P. v. R. Schleyer, *Carbonium Ions*, Wiley, New York, 1970.
18. H. Mayr, *Angew. Chem.*, *29*:1371 (1990).
19. T. H. Lowry and K. S. Richardson, *Mechanism and Theory in Organic Chemistry*, Harper & Row, New York, 1987.
20. M. Szwarc, *Carbanions, Living Polymers and Electron Transfer Processes*, Interscience Publishers, New York, 1968.
21. M. Szwarc, *Ions and Ion Pairs in Organic Reactions*, Wiley, New York, 1974.
22. M. Szwarc and M. Van Beylen, *Ionic Polymerization and Living Polymers*, Chapman & Hall, New York, 1993.

23. B. Vollmert and J. Huang, *Makromol. Chem., Rapid Commun.*, *2*:467 (1981).
24. B. Vollmert and J. Huang, *Makromol. Chem., Rapid Commun.*, *1*:133 (1980).
25. G. Hild, C. Strazielle and P. Rempp, *Eur. Polym. J.*, *19*:721 (1983).
26. G. Hild, A. Kohler and P. Rempp, *Eur. Polym. J.*, *16*:525 (1980).
27. D. Geiser and H. Hocker, *Macromolecules*, *13*:653 (1980).
28. Y. Gan, J. Zoller, and T. E. Hogen-Esch, *ACS Polym. Preprints*, *34(1)*:69 (1993).
29. S. Penczek and K. Matyjaszewski, *J. Polym. Sci., Polym. Symp.*, *56*:255 (1976).
30. V. Percec, C. G. Cho, C. Pugh, and D. Tomazos, *Macromolecules*, *25*:1164 (1992).
31. V. Percec, C. G. Cho, and C. Pugh, *Macromolecules*, *24*:3227 (1991).
32. V. Percec, C. G. Cho, and C. Pugh, *J. Mater. Chem.*, *1*:217 (1991).
33. R. W. Lenz, *Makromol. Chem. Suppl.*, *4*:47 (1981).
34. O. Nuyken, M. B. Leitner, and G. Maier, *Makromol. Chem*, *193*:3083 (1992).
35. K. Matyjaszewski, C.-H. Lin, and C. Pugh, *Macromolecules*, *26*:2649 (1993).
36. H. Mayr, R. Schneider, B. Irrgang, and C. Schade, *J. Am. Chem. Soc.*, *112*: 4454 (1990).
37. J. Brandrup and E. H. Immergut, *Polymer Handbook*, Wiley, New York, 1989.
38. H. W. McCormick, D. J. Worsfold, and S. Bywater, *J. Polym. Sci.*, *25*:488 (1957).
39. H. Sawada, *Thermodynamics of Polymerization*, Dekker, New York, 1976.
40. D. J. Worsfold and S. Bywater, *J. Polym. Sci.*, *26*:299 (1957).
41. J. Sebenda, *Pure Appl. Chem.*, *48*:329 (1976).
42. R. Huisgen, H. Brade, H. Waltz and I. Glogger, *Chem. Ber.*, *90*:1437 (1956).
43. K. J. Ivin and J. Leonard, *Eur. Polym. J.*, *6*:331 (1970).
44. E. J. Goethals, P. Bossaer, and R. Devaux, *Polym. Bull.*, *6*:121 (1980).
45. S. Winstein, E. Clippinger, A. H. Fainberg, R. Heck, and G. C. G. C. Robinson, *J. Am. Chem. Soc.*, *78*:328 (1956).
46. E. Gold, *Pure Appl. Chem.*, *55*:1281 (1983).
47. W. H. Powell, *Pure Appl. Chem.*, *65*:1357 (1993).
48. K. Matyjaszewski, *Makromol. Chem., Macromol. Symp.*, *13/14*:433 (1988).
49. K. Matyjaszewski, *Eur. Polym. J.* *19*:787 (1983).
50. N. Kalfoglou and M. Szwarc, *J. Phys. Chem.*, *72*:2233 (1968).
51. P. M. Bower, A. Ledwith, and D. C. Sherrington, *J. Chem. Soc. (B)*, 1511 (1971).
52. W. Gogolczyk, S. Slomkowski, and S. Penczek, *J. Chem. Soc. Perkin II*, 1729 (1977).
53. R. Schneider, H. Mayr, and P. H. Plesch, *Ber. Bunsenges. Phys. Chem.*, *91*:1369 (1987).
54. P. H. Plesch and V. T. Stannett, *J. Macromol. Sci. Chem.*, *A18*:425 (1982).
55. J. M. Lehn, K. Mislow, and G. Wagner, *J. Am. Chem. Soc.*, *92*:4050 (1970).
56. K. Mislow, *Angew. Chem.*, *9*:400 (1970).
57. J. B. Lambert and D. H. Johnson, *J. Am. Chem. Soc.*, *90*:1349 (1968).

58. J. V. Crivello and J. H. W. Lam, *J. Polym. Sci., Polym. Symp.*, *56*:383 (1976).
59. K. Matyjaszewski, *J. Macromol. Sci. Rev.*, *26*:1 (1986).
60. J. S. Hrkach and K. Matyjaszewski, *Macromolecules*, *23*:4042 (1990).
61. *CRC Handbook of Physics and Chemistry*, 67th ed., CRC Press, Boca Raton, 1986–1987, p. F156.
62. M. Bjoroy, B. B. Saunders, S. Esperas, and J. Songstad, *Phosphorus*, *83*: (1976).
63. S. F. Florquin and E. J. Goethals, *Makromol. Chem.* *82*:3371 (1981).
64. V. Gutmann and V. Hampel, *Monatsh. Chem.*, *92*:582 (1961).
65. C. B. Kim and K. T. Leffek, *Can. J. Chem.*, *53*:3408 (1975).
66. S. Penczek and R. Szymanski, *Polymer J.*, *12*:617 (1980).
67. K. Matyjaszewski, *J. Polym. Sci., Polym. Chem. Ed.*, *22*:29 (1984).
68. M. Miyamoto, K. Aoi, and T. Saegusa, *Macromolecules*, *24*:11 (1991).
69. Y. Yamashita, *J. Polym. Sci. Polym. Symp.*, *56*:447 (1976).
70. J. E. Chandler, B. H. Johnson, and R. W. Lenz, *Macromolecules*, *13*:377 (1980).

2

Fundamentals of the Reactions of Carbocations with Nucleophiles

HERBERT MAYR Institute for Organic Chemistry, Technical University of Darmstadt, Darmstadt, Germany

I. INTRODUCTION

Carbocations are molecules with a formal positive charge at carbon and an even number of electrons. Although the first examples of such ions were reported in 1902, when Baeyer [1,2] recognized the salt-like character of the compounds formed from triphenylmethanol and sulfuric acid [3], the concept of carbocation chemistry essentially was developed by Meerwein and Ingold in the 1920s [4]. Meerwein rationalized the rearrangement of camphene hydrochloride to isobornyl chloride by suggesting the ionization of the C—Cl bond and successive rearrangement of the cationic intermediate [5]. Other molecular rearrangements were interpreted analogously [6]. Kinetic investigations led Ingold to the conclusion that nucleophilic aliphatic substitutions follow two different mechanisms, one of which (the so-called S_N1 mechanism) involves the intermediacy of carbocations [7]. Winstein's kinetic and stereochemical studies of solvolytic displacement reactions revealed the importance of ion pairing [8] and anchimeric assistance [9] in reactions with a stepwise mechanism. These investigations furthermore led to the formulation of nonclassical carbocations, i.e., carbocations which cannot be properly described by two-electron two-center bonds [10].

Until the early 1960s information about carbocations has been obtained almost exclusively from indirect evidence. A major change occurred in 1962 when Olah reported the generation and direct observation of carbocations as long-lived species in solvents of low nucleophilicity (superacidic conditions) [11,12]. Many types of carbocations have since

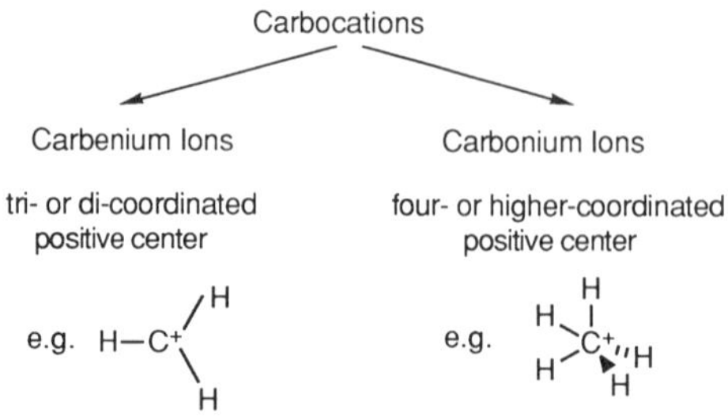

Scheme 1

then been investigated by a combination of these methods, including quantum chemical calculations [13], and nowadays carbocations belong to the best characterized reactive intermediates [14].

The nomenclature—carbonium ions, carbyl ions, carbenium ions, carbocations—has undergone several changes during their short history until Olah's suggestion [11] was finally accepted [15]. One now differentiates carbenium ions, i.e., cations with tri- or di-coordinated positive carbon (formally protonated carbenes) from carbonium ions, in which the coordination of the formally positive carbon is four and greater (e.g., protonated alkanes, see Scheme 1). According to recent investigations, many species, which have previously been considered to be nonbridged "classical" carbocations (i.e., carbenium ions), show some degree of bridging [16–18]. Since the differentiation between carbenium and carbonium ions becomes difficult in those cases, one should consequently use the general term carbocations whenever there is any ambiguity.

II. STABILITY SCALES FOR CARBOCATIONS

A. General

"Stability scales" compare isomeric and nonisomeric carbocations with respect to a certain property. As there is more than one reasonable property, there is none unique, "the correct stability scale," but there are several of them, each of which can conceptionally be justified.

To avoid ambiguity by intermixing different stability scales, it would be preferable to completely abandon the term "carbocation stability" and to clearly designate the properties that are going to be compared. Nevertheless, the term is in current use in discussions of chemical reactivity, and because the scales employed in this contribution are roughly correlated with each other, we will continue to talk about "carbocation stabilities" in a qualitative sense.

As discussed elsewhere [19], there are two main categories of "stability scales," which relate either the Lewis- [Eqs. (1a, b)] or the Bronsted-acidities [Eq. (1c)] of carbocations. In this contribution we will only refer to the Lewis acidity scales, i.e., scales that compare kinetic or thermodynamic features of the forward or backward reactions (1a, b).

$$R^+ + X^- \rightleftharpoons R\text{-}X \tag{1a}$$

$$R^+ + Y \rightleftharpoons R\text{-}Y^+ \tag{1b}$$

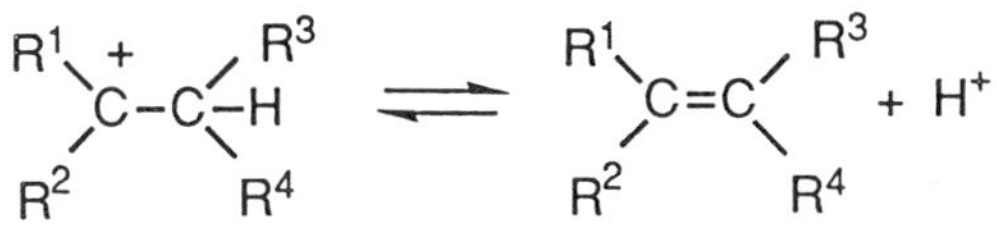
(1c)

The correlations between these scales, i.e., the justification for the further use of the term "stability," will be discussed in Section F.

B. Hydride Affinity Scale for the Gas Phase

Thermochemical properties referring to the gas phase are not obscured by solvent effects and are, from a fundamentalist's point of view, the ideal basis for stability scales.

Heats of formation of carbocations, which are the basis data for these scales have been derived from the heats of formation of the neutral precursors (e.g., radicals) and the adiabatic ionization energies [= ionization potentials, Eq. (2)] [20].

$$R\cdot \xrightarrow{IP_a} R^+ + e \tag{2}$$

The data thus obtained have been supplemented by relative heats of formation obtained by the study of proton, hydride, and halide transfer equilibria [Eqs. (3, 4)] in a high-pressure mass spectrometer, flow tube, or ion cyclotron resonance spectrometer [20].

$$AH^+ + B \rightleftharpoons A + BH^+ \tag{3}$$

$$R_A^+ + R_B X \rightleftharpoons R_A X + R_B^+ \tag{4}$$

The heats of formation of the carbocations are then combined with the heats of formation of the corresponding hydrocarbons [21] and ΔH_f^0 of H^- (139 kJ mol^{-1}) [23] to yield the hydride affinity scale of Scheme 2. Although hydride affinities of larger carbocations (molecular formula $\gtrsim$

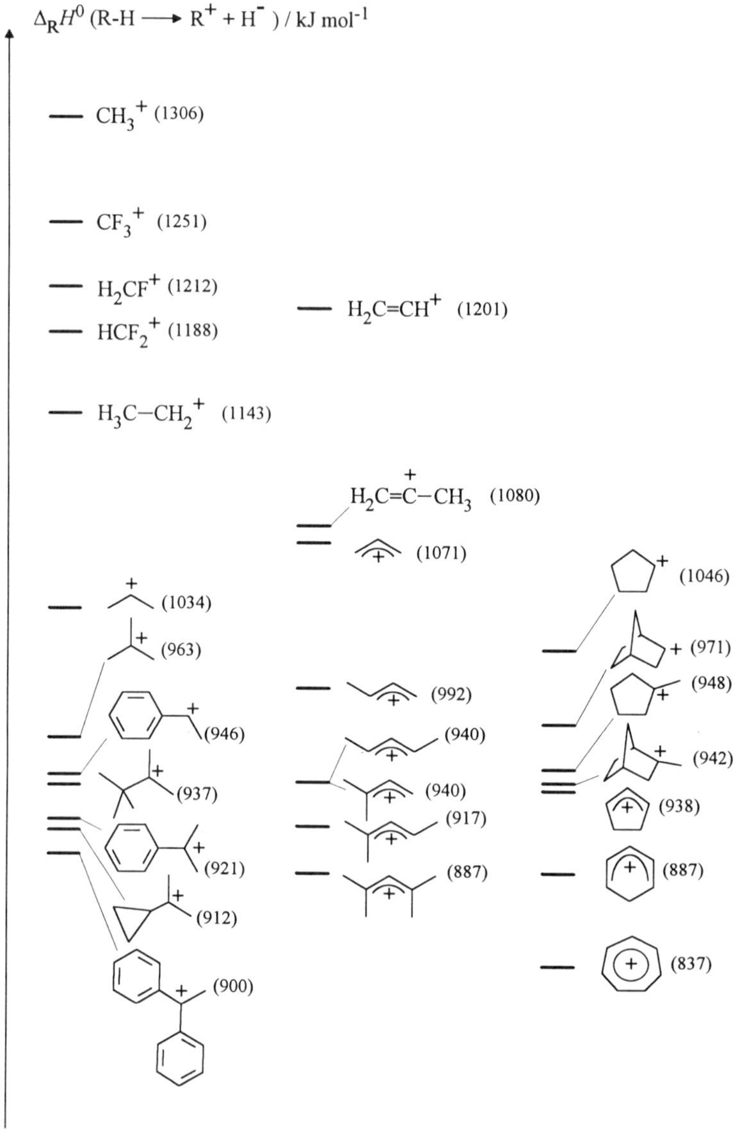

Scheme 2 Hydride affinity scale of carbocations. (data from Ref. 23.) With $\Delta H_f^0(H^-) = 145$ kJ mol^{-1} [20] the calculated hydride affinities are 6 kJ mol^{-1} greater than listed in this scheme.

C_6), correlate well with solution properties (see Section F), hydride affinities of smaller cations strongly depend on the number of heavy atoms [22] and are, therefore, of limited value for the chemistry in solution.

C. Heats of Ionization

Arnett and co-workers investigated the heats of reactions of alkyl chlorides with SbF_5/SO_2ClF [Eq. (5)] and of alcohols in $SbF_5/HSO_3F/SO_2ClF$ [Eq. (6)] [24–26].

$$R\text{—}Cl \xrightarrow[-50 \text{ to } -120^\circ C]{SbF_5/SO_2ClF} R^+; \qquad \Delta H_i \tag{5}$$

$$R\text{—}OH \xrightarrow[SO_2ClF,\ -40^\circ C]{SbF_5/HSO_3F} R^+; \qquad \Delta H_{rxn} \tag{6}$$

By these methods, solutions of highly stabilized (e.g., trityl cations) as well as of relatively unstable carbocations (e.g., sec-alkyl cations) have been produced. Although the precision of the calorimetric measurements is smaller than that of most equilibrium determinations, it is an advantage of Arnett's approach that very different types of carbocations can be studied by the same method (Scheme 3). Error propagations, which may be introduced when a series of equilibrium constants or overlapping scales are connected, are thus eliminated.

D. Equilibria in Solution

Because of the high tendency of carbocations to react with neutral molecules in various ways [11,27,28] solution equilibria as expressed by Eqs. (3) and (4) have only been determined for some relatively stable cations. A well-known carbocation scale of this category is the pK_{R^+} scale that is based on the equilibrium (7) [29,30].

$$R^+ + H_2O \overset{K_{R^+}}{\rightleftharpoons} R\text{—}OH + H^+ \tag{7}$$

$$pK_{R^+} = H_R + \lg \frac{[R^+]}{[ROH]} \tag{8}$$

According to Eq. (8), the pK_{R^+} value equals the acidity function (H_R) of that solution, in which equal concentrations of carbocation and nonionized alcohol coexist. The acidity function H_R has been derived by Deno, studying equilibria as described by Eq. (7); it approaches pH in dilute aqueous solution. For the reasons discussed above, the application of this method is limited and has been used mostly to characterize aryl-substituted carbocations (Scheme 4) which do not react with their precursor alcohols.

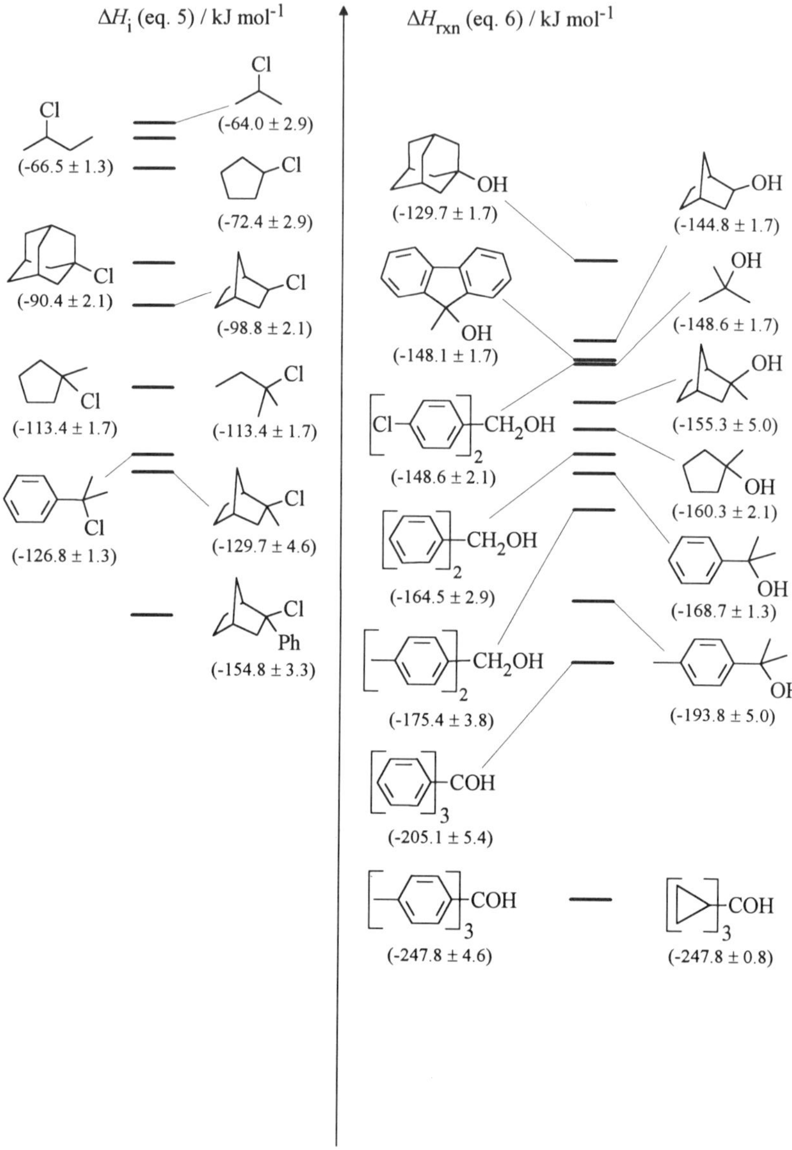

Scheme 3 Heats of ionization of alkyl chlorides and of alcohols under superacidic conditions. (data from Ref. 26.)

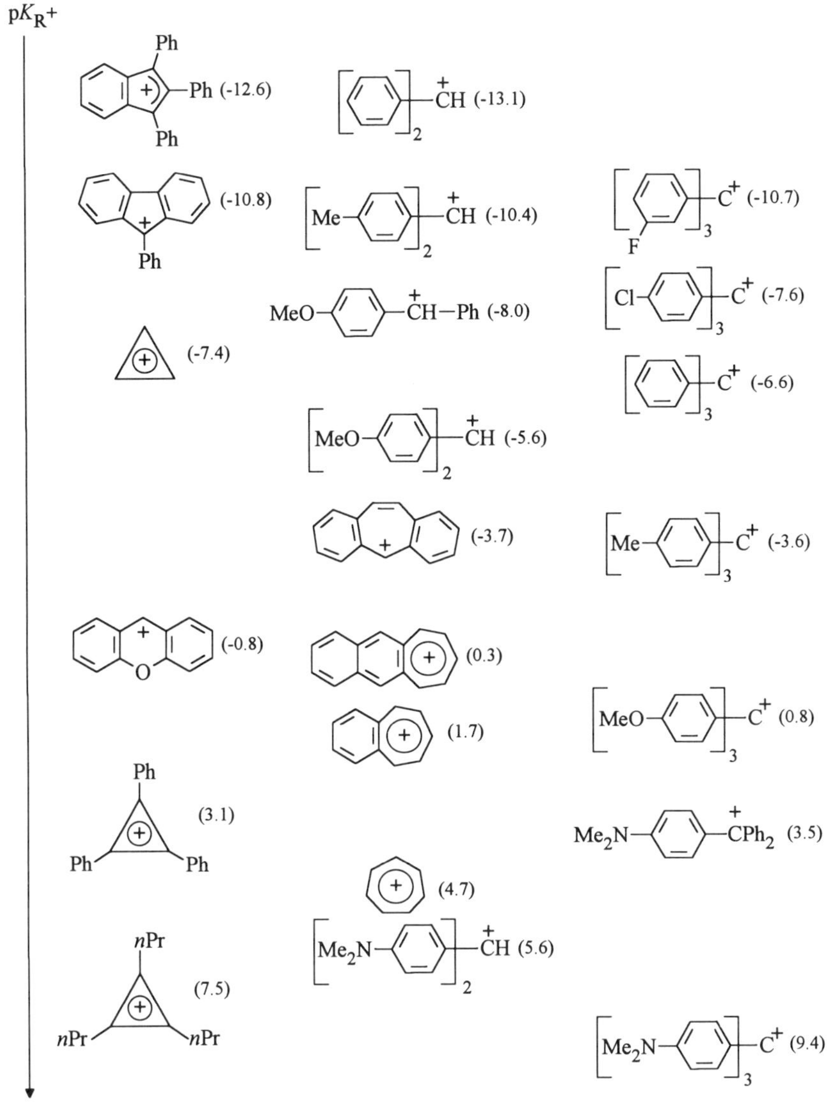

Scheme 4 pK_{R^+} scale for carbocations (data from Refs. 19, 31–33).

Chloride transfer equilibria of triarylmethyl- [34] and diarylmethyl cations [35] have been determined by ^{1}H NMR spectroscopy; they have been combined to give a chloride affinity scale (Scheme 5).

E. S_N1-Solvolysis Rates

Traditionally, relative "stabilities" of carbocations have been derived from the comparison of the rates of solvolysis reactions following the S_N1 mechanism, for which the designation $D_N + A_N$ has recently been proposed [36]. The comparison of solvolytic rate constants for substrates of a large structural variety is hampered by the fact that the published solvolysis rates refer to different solvents, different temperatures, and precursors with different leaving groups. Dau-Schmidt has, therefore, converted solvolysis rates of a manifold of alkyl chlorides and bromides to standard conditions, i.e., k_{solv} of RCl in 100% EtOH at 25° C (Scheme 6) [37]. Although from a theoretical point of view, ethanol is not an ideal solvent for observing unassisted S_N1-type reactions (nucleophilic solvent participation), it has been selected as the reference solvent because most available experimental data have been collected in solvents of comparable nucleophilicity, a fact which made conversions to 100% ethanol relatively unproblematic [38].

$$\text{Aryl}_2\text{CHCl} + \text{Aryl}'_2\text{CH}^+ \underset{}{\overset{K}{\rightleftharpoons}} \text{Aryl}_2\text{CH}^+ + \text{Aryl}'_2\text{CHCl}$$

K_{203}			Aryl_2CH^+ / Aryl_2CHCl	$\Delta\Delta G^0$ / kJ mol^{-1}
			An_2CH	0.0
		0.014		
			An(PhOPh)CH	7.2
		0.049		
			AnTolCH	12.3
		0.025		
	0.0098		AnPhCH	18.5
		0.50		
			(PhOPh)TolCH	19.7
		0.51		
	0.026		An(ClPh)CH	20.9
0.014			(PhOPh)PhCH	25.9
		0.26		
			Tol_2CH	28.1

An = p-$MeOC_6H_4$, Tol = p-MeC_6H_4, PhOPh = p-H_5C_6O-C_6H_4, ClPh = p-ClC_6H_4

Scheme 5 ^{1}H NMR spectroscopic determination of chloride transfer equilibria (−70° C, CD_2Cl_2). (From Ref. 35.)

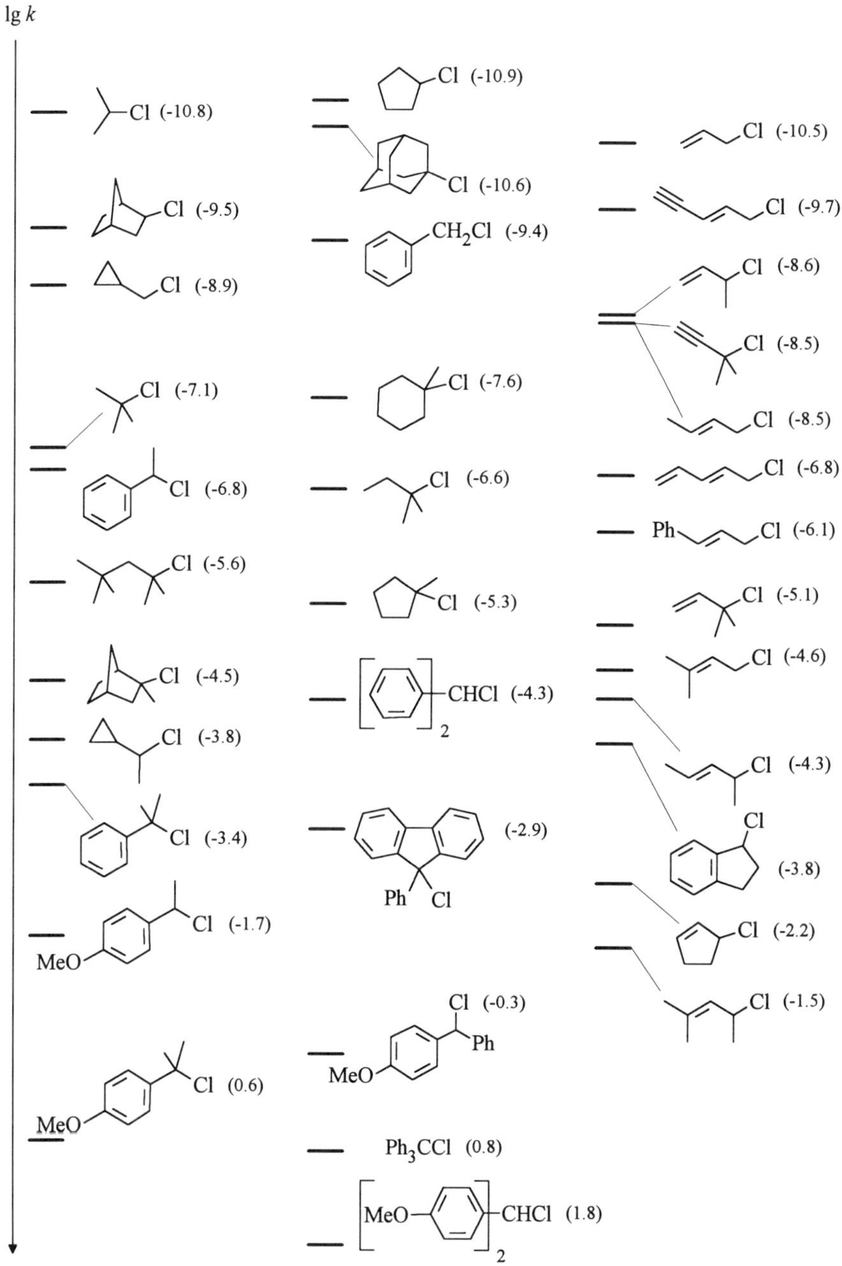

Scheme 6 Solvolysis rate constants of alkyl chlorides converted to standard conditions (RCl, 100% EtOH, 25° C). (Data from Ref. 37.)

F. Correlations Between the Various "Stability" Scales

As indicated in Section A, all carbocation scales reported in Sections B through E refer to the kinetics or thermodynamics of the forward or backward reaction [Eqs. (1a, b)]. Because substituent effects on carbocations are greater than those on the neutral counterparts, it is not surprising that all these scales correlate with each other. The correlations (9–12) have been reported by Arnett and Hofelich [26].

ΔH_{rxn}(ROH, $SbF_5/HSO_3F/SO_2ClF$) vs. ΔH_i(RCl, SbF_5/SO_2ClF); r = 0.97 for 9 points.

$$\Delta H_{rxn} = 1.03\ (\Delta H_i) - 41.0 \tag{9}$$

ΔH_{rxn} vs. pK_{R^+}; $r = 0.95$ for 12 points.

$$\Delta H_{rxn} = -6.90\ pK_{R^+} - 250.2 \tag{10}$$

ΔH_i vs. log k for ethanolysis of RCl at 25° C (correction for nucleophilic solvent participation); $r = 0.98$ for 10 points.

$$\Delta H_i = -6.78 \log k_{solv} - 156.5 \tag{11}$$

ΔH_i vs. gas-phase hydride affinity HIA; $r = 0.97$ for 15 points.

$$\Delta H_i = 0.600\ \text{HIA} - 682 \tag{12}$$

All energy values are expressed in kJ mol^{-1}.

Mutual interchanges between pK_{R^+} values, gas-phase hydride affinities, and ethanolysis rate constants can, in principle, be derived from Eqs. (9–12). Because of the use of more extended data sets and of uncorrected ethanolysis rate constants, the correlations presented in Ref. 19 are slightly different.

G. Ionization Equilibria of Alkyl Halide/Metal Halide Mixtures: Lewis Acidities of Metal Chlorides

Many synthetic reactions, that proceed via carbocations, produce these intermediates from mixtures of alkyl halides and Lewis acidic metal halides. The concentration of carbocations produced under these conditions depends on the tendency of the alkyl halides to ionize ("carbocation stability") and the strengths of the Lewis acids in a certain solvent. Because the ionizing abilities of alkyl halides can be derived from the schemes and correlations given in Sections B through F, we shall now concentrate on the relative halide affinities of the Lewis acidic metal halides. For this purpose, we consider Eq. (13) which describes the exchange of a chloride ion between the Lewis acids R^+ and MCl_n.

$$\text{R-Cl} + MCl_n \rightleftharpoons R^+ + MCl_{n+1}^- \tag{13}$$

Let us first ignore ion-pairing phenomena. With this assumption, the chloride transfer equilibria (13) correspond to the chloride transfer equilibria between two carbocations which were described in Scheme 5 and thus provide a comparison of the chloride affinities of metal halides and of carbocations. One would expect the right side of this equilibrium to be favored if MCl_n is the stronger Lewis acid and the left side when R^+ is the stronger Lewis acid.

The relative chloride affinities of the benzhydryl cations from Scheme 5 are now graphically displayed in Scheme 7 (bottom, right). The correlation between the chloride affinities of diarylcarbenium ions and the ethanolysis rate constants of the corresponding diarylmethyl chlorides [39] allows the chloride affinity scale of the diarylmethyl cations to extend to the less stabilized systems (Scheme 7, bottom left). In the ΔG^0 scale shown in Scheme 7, 10 kJ mol^{-1} corresponds to $K = 374$.

If the chloride transfer from diarylmethyl chlorides to metal chlorides MCl_n followed Eq. (13), one might derive relative chloride affinities of MCl_n and $Aryl_2CH^+$ from the same type of NMR experiments, which led to the relative chloride affinities of various $Aryl_2CH^+$ species. In this way one would also obtain a chloride affinity scale of various metal halides. The situation is more complex, however!

Let us first look at the chloride transfer from $Aryl_2CHCl$ to the well-behaved Lewis acid BCl_3. This Lewis acid is monomeric in solution, accepts only one chloride ion to give BCl_4^-, and this anion does *not* coordinate with a second molecule of BCl_3 to give binuclear complexes. It has

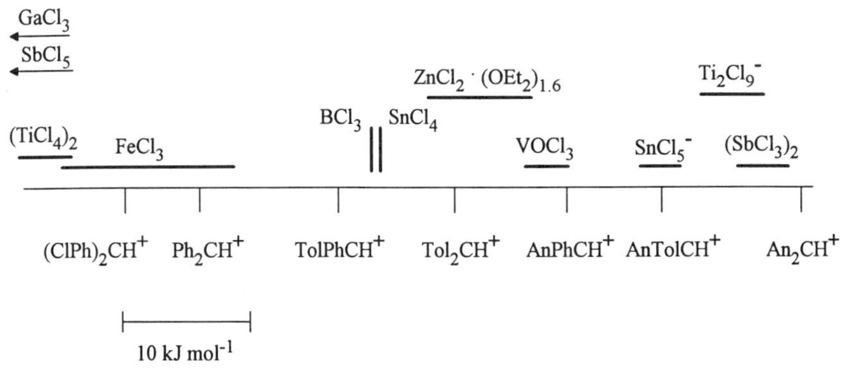

Scheme 7 Relative chloride affinities of diarylcarbenium ions and of metal chlorides in CH_2Cl_2 ($-70°$ C). (From Ref. 40.)

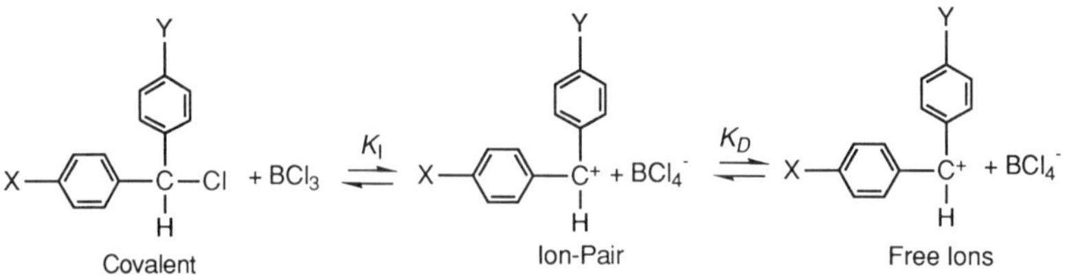

$$K_I = \frac{[\text{Aryl}_2\text{CH}^+\ \text{BCl}_4^-]}{[\text{ArylCHCl}][\text{BCl}_3]}\ ; \qquad K_D = \frac{[\text{Aryl}_2\text{CH}^+][\text{BCl}_4^-]}{[\text{Aryl}_2\text{CH}^+\ \text{BCl}_4^-]}\ ;$$

Scheme 8

been reported that mixtures of diarylmethyl chlorides and BCl_3 can be described adequately by the formalism shown in Scheme 8 [41]. The benzhydryl chlorides and BCl_3 react with formation of ion pairs (ionization constant, K_I) which dissociate to give the free ions (dissociation constant, K_D). Because paired and free diarylcarbenium ions show only slightly different UV-visible spectra, [41], spectrophotometric measurements allow the determination of the total carbocation concentration. On the other hand, only free ions are detected by conductometric analysis, and a combination of both methods allows the determination of K_I and K_D using the theory of binary ionogenic equilibria [42,43].

Figure 1 shows that conductance increases instantaneously, when dianisylmethyl chloride is titrated with BCl_3. A limiting value is reached after the addition of 1 equivalent of BCl_3 indicating a large value of K_I, the magnitude of which cannot be derived from this titration experiment. This curve qualitatively shows that BCl_3 is a much stronger chloride acceptor than An_2CH^+. As shown in Figure 1, the titration curves flatten as the electron-releasing abilities of the *p*-substituents in the diarylmethyl systems decrease, and for the ionization of Tol_2CHCl and TolPhCHCl with BCl_3, ionization constants, K_I, of 64 and 0.17, respectively, have been determined.

In contrast to K_I, which strongly depends on the substitution pattern of the carbocations, the dissociation constant K_D, which could be obtained for all systems shown in Fig. 1, was found to be almost constant, $(1.9–2.9) \times 10^{-4}$ mol L^{-1} for CH_2Cl_2 at $-70°$ C, independent of the stabilization of the carbocations. Accordingly, similar K_D values have been reported for other carbocation salts with complex counter ions [41].

These numbers show that, except in highly dilute solutions, carbocations are predominantly paired in CH_2Cl_2, and chloride affinities can

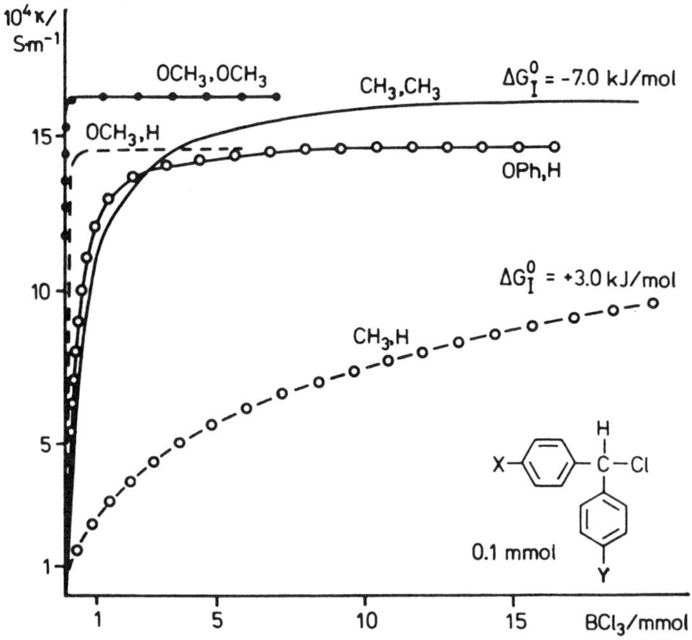

Figure 1 Conductance of 1.1×10^{-3} *M* diarylmethyl chloride solutions in the presence of variable amounts of BCl_3 (CH_2Cl_2, $-70°$ C). (From Ref. 41, reprinted with permission of VCH Verlagsgesellschaft.) Substituents *X*, *Y* are indicated beside the graphs.

be based on the K_I values. From the quoted K_I values for ionization equilibria with BCl_3, one can derive that BCl_3 is a stronger chloride acceptor than Tol_2CH^+ and a weaker chloride acceptor than $TolPhCH^+$, so that the position of BCl_3 in Scheme 7 can be fixed.

Experiments similar to those described in Fig. 1, and titrations of $Aryl_2CH^+MCl_{n+1}^-/MCl_n$ mixtures with $NR_4^+Cl^-$ solutions have been performed to characterize the relative chloride affinities of other Lewis acids MCl_n [40]. It turned out, however, that the well-defined behavior of BCl_3 could not be observed with other systems. The positions of all other metal chlorides in Scheme 7, therefore, have only qualitative character. Relatively small concentrations of $SbCl_5$ and $GaCl_3$ were sufficient to completely ionize $(p\text{-}ClC_6H_4)_2CHCl$, the weakest chloride donor of this series. The values of these ionization constants could not be determined, and only a limiting value of the chloride affinities of these two Lewis acids can be given.

$TiCl_4$ also acts as a very strong chloride acceptor, when 2 equivalents of $TiCl_4$ per Cl^- are available. The $Ti_2Cl_9^-$ ion thus formed has been characterized by IR [44] and x-ray crystallography [45]. As $Ti_2Cl_9^-$ is a relatively weak chloride acceptor, in between $AnTolCH^+$ and An_2CH^+, the high chloride affinity of $TiCl_4$ only becomes effective, when it is used in high concentration. The formation of $Ti_2Cl_{10}^{2-}$ ions has been observed when Ph_3CCl was treated with $TiCl_4$ in PhCOCl, $POCl_3$, or $PhPOCl_2$ [46–48]. At higher $[Cl^-]/[TiCl_4]$ ratio, $TiCl_6^{2-}$ ions are produced [49].

Very small concentrations of $FeCl_3$ were sufficient to completely ionize Tol_2CHCl, indicating a high chloride affinity of $FeCl_3$. More accurate comparisons with less stabilized carbenium ions failed because of the low solubility of $FeCl_3$.

Tin(IV)chloride can accept two Cl^- ions per molecule. From the ionization equilibrium with Tol_2CHCl one can derive that the first chloride affinity resembles that of BCl_3, whereas $SnCl_5^-$ is a somewhat weaker chloride acceptor comparable to $AnTolCH^+$. In accord with the relatively small difference of the chloride affinities of $SnCl_4$ and $SnCl_5^-$, the Lewis acidity of $SnCl_4$ was found even to be increased by coordination with a neutral ligand: When $SnCl_4$ and carbonyl compounds have been mixed in a 1:1 ratio, 2:1 complexes and free $SnCl_4$ have been observed by NMR, while the expected 1:1 complexes have not been detectable [50].

Zinc chloride has been investigated as an ether complex, which is soluble in CH_2Cl_2 [51]. The $ZnCl_2 \cdot (OEt_2)_{1.6}$ mixture (50 equivalents) is able to completely ionize AnPhCHCl and to partially ionize Tol_2CHCl. Because of the complex coordination equilibria with ether, a detailed description of the system does not appear to be possible.

The liquid Lewis acid $VOCl_3$ was found to completely ionize AnTolCHCl but not Tol_2CHCl; a detailed analysis of the titration curves as in the case of BCl_3 also failed.

Antimony(III) chloride, a relatively weak Lewis acid, shows some resemblance to the behavior of $TiCl_4$, as its full chloride affinity can only become effective when two molecules of $SbCl_3$ per Cl^- can be provided. With an excess of $SbCl_3$, An_2CHCl can be ionized almost completely (probably formation of $Sb_2Cl_7^-$). The resulting position of $SbCl_3$ in Scheme 7 is thus in qualitative agreement with Penczek's investigation [52a] which yielded $\Delta G_I^0 = -3.2$ kJ mol^{-1} for the reaction $Ph_3CCl + SbCl_3 \rightleftarrows Ph_3C^+SbCl_4^-$ in CH_2Cl_2 at 25° C (compare pK_{R^+} of Ph_3C^+ and An_2CH^+ in Scheme 4).

Although the chloride affinities of most metal chlorides thus determined are only approximate, comparison with an analogous scale determined by Gutmann in CH_3CN [53,54] shows distinct deviations. Because of coordination of the strong Lewis acids with the donor solvent, a leveling of the strengths of these Lewis acids in CH_3CN is observed. The coordination of carbocations and other Lewis acids with *n*-donor molecules has been studied in detail [52b,150,163].

III. REACTIONS OF CARBOCATIONS WITH ALKENES

A. Selective Formation of [1:1] Products

Mixtures of alkyl halides and Lewis acids are well-known initiating systems for the polymerizations of alkenes, and the mechanism suggested for these reactions by Kennedy [55] appears to be generally accepted (Scheme 9), although the importance of the chain transfer step from initiator has been questioned [56].

If the propagation step of the sequence described in Scheme 9 could be suppressed, the gross reaction (14) would result, and this reaction would allow mechanistic details of the reactions of carbocations with alkenes to be investigated.

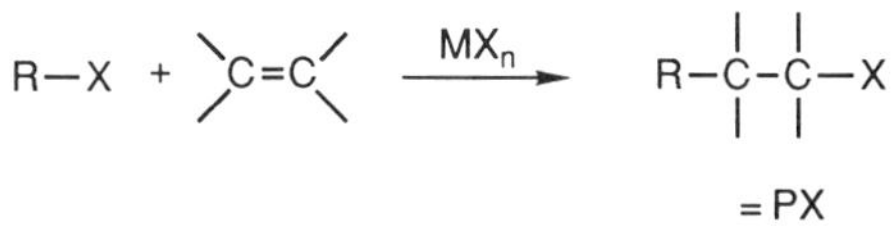

(14)

Because reaction (14) generates a product (P*X*), which may exhibit similar electrophilic properties as the reactant R*X*, it will depend on the relative

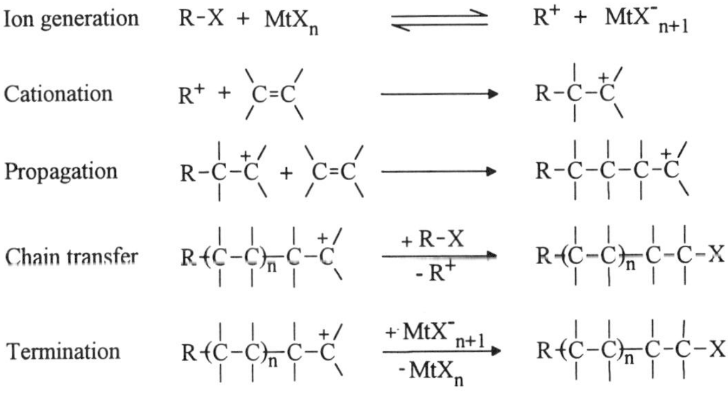

Scheme 9 Mechanism of alkyl halide/Lewis acid-initiated polymerizations of alkenes. (From Ref. 57.)

electrophilicities of reactant and product, whether the reaction will produce polymers or terminate at the [1:1] product stage. The competition situation described in Scheme 10 is encountered, and in order to avoid successive reactions of the product P*X*, one has to find conditions, under which R*X*/R^+ is more reactive than P*X*/P^+.

A qualitative analysis of this problem can be based on Scheme 11, which shows energy profiles for the additions of various alkyl halides *to the same alkene*. In accord with experimental results (see Section III.D.5. and Ref. 58), the Gibbs free energy difference between R*X* and R—C—C^+ (= P^+) in Scheme 11 is kept constant, since the stabilization of R—C—C^+ is almost exclusively determined by the substituents at the new carbocationic center and only to a very small degree by the nature of R. According to the Bell-Evans-Polanyi principle [59], the height of the transition state of the rate-determining step (R^+ + C=C) will decrease, as the energy of R^+ is lowered. As long as the reactants are predominantly covalent ($R_A X$. . .$R_C X$), stabilization of R^+ will be associated with a decrease of the activation free enthalpy of the overall reaction ($\Delta G^\ddagger$). In contrast, $\Delta G^\ddagger$ will grow with increasing stabilization of R^+, if the reactants are predominantly ionic (R_{C^+}. . .R_{E^+}).

On the basis of these considerations it has been concluded that under given reaction conditions (Lewis acid/solvent) the reactivity maximum is found for an alkylating system (R*X*/R^+) that is approximately half-ionized [60,61]. Scheme 11 suggests that the electrophilic reactivity of R*X* increases with increasing stabilization of R^+ *if only small equilibrium concentrations of carbocations are involved*. In accord with this analysis, the relative alkylating abilities of alkyl chlorides have been found to be proportional to their ethanolysis rates (Fig. 2) [62]. The only compound that deviates from this correlation is trityl chloride which alkylates considerably more slowly than expected from its solvolysis rate.

Figure 2 provides the experimental justification for the long-standing rule that alkylations as shown in Eq. (14) (conditions of *incomplete* ionization) will only yield 1:1 products selectively, if the reactants R*X* solvolyze

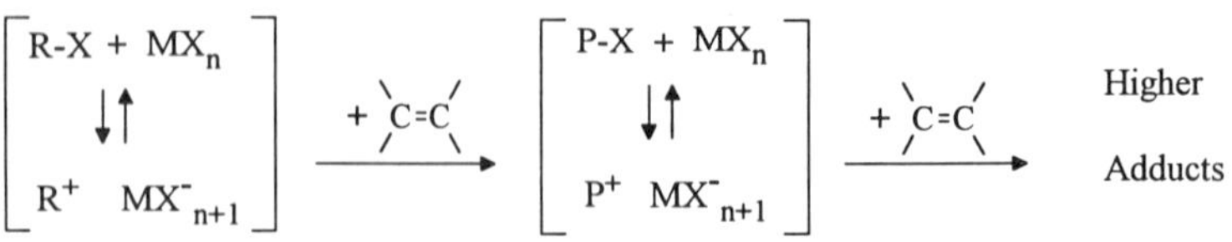

Scheme 10 Chemoselectivity control in Lewis acid promoted reactions of alkyl halides with alkenes.

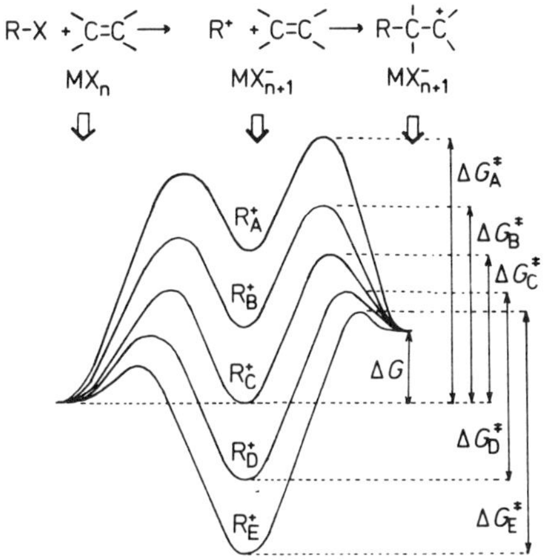

Scheme 11 Schematic energy profiles for the Lewis acid-initiated additions of various alkyl halides to a certain alkene. (From Ref. 60, reprinted with permission of Hüthig & Wepf Verlag, Zug, Switzerland.)

faster than the products P*X* [63]. Preparative investigations had demonstrated the practicability of this working hypothesis (Scheme 12) [64,65]. If alkyl halides are ordered according to increasing solvolysis rates from top to bottom, and alkenes are arranged in a way that the solvolysis rates of the products increase from left to right, only those combinations of alkyl halides and alkenes which are located in the lower left section of Scheme 12 fulfill the quoted criterion for selective formation of 1:1 products, and in fact 1:1 products from these combinations were isolable. Empty boxes in Scheme 12 indicate that oligomerization (formation of higher adducts) is predicted, but has not been examined.

B. Structures of the Addition Products

1. *Regio- and Stereoselectivity of the Addition Reactions*

Like proton-induced H*X* additions [66–68], additions of carbocations to alkenes proceed with strict regioselectivity, the orientation being determined by the stabilities of the intermediate carbocations (Markovnikov rule). In this respect, carbocation additions differ from other electrophilic additions, as sulfenylations or selenylations, where the orientation is controlled by the nucleophilic attack at the bridged cationic intermediate (Scheme 13) [67, p. 860].

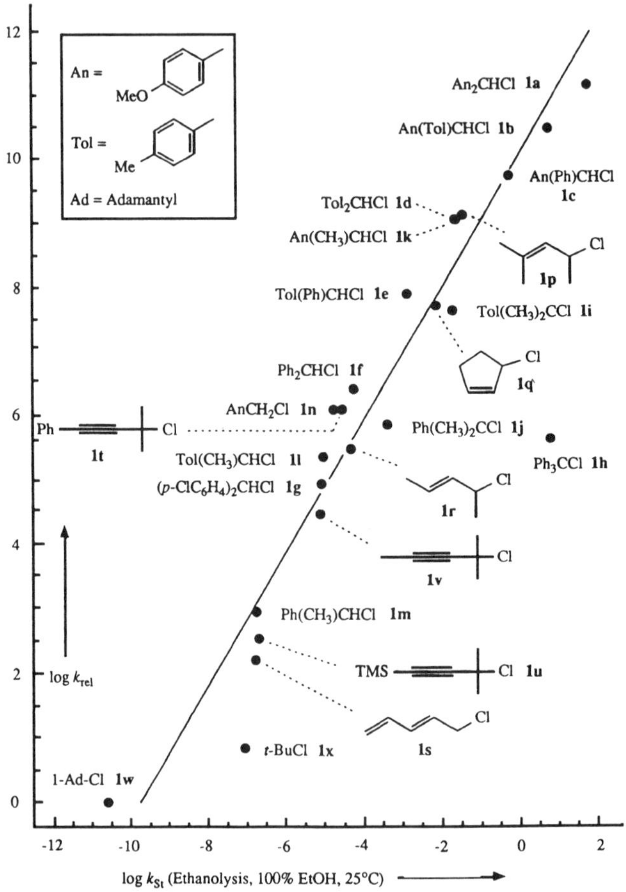

Figure 2 Correlation of the relative reactivities of alkyl chlorides **1a-x** toward allyltrimethylsilane (CH_2Cl_2, $-70°$ C) with their ethanolysis rate constants (25° C). The value for Ph_3CCl has not been used for calculating the correlation equation $\log k_{rel} = 1.036 \log k(EtOH) + 10.1$ ($r = 0.971$). (From Ref. 62, reprinted with permission of VCH Verlagsgesellschaft.)

Even the sterically shielded 2,4,4-trimethyl-2-pentene is selectively attacked at the tert-butyl-substituted olefinic position (Scheme 14) [82], in accord with electronic control of the electrophilic attack.

Halogenated ethylenes react with carbocations at the nonhalogenated or least-halogenated position [70–77], not because the α-halogenated car-

Solvolysis Rates of 1:1 Products →

Solvolysis Rates of Reactants ↓

				Ph		Ph	OEt
$(CH_3)_2CHCl$	—	—	—		—		
$(CH_3)_3CCl$	41%	35%	—	—	—		
$Ph{-}CH(CH_3){-}Cl$	52%	50%	71%	72%	—	—	
$(CH_3)_2C{=}CH{-}CH_2Cl$	32%	27%	65%	75%	10%	—	—
$Ph{-}C{\equiv}C{-}C(CH_3)_2Cl$	67%	46%	93%	91%	67%	—	—
$Ph{-}C(CH_3)_2Cl$	58%	65%	71%	71%	64%	58%	—
Ph_2CHCl	92%	85%	97%	88%	82%	75%	—
$CH_3O{-}CH_2Cl$	47%	70%	60%	75%	64%	37%	—
$CH_3O{-}CH(Ph){-}Cl$	78%	65%	90%	57%	52%	69%	68%

Scheme 12 Yields of 1:1 products from Lewis acid-catalyzed reactions of alkyl halides with alkenes. ($R{-}X + {>}C{=}C{<} \longrightarrow R{-}\overset{|}{\underset{|}{C}}{-}\overset{|}{\underset{|}{C}}{-}X$). (Reprinted with permission from Ref. 64. Copyright 1983 American Chemical Society.)

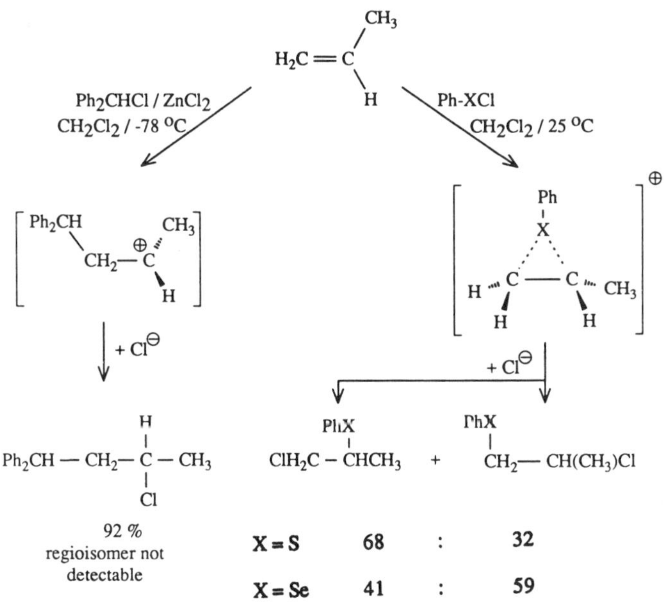

Scheme 13

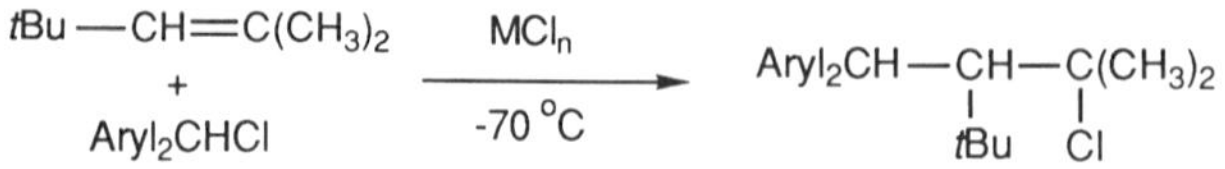

Scheme 14

bocations thus produced are very much stabilized by the halogen but because the alternative orientation would yield a carbocation that is destabilized by halogen in β-position (Scheme 15) [78].

In accord with the Hammett σ_p-value of CH_2Br, which is close to zero (0.14) [79], allyl bromide has been reported to be attacked at both termini of the double bond to give a mixture of two regioisomers (Scheme 16) [80].

Alkyl-substituted 1,3-dienes react so that the allyl cation with the highest possible number of terminal alkyl groups is formed, leading to the regioselectivity indicated by the arrows in Scheme 17 [81–83]. With highly reactive carbocations the selectivity is reduced, however, and the prenyl cation has been reported to undergo approximately 20% of 4-attack at isoprene besides 80% of the regular 1-attack [83c,d]. 1-Buten-3-yne is

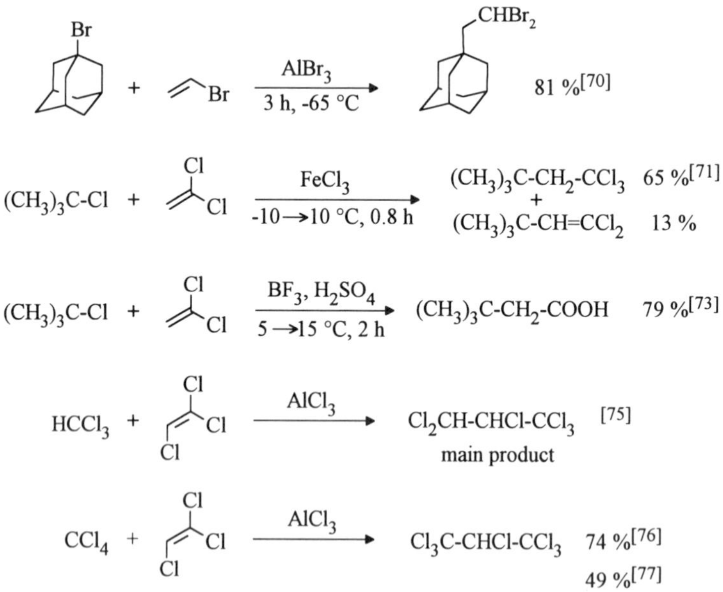

Scheme 15 (Data from Refs. 70, 71, 73–77.)

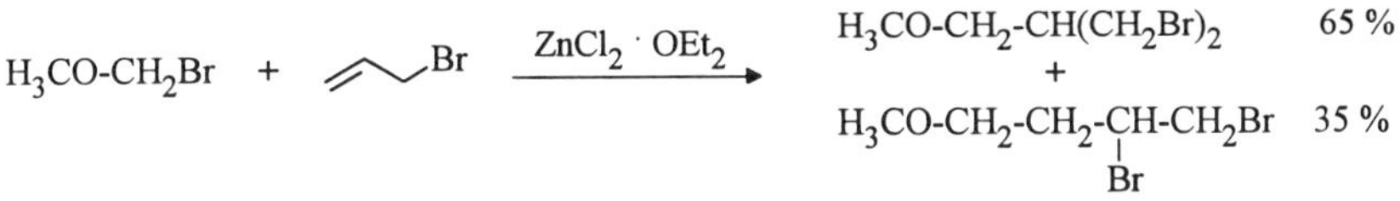

Scheme 16

attacked by carbon electrophiles at the 4-position, and the reversal of the direction of addition in 1-hexene-3-yne (Scheme 17) can be explained by the electronic effect of the extra ethyl group on the cationic intermediate.

According to quantum chemical calculations, additions of carbocations to alkenes are expected to proceed via π-complexes [84, 85]. The high antistereoselectivity of carbocationic cyclizations had been explained by this hypothesis [86–88].

Stereochemical investigations of intermolecular additions showed that the syn/anti-adduct ratio was dependent on the nature of the Lewis acid. In reactions of diarylmethyl chlorides to (*Z*)- and (*E*)-2-butene the formation of the antiadduct was generally favored (Scheme 18) [89]. Because the different product mixtures have been obtained from the stereoisomeric 2-butenes, the intermediacy of long-lived, freely rotating *sec*-alkyl cations has been excluded.

Ion-pair effects cannot account for the stereoselectivities described in Scheme 18 as their operation should give rise to syn selectivity [90]. The observation that the addition rates of diarylcarbenium salts to alkenes are independent of the nature and the concentration of the negative counterions (see Section III.D.3) excludes the simultaneous attack of the carbocation and its counterion at the π-bond of 2-butene.

It was possible to explain the stereochemical phenomena of Scheme 18 by assuming the intermediacy of partially bridged cations **3** [89]. A detailed analysis of the transition states, which explains the simultaneous formation of rearranged products, will be developed in Section III.B.2.

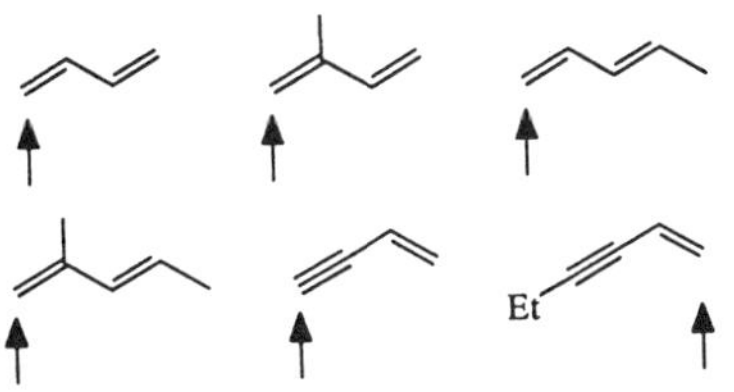

Scheme 17 Sites for attack by carbenium ions on conjugated dienes and enynes. (From Ref. 69, reprinted with permission of VCH Verlagsgesellschaft.)

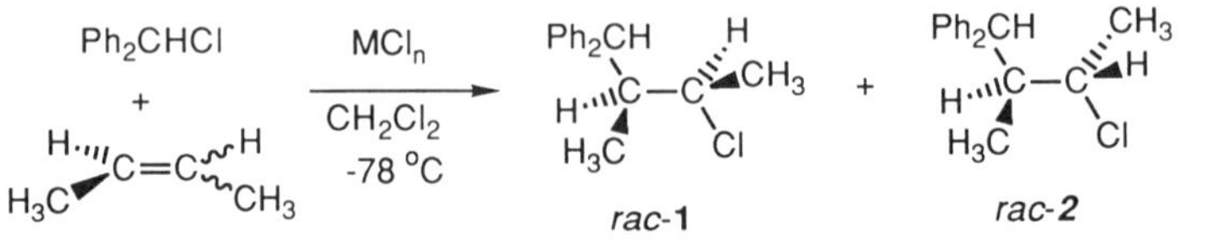

(*Z*)-2-butene	$ZnCl_2$	86	:	14
	BCl_3	84	:	16
	$SnCl_4$	91	:	9
	$TiCl_4$	96	:	4
(*E*)-2-butene	$ZnCl_2$	4	:	96
	BCl_3	12	:	88
	$SnCl_4$	8	:	92
	$TiCl_4$	< 1	:	> 99

Scheme 18 Diastereoselectivities of the Lewis acid catalyzed reactions of benzhydryl chloride with (*E*)- and (*Z*)-2-butene. Further products, arising from hydride shifts, are described in Scheme 21. (From Ref. 19.)

Interaction of the carbenium center with one of the phenyl rings, as shown in formula **4** (Scheme 19) has been suggested as an alternative explanation for the observed stereoselectivities [91].

Increasing electron-releasing ability of the substituents at the newly developing carbocationic center reduces the amount of bridging. Accordingly, the Lewis acid promoted reactions of diphenylmethyl chloride with both (*E*)- and (*Z*)-1-phenyl-1-propene yield the same stereoisomer (**5**) as the major product, indicating a more rapid rotation of the intermediate benzyl cation [89]. Analogous changes of stereoselectivity have been reported for bromine additions to butenes and styrenes [92–94].

We are not aware of any systematic investigations on the diastereoselectivities of the CC-bond-forming step in intermolecular reactions of carbocations with alkenes. Generally, we observed only low stereoselectivities in such cases, as illustrated for the Lewis acid catalyzed addition of 4-chloro-2-pentene to (*Z*)-2-butene (Scheme 19). The *si,re* transition state **6** is slightly favored (75:25) over the *si,si* (and *re,re*) transition state **7**, and for the corresponding addition to (*E*)-2-butene, the advantage of *si,re* over *si,si* (or *re,re*) sinks to 57:43 [95].

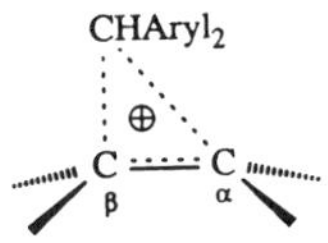

3

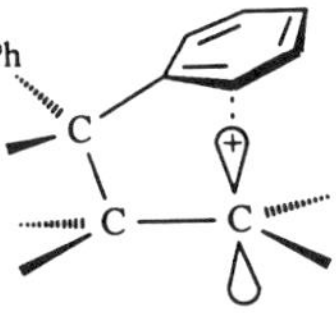

4

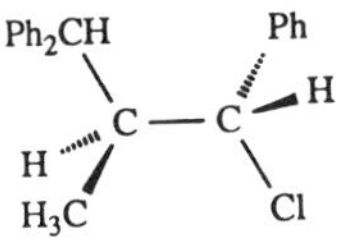

5

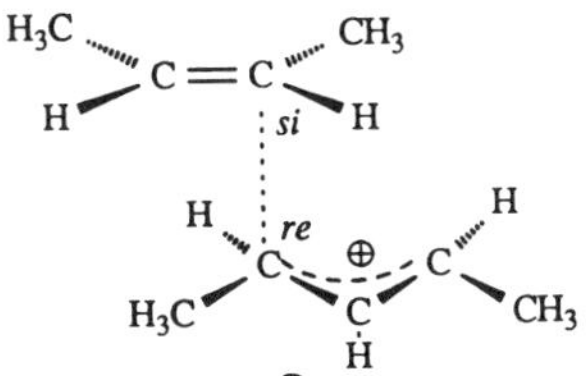

6

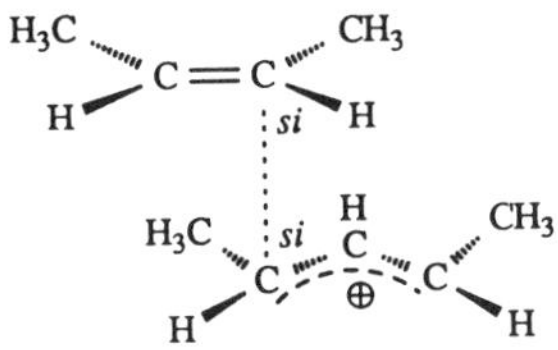

7

Scheme 19 Diastereomeric transition states for the reaction of the 1,3-dimethylallyl cation with (*E*)-2-butene. (From Ref. 69.)

2. *Reactions of the Initial Carbocationic Adducts*

The carbocations **8** formed by electrophilic addition to an alkene may immediately be trapped by an anion to give the ordinary addition products **9** (Scheme 20, a) as discussed in Section III.B.1, or eliminate a proton to give the substitution product **10** (Scheme 20, b). Usually proton eliminations, which can also occur at other positions than shown in Scheme 20, play a subordinate role when the reactions are carried out at low temperature ($-70°$ C). Alternatively, 1,2-hydride or alkyl migrations (Scheme 20, c and d) may take place, and if there is a nucleophilic group in **8** (usually C=C, C≡C or arene ring), cyclization may occur [96]. The cations **11–13** thus produced, may then be trapped by X^-, eliminate H^+ or undergo further rearrangements (c, d, e) before an uncharged molecule is formed. In view of the manifold of possible reactions, only one of the possible hydride and alkyl shifts is shown in Scheme 20, and the fact that many of these reactions possess very low activation energies, it is astonishing that very often the ordinary adducts **9** can be isolated in high yield; sometimes they are the exclusive products. An explanation for this behavior may be obtained from Scheme 21, which analyzes the Lewis acid-catalyzed reactions of diphenylmethyl chloride with the (*E*,*Z*)-isomeric 2-butenes.

As discussed in Section III.B.1, these reactions are assumed to proceed via partially bridged cations. Attack of Ph_2CH^+ at (*Z*)- and (*E*)-2-

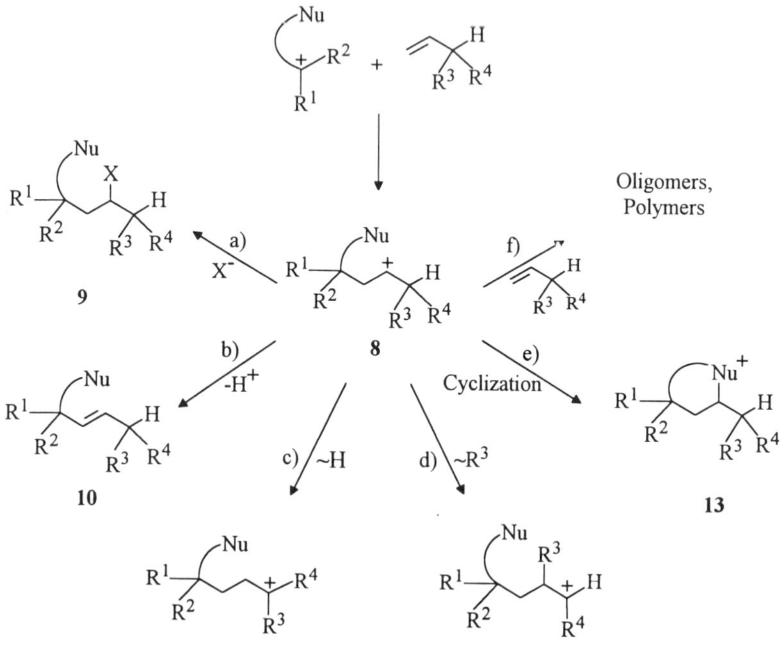

Scheme 20

butene initially yields the rotational isomers **14a** and **14b** (Scheme 21). The thermodynamically favorable 1,2-hydride shift, that converts the secondary cations **14a**, **b** into the tertiary cation **16** can only occur after a 60° rotation of **14a** and **14b** to give the rotational isomers **15a** and **15b**, where the C—H bonds have the proper orientation to undergo the activationless hydride migration. One can easily see that more conformational strain is built up during the rotation **14b** → **15b** than during the corresponding rotation **14a** → **15a**. The higher antiselectivity (back-side attack at **14b** → *rac*-**2**) and the lower amount of rearranged products in reactions with (*E*)-2-butene (compared with those with (*Z*)-2-butene) can thus be explained. In both reaction series, the highest stereoselectivity and lowest percentage of rearranged products was obtained with $TiCl_4$ as catalyst. This behavior is unusual because, in additions to most other π-systems, an optimum of chemoselectivity has been observed with $ZnCl_2{\cdot}OEt_2$ as catalyst.

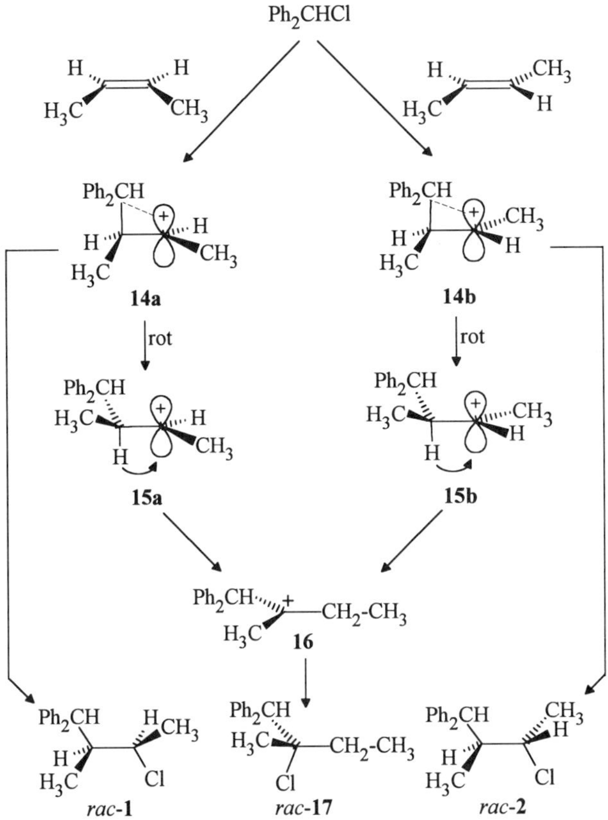

	$ZnCl_2$	3.3	14.6	72.6
	BCl_3	9.0	21.3	64.1
	$SnCl_4$	6.4	21.0	69.4
	$TiCl_4$	-	3.9	96.1
	$ZnCl_2$	42.6	42.3	6.8
	BCl_3	29.9	61.2	5.9
	$SnCl_4$	30.8	65.0	2.9
	$TiCl_4$	75.6	20.4	2.9

Scheme 21 Mechanism of the Lewis acid catalyzed reactions of benzhydryl chloride with (*E*)- and (*z*)-2-butene. Ref. 89 describes the formation of some additional rearranged products in low yields, which do not affect the general conclusions, drawn in this discussion.

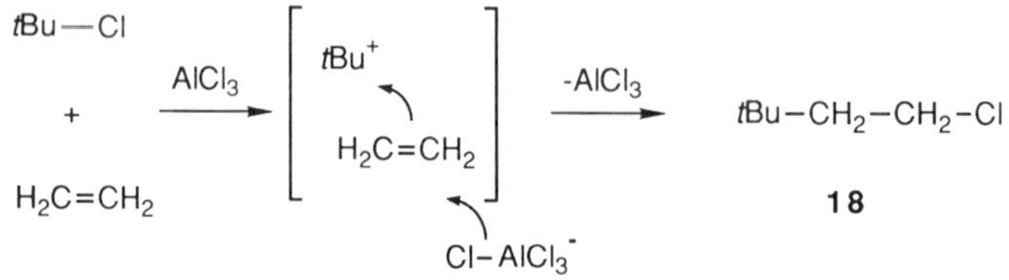

18

Scheme 22

In analogy to these observations, alkylations of cyclopentene and cyclohexene, which also represent (*Z*)-1,2-dialkylated ethylenes, afforded complex product mixtures due to extensive rearrangements.

In view of these results, it is surprising that no rearrangement occurs during the aluminium chloride-catalyzed reaction of *tert*-butyl chloride with ethylene [97]. The intermediacy of a primary carbocation can, therefore, be ruled out and the selective formation of **18** may be rationalized by assuming the intermediacy of a bridged cation (*tert*-butyl-bridged ethylene). Alternatively, the attack of the *tert*-butyl cation at ethylene may be nucleophilically assisted by the $AlCl_4^-$ ion because of the low stability of the 3,3-dimethyl-1-butyl cation (Scheme 22) [98].

Rearrangements that result in little or no stabilization of the carbenium ion (e.g., tertiary → tertiary or secondary → secondary) are observed occasionally, but can be suppressed partly or totally by increasing the rate of ionic recombination (v_1, Scheme 23).

Because rearrangements ($R^+ \rightarrow R'^+$) can occur in paired or unpaired ions, their rate ($v_2 + v_3$) depends little ($v_2 \approx v_3$) or not at all ($v_2 = v_3$) on the degree of ion pairing. The rate of ion combination (v_1, v_4), on the

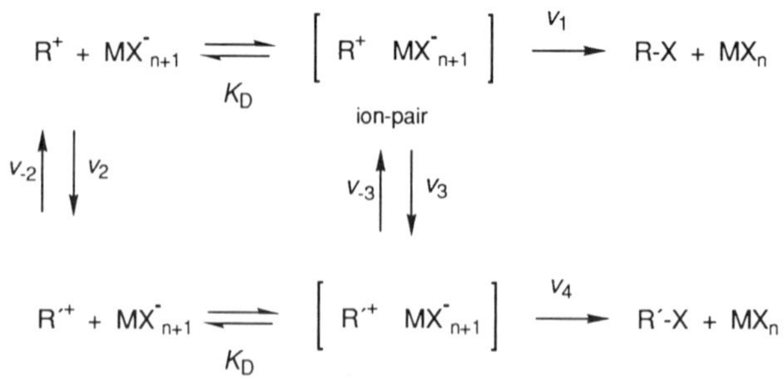

Scheme 23

other hand, is proportional to the concentration of ion pairs and can, for instance, be increased by addition of a quaternary ammonium salt ($R_4N^+MX_n^-$). Scheme 24 shows that the yield of **21** relative to that of the cyclization product **22** increases with increasing BCl_4^- concentration. The **21**:**22** ratio remains essentially constant, however, once the majority of the ions are paired [58].

The fact that the $ZnCl_2$-catalyzed reaction of the allyl chloride **23** with styrene yields a 65:35 mixture of **27** and **28**, whereas pure **27** is obtained under the same conditions in the presence of benzyltriethylammonium trichlorozincate, can be explained in the same way [99,100]. The 1,3-vinyl

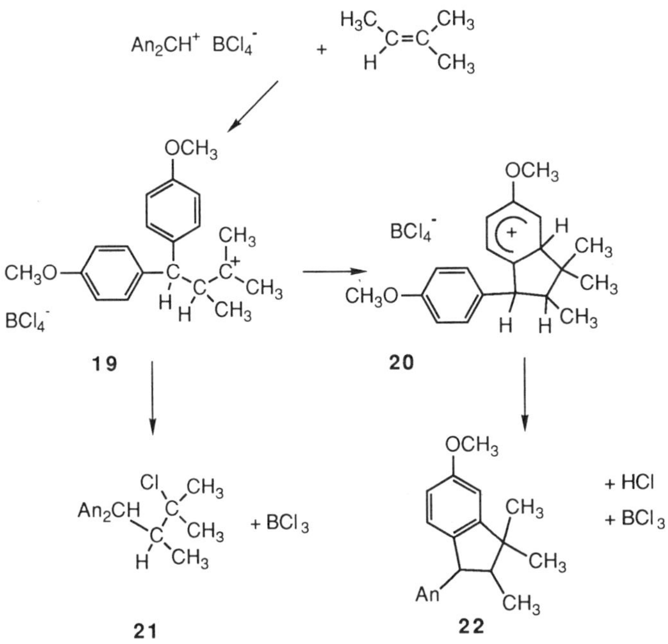

$[BCl_4^-]$ / mmol L^{-1}	0.055	0.56	3.4	15	49
21/22	0.7	2.1	4.7	6.1	5.3

Scheme 24 The influence of benzyltriethylammonium tetrachloroborate on the product ratio observed in the reaction of bis(*p*-anisyl)carbenium tetrachloroborate (0.055 mmol L^{-1}) with 2-methyl-2-butene in CH_2Cl_2 at $-70°$ C. (From Refs. 58, 69.)

shift **24** → **25** → **26** is suppressed at a higher $ZnCl_3^-$ concentration which accelerates the transfer of a chloride ion to **24** (Scheme 25).

Low reaction temperatures also favor the isolation of the initial addition products (Scheme 26). Although the $ZnCl_2 \cdot OEt_2$-catalyzed reactions of the propargyl chlorides **32** (R=CH_3, Ph) with isobutene and isoprene at −78° C gave the acyclic addition products **29** and **33**, respectively, the cyclic vinyl chlorides **30**, **31**, and **34** were obtained at 0° C [81,101,102].

As already shown for hydride shifts, there are also cases, in which consecutive π-cyclizations cannot be suppressed. For none of the reactions, described in Scheme 27, conditions could be found, under which acyclic products were isolable [99,103–105]. The gem-dialkyl effect [9] accelerated the cyclizations so much that these reactions afforded cyclized products exclusively.

Analogously, rearrangements could not be suppressed in reactions with methylenecyclopropane, vinylcyclopropane or norbornene (Scheme 28). The intermediate carbocations **35–37** are not intercepted, and the rearranged products **38–40** have been isolated instead [106–108].

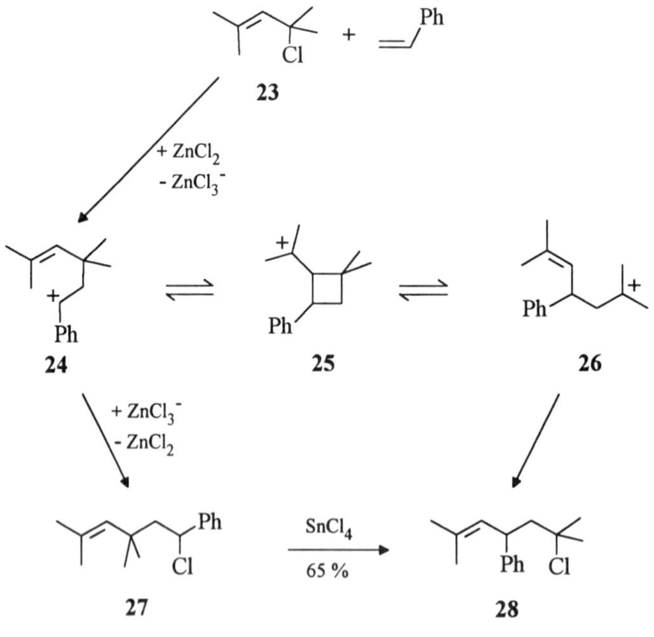

Scheme 25 (From Ref. 69.)

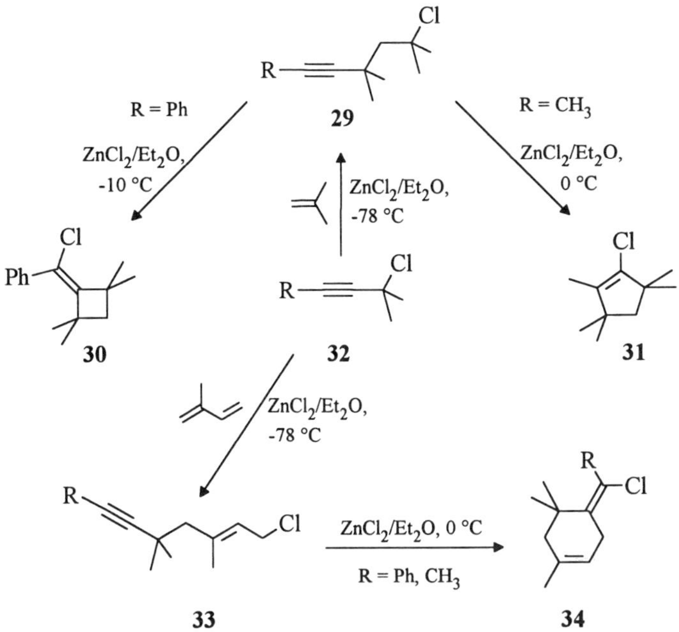

Scheme 26

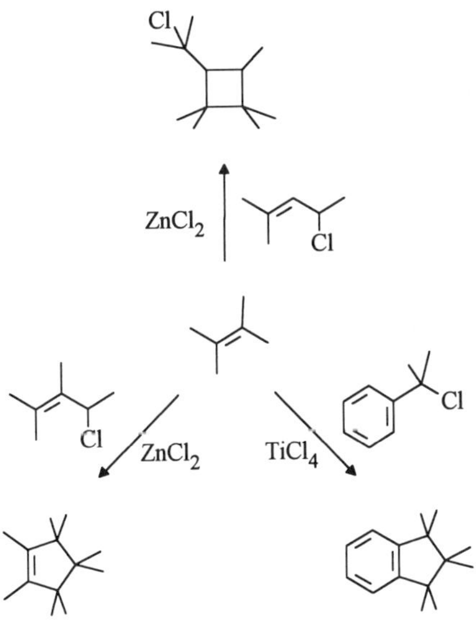

Scheme 27

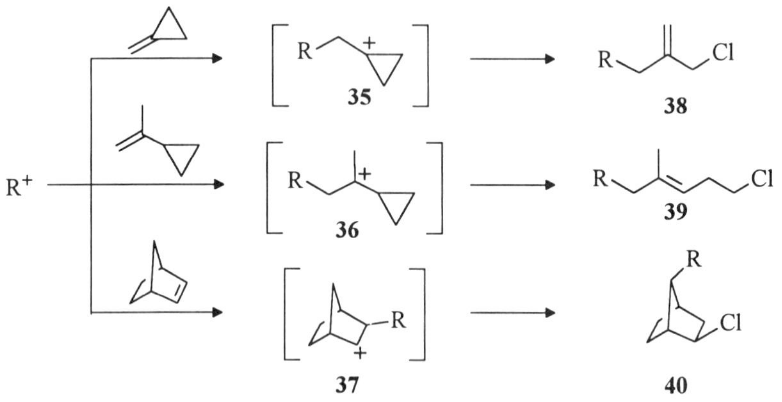

Scheme 28

C. Thermodynamics of Addition Reactions: Addition versus Grob-Fragmentation

To treat the thermodynamics of addition reactions, we consider four extreme cases that differ in the degree of ionization of reactants and products: (a) reactants and products are covalent; (b) reactants and products are ionic; (c) reactants are covalent, products ionic; and (d) reactants are ionic, products covalent. The classifications a–d are also used in the subscripts ($\Delta G_a^0 \ldots \Delta G_d^0$) in Scheme 29.

Case a. If carbocations exist only in small equilibrium concentrations (i.e., the reactants and products are predominantly covalent), ΔH_a^0

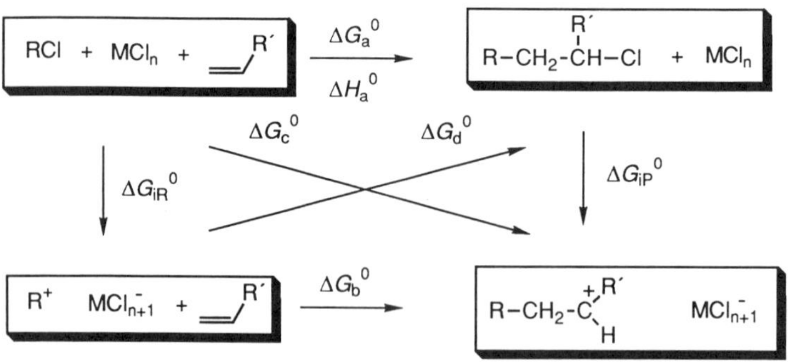

Scheme 29 (From Ref. 69.)

is given by the energy gain for the conversion of a π_{cc} into a σ_{cc} bond. This energy is estimated from mean bond energies [109] to be $\Delta H_a{}^0 = -84$ kJ mol^{-1}, in fair agreement with the calorimetrically determined heats of reaction [35] for the addition of the *p*-methoxy-substituted benzhydryl chloride to 2-methyl-1-pentene (−84.5), styrene (−83.4), 2-methyl-2-butene (−78.9), and isoprene (−74.2).

Changes in the ground state energy of the π-systems and different degrees of steric strain in the addition products have little effect on the magnitude of the enthalpy of addition. Using reaction entropies of −150 to −175 J mol^{-1} K^{-1} determined from increments [110], we obtain $\Delta G_a{}^0$ (−70° C) ≈ −50 kJ mol^{-1}. The magnitude of $\Delta G_a{}^0$, which is independent of the Lewis acid, only varies significantly if considerable steric strain is present in the addition products. However, as even the strongly hindered tris(*p*-chlorophenyl)methyl chloride **41** reacts with isobutene at −10° C to give the adduct **42** [111], we can conclude that $\Delta G_a{}^0$ only becomes positive in extreme cases.

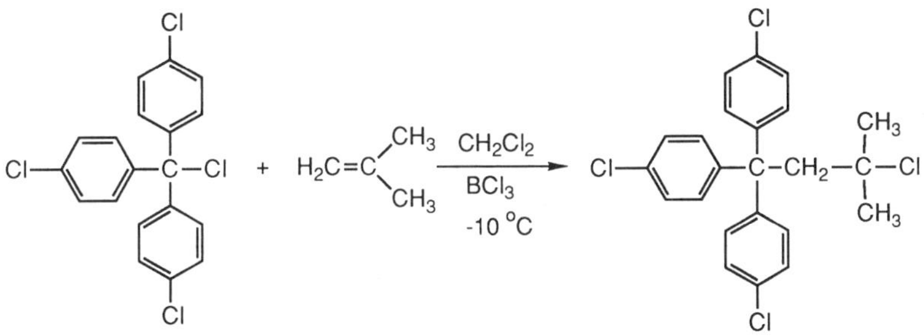

41 **42** [*1*]

Case b. If the reactants and products are ionic, i.e., ΔG^0_{iR}, $\Delta G^0_{iP} < 0$, the standard free energy of reaction according to Scheme 29 is given by Eq. (15).

$$\Delta G_b{}^0 = \Delta G_a{}^0 - (\Delta G^0_{iR} - \Delta G^0_{iP}) \tag{15}$$

Equation (15) shows that $\Delta G_a{}^0$ must be combined with the difference in the free energies of ionization of reactants and products in this case. To have a negative $\Delta G_b{}^0$, the difference in carbocationic stabilization $(\Delta G^0_{iR} - \Delta G^0_{iP})$ must not exceed the energy gain by conversion of a π_{cc} into a σ_{cc} bond ($\Delta G_a{}^0 \approx -50$ kJ mol^{-1}, see Case a.).

Case c. If ionic products (i.e., $\Delta G^0_{iP} < 0$) are to be formed from covalent reactants (i.e., $\Delta G^0_{iR} > 0$), the standard free energy of reaction is given by Eq. (16).

$$\Delta G_c{}^0 = \Delta G_a{}^0 + \Delta G^0_{iP} \tag{16}$$

Since now $\Delta G_a{}^0$ and ΔG^0_{iP} are negative, $\Delta G_c{}^0$ must also adopt a negative value. It is therefore expected that all addition reactions of type c should be favored thermodynamically.

Case d. If covalent products (i.e., $\Delta G^0_{iP} > 0$) are to be formed from ionic reactants ($\Delta G^0_{iR} < 0$), the standard free energy of reaction is given by Eq. (17).

$$\Delta G_d{}^0 = \Delta G_a{}^0 - \Delta G^0_{iR} \tag{17}$$

According to Eq. (17), $\Delta G_d{}^0$ is negative if $\Delta G_a{}^0 < \Delta G^0_{iR}$. A value of -50 kJ mol^{-1} was estimated for $\Delta G_a{}^0$ ($-70°$ C) above, so that carbocation salts can only add to olefins to form covalent products if $\Delta G^0_{iR} > -50$ kJ mol^{-1}. As ΔG^0_{iR} depends both on the structure of R^+ and on the Lewis acidity of MCl_n, we can conclude that the thermodynamic driving force for Case d increases with decreasing stabilization of R^+ and decreasing chloride-ion affinity of MCl_n. Because the magnitude of ΔG^0_{iR} can be estimated from the relative chloride affinities of carbocations and Lewis acidic metal halides in Scheme 7 (Section II.G), one can derive which carbocationic salts might add to alkenes with formation of covalent products.

The reverse of the addition reactions discussed here is the Grob fragmentation [112–114]. In agreement with the considerations for Case d, fragmentation is observed if the CC bond cleavage yields a well-stabilized carbenium ion, as for instance in the ionization of β-amino-substituted alkyl derivatives (Scheme 30).

On the other hand, Overman described that additions of iminium ions to alkynes can be achieved when nucleophilic anions are added to the reaction mixture [115,116].

Analogously, fragmentation to form the aromatic tropylium ion ($pK_{R^+} = +4.76$) [33] is observed when **43** (R=H) is treated with perchloric acid [117]. To demonstrate the counterion effect discussed above, we combined tropylium tetrafluoroborate (**44**·$BF_4{}^-$) with isobutene in methanol and isolated **43** (R = CH_3) in 45% yield [118].

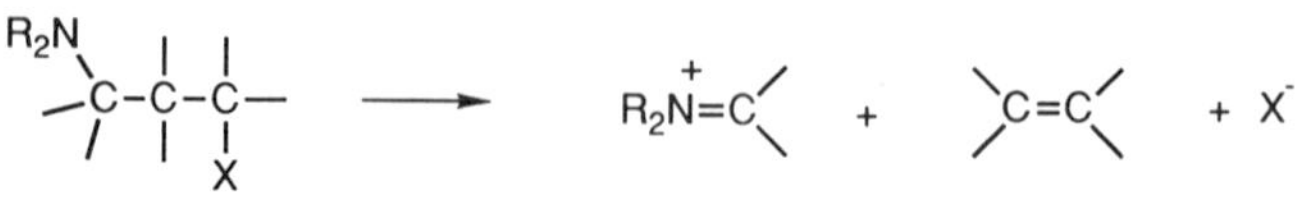

Scheme 30

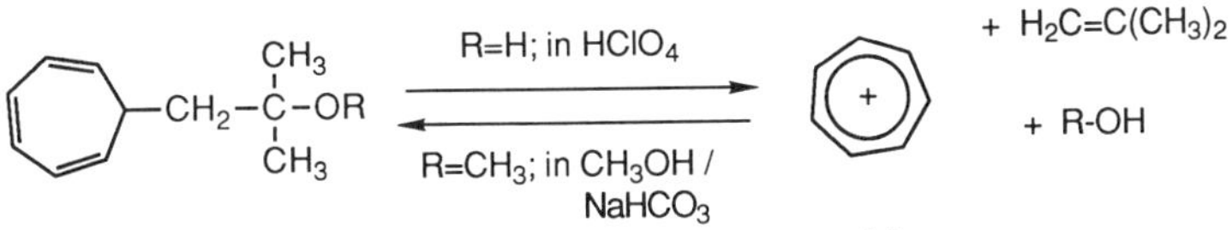

43 44 [2]

D. Kinetic Investigations

1. General

Until recently, knowledge about absolute and relative rates of reaction of alkenes with carbocations was very limited and came almost exclusively from studies of carbocationic polymerizations [119–125]. The situation changed, when it became obvious that reactions of carbocations with alkenes do not necessarily yield polymers, but terminate at the 1:1 product stage under appropriately selected conditions (see Section III.A). Three main sources for kinetic data are now available: Relative alkene and carbocation reactivities from competition experiments, absolute rates for reactions of stable carbocation salts with alkenes, and absolute rates for the reactions of Laser-photolytically generated carbocations with alkenes. All three sets of data are in perfect mutual agreement, i.e., each of these sets of data is supported by two independent data sets.

Prior to the quoted investigations, Dorfman had reported rate constants for the reactions of radiolytically generated benzyl- and benzhydryl cations [126] with some alkenes and dienes [127]. Although Dorfman's data on benzhydryl cations agreed well with the results of the other methods, two independent groups came to the conclusion that the corresponding benzyl cation data must be erroneous [128,129]. Because only a very limited number of compounds had been studied by the radiolytic method, details of this technique will not be presented here.

2. Kinetic Methods

a. Determination of Relative Reactivities by Competition Experiments. Relative reactivities of nucleophiles can be determined by treating carbocation precursors R-Cl with a metal chloride, MCl_n, in presence of an excess of two competing olefins (or other nucleophiles). If reaction conditions are selected, which exclude consecutive reactions of the [1:1] products with the excess alkenes (Section III.A), the relative reactivities of the two alkenes can be derived from the ratio of the addition products **45** and **46** (Scheme 31) [107,130–132] using the known algorithms [133,134]. The competition constants thus determined will only reflect the relative

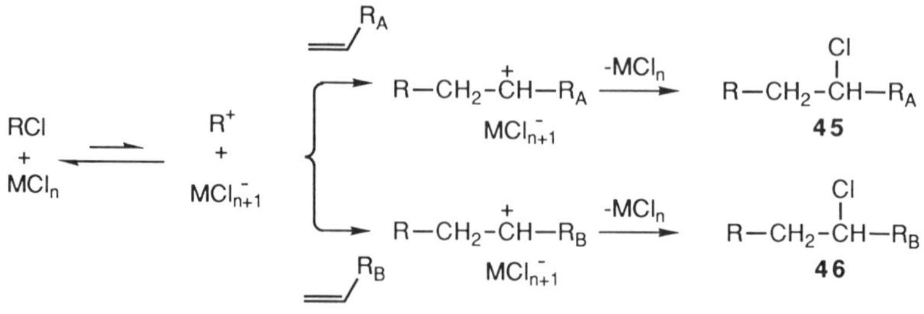

45
46

Scheme 31

alkene reactivities if the attack of the carbocation R^+ at the olefin is irreversible. The cross-over experiment described in Scheme 32 proved this behavior [130].

Phenyltolylmethyl chloride and olefin **47** reacted in presence of $ZnCl_2$ to give the triphenyltolyl-substituted compound **50** as the only product. The same adduct was produced exclusively by combination of diphenylmethyl chloride with olefin **49**. If the intermediate cation **48** would undergo fragmentation under these conditions, the symmetrical diphenylditolyl and the tetraphenyl homolog of **50** should be produced along with the unsymmetrical adduct **50**. Because none of these adducts was observable, irreversible formation of **48** has been proven. In the same way, the irreversibility of the CC-bond-forming step in reactions of $An_2CH^+BCl_4^-$ with 1,1-dialkylethylenes has been confirmed [58].

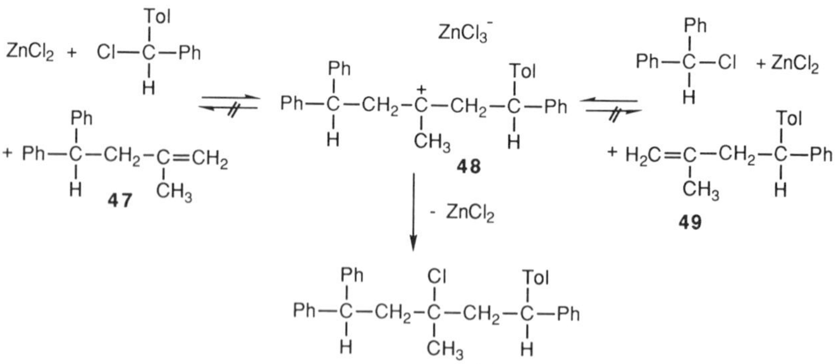

47
48
49
50

Scheme 32

Analogous competition experiments can be performed to derive the relative reactivities of carbocations (Scheme 33). When two alkyl chlorides R_ACl and R_BCl are completely ionized by an excess of MCl_n, the relative reactivities of R_A^+ and R_B^+ can be derived from the product ratio **51**:**52**. The more complex situation which is encountered, when R_ACl and R_BCl are only partially ionized, will be discussed in Section III.D.5.

b. Determination of Absolute Rate Constants Using Stable Carbocation Salts. Most kinetic data on carbocation reactivities presently available have been obtained with the work station depicted in Figure 3. In a typical experiment, the carbocation precursor (e.g., diarylmethyl chloride) is added dropwise to a solution of a Lewis acid (e.g., BCl_3) which must be strong enough (see Section II.G) to rapidly and completely ionize the carbocation precursor. Conductance and absorbance (fiber optics) are measured after addition of each portion of $Aryl_2CHCl$, and because the total concentration of the carbocation is known from the amount of precursor ($Aryl_2CHCl$) added, calibration curves are obtained that relate conductance and absorbance to the total concentration of the carbocations (free + paired!).

When an appropriate alkene is added, color and conductance disappear due to the formation of covalent adducts (Scheme 34). For the success of the method it is essential that the chloride transfer from the complex anion to the new carbocation is fast and complete, because only then the selective formation of 1:1 products and the controlled decay of absorbance and conductance is warranted. Ideally, both quantities (absorbance and conductance) yield the same dependence of carbocation concentration on time, and the second-order rate law (18) is generally obeyed.

$$-d[\mathrm{Aryl_2CH^+}]/dt = k_2[\mathrm{Aryl_2CH^+}][\mathrm{alkene}] \qquad (18)$$

A rationalization for this unexpectedly simple rate law, where $[R^+]$ represents the total concentration of carbocations (i.e., free and paired cations) is given in Section III.D.3.

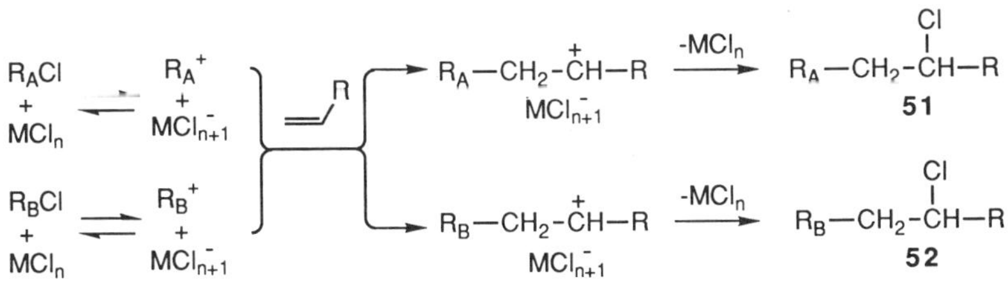

Scheme 33

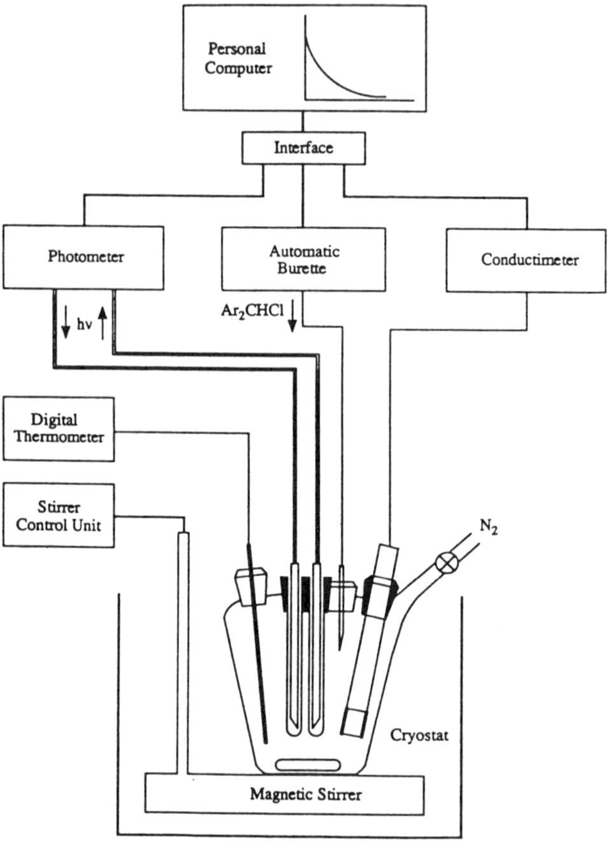

Figure 3 Work station for determining reactivities of diarylcarbenium ions. (Reprinted with permission from Ref. 58.)

c. **Kinetic Studies with Laser-Flash Photolytically Generated Carbocations.** Although irradiation of diarylmethyl chlorides with 248 nm light in CH_2Cl_2, THF, and cyclohexane almost exclusively led to homolytic cleavage of the C—Cl bond, homolytic and heterolytic cleavage takes place in acetonitrile [135]. The quantum yields for homolysis (0.2–0.4) were found to be rather independent of the nature of the substituent on the benzene ring, but the quantum yields for heterolysis increase with increasing electron-donor strength from 0.05 for $(p\text{-ClC}_6H_4)_2CHCl$ to 0.3 for An_2CHCl, as derived from the UV-visible signals. In the presence of

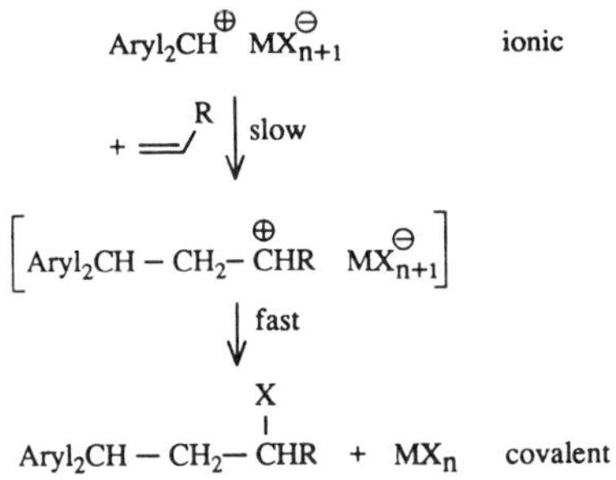

Scheme 34 (From Ref. 69, reprinted with permission of VCH Verlagsgesellschaft.)

nucleophiles, the decay of the transient carbocations follows a pseudo-first-order rate law [[Nuc]$\gg$[$Aryl_2CH^+$], Eq. (19)], and the second-order rate constants k_2 can be derived from the slopes of k_{obs} vs. [Nuc] plots [Eq. (20)].

$$-d[Aryl_2CH^+]/dt = k_{obs}[Aryl_2CH^+] \tag{19}$$

$$k_{obs} = k_0 + k_2[Nuc] \tag{20}$$

The intercept k_0 accounts for the reaction of $Aryl_2CH^+$ with the solvent acetonitrile and for the recombination with Cl^-. Since the magnitude of k_0 ranges from 3×10^5 to 3×10^6 sec^{-1} for the carbocations under concern, only reactive nucleophiles with $k_2 \cdot [Nuc] \geq 10^7$ sec^{-1} can be studied with this method [136].

3. *Counterion, Ion Pairing, and Solvent Effects*

When the carbocations are generated by Laser flash photolysis, the ion pair collapse with the nucleophilic counterion Cl^- is so fast [136] that the decay cannot be followed with the instrumentation used for these experiments, i.e., only those carbocations which manage to escape from the [$Aryl_2CH^+Cl^-$] ion pair can be observed. Consequently, all rate constants determined for the Laser photolytically produced carbocations refer to the reactions of the nonpaired entities.

Paired and nonpaired carbocations exist in solutions of carbocations with complex counterions [41]. With the dissociation constant $K_D = 1.9 \times 10^{-4}$ mol L^{-1} (see Section II.G) one calculates 83% free ions for a 4.5×10^{-5} M solution and 19% free ions for a 4×10^{-3} M solution of $AnPhCH^+BCl_4^-$ in CH_2Cl_2 at $-70°$ C (Table 1). Yet, according to Table 1, the observed rate constants are exactly the same in both solutions,

Table 1 Independence of the Second-Order Rate Constants (CH_2Cl_2, $-70°$ C) of the Degree of Ion-Pairing

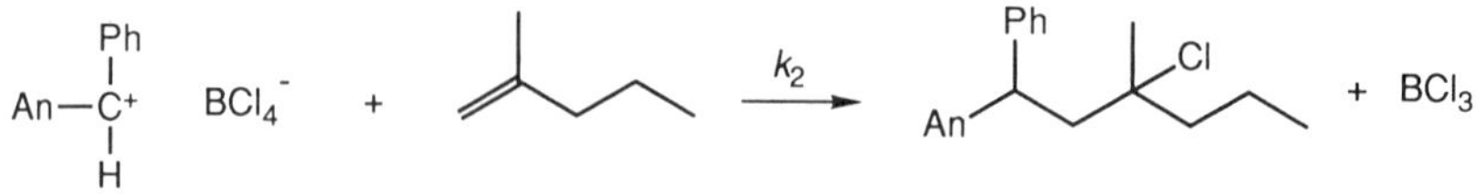

$[AnPhCH^+]_0$ / mol L^{-1}	% free ions[a]	k_2 / L mol^{-1} s^{-1}
4.5×10^{-5}	83 - 96	27.8
8.8×10^{-5}	74 - 95	26.0
2.1×10^{-4}	60 - 89	26.2
4.1×10^{-3}	19 - 38	26.8
1.0×10^{-4} [b]	≈ 13	25.8

a) percentage of free ions at t = 0 and after 80 - 88 % conversion i. e. at the end of the evaluated range; for $K_D = 1.9 \times 10^{-4}$ mol L^{-1}

b) In the presence of 1.0×10^{-2} mol L^{-1} benzyltriethylammonium tetrachloroborate

Source: Ref. 58.

indicating identical reactivities of free and paired ions. This conclusion can also be derived from the fact that these reactions follow second-order kinetics [Eq. (18)]. If free ions and ion pairs react with different rate constants, Eq. (18) should not hold for a large degree of conversion, because the ratio free/paired ions increases during the reaction as the ions are consumed. Table 1, entry 5 shows that even in the presence of benzyltriethylammonium salts (10^{-2} mol L^{-1}), which keeps the percentage of free ions at ~13% throughout the reaction, the same k_2 value is observed. A similar behavior has previously been reported for onium ions [91b].

How can this behavior be explained? In Section II.G, it was reported that the ionization constants (K_I) of $Aryl_2CHCl/MCl_n$ mixtures strongly depend on the substituents in the benzene ring, whereas the dissociation constants K_D are essentially independent of the nature of the cations. A literature search showed that K_D values of about 10^{-4} mol L^{-1} in dichloromethane were found for all combinations of the cations $Aryl_2CH^+$, $Aryl_3C^+$, tropylium, trimethylpyrylium, and tetraethylammonium with the anions BCl_4^-, $SbCl_6^-$, SbF_6^-, and AsF_6^- [41]. This behavior is expected from the "sphere in continuum" model for ions of comparable effective size in a certain solvent [137].

With $\Delta G^0 = -RT \ln K_D$ one can calculate that the standard Gibbs enthalpy of all these ion pairs is approximately 14 kJ mol^{-1} lower than that of the free ions (CH_2Cl_2, −70° C). As K_D was found to be similar for cations of substantially different structure, the same order of magnitude of K_D can also be expected for the cations $Aryl_2CH$-CH_2-CHR^+ and for the activated complexes preceding these cations, as indicated in Figure 4. A constant population ratio between paired and nonpaired ions in ground and transition states can thus be expected, and the independence of the reaction rates of the degree of ion pairing can be rationalized.

The situation is quite different for carbanionic systems, for which large reactivity differences between free ions and ion pairs are well established [138]. What makes the difference? Scheme 35 outlines that the term "ion pairing" has a different meaning in carbocationic and carbanionic chemistry [139].

The left column of Scheme 35 repeats the ionization-dissociation scheme discussed in Section II.G. If carbon is connected to an electronegative element, one speaks of a covalent compound with a polarized C—X bond. This treatment is justified as there is an approximately tetrahedral environment of the corresponding carbon center. Diphenylmethyl chloride, for example, is never termed a contact ion pair. A well-defined ionization step, which was discussed in Section II.G, generates a carbocation

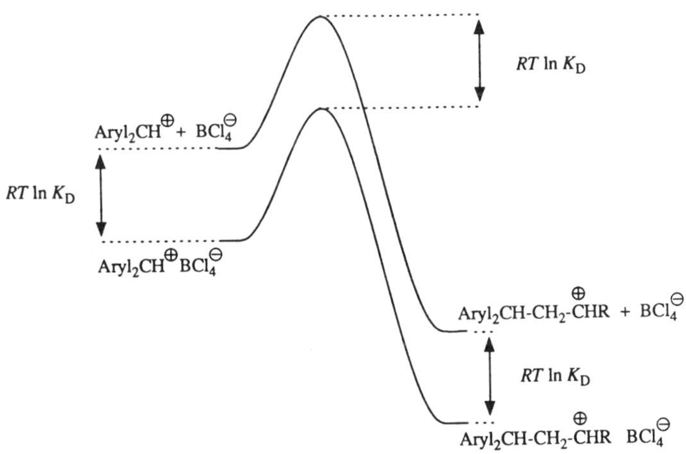

Figure 4 Energy profiles for the rate-determining steps of the additions of paired and unpaired carbenium ions to alkenes. (From Ref. 69, reprinted with permission of VCH Verlagsgesellschaft.)

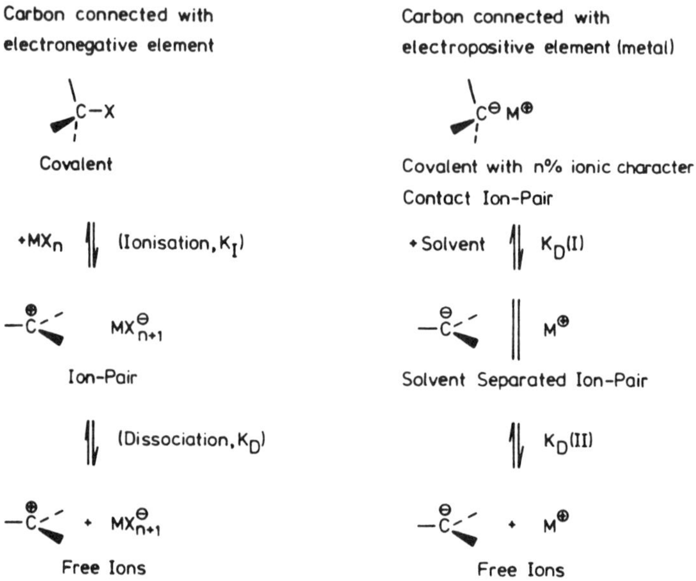

Scheme 35 Carbocationic and carbanionic species which are relevant in polymerizations of alkenes. (From Ref. 139, reprinted with permission of Hüthig & Wepf Verlag, Zug, Switzerland.)

with a planar carbenium center, a process which is associated with large changes in NMR and UV-visible spectra.

Electrostatic interactions restrain the partners in the ion pair. Covalent interactions do not play a role, and there is almost no structural change when the complex counterion is completely removed, i.e., when the free ions are produced. As a consequence, the dissociation step is associated with very minute spectroscopic changes [41], and it is not surprising that entities that are almost identical in their spectroscopic properties also show almost identical reactivity.

Whereas ionization and dissociation are clearly defined processes on the left side of Scheme 35, the situation is more complicated for carbanionic systems (Scheme 35, right). Organic alkali metal compounds, for example, which often exist as aggregates, are often described as covalent species with a certain percentage of ionic character [140–142]. If the formal carbanion is a resonance-stabilized species (e.g, diphenylmethyl lithium or sodium), the species with the closest interaction between the organic fragment and the metal is usually called a contact ion pair. In contrast to the situation on the left of Scheme 35, the interactions between carbanion and metal cation are not purely electrostatic; a covalent or polarization term is also involved. Consequently, considerable spectral changes in NMR and UV-visible spectra can be observed when this interaction is abandoned [143–145] and it is hardly surprising that species with significantly different spectral properties also show different reactivity.

The dissociation constant K_D, which is often derived from spectral properties in carbanion chemistry, therefore includes a covalent term that corresponds to K_I in carbocationic chemistry. As one would not expect equal reactivity of benzhydryl chloride and benzhydryl cations, one also should not expect equal reactivity for benzhydryl lithium and benzhydryl anions. As one realizes that the terms "contact ion-pair" and "dissociation" have a different meaning in carbocation and carbanion chemistry, the apparent discrepancies quoted above, will disappear.

After establishing that free and paired carbocations show the same reactivity, it is not surprising that the reaction rates are also independent of the nature of the counterions (Table 2). The slightly reduced k_2 value for the reaction of the $SnCl_5^-$ salt has been explained by a small degree of reversibility of this reaction; because of the low solubility of this salt, only a very small anion concentration could be employed.

The reaction of the bis(*p*-methoxyphenyl)carbenium tetrachloroborate with 2-methyl-1-pentene has finally been used to study the influence of solvent polarity on reaction rates [58]. Because the rate-determining step involves the formation of a monopositively charged ion from another monopositively charged ion and a neutral component, so that charge is neither generated nor destroyed, the rate constants increase only slightly with increasing solvent polarity ($k_{CH_3NO_2}/k_{CHCl_3} = 4.8$). While the correlation between log k_2 and ϵ is rather poor, a fair correlation with the solvent polarity parameter $E_T(30)$ [146] was found (Fig. 5).

Table 2 Rate Constants for the Reactions of Various Bis(*p*-methoxyphenyl) Carbenium Salts with 2-Methyl-1-pentene in CH_2Cl_2 at $-70°$ C

$[Anion]_0$/mmol L^{-1}	k_2/L mol^{-1} sec^{-1}
$BCl_4^-/4.60 \times 10^{-4}$	2.64×10^{-2}
$BClBr_3^-/8.70 \times 10^{-4}$	2.70×10^{-2}
$B(OCH_3)Cl_3^-/7.01 \times 10^{-4}$	2.81×10^{-2}
$SnCl_5^-/4.22 \times 10^{-5}$	2.48×10^{-2}

Source: Ref. 58.

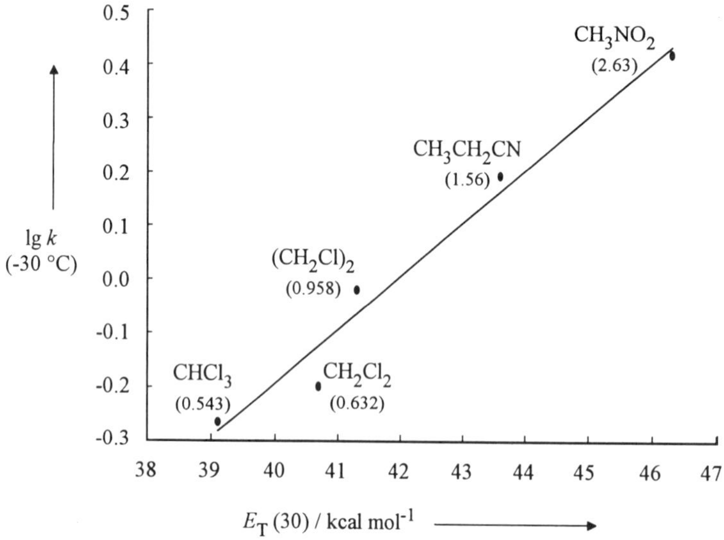

Figure 5 Rate constants for the reaction of $An_2CH^+BCl_4^-$ with 2-methyl-1-pentene as a function of the solvent polarity parameter $E_T(30)$. (From Ref. 58.) $\log k = 0.0995\ E_T(30) - 4.175$ ($r = 0.986$); k_2 values ($-30°$ C) in parentheses.

4. *Variation of the Alkenes*

a. Overview. Scheme 36 shows rate constants of a series of representative hydrocarbons toward the *p*-methoxy-substituted benzhydryl cation, the electrophile that had been used for most kinetic investigations [82]. Several features can be realized: In the left column, there are two blocks of olefins, a lower one with compounds that give secondary and an upper one with compounds that give tertiary carbocations. The fact that isobutene is 2×10^4 times more reactive than the (*E*,*Z*)-isomeric 2-butenes demonstrates the unequal effect of methyl groups at the two termini of the double bond, in accord with an unsymmetrical transition state. Norbornene sits between these two blocks as expected from the location of the norbornyl cation in the stability scales (cf., Section II).

The conjugated π-systems in the right column of Scheme 36 show reactivities similar to those of the alkyl-substituted double bonds, and there is no evidence that conjugated double bonds are generally more easily attacked by electrophiles.

In the following discussion it is usually not necessary to take into account the method or the reference carbocation which has been em-

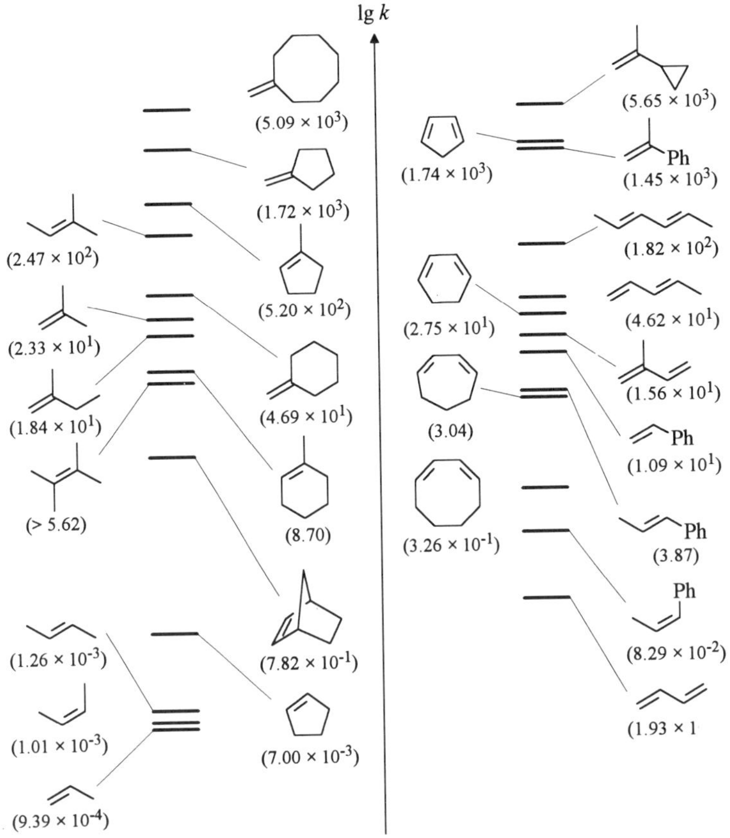

Scheme 36 k_2 (L mol^{-1} sec^{-1}) for $AnPhCH^+$ + alkenes and dienes (CH_2Cl_2, −70° C; data from Refs. 82 and 106).

ployed for comparing certain alkenes. The arbitrarily selected couples in Table 3 prove that the same conclusions can be derived from relative alkene reactivities determined by the competition method (with $Tol_2CHCl/ZnCl_2 \cdot OEt_2$) or from direct rate measurements (with $AnPhCH^+BCl_4^-$). It has been found, however, that the relative reactivities of alkenes with different substitution at the position of electrophilic attack do depend on the electrophile [128]. As a consequence, these comparisons have to be more sophisticated.

The consistency of the rate constants determined for the reaction of methylene-cyclopentane with stable carbocation salts in CH_2Cl_2 and with

Table 3 Comparison of Relative Alkene Reactivities Determined by Competition Experiments and Measurements of Absolute Rate Constants (CH_2Cl_2/−70° C)

Comparison	Tol_2CH^{+} [a]	$AnPhCH^{+}$ [b]
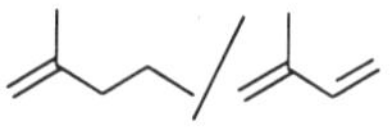	1.79	1.65
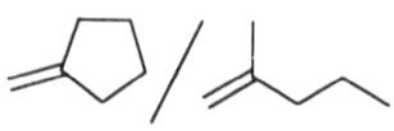	95.1	67.8
	82.3	63.8
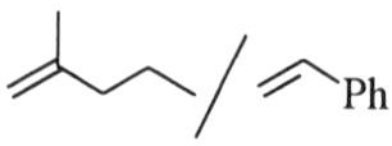	3.97	2.37
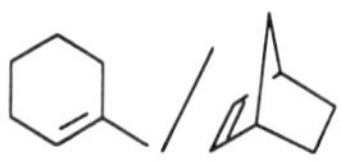	13.6	11.1
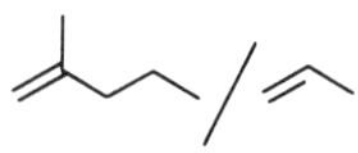	3.16×10^4	2.75×10^4

[a] Competition method: Tol_2CHCl, $ZnCl_2 \cdot OEt_2$, CH_2Cl_2, −70° C. *Source*: Refs. 107 and 130.
[b] Direct rate measurements: $AnPhCH^{+}BCl_4^{-}$, CH_2Cl_2, −70° C. *Source*: Refs. 82 and 128.

Laser flash photolytically generated carbocations in CH_3CN is demonstrated by Fig. 6. Because of the small solvent effects discussed above, solvent corrections have not been performed for this correlation. One has to be aware of leveling effects, however, when rate constants $>5 \times 10^7$ L mol^{-1} sec^{-1} are considered [136].

b. Structure and Reactivity of Alkenes and Dienes. Substituent effects are usually discussed in terms of steric and electronic effects. Steric effects are usually negligible when groups far remote from the reaction center are exchanged. Figure 7, which compares the relative reactivities of *p*- and *m*-substituted styrenes toward Tol_2CH^{+}, shows a linear correlation of log k with σ^{+} (correlation coefficient $r = 0.993$); the corresponding correlation with σ is of somewhat lower quality ($r = 0.984$). The Hammett ρ^{+}-value of −5.0, derived from Fig. 7 indicates a transition state with a high positive charge density at the benzylic carbon of styrene [107].

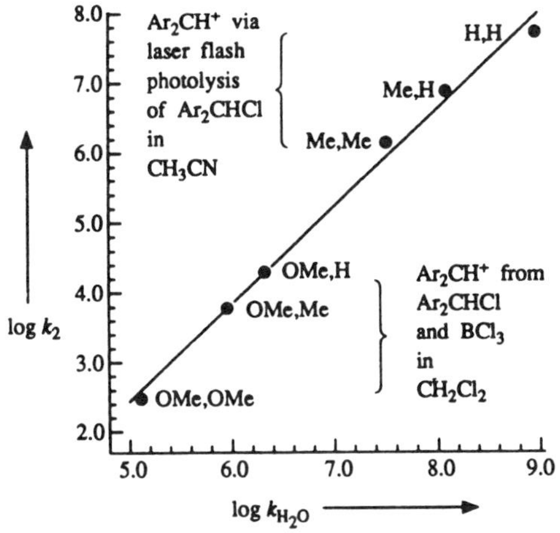

Figure 6 Correlation of the reactivities of benzhydryl cations (parasubstituents in graph) toward methylenecyclopentane at 20° C with the corresponding reactivities toward H_2O (k_{H_2O} in H_2O/acetonitrile = 2/1; sec^{-1}); log k_2 = 1.397 log k_{H_2O} − 4.55. (Reprinted with permission from Ref. 136. Copyright 1991 American Chemical Society.)

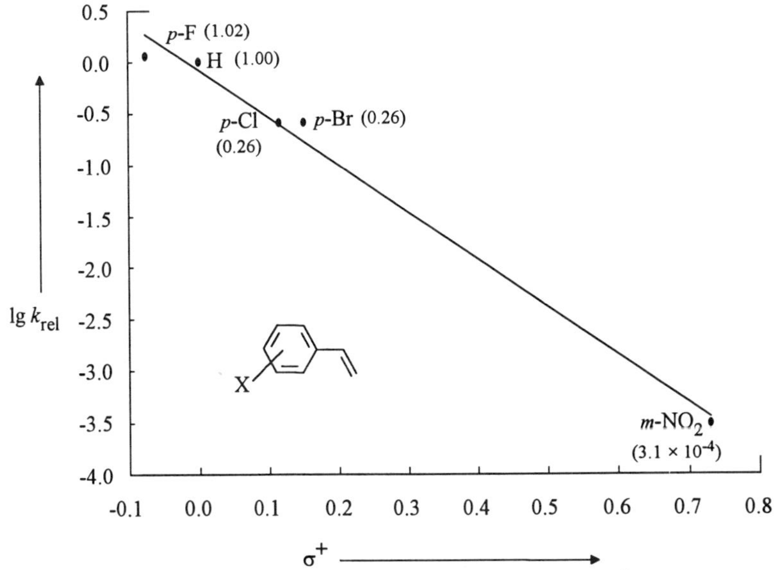

Figure 7 Relative reactivities of substituted styrenes toward $Tol_2CHCl/ZnCl_2/Et_2O$ (CH_2Cl_2, −70° C). (From Ref. 107.) k_{rel} values in parentheses.

Let us now keep electronic effects almost constant and change the size of alkyl groups. According to Scheme 37 (left column), successive replacement of one methyl group in isobutene by ethyl, isopropyl, and tert-butyl leads to a moderate decrease of reactivity, showing a small influence of steric effects at the position of the new carbocation center. When branching occurs in homoallylic position (Scheme 37, right), the hyperconjugative acceleration even overcompensates the steric retardation.

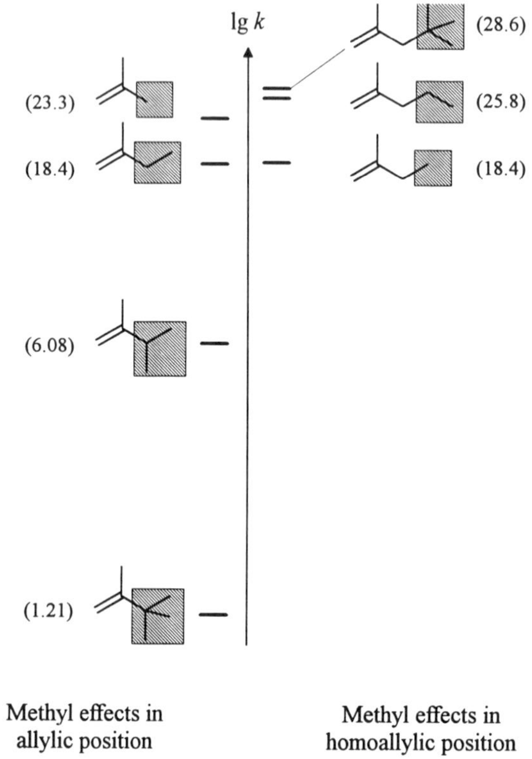

Scheme 37 k_2(L mol^{-1} sec^{-1}) for AnPhCH$^+$ + $H_2C{=}C(CH_3)R$ (CH_2Cl_2, $-70°C$). (From Ref. 82.)

In contrast, dramatic steric effects are found, when a tert-butyl group is introduced at the position of electrophilic attack. Scheme 38 shows, that the tert-butyl-substituted compound reacts at least 10^3 times more slowly than the corresponding methyl-substituted compound. The determination of the accurate rate constant for 2,4,4-trimethyl-2-pentene failed, as this

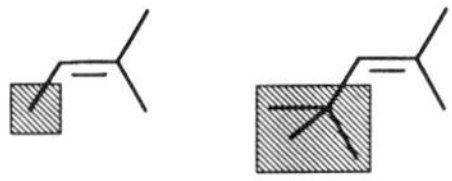

(247) (< 0.247)

Scheme 38 k_2 (L mol^{-1} sec^{-1}) for $AnPhCH^+$ + $RCH{=}C(CH_3)_2$ (CH_2Cl_2, −70° C). (From Ref. 82.)

olefin underwent acid-catalyzed rearrangement into the more reactive 2,4,4-trimethyl-1-pentene during the course of the kinetic experiment [82]. From these data one can derive that the substituent effects at the new carbocationic center, which are discussed in Scheme 39, do predominantly have an electronic origin. The left column of Scheme 39 shows the familiar substituent order $CH_3 < CH{=}CH_2 < Ph$ for stabilization of the developing carbenium center. Depending on its configuration, (*Z*) or (*E*), the effect of the 1-propenyl group is somewhat smaller or larger than that of phenyl [82,147]. The phenyl/methyl ratio of 10^4 (styrene/propene; Scheme 39, left) decreases to 62 (α-methylstyrene/isobutene; Scheme 39, right) when the electron demand of the new carbocation center is reduced by the presence of an extra methyl group. Both, vinyl and methyl group now accelerate by about four orders of magnitude compared with hydrogen (Scheme 39, right). The ethynyl group even desactivates relative to hydrogen; as 2-methyl-1-butene-3-yne was reported to be inert under conditions in which propene reacts smoothly with electrophiles [107], one can estimate that this enyne is at least one order of magnitude less reactive than propene.

Scheme 39 further shows that the introduction of a phenyl group into an allylic position of isobutene causes a rate reduction of 20. The reactivity of the benzyl-substituted double bond is obviously diminished by the negative inductive effect of phenyl, and phenonium stabilization of the intermediate carbocation does not take place [148].

Comparison of the 1,3-dienes in both columns of Scheme 39 shows that 1,3-butadiene is the least reactive compound, whereas all other 1,3-dienes in this scheme show similar reactivity. This behavior reflects the fact that 1,3-butadiene yields a terminally monoalkylated allyl cation whereas the other dienes give terminally dialkylated allyl cations (cf., Section III.B.1., Scheme 17). The additional methyl group in the central allylic position, which is present in the allyl cation generated from 2,3-dimethylbutadiene, does not contribute to the stabilization.

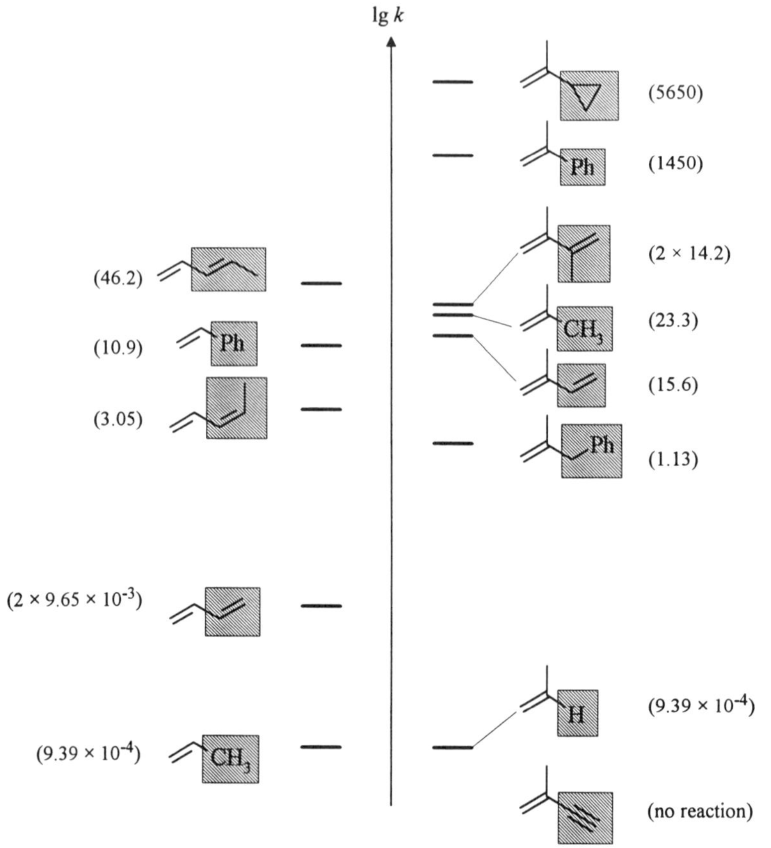

Scheme 39 k_2 (L mol^{-1} sec^{-1}) for $AnPhCH^+$ + $H_2C{=}CHR$ (left) and $H_2C{=}C(CH_3)R$ (right); (CH_2Cl_2, −70° C). (From Refs. 82, 147, 148.)

Terminally disubstituted allyl cations are also produced from the dienes listed in Scheme 40. Accordingly, 2(*E*),4(*E*)-hexadiene shows a similar reactivity as (*E*)-1,3-pentadiene (see Scheme 39). The strong dependence of the diene reactivity on ring size is obvious. Cyclopentadiene is considerably more reactive than its acyclic analogon, which may be explained by the "aromatic stabilization" of the 2-cyclopentenyl cation [149]. Increasing ring size disturbs the coplanarity of the 1,3-diene fragment and of the intermediate allyl cation, and the reactivity decreases by two orders of magnitude from 1,3-cyclohexadiene to 1,3-cyclooctadiene.

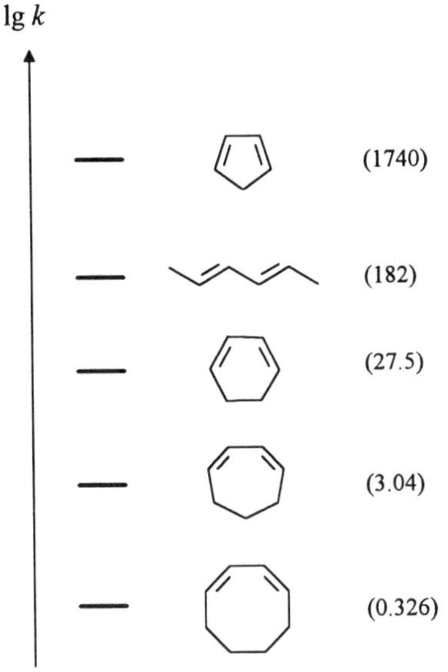

Scheme 40 k_2 (L mol^{-1} sec^{-1}) for AnPhCH$^+$ + cycloalka-1,3-dienes (CH_2Cl_2, −70° C). (From Ref. 82.)

In nonconjugated dienes, the negative inductive effect of the extra double bond reduces the nucleophilicity of the other π-system (Fig. 8) [148]. There is no evidence for anchimeric assistance by the additional double bond, as observed in certain solvolytic reactions [10,151–153]. When the two double bonds are separated by four methylene groups, they behave as isolated double bonds, and their reactivity equals that of an analogous monoene.

A remarkable dependence of the reactivity on ring size has been found in the series of methylenecycloalkanes (Fig. 9) [106]. The exceptionally low rate constant for methylenecyclopropane indicates that the low solvolysis rates of cyclopropyl derivatives [154] are not only caused by the unfavorable change of hybridization of one ring carbon in cyclopropane but also by the low stability of the cyclopropyl cation relative to a compound with the same hybridization (methylenecyclopropane). The destabilization of the cyclopropyl cation must actually be greater than indicated by the numbers in Fig. 9 as the transition state of the electrophilic attack may already profit from the stabilizing ring-opening process (cf., Section III.B.2).

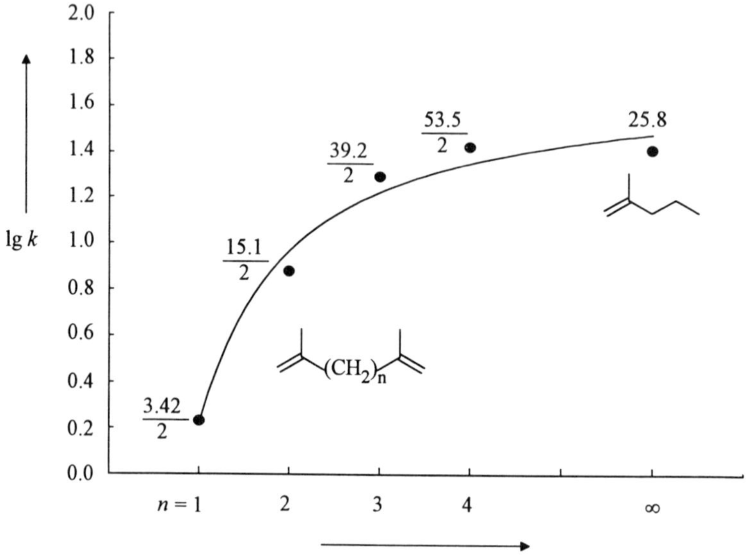

Figure 8 Rate constants (L mol^{-1} sec^{-1} per double bond) for the reactions of $AnPhCH^+$ with nonconjugated dienes (CH_2Cl_2, $-70°$ C). (From Ref. 148.)

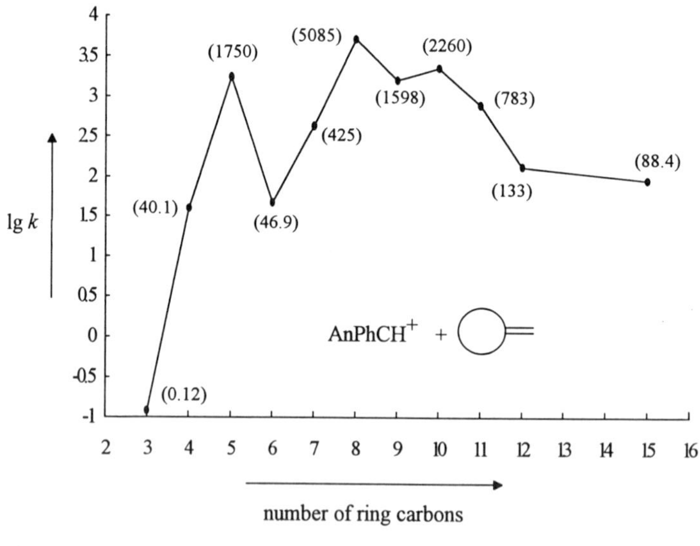

Figure 9 Second-order rate constants (L mol^{-1} sec^{-1}) for the reactions of methylenecycloalkanes with $AnPhCH^+$(CH_2Cl_2, $-70°$ C). (From Ref. 106.)

Spectacular reactivity differences have also been observed for the larger ring systems in which electronic effects are less obvious. Methylenecyclooctane, for example, is 10^2 times more reactive than methylenecyclohexane. It has been reported that the rates of carbocation additions to methylenecycloalkanes correlate with the solvolysis rates of the corresponding cycloalkyl derivatives [106], which have previously been rationalized by different changes of strain during the rehybridization in the rate-determining step [155–157]. Because none of the ring carbons of the methylenecycloalkanes changes hybridization in the rate-determining step of the electrophilic additions, Brown's (I)-strain theory of cycloalkyl solvolyses [158] has been questioned.

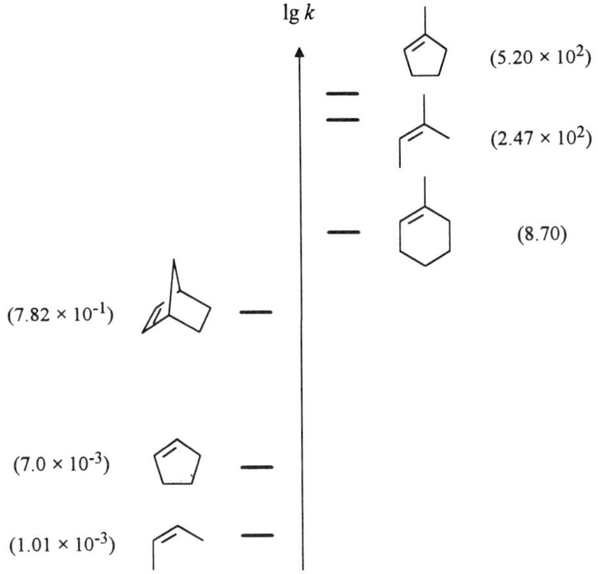

Scheme 41 k_2 (L mol^{-1} sec^{-1}) for reactions with $AnPhCH^+$ (CH_2Cl_2, $-70°$ C). (From Ref. 82.)

The few available data for carbocation additions to cycloalkenes (Scheme 41) show an analogous reactivity order: Cyclopentenes are more reactive than the acyclic analogs, and the only cyclohexene derivative shown in Scheme 41 is less reactive. Because of the paucity of data, this analogy should not be overinterpreted. The location of norbornene between the compounds which give secondary and tertiary carbocations has already been mentioned (Scheme 36 in Section III.D.4.a.).

Although an interplay of enthalpic and entropic effects has been shown to be responsible for the reactivity order of the methylenecycloalkanes, one can summarize that generally the entropy effects caused by substituent variation at the developing carbenium center are small compared with the enthalpic effects.

The effect of substituents at the position of attack of the electrophile is quite different, however. In general, one can find that the introduction

Table 4 Methyl Effects in Alkenes at the Position of Electrophilic Attack (Reactions with the *p*-Methoxy-Substituted Benzhydryl Cation, $An(Ph)CH^+$, CH_2Cl_2, $-70°$ C)

	$\Delta H^‡$ [kJ mol^{-1}]	$\Delta S^‡$ [J mol^{-1} K^{-1}]	k_2 [L mol^{-1} s^{-1}]	k_{rel}	$\Delta\Delta S^‡$ (rel.)
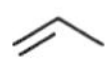	32.6	-139	9.39×10^{-4}	1.0	0
	31.4	-145	1.01×10^{-3}	1.1	-6
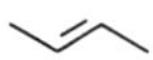	29.9	-150	1.26×10^{-3}	1.3	-11
	21.1	-112	23.3	1.0	0
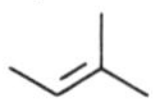	7.5	-159	247	10.6	-47
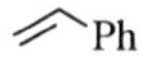	19.3	-127	10.9	1.0	0
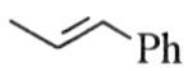	15.5	-154	3.87	0.36	-27
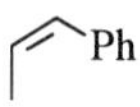	26.1	-134	0.083	0.0076	-7
	22.4	-99	46.2	1.0	0
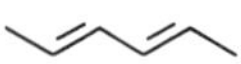	15.1	-124	182	3.9	-25
	15.5	-122	187	1.0	0
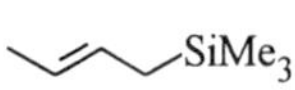	8.6	-130	4190	22.4	-8

Source: Refs. 69, 82, and 159.

of a methyl group at the olefinic position which is attacked by the electrophile leads to a more negative value of $\Delta S^{\ddagger}$ and to a simultaneous decrease in the enthalpy of activation, $\Delta H^{\ddagger}$. This effect is observed analogously in the allylsilane series [159] (Table 4) and only *cis*-1-phenyl-1-propene shows an enthalpic effect that deviates from this rule. Because of the counteracting enthalpic and entropic effects, a methyl group at the position of the electrophilic attack may thus either accelerate or retard the reaction [69].

Figure 10 shows that the same relative reactivities of terminal π-systems, which have been determined with respect to $AnPhCH^+$ as the reference electrophile, can also be observed with respect to An_2CH^+, which is 3 orders of magnitude less electrophilic, or with Tol_2CH^+, which is 2 orders of magnitude more electrophilic than $AnPhCH^+$, i.e., Ritchie's constant selectivity relationship [160] is also observed for this type of reactions.

The relative π-nucleophilicities are not generally electrophile-independent, however, as already mentioned in Section III.D.4.a. Figure 11 shows that π-systems with a methyl group at the position of the electrophilic attack (more general: systems with a more negative entropy of acti-

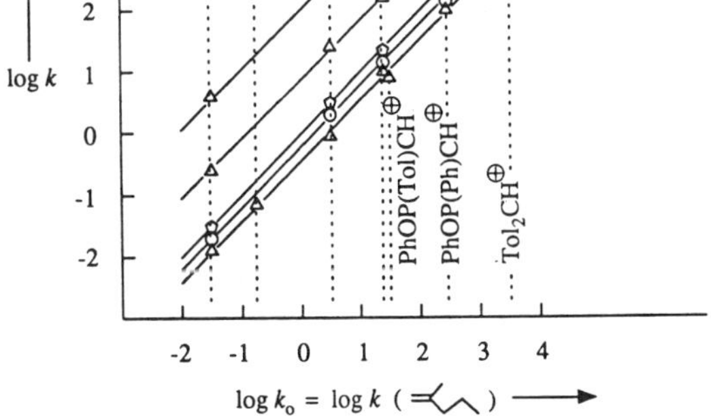

Figure 10 Reactivities of terminal alkenes toward diarylcarbenium ions (−70° C, CH_2Cl_2, reference reaction: $Aryl_2CH^+$ + 2-methyl-1-pentene). (Reprinted with permission from Ref. 128. Copyright 1990 American Chemical Society.)

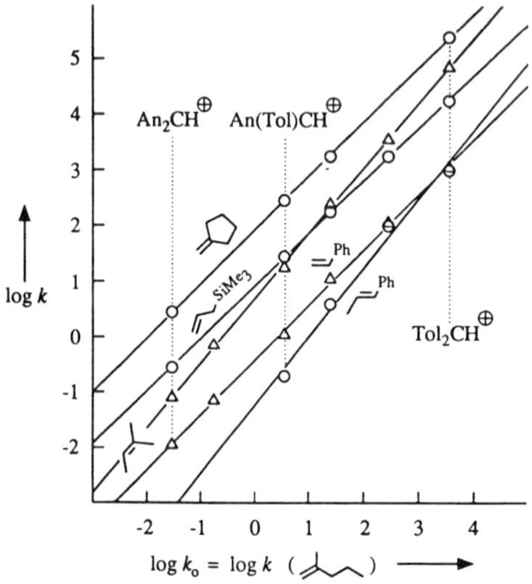

Figure 11 Reactivities of olefins toward diarylcarbenium ions (−70° C, CH_2Cl_2, reference reaction: $Aryl_2CH^+$ + 2-methyl-1-pentene). (Reprinted with permission from Ref. 128. Copyright 1990 American Chemical Society.)

vation) also follow linear free energy relationships, but with larger slopes. Crossing of some correlation lines occurs with the consequence that styrene is 5.4 times more reactive than *trans*-β-methylstyrene toward An-TolCH$^+$ while these two nucleophiles show equal reactivity toward Tol_2CH^+ [128]. As expected from the extension of the correlation lines in Fig. 11, *trans*-β-methylstyrene was found to be 4.4 times more reactive than styrene in competition experiments with $Ph_2CHCl/ZnCl_2$ [107]. The latter examples show that simple structure-reactivity discussions are not any longer possible if variations of the direct environment of the reaction centers are included.

c. Origin of the Substituent Effects and Comparison with Other Electrophilic Additions. Symmetrically bridged and open transition states as shown in Scheme 42 are the two extreme pathways electrophilic additions

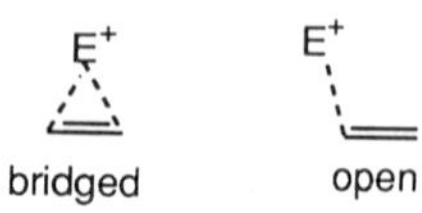

Scheme 42

to CC-double bonds can adopt. Although strongly bridged transition states are usually encountered in electrophilic brominations, chlorinations, sulfenyl- and selenyl halide additions [66,67,161,162,164,165], proton additions are assumed to proceed through open transition states [68,166–170]. Ruasse et al. [171] have employed methyl effects to differentiate the two modes of attack. Whereas bridged transition states are stabilized by methyl groups at both termini of the double bond, open transition states are stabilized predominantly by methyl groups at the developing carbenium center.

Accordingly, Table 5 shows the reactivity order

[3]

for the bridging electrophiles (entries 1–3) and the order

[4]

for proton additions (entry 4). Obviously, there is no resemblance between the substituent effects for halogenations and sulfenylations on one side and for carbocation additions on the other side. Some similarities between the substituent effects for proton and carbocation additions (entries 4 and 5) trigger a closer comparison.

Figure 12 reveals a very moderate correlation between the two sets of data. It is evident that conjugated π-systems (1,3-dienes, styrenes) all lie above the correlation line, i.e., these compounds react much faster with carbocations than expected from their reactivity toward protons. If one assumes that the transition states of both types of reactions are controlled by the stabilization of the newly developing carbocations, one has to conclude that the deviations arise because proton additions to conjugated π-systems are unusually slow and not because carbocation additions are unusually fast. The similar reactivity of isobutene and styrene toward carbocations agrees well with the rule of thumb that in simple carbocations the stabilizing effect of one phenyl group is equivalent to that of two methyl groups [172]. Higher resonance stabilization of the ground state of styrene has been suggested to explain why styrene reacts 10^3 times more slowly with Bronsted acids than isobutene [170]. This explanation is not acceptable, however, because the heats of hydrogenation of isobutene and styrene are identical within experimental error [21]. Possibly, differences in solvation between aryl, vinyl, and alkyl groups adjacent to the carbocationic center [173,174] in water (solvent for H_3O^+ additions)

Table 5 Relative Reactivities of Alkenes Toward Various Electrophiles

	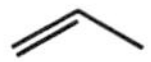		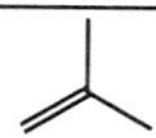	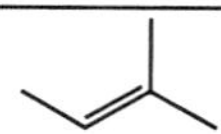	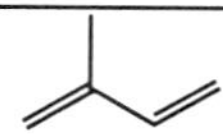	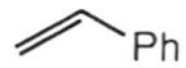	ref
Cl_2(without solvent, -9 and 25 °C)	1.00[a]	50	58	1.1×10^4			[161]
Br_2(CH_3OH, 25 °C)	1.00	32.3	92.3	3.2×10^3		3.80	[171, 165]
2,4-$(O_2N)_2C_6H_3SCl$ (CH_3CO_2H, 30 °C)	1.00[a]	2.08	3.39	15.4		0.66	[164]
H_3O^+(H_2SO_4, 25 °C)	1.00	14.8	1.56×10^5	9.03×10^4	1.34×10^4	1.01×10^2	[166, 167, 169]
$AnPhCH^+$(CH_2Cl_2, -70 °C)	1.00	1.34	2.48×10^4	2.63×10^5	1.66×10^4	1.16×10^4	[82]

[a]1-Butene. *Source*: Refs. 82, 161, 164–167, 169, 171.

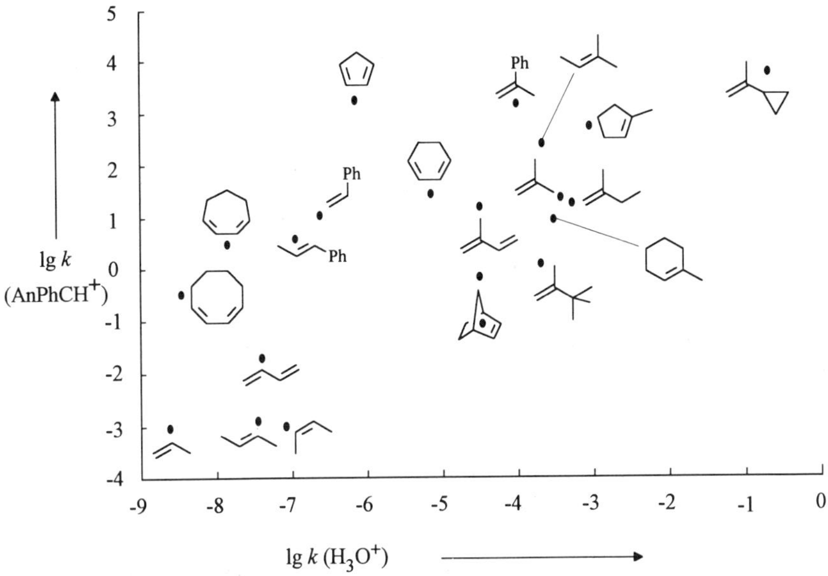

Figure 12 Correlation of the reactivities of alkenes and dienes toward H_3O^+(H_2O, 25° C) and the carbenium ion $AnPhCH^+$(CH_2Cl_2, −70° C). (From Ref. 82.)

and organic solvents (R^+ additions) account for the unexpectedly low protonation rates of conjugated double bonds. A definitive answer is still elusive.

In summary, there is some resemblance between the rates of addition of protons and carbocations (Fig. 12), but the correlation is not good enough to make useful predictions for one set of data from the other.

According to Section III.C [Scheme 29 and Eq. (15)], the standard Gibbs enthalpy for reaction (21) can be calculated from Eq. (22).

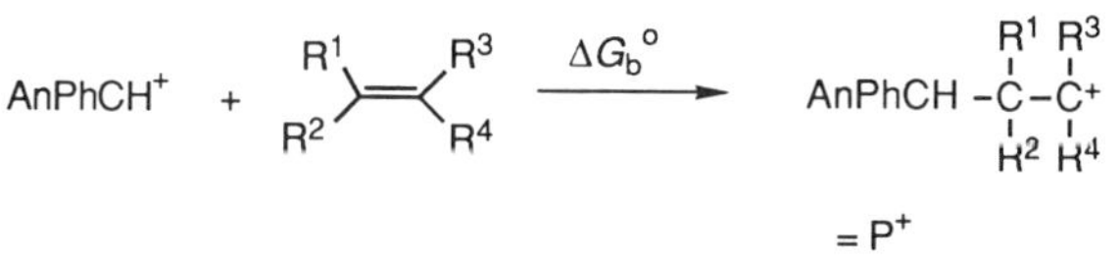

(21)

$$\Delta G_b^0 = \Delta G_a^0\,(\pi_{cc} \rightarrow \sigma_{cc}) - \Delta G_i^0\,(\mathrm{AnPhCHCl} \rightarrow \mathrm{AnPhCH^+}) + \Delta G_i^0\,(\mathrm{PCl} \rightarrow \mathrm{P^+}) \quad (22)$$

As the energy for the conversion of a π- into a σ-bond is fairly constant (exceptions have been discussed), the differences in $\Delta G_b{}^0$ are mostly dependent on the ionization free enthalpy of the products, ΔG_i^0 (PCl → P^+). According to Section II.F, this quantity correlates with several other carbocation stability parameters, e.g., with the solvolysis rate constants of the addition products PCl.

In agreement with these considerations, a fairly good correlation between the rate constants for reaction (Eq. 21) and the solvolysis rates of the addition products has been observed (Fig. 13). Taking into account that approximately 90% of carbocationic character is manifest in the solvolysis transition states [175], the slope of the correlation in Figure 13 (slope × 2.3 RT = 0.45) indicates that roughly half of the character of the new carbocations is developed in the transition states of the addition reactions.

5. *Stability and Reactivity of Carbocations*

The relationship between structure and reactivity of diarylcarbenium ions has already been mentioned in the discussion of Figures 10 and 11 (Section III.D.4.b). Some numbers on which these figures are based, are given in Table 6. They show that the ditolylcarbenium ion (Tol_2CH^+) reacts 10^5 times faster with 2-methyl-1-pentene and 10^6 times faster with 2-methyl-2-butene than the better stabilized dianisylcarbenium ion (An_2CH^+). It

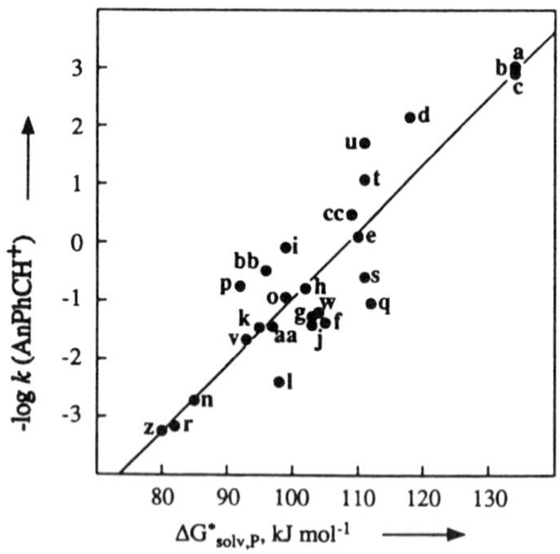

Figure 13 Correlation of the reactivities of alkenes and dienes toward the carbenium ion $AnPhCH^+$ (CH_2Cl_2, −70° C) with the free energies of activation for ethanolysis of the addition products P-Cl. (Reprinted with permission from Ref. 82. Copyright 1990 American Chemical Society.) Letters refer to compound symbols in Ref. 82.

Table 6 Rate Constants and Activation Parameters for the Additions of Diarylcarbenium Ions to 2-Methyl-1-pentene and 2-Methyl-2-butene in Dichloromethane

	2-Methyl-1-pentene			2-Methyl-2-butene		
Carbenium Ion	$k_2(-70°\ C)$ (L mol^{-1} sec^{-1})	$\Delta H‡$ (kJ mol^{-1})	$\Delta S‡$ (J mol^{-1} K^{-1})	$k_2(-70°\ C)$ (L mol^{-1} sec^{-1})	$\Delta H‡$ (kJ mol^{-1})	$\Delta S‡$ (J mol^{-1} K^{-1})
An_2CH^+	2.92×10^{-2}	29.7	−125	8.38×10^{-2}	21.9	−155
$An(p\text{-}PhOC_6H_4)CH^+$	1.69×10^{-1}			7.5×10^{-1}		
$An(Tol)CH^+$	3.38	22.7	−120	1.83×10^{1}	13.6	−151
$An(Ph)CH^+$	2.58×10^{1}	19.5	−119	2.47×10^{2}	7.5	−159
$p\text{-}PhOC_6H_4(Tol)CH^+$	3.30×10^{1}					
$p\text{-}PhOC_6H_4(Ph)CH^+$	2.86×10^{2}	15.3	−119	3.92×10^{3}	5.3	−147
Tol_2CH^+	3.40×10^{3}	11.6	−117	6.79×10^{4}	−1.4	−156

Source: Refs. 58 and 128.

can be seen that these variations of reactivity are entirely caused by changes in $\Delta H^‡$. The activation entropy, $\Delta S^‡$, remains constant within a reaction series, in analogy to the observation that structural variations of the alkenes at positions remote from the reaction center predominantly affect $\Delta H^‡$ while $\Delta S^‡$ remains unchanged.

When $\Delta G^‡$ of these reactions is plotted against the free energy of ionization (ΔG_i^0) (see Section II.D) linear relationships with gradients of 0.67 and 0.79 are obtained (Fig. 14). Analogous rate-equilibrium relationships for other olefins gave slopes between 0.64 and 0.94. These numbers indicate that in the transition states more than two-thirds of the positive charge has been removed from the diarylmethyl fragment [176]. Because it had just been discussed that about one-half of a positive charge unit has arrived at the new carbocation center (Section III.D.4.c.), some charge is missing. Partial bridging, which locates partial charges at the other olefinic carbon or nonperfect synchronization [177] may account for the deviation. Because ΔG_i^0 is linearly correlated with other carbocation stability parameters, as pK_{R^+} or solvolysis rate constants (Section II.F), carbocation reactivities can also be derived from these quantities, and the corresponding correlation equations have been published [128]. Although in some cases evidence against the relationship between reactivity and thermodynamic stability of carbocations has been presented [178–181], the majority of data indicate that rate constants for various types of carbocations can be calculated from stability parameters as pK_{R^+} [69]. Correlations as shown in Figure 14 are usually linear up to rate constants of 5×10^7 L mol^{-1} sec^{-1} [136]. Only beyond that border, when the diffusion limit ($\approx$3

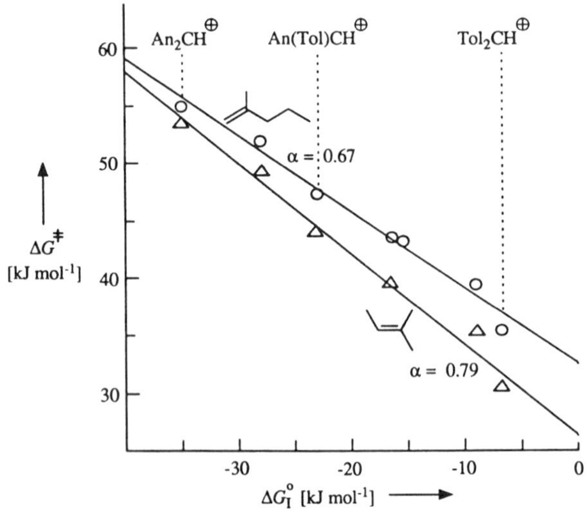

Figure 14 Correlation between $\Delta G^{\ddagger}$ for the reactions of diarylcarbenium ions with alkenes and ΔG_i^0 for the ionization of the corresponding diarylmethyl chlorides with BCl_3. (From Ref. 69, reprinted with permission of VCH Verlagsgesellschaft.)

$\times$ 10^9 L mol^{-1} sec^{-1}) is approached, deviations from these correlations become noticeable [136,182].

Rate and equilibrium constants have been assembled to construct the complete energy profiles for the reactions of various benzhydryl chlorides with 2-methyl-1-pentene (Fig. 15) [58,69]. Although kinetic and mechanistic details about the ionization step are known in some cases [58], this part has been omitted in Fig. 15, as it does not affect the further conclusions. Figure 15, in which the energies of the non-ionized benzhydryl chlorides are set at the same level, shows clearly that the order of stability of the benzhydryl cations is retained in the transition states for the additions, but that the separations of the energy levels are much smaller than in the ground states. Only one-third of the positive charge still resides in the benzhydryl fragments. An important consequence for the control of relative electrophilicities results.

Figure 16 describes the results of competition experiments, in which mixtures of the benzhydryl chlorides AnPhCHCl and AnTolCHCl in presence of variable amounts of BCl_3 are treated with a small amount of

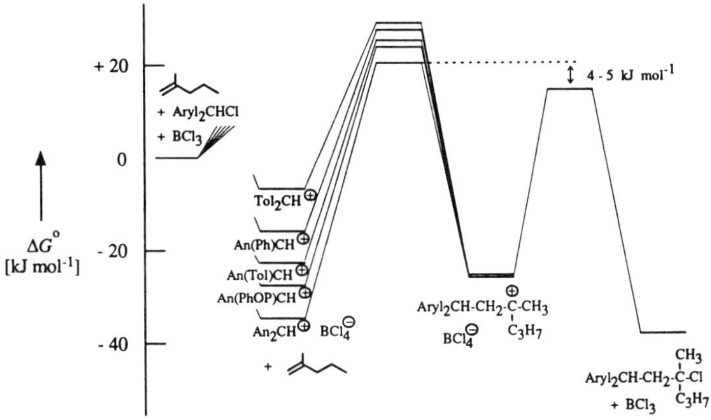

Figure 15 Free-energy profiles for the boron trichloride-induced additions of diarylmethyl chlorides to 2-methyl-1-pentene. (From Ref. 69, reprinted with permission of VCH Verlagsgesellschaft.)

2-methyl-1-pentene [61,183]. In presence of 5 equivalents of BCl_3, $AnPhCH^+$ reacts 7.2 times faster than $AnTolCH^+$, the same ratio as derived from the directly measured rate constants of these two ions (Table 6). When the concentration of BCl_3 is reduced, the reactivity ratio is reversed. In presence of catalytic amounts of BCl_3, a low concentration of the more reactive carbocation ($AnPhCH^+$) competes with a higher concentration of the less reactive carbocation ($AnTolCH^+$). A rapid ionization preequilibrium is realized, and Fig. 17, a section of Fig. 15, illustrates the Curtin-Hammett situation [184–186] encountered. If small concentrations of the cations $AnPhCH^+$ and $AnTolCH^+$ are reversibly generated from the corresponding benzhydryl chlorides, the observed reactivity ratio can be derived from the relative heights of the two activated complexes. From $\Delta\Delta G^{\ddagger} = 2.7$ kJ mol^{-1} one calculates the reactivity of AnPhCHCl to be 0.18 times that of AnTolCHCl, i.e., the same ratio as observed in the competition experiment with catalytic amounts of Lewis acid.

The same line of arguments rationalizes why the reactivity ratio of two electrophiles is growing with increasing difference of their standard ionization free enthalpies. Under conditions of almost complete ionization, the *p*-phenoxy-substituted benzhydryl cation is 5400 times more reactive than the bis(*p*-methoxy)-substituted analog, whereas in presence of catalytic amounts of Lewis acid, the relative reactivity of the two compounds is 0.016 (Fig. 18).

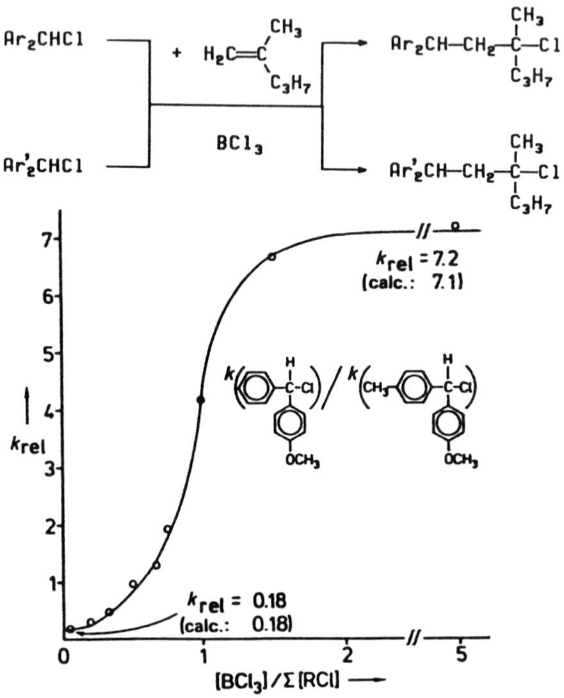

Figure 16 Relative reactivities of two diarylmethyl chlorides toward 2-methyl-1-pentene in CH_2Cl_2 at $-70°$ C as a function of the Lewis acid concentration. (From Ref. 183, reprinted by permission of Kluwer Academic Publishers.)

Of course, the relative reactivities do not only depend on the quantity but also on the nature of the Lewis acid. Figure 19 shows that different selectivity graphs are obtained for different ionizing systems [60,183]. The left part of the BCl_3 graph is identical with the corresponding graph in Fig. 14; it shows the increasing reactivity of diarylcarbenium ions with decreasing thermodynamic stability. The reactivity maximum is reached for the ditolyl system and further reduction of the stabilization of the carbocations reduces the reactivity, because these systems are only partially ionized by BCl_3.

The dimethoxy-substituted benzhydryl chloride is also fully ionized in a $SnCl_4$/EtOAc/CH_2Cl_2 solution [60], and the reactivity toward 2-methyl-1-pentene is identical with that in the BCl_3/CH_2Cl_2 solution (Fig. 19). AnPhCHCl is not fully ionized under these conditions, however, and

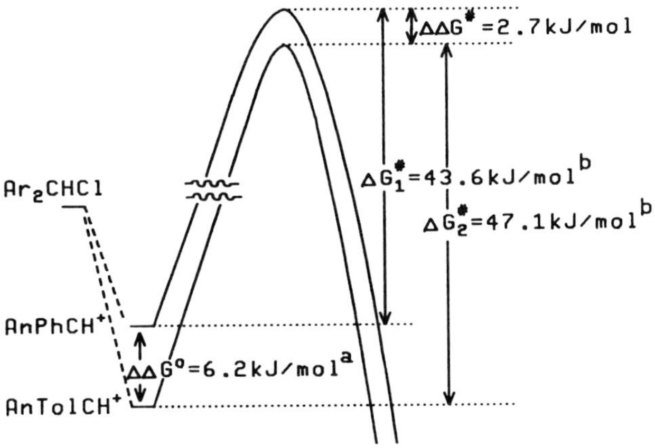

Figure 17 Energy profiles for the reactions of *p*-methoxy and of *p*-methoxy-*p′*-methyl-benzhydryl chloride with 2-methyl-1-pentene (−70° C). (a) From Scheme 5; (b) from Table 6. (From Ref. 183, reprinted by permission of Kluwer Academic Publishers.)

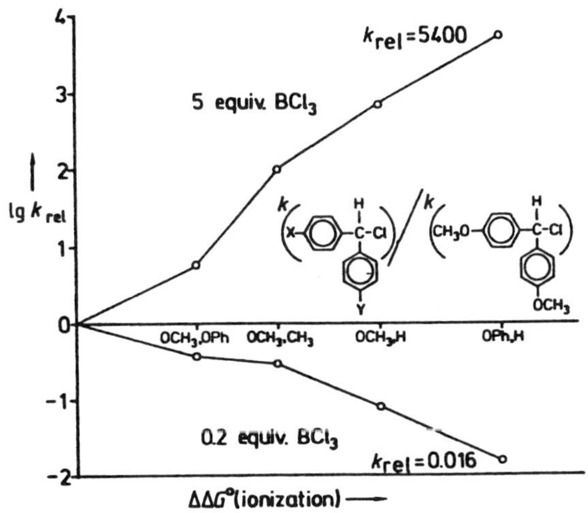

Figure 18 Relative reactivities of para-substituted diarylmethyl chlorides toward 2-methyl-1-pentene in presence of catalytic amounts of Lewis acid (bottom) and under conditions of complete ionization (top) (CH_2Cl_2/−70° C). (From Ref. 183, reprinted by permission of Kluwer Academic Publishers.)

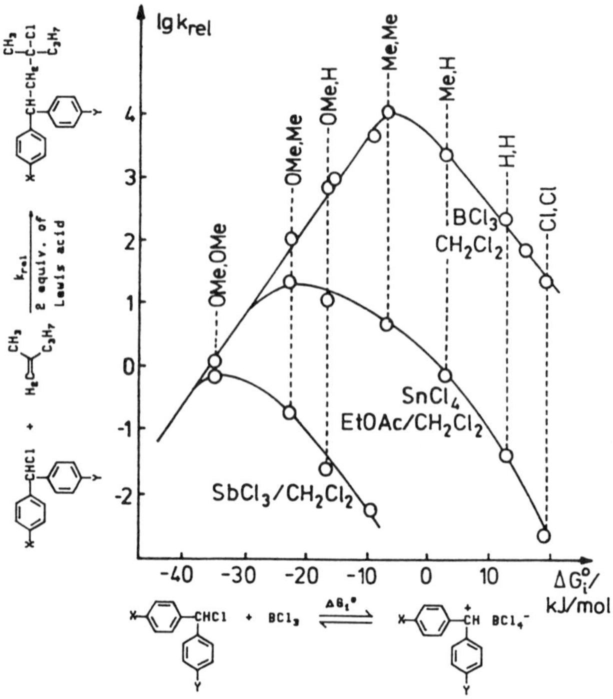

Figure 19 Relative reactivities of diarylmethyl chlorides in presence of 2 equivalents of Lewis acid. (From Ref. 183, reprinted by permission of Kluwer Academic Publishers.)

further decrease of the carbocation stabilization causes the reactivity in $SnCl_4$/EtOAc/CH_2Cl_2 to decrease because of the reduction of the carbocation concentration.

The weak Lewis acid $SbCl_3$ does not even fully ionize the dianisylmethyl chloride, and only one branch of the $SbCl_3$/CH_2Cl_2 graph can be seen in Fig. 19. It has been concluded that any Lewis acid/solvent system can be represented by a characteristic graph as shown in Fig. 19 [183]. An increase of the Lewis acid strength (see Scheme 7) will give rise to an increase of the reactivity maximum and its shift toward less stabilized carbocations. Because the reactivity maximum will be found for that RCl/R^+ system, which is approximately 50% ionized by the given Lewis acid, the most reactive RCl/R^+ couples in a CH_2Cl_2/Lewis acid mixture will be those that are located close to the corresponding metal halides in

Scheme 7. Figure 19 finally provides the quantitative basis for the qualitative model (Scheme 11), which has been employed for deriving conditions that allow the selective synthesis of 1:1 products by Lewis acid induced reactions of alkyl halides with alkenes.

As these considerations also apply to initiations of carbocationic polymerizations and to initiations and propagations in living carbocationic polymerizations, the choice of Lewis acids for these purposes can now be designed systematically.

IV. REACTIONS OF CARBOCATIONS WITH OTHER π-NUCLEOPHILES

A. Allylsilanes, Allylgermanes, and Allylstannanes

In organic syntheses allylsilanes and allylstannanes have been used extensively as allyl anion equivalents during the last two decades [187–190]. The regioselective attack of electrophiles, which finally yields products with allylic inversion (Scheme 43), has been explained by the hyperconjugative stabilization of carbenium centers by the carbon-silicon or carbon-tin bond in the β-position [191–196], which has initially been derived from solvolytic experiments [197–199].

The electrophile E^+ in Scheme 43 may correspond to one of the benzhydryl cations used for the kinetic studies with alkenes. The reactions of $Aryl_2CH^+MCl_{n+1}^-$ with allylsilanes have been shown to proceed via rate-determining attack of $Aryl_2CH^+$ at the CC-double bond, followed by rapid demetallation, and the same kinetic methods as described for alkenes in Section III.D.2, can be applied for determining the nucleophilicities of the allylelement compounds. Again, all three methods gave consistent results. Care has to be taken, however, in experiments with allylstannanes, as these undergo rapid transmetallation reactions with a variety of Lewis acids [159 and references quoted therein]. Benzhydryl triflates, which may be produced from benzhydryl chlorides and trimethylsilyl triflate, have therefore been employed for investigating the allylic tin compounds, as trimethylsilyl triflate was proven not to react with allylstannanes under the conditions of the kinetic experiments [159].

The compounds shown in Scheme 44 cover such a wide range of reactivities that a single electrophile is not sufficient for the kinetic com-

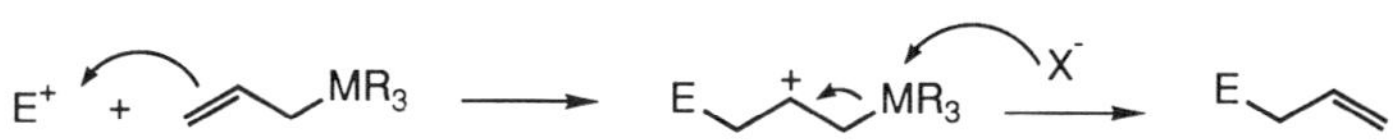

Scheme 43

parison of all nucleophiles. As the relative reactivities of terminal CC-double bonds are almost independent of the electrophilicities of their reaction partners (see Section III.D.4.b), the rate constants obtained with different benzhydryl cations can be combined to give the reactivity order shown in Scheme 44.

The comparison of propene, allyltrimethylsilane, and isobutene indicates, that introduction of a trimethylsilyl group in β-position of the developing carbenium center activates more than a methyl group in α-position. Both series of triphenyl element compounds (left and right column Scheme 44) show the reactivity pattern Si < Ge ≪ Sn, but variation of the substituents at silicon and tin was found to largely affect the reactivity of the double bond. While in the allyl series (right column), the trialkylsilanes and -stannanes are 2 to 3 orders of magnitude more reactive than the

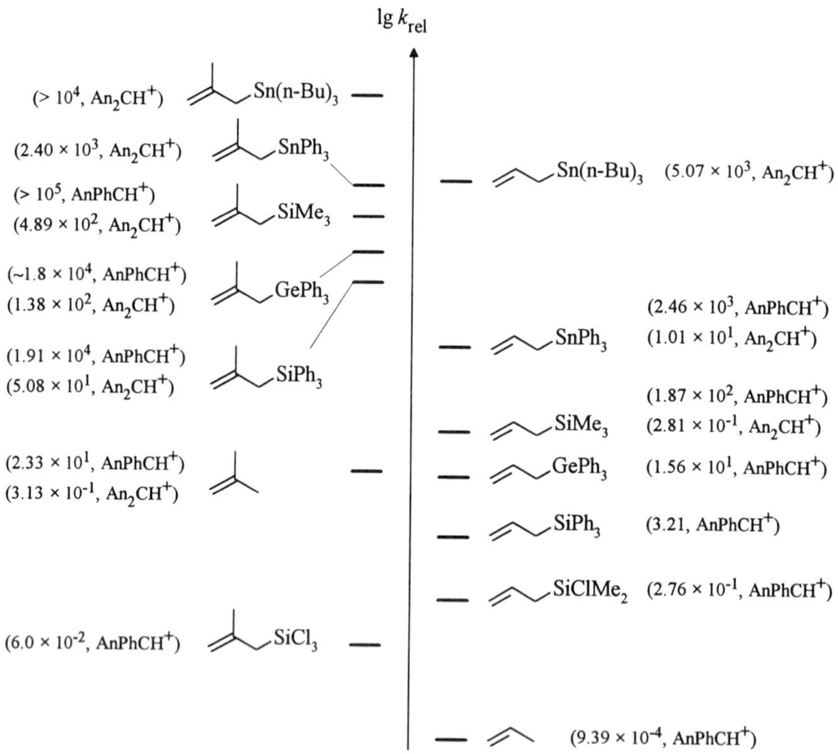

Scheme 44 k_2 (L mol^{-1} sec^{-1}) for reactions with An_2CH^+ or $AnPhCH^+$ (CH_2Cl_2, $-70°$ C). (From Ref. 159.)

triphenyl analogs, the reactivity ratio is reduced to 1 order of magnitude in the methallyl series (left column) because of the reduced electron demand of the developing carbocation center.

Even greater effects on rates are observed when a chlorine substituent at silicon is introduced. Although a trimethylsilyl group activates the π-bond of propene by a factor of 2×10^5, the acceleration by a chlorodimethylsilyl group is only 3×10^2. The left column shows that an allylic trichlorosilyl group even desactivates the double bond by a factor of 4×10^2, and one can interpolate that the substituent effect of a dichloromethylsilyl group is similar to that of hydrogen.

Methyl groups at the position of electrophilic attack exert exactly the same enthalpic and entropic effects as in the alkene series (Table 4), and one can summarize that the attack of carbocations at alkenes and at allylsilanes, allylgermanes, and allylstannanes follows the same mechanism. The differences between these classes of nucleophiles are encountered after the rate-determining step: While ordinary carbocations (produced from alkenes) usually accept a chloride ion to give addition products, the β-metal-substituted carbocations are generally demetalated to yield the S_E2' products. It has been reported, however, that β-silyl-substituted carbenium ions with bulky substituents at silicon may also act as chloride acceptors with the consequence that in these cases allylsilanes yield addition products in the same way as ordinary alkenes do [159].

B. Alkyl Enol Ethers and Silyl Enol Ethers

Whereas alkyl enol ethers easily polymerize upon treatment with electrophilic reagents [202], silylated enol ethers are rapidly desilylated after the electrophilic attack, and regeneration of the carbonyl group prevents polymerization. Therefore, silyl enol ethers are frequently used reagents for synthesizing α-substituted carbonyl compounds (Scheme 45) [200,201,203–205].

Because of the high nucleophilicity of the donor-substituted double bonds, rate constants for the attack of electrophiles at silyl enol ethers have been determined with the weak electrophile $(p\text{-}Me_2N\text{-}C_6H_4)_2CH^+$ (Scheme 46) [206]. One can extract from these data that the nucleophilicity of silyl enol ethers is in between that of analogously substituted allylsilanes and allylstannanes (Scheme 47).

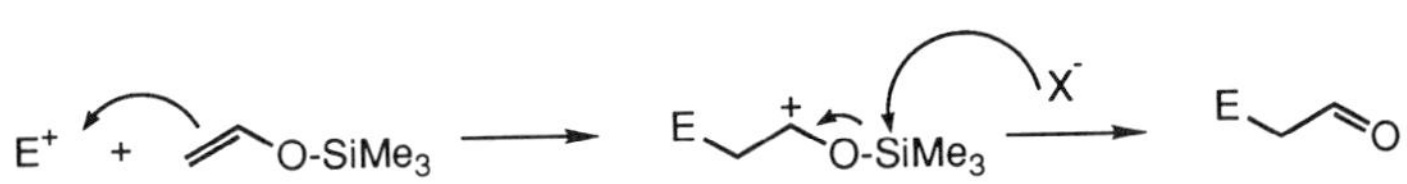

Scheme 45

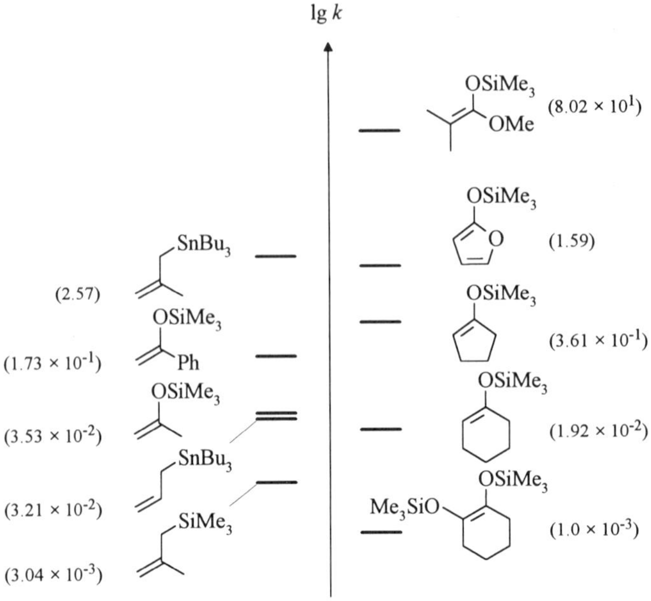

Scheme 46 k_2 ($L\ mol^{-1}\ sec^{-1}$) for reactions with $(p\text{-}Me_2NC_6H_4)_2CH^+$ (CH_2Cl_2, 20° C). (From Refs. 206, 182.)

Silylated ketene acetals are more reactive than silyl enol ethers (Scheme 46), and the higher reactivity of cyclopentenes compared to cyclohexenes, which has already been reported for the hydrocarbon series (Scheme 41), is also observed for this class of compounds. The negative inductive effect of oxygen, which operates at the position of electrophilic attack, makes the bisenol ether (Scheme 46, right column, bottom) 20 times less reactive than the structurally analogous monoenol ether.

Alkyl enol ethers polymerize under these conditions, and their reactivity, as that of other strong nucleophiles, has been determined with the LASER flash method (Table 7). Because this method of carbocation generation produces the nucleophilic counterions Cl^-, eventual polymeriza-

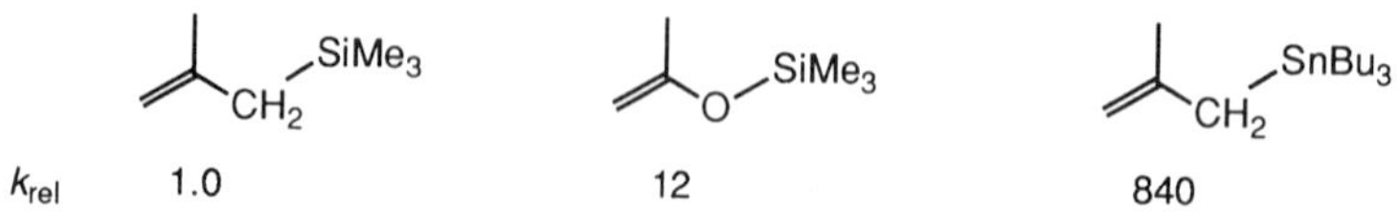

Scheme 47

tion cascades terminate after several chain growth steps by combination of the carboxonium ions with Cl^-. As the enol ethers are used in high excess over $Aryl_2CH^+$, pseudo-first-order kinetics is observed, also if more than one enol ether molecule per benzhydryl cation is consumed.

Scheme 48, which compares silyl and alkyl enol ethers of analogous structure, shows that the two classes of compounds have similar reactivities. Although the rate constants for the bis(*p*-chlorophenyl)-carbenium

Table 7 Second-Order Rate Constants (L mol^{-1} sec^{-1}) for the Reactions of Benzhydryl Cations with π-Nucleophiles in Acetonitrile at 20° C

Nucleophile	$(p\text{-ClC}_6H_4)_2CH^+$	$(p\text{-CH}_3C_6H_4)_2CH^+$
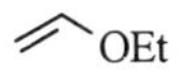	1.7×10^8	6.3×10^6
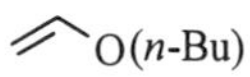	2.0×10^8	
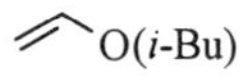	1.9×10^8	
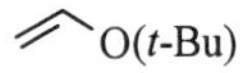	4.2×10^8	
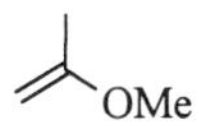	7.4×10^8	3.7×10^7
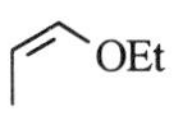	7.7×10^8	
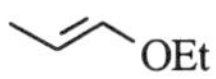	1.3×10^9	
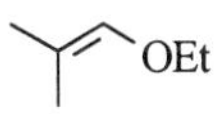	2.2×10^9	2.2×10^8
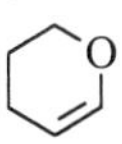	2.7×10^8	
	1.3×10^9	
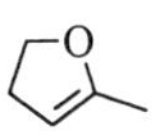	3.5×10^9	1.2×10^8
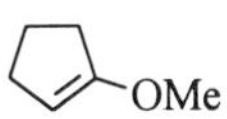	1.9×10^9	2.2×10^8
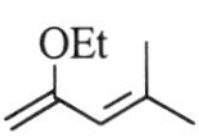	$\sim 4 \times 10^9$	

Source: Ref. 136.

$(p\text{-ClC}_6\text{H}_4)_2\text{CH}^+$	1.5×10^9	2.2×10^9	2.5×10^9	1.9×10^9
$(p\text{-H}_3\text{CC}_6\text{H}_4)_2\text{CH}^+$	4.2×10^7	2.2×10^8	6.0×10^8	2.2×10^8

Scheme 48 k_2 (L mol^{-1} sec^{-1}, 20° C, CH_3CN). (From Ref. 136.)

ion do not give much information because of the proximity of the diffusion limit, the data obtained with the ditolylcarbenium ion indicate that silyl enol ethers can be somewhat more or less nucleophilic than the corresponding alkyl enol ethers, depending on structure.

The proximity of the diffusion limit also inhibits a detailed discussion of the data in Table 7, but a significant difference to the substituent effects discussed in Section III.D.4 is obvious. Whereas the reactivities of terminal alkenes, dienes, and styrenes toward $AnPhCH^+$ correlate with the stabilities of the new carbenium ions and not with the ionization potentials of the π-nucleophiles [69], the situation is different for the reactions of enol ethers with $(p\text{-ClC}_6\text{H}_4)_2\text{CH}^+$ [136]. In this reaction series, methyl groups at the position of electrophilic attack activate the enol ether double bonds more than methyl groups at the new carbocationic center, i.e., the relative activation free enthalpies are not controlled any longer by the stabilities of the intermediate carbocations but by the ionization potentials of the enol ethers (Fig. 20). An interpretation of the correlation in Fig. 20 has not yet been given, but one can alternatively discuss early transition states which are controlled by frontier orbital interactions or the involvement of outer sphere electron transfer processes [220].

C. Alkynes

Lewis acid-catalyzed reactions of readily ionizable alkyl halides with alkynes yield vinyl halides [207]. The regioselectivities of these additions can be rationalized by the relative stabilities of the intermediate vinyl cations. Unlike the situation described for additions to alkenes, there is no preference for anti-additions, and the stereoselectivities can be explained by the intermediacy of nonbridged species [208]. The site of nu-

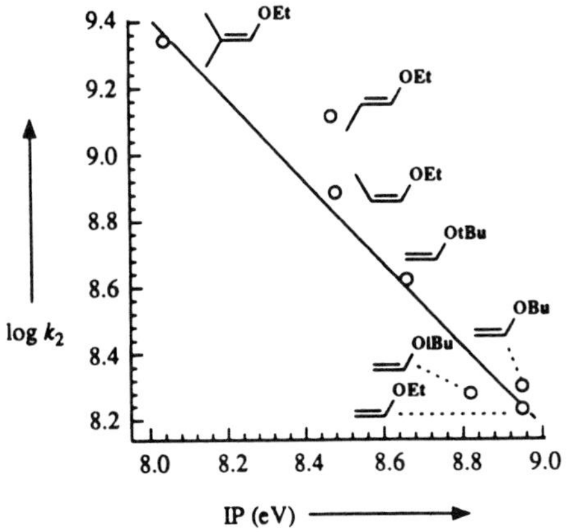

Figure 20 Correlation of the reactivities of various enol ethers toward $(p\text{-}ClC_6H_4)_2CH^+$ at 20° C in acetonitrile [L mol^{-1} sec^{-1}] with their ionization potentials. (Reprinted with permission from Ref. 136. Copyright 1991 American Chemical Society.)

cleophilic attack on the intermediate vinyl cation is determined by the relative sizes of the substituents R and R′ (Scheme 49).

This interpretation is supported by the fact that a 1:1 mixture of stereoisomeric vinyl chlorides is isolated from the benzylation of 1,3-diphenylpropyne with labelled benzyl chloride (Scheme 50) [208].

Whereas electrophiles with strong bridging tendency (e.g., halogens) react considerably faster with alkenes than with analogously substituted alkynes, protons [209,210] and carbocations [211] have been reported to attack analogously substituted double and triple bonds with similar rates [212].

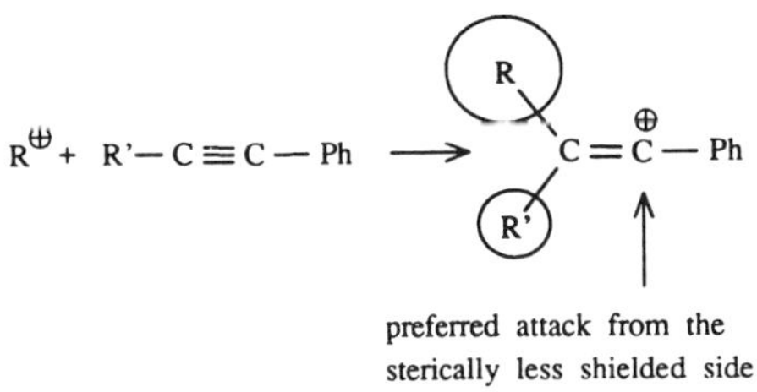

Scheme 49

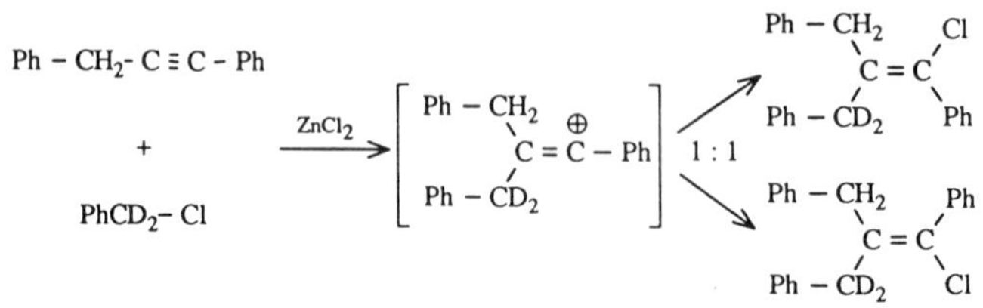

Scheme 50

As the reactions of carbocations with alkynes have higher activation enthalpies and less negative activation entropies than analogous reactions with alkenes, the relative reactivities of styrene and phenylacetylene have been found to strongly depend on temperature (Table 8) [213].

According to Scheme 51, at −70° C alkynes are somewhat less reactive toward carbocations than alkenes with analogous substitution, but the reactivity scales for alkynes and alkenes overlap, and it is certainly incorrect to say that alkynes are generally weaker nucleophiles than alkenes.

D. Arenes

Rates for the key step of Friedel-Crafts alkylations, i.e., the attack of carbocations on arenes have only recently been reported [214,215]. Problems arising from the reversibility of the CC-bond-forming step have been overcome by performing experiments in presence of $R_4N^+MCl_{n+1}^-$ salts, where MCl_{n+1}^- acts as a base for the rapid deprotonation of the intermediate benzenium ions (Scheme 52).

According to Fig. 21, the rate constants correlate with the basicities of the arenes. From the gradients of these correlations (0.94–1.05) which

Table 8 Relative Reactivities of Styrene and Phenylacetylene Toward Tol_2CH^+ at Various Temperatures

T/° C	$k(PhCH{=}CH_2)/k(PhC{\equiv}CH)$
−70	164
−40	62.7
−10	27.6
20	15.7

Source: Ref. 213.

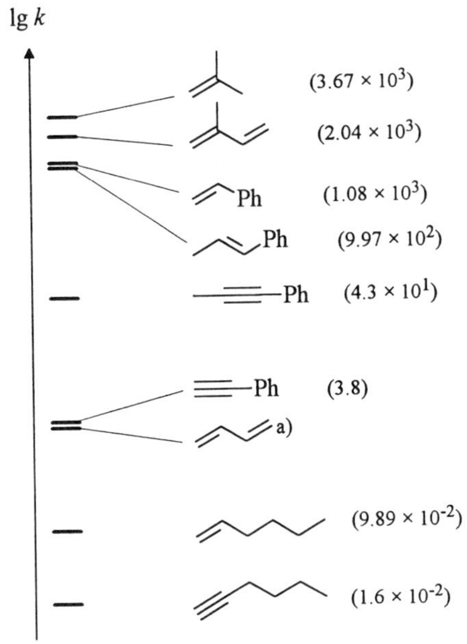

a) Extrapolated value.

Scheme 51 k_2 (L mol^{-1} sec^{-1}) for reactions with Tol_2CH^+ (CH_2Cl_2, $-70°$ C, based on absolute rate measurements). (From Refs. 213, 128.)

reduce to 0.63–0.80, when the different reference temperatures of rate and equilibrium measurements are considered, one can derive a transition state with highly developed benzenium ion character [215].

The manifold of data on relative nucleophilic reactivities of arenes, which are available from competition experiments [27,28,79,216], can now be combined with the few absolute rate constants to directly compare the reactivities of the large variety of aromatic and aliphatic π-systems.

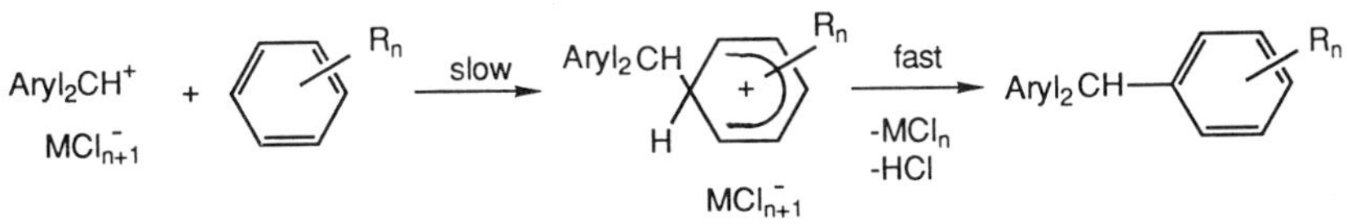

Scheme 52

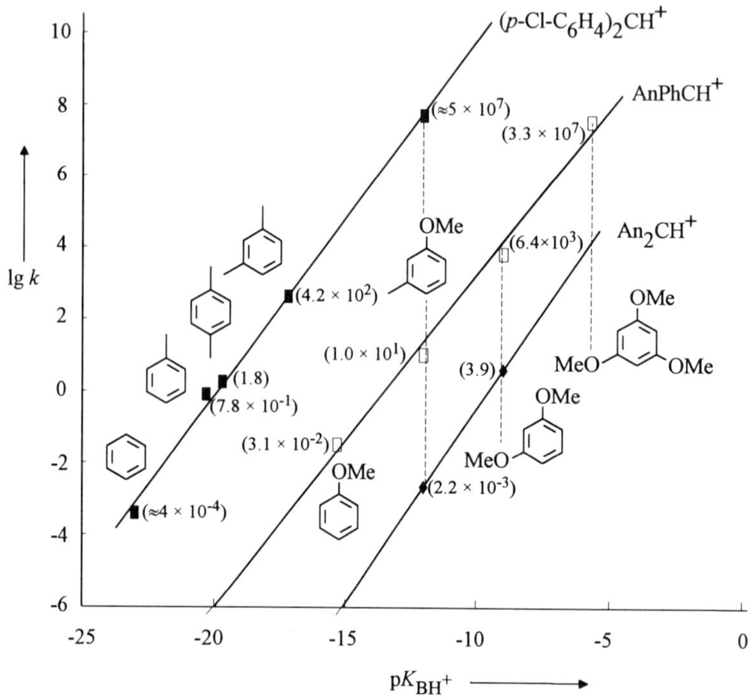

Figure 21 Relationship between the basicity of arenes and their reactivity [L mol^{-1} sec^{-1}] to benzhydryl cations (CH_2Cl_2, $-70°$ C). (From Ref. 215.)

V. SUMMARY: A REACTIVITY SCALE FOR π-NUCLEOPHILES

Since Swain's and Scott's efforts [217] to quantify the kinetic term "nucleophilicity," chemists have continued to search for a quantitative concept of nucleophilic reactivity [218]. Most of this work has dealt with S_N2 type reactions, however, and the marked dependence of the relative strengths of nucleophiles on the nature of the electrophile and the polarity of the solvent has become textbook knowledge.

As already established for combinations of cations with *n*-nucleophiles [33,160,219], the situation is less complicated for the reactions of carbocations with π-systems. Solvent polarity plays only a minor role (Section III.D.3) and, for many π-systems, the relative reactivity has been found to be electrophile-independent (Fig. 10, Section III.D.4.b). Also for these systems, the construction of a universal nucleophilicity scale is not unproblematic, however. Remember Fig. 11, which shows that An_2CH^+ reacts 3.4 times faster with allyltrimethylsilane than with 2-methyl-2-bu-

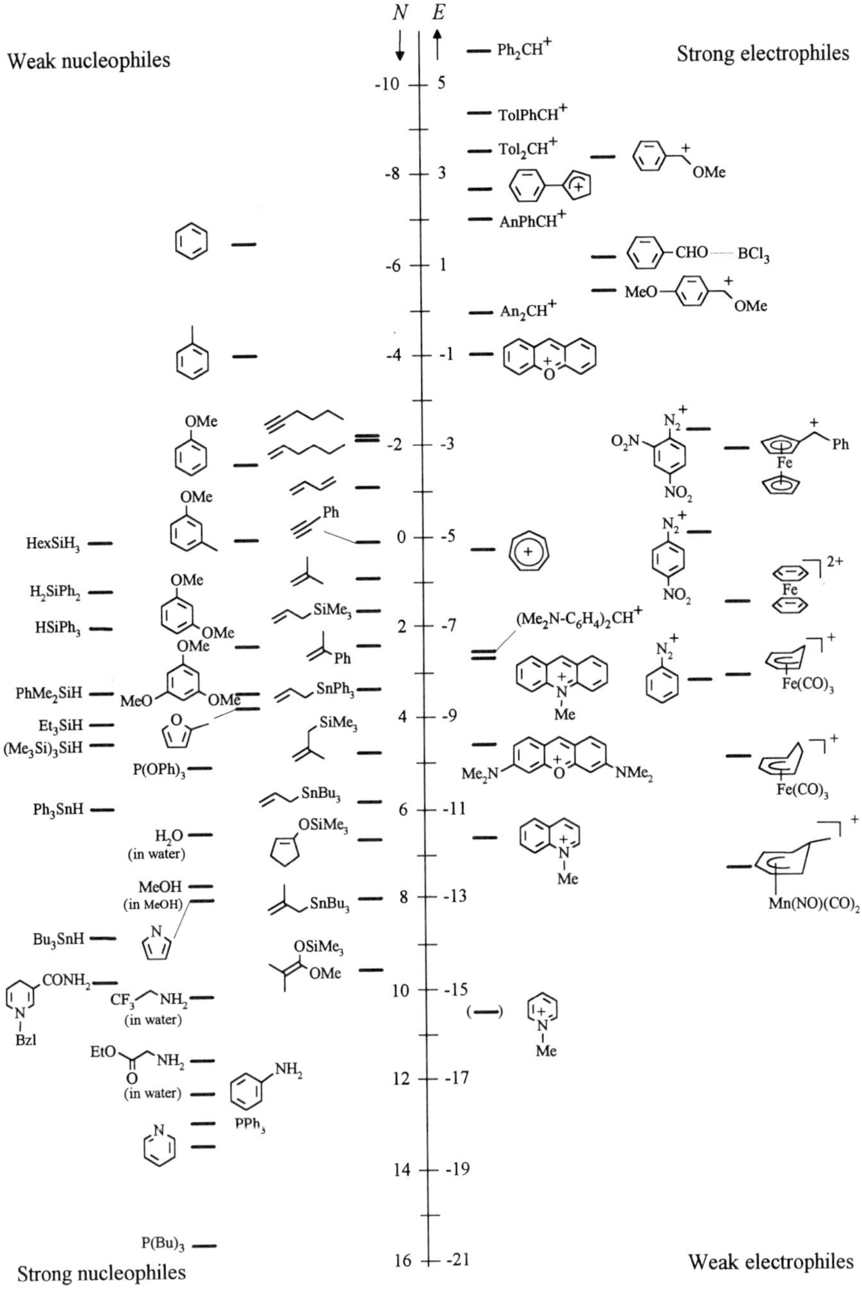

Scheme 53 Electrophilicity and nucleophilicity parameters according to Eq. (23). (From Ref. 182, reprinted with permission of VCH Verlagsgesellschaft.)

tene, whereas Tol_2CH^+ shows the inverse reactivity order for these two compounds (ratio 0.26) [128]. Also temperature plays an important role: While above −57° C, $AnPhCH^+$ reacts faster with allyltrimethylsilane than with 2-methyl-2-butene, the reactivity order is opposite below −57° C [128]. Should the attractive idea of constructing a universal reactivity scale for π-nucleophiles therefore be abandoned?

$$\lg k = s\,(N + E) \qquad \text{for } 20°\text{ C} \tag{23}$$

Mayr and Patz have recently evaluated 56 reaction series, mostly for reactions as described in this article, and derived Eq. (23), in which carbocations are characterized by the electrophilicity parameter E, whereas nucleophiles are characterized by the nucleophilicity parameter N and the slope parameter s [182]. The latter quantity, s, which basically describes the slopes of plots as shown in Figs. 10 and 11, ranges from 0.8 to 1.2 for 91 % of the π-nucleophiles investigated. The mathematical form of Eq. (23) implies that the exact value of s will usually only be of importance when rate constants, which strongly deviate from 1 (e.g., $|\log k| > 5$), are considered. Some of the characterized nucleophiles and electrophiles are listed in Scheme 53, where the two scales are arranged in such a way that electrophiles and nucleophiles which are located at the same level are predicted to combine with rate constants of $\lg k = -5\,s$. With $s \approx 1$ one expects slow combinations for electrophile-nucleophile pairs at the same level, whereas reactions of nucleophiles with electrophiles located below them are expected to be very slow or not to occur at all at 20° C.

VI. EPILOGUE

A large part of the discussions in this chapter was concerned with carbocationic [1:1]-telomerizations, i.e., Lewis acid-promoted additions of alkyl halides and related compounds to CC-double bonds with formation of 1:1 products. The relevance of these studies for the initiation of carbocationic polymerizations, as described in Scheme 9 (Section III.A), is obvious.

Ion generation and cationation are identical in [1:1]-telomerizations and polymerizations, and the rules derived for identifying the most reactive RX/MX_n combinations (Schemes 11 and Fig. 19) apply analogously to polymerizations. Most living cationic polymerizations presently known proceed via dormant species, i.e., the growing chain is intercepted by an anion or n-nucleophile after each propagation step to give an ester or onium ion, which undergo heterolysis before the next monomer unit is attacked. Because initiation should be faster than propagation to obtain a narrow molecular weight distribution, the treatment in Section III.A (selective formation of 1:1 products) provides a guide how to select initiator and Lewis acid for realizing an optimal termination reactivation sequence.

The rearrangement processes, which can easily be studied in Lewis acid-catalyzed additions of alkyl halides to alkenes (Section III.B.2), give information about possible rearrangements in a growing chain produced from the corresponding monomer.

Comparison of the relative alkene reactivites reported in Section III.D.4 with known copolymerization parameters [57] indicates qualitative agreement between the two sets of data. It is thus possible to use the reactivity scales in Section III.D.4, for predicting approximate copolymerization parameters, but the calculation of accurate rate ratios in copolymerization suffers from the fact that alkene reactivities are not completely independent of the nature of the attacking carbocation. This is especially true if the propagation rates get close to the diffusion limit. It is not possible to directly measure propagation rate constants by investigating cationic telomerizations, but extrapolations to such data on the basis of Eq. (23) in Section V are conceivable.

The investigations described in this chapter show that nature and concentration of the negative counterion has a great influence on the outcome of carbocationic telomerizations because of the different rates of ion-pair collapse and because of reionization of eventually produced 1:1 adducts from alkyl halides and alkenes. On the other hand, it was found that the rate of attack of a carbocation at an alkene is generally independent of the nature of the complex counterion and that free and paired carbocations react with equal rates. If this conclusion also holds for carbocationic polymerizations, a reinterpretation of many polymerization kinetics becomes necessary.

ACKNOWLEDGMENT

Support of this work by the *Deutsche Forschungsgemeinschaft* and the *Fonds der Chemischen Industrie* is gratefully acknowledged. I thank Dr. B. Dogan, Mrs. J. Bott, and Mr. M. Roth for their help in preparing this manuscript.

Set processes have been excluded recently: M. Patz, H. Mayr, J. Maruta, and S. Fukuzumi, *Angew. Chem. Int. Ed. Engl. 34*: 1225–1227 (1995).

REFERENCES AND NOTES

1. A. Baeyer and V. Villiger, *Chem. Ber. 35*: 3013–3033 (1902).
2. A. Baeyer and V. Villiger, *Chem. Ber. 35*: 1189–1201 (1902).
3a. F. Kehrmann and F. Wentzel, *Chem. Ber. 34*: 3815–3819 (1901).
3b. F. Kehrmann and F. Wentzel, *Chem. Ber. 35*: 622 (1902).

4. C. D. Nenitzescu, Historical outlook, Carbonium Ions (G. A. Olah and P. v. R. Schleyer; eds.), Wiley-Interscience, New York, 1968, Vol. 1, Chap. 1.
5. H. Meerwein and K. v. Emster, *Chem. Ber. 55*: 2500–2528 (1922).
6. M. Saunders, J. Chandrasekhar, and P. v. R. Schleyer, Rearrangements of carbocations, *Rearrangements in Ground and Excited States* (P. de Mayo, ed.) Academic Press, New York, 1980, Vol. 1, pp. 1–53.
7. C. K. Ingold, *Structure and Mechanism in Organic Chemistry*, 2nd ed, Cornell University Press, Ithaca, 1969, Chap. 7.
8. *Ions and Ion Pairs in Organic Reactions* (M. Szwarc, ed.), John Wiley & Sons, New York, 1972, Vols. 1–2.
9. B. Capon and S. P. McManus, *Neighboring Group Participation*, Plenum, New York, 1976, Vol. 1; p. 58.
10. S. Winstein, Nonclassical ions and homoaromaticity, *Carbonium Ions* (G. A. Olah and P. v. R. Schleyer, eds.), Wiley-Interscience, New York, 1972, Vol. 3, Chap. 22.
11a. G. A. Olah, *Angew. Chem. 85*: 183–225 (1973).
11b. G. A. Olah, *Angew. Chem. Int. Ed. Engl. 12*: 173–212 (1973).
12. G. A. Olah, G. K. S. Prakash, and J. Sommer, *Superacids*, John Wiley & Sons, New York, 1985.
13. W. J. Hehre, L. Radom, P. v. R. Schleyer, and J. A. Pople, *Ab Initio Molecular Orbital Theory*, John Wiley & Sons, New York, 1986.
14. P. Vogel, *Carbocation Chemistry*, Elsevier, Amsterdam, 1985.
15. W. H. Powell, *Pure Appl. Chem. 65*: 1357–1455 (1993).
16. P. v. R. Schleyer, J. W. de M. Carneiro, W. Koch, and K. Raghavachari, *J. Am. Chem. Soc. 111*: 5475–5477 (1989).
17. P. v. R. Schleyer, W. Koch, B. Liu, and U. Fleischer, *J. Chem. Soc. Chem. Commun.* 1098–1099 (1989).
18. J. W. de M. Carneiro and P. v. R. Schleyer, *J. Am. Chem. Soc. 112*: 4064–4066 (1990).
19. H. Mayr, G. Hagen, J.-P. Dau-Schmidt, and T. W. Bentley, *Chem. Rev.*, in preparation.
20. S. G. Lias, J. E. Bartmess, J. F. Liebmann, J. L. Homes, R. D. Levin, and W. G. Mallard, *J. Phys. Chem. Ref. Data 17, Suppl. 1* (1988).
21. J. B. Pedley, R. D. Naylor, and S. P. Kirby, *Thermochemical Data of Organic Compounds*, 2nd ed., Chapman and Hall, London, 1986.
22. J. L. Holmes and F. P. Lossing, *Can. J. Chem. 60*: 2365–2371 (1982).
23. D. H. Aue and M. T. Bowers, Stabilities of positive ions from equilibrium gas-phase basicity measurements, *Gas Phase Ion Chemistry* (M. T. Bowers, ed.), Academic Press, New York, 1979, Vol. 2, Chap. 9.
24. E. M. Arnett and C. Petro, *J. Am. Chem. Soc. 100*: 2563–2564 (1978).
25. E. M. Arnett and C. Petro, *J. Am. Chem. Soc. 100*: 5408–5416 (1978).
26. E. M. Arnett and T. C. Hofelich, *J. Am. Chem. Soc. 105*: 2889–2895 (1983).
27. G. A. Olah, *Friedel-Crafts Chemistry*, Wiley, New York, 1973.
28. G. A. Olah, *Friedel-Crafts and Related Reactions*, Wiley, New York, 1963–1965, Vol. 1–4.

29. N. C. Deno, J. J. Jaruzelski, and A. Schriesheim, *J. Am. Chem. Soc. 77*: 3044–3051 (1955).
30. N. C. Deno and A. Schriesheim, *J. Am. Chem. Soc. 77*: 3051–3054 (1955).
31. K. Okamoto, K. Takeuchi, K. Komatsu, Y. Kubota, R. Ohara, M. Arima, K. Takahashi, Y. Waki, and S. Shirai, *Tetrahedron 39*: 4011–4024 (1983).
32. H. H. Freedman, *Carbonium Ions* (G. A. Olah, and P. v. R. Schleyer, eds.), Wiley, New York, 1973, Vol. 4; pp. 1501–1578.
33. C. D. Ritchie, *Can. J. Chem. 64*: 2239–2250 (1986).
34. S. V. McKinley, J. W. Rakshys, Jr., A. E. Young, and H. H. Freedman, *J. Am. Chem. Soc. 93*: 4715–4724 (1971).
35. C. Schade, H. Mayr, and E. M. Arnett, *J. Am. Chem. Soc. 110*: 567–571 (1988).
36. R. D. Guthrie, *Pure Appl. Chem. 61*: 23–56 (1989).
37. J. P. Dau-Schmidt, Ph.D. thesis, Medizinische Universität zu Lübeck, 1992.
38. T. W. Bentley, J.-P. Dau-Schmidt, G. Llewellyn, and H. Mayr, *J. Org. Chem. 57*: 2387–2392 (1992).
39. C. Schade and H. Mayr, *Tetrahedron 44*: 5761–5770 (1988).
40. C. Schade, E. Bäuml, and H. Mayr, preliminary results.
41. R. Schneider, H. Mayr, and P. H. Plesch, *Ber. Bunsenges. Phys. Chem. 91*: 1369–1374 (1987).
42. D. W. Grattan and P. H. Plesch, *J. Electroanal. Chem. 103*: 81–94 (1979).
43a. G. E. Holdcroft and P. H. Plesch, *J. Chem. Res.* (*S*) 36–37 (1985).
43b. G. E. Holdcroft and P. H. Plesch *J. Chem. Res.* (*M*) 401–450 (1985).
44. C. S. Creaser and J. A. Creighton, *J. Chem. Soc. Dalton Trans.* 1402–1405 (1975).
45. T. J. Kistenmacher and G. D. Stucky, *Inorg. Chem. 10*: 122 (1971).
46. M. Baaz, V. Gutmann, and J. R. Masaguer, *Monatsh. Chem. 92*: 582–589 (1961).
47. M. Baaz, V. Gutmann, and J. R. Masaguer, *Monatsh. Chem. 92*: 590–599 (1961).
48. V. Gutmann and G. Hampel, *Monatsh. Chem. 92*: 1048–1054 (1961).
49. V. Gutmann and F. Mairinger, *Monatsh. Chem. 92*: 720–724 (1961).
50. S. E. Denmark and N. G. Almstead, *J. Am. Chem. Soc. 115*: 3133–3139 (1993).
51. H. Mayr and W. Striepe, *J. Org. Chem. 50*: 2995–2998 (1985).
52a. S. Penczek and P. Kubisa, *Makromol. Chem. 165*: 121–135 (1973).
52b. S. Slomkowski and S. Penczek, *J. Chem. Soc. Perkin Trans 2*: 1718–1722 (1974).
53. M. Baaz, V. Gutmann, and O. Kunze, *Monatsh. Chem. 93*: 1162–1175 (1962).
54. V. Gutmann and R. Schmid, *Monatsh. Chem. 102*: 798–805 (1971).
55. J. P. Kennedy and E. Maréchal, *J. Polym. Sci. Macromol. Rev. 16*: 123–197 (1981).
56. H. Mayr and C. Schade, *Makromol. Chem. Rapid Commun. 9*: 483–488 (1988).

57. J. P. Kennedy and E. Maréchal, *Carbocationic Polymerization*, Wiley, New York, 1982.
58. H. Mayr, R. Schneider, C. Schade, J. Bartl, and R. Bederke, *J. Am. Chem. Soc. 112*: 4446–4454 (1990).
59. M. J. S. Dewar and R. C. Dougherty, *The PMO Theory of Organic Chemistry*, Plenum Press, New York, 1975, pp. 218–219.
60. H. Mayr and R. Schneider, *Makromol. Chem. Rapid Commun. 5*: 43–46 (1984).
61a. H. Mayr, C. Schade, M. Rubow, and R. Schneider, *Angew. Chem. 99*: 1059–1060 (1987).
61b. H. Mayr, C. Schade, M. Rubow, and R. Schneider, *Angew. Chem. Int. Ed. Engl. 26*: 1029–1030 (1987).
62. J.-P. Dau-Schmidt and H. Mayr, *Chem. Ber. 127*, 205–212 (1994).
63a. H. Mayr, *Angew. Chem. 93*: 202–204 (1981).
63b. H. Mayr, *Angew. Chem. Int. Ed. Engl. 20*: 184–186 (1981).
64. H. Mayr and W. Striepe, *J. Org. Chem. 48*: 1159–1165 (1983).
65. E. Muks, *Proc. Estonian Acad. Sci. Chem. 39*: 94–102 (1990).
66. P. B. D. De la Mare and R. Bolton, *Electrophilic Additions to Unsaturated Systems*, 2nd ed., Elsevier, Amsterdam, 1982, Vol. 3.
67. G. H. Schmid and D. G. Garratt, Electrophilic additions to carbon-carbon double bonds; The chemistry of double-bonded functional groups, *The Chemistry of Functional Groups* (S. Patai, ed.), Wiley, London, 1977, Suppl. A, pp. 725–912.
68. V. J. Nowlan and T. T. Tidwell, *Acc. Chem. Res. 10*: 252–258 (1977).
69a. H. Mayr, *Angew. Chem. 102*: 1415–1428 (1990).
69b. H. Mayr, *Angew. Chem. Int. Ed. Engl. 29*: 1371–1384 (1990).
70. H. Stetter and P. Goebel, *Chem. Ber. 95*: 1039–1042 (1962).
71. L. Schmerling, *J. Am. Chem. Soc. 71*: 698–701, 701–703 (1949).
72. L. Schmerling, Condensation of haloalkanes with alkenes and haloalkenes, *Friedel-Crafts and Related Reactions*, Wiley, New York, 1964 Vol. 2, Chap. 26, pp. 1133–1173.
73a. K. Bott and H. Hellmann, *Angew. Chem. 78*: 932–936 (1966).
73b. K. Bott and H. Hellmann *Angew. Chem. Int. Ed. Engl. 5*: 870–874 (1966).
74a. K. Bott, *Angew. Chem. 92*: 169–176 (1980).
74b. K. Bott, *Angew. Chem. Int. Ed. Engl. 19*: 171–178 (1980).
75. H. J. Prins, *J. Prakt. Chem. 89*: 414–451 (1914).
76. H. J. Prins, *Rec. Trav. Chim. 69*: 1003–1006 (1950).
77. A. L. Henne and E. C. Ladd, *J. Am. Chem. Soc. 60*: 2491–2495 (1938).
78. This interpretation is based on kinetic investigations of electrophilic additions to haloalkenes. H. Mayr and M. Glaser, unpublished results.
79. R. Taylor, *Electrophilic Aromatic Substitution*, Wiley, Chichester, 1990, Chap. 11, pp. 455–496.
80. A. Z. Shikhmamedbekova and F. A. Mamedov, *Uch. Zap. Azerb, Gos. Univ. Ser. Khim. Nauk.* 54–63 (1966). *Chem. Abstr. 67*: 81763j (1967).
81. H. Mayr and H. Klein, *Chem. Ber. 115*: 3528–3546 (1982).
82. H. Mayr, R. Schneider, B. Irrgang, and C. Schade, *J. Am. Chem. Soc. 112*: 4454–4459 (1990).

83a. H. Gross and E. Höft, *Angew. Chem. 79*: 358–378 (1967).
83b. H. Gross and E. Höft, *Angew. Chem. Int. Ed. Engl. 6*: 335–355 (1967).
83c. K. V. Lääts and E. A. Muks, *Zhur. Org. Khim. 10*: 162–164 (1974).
83d. K. V. Lääts, T. A. Kaal, I. A. Kalja, and E. A. Muks, *Zhur. Org. Khim. 10*: 159–161 (1974).
84. M. J. S. Dewar and C. H. Reynolds, *J. Am. Chem. Soc. 106*: 1744–1750 (1984).
85. P. Hallpap, M. Bölke, G. Heublein, and C. Weiss, *Makromol. Chem., Macromol. Symp. 3*: 47–63 (1986).
86. P. A. Bartlett, Olefin cyclization processes that form carbon-carbon bonds; Olefin cyclization processes that form carbon-heteroatom bonds, *Asymmetric Synthesis*, (J. D. Morrison, Ed.), Academic Press, New York, 1984, Vol. 3, Part B, Chap. 5 and 6, pp. 341–454.
87a. W. S. Johnson, *Angew. Chem. 88*: 33–49 (1976).
87b. W. S. Johnson, *Angew. Chem. Int. Ed. Engl. 15*: 9–17 (1976).
88. W. S. Johnson, *Bioorg. Chem. 5*: 51–98 (1976).
89. R. Pock, H. Mayr, M. Rubow, and E. Wilhelm, *J. Am. Chem. Soc. 108*: 7767–7772 (1986).
90a. M. J. S. Dewar and R. C. Fahey, *Angew. Chem. 76*: 320–325 (1964).
90b. M. J. S. Dewar and R. C. Fahey, *Angew. Chem. Int. Ed. Engl. 3*; 245–249 (1964).
91a. K. Matyjaszewski, *Comprehensive Polymer Science*, (G. Allen and J. C. Bevington, eds.), Pergamon Press, Oxford, 1989, Vol. 4, Chap. 41.
91b. K. Matyjaszewski, *Makromol Chem., Macromol. Symp. 54/55*: 51–71 (1992).
92. J. H. Rolston and K. Yates, *J. Am. Chem. Soc. 91*: 1469–1476 (1969).
93. J. H. Rolston and K. Yates, *J. Am. Chem. Soc. 91*: 1477–1483 (1969).
94. M.-F. Ruasse, *Acc. Chem. Res. 23*: 87–93 (1990).
95. E. von der Brüggen, Ph.D. thesis, Medizinische Universität zu Lübeck, 1988.
96. E. Muks, *Proc. Estonian Acad. Sci. Chem. 39*: 103–107 (1990).
97. L. Schmerling, *J. Am. Chem. Soc. 67*; 1152–1154 (1945).
98. W. P. Jencks, *Acc. Chem. Res. 13*: 161–169 (1980).
99. H. Mayr, H. Klein, and G. Kolberg, *Chem. Ber. 117*: 2555–2579 (1984).
100. E. Bäuml, G. Kolberg, and H. Mayr, *Tetrahedron Lett. 28*: 387–390 (1987).
101. H. Mayr, B. Seitz, and I. K. Halberstadt-Kausch, *J. Org. Chem. 46*: 1041–1043 (1981).
102. H. Mayr and H. Klein, *J. Org. Chem. 46*: 4097–4100 (1981).
103a. H. Klein and H. Mayr, *Angew. Chem. 93*: 1069–1070 (1981).
103b. H. Klein and H. Mayr, *Angew. Chem. Int. Ed. Engl. 20*: 1027–1029 (1981).
104a. H. Klein, G. Freyberger, and H. Mayr, *Angew. Chem. 95*: 62–63 (1983).
104b. H. Klein, G. Freyberger, and H. Mayr, *Angew. Chem. Int. Ed. Engl. 22*: 49 (1983).
105. J. Baran and H. Mayr, *J. Org. Chem. 53*: 4626–4628 (1988).
106. M. Roth, C. Schade, and H. Mayr, *J. Org. Chem. 59*: 169–172 (1994).
107. R. Pock and H. Mayr, *Chem. Ber. 119*: 2497–2509 (1986).
108. R. Pock, H. Klein, and H. Mayr, *Chem. Ber. 119*: 929–942 (1986).

109. A. Streitwieser Jr. and C. H. Heathcock, *Introduction to Organic Chemistry*, 3rd ed., Macmillan, New York, 1985, p. 1153.
110. S. W. Benson, *Thermochemical Kinetics*, 2nd ed., Wiley-Interscience, New York, 1976.
111. C. Schade and H. Mayr, *Makromol. Chem., Rapid. Commun. 9*: 477–482 (1988).
112a. C. A. Grob and P. W. Schiess, *Angew. Chem. 79*: 1–27 (1967).
112b. C. A. Grob and P. W. Schiess, *Angew. Chem. Int. Ed. Engl. 6*: 1–15 (1967).
113a. C. A. Grob, *Angew. Chem. 81*: 543–554 (1969).
113b. C. A. Grob, *Angew. Chem. Int. Ed. Engl. 8*: 535–546 (1969).
114. K. B. Becker and C. A. Grob, The formation of unsaturated groups by heterolytic fragmentation, *The Chemistry of Functional Groups* (S. Patai, ed.) Wiley, London, 1977, Suppl. A, pp. 653–723.
115. L. E. Overman and M. J. Sharp, *J. Am. Chem. Soc. 110*: 612–614 (1988).
116. L. E. Overman and A. K. Sarkar, *Tetrahedron Lett. 33*: 4103–4106 (1992).
117. K. Conrow, *J. Am. Chem. Soc. 81*: 5461–5467 (1959).
118. E. Bäuml, C. Schell, and H. Mayr, unpublished results.
119. P. H. Plesch, *Prog. Reaction Kinetics 18*: 1–62 (1993).
120. J. P. Vairon, A. Rives, and C. Bunel, *Makromol. Chem. Macromol. Symp. 60*: 97–105 (1992).
121. R. B. Taylor and F. Williams, *J. Am. Chem. Soc. 91*: 3728–3732 (1969).
122. D. C. Pepper, *J. Polymer Sci. Symp. 56*: 39–46 (1976).
123. F. Subira, J.-P. Vairon, and P. Sigwalt, *Macromolecules 21*: 2339–2346 (1988).
124. T. Kunitake and K. Takarabe, *Macromolecules 12*: 1061–1067 (1979).
125. T. Kunitake and K. Takarabe, *Macromolecules 12*: 1067–1071 (1979).
126. L. M. Dorfman, R. J. Sujdak, and B. Bockrath, *Acc. Chem. Res. 9*: 352–357 (1976).
127. Y. Wang and L. M. Dorfman, *Macromolecules 13*: 63–65 (1980).
128. H. Mayr, R. Schneider, and U. Grabis, *J. Am. Chem. Soc. 112*: 4460–4467 (1990).
129a. R. A. McClelland, C. Chan, F. Cozens, A. Modro, and S. Steenken, *Angew. Chem. 103*: 1389–1391 (1991).
129b. R. A. McClelland, C. Chan, F. Cozens, A. Modro, and S. Steenken, *Angew. Chem. Int. Ed. Engl. 103*: 1337 (1991).
130. H. Mayr and R. Pock, *Chem. Ber. 119*: 2473–2496 (1986).
131. E. Muks, M. Railjan, and A. Erm, *Proc. Estonian Acad. Sci. Chem. 40*: 93–97 (1991).
132. E. Muks, M. Railjan, I. Loiveke, and M. Heinväli, *Proc. Estonian Acad. Sci. Chem. 41*: 26–29 (1992).
133a. R. Huisgen, *Angew. Chem. 82*: 783–794 (1970).
133b. R. Huisgen, *Angew. Chem. Int. Ed. Engl. 9*: 751–762 (1970).
134. C. K. Ingold and F. R. Shaw, *J. Chem. Soc.* 575–581 (1949).
135. J. Bartl, S. Steenken, H. Mayr, and R. A. McClelland, *J. Am. Chem. Soc. 112*: 6918–6928 (1990).

136. J. Bartl, S. Steenken, and H. Mayr, *J. Am. Chem. Soc. 113*: 7710–7716 (1991).
137. J. T. Denison and J. B. Ramsey, *J. Am. Chem. Soc. 77*: 2615–2621 (1955).
138. M. Szwarc, Ionic polymerization, *Ions and Ion Pairs in Organic Reactions* (M. Szwarc, ed.), John Wiley & Sons, New York 1972 Vol. 2, Chap. 4.
139. H. Mayr, R. Schneider, and C. Schade, *Makromol. Chem. Macromol. Symp. 13/14*: 43–59 (1988).
140. J. B. Collins and A. Streitwieser Jr., *J. Comput. Chem. 1*: 81 (1980).
141. A. E. Reed, R. B. Weinstock, and F. Weinhold, *J. Chem. Phys. 83*: 735–746 (1985).
142. R. F. W. Bader and P. J. MacDougall, *J. Am. Chem. Soc. 107*: 6788–6795 (1985).
143. D. H. O'Brien, C. R. Russell, and A. J. Hart, *J. Am. Chem. Soc. 98*: 7427–7429 (1976).
144. D. H. O'Brien, C. R. Russell, and A. J. Hart, *J. Am. Chem. Soc. 101*: 633–639 (1979).
145. J. Smid, Spectrophotometric studies of ion-pair equilibria, *Ions and Ion Pairs in Organic Reactions* (M. Szwarc, ed.), John Wiley & Sons, New York, 1972, Vol. 1, Chap. 3.
146. C. Reichardt, *Solvents and Solvent Effects in Organic Chemistry*, 2nd ed., VCH Verlagsgesellschaft, Weinheim, 1988.
147. M. Hartnagel, Diplomarbeit, Technische Hochschule Darmstadt, 1993.
148. B. Irrgang and H. Mayr, *Tetrahedron 47*: 219–228 (1991).
149. T. S. Sorensen, *J. Am. Chem. Soc. 91*: 6398–6403 (1969).
150. D. P. N. Satchell and R. S. Satchell, *Chem. Rev. 69*: 251–278 (1969).
151. P. R. Story and B. C. Clark Jr., Homoallylic and homoaromatic cations, *Carbonium Ions* (G. A. Olah and P. v. R. Schleyer, eds.), Wiley, New York, 1972, Vol. 3, Chap. 23.
152. S. Winstein, M. Shatavsky, C. Norton, and R. B. Woodward, *J. Am. Chem. Soc. 77*: 4183–4184 (1955).
153. S. Winstein and M. Shatavsky, *J. Am. Chem. Soc. 78*: 592–597 (1956).
154. P. v. R. Schleyer, W. F. Sliwinski, G. W. Van Dine, U. Schöllkopf, J. Paust, and K. Fellenberger, *J. Am. Chem. Soc. 94*: 125–133 (1972).
155. H.-J. Schneider, N. Becker, G. Schmidt, and F. Thomas, *J. Org. Chem. 51*: 3602–3607 (1986).
156. H.-J. Schneider and F. Thomas, *J. Am. Chem. Soc. 102*: 1424–1425 (1980).
157. H.-J. Schneider, G. Schmidt, and F. Thomas, *J. Am. Chem. Soc. 105*: 3556–3563 (1983).
158. H. C. Brown and M. Borkowski, *J. Am. Chem. Soc. 74*: 1894–1902 (1952).
159. G. Hagen and H. Mayr, *J. Am. Chem. Soc. 113*: 4954–4961 (1991).
160. C. D. Ritchie, *Acc. Chem. Res. 5*: 348–354 (1972).
161. M. L. Poutsma, *J. Am. Chem. Soc. 87*: 4285–4293 (1965).
162. F. Freeman, *Chem. Rev. 75*: 439–490 (1975).
163. J.-F. Gal and P.-C. Maria, *Prog. Phys. Org. Chem. 17*: 159–238 (1990).
164. G. Collin, U. Jahnke, G. Just, G. Lorenz, W. Pritzkow, M. Röllig, L. Win-

guth, P. Dietrich, C.-E. Döring, H. G. Hauthal, and A. Wiedenhöft, *J. Prakt. Chem. 311*: 238–255 (1969).
165. E. Bienvenue-Goëtz and J.-E. Dubois, *Tetrahedron 34*: 2021–2026 (1978).
166. W. K. Chwang, V. J. Nowlan, and T. T. Tidwell, *J. Am. Chem. Soc. 99*: 7233–7238 (1977).
167. P. Knittel and T. T. Tidwell, *J. Am. Chem. Soc. 99*: 3408–3414 (1977).
168. K. Oyama and T. T. Tidwell, *J. Am. Chem. Soc. 98*: 947–951 (1976).
169. W. K. Chwang, P. Knittel, K. M. Koshy, and T. T. Tidwell, *J. Am. Chem. Soc. 99*: 3395–3401 (1977).
170. J. L. Jensen and V. Uaprasert, *J. Org. Chem. 41*: 649–654 (1976).
171. M. F. Ruasse, J. E. Dubois, and A. Argile, *J. Org. Chem. 44*: 1173–1175 (1979).
172. A. Streitwieser Jr., *Solvolytic Displacement Reactions*, McGraw-Hill, New York, 1962, p. 43.
173. K.-T. Liu and H.-C. Sheu, *J. Org. Chem. 56*: 3021–3025 (1991).
174. T. W. Bentley, I. S. Koo, and S. J. Norman, *J. Org. Chem. 56*: 1604–1609 (1991).
175. E. M. Arnett, C. Petro, and P. V. R. Schleyer, *J. Am. Chem. Soc. 101*: 522–526 (1979).
176. S. Thea and A. Williams, *Chem. Soc. Rev. 15*: 125–140 (1986).
177. C. F. Bernasconi, *Acc. Chem. Res. 20*: 301–308 (1987).
178. J. P. Richard, T. L. Amyes, L. Bei, and V. Stubblefield, *J. Am. Chem. Soc. 112*: 9513–9519 (1990).
179. T. L. Amyes and J. P. Richard, *J. Chem. Soc. Chem. Commun.* 200–202 (1991).
180. L. J. Johnston, P. Kwong, A. Shelemay, and E. Lee-Ruff, *J. Am. Chem. Soc. 115*: 1664–1669 (1993).
181. T. L. Amyes, I. W. Stevens, and J. P. Richard, *J. Org. Chem. 58*: 6057–6066 (1993).
182a. H. Mayr and M. Patz, *Angew. Chem. 106*: 990–1010 (1994).
182b. H. Mayr and M. Patz, *Angew. Chem. Int. Ed. Engl. 33*: 938–957 (1994).
183. H. Mayr, Control of electrophilicity in aliphatic Friedel-Crafts reactions, *Selectivities in Lewis Acid Promoted Reactions* (D. Schinzer, ed.), NATO ASI Series C, Vol. 289, 1989, pp. 21–36.
184. D. Y. Curtin, *Rec. Chem. Prog. 15*: 111 (1954).
185. L. P. Hammett, *Physical Organic Chemistry*, McGraw-Hill, New York, 1970.
186. J. I. Seeman, *Chem. Rev. 83*: 83–134 (1983).
187. E. W. Colvin, *Silicon in Organic Synthesis*, Butterworths, London, 1981.
188. W. P. Weber, *Silicon Reactions for Organic Synthesis*, Springer, Berlin, 1983.
189. I. Fleming, Organic Silicon Chemistry, *Comprehensive Organic Chemistry*, (D. H. R. Barton and W. D. Ollis, eds.), Pergamon Press, Oxford, 1979, Vol. 3, pp. 541–686.

190. M. Pereyre, J.-P. Quintard, and A. Rahm, *Tin in Organic Synthesis*, Butterworths, London, 1987.
191. S. G. Wierschke, J. Chandrasekhar, and W. L. Jorgensen, *J. Am. Chem. Soc. 107*: 1496–1500 (1985).
192. J. C. White, R. J. Cave, and E. R. Davidson, *J. Am. Chem. Soc. 110*: 6308–6314 (1988).
193. M. R. Ibrahim and W. L. Jorgensen, *J. Am. Chem. Soc. 111*: 819–824 (1989).
194. C. Eaborn, *J. Chem. Soc., Chem. Commun.* 1255 (1972).
195. J. B. Lambert, *Tetrahedron 46*: 2677–2689 (1990).
196. Y. Apeloig, Theoretical aspects of organosilicon compounds; The chemistry of organic silicon compounds, Part I, *The Chemistry of Functional Groups* (S. Patai and Z. Rappoport, eds.), Wiley, Chichester, 1989, Chap. 2, pp. 57–225.
197. J. B. Lambert, G.-t. Wang, R. B. Finzel, and D. H. Teramura, *J. Am. Chem. Soc. 109*: 7838–7845 (1987).
198. J. B. Lambert and G.-t. Wang, *Tetrahedron Lett. 29*: 2551–2554 (1988).
199. J. B. Lambert, G.-t. Wang, and D. H. Teramura, *J. Org. Chem. 53*: 5422–5428 (1988).
200a. M. T. Reetz, *Angew. Chem. 94*: 97–109 (1982).
200b. M. T. Reetz, *Angew. Chem. Int. Ed. Engl. 21*: 96–108 (1982).
201. T. Mukaiyama and M. Murakami, *Synthesis* 1043–1054 (1987).
202. L. S. Povarov, *Russ. Chem. Rev. (Engl. Transl.) 34*: 639–656 (1965).
203. D. Caine, Alkylations of enols and enolates, *Comprehensive Organic Synthesis* (B. M. Trost, I. Fleming, and G. Pattenden, eds.), Pergamon Press, Oxford, 1991, Vol. 3; Chap. 1.1, p. 25.
204. I. Paterson *Tetrahedron 44*: 4207–4219 (1988).
205. P. Brownbridge, *Synthesis* 85–104 (1983).
206. M. Patz and H. Mayr, *Tetrahedron Lett. 34*: 3393–3396 (1993).
207. P. J. Stang, Z. Rappoport, M. Hanack, and L. R. Subramanian, *Vinyl Cations*, Academic Press, New York, 1979.
208. F. Marcuzzi and G. Melloni, *J. Am. Chem. Soc. 98*: 3295–3300 (1976).
209. K. Yates, G. H. Schmid, T. W. Regulski, D. G. Garratt, H.-W. Leung, and R. McDonald, *J. Am. Chem. Soc. 95*: 160–165 (1973).
210. G. Modena, F. Rivetti, G. Scorrano, and U. Tonellato, *J. Am. Chem. Soc. 99*: 3392–3395 (1977).
211. F. Marcuzzi, G. Melloni, and G. Modena, *J. Org. Chem. 44*: 3022–3028 (1979).
212. G. Melloni, G. Modena, and U. Tonellato, *Acc. Chem. Res. 14*: 227–233 (1981).
213. H. Mayr, J. L. Gonzalez, and K. Lüdtke, *Chem. Ber. 127*: 525–531 (1994).
214a. F. Cozens, J. Li, R. A. McClelland, and S. Steenken, *Angew. Chem. 104*: 753–755 (1992).
214b. F. Cozens, J. Li, R. A. McClelland, and S. Steenken *Angew. Chem. Int. Ed. Engl. 31*: 743–745 (1992).

215a. H. Mayr, J. Bartl, and G. Hagen, *Angew. Chem. 104*: 1689–1691 (1992).
215b. H. Mayr, J. Bartl, and G. Hagen, *Angew. Chem. Int. Ed. Engl. 31*: 1613–1615 (1992).
216. A. R. Katritzky and R. Taylor, *Adv. Heterocyc. Chem. 47*: 1–467 (1990).
217. C. G. Swain and C. B. Scott, *J. Am. Chem. Soc. 75*: 141–147 (1953).
218. J. M. Harris and S. P. McManus, eds., *Nucleophilicity: Advances in Chemistry Series Nr. 215*, ACS, Washington, DC, 1987.
219. C. D. Ritchie and M. Sawada, *J. Am. Chem. Soc. 99*: 3754–3761 (1977).

3
Mechanistic Aspects of Cationic Polymerization of Alkenes

KRZYSZTOF MATYJASZEWSKI Carnegie Mellon University, Pittsburgh, Pennsylvania

COLEEN PUGH The University of Michigan, Ann Arbor, Michigan

I. INTRODUCTION

Because a polymerization comprises several elementary reactions, the full mechanistic description of cationic polymerizations of alkenes must discuss the mechanisms of initiation, propagation, transfer, termination, and other reactions such as isomerization and rearrangements. These elementary reactions correspond to generation of active sites, monomer consumption, transfer of activity from one chain end to a transfer agent such as monomer or counterion, and termination of the kinetic chain, respectively. Carbenium ions are involved in nearly all of these reactions and therefore will be discussed first. This discussion supplements the general information provided in Chapter 2, which is devoted to reactions of carbenium ions with various nucleophiles. The vast mechanistic information provided by living/controlled carbocationic polymerizations is covered in Chapter 4 and will not be discussed here. Our discussion focuses on systems which often do not provide well-defined polymers because of slow initiation, transfer and/or termination. Nevertheless, analysis of the chemistry and kinetics of these elementary reactions has elucidated the advantages and disadvantages of various initiating systems, which has subsequently led to the development of new "living" systems for the synthesis of well-defined polymers.

II. CARBENIUM IONS

As discussed in Chapters 1 and 2, carbenium ions are species with a formal positive charge on trivalent carbon, in contrast to pentavalent carbonium ions. They are very reactive species which must be stabilized by suitable substituents to participate in polymerization. The stability of carbenium ions increases with increasing substitution and with the substituents' electron-donating ability [1–4]. The reactivity of carbenium ions decreases in the order: primary ($CH_3^+ > CH_3CH_2^+$) > secondary (Me_2CH^+) > tertiary (Me_3C^+). For example, the unsubstituted methylium ion is extremely reactive and has not been observed directly. Although it can not decompose (except to H^+ and carbene), it reacts even with very weak nucleophiles such as alkyl halides to form halonium ions [5].

Similarly, benzylium ions ($PhCH_2^+$) are more reactive than carbenium ions with two phenyl groups (Ph_2CH^+), which are more reactive than trityl cations (Ph_3C^+). For example, cumyl and styryl cations readily decompose even at low temperatures when they are generated in typical polymerization solvents such as CH_2Cl_2, CH_3Cl, $CHCl_3$, and $C_6H_5CH_3$. These carbenium ions can deprotonate to form monomers that can oligomerize and participate in Friedel-Crafts aromatic alkylation. However, although unsubstituted benzhydryl ions are stable only at low temperatures, benzhydryl ions with two *p*-methoxy groups are stable at room temperature for limited periods as discussed in Chapter 2. Triarylmethylium and tropylium salts are the only two carbenium ions that have lifetimes sufficiently long at room temperature to be commercially available. The stability of these carbenium ions depends on the structure of their counteranions; those based on SbF_6^- and AsF_6^- are the most stable. Aryl group stabilization is most pronounced when the aryl substituents are coplanar. However, ^1H NMR shielding demonstrates that steric hindrance, especially by ortho hydrogen atoms or substituents, causes the phenyl groups in triarylmethylium ions to twist considerably out of the plane [6,7]. This results in a "propeller-like" shape. Phenyl and vinyl substituents stabilize carbenium ions through resonance [Eq. (1)].

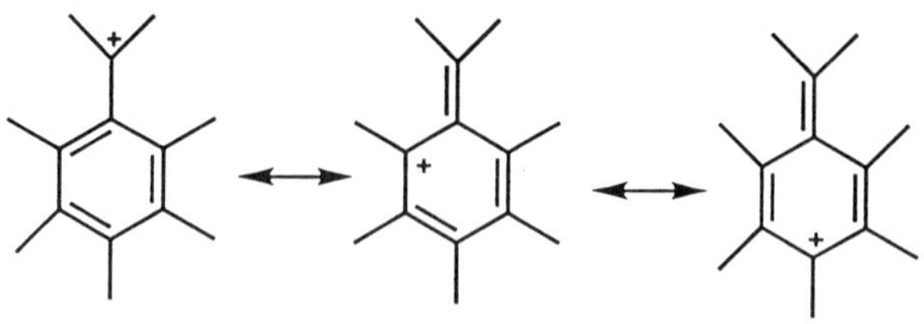

(1)

Electron-donating substituents in ortho and especially in para positions additionally stabilize carbenium ions. Therefore, *p*-methyl and *p*-methoxystyrene readily polymerize cationically.

Alkyl groups stabilize carbenium ions by weak electron donation and by hyperconjugation [Eq. (2)].

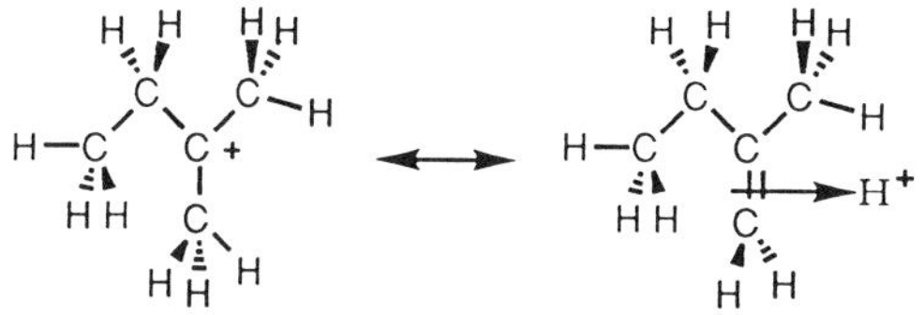
(2)

The tertiary propagating carbenium ions in isobutene polymerizations are stabilized by eight β-H atoms through hyperconjugation. Therefore, 1,1-disubstituted alkenes polymerize cationically much more readily than monosubstituted α-olefins.

Substituents with α-heteroatoms stabilize carbenium ions via resonance, which leads formally to oxonium ions [Eq. (3)].

$$CH_3-O-CH_2^+ \rightleftharpoons CH_3-\overset{+}{O}=CH_2$$
$$CH_3-\overset{+}{C}=O \rightleftharpoons CH_3-C\equiv O^+ \qquad (3)$$

The most reactive monomers in cationic polymerizations, *N*-vinyl carbazole, vinyl ethers, and *p*-methoxy-α-methylstyrene provide the most stable carbenium ions. Carbenium ions generated from the first two monomers are stabilized by α-heteroatoms, whereas the latter monomer generates a tertiary carbenium ion with a strongly electron donating *p*-substituent. Carbenium ions with two alkoxy groups, such as 1,3-dioxolane-2-ylium cations and their acyclic analogs (Chapter 6) are stable at room temperature in the absence of moisture.

Carbenium ions are sp^2 hybridized with the empty orbital perpendicular to the plane containing the three substituents. Calculations confirm both the flat structure of carbenium ions and the shorter linkage with aryl substituents due to the partial double-bond character. As discussed previously, the positive charge is not localized just on the sp^2-hybridized carbon of the carbenium ion, but is dispersed onto neighboring substituents. Table 1 presents the charge distribution in model carbenium ions formed by protonation of propene, isobutene, styrene, α-methylstyrene, methyl vinyl ether, and methyl propenyl ether.

These calculations demonstrate that up to 30% of the charge is located at each β-hydrogen. β-Protons therefore are easily abstracted by any basic

Table 1 Charge Distribution on β-H Atoms and sp^2-Hybridized C^+ Atoms Calculated by RHF/6-31G* for Carbocations Derived from Various Alkenes

Parent alkene	Proportion of charge on β-H	Proportion of charge on C^+	Ratio of charges β-H/C^+
Propene	0.28989	0.17708	1.6371
Isobutene	0.27404	0.30883	0.88735
Styrene	0.24899	0.18738	1.3288
α-Methylstyrene	0.24638	0.19715	1.2497
Me Vinyl Ether	0.27600	0.38800	0.71134
Me Propenyl Ether	0.26598	0.58817	0.45222

component present in the system. Thus, the major transfer reaction in cationic polymerizations of alkenes is β-proton elimination. The results of the RHF/6-31G* *ab initio* calculations presented in Table 1 show higher charge distribution on β-hydrogens than calculated previously using lower basis set method (STO-3G) [67]. The ratio of charges on β-H atoms and C^+ may be roughly correlated with the ability to transfer in polymerization. Tertiary cations apparently have a lower tendency to transfer than secondary cations do in spite of more β-hydrogens. An alkoxy substituent decreases the probability of transfer more than methyl or phenyl groups.

A. Identification of Carbenium Ions by NMR

A number of stable carbenium ions can be generated in superacid media; most NMR data on chemical shifts and coupling constants of carbenium ions have been reported in superacid media [7–10]. However, superacids cannot be used for polymerization studies because excess superacid not only stabilizes the ionic species, but also protonates any alkene which would otherwise be available for oligomerization/polymerization.

Only the most stabilized carbenium ions such as those derived from vinyl ethers can be studied directly by NMR at low temperatures in organic solvents typically used for polymerizations. As discussed in detail in Chapter 4, the α-methine proton of α-isobutoxyethyl carbenium ion is strongly deshielded, and absorbs at $\approx$10 ppm [Eq. (4)].

$$\underset{\approx 5\text{ to }6\text{ ppm}}{CH_3\text{-}CH(OR)\text{-}X} + LA \rightleftharpoons \underset{\approx 10\text{ ppm}}{CH_3\text{-}C^+H(OR),\ LAX^-} \quad (4)$$

The ^1H NMR spectra also indicate that the methylene protons in the isobutoxy group are equivalent in the carbenium ion [11], in contrast to their magnetic inequivalence in the covalent precursor. That is, addition of Lewis acid to the alkyl chloride leads to rapid ionization as shown by one broad signal at 3.6 ppm for the methylene group. This resolves into a doublet if ionization is sufficiently fast. Two AB doublets are observed if exchange is slow. Note that the methylene carbon atom and carbocationic center are connected through an oxygen atom in the Newman projection shown in Eq. (5).

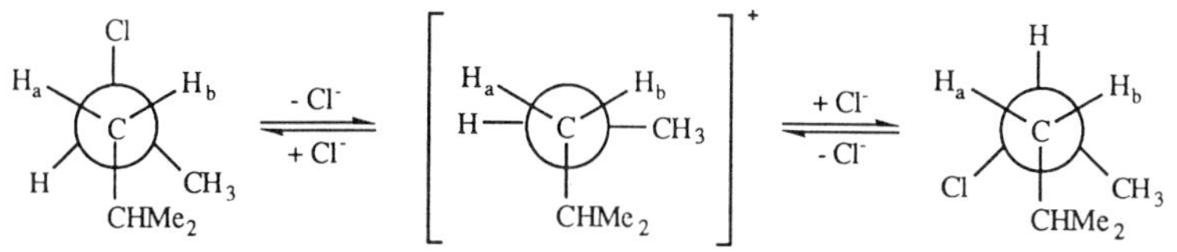

(5)

However, Lewis acid not only ionizes the alkyl halide, but may also complex with a nucleophilic oxygen atom. This reaction is important in the presence of both excess vinyl ether (monomer) and excess polymer. We will return to this reaction in Section IV.C.2.

Carbenium ions can be generated not only by ionization of alkyl halides with Lewis acids, but also by protonation of double bonds. Unfortunately, addition of excess triflic acid to a vinyl ether leads to polymer rather than to the monomeric cation [Eq. (6)].

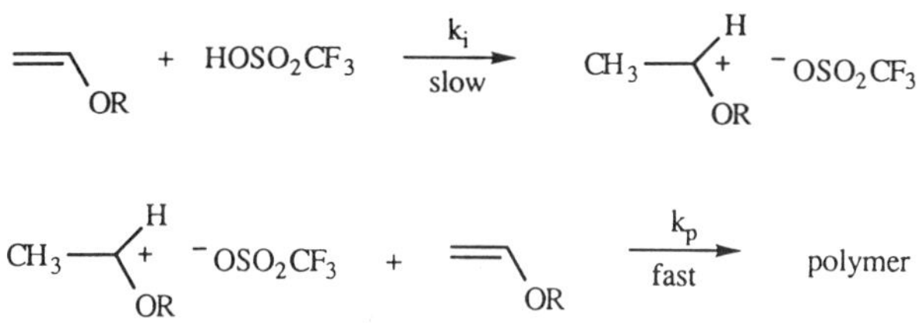

(6)

Initiation is apparently slower than propagation. That is, the nucleophilicity of vinyl ethers is higher than their basicity. Other monomers such as *p*-methoxy-α-methylstyrene are apparently more basic and react rapidly with acid. In addition, the equilibrium monomer concentrations of α-methylstyrenes are relatively high ($[M]_\infty \approx 0.2$ mol/L at $-30°$ C). Because they can not polymerize at low concentration, they are ideal monomers for model studies [12,13]. The equilibrium constants of dimerization and trimerization are much larger than that for the formation of high polymer. Therefore, dimers and trimers can be formed below $[M]_\infty$ although high polymers cannot.

Figure 1 shows the ^1H NMR spectrum of the monomeric carbenium ion formed at $-63°$ C in CH_2Cl_2 by reaction of p-methoxy-α-methylstyrene with excess triflic acid. A deshielded singlet is observed for the α-methyl group when p-methoxy-α-methylstyrene is protonated. Two α-methyl groups absorb at 3.04 ppm, which is 0.89 ppm downfield from that of monomer. The positive charge is delocalized substantially onto the aromatic ring and the p-methoxy group. The p-methoxy protons absorb at 4.21 ppm, which is 0.38 ppm downfield from that of monomer. The chemical shifts of protons ortho and meta to the carbocatonic center are 8.60 and 7.31 ppm, respectively.

Triflic acid homoconjugates with triflate anion, and two equivalents of acid are therefore necessary per cation. If only an equimolar amount of acid is used relative to p-methoxy-α-methylstyrene, then a dimeric cation M_2^+ is formed, along with small amounts of monomeric M_1^+ cation and trimeric M_3^+ cation according to Eq. (7). No higher oligomers ($n >$ 3) can be formed due to the low $[M]_0$ ($[M]_0 < M]_\infty$).

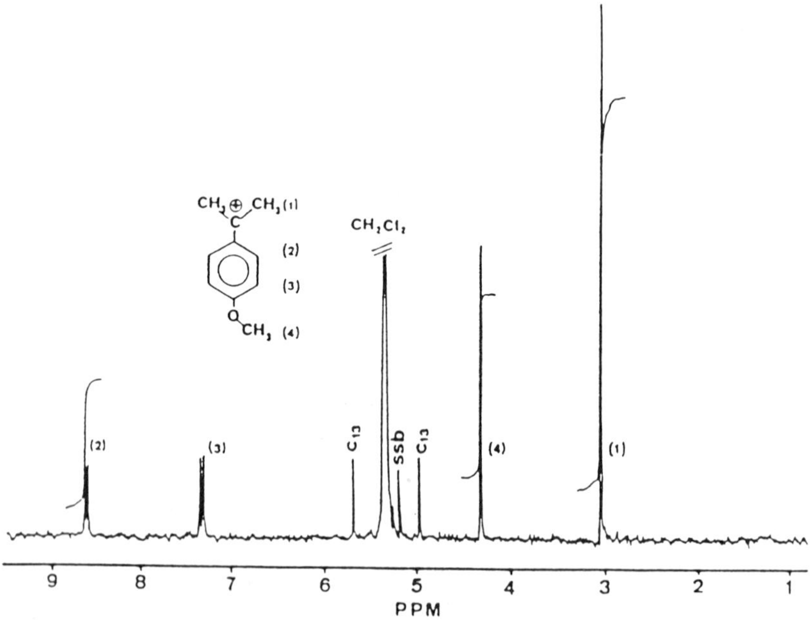

Figure 1 ^1H NMR spectrum of the reaction mixture of $[p\text{MeO}\alpha\text{MeSt}]_0 = 0.02$ M and $[\text{HOTf}]_0 = 0.05$ M in CD_2Cl_2 at $-63°$ C (From Ref. 13).

$$\begin{aligned} (HA)_2 + M &\rightleftharpoons M_1^+, HA_2^- \\ M_1^+, HA_2^- + M &\rightleftharpoons M_2^+, HA_2^- \\ M_2^+, HA_2^- + M &\rightleftharpoons M_3^+, HA_2^- \\ M_3^+, HA_2^- + M &\rightleftharpoons M_4^+, HA_2^- \end{aligned} \tag{7}$$

The ^{1}H NMR spectrum of the oligomeric carbenium ions is shown in Fig. 2. The methoxy group from the charged aromatic ring in the "dimeric cation" absorbs at 4.26 ppm, which is 0.05 ppm upfield from that in the monomeric cation; the second methoxy group absorbs at 3.75 ppm. The

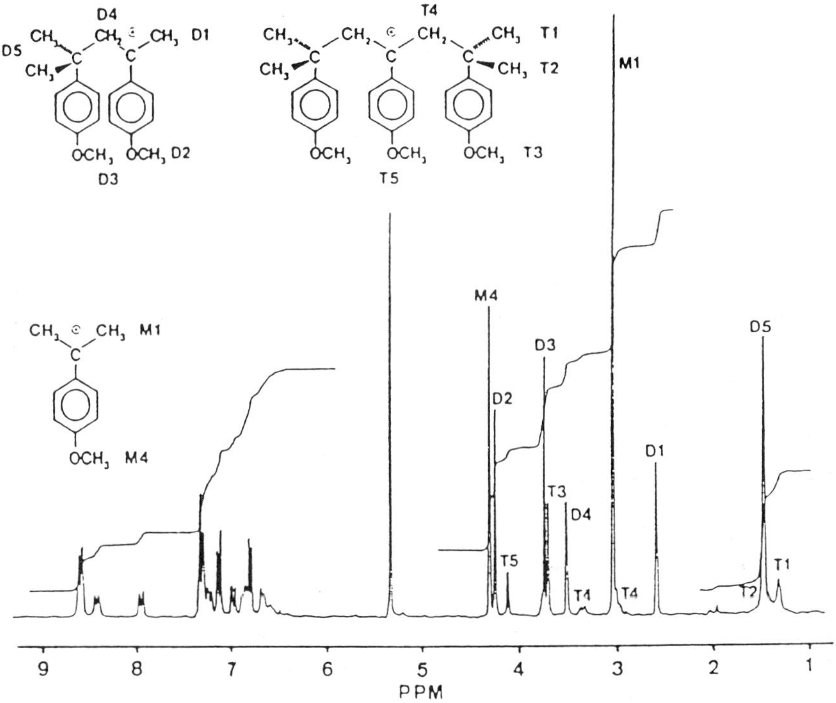

Figure 2 ^{1}H NMR spectrum of the reaction mixture of $[p\text{MeO}\alpha\text{MeSt}]_0 = 0.03$ M and $[\text{HOTf}]_0 = 0.037$ M in CD_2Cl_2 after reaching equilibrium at $-65°$ C (From Ref. 13).

α-methylene and α-methyl groups absorb at 3.52 ppm and 2.56 ppm, respectively, whereas both terminal α-methyl groups absorb at 1.50 ppm.

The aromatic protons from the second ring absorb in a typical AB pattern (ortho-H at 7.13 ppm, meta-H at 6.80 ppm, J_{AB} = 8.7 Hz). However, all four protons in the charged ring absorb at different chemical shifts, with the signals at 8.42 ppm and 7.96 ppm assigned to the ortho protons, and those at 7.23 ppm and 6.98 ppm assigned to the meta protons. The coupling constant between ortho protons is weaker (1.5 Hz) than that between meta protons (2.5 Hz). The nonequivalency of aromatic protons in the first ring demonstrates that rotation is slow around the carbocationic center and the aromatic nucleus, which indicates that there is partial double-bond character in the bond connecting these two groups. On the other hand, the equivalency of the two terminal methyl groups indicates that rotation is fast around the corresponding C-Ar bond [Eq. (8)].

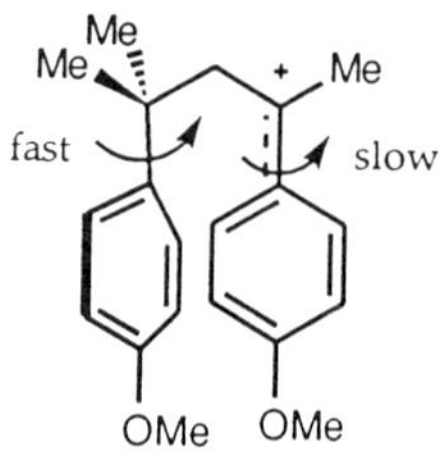

(8)

Small signals due to the trimeric cation can also be detected in Fig. 2. Figure 3 demonstrates that these signals are much easier to detect in the presence of excess monomer. Surprisingly, the charge in the trimeric cation is on the central carbon atom rather than on the terminal carbon atom; the isomerized $M_{3is}{}^+$ cation must be thermodynamically more stable than the original $M_3{}^+$ [Eq. (9)]. The signal of the central methoxy group appears at 4.10 ppm, which is 0.16 ppm upfield from $M_2{}^+$ and 0.11 ppm upfield from $M_1{}^+$. The methylene protons absorb in an AB pattern at 3.35 ppm and 3.00 ppm (J_{AB} = 10 Hz), apparently due to slow rotation of the alkyl groups in the isomerized cation. This is confirmed by the nonequivalency of the methyl groups absorbing at 1.49 and 1.32 ppm. Slow rotation indicates that there are strong interactions between the carbocationic center and the two neighboring aromatic rings. The deep red color of the reaction mixture at this stage indicates that the interaction may involve charge transfer between the carbocationic center and neighboring aromatic rings.

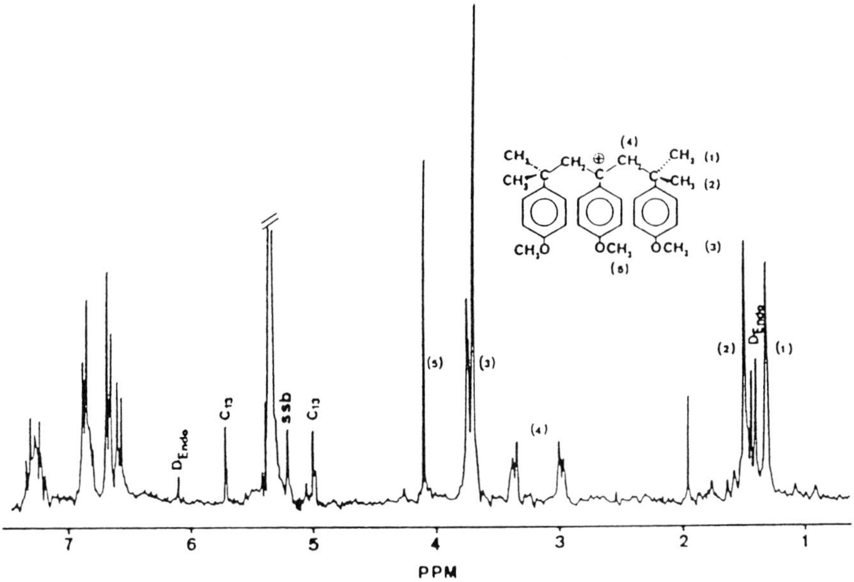

Figure 3 ^{1}H NMR spectrum of the reaction mixture of $[p\text{MeO}\alpha\text{MeSt}]_0 = 0.016$ M and $[\text{HOTf}]_0 = 0.008$ M in CD_2Cl_2 after reaching equilibrium at $-65°$ C. (From Ref. 13).

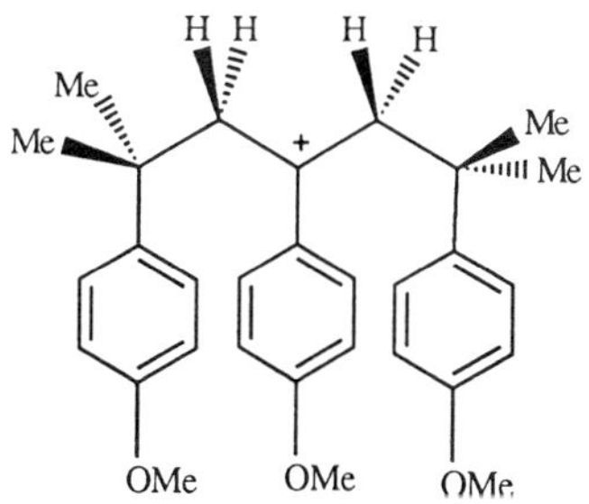

(9)

The exact mechanism of M_3^+ isomerization has not been established yet. It may involve a 1,3-methide shift, or reaction of the unsaturated exo-dimer with monomeric cation [Eq. (10)]. The unsaturated dimer is formed by deprotonation of the dimeric cation.

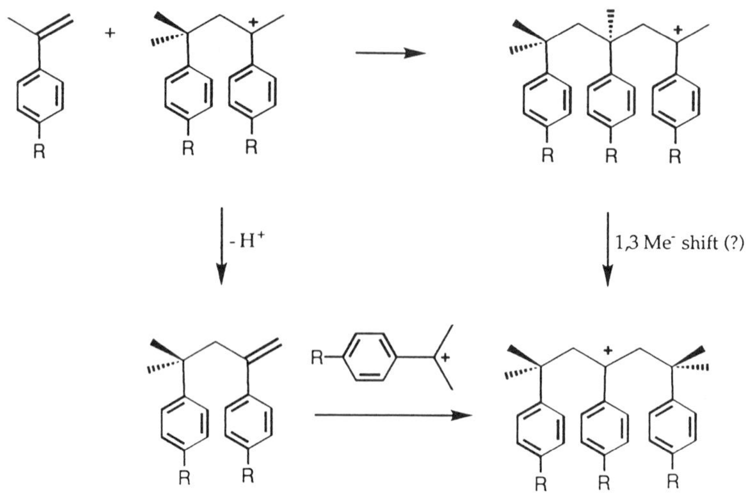

(10)

The timescale of NMR experiments is usually at least 1 min, which is long enough for carbenium ions to react with thousands of monomer molecules. Because isomerized structures are not observed in polymers formed from α-methylstyrenes, isomerization must occur over a much longer time than that required for complete polymerization.

Other reactions occur at even longer times. Indans and indanyl cations form first; these are the final products for alkyl-substituted α-methylstyrenes. However, further reactions of the p-methoxy derivative leads to formation of spirobiindan and anisol [Eq. (11)].

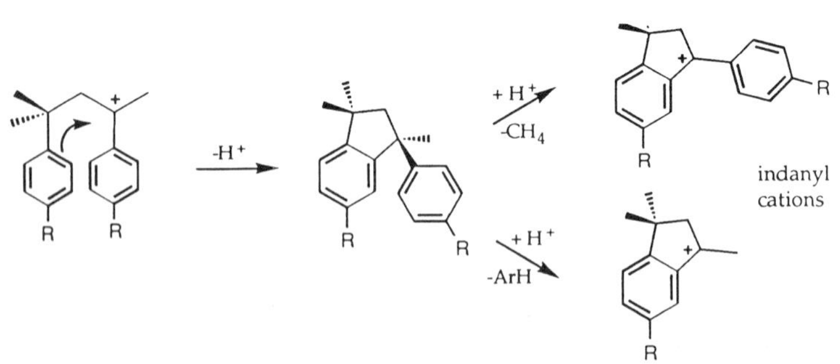

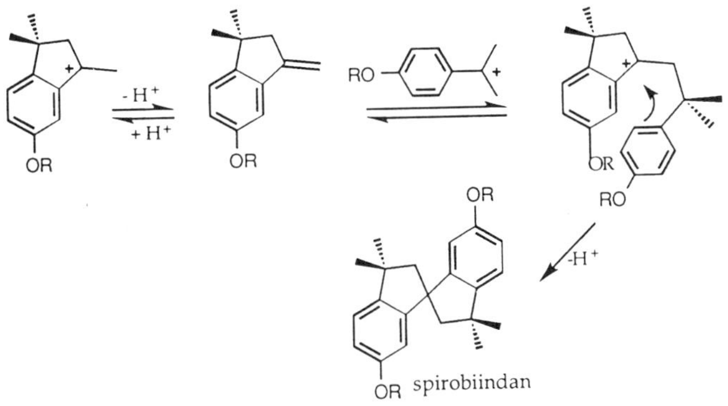

(11)

The ^{1}H NMR chemical shifts of these carbenium ions, unsaturated dimers, and indan derivatives are summarized in Tables 2–5.

B. Identification of Carbenium Ions by UV-Visible Spectroscopy

UV-visible spectroscopy is very useful for studying low concentrations of carbenium ions. Unfortunately, it is limited to carbenium ions α-substituted with aryl groups. Styryl and cumyl cations absorb at $\lambda_{max} = 330$–400 nm depending on the substituents in the aromatic ring, with extinction coefficients $\epsilon = 10{,}000$–$30{,}000\ mol^{-1}\cdot L\cdot cm^{-1}$. The latter value is probably more reliable because it was obtained from a well-defined oligomerization system [14]; the lower value was obtained in highly dilute superacid media [15]. The absorption maxima of carbenium ions in CH_2Cl_2 observed by stopped-flow experiments of styrene and α-methylstyrene polymerizations were $\lambda_{max} \approx 335 \pm 10$ nm [16–18]. The lifetime of carbenium ions depends on the counteranion and temperature. For example, the lifetime of the styryl cation is less than 1 sec at room temperature when triflate is the anion, but increases to nearly 1 hr at $-70°$ C [18]. These carbenium ions decompose by intramolecular cyclization to form polymers with indanyl end groups; more stable cations are formed at longer times by hydride transfer as shown in Eq. (11). The UV spectra of carbenium ions derived from various styrenes are summarized in Table 6.

Monomeric styryl cations were observed recently at 315 and 325 nm during flash photolysis of styrene and α-methylstyrene, respectively, in trifluoroethanol [20]. Absorbance of these model compounds at wavelengths lower than those of the propagating species may be due to the

Table 2 Chemical Shifts and Coupling Constants of the Carbocations and Unsaturated Species for pMeO-α-MeSt

compd	structure	chem shift	coupling const, Hz	compd	structure	chem shift	coupling const, Hz
M_1^+	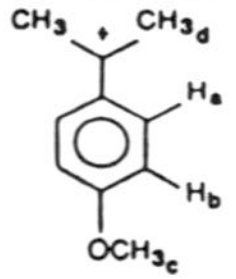	8.60 (H_a), 7.31 (H_b) 4.31 (H_c), 3.04 (H_d)	J_{AB} = 9.2	M	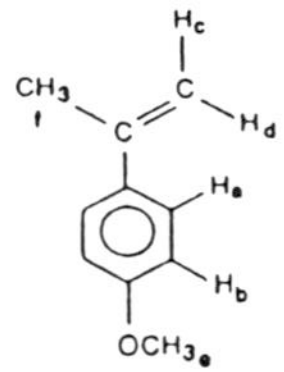	7.44 (H_a), 6.89 (H_b) 5.31 (H_c) 5.01 (H_d) 3.835 (H_e) 2.155 (H_f)	J_{AB} = 8.9 J_{CD} = 1.5 J_{DF} = 1.5 J_{CF} = 0.8
M_2^+	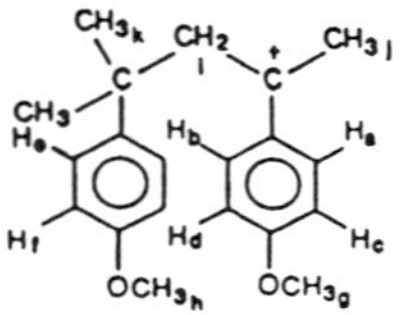	8.42 (H_a), 7.96 (H_b) 7.23 (H_c), 6.98 (H_d) 7.13 (H_e), 6.80 (H_f) 4.26 (H_g), 3.75 (H_h) 3.52 (H_i), 2.56 (H_j) 1.50 (H_k)	J_{AB} = 1.5 J_{CD} = 2.5 J_{EF} = 8.7 J_{AC} = 9.7 J_{BD} = 9.7	D_E(endo) (trans)	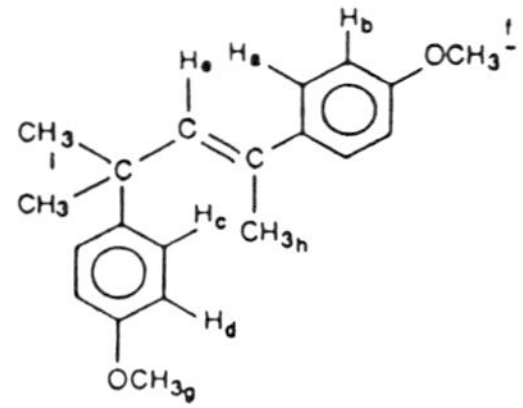	7.38 (H_a–H_c) 6.86 (H_b) 6.09 (H_e) 3.85 (H_f) 3.845 (H_g) 1.625 (H_h) 1.545 (H_i)	J_{AB} = 8.8 J_{CD} = 8.6 J_{EH} = 1.2
M_{3l}^+	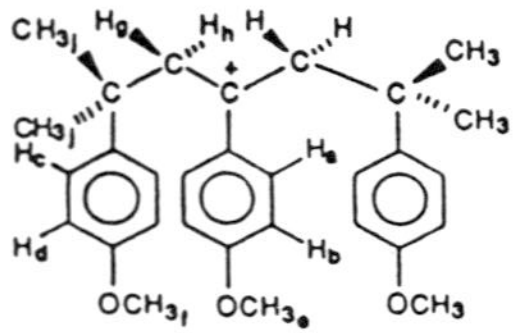	7.25 (H_a), 6.60 (H_b) 6.86 (H_c), 6.65 (H_d) 4.10 (H_e), 3.70 (H_f) 3.35 (H_g), 3.00 (H_h) 1.49 (H_i), 1.32 (H_j)	J_{CD} = 8.6 J_{GH} = 9	D_E(exo)	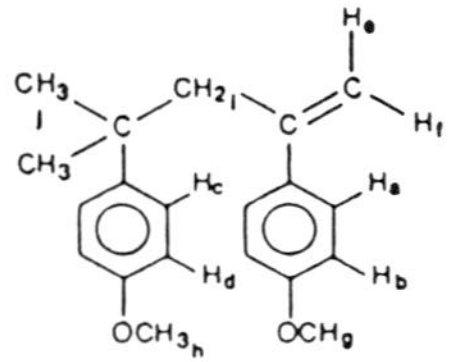	7.25 (H_a) 7.22 (H_c) 6.78 (H_b–H_d) 5.15 (H_e) 4.77 (H_f) 3.83–3.82 (H_g–H_h) 2.82 (H_i) 1.262 (H_j)	J_{AB} = 8.8 J_{CD} = 8.8 J_{EF} = 1.8

Table 3 ^1H NMR Chemical Shifts of Monomeric Cations (M_1^+) in CD_2Cl_2 at −70° C

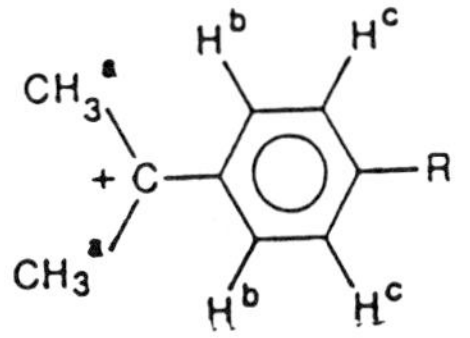

R	H^a	H^b	H^c	H(R)	J_{bc}, (Hz)
H	3.62	8.78	8.02	8.62	8.4
CH_3	3.44	8.63	7.79	2.80	8.6
$C(CH_3)_3$	3.46	8.67	7.97	1.44	8.9
OCH_3	3.04	8.60	7.31	4.31	9.2

absence of penultimate aromatic rings, and/or to the solvent used. For example, the monomeric cation derived from *p*-methoxy-*α*-methylstyrene and triflic acid absorbs at 368 nm in CH_2Cl_2 [13] and at 360 nm in CF_3CH_2OH [20]. As shown in Fig. 4, the cumyl carbenium ion absorbance also involves a low-energy shoulder. Similar shoulders are observed with 1-phenylethyl derivatives.

Table 4 ^1H NMR Chemical Shifts of Indans in CD_2Cl_2 at −70° C

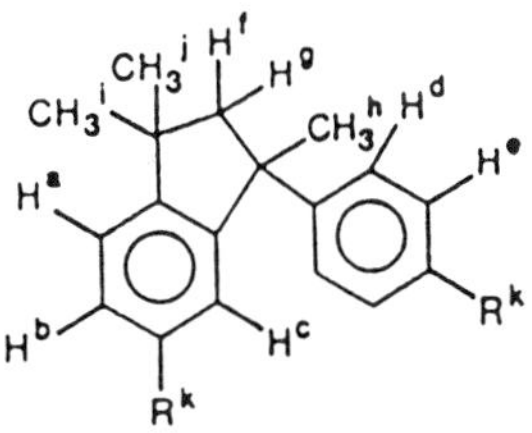

R	H^a	H^b	H^c	H^d	H^e	H^f	H^g	H^h	H^i	H^j	H^k		J_{AB} (Hz)	J_{BC} (Hz)	J_{DE} (Hz)	J_{FG} (Hz)
H	*m*	*m*	*m*	*m*	*m*	2.38	2.17	1.63	1.29	0.92	*m*	*m*				13.0
CH_3	7.08	7.08	6.87	7.04	7.04	2.34	2.12	1.58	1.26	0.93	2.30	2.2				13.0
$C(CH_3)_3$	7.12	7.29	7.17	7.03	7.22	2.32	2.13	1.60	1.22	0.85	1.27	1.2	8	1.5	8.4	12.9
OCH_3	7.12	6.86	6.65	7.08	6.79	2.32	2.15	1.60	1.26	0.91	3.83	3.7	11	2.5	8.7	12.5

m Multiplet in the region 7.30–7.10 ppm.

Table 5 ^{1}H NMR Chemical shifts of indanyl cations in CD_2Cl_2 at $-70°$ C

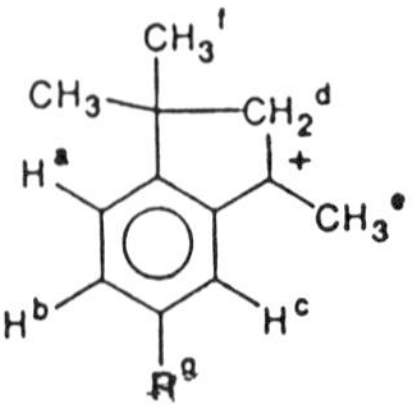

R	H^a	H^b	H^c	H^d	H^e	H^f	H^g
H	7.90	8.60	8.50	3.87	3.58	1.60	7.90
CH_3	7.92	8.51	8.30	3.85	3.44	1.58	2.63
$C(CH_3)_3$	7.87	8.67	8.18	3.76	3.37	1.47	1.36

The unsubstituted benzhydryl cation absorbs at longer wavelengths (≈440 nm) than cumyl and styryl derivatives due to more pronounced charge delocalization by the additional aromatic ring. In contrast, trityl carbenium ions absorb at only 400–430 nm in spite of having three phenyl groups; this is probably due to their propeller-like structure. Electron-donating and resonance-stabilizing substituents such as alkyl, alkoxy and amino, especially in para positions, lead to stronger delocalization and lower energy absorptions. For example, the trityl salt with three *p*-(dimethylamino) substituents is used as a dye (crystal violet, $\lambda_{max} = 588$ nm).

The absorption at UV-visible wavelengths and the accompanying color may be due not only to propagating carbenium ions, but also to inactive species. As discussed in the previous section, growing carbenium ions can isomerize to less reactive species with more delocalized charge. For example, the absorption at 380 nm in *p*-methoxystyrene polymerizations is probably due to isomerized species rather than to growing carbenium ions (≈350 nm) [29] (cf. Fig. 4). Indanyl cations, such as 1-phenylindan-1-ylium, absorb at 414 nm with $\epsilon = 32{,}000\ mol^{-1} \cdot L \cdot cm^{-1}$ [26]. Thus, a combination of UV and NMR data is ideal for elucidating both the structure and reactivities of carbenium ions. Although NMR spectroscopy is superior to UV-visible spectroscopy for determining structure, the low concentrations used in UV-visible spectroscopy are more typical of the concentrations of carbenium ions present in real polymerization systems. Stopped-flow UV-visible spectroscopy is especially useful in detecting transient reactive carbenium ions.

Table 6 Maxima of absorption and extinction coefficients of model and growing carbenium ions

Monomer	Cation	λ_{max} (nm)	ϵ ($mol^{-1} \cdot L \cdot cm^{-1}$)	Comments	Ref.
St	$—CH_2CHPh^+$	340	10^4	1a	[19]
St	$MeCHPh^+$	315		1b	[20]
α-MeSt	$—CH_2CMePh^+$	336	10^4	1a; $DP_n < 2$	[21]
α-MeSt	$—CH_2CMePh^+$	350	10^4	1a; polymer	[19]
α-MeSt	Me_2CPh^+	333	$2.6 \cdot 10^4$		[14]
α-MeSt	Me_2CPh^+	326	$1.1 \cdot 10^4$	2	[15]
α-MeSt	Me_2CPh^+	325		1b	[20]
p-MeSt	$—CH_2CHAr^+$	334			[22]
*p*MeSt	$MeCHAr^+$	325		1b	[20]
p-MeOSt	$—CH_2CHAr^+$	340			[19]
p-MeOSt	$—CH_2CHAr^+$	380	$2.8 \cdot 10^4$	2	[23]
p-MeOSt	$MeCHAr^+$	348	$2.8 \cdot 10^4$	2	[24]
p-MeOSt	$MeCHAr^+$	340		1b	[20]
p-ClSt	$—CH_2CHAr^+$	325			[19]
2,4,6-MeSt	$—CH_2CHAr^+$	325			[19]
p-Me-α-MeSt	Me_2Ar^+	345	$>2.5 \cdot 10^4$	4	[14]
p-Me-α-MeSt	Me_2Ar^+	340		1b	[20]
p-Me-α-MeSt	$—CH_2CMeAr^+$	358	$>2.2 \cdot 10^4$	3	[14]
p-iPr-α-MeSt	$—CH_2CMeAr^+$	362	$>1.6 \cdot 10^4$	3	[25]
p-*t*-Bu-α-MeSt	Me_2Ar^+	350	$3.5 \cdot 10^4$	4	[14]
p-*t*-Bu-α-MeSt	$—CH_2CMeAr^+$	363	$>2.2 \cdot 10^4$	3	[14]
p-MeO-α-MeSt	Me_2Ar^+	368	$2.9 \cdot 10^4$	4	[14]
p-MeO-α-MeSt	Me_2Ar^+	360		1b	[20]
p-MeO-α-MeSt	$—CH_2CMeAr^+$	382	$>2.5 \cdot 10^4$	3	[14]
IN	$—CHRCHAr^+$	404			[22]
3-iPrIN	$—CHRCHAr^+$	320	$2.9 \cdot 10^4$		[26]
3-MeIN	$—CHRCHAr^+$	312	$3 \cdot 10^4$		[26]
3-PhIN	$—CHRCHAr^+$	412	$3.2 \cdot 10^4$		[26]
1,1-DPE	$MeCPh^+{}_2$	435	$2.6 \cdot 10^4$	5	[22,27,28]

1a. ϵ assumed from the model tertiary alcohol in superacid media; 1b. cations generated by flash photolysis in trifluoroethanol; additional low-energy shoulders were observed for *p*-H and *p*-Me but not for *p*-MeO derivatives.

2. ϵ assumed quantitative formation of ions.

4. Confirmed by NMR.

3. Assumed A^- structure, ϵ should be higher for HA_2^- counterions.

5. Additional bands were observed at 520 nm (and 395 nm) and at 456 nm (and 330 nm) and were ascribed to the cation solvated internally by the neighboring aromatic ring or by monomer.

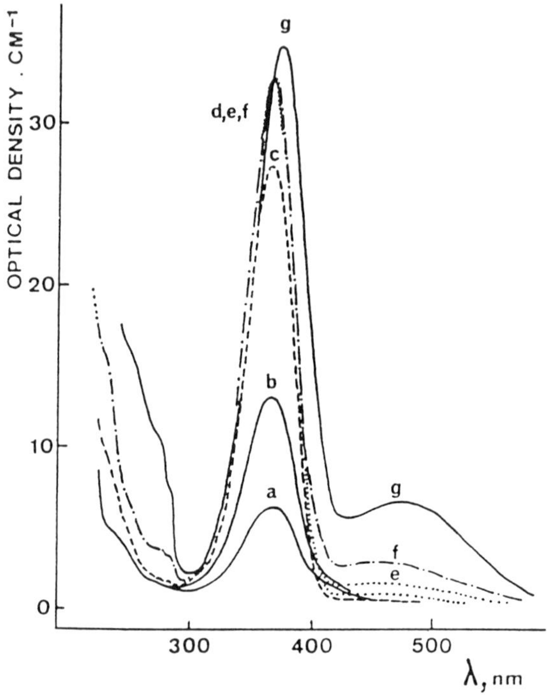

Figure 4 UV spectra of the reaction mixtures of *p*-methoxy-α-methylstyrene with CF_3SO_3H in CH_2Cl_2 at −72° C $[HOTf]_0$ = 0.0021 *M* (6.84 × 10^{-5} mol). $[pMeO\alpha MeSt]_0/[HOTf]_0$: (a) 0.1; (b) 0.266; (c) 0.45; (d) 0.614; (e) 0.79; (f) 0.96; (g) 2.23 (From Ref. 13).

C. Identification of Carbenium Ions by Other Spectroscopic Techniques

There is limited structural information on spectroscopy of carbenium ions from other techniques such as Raman [30], IR [30], photoelectron spectroscopy (ESCA) [31], circular dichroism (CD) [32], magnetic CD [33], polarography and voltametry [3,31]; none of these techniques have been applied to polymerization systems successfully yet.

D. Identification of Carbenium Ions by Conductometric Measurements

Conductometry is used to determine the concentration or fraction of dissociated ions that conduct electricity (-C^+ and X^-). However, it is only reliable when studying the dissociation of stable carbenium ions such as trityl and benzhydryl ions. Less stable carbenium ions isomerize, rearrange and/or decompose to more stable cationic species which may also conduct. That is, conductometry provides no structural information and therefore must be supplemented by other characterization techniques such as UV and NMR, which can distinguish between covalent (-C-X) and ionic species (-C^+, X^- plus -C^+), and between carbenium and onium structures. However, spectroscopy usually cannot determine the degree of dissociation because the spectra of ions and ion pairs are nearly identical. For example, conversion of free dianisylmethylium tetrachloroborate ions to the corresponding ion pair using salts with common ions results in a blue shift of less than 2 nm [34].

Thus, spectroscopic analysis and conductometry can be combined to determine both equilibrium constants K_I and K_D [Eq. (12)].

$$\text{...-C-X} \underset{}{\overset{K_I}{\rightleftharpoons}} \text{...-C}^+\text{, X}^- \underset{}{\overset{K_D}{\rightleftharpoons}} \text{...-C}^+ + \text{X}^- \tag{12}$$

The dissociation constants of trityl and benzhydryl salts are $K_D \approx 10^{-4}$ mol/L in CH_2Cl_2 at 20° C, which corresponds to 50% dissociation at $2 \cdot 10^{-4}$ mol/L total concentration of carbocationic species (cf. Table 7) [34]. The dissociation constants are several orders of magnitude higher than those in analogous anionic systems, which are typically $K_D \approx 10^{-7}$ mol/L [12]. As discussed in Section IV.C.1, this may be ascribed to the large size of counterions in cationic systems (e.g., ionic radius of $SbCl_6^- \approx 3.0$ Å) compared with those in anionic systems (e.g., ionic radius of $Li^+ \approx 0.68$ Å), and to the stronger solvation of cations versus anions. However, the dissociation constants estimated by the common ion effect in cationic polymerizations of styrene with perchlorate and triflate anions are similar to those in anionic systems ($K_D \approx 10^{-7}$ mol/L) [16,17]. This may be because styryl cations are secondary rather than tertiary ions. For example, the dissociation constants of secondary ammonium ions are 100 times smaller than those of quaternary ammonium ions [39].

The dissociation constants of spherical ions should increase with increasing temperature (T) and increasing dielectric constant of the medium (ϵ) according to Eq. (13),

$$K_D = K_D^{o} \cdot \exp\{-e^2 \cdot z^2/[(r_+ + r_-)\epsilon kT]\} \tag{13}$$

where e is the electron charge, z is the ions' valency (usually 1), k is Boltzmann's constant and $(r_+ + r_-)$ is the sum of the anion's and cation's radii [40]. $K_D{}^0$ is the dissociation constant of a fictitious "uncharged ion

Table 7 Molar conductivities, ion pair dissociation constants, and thermodynamic parameters for the dissociation of carbenium ion salts in dichloromethane

Salt[a]	T (° C)	$\Lambda \cdot 10^3$ ($S \cdot m^2 \cdot mol^{-1}$)	$10^4 K_D$ ($mol \cdot L^{-1}$)	ΔH^0 ($kJ \cdot mol^{-1}$)	ΔS^0 ($J \cdot mol^{-1} \cdot K^{-1}$)	Ref.
B-OCH_3,OCH_3	−70	3.50 ± 0.14	2.2 ± 0.2	−2.3	−81	[34]
B-OCH_3,OCH_3	−20	8.42	1.6			
B-OCH_3,OPh	−70	3.15 ± 0.06	2.1 ± 0.4			[34]
B-OCH_3,CH_3	−70	3.49 ± 0.14	2.4 ± 0.3	−3.8	−88	[34]
B-PPh,CH_3	−70	2.99 ± 0.17	2.9 ± 0.5			[34]
B-OCH_3,H	−70	3.37 ± 0.12	1.9 ± 0.2	−7.0	−105	[34]
B-CH_3,CH_3	−70	3.31 ± 0.25	1.9 ± 0.1			[34]
$Tr^+SbCl_6^-$	−45	6.17	0.73	−10.5	−126	[35]
$Pyr^+SbCl_6^-$	−45	5.95	0.88	−13.0	−138	[35]
$Et_4N^+SbCl_6^-$	−45	5.53	1.2	−6.3	−100	[35]
$Ph_3C^+SbCl_6^-$	−45	4.13	2.9	−3.3	−71	[36]
$Ph_3C^+SbCl_6^-$	−45	4.43	5.3	−8.4	−96	[35]
$Ph_3C^+SbCl_6^-$	25	10.8	1.9			
$Ph_3C^+SbF_6^-$	0	9.71	2.4	−9.2	−105	[37]
$Ph_3C^+AsF_6^-$	0	9.69	1.9	−9.2	−105	[37]
$Ph_3C^+Cl^-$	0			0	−152	[38]

[a] B, benzhydryl tetrachloroborate ($Ar_2CH^+BCl_4^-$) with the corresponding p-substituents; Tr, tropylium; Pyr, 2,4,5-trimethylpyrylium; Et, Ethyl; Ph, Phenyl.

pair,'' i.e., a species whose structure is identical to that of the real ion pair but bearing no charge [12]. According to this equation, dissociation should be endothermic because of the ion's higher mobility at higher temperatures. However, dissociation is weakly exothermic ($\Delta H_D \approx -10$ to -2 kJ/mol) as shown in Table 7.

This exothermicity is apparently due to solvation being exothermic; free ions are more strongly solvated than ion pairs. In addition, the dielectric constant of a medium increases with decreasing temperature; $d\ln\epsilon/dT$ is generally less than -1 [12]. Nevertheless, temperature has only a minor effect on dissociation. For example, the dissociation constant of dianisylmethylium tetrachloroborate increases from $K_D = 1.6 \cdot 10^{-4}$ mol/L at $-20°$ C, to $K_D = 2.2 \cdot 10^{-4}$ mol/L at $-70°$ C in CH_2Cl_2 [34]. Changing the dielectric constant independent of temperature has a more pronounced effect; the dissociation constant of trityl salt increases 100-fold on going from CH_2Cl_2 to nitrobenzene. Dissociation is almost completely suppressed in less polar solvents such as benzene, toluene, and other hydrocarbons.

E. Solvation of Carbenium Ions

Macroscopic solvent effects can be described by the dielectric constant of a medium, whereas the effects of polarization, induced dipoles, and specific solvation are examples of microscopic solvent effects. Carbenium ions are very strong electrophiles that interact reversibly with several components of the reaction mixture in addition to undergoing initiation, propagation, transfer, and termination. These interactions may be relatively weak as in dispersive interactions, which last less than it takes for a bond vibration ($<10^{-14}$ sec), and are thus considered to involve "sticky" collisions. Stronger interactions lead to long-lived intermediates and/or complex formation, often with a change of hybridization. For example, onium ions are formed with *n*-donors. Even stable trityl ions react very rapidly with amines to form ammonium ions [41], and with water, alcohol, ethers, and esters to form oxonium ions. Onium ion formation is reversible, with the equilibrium constant depending on the nucleophile, cation, solvent, and temperature (cf., Section IV.C.3).

Electron-deficient carbenium ions interact not only with *n*-donors but also with other electron-rich compounds, including alkenes, alkynes, and aromatic rings. For example, vinyl monomers and polymer chains may complex carbenium ions. Electrophilic addition (propagation) may proceed by *Π*-complexation as in the asymmetric complex (C*) shown in Eq. 14.

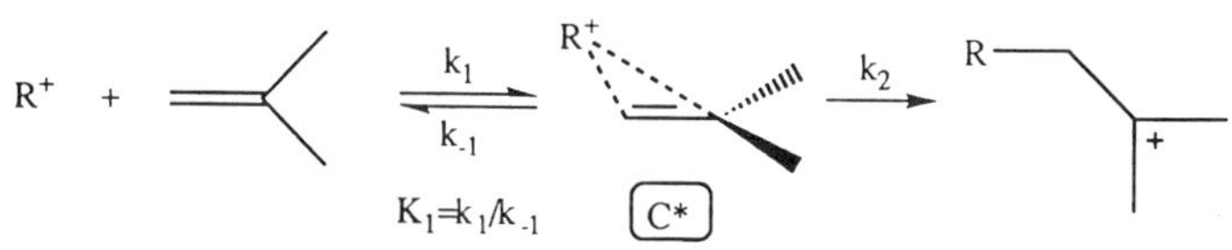

(14)

If complexation is fast and reversible, the apparent rate constant of propagation is the product of the complexation constant and the rate of rearrangement of the complex (K_1k_2). If rearrangement is fast but complexation is slow, the apparent rate constant will be equal to the rate of complex formation (k_1). In either case, the resulting kinetics should be first order in monomer. However, very strong complexation in which nearly all of the active sites are complexed by monomer followed by slow unimolecular rearrangement may lead to zero-order kinetics in monomer [Eq. (15)]. Complexation of Lewis acid by monomer and slow generation of carbenium ions from dormant species may also result in zero-order kinetics in monomer.

weak but fast complexation

$$-d[M]/dt = K_1 \cdot k_2 \cdot [M] \cdot [R^+]$$

weak and slow complexation

$$-d[M]/dt = k_1 \cdot [M] \cdot [R^+] \qquad (15)$$

*strong complexation, slow rearrangement of the complex C**

$$-d[M]/dt = k_2 [C^*]$$

There is some spectroscopic evidence that aromatic compounds complex carbenium ions [42]. For example, the complexation equilibrium constant between trityl ions and hexamethylbenzene is $K = 68 \text{ mol}^{-1} \text{ L}$ at 0° C [43]. Complexation should be stronger with more electrophilic carbenium ions such as those derived from styrene and α-methylstyrene. On the other hand, the monoalkyl-substituted phenyl rings attached to the polymer chain are weaker nucleophiles than hexamethylbenzene. A complexation constant $K = 4 \text{ mol}^{-1} \text{ L}$ was reported for trityl cation and styrene [43]. Similar complexes have been proposed to explain the red color observed in inifer systems based on 1,4-bis(1-chloro-1-methylethyl)benzene and BCl_3 in CH_2Cl_2 at low temperature [44].

Weak interactions of aromatic rings with growing carbenium ions could affect the overall polymerization rate and possibly the stereochemistry of addition. For example, the syndiotacticity of poly(α-methylstyrene) prepared cationically may be due to interaction of propagating styryl cations with penultimate aromatic rings [45]. If these interactions are strong enough, intramolecular Friedel-Crafts alkylation occurs, resulting in indanyl end groups as shown in Eq. (11).

The reactivities of carbenium ions may conceivably depend on their solvation state. The trimodal molecular weight distribution observed in p-methoxystyrene polymerizations has been assigned to three independently growing species complexed with either monomer, solvent (CH_2Cl_2), or the penultimate aromatic ring [46], each with different reactivities. However, it is unlikely that the lifetime of each species would be long enough for such long polymer chains to grow before exchanging with one another. This system will be discussed further when considering the mechanism of propagation in Section IV.D.

F. Reversible Conversion of Carbenium Ions and Dormant Species

Carbenium ions often exist in dynamic equilibrium with dormant species formed by reaction with a nucleophile [Eq. (16)].

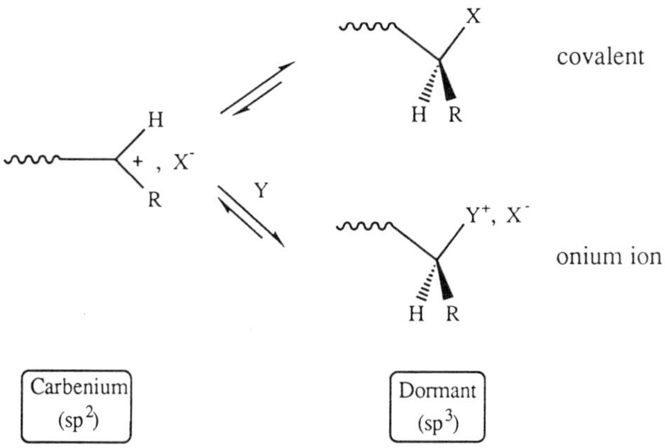

(16)

Covalent species are generated when the nucleophile is an anion, whereas onium ions are produced when the nucleophile is not charged. Formation of either dormant species changes the hybridization at carbon from sp^2 to sp^3 with its accompanying energy barrier. These dormant species generally do not react directly with alkenes, and must therefore ionize first; only the resulting carbenium ions react with double bonds. In some systems, only low concentrations of carbenium ions exist for a short time. Although their lifetime may be shorter than 1 μsec, carbenium ions can be considered as individual chemical entities if their lifetime is longer than a bond vibration ($\approx 10^{-14}$ sec). A low stationary concentration of carbenium ions is maintained when they form reversibly. The stationary concentration of carbenium ions may be as low as parts per million relative to the dormant species. Because such low concentrations of carbenium ions are difficult to detect directly, propagation has sometimes been erroneously proposed to proceed by direct reaction of monomer with covalent species.

If concentrations of carbenium ions are too low to be observed directly, they must be detected indirectly in kinetic studies of the racemization of optically active dormant species, ligand exchange and/or detailed studies of the effect of substituents, solvent and salts. Some of the most convincing and elegant work in this area was presented in Chapter 2 using primarily benzhydryl derivatives. As discussed in the next section, correlations between ionization rates and equilibrium constants, rates of solvolysis and rate constants of electrophilic addition can be interpolated and in some cases extrapolated to cationic polymerizations of alkenes to evaluate the reactivities of various active species and the dynamics of their isomerization.

1. Covalent Species

Model racemization, exchange, and solvolysis studies were performed using optically active and radiolabeled compounds. These reactions may be spontaneous or catalyzed by protonic (e.g., RCO_2H) or Lewis acids (e.g., $HgCl_2$). For example, these reactions have been followed using optically active 1-phenylethyl acetate or chloride and radiolabeled $HgCl_2$ as shown in Eq. (17). Alternatively, the acids and leaving groups may contain radiolabeled Cl or O atoms.

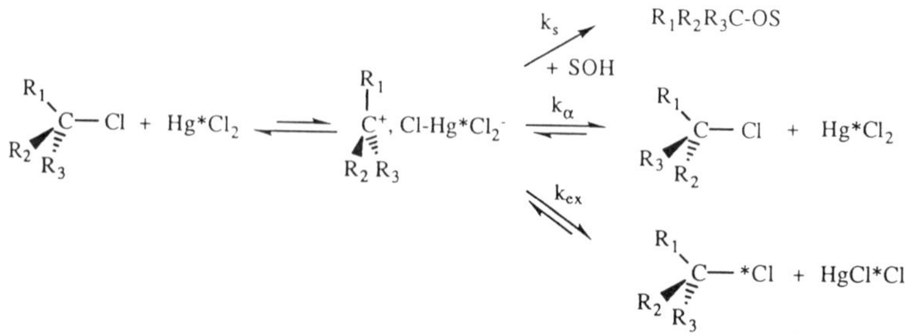

(17)

The overall consumption rate of the covalent precursor (k_S) is determined by HPLC and/or titration measurements; this correlates with monomer consumption in propagation. The rate of racemization of optically active 1-phenylethyl chloride (k_α) is determined by polarimetric measurements. Racemization is usually faster than solvolysis, confirming that activation is reversible and that internal return may occur before the carbenium ion reacts with an external nucleophile. Racemization requires not only that the C—Cl bond of the covalent precursor is broken, but that the lifetime of the ion pair is long enough for the flat carbenium ion to rotate, such that both sides of the carbenium ion are completely equivalent as shown in Eq. 18.

(18)

Cleavage of the C—Cl bond is most easily followed by exchange reactions (k_{ex}), such as exchange with radiolabeled *Cl atoms from the activator ($H_g{}^*Cl_2$) or an added salt ($NBu_4{}^+$, $^*Cl^-$). Exchange is usually faster than racemization.

1-Phenylethyl chloride ionizes and therefore racemizes in nitromethane both spontaneously and in the presence of various acids (e.g., HCl) [47]. HCl elimination (dehydrochlorination) also occurs. However, unimolecular racemization is four times faster than this side reaction at 99° C;

$k_\alpha^1 = 1.2 \times 10^{-5}\ sec^{-1}$ and $k_D^1 = 0.3 \times 10^{-5}\ sec^{-1}$. The rate of dehydrochlorination is exactly the same as the rate of Cl atom exchange with radiolabeled HCl. Both reactions pass through the same intermediate and simply describe the rate of capture of the 1-phenylethyl carbenium ion; both rates are slower than racemization (internal return).

Spontaneous racemization is very slow in nonpolar solvents at ambient temperature, but is greatly accelerated by protonic and Lewis acids. Racemization is first order in both 1-phenylethyl chloride and acid. Racemization catalyzed by $SnCl_4$ in CCl_4 at 25° C proceeds with a rate constant $k_\alpha = 1.5 \times 10^{-2}\ mol^{-1} \cdot L \cdot sec^{-1}$ [48]. Because styrene and 1-phenylethyl chloride consumption is 90 times slower than racemization, the rate of racemization is not affected by adding styrene to the system. That is, the efficiency of ion capture by styrene is low, whereas the ion pair collapse must be very fast. Racemization of 1-phenylethyl chloride with $SnCl_4$ is nearly 100 times faster in benzene than in CCl_4, $k_\alpha = 1.3\ mol^{-1} \cdot L \cdot sec^{-1}$ at 25° C [49], with activation parameters $\Delta H\ddagger = 35\ kJ \cdot mol^{-1}$ and $\Delta S\ddagger = 120\ J \cdot mol^{-1} \cdot K^{-1}$.

Racemization was also studied using other Lewis acids ($AlCl_3$, $GaCl_3$, $HgCl_2$, $ZnCl_2$, etc.), but in strongly nucleophilic solvents such as acetone and diethyl ether [50]. Lewis acids interact with these solvents to form complexes [51] which are much less reactive than "naked" Lewis acids. Because the strongest Lewis acids form the tightest and least reactive complexes, the reported relative rates of racemization are less useful than those determined in solvents such as carbon tetrachloride and benzene which interact only weakly with $SnCl_4$ [52]. Nevertheless, the relative orders of reactivity $ZnI_2 > ZnBr_2 > ZnCl_2 > HgI_2 > HgBr_2 > HgCl_2$ observed in acetone and $GaCl_3 > AlCl_3 > BF_3$ in ether correspond well to the intuitive order of Lewis acid strength.

Both *p*-chlorobenzhydryl chloride and cumyl chloride ionize much more readily than 1-phenylethyl chloride. Thus, optically active *p*-chlorobenzhydryl chloride racemizes at lower temperatures. This is still slow in nonpolar solvents, but can be accelerated significantly by Lewis acids [53]. Table 8 compares the rates of racemization (k_α^2) with the rates of chlorine exchange (k_{ex}^2) in three solvents using radiolabeled $Hg^{36}Cl_2$. The ratio $k_\alpha^2/k_{ex}^2 = 1.5$ observed in both acetone and 80% aqueous acetone indicates that RCl is regenerated from racemic ion pairs in which all three chlorine atoms in $HgCl_3^-$ have equal reactivity because these counterions always contain two Cl atoms from the original $HgCl_2$ and one from RCl. The equivalence of Cl atoms in the ion pair and faster racemization than solvolysis in aqueous acetone ($k_\alpha^2/k_s^2 = 1.6$) indicate that the reactive intermediates are truly ionic rather than activated complexes in which the original chlorine atom returns preferentially back to RCl.

Table 8 Rate constants of racemization and ^{36}Cl exchange of *p*-chlorobenzydryl chloride at 25° C

Solvent	k_α^2, × 10^3 ($mol^{-1} \cdot L \cdot sec^{-1}$)	k_{ex}^2, × 10^3 ($mol^{-1} \cdot L \cdot sec^{-1}$)	k_α^2/k_{ex}^2
Acetonitrile	110	110	1.0
Acetone	18.7	12.4	1.5
Aq. acetone	107	30	1.6
Benzene	1.04	4.37	0.24

Source: Ref. 53.

The rate constants k_α^2 and k_{ex}^2 are equal in acetonitrile (the most polar solvent studied) indicating that the carbenium ion returns to the covalent species by reaction with $HgCl_3^-$ anions in which all chlorine atoms are identically labeled. Thus, racemic RCl is regenerated by $HgCl_3^-$ anions which have fully equilibrated with the original chlorine atom from RCl and the total chlorine atom content. This suggests that the ion pairs are either loose or dissociated with long lifetimes. In contrast, $k_\alpha^2/k_{ex}^2 = 0.24$ in benzene (the least polar solvent considered). In this case, the chlorine atoms in $HgCl_3^-$ in the ion pair equilibrate six times faster than the faces of the benzhydryl cation in the same ion pair, assuming that ionization is 1.5 times faster than exchange. This suggests that the ion pair is extremely tight. Nevertheless, ionization and exchange occur quite rapidly even in nonpolar benzene.

Similar studies were conducted with 1-phenylethyl benzoate derivatives labeled with ^{18}O (radioactive) or ^{17}O (NMR active) carbonyl groups. Substituents in the alkyl groups and in the leaving group affect both the rates and the nature of the carbenium ion intermediates. For example, 1-phenylethyl *p*-nitrobenzoate undergoes solvolysis [Eq. (19)] at 60° C in 70% aqueous acetone 30,000 times slower than the corresponding 1-(*p*-methoxyphenyl)ethyl derivative [54].

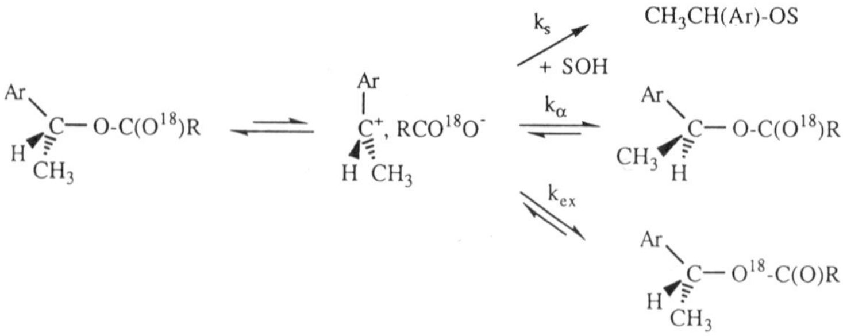

(19)

Experiments with optically active and ^{18}O-labeled benzoates have shown that the ratio of the rate constants of exchange of the labeled atoms and solvolysis are not sensitive to substituents ($k_{ex}/k_s = 0.7$). In contrast, the ratio of the rate constants of racemization and exchange does depend on the substituents. Although $k_\alpha/k_{ex} = 0.7$ for the anisyl derivative, no racemization was detected for the 1-phenylethyl p-nitrobenzoate ($k_\alpha/k_{ex} < 0.02$). Both this result and the observed salt effects were explained by the formation of tight ion pairs for the p-nitrobenzoate derivative. Bond breaking and randomization of labeled benzoate groups occurred without racemization, which would require rotation of the substituents in the carbenium ion or attack from the other side of the planar cation. On the other hand, the lifetime of the carbenium ions derived from 1-(p-methoxphenyl) ethyl p-nitrobenzoate is sufficient for full rotation of the cation, resulting in similar rates of randomization and racemization, and demonstrating that the ion pairs are loose (solvent separated). Randomization and racemization follow first-order kinetics. Similar results were obtained by ^{17}O NMR with ^{17}O-labeled compounds [55].

Dynamic NMR can be used to determine the kinetics of ionization if it is slow enough and if no side reactions occur. The rate of spontaneous ionization of trityl derivatives depends on the leaving group, the substituents on the trityl moiety, the solvent, and the temperature [42]. The rate constants of ionization of trityl chloride in 10:1 SO_2/CD_2Cl_2 are $k_i = 270$ sec^{-1} at $-53°$ C and $k_i = 35$ sec^{-1} at $-80°$ C; the rates of recombination of counterions within the ion pair are $k_r = 320$ sec^{-1} at $-53°$ C and $k_r = 8$ sec^{-1} at $-80°$ C. This also indicates that ionization is exothermic in this system.

Spontaneous ionization requires both good leaving groups and that the resulting carbenium ions are sufficiently stable. For example, although primary triflates are very stable covalent species which do not self-ionize, secondary triflates with phenyl substitents are very reactive and spontaneously ionize. The ionization equilibrium of styryl triflate could not be established because of side reactions such as Friedel-Crafts alkylation [56]. On the other hand, methoxymethylium triflate is partially ionized with equilibrium constants $K_I = 5 \cdot 10^{-4}$ at 10° C and $K_I = 2 \cdot 10^{-4}$ at $-70°$ C in SO_2 [57]. In this system, ionization is endothermic. Secondary triflates with alkoxy substituents, such as those in polymerizations of vinyl ethers, are apparently more strongly ionized than their primary counterparts [58,59].

2. *Onium Ions*

Carbenium ions react with a variety of nucleophiles, including both anions and neutral species. For example, ethers and acetals form oxonium ions, sulfides generate sulfonium ions, amines form ammonium ions, and phosphines generate phosphonium ions. Formation of onium ions is exothermic, with the enthalpy of equilibrium depending on the structures of carbocation and nucleophile. For example $\Delta H \approx -40$ kJ/mol for the reaction of trityl salts with ethers [60]. UV spectroscopy is particularly useful for following the formation of onium ions by reaction of n-donors with carbenium ions α-substituted with aromatic groups because only the carbenium ions absorb at UV-visible wavelengths. NMR can be used to follow onium ion formation when the carbenium ions do not contain aromatic groups, although this requires higher concentrations of carbenium ions and longer spectra acquisition times.

The dynamics of reversible onium ion formation has been studied by generating carbenium ions in the presence of nucleophiles using pulse radiolysis or flash photolysis, and following the rate of disappearance of the carbenium ions by UV. As discussed in Chapter 2, the kinetics of reaction of various electrophiles with nucleophiles obey a general reactivity/selectivity relationship. The rates of reaction of various nucleophiles with carbenium ions are summarized in Table 9. These rates often approach diffusion controlled limits ($k \approx 10^{10}$ $mol^{-1} \cdot L \cdot sec^{-1}$). The rates are slower for less nucleophilic and less electrophilic compounds, and are particularly slow with sterically hindered amines such as lutidine (2,6-dimethylpyridine) [63]. Solvent effects are minimal when the reactions are diffusion controlled, although tributyl amines react slower with carbenium ions in more nucleophilic dichloroethane than in methylene chloride.

If onium ion formation is reversible, the position of the equilibrium shown in Eq. (20) is defined by the rate constants of association and dissociation and therefore by the electrophilicity of the carbenium ion and the nucleophilicity of the nucleophile.

$$CH_3{-}\overset{+}{C}H(R) \;+\; Nu\langle \;\underset{k_d}{\overset{k_a}{\rightleftharpoons}}\; CH_3{-}CH(R){-}\overset{+}{N}u\langle \qquad (20)$$

As demonstrated by the association rate constants listed in Table 10, association is relatively fast and has low activation energies. Table 10 also tabulates the equilibrium constants for reaction of a variety of nucleophiles with carbenium ions. Most of the equilibrium constants involving trityl carbenium ions were obtained from UV studies, whereas those of methoxymethylium carbenium ions with both dimethyl ether and methylal were calculated using dynamic NMR [64]. Values for isobutoxy alkyl derivatives have been estimated from polymerization kinetics. The data presented in Table 10 demonstrate that the equilibrium constants are lower for weaker nucleophiles and more stable carbenium ions. For example, carbenium ions react faster with diethyl ether than with the less nucleo-

Table 9 Rate constants of the reaction of carbocations with various nucleophiles

Carbocation	Nucleophile	Solvent[a]	T (° C)	k_a ($mol^{-1} \cdot L \cdot sec^{-1}$)	Ref.
Ph_2CH^+	I^-	DCE	24	$4.6 \cdot 10^{10}$	[61]
Ph_2CH^+	Br^-	DCE	24	$5.9 \cdot 10^{10}$	[61]
Ph_2CH^+	Cl^-	DCE	24	$8.5 \cdot 10^{10}$	[61]
Ph_2CH^+	$(C_2H_5)_3N$	MC	24	$2.9 \cdot 10^9$	[41]
Ph_2CH^+	$(C_3H_7)_3N$	MC	24	$1.5 \cdot 10^9$	[41]
Ph_2CH^+	$(C_4H_9)_3N$	MC	24	$1.1 \cdot 10^9$	[41]
Ph_2CH^+	$(C_4H_9)_3N$	DCE	24	$5 \cdot 10^8$	[41]
Ph_2CH^+	$(C_4H_9)_3N$	TCE	24	$8.5 \cdot 10^8$	[41]
Ph_2CH^+	NH_3	DCE	24	$4.3 \cdot 10^9$	[61]
Ph_2CH^+	H_2O	DCE	24	$1.3 \cdot 10^6$	[61]
Ph_2CH^+	$(H_2O)_2$	DCE	24	$1 \cdot 10^8$	[61]
Ph_2CH^+	F^-	CH_3CN	20	$2 \cdot 10^{10}$	[62]
Ph_2CH^+	Cl^-	CH_3CN	20	$2 \cdot 10^{10}$	[62]
Ph_2CH^+	Br^-	CH_3CN	20	$2.1 \cdot 10^{10}$	[62]
Ph_2CH^+	I^-	CH_3CN	20	$1.9 \cdot 10^{10}$	[62]
Ph_2CH^+	SCN^-	CH_3CN	20	$2.7 \cdot 10^{10}$	[62]
Ph_2CH^+	SH^-	CH_3CN	20	$2.4 \cdot 10^{10}$	[62]
Ph_2CH^+	$H_2PO_4^-$	CH_3CN	20	$2.1 \cdot 10^{10}$	[62]
Ph_2CH^+	NO_2^-	CH_3CN	20	$2.2 \cdot 10^{10}$	[62]
Ph_2CH^+	NO_3^-	CH_3CN	20	$2.1 \cdot 10^{10}$	[62]
(p-anisyl)$_2CH^+$	F^-	CH_3CN	20	$1.4 \cdot 10^{10}$	[62]
(p-anisyl)$_2CH^+$	Cl^-	CH_3CN	20	$1.5 \cdot 10^{10}$	[62]
(p-anisyl)$_2CH^+$	Br^-	CH_3CN	20	$1.6 \cdot 10^{10}$	[62]
(p-anisyl)$_2CH^+$	SCN^-	CH_3CN	20	$1.4 \cdot 10^{10}$	[62]
(p-anisyl)$_2CH^+$	$H_2PO_4^-$	CH_3CN	20	$1.3 \cdot 10^{10}$	[62]
(p-anisyl)$_2CH^+$	NO_3^-	CH_3CN	20	$2.5 \cdot 10^9$	[62]
Ph_3C^+	I^-	DCE	24	$4.6 \cdot 10^{10}$	[61]
Ph_3C^+	Br^-	DCE	24	$5.9 \cdot 10^{10}$	[61]
Ph_3C^+	Cl^-	DCE	24	$8.5 \cdot 10^{10}$	[61]
Ph_3C^+	$(C_2H_5)_3N$	MC	24	$1.8 \cdot 10^8$	[41]
Ph_3C^+	$(C_3H_7)_3N$	MC	24	$5 \cdot 10^7$	[41]
Ph_3C^+	$(C_3H_7)_3N$	DCE	24	$1.5 \cdot 10^7$	[41]
Ph_3C^+	$(C_4H_9)_3N$	DCE	24	$4 \cdot 10^6$	[41]
Ph_3C^+	NH_3	DCE	24	$2.4 \cdot 10^7$	[41]
Ph_3C^+	Lutidine	DCE	24	10^3	[63]
Ph_3C^+	Py	DCE	24	10^7	[63]

[a] MC, methylene chloride, DCE, 1,2-dichloroethane, TCE, 1,1,2-trichloroethane.

Table 10 Equilibrium and rate constants of the onium ions formation

Carbocation	Nucleophile	Solvent	T (° C)	k_a ($mol^{-1} \cdot L \cdot sec^{-1}$)	k_d (sec^{-1})	K_e ($= k_a/k_d$) ($mol^{-1} \cdot L$)	k_{exch} ($mol^{-1} \cdot L \cdot sec^{-1}$)	Ref.
Ph_3C^+	H_2O	CH_2Cl_2	25			$6.2 \cdot 10^1$		[64]
Ph_3C^+	H_2O	CH_2Cl_2	−73			$1.4 \cdot 10^5$		[64]
Ph_3C^+	THF	CH_2Cl_2	25			$1.6 \cdot 10^0$		[64]
Ph_3C^+	THF	CH_2Cl_2	−73			$1.8 \cdot 10^5$		[64]
Ph_3C^+	$(C_2H_5)_2O$	CH_2Cl_2	25			$2.3 \cdot 10^{-1}$		[64]
Ph_3C^+	$(C_2H_5)_2O$	CH_2Cl_2	−73			$2.9 \cdot 10^5$		[64]
Ph_3C^+	1,3-Dioxolane	CH_2Cl_2	25			$6.4 \cdot 10^{-2}$		[64]
Ph_3C^+	1,3-Dioxolane	CH_2Cl_2	−73			$2.6 \cdot 10^1$		[64]
Ph_3C^+	Methylal	CH_2Cl_2	25			$4.4 \cdot 10^{-2}$		[64]
Ph_3C^+	Methylal	CH_2Cl_2	−73			$4.4 \cdot 10^{-1}$		[64]
Ph_3C^+	$(C_6H_5)_3P$	CH_3CN	25			$7.1 \cdot 10^7$		[65]
Ph_3C^+	$(C_6H_5)_3P$	CH_2Cl_2	25			$1.6 \cdot 10^7$		[65]
Ph_3C^+	$(p\text{-}ClC_6H_4)_3P$	CH_3CN	25			$5 \cdot 10^4$		[65]
$CH_3OCH_2^+$	Methyl-2-chloroethyl ether	SO_2	−70			$6 \cdot 10^3$		[64]
$CH_3OCH_2^+$	Di-2-chloroethyl ether	SO_2	−70			$1.8 \cdot 10^2$		[64]
$\sim CH_2CH^+$ (OiBu)	THT	CH_2Cl_2	−30			10^3		[59]
$\sim CH_2CH^+$ (OiBu)	$(p\text{-}ClC_6H_4)_3P$	CH_2Cl_2	RT			10^{10}		[66]
$CH_3OCH_2^+$	$(CH_3O)_2CH_2$	SO_2	−70	$2 \cdot 10^6$	$7 \cdot 10^2$	$3 \cdot 10^3$	$2 \cdot 10^4$	[64]
$CH_3OCH_2^+$	$(CH_3O)_2CH_2$	SO_2	−50	$2 \cdot 10^6$	$2 \cdot 10^3$	$1 \cdot 10^3$	$3 \cdot 10^4$	[64]
$CH_3OCH_2^+$	$(CH_3)_2O$	SO_2	−70	$1.2 \cdot 10^7$	4.0 ± 0.7	$2.9 \cdot 10^6$	$2 \cdot 10^2$	[57]
$CH_3OCH_2^+$	$(CH_3)_2O$	SO_2	−50	$1.9 \cdot 10^7$	$5 \cdot 10^1$	$4 \cdot 10^5$	$1.5 \cdot 10^1$	[57]
$CH_3OCH_2^+$	$(CH_3)_2O$	SO_2	−30	$3.2 \cdot 10^6$	$7 \cdot 10^1$	$4.5 \cdot 10^4$	$5 \cdot 10^3$	[57]
$CH_3OCH_2^+$	$(CH_3)_2O$	SO_2	−10	$8.6 \cdot 10^6$	$4 \cdot 10^2$	$2.1 \cdot 10^4$	$2 \cdot 10^4$	[57]

philic acetal, whereas dissociation of the acetal-based oxonium ions back to carbenium ions is much faster than dissociation of the oxonium ions based on diethyl ether. The equilibrium constants are therefore much higher for ethers than for acetals.

In addition to reaction with carbenium ions as in Eq. (20), nucleophiles may react directly with onium ions in an exchange reaction [Eq. (21)]. Although this exchange reaction is often faster than dissociation of onium ions [57], the latter route dominates with more stable secondary carbenium ions [67].

(21)

III. INITIATION

A. Chemistry of Initiation

Initiation is a prerequisite to any chain polymerization; it generates the active sites capable of propagation. Cationic initiation of alkenes is often a complex process involving more than one step in addition to the ubiquitous electrophilic addition of initiating species to the double bond of the monomer to generate carbenium ions. It generally occurs by either protonation of an olefinic monomer with a protonic acid, or by cationation with a cationogenic compound. These protonic acids and cationogens are often supplemented by other components of a complex initiating system. Whereas the initiator is the compound (protonic acid, carbenium ion, ester, alcohol, water) which reacts directly with monomer to produce the head group, an activator is a compound such as a Lewis acid which regulates both the initiation and polymerization rates, but which does not react directly with monomer. If activation is too strong, a deactivator may also be required to control the polymerization. Deactivators such as salts and nucleophiles interfere with the growing species and are not directly involved in initiation.

Thus, the terms initiator and coinitiator, as well as catalyst and cocatalyst, must be clearly distinguished. As proposed earlier [68], an initiator is consumed in the initiation process whereas a catalyst remains unchanged during the polymerization. In the polymerization of alkenes initiated directly by Lewis acids (e.g., iodine initiated polymerization of vinyl ethers) [69], the Lewis acid plays both roles. Nevertheless, Lewis acids usually act only as catalysts rather than as initiators, with protonogenic compounds such as adventitious moisture being the initiator.

Polymerization may also be induced by oxidation and formation of radical cations that recombine to form dications, or rearrange to form radicals and cations that initiate polymerization. Radical cations are also generated by other methods such as field ionization and emission, electrochemistry and photochemical initiation. Although the actual initiating species may not be directly observable, their fate and the mechanism of initiation can be studied by a combination of kinetic measurements, molecular weight determination, and end group analysis. End group analysis requires special labeling of the initiator if the polymers generated are of high molecular weight. For example, end groups can be studied by ^{1}H NMR if a nondeuterated initiator initiates polymerization of a deuterated monomer; labeling the initiator with ^{13}C enhances ^{13}C NMR analysis of end groups. Alternatively, the fate of radiolabeled compounds can be followed. However, the initiation mechanism resulting from electric field and ionization methods remain obscure, although the nature of the growing species can be established by polymerization in the presence of radical and/or cationic scavangers.

Section III of this chapter covers the fundamentals of carbocationic initiation. Further information can be obtained from two excellent reviews [70,71].

1. Overview of Cationic Initiators

Several classes of compounds initiate cationic polymerizations of alkenes, including protonic acids, Lewis acids (usually in combination with a cation or proton source), stable carbenium ions, oxidizing reagents, and other strong electrophiles. This section attempts to explain the mechanism of initiation with quantitative information when available; physical means of initiation (electric current, γ-rays, field ionization and emission, nuclear chemical initiation) will not be discussed.

The initiator influences the polymerization in several ways. For example, the relative rates of initiation and propagation, and therefore the efficiency of initiation, influence the molecular weight distribution. Complex initiating systems have been developed to control both the overall rate of polymerization and the molecular weight distribution (Chapter 4). The structure of the initiator is also very important because it may provide a site for side reactions both before and after being incorporated as the polymer's end group. As discussed in Chapter 5, initiators with stable functional groups are used to prepare macromonomers and telechelics. Thus, although the concentration of the end group introduced by the initiator is negligible relative to that of the many monomer repeat units along the polymer backbone, correct selection of the initiating system is extremely important for controlling the molecular weight, molecular weight distribution, and topology of the polymer.

New initiators should provide as much control as possible over the polymerization and the resulting polymer's structure. The ideal initiator should be stable at room temperature for an infinite amount of time and should generate the growing species quantitatively. The growing species should also remain active throughout the entire polymerization. In order to provide sufficient time for synthetic manipulations such as forced termination by a functional reagent or addition of a second comonomer, the reaction half-lifetime should be in the range of minutes to hours. If the reaction time is too short, spontaneous termination may occur before further synthetic manipulations can be performed. On the other hand, reaction times longer than a few days are also obviously inconvenient.

The initiator stability must therefore be balanced with its required reactivity. For example, stable carbenium ions such as tropylium or trityl react very slowly with alkenes and are therefore not optimal initiators. Protonic acids may be sufficiently reactive but often have anions that are basic and cause transfer by β-proton abstraction from the growing carbenium ion; anions resulting from Lewis acid initiators are less basic. Many of the new efficient initiators are prepared by mixing two stable components, such as an alkyl halide and a Lewis acid, which provide reversibly active carbenium ions.

2. Protonic Acids

There are a number of factors that determine whether a protonic acid can initiate polymerization of alkenes. Their acidity (pK_a), and therefore the basicity of the resulting counteranion, determines the efficiency of initiation. Although reliable pK_a values of acids stronger than sulfuric or hydroiodic ($pK_a < -9$) are difficult to obtain in aqueous solutions due to their nearly complete dissociation, the pK_a values of acetic acid (4.75) and trichloroacetic acid (0.7) in water provide useful references. Conductometric and potentiometric estimates of the pK_a values of selected protonic acids in various organic solvents are summarized in Table 11 in descending acid strength. These values are not very precise, however, because the amount of moisture in each system was not monitored precisely.

The lower pK_a values in acetonitrile indicate that dissociation is stronger in this solvent. However, cationic initiation is not very efficient in acetonitrile because some ionized protons react with solvent. The relative acid strength is similar in most of the solvents shown in Table 11, with some fluctuations in acetic acid. For example, the relative strengths of perchloric and triflic acids are reversed in acetic acid. Although the higher pK_a values of triflic acid in most solvents indicate that it is more basic, both the kinetics of ring-opening polymerization of cyclic ethers [80] and the rates of some organic reactions demonstrate that triflate anion is less nucleophilic than perchlorate.

Both the nucleophilicity and basicity of the counteranion have important consequences on the mechanism of carbocationic polymerizations.

Table 11 pK_a Values of selected protonic acids in organic solvents

Acid	CH_3CN	Ref.	CH_3NO_2	Ref.	$C_2H_4Cl_2$	Ref.	CH_3COOH	Ref.
ClO_4H	1.6	[72]	2.0	[72]	3	[73]	4.9	[74]
CF_3SO_3H	2.6	[72]	3.0	[72]	7.3	[75]	4.7	[74]
FSO_3H	3.4	[72]	4.3	[72]			6.1	[74]
HI					7.9	[75]	5.8	[74]
HBr	5.5	[76]			8.7	[75]	5.6	[74]
H_2SO_4	7.3	[76]					7.0	[74]
CH_3SO_3H	8.4	[72]	6.0	[72]			8.6	[72]
HCl	8.9	[76]			10.8	[75]	8.4	[74]
CF_3COOH	10.6	[77]			7	[78]	11.4	[79]
CCl_3COOH	12.7	[77]					12.2	[79]
CCl_2HCOOH	13.2	[77]						
$CClH_2COOH$	15.3	[77]						
CH_3COOH	22.5	[77]					12.8	[79]

For example, styryl cations react with perchlorate to form covalent esters that are relatively stable at low temperatures [81], whereas the more basic triflate anion tends to abstract β-protons from carbenium ions in a transfer reaction [56]. The basicity of the counteranion determines the contribution of β-proton elimination relative to propagation, and therefore the limiting molecular weight in a polymerization.

In addition to the influence of moisture, conductometric data are affected by dimerization of the acid and by homoconjugation. Homoconjugation of perchloric acid is shown in Eq. (22) [82,83].

$$2\ ClO_4H \rightleftarrows ClO_4^- + \{ClO_4H_2\}^+ \tag{22}$$

Although dimerization of weak carboxylic acids via hydrogen bonding is well documented [84–90], dimerization of acids such as triflic and perchloric acid is less certain [91]. The dimerization constant for triflic acid was initially estimated to be 10^3 mol/L at room temperature in polar solvents such as acetonitrile [92]; however, this could not be confirmed [71]. Kinetic measurements support aggregation in chlorinated solvents such as CH_2Cl_2 [18]. That is, the initial external orders in triflic acid are as high as 3.3 at $-23°$ C and 3.8 at $-32°$ C in styrene polymerizations. On the other hand, the rate of propagation is proportional to $[CF_3SO_3H]_0^2$ once the stationary state is reached, indicating that the dimeric acid is involved in reinitiation. Although trimers and/or tetramers of the acid may form at higher concentrations, competing reinitiation by dimers precludes formation of higher aggregates in polymerizing systems [18].

Although the strength of a protonic acid and/or the basicity of its counteranion may prevent it from acting as an efficient initiator on its own, such protonic acids are often used to preform alkyl halides or ester adducts with the alkene, which can then be used to initiate polymerization after activation with a Lewis acid [Eq. (23)].

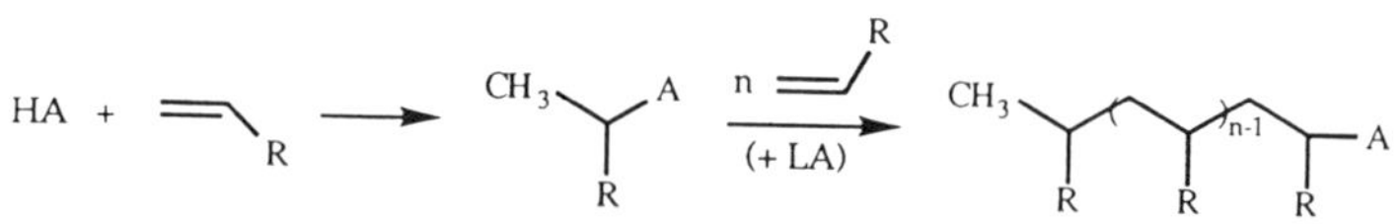

(23)

Polymerization then proceeds via reversible formation of carbenium ions, which subsequently react with monomer. Nonopolar solvents and excess acid are typically used to prepare the first adduct in high yield. If the acid is a gas, it can be bubbled through the monomer solution, with excess acid removed under reduced pressure.

a. Most Important Protonic Acids as Initiators for Cationic Polymerization. The acids reviewed in this section are grouped into the four classes of carboxylic acids, hydrogen halides, other moderately strong acids with $pK_a \approx 5$–10 in AcOH (phosphoric, sulfuric, methanesulfonic, fluorosulfonic), and very strong acids with $pK_a < 5$ (triflic and perchloric).

1. Carboxylic Acids. As shown previously in Table 11, carboxylic acids are weaker acids than hydrogen halides. Acetic acid polymerizes only the most reactive monomer: *N*-vinyl carbazole [93]. In this case, reactions carried out in nonpolar solvents yield the corresponding alkyl acetate, whereas more polar solvents and higher acid concentration favors polymer formation [93,94]. Monochloro-, dichloro-, and trichloroacetic acids are much stronger acids than unsubstituted acetic acid, but weaker than hydrogen halides. Nevertheless, they are more efficient initiators for carbocationic polymerizations than hydrogen halides. Dichloroacetic and trichloroacetic acids dimerize 1,1-diphenylethylene and initiate polymerization of α-methylstyrene, styrene, and cyclopentadiene [95–99]. Trifluoroacetic acid is the strongest acid among this group and has been studied extensively as an initiator for polymerization of styrene and other alkenes [100–103].

Styrene polymerization initiated by trifluoroacetic acid can be used as a standard in demonstrating the effects of solvent and concentration. Polymerization is favored in more polar solvents and at high acid concentration [101]. Only the 1:1 adduct is formed when acid is slowly added to styrene [100]. However, fast polymerization occurs when styrene is added to the concentrated acid [Eq. (24)].

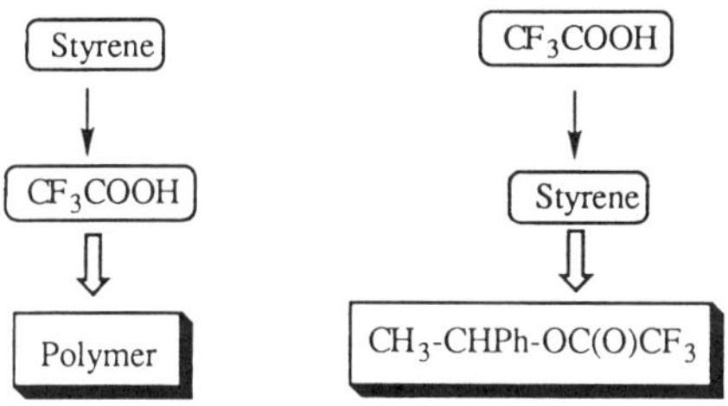

(24)

Acid adds slowly to styrene to form covalent esters, which have been detected directly by ^{19}F NMR [101]. Carbenium ions capable of propagation are then generated by ionization of the covalent esters with excess acid. However, the concentration of carbenium ions is too low to detect directly by spectroscopy, and propagation was initially proposed to occur by a nonionic mechanism.

The high reaction orders in acid observed in these polymerizations further support their role as both activator and initiator [101,102]. In addition, because acid adds only slowly to styrene to form 1-phenylethyl trifluoroacetate, the polymerization rate increases in the presence of preformed 1-phenylethyl trifluoroacetate initiator, which is then activated by excess acid. When optically active 1-phenylethyl trifluoroacetate is used as the initiator, its rate of incorporation into the polymer backbone is five times slower than its rate of racemization [103]. This confirms that carbenium ion intermediates are generated in the polymerization, and that not all ionization steps are followed successfully by monomer addition.

Although carboxylic acids generally form 1:1 adducts with alkenes, the resulting esters are easily ionized in the presence of either Lewis or protonic acids. The higher efficiency of chlorinated acetic acids relative to hydrogen halides is ascribed to the ability of their 1:1 adducts to coordinate with excess acid. Alkyl halides are eventually formed when carboxylic acids are used to initiate polymerization in the presence of a Lewis acid due to migration of the carboxylate moiety to the Lewis acid [Eq. (25)]. Similarly, styrene and isobutene polymerizations initiated by preformed alkyl acetate adducts in the presence of BCl_3 always produce Cl-terminated chains [104,105].

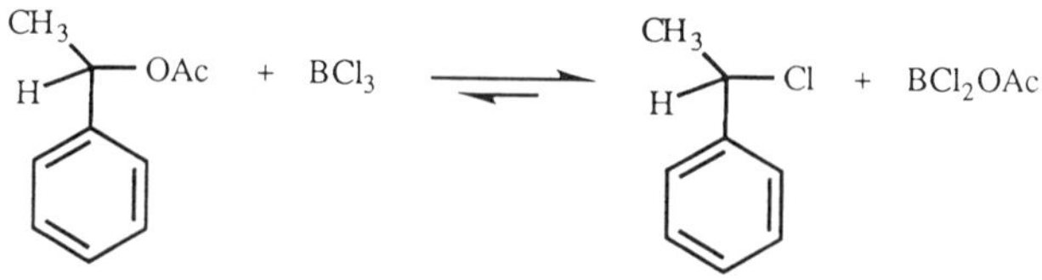

(25)

The 1:1 adducts of various carboxylic acids and styrene, vinyl ethers, and isobutene have been isolated and used as initiators in the presence of Lewis acid activators. The polymerization rates correlate with the basicity of the leaving groups. However, isobutene polymerizes $\approx 10^3$ times slower when initiated by pivalates and isobutyrates in the presence of BCl_3 than when initiated by acetates, even though they have similar pK_a values [106]. Coordination of the covalent adducts with BCl_3 is evidently hindered when the alkyl substituents are bulkier. More detailed studies on vinyl ether polymerizations using a series of substituted benzoates demonstrate that the pK_a values of the parent acid affects both the initiation rate and dynamics of ionization, and therefore the ability to prepare well-defined polymers [107].

2. Hydrogen Halides. Hydrogen halides add to double bonds to form alkyl halides in high yield [71]. This spontaneous addition is usually second order in acid, suggesting that either dimers of the acid are involved or halide anions react with the acid/alkene adduct by an Ad_E3 mechanism. Depending on the acid and nucleophile involved, either syn or anti addition is possible [Eq. (26)].

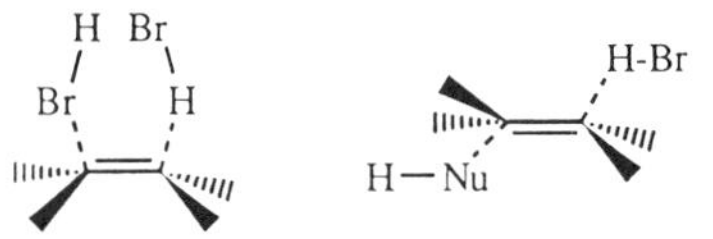

(26)

DBr, DCl, and CH_3SO_3D additions to E- and Z-2-butene proceed without diastereoisomerization, H/D exchange, or positional isomerization [108,109]. Although this suggests that carbenium ions may not develop completely, carbenium ion intermediates are apparently involved when the reaction is catalyzed by triflic acid. That is, triflic acid catalysis greatly increases the rate, and both stereo- and positional isomerization occur in its presence [110].

Most alkenes do not polymerize in the presence of only hydrogen halides. One noticeable exception is *N*-vinyl carbazole, the most nucleophilic alkene, which is successfully polymerized by HI, HBr, and HCl [111]. Polymerization by HI produces well-defined polymers [112]. Cyclohexyl vinyl ether also produces well-defined polymers when initiated by HI, although the polymerization is slow [113]. Other vinyl ethers form 1:1 α-alkoxyiodoethanes adducts with hydrogen iodide rather than polymer, especially in nonpolar solvents [Eq. (27)].

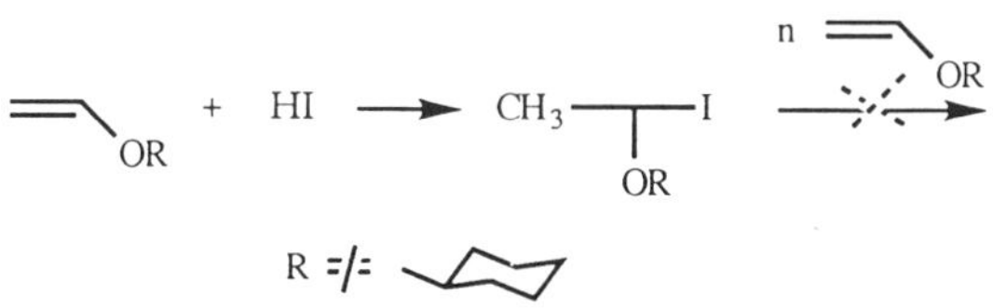

(27)

However, ionization of the adducts should be more pronounced in more polar solvents and at lower temperatures if ionization is exothermic. Most vinyl ethers polymerize under these conditions [114]. Nevertheless, traces of iodine may catalyze polymerization, because Lewis acids act as coinitiators. Moreover, even styrene oligomerizes in the presence of high concentrations of dry HCl in polar solvents at $-78°$ C [115].

3. Other Acids of Moderate Strength. Phosphoric, sulfuric, methanesulfonic, and fluorosulfonic acids are moderately strong acids ($pK_a \approx$ 5–10 in AcOH) which have been used to initiate cationic polymerizations. However, sulfuric acid is rarely used anymore due to its low solubility in organic solvents and its high oxidizing power, even though it was the first protonic acid used for carbocationic polymerizations [116]. In this case, inactive bisulfates are formed [117–120] which can be reactivated by other acids [121]. Nitric acid is also a strong oxidant and is rarely used for polymerizations, although it does successfully polymerize *N*-vinyl carbazole [122]. Methanesulfonic acid polymerizes styrene in more polar CH_3NO_2 and CH_2Cl_2 [123], and *p*-methoxystyrene even in less polar solvents [124]. In the latter case, carbenium ion intermediates were detected by UV spectroscopy in a stopped-flow system.

Stoichiometric adducts of organophosphoric acids and monomer have been used to polymerize vinyl ethers in the presence of Lewis acid. Phosphinic ($R^1R^2PO_2H$; $R^1 = R^2 =$ PhO, BuO, or Ph; $R^1 =$ Ph, $R^2 =$ H) and phosphoric acid derivatives polymerize isobutyl vinyl ether in toluene at <0° C in the presence of $ZnCl_2$. The molecular weights of the resulting polymers are directly proportional to monomer conversion and correspond to the calculated values based on formation of one polymer chain per $R^1R^2PO_2H$ molecule. Cationic polymerizations are also induced by the diprotic acid, $PhOP(O)(OH)_2$, coupled with $ZnCl_2$; in this case, one mole of the acid generates two polymer chains. ^{31}P and ^1H NMR analysis of model reactions and polymerizations initiated by $(PhO)_2PO_2H/ZnCl_2$ demonstrate that the 1:1 adduct forms first, which acts as the true initiator. Ionic intermediates are detected indirectly by exchange of the terminal phosphate group with chlorines from the Lewis acid.

4. Strong Acids. Perchloric and triflic acids are the two strongest protonic acids with noncomplex anions which are commercially available. Perchloric acid is sold as an aqueous solution; safe methods for preparing and handling the anhydrous acid are described in the literature [125]. Triflic acid is available in neat form and is considered the standard protonic acid for testing the cationic polymerizability of potential monomers. This is because the nucleophilicity of the triflate anion is very low, and therefore it generates a higher proportion of ionic species relative to covalent dormant species than other oxyacids do. Unfortunately, triflate anions are relatively basic and often abstract β-protons, thereby limiting the molecular weights possible.

Even very small amounts of anhydrous perchloric and triflic acids ($<10^{-3}$ *M*) polymerize styrene rapidly and quantitatively [123,126–130]. Carbenium ions were initially not detected in these systems, and they were therefore proposed to proceed by a pseudocationic mechanism in which covalent esters react directly with styrene in a concerted muticenter rearrangement [128]. However, short-lived carbenium ions have since been detected directly by stopped-flow UV [17–19,131]. The mechanism of the propagation step in these systems is discussed in more detail in Section IV.D.2.a.

Perchloric and triflic acids polymerize nearly all cationically polymerizable monomers, except α-olefins and isobutene. Unsaturated dimers and trimers are formed preferentially by β-proton elimination due to the relatively high basicity of the noncomplex counteranions [cf. Eq. (20)] [132]. In general, triflic and perchloric acids do not generate well-defined polymers. Their acidity is too strong, and they therefore generate carbenium ions in too high concentration, resulting in polymerizations which are too fast. The concentration of carbenium ions can be reduced by equilibration with onium ions, as in vinyl ether polymerizations in the presence of sulfides [58,133]. However, unsaturated end groups are still obtained in most systems initiated with triflic acid.

b. Kinetics. Initiation with protonic acids is usually first order in monomer and second or third order in acid. These higher orders in acid indicate that initiation involves either initiation by dimer and/or higher aggregates of the acid, an Ad_E3 mechanism, or activation of the covalent ester adducts by additional acid. Initiation is usually slower than propagation, with a relatively large amount of acid left unreacted at the end of the polymerization. For example, the yield of carbenium ions in styrene polymerizations initiated by triflic acid is usually less than 10%, but increases to 30% at low temperatures ($< -60°$ C) [18]. Acids are sometimes regenerated by β-proton abstraction; this indicates that protonation of monomers and/or linear oligomers is reversible.

The rate constants of initiation of styrene by triflic acid are estimated to be roughly $k_i \approx 10\ M^{-1}\cdot sec^{-1}$ at 0° C in CH_2Cl_2; this is 10,000 times smaller than the corresponding propagation rate constant under similar conditions [17]. However, it does not take into account the higher order kinetics in acid [134]. Initiation of more nucleophilic monomers is faster, with $k_i = 10^3\ M^{-1}\cdot sec^{-1}$ for α-methylstyrene [21] and $k_i = 5\cdot 10^4\ M^{-1}\cdot sec^{-1}$ for p-methoxystyrene [23], as determined by stopped-flow methods at ambient temperature in CH_2Cl_2 and $C_2H_4Cl_2$.

3. *Lewis Acids*

In addition to protonic acids, Lewis acids are the most common initiators of carbocationic polymerizations. Two mechanisms are possible. Direct initiation is rare and usually slow. The more prevalent mechanism is by cocatalysis in binary systems, with the Lewis acid acting as a coinitiator or catalyst rather than as initiator. Cationating or protonating species are the true initiators, which are therefore the species incorporated at the polymer's end group. The most common initiator is adventitious water in insufficiently dried systems. Thus, mechanistic studies should be performed under stringently dry conditions or in the presence of proton traps such as hindered pyridines. In addition to water, the protonating reagent may be an alcohol, carboxylic acid, amine, or amide [Eq. (28)].

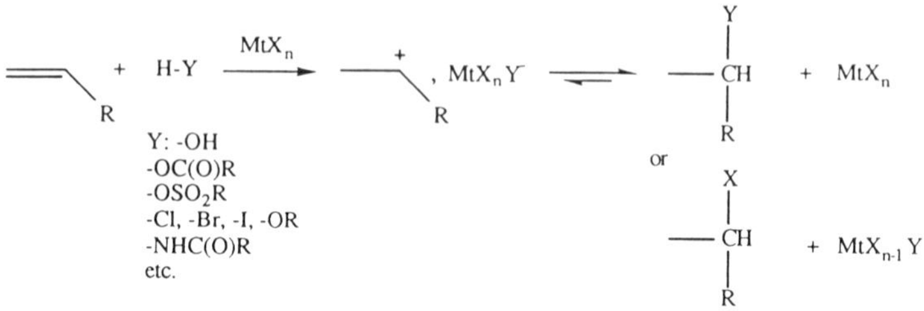

(28)

Cationating reagents include alkyl halides, esters, ethers, and anhydrides. These compounds do not react with alkenes without prior ionization or activation by a Lewis or protonic acid as shown in Eq. (29).

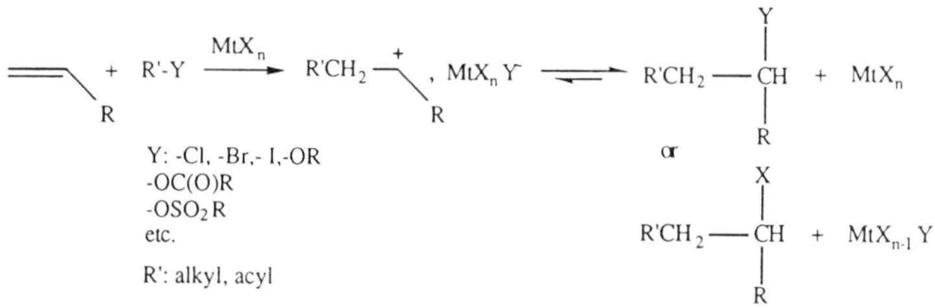

(29)

Some Lewis acid-initiated polymerizations have been proposed to proceed by direct addition of the Lewis acid to the monomer's double bond. However, this is usually an exception, and has been clearly proven only for iodine [69,135] and boron halide [136,137] initiated systems. Iodination and haloboration are reversible processes which produce deactivated alkyl halides due to the electron-withdrawing substituent at the neighboring carbon [Eq. (30)].

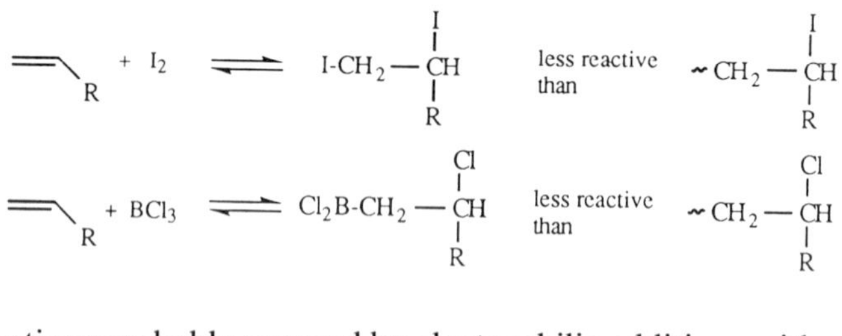

(30)

Both reactions probably proceed by electrophilic additions without carbenium ion intermediates, possibly through iodonium ions or concerted additions resembling hydroboration. The reversible iodine/*n*-butyl vinyl ether addition has an equilibrium constant $K = 10$ mol/L in hexane at 0° C. It is exothermic ($\Delta H = -32$ kJ/mol) [69] and quite fast, with complete addition in <1 min.

Aluminum chloride and bromide were proposed to self-ionize and add to monomer according to Eq. (31) [138–142].

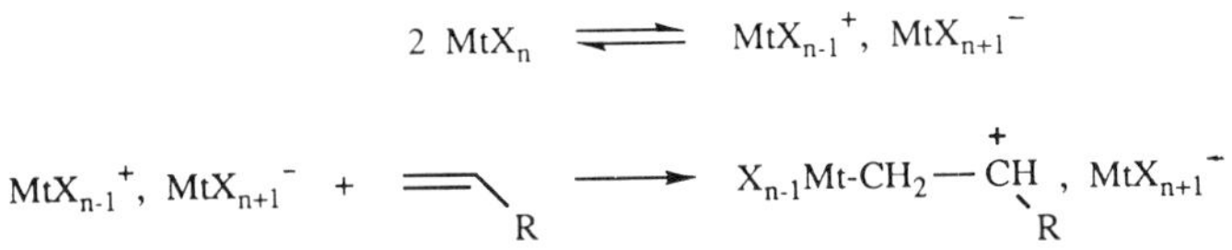

(31)

Other Lewis acids have been proposed to initiate polymerization after self-ionization. However, most Lewis acids do not initiate polymerization under stringently dry conditions, which strongly indicates that the initiating systems are binary.

Alternatively, Lewis acids such as $SbCl_5$ may initiate oligomerization directly by electron transfer from extremely reactive alkenes such as 1,1-diphenylethylene and 1,1-di(*p*-methoxyphenyl)ethylene [28,143,144]. The dimeric tail-to-tail carbenium ion of 1,1-diphenylethylene shown in Eq. (32) was observed, and its formation explained by a radical cation intermediate. Because 1,1-diarylethylenes can not polymerize, only oligomerization was observed.

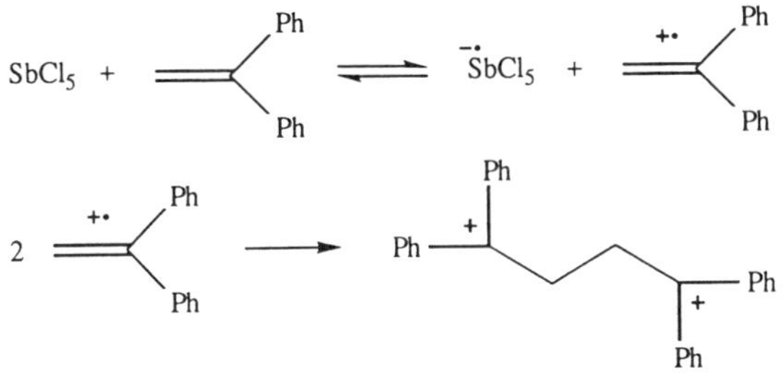

(32)

$SbCl_5$ also chlorinates alkenes and is reduced to the much less reactive $SbCl_3$. This reaction probably prevents formation of high molecular weight polystyrenes [145].

Direct initiation by monomeric Lewis acids could also occur by a zwitterionic mechanism [146,147]. The initially formed zwitterion could then rearrange to a species identical to that formed by disproportionation [Eq. (33)].

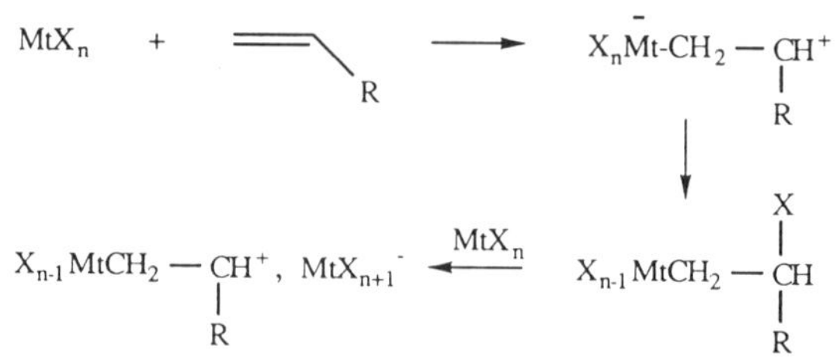

(33)

However, zwitterions have never been observed and the exact initiation mechanism is not known. It may occur by a one-step mechanism similar to haloboration [137].

a. Review of Lewis Acids. As discussed in the previous section, Lewis acids are most often used as one component of a binary system. This results in faster and therefore more efficient initiation. It also provides better control of the polymerization rate, the number of chains generated and their molecular weight. Direct initiation is slow and rarely occurs. Even if direct initiation does occur, cocatalysis with traces of protonogenic (moisture) or cationogenic species in the system will always compete. Cocatalysis can only be suppressed completely under super dry conditions and/or in the presence of proton traps. Therefore, Lewis acids are used as activators or coinitiators in most carbocationic polymerizations. The true initiating species is either an alkyl ester or halide, or a protonating agent such as adventitious or a controlled amount of water.

The polymerization rate depends on the equilibrium position between carbenium ions and the corresponding covalent adduct. The molecular weight distribution is determined by the dynamics of this equilibrium, particularly the rate of deactivation of active to dormant species. If this exchange is slow in comparison to propagation, the distribution will be broad or sometimes even polymodal. In contrast, nearly Poisson distributions ($\overline{M}_w/\overline{M}_n \approx 1$) can be obtained if exchange is faster than propagation. Thus, the lability of the Lewis acid's ligands and their relative nucleophilicities and basicities are important in controlling the polymerization. Ligands may exchange with the leaving group of the original initiator as shown previously in Eqs. (28) and (29).

The order of Lewis acid strengths was discussed in Chapter 2. The following three general rules correlate their structure and reactivity [51].

1. Ligands increase the Lewis acid strength in the order: aryl, alkyl $\ll$ carboxyl $<$ alkoxy $<$ I $<$ Br $<$ Cl $<$ F.
2. Lewis acid strength generally increases with increasing atomic number in the same group: $BX_3 < AlX_3 < GaX_3$; $SiCl_4 < GeCl_4 < SnCl_4$;

$PCl_5 < AsCl_5 < SbCl_5$. (This rule does not apply to Lewis acids based on some transition metals.)

3. Lewis acidity increases with increasing number of ligands and therefore with the oxidation state of the central atom: $SnCl_2 \ll SnCl_4$; $PCl_3 \ll PCl_5$; $SbF_3 \ll SbF_5$.

Thus, SbF_5 and $GaCl_3$ are some of the strongest Lewis acids. However, very strong Lewis acids will not be optimal if they lead to exceedingly fast and uncontrolled polymerizations. Moreover, collapse of the ion pair involving Lewis acids with fluorine ligands is irreversible due to the very strong C—F bond, resulting in termination. Thus, the overall polymerization rates in some systems may be lower with stronger Lewis acids than with weaker Lewis acids (e.g., BF_3 vs. BCl_3 or BBr_3), whereas the opposite is true in other systems. In addition, very strong Lewis acids may form complexes even with weak nucleophiles such as solvent, monomer, or an additive. These complexes are relatively inactive, leading to overall rates of polymerization that are lower than those observed with weaker Lewis acids which do not form such complexes.

The chemistry of Lewis acids is quite varied, and equilibria such as those shown in Eqs. (28) and (29) should often be supplemented with additional possibilities. Some Lewis acids form dimers that have very different reactivities than those of the monomeric acids. For example, the dimer of titanium chloride is much more reactive than monomeric $TiCl_4$ (cf., Chapter 2). Alkyl aluminum halides also dimerize in solution, whereas boron and tin halides are monomeric. Tin tetrachloride can complex up to two chloride ligands to form $SnCl_6^{2-}$. Therefore, $SnCl_5^-$ can also act as a Lewis acid, although it is weaker than $SnCl_4$ [148]. Transition metal halides based on tungsten, vanadium, iron, and titanium may coordinate alkenes, and therefore initiate polymerization by either a coordinative or cationic mechanism. Other Lewis acids add to alkenes; this may be slow as in haloboration and iodine addition, or faster as with antimony pentachloride.

The most frequently used Lewis acids based on boron, aluminum, tin, and titanium halides are discussed in more detail in the next section. These acids are used most frequently because of their availability and relatively low cost. Strong Lewis acids such as $GaCl_3$ and the halides of antimony and arsenic are used less often.

1. Iodine. Iodine successfully initiates or catalyzes polymerization of *N*-vinyl carbazole, vinyl ethers, and styrenes. It adds to double bonds to form 1,2-diiodosubstituted ethanes which are subsequently ionized in the presence of excess iodine [69,149–151]. Stopped-flow UV studies demonstrated unambiguously that carbenium ions are present in *p*-methoxystyrene polymerizations initiated by I_2 [124]. Alternatively, I_2-initiated polymerizations of styrene have been proposed to proceed by HI elimination of the adduct, which subsequently reacts with styrene to form the more reactive 1-phenylethyl iodide adduct [Eq. (34)] [135].

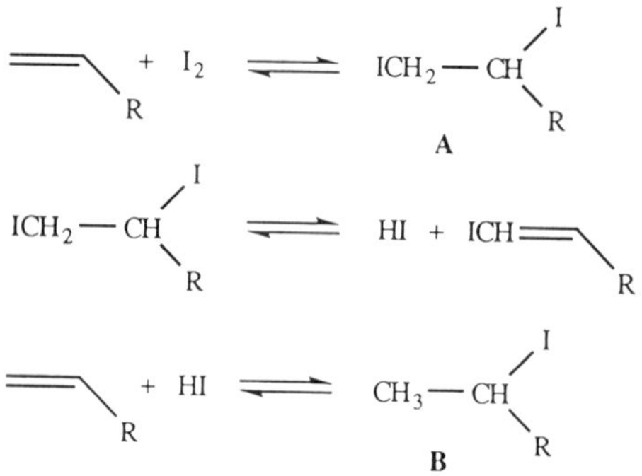

(34)

B much more reactive than A; no β-electron withdrawing group

Initiation by iodine alone is slow. However, iodine activates initiation of HI/monomer adducts. Well-defined polymers are produced because the rates of initiation and propagation are similar [114].

2. *Boron Halides.* Although BF_3 is one of the most frequently used Lewis acids for carbocationic polymerizations, it is generally not the best choice. Boron trifluoride is often used as a complex with diethyl ether, which is much less reactive than the pure Lewis acid. Boron trifluoride requires a cationogen or a protonogen to initiate polymerization. In the presence of water, BF_3 polymerizes nearly all monomers, including isobutene, styrenes, vinyl ethers, and *N*-vinyl carbazole. Nevertheless, initiation of isobutene by BF_3 in the presence of ^3H_2O at $-78°$ C results in only 1% of the chains being radiolabeled [152]. This indicates that initiation with "H^+ BF_3OH^-" is slow; polymerization must therefore be dominated by transfer. In other cases, polymerization is incomplete due to termination by F^- to generate inactive alkyl-fluoride end groups.

Boron trichloride and tribromide successfully polymerize styrenes and isobutene. These Lewis acids are typically used in combination with water or alkyl chlorides, acetates, ethers, and alcohols [105,153]. In contrast to earlier reports, BCl_3 can self-initiate polymerization of styrene and isobutene [137] by haloboration, and subsequent activation of the resulting alkyl chlorides by excess Lewis acid. Direct initiation was confirmed by the formation of lower molecular weight polymers than predicted from the ratio of monomer to initiator, even when transfer was low. However, the —BCl_2 substituent at the α-carbon reduces the reactivity of the alkyl chloride. Thus, self-initiation is negligible in the presence of powerful initiators.

Boron trichloride and tribromide exchange ligands when used with alkyl halides and acetates initiators. For example, as shown in Eq. (35), 1-phenylethyl bromide and BBr_2Cl are formed rapidly and quantitatively when 1-phenylethyl chloride is mixed with an equimolar amount of boron tribromide [59,67].

$$CH_3CH(Ph)\text{-}Cl + BBr_3 \rightleftharpoons \left[CH_3CH^+(Ph),\ {}^-BClBr_3\right] \rightleftharpoons CH_3CH(Ph)\text{-}Br + BClBr_2 \quad (35)$$

This is apparently due to boron having a higher affinity, in comparison to carbon, for chlorine than bromine.

3. Aluminum Halides and Organohalides. Aluminum trichloride is often used in reactions involving cationic intermediates, especially in Friedel-Crafts alkylations and acylations, and also in polymerizations. However, its solubility is low in nonpolar organic solvents, making quantitative studies of initiation difficult. Aluminum chloride alone does not initiate polymerization of dry styrene [154], although $AlBr_3$ self-initiates polymerization of isobutene [140]. In general, aluminum halides are used in the presence of water or hydrogen halides [155,156]. Model studies using nonpolymerizable 1,1-diphenylethylene have demonstrated that the concentration of carbenium ions generated is proportional to $[AlCl_3]_0$ when $[AlCl_3]_0/[H_2O]_0 \leq 2$, but is independent of water concentration at higher ratios. One molecule of water produces 1.3 carbenium ions, suggesting that alumoxanes are formed [157]. Hindered pyridines do not inhibit initiation in this system [158], which suggests that initiation occurs by a concerted mechanism rather than through free protons.

Aluminum alkyls ($AlEt_2Cl$ and $AlEtCl_2$) may initiate polymerization directly by addition of an ethyl group across the monomer's double bonds. For example, ethylaluminum dichloride initiates isobutene and styrene polymerizations in the absence of proton donors [159]. However, cationogenic or protonogenic compounds are added to most systems [70,71], and diethylaluminum chloride does not initiate styrene and vinyl ether polymerizations on its own. Surprisingly, highly stereoregular poly(isobutylvinyl ether) is produced in the presence of oxygen [160].

4. Tin Halides. Tin tetrachloride was the first Lewis acid used for carbocationic polymerizations [161]. It does not initiate polymerization of alkenes in the absence of cationogens and protonogens, and it is used most often in combination with alkyl chlorides and esters preformed by reaction of the corresponding protonic acid with monomer. In the presence of water, the polymerization rate increases up to a certain ratio of $[H_2O]_0/[SnCl_4]_0$, and then decreases [162]. The rate increase is simply due to a greater number of propagating chains due to a higher concentration of initiating water. The rate then decreases with greater concentrations of water due to termination by water to form inactive oxonium and hydronium ions. The ratio of $[H_2O]_0/[SnCl_4]_0$ at which the maximum rate occurs depends on the monomer, solvent, and temperature; it ranges from approximately 1 to 200% and is lowest in less polar solvents (CCl_4) and with weakly nucleophilic monomers such as styrene. The rate of monomer conversion maximizes at high ratios of $[H_2O]_0/[SnCl_4]_0$ in more polar chlorinated solvents using more reactive "monomers" such as 1,1-diphenylethylene [162–164]. Values greater than 100% indicate that either hexacoordinated stannate dianions are involved, or termination by water is inefficient.

Tin tetrabromide is a much weaker Lewis acid than $SnCl_4$. However, its ligands are more labile and it therefore successfully polymerizes α-methylstyrene to generate well-defined polymers [165,166].

5. Titanium Halides. Titanium tetrachloride is frequently used in cationic polymerizations due to its relatively low cost and high Lewis acidity; it is often used in combination with protonating agents such as hydrogen halides and carboxylic acids [167]. In the latter case, it was suggested that carboxylate replaces a chloride ligand to form the weaker Lewis acid, titanium trichloroacetates [168]. Similarly, titanium chlorides with alkoxy ligands are generated in situ in indene polymerizations using arylalkyl ethers [Eq. (36)] [169,170]. They are very efficient initiators for cyclopentadiene to form soluble linear polymers [171].

$$\text{R-OR}' + \text{TiCl}_4 \rightleftharpoons \text{R}^+, {}^-\text{TiCl}_4\text{OR}' \rightleftharpoons \text{R-Cl} + \text{TiCl}_3\text{OR}' \qquad (36)$$

As outlined in Section III.A.3.a, the strength of the Lewis acid with mixed chloride and alkoxy derivatives decreases as the number of chloride ligands are replaced with alkoxy groups. Titanium chloride with one alkoxy group polymerizes styrene and α-methylstyrene; Lewis acid with two alkoxy groups is too weak to initiate polymerization of styrene, but will initiate polymerization of α-methylstyrene and vinyl ethers. The Lewis acidity of titanium chloride derivatives with three alkoxy groups are so low that only vinyl ether polymerizations reach reasonable conversions.

Initiation is complicated by dimerization of the Lewis acid and by complexation of the Lewis acid with olefinic double bonds of the monomer. The mechanism of initiation is not clear, although partial monomer conversion occurs even in stringently dried systems in a sealed apparatus [172,173]. The polymerization yield increases as temperature decreases, suggesting that $TiCl_4$ addition to double bonds is exothermic and reversible. Indeed, mixtures of titanium halides with alkenes are usually colored, apparently due to weak complexation. However, complexation constants have not been determined. Alternatively, Π-complexation with double bonds may be greater at higher temperatures, albeit nonproductive in monomer addition. This would then reduce the concentration of Lewis acid available for initiation and formation of more reactive dimers. As discussed in Chapter 2, the extent of Friedel-Crafts alkylation of aromatic rings is higher in reactions using $TiCl_4$, than in similar reactions using other Lewis acids.

Lewis acids based on titanium tend to aggregate and form dimers which are usually more reactive than their monomeric precursors (cf., Chapter 2). The degree of aggregation depends on the solvent, temperature, and the ligands attached to titanium; no dimerization was detected by cryoscopy at $-95°$ C in CH_2Cl_2 [174]. However, kinetic measurements of isobutene and styrene polymerizations indicate that polymerization is second order in titanium chloride [175,176], perhaps due to formation of a low concentration of the more reactive dimer or more stable $Ti_2Cl_9^-$ anions. However, polymerizations performed at lower [$TiCl_4$] were reported to be first order in titanium chloride [105].

6. *Other Lewis Acids.* Several relatively weak Lewis acids such as zinc halides and mercury halides initiate polymerization of the most reactive monomers such as *N*-vinyl carbazole, vinyl ethers, and alkoxystyrenes. Many of these acids have poor solubility in hydrocarbons and halogenated hydrocarbons and are therefore used as acetone or ether solutions. However, such solvents act as nucleophiles, and therefore decrease the acids' Lewis acidity.

Strong Lewis acids such as antimony pentachloride tend to undergo side reactions such as electron transfer reactions, formation of strong complexes with ethers, and direct addition to alkenes to produce 1,2-dichloroadducts and much weaker Lewis acids $SbCl_3$. Although antimony pentafluoride is a stronger Lewis acid than $SbCl_5$, it apparently does not undergo these side reactions due to its higher affinity for halides to form SbF_5Cl^- anions. However, it reacts with glass walls, especially in incompletely dried systems, and is therefore not used very often.

b. Kinetics. Initiation can be followed by consumption of the initiator and monomer. However, the initiator is often present at low concentrations and can not be easily detected. In some systems, it is still possible to study the kinetics of initiation if there is an induction period due to propagation being much faster.

The overall rates of polymerization initiated by alkyl esters and halides in the presence of Lewis acids is usually first order in monomer (M), initiator (RX) and Lewis acid (LA) [Eq. (37)].

$$R_p = -d[\mathrm{M}]/dt = k_p[\mathrm{M}]\cdot[\mathrm{R}X]_0[\cdot\mathrm{LA}]_0 \tag{37}$$

If initiation is faster or comparable to propagation and termination is negligible, kinetic plots are straight in semilogarithmic coordinates. Initiation is faster than propagation and not kinetically detectable in polymerizations of isobutene and styrene initiated by cumyl derivatives because the initiator is more easily ionized than the propagating species. However, if the initiator is less easily ionized than the propagating species as in α-methylstyrene polymerizations initiated by cumyl derivatives, and in isobutene polymerizations initiated by t-butyl derivatives (cf., also Section III.A.5), then initiation may be incomplete and the overall polymerization rate will increase continuously.

Polymerizations of vinyl ethers initiated by HI/I_2 in hexane are zero order in monomer, but first order in monomer in more polar solvents. Several groups [177–179] have proposed that this zero-order dependence is due to monomer complexing with iodine. Alternatively, polymerization will be zero order in monomer if ionization is the rate-determining step, followed by faster addition of monomer to carbenium ions [180]. In this case, the concentration of monomer will not affect the polymerization rate.

Polymerizations which are second order in Lewis acid have also been observed, including self-initiated polymerizations with $AlBr_3$ [181], and isobutene polymerizations initiated by alkyl esters and halides activated by $TiCl_4$ [175,182].

Kinetic analysis of isobutene polymerizations initiated by BCl_3 was recently used to distinguish between haloboration and self-ionization of the Lewis acid [Eq. (38)].

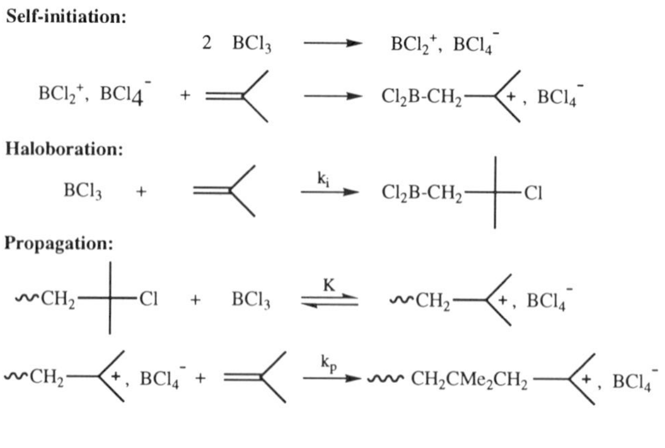

(38)

A linear increase in molecular weight as a function of $p^{1/2}$ (p = conversion) and a linear dependence of $\ln[(1 + p^{1/2})/(1 - p^{1/2})$ vs $[BCl_3]$ indicate that the slow initiation proceeds by direct haloboration ($k_i \approx 10^{-5}$ $mol^{-1}{\cdot}L{\cdot}sec^{-1}$ at $-50°$ C), which is much slower than the overall rate of polymerization ($k_p^{app} \approx k_p{\cdot}K \approx 10^{-1}$ $mol^{-1}{\cdot}L{\cdot}sec^{-1}$). Surprisingly, neither rate constant varies with temperature in the range $-25°$ C to $-60°$ C, indicating that haloboration has a low enthalpy of activation and that the exothermicity of ionization is similar to the activation energy for propagation in carbocationic polymerizations of isobutene. Low activation energies are probably not compatible with low values of the rate constant of initiation and may indicate that haloboration is reversible and that the rate-limiting step is activation of the adduct deactivated by BCl_2 group at the β-position [cf., Eq. (30)].

4. *Carbenium Ions*

Trityl (triphenylmethylium) and tropylium cations are two commercially available carbenium ions. They react with alkenes by either direct addition or by hydride abstraction; tropylium ions are usually less reactive than trityl. As shown in Eq. (39), trityl ions initiate styrene polymerization by direct addition.

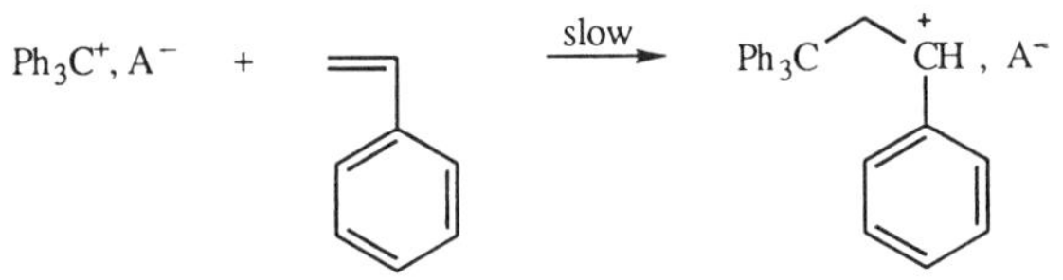

(39)

Hydride abstraction occurs when the monomer is sterically crowded, as with α-methylstyrene [Eq. (40)]. This generates a carbenium ion, which then adds to additional monomer.

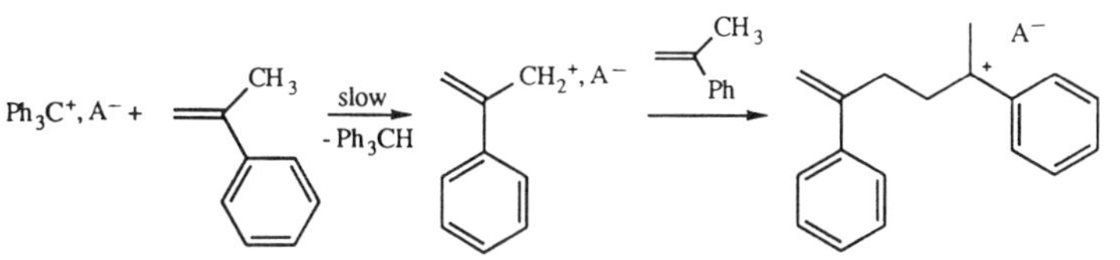

(40)

Electronic effects are also important. For example, unsubstituted trityl salts do not add to isobutene, although tris(p-chlorophenyl)methylium salt reacts readily [183].

Trityl salts have often been used to study the kinetics of initiation by following the disappearance of the salt and the monomer by UV spectroscopy. For example, trityl salts absorb strongly ($\epsilon \approx 40{,}000$ $M^{-1}{\cdot}cm^{-1}$) at 410–440 nm, in contrast to either their adducts or triphenylmethane.

Table 12 lists the rate constants of initiation in numerous systems calculated from spectrometric measurements. These rate constants are 10^3–10^6 times smaller than the rate constants of propagation ($k_p^+ \approx 10^5$ $mol^{-1}·L·sec^{-1}$) under similar conditions, with the largest differences occurring with less reactive monomers such as styrenes and with sterically hindered monomers such as α-methylstyrene.

Trityl hexachloroantimonate polymerizes styrene only at relatively high concentrations ($>10^{-2}$ *M*) [145]. This is apparently due to rapid formation of the covalent chloro adducts, which are relatively inactive in the presence of only a small amount of Lewis acid. Moreover, the $SbCl_5$ generated in this reaction can add to styrene to form a 1,2-dichloroadduct and inactive $SbCl_3$. In contrast, polymerization is fast with trityl hexafluoroantimonate [145], demonstrating the SbF_6^- counterion does not decompose to SbF_5 and F-terminated chains.

Carbenium ions can also be generated in situ from, for example, Lewis acids and alkyl esters or halides. This involves reversible ion formation as discussed in detail in the next section. Alternatively, carbenium ions can be generated by reaction of alkyl halides with silver salts with complex anions such as $AgSbF_6$ and $AgBF_4$. This method is useful for generating perfluorinated complex anions because most alkyl fluorides are not ionized by Lewis acids.

Table 12 Initiation Rate Constants of Vinyl Monomers by Trityl Salts

Monomer[a]	Counterion	Solvent	T (° C)	k_i ($mol^{-1}·L·sec^{-1}$)	E_i ($kJ·mol^{-1}$)	Ref.
Styrene	$SnCl_5^-$	$(CH_2Cl)_2$	30	0.0009	28.1	[184]
Styrene	$SbCl_6^-$	CH_2Cl_2	30	0.0009		[145]
α-MeSt	$SnCl_5^-$	$(CH_2Cl)_2$	30	0.002	21.0	[185]
p-MeSt	$SnCl_5^-$	$(CH_2Cl)_2$	30	0.06		[185]
p-MeOSt	$SbCl_6^-$	CH_2Cl_2	20	0.6	55.9	[186]
Indene	$SnCl_5^-$	CH_2Cl_2	20	0.0007	77.3	[187]
CPD	$SbCl_6^-$	CH_2Cl_2	22	0.001	9.2	[188]
MVE	$SbCl_6^-$	CH_2Cl_2	0	0.6	36.5	[189]
EVE	$SbCl_6^-$	CH_2Cl_2	0	2.3	28.1	[189]
IBVE	$SbCl_6^-$	CH_2Cl_2	0	5.4	39.1	[189]
IBVE	$SbCl_6^-$	CH_2Cl_2	20	16	39	[190]
IPVE	$SbCl_6^-$	CH_2Cl_2	0	15	23.5	[189]
NVC	SbF_6^-	CH_2Cl_2	20	130	44.5	[191]

[a] CPD, cyclopentadiene; MVE, methyl vinyl ether; EVE, ethyl vinyl ether; IBVE, isobutyl vinyl ether; IPVE, isopropyl vinyl ether; NVC, *N*-vinyl carbazole.

5. Covalent Esters and Halides

Covalent esters and halides do not initiate carbocationic polymerizations directly, and must therefore be ionized first. This may be due to the high proportion of charge which must be transferred to the alkene in the transition state (cf., Chapter 2). Spontaneous ionization is very rare and occurs only in polar solvents at low temperatures if the leaving group is sufficiently good and the substituents are strongly stabilizing. Nevertheless, handling initiators which spontaneously ionize is problematic because most carbenium ions rapidly decompose. It is generally preferable to prepare these reactive species in situ, which may be as simple as transferring the covalent precursor from a nonpolar to a more polar solvent or adding Lewis acid.

Lewis acids are typically used to activate dormant covalent esters and halides, although protonic acids are also used. Lewis acids that self-initiate should not be used if well-defined polymers are desired. If initiation is fast, the number of growing chains equals the number of covalent initiating molecules. Fast initiation occurs when the covalent initiator is ionized more easily than, or comparable to, the macromolecular covalent species involved in propagation, unless it is sterically hindered or strongly stabilized as with trityl and tropylium species. The rate of initiation is faster than propagation if the initiator contains a better leaving group than is present in the growing species, as in the alkyl acetate/BCl3 initiating systems discussed earlier [cf., Eq. (21)] [104].

The rates of initiation and propagation are comparable when the covalent initiator and dormant chain ends have similar structures. Therefore, 1-phenylethyl precursors are useful initiators for styrene polymerizations, but are poor initiators for α-methylstyrene and vinyl ether polymerizations. Similarly, cumyl derivatives are good initiators for isobutene and styrene, but are poor initiators for vinyl ethers; their initiation of α-methylstyrene is apparently slow [165]. 1-Alkoxyethyl derivatives are successful initiators for vinyl ethers, styrenes, and presumably isobutene polymerizations [165,192]. *t*-Butyl derivatives initiate polymerization of isobutene slowly [105]. This is mirrored in model studies that show that *t*-butyl chloride undergoes solvolysis approximately 30 times slower than 2-chloro-2,4,4-trimethylpentane [193]. This may be due to insufficient B-strain in monomeric tertiary precursors [194]. In contrast, monomeric and dimeric or polymeric structures of secondary esters and halides apparently have similar reactivity.

Because the covalent initiator is incorporated into the polymer chain as an end group, end-functionalized polymers can be prepared using functionalized initiators. This has been used extensively in vinyl ether polymerizations and is reviewed in Chapter 5. However, most functional groups are capable of reacting with the active species and must therefore be protected. Hydroxy groups can be introduced using silyl ether-protected initiators; amino groups are introduced via phthalimide derivatives. Aromatic groups within the initiator may also undergo side reactions. For example, indanyl end groups are generated in isobutene polymerizations initiated by cumyl derivatives [Eq. (41)].

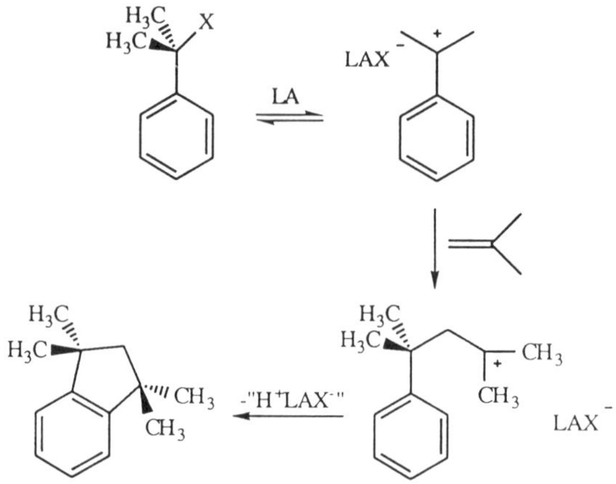

(41)

Intramolecular cyclization is prevented by substituents at either the ortho or meta positions [Eq. (42)] [105].

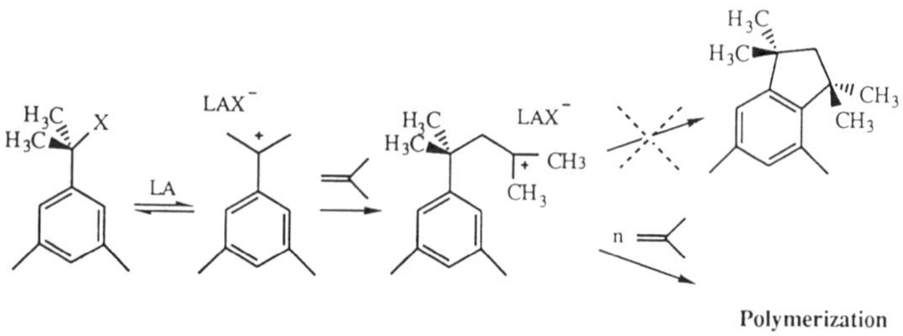

(42)

B. "Special" Initiators

In addition to protons, alkyl and acyl groups, silyl groups are sufficiently electropositive to initiate carbocationic polymerizations. Silicon is more electropositive than carbon, and therefore reacts with many nucleophiles, especially those based on oxygen. The resulting silicon-heteroatom bonds are relatively labile; trimethylsilyl is thus often referred to as a "bulky proton." Trimethylsilyl triflate (TMSOTf) has been used to initiate cationic polymerizations of alkenes and heterocycles, although direct initiation is generally not observed. Instead, TMSOTf hydrolyses to generate triflic acid, which is the true initiator. Therefore, no polymerization occurs in the polymerization of styrene initiated by TMSOTf in the presence of a hindered pyridine which acts as a proton trap [195]. Polymerization begins when moisture is allowed to enter the system, and the hindered pyridine is consumed by triflic acid formed by hydrolysis of TMSOTf. Vinyl ethers are polymerized by cationation using TMSOTf, silyl iodide and silyl bromide only in the presence of compounds with carbonyl groups. As shown in Eq. (43), ketones, aldehydes, and acetals react with these silyl reagents to generate siloxy carbenium ions, which then initiate polymerization [196–199].

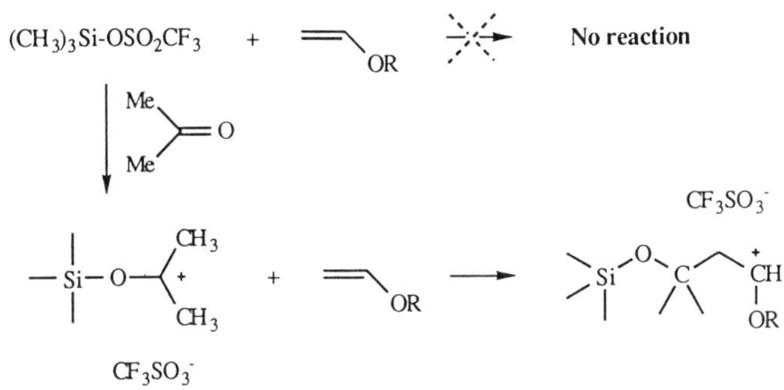

(43)

Oligosilanes and polysilanes initiate multidirectional growth. For example, poly(*N*-vinyl carbazole) and poly(vinyl ether) side chains have been initiated from polysilanes containing several triflate groups to prepare comb-like copolymers [200–202].

Initiators based on halonium and sulfonium salts are used commercially in various microlithographic processes and in the coating industry. Onium salts were developed commercially as photoinitiators due to the lower sensitivity of cationic polymerizations to oxygen compared to radical polymerizations. Aromatic halonium and sulfonium salts with complex anions such as SbF_6^-, AsF_6^- and BF_4^- do not initiate cationic polymerizations spontaneously, but must be activated by UV irradiation.

Photoinitiators which produce species-inducing cationic polymerization of alkenes and heterocycles are usually based on diarylhalonium salts and triarylsulfonium salts [203]. The mechanism of the direct photolysis of these salts is quite complex and involves formation of excited singlet species which can reversibly convert by intersystem crossing (ISC) to the excited triplet. The triplet decays to the ground state or produces a triplet aryl radical and haloarene radical cation [Eq. (44)] [204].

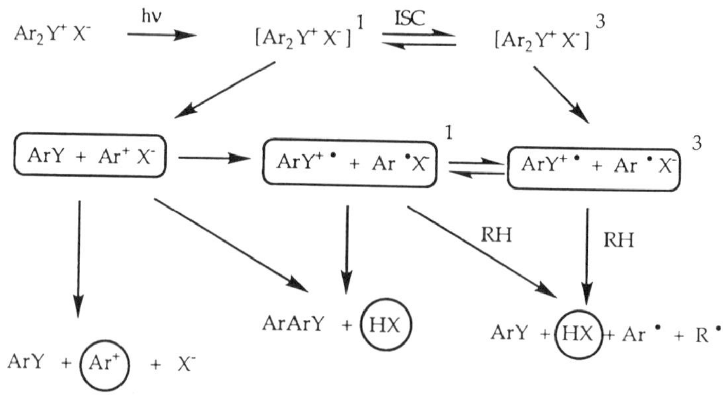

(44)

$$Ar^+ + M \longrightarrow ArM^+$$

$$Ar^+ + ArY \longrightarrow ArArY + H^+$$

$$Ar^+ + RH \longrightarrow ArR + H^+$$

$$ArY^{+\bullet} + M \longrightarrow {}^{\bullet}YArM^+$$

$$ArY^{+\bullet} + Ar^{\bullet} \longrightarrow ArArY + H^+$$

$$H^+ + M \longrightarrow HM^+$$

The caged species may escape geminate recombination and produce various species that can initiate cationic polymerization. Solvent (RH) often participates in these reactions producing protonic acids. As shown in Eq. (44), protonic acids are also formed by reaction of radical cations with aryl radicals or by Friedel-Crafts arylation. Up to 70% of the protonic acid is formed upon photolysis of diaryliodonium salts [205]. In addition to initiation by protons, arenium cations and haloarene radical cations can react directly with monomer. The efficiency of these salts as cationic initiators depends strongly on the counterions. Those with complex anions such as hexafluoroantimonate, hexafluorophosphate, and triflate are the most efficient.

Cationic polymerization can also be initiated by some highly acidic inorganic compounds such as heteropolyacids and acidic clays. Clays may find some commercial applications due to their relatively low cost and facile catalyst removal. Polymerization within the clays' layered structures may also enable control of the microstructure and perhaps suppress side reactions. We expect this field to develop rapidly in the near future.

IV. PROPAGATION

Statistically, propagation is the most important elementary reaction of a polymerization because it results in formation of the entire macromolecule except the end groups. Thus, the polymerization rate, molecular weight, and molecular weight distribution may be regulated if the propagation mechanism is elucidated. This information may also be used to increase the regioselectivity and stereoselectivity of propagation, thereby yielding more controlled microstructures. Higher molecular weight polymers can be prepared with well-defined structures if the chemoselectivity is increased. Ultimately, a complete mechanistic picture for all of the alkene polymerizations will enable the synthesis of well-defined copolymers with predictable compositions and microstructures by controlled copolymerizations.

This section presents the mechanistic aspects of carbocationic propagation; ''living'' systems will be described in more detail in Chapter 4. To completely understand the propagation mechanism, the structure of both the monomer and growing chain ends must be known. However, there are a number of factors which complicate establishing the structure of the growing chain end, including its often short lifetime. Ionic polymerizations usually involve several different active species, each with different reactivities toward monomer and other reagents [Eq. (45)].

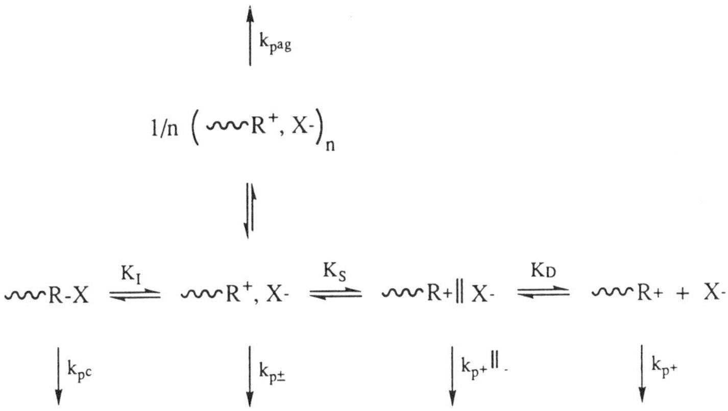

(45)

In carbanionic systems, the reactivity of these active centers generally follows their degree of ionization: covalent species ≪ tight ion pairs ≪ loose ion pairs ≈ free ions. However, in carbocationic systems the reactivities of carbenium ions are similar regardless of the degree of association, and covalent species are not active at all. Ionic clusters are formed by aggregation of ion pairs in nonpolar media and at higher concentrations, even in the presence of added salts.

In addition to covalent species and carbenium ions, the equilibria may involve onium ions, which are formed by reaction of carbenium ions with noncharged nucleophiles [Eq. (46a)]. This decreases the carbenium ions' lifetime, and therefore the time available for isomerization to more stable and less reactive carbenium ions via hydride and alkyl anion shifts [Eq. (46b)]. Decreasing the probability of rearrangements by decreasing the carbenium ions' lifetime is especially useful because such rearrangements can not be prevented by decreasing the polymerization temperature.

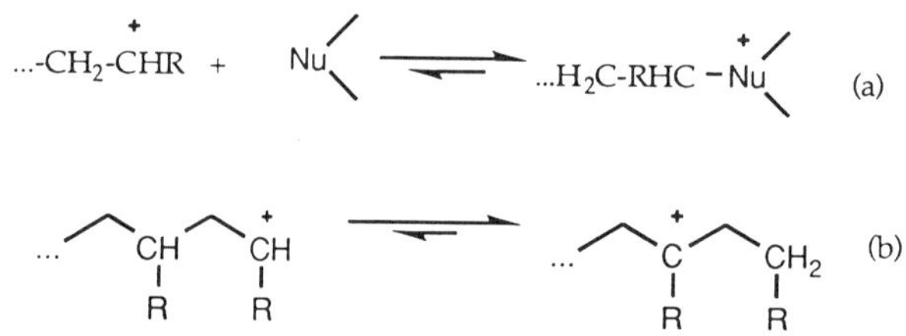

(46)

Propagation and other elementary reactions involving carbenium ions experience significant solvent effects. Because solvation is exothermic, temperature effects are also considerable. However, each elementary reaction is affected by temperature differently, which sometimes results in negative overall activation energies [186]. The activation energies of side reactions such as transfer are usually higher than those of propagation. Therefore, it is necessary to work at low temperature to produce high molecular weight polymer by carbocationic polymerizations ($< -100°$ C for butyl rubber).

In order to prevent erroneous correlations of structure and reactivity, only active species of the same type should be compared. For example, carbenium ions with electron-donating substituents are more stable and less reactive than those with electron-withdrawing substituents, whereas covalent species with electron-donating substituents ionize much easier than those with electron-withdrawing substituents. Both factors result in a higher concentration of the more stabilized carbenium ion, which may lead to a higher apparent or overall rate constant for electrophilic addition of carbenium ions to a standard alkene, contrary to the order of reactivity of the carbenium ions themselves. Thus, absolute rather than apparent rate constants should be used to establish structure-reactivity relationships.

A. Thermodynamics

Thermodynamics determines whether or not a monomer will polymerize, to what extent it polymerizes, and what conditions such as solvent, temperature, and concentrations are required. As discussed in Chapter 1, the thermodynamic polymerizability of a monomer is independent of the mechanism and is therefore identical for radical, anionic, cationic, and coordinative mechanisms if structurally identical polymers are obtained. Although this requires that both the end groups and the microstructure are the same, the influence of regioselectivity and stereoselectivity on the enthalpy and/or entropy of polymerization has not been confirmed experimentally yet.

The reversibility of propagation, or more specifically, the position of the equilibrium as determined by the ratio of the rate constants of propagation and depropagation is also independent of the mechanism. The equilibrium monomer concentration of monosubstituted alkenes such as styrenes and vinyl ethers are so low ($[M]_\infty < 10^{-6}$ mol/L) at temperatures used for carbocationic polymerizations that the reversibility of polymerization can be neglected.

However, the equilibrium monomer concentrations of disubstituted alkenes is measurable. The equilibrium constants for dimerization, trimerization, and polymerization of α-methylstyrene have been determined as a function of temperature under anionic conditions [12]; similar values should be obtained under cationic conditions. Unfortunately, the equilibrium position can't be determined directly under cationic conditions due to the irreversible side reactions of isomerization and indan and spirobiindan formation (Section II.A). The equilibrium monomer concentrations of isobutene and isopropenyl vinyl ethers should also be relatively high, albeit lower than those of α-methylstyrenes. However, the true equilibrium can't be reached with these monomers due to irreversible side reactions, and reliable data are therefore not available. Nevertheless, the ceiling temperature of isobutene polymerization is apparently between 50 and 150° C.

As discussed in more detail in subsequent sections, growing carbenium ions are highly susceptible to transfer reactions by β-proton elimination, to generate chains with unsaturated end groups. Formally, these polymers with unsaturated end groups are macromonomers, which can be polymerized and/or copolymerized to form either star-like or comb-like structures. However, sterically hindered 1,2-disubstituted alkenes are usually formed, which do not readily homopolymerize, but may copolymerize if any unreacted monomer remains in the reaction mixture. Nevertheless, the significant increase in molecular weight at nearly complete conversion in polymerizations of styrenes [182,206] may be explained by copolymerization of the macromonomers; intermolecular Friedel-Crafts alkylation of phenyl groups along another polymer backbone is unlikely to compete with unimolecular indan formation.

Transfer in α-methylstyrene and isobutene polymerizations occurs with β-proton elimination from either the endo- or exoposition [Eq. (47)] [14,132]. The latter is not only favored kinetically, but produces a more reactive alkene which may copolymerize. In contrast, the thermodynamically more stable internal alkene should be too sterically hindered to copolymerize.

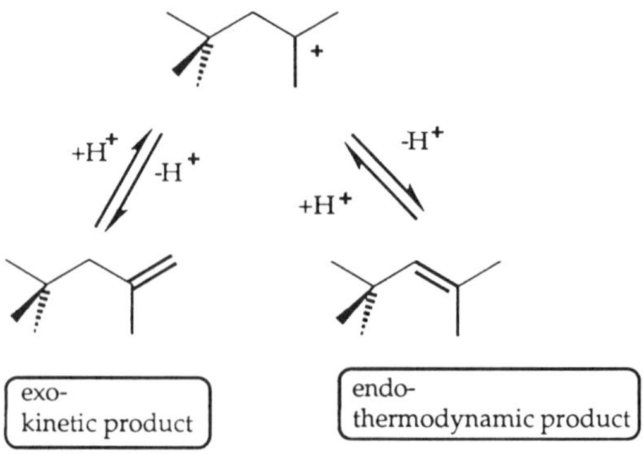

(47)

B. Kinetics

The degenerative nature of propagation results in reformation of the same active species, but with monomer consumption and chain growth. Although the monomer's thermodynamic polymerizability is independent of the mechanism, the mechanism and structure of the active species determines the rate of monomer conversion. The structure of the active species involved in carbocationic polymerizations was discussed in Section II; detailed information on the reactivities of model species was presented in Chapter 2, with the conclusion that covalent precursors do not react directly with alkenes, but must first ionize to sp^2-hybridized carbenium ions. Only the resulting carbenium ions can add to double bonds.

Kinetics is one of the most fundamental tools used to elucidate a reaction mechanism. Kinetic measurements require determination of the concentrations of all reagents and active species as a function of time. The rate of polymerization is defined as the rate of monomer consumption, which depends on the concentration of monomer and active centers. If polymerization is internally first order in monomer, the slopes of the semilogarithmic anamorphoses are equal to the product of the rate constant and the concentration of growing species. If a variety of species [cf., Eq. (45)] such as free ions (P^+), ion pairs ($P^{\pm}$), covalent adducts (P^c), and aggregates (P^{ag}) are involved, the rate of polymerization is described by Eq. (48), and the slope is defined by Eq. (49).

$$R_p = -d[\mathrm{M}]/dt = k_p \cdot [\mathrm{M}] \cdot [\mathrm{P}^*] \tag{48}$$

$$\text{Slope} = -d\ln[\mathrm{M}]/dt = k_p{}^{+} \cdot [\mathrm{P}^+] + k_p{}^{\pm} \cdot [\mathrm{P}^{\pm}] + k_p{}^{\mathrm{c}} \cdot [\mathrm{P^c}] + k_p{}^{\mathrm{ag}} \cdot [\mathrm{P^{ag}}] + \ldots = \sum_{n=1}^{\infty} k_p{}^{n} \cdot [\mathrm{P}^{*n}] \tag{49}$$

Thus, the concentration of each species must be determined to determine the absolute rate constant(s) of propagation. If not, only the apparent rate constant of propagation is obtained.

Earlier work assumed that initiation was instantaneous and quantitative, such that the concentration of growing carbenium ions was equal to the concentration of initiator used [Eq. (50)].

$$-d\ln[\mathrm{M}]/dt = k_p \cdot [\mathrm{P}^*] \approx k_p{}^{\mathrm{app}} \cdot [\mathrm{I}]_0 \tag{50}$$

However, the low yield of carbenium ions ($\ll 1\%$) and often incomplete initiation in these classic cationic systems resulted in very low apparent rate constants of propagation ($k_p{}^{\mathrm{app}} \approx 1$ mol^{-1}·L·sec^{-1}). In the past 20 years, three methods have been developed to obtain more accurate rate constants for carbocationic propagation. They are based on direct measurement of the concentration of growing species by rapid spectroscopic techniques, on systems initiated by γ-irradiation, and on systems initiated by trityl salts which are consumed slowly. The latter method follows the concentration of unreacted initiator spectroscopically. Although all three methods have some disadvantages, they provide consistent carbocationic propagation rate constants ($k_p \approx 10^{5\pm1}$ mol^{-1}·L·sec^{-1} at $\approx 0°$ C) [207].

Monomer consumption is easily followed by several techniques, including spectroscopy (NMR, UV, IR), chromatography, dilatometry, calorimetry, and gravimetry. However, the concentration of growing species in typical polymerizations is much lower than that of monomer and polymer, and is not easily detected spectroscopically. As discussed in Section II, NMR is used routinely to study model compounds, but has not been extended to polymerization studies yet. This is because the intensity of resonances due to the growing carbenium ions are much weaker than those from monomer and polymer, and because the propagating carbenium ions are relatively unstable. Phosphines have been added to cationic polymerizations to study the terminated chain ends in their dormant form by ^{31}P NMR [66,208]. However, all other electrophilic species such as covalent adducts and onium and isomerized carbenium ions also react with these strong nucleophiles, thus preventing determination of the concentration of only the growing carbenium ions. Similar reactions have been performed with malonate anions [209] and naphthoxides (UV studies) [210] to establish the concentration of all potentially reactive chains.

Carbenium ions with α-aromatic substituents (e.g., styrene derivatives) absorb at very different UV-visible wavelengths compared to the corresponding monomers and polymer chains. In such cases, UV-visible spectroscopy can be used to quantitatively determine the concentration of growing carbenium ions. Because the extinction coefficients of these carbenium ions are typically $\epsilon \approx 10{,}000$ to $30{,}000\ mol^{-1} \cdot L \cdot cm^{-1}$, they can be detected only at concentrations higher than $[C^+] \approx 10^{-6}\ mol^{-1} \cdot L$. In some of the systems discussed in this chapter, the concentration of carbenium ions are below the detection limit of UV. On the other hand, those systems with carbenium ion concentrations $[C^+] \approx 10^{-5}\ mol^{-1} \cdot L$ will propagate very rapidly because $k_p \approx 10^5\ mol^{-1} \cdot L \cdot sec^{-1}$. The resulting half-lifetimes ($\tau_{1/2} \approx 1$ sec) will require rapid detection techniques, such as stopped-flow, to follow concentration changes in monomer and especially carbenium ions.

1. *Stopped-Flow Studies*

Currently, the only technique available to observe the evolution of the concentration of growing carbenium ions directly during a polymerization is by UV spectroscopy of monomers with α-aryl substituents. This requires rapid scanning because the lifetime of growing carbenium ions is very short and the polymerization is very fast. Although stopped-flow is the most successful rapid scanning technique in studies of carbocationic polymerizations [16–19,23,211], flash photolysis [20], and pulse radiolysis [41] used for model studies could also potentially be applied to polymerizations. The limiting factor in obtaining quantitative data from stopped-flow studies is the reliability of the extinction coefficient used to calculate the concentration of growing carbenium ions. Accurate values of monomer conversion may also be a problem because monomer often absorbs too strongly to be observed simultaneously with the low concentration of carbenium ions.

As discussed in Section II, results obtained by either UV or conductometry alone may be misleading because neither establish the precise structure of the absorbing and/or conducting species. For example, the 380-nm absorption observed in *p*-methoxystyrene polymerizations is due to isomerized carbenium ions which do not propagate [29]. Similarly, both carbenium and dormant onium ions conduct. Nevertheless, these techniques can be combined with model studies and kinetic analysis to determine the structure of the absorbing and/or conducting species. Such techniques confirm that the 330-nm absorption in protonic acid-initiated polymerizations of styrene is due to growing carbenium ions [16], although these polymerizations were previously assumed to propagate by a pseudocationic mechanism without carbenium ion intermediates [212].

Table 13 tabulates the rate constants of propagation of styrene monomers calculated from UV data assuming that $\epsilon = 10{,}000\ mol^{-1}\cdot L\cdot cm^{-1}$ for growing styrene and α-methylstyrene carbenium ions, and $\epsilon = 28{,}000\ mol^{-1}\cdot L\cdot cm^{-1}$ for carbenium ions based on *p*-methoxystyrene derivatives. The former extinction coefficient was determined in superacid media assuming quantitative formation of carbenium ions, and is probably

Table 13 Rate Constants of the Cationic Propagation of Various Styrenes Measured by Stopped-Flow Method[a]

Monomer, $[M]_0$ ($mol\cdot L^{-1}$)	Initiator, $[I]_0$ ($mol\cdot L^{-1}$)	Solvent	T (° C)	$k_p\cdot 10^{-3}$ ($mol^{-1}\cdot L\cdot sec^{-1}$)	$k_p^+/k_p^{\pm}$	Comment	Ref.
Styrene, 0.2	$HClO_4$, 10^{-2}	CH_2Cl_2	−80	10			[19]
Styrene, 0.2	$HClO_4$, $6\cdot 10^{-3}$	CH_2Cl_2	−80	5	20	b	[213]
Styrene, 0.2	$HClO_4$, $6\cdot 10^{-3}$	CH_2Cl_2	−60	15	8	b	[213]
Styrene, 0.01	CF_3SO_3H, $2\cdot 10^{-3}$	CH_2Cl_2	−62	1			[18]
Styrene, 0.01	CF_3SO_3H, $2\cdot 10^{-3}$	CH_2Cl_2	−10	100			[18]
Styrene, 0.4	CF_3SO_3H, $3\cdot 10^{-3}$	$C_2H_4Cl_2$	−1	50	6	b	[17]
Styrene, 0.4	CF_3SO_3H, $5\cdot 10^{-3}$	$C_2H_4Cl_2$	30	200	24	b	[17]
α-MeSt, 2.10^{-4}	CF_3SO_3H, 10^{-2}	$C_2H_4Cl_2$	30	30	–	c	[21]
α-MeSt, 3.10^{-4}	H_2SO_4, 10^{-2}	$C_2H_4Cl_2$	30	3,000	–	c	[21]
p-MeOSt, 5.10^{-3}	CH_3COClO_4, 10^{-4}	$C_2H_4Cl_2$	30	70	–	d	[23]
p-MeOSt, 5.10^{-3}	CF_3SO_3H, 10^{-4}	$C_2H_4Cl_2$	30	100	–	d	[23]
p-MeOSt, 5.10^{-3}	$SnCl_4$, 10^{-4}	$C_2H_4Cl_2$	30	10	–		[23]
p-iPr-α-MeSt, 0.02	CF_3SO_3H, 10^{-4}	CH_2Cl_2	−58	8	–	c	[25]
p-iPr-α-MeSt, 0.11	CF_3SO_3H, 10^{-4}	CH_2Cl_2	−56	60	–	c	[25]

[a] Rate constants were calculated by assuming the extinction coefficient of growing macromolecular styryl cation to be $\epsilon = 10{,}000\ mol^{-1}\cdot L\cdot cm^{-1}$. Values of k_p might be three times larger, if $\epsilon \approx 30{,}000\ mol^{-1}\cdot L\cdot cm^{-1}$

[b] Relative reactivities of ions and ion pairs estimated from common ion effect.

[c] Polymerization impossible ($[M]_0 < [M]_e$); rate coefficient gives lower limit of the rate constant of propagation and may be equal to the rate constant of transfer. Note strong increase of the rate coefficient when $[M]_0 > [M]_e$ (=0.05 mol/L at −60° C).

[d] $\epsilon = 28{,}000\ mol^{-1}\cdot L\cdot cm^{-1}$ observed, yield of cations between 100% and 50%.

only one-third of the actual value [15]. The second extinction coefficient is much more reliable although the absorption maximum at 380 nm is apparently not due to the growing carbenium ions because they typically absorb at 340 nm [19,20]; the isomerized carbenium ions [13] shown in Eq. (8) absorb at 380 nm.

The most important conclusion from stopped-flow studies is that the rate constants of propagation of several styrene derivatives are approximately $k_p \approx 10^{5\pm1}$ mol^{-1}·L·sec^{-1} at 0° C, which is relatively high compared with those of radical and anionic systems (average $k_p \approx 10^2$ mol^{-1}·L·sec^{-1} at 0° C). Solvent effects are noticeable, with propagation slower in more nucleophilic 1,2-dichloroethane [17] than in CH_2Cl_2 [18] under comparable conditions. That is, the carbenium ion reactivity is apparently reduced by interaction with more nucleophilic solvents. However, such interactions do not result in formation of chloronium ions, whose spectra would be very different compared to those of the corresponding carbenium ions.

As discussed in Chapter 2 and later parts of this section, the reactivities of carbenium ions and ion pairs are similar, and independent of the structure of the counteranion. Although the rate constants of α-methylstyrene "polymerizations" initiated by triflic acid are significantly different than those initiated by sulfuric acid, the reactions were performed at low monomer concentrations ($[M]_0 < [M]_e$), which thermodynamically prevent polymerization [214]. Monomer is converted to linear unsaturated dimers. However, dimerization requires β-proton elimination. Thus, the faster apparent rate of "polymerization" with sulfuric acid may be due to faster β-proton elimination by transfer to the more basic sulfate counteranion according to Eq. (51).

$$
\begin{aligned}
&\mathrm{HA + M} \quad \overset{k_i}{\rightleftharpoons} \quad \mathrm{M_1^+, A^-} \\
&\mathrm{M_1^+, A^- + M} \quad \overset{k_p^1}{\rightleftharpoons} \quad \mathrm{M_2^+, A^-} \\
&\mathrm{M_2^+, A^- + M} \quad \overset{k_p^2}{\rightleftharpoons} \quad \mathrm{M_3^+, A^-} \\
&\mathrm{M_1^+, A^-} \quad \overset{k_{tr}^A}{\rightleftharpoons} \quad \mathrm{M_2^= + HA}
\end{aligned}
\tag{51}
$$

Propagation by reaction of M_3^+ and M_n^+ is not possible at $[M]_0 < [M]_e$. Therefore, the rate-limiting step in dimerization may be transfer to counteranion. The true rate constants of propagation for α-methylstyrene may therefore be even higher than reported with either counteranion. This is supported by the much higher rates of monomer consumption in p-isopropyl-α-methylstyrene polymerizations at $[M]_0 > [M]_e$ [25], in which propagation is possible. Only linear oligomers with DP = 19 were formed in this experiment, although dimers and trimers are formed at lower $[M]_0$.

The ratio of the reactivities of ions and ion pairs ($k_p^+/k_p^\pm$) are also included in Table 13. They were determined from kinetic studies of the apparent rate constants at either different acid concentrations which vary the extent of dissociation into free ions, or in the presence of tetrabutylammonium salts with common counteranions such as perchlorates and triflates. This results in ratios of the reactivities of ions and ion pairs of approximately 6 to 24. However, addition of an equimolar amount of salt to triflic acid may lead conjugation of acid with anions [215], with complete deactivation of the system. Therefore, the lower rate constants of propagation for ion pairs may be partially due to removal of the acid from the system. Thus, the values reported in Table 13 can be considered the upper limit of $k_p^+/k_p^\pm$. The true ratio might be lower, with very similar reactivities for ions and ion pairs as in model systems [4]. Miscalculations of the ratio of reactivities of ions and ion pairs has led to unrealistic values of activation parameters calculated for propagation by ions ($E_p^\ddagger = 51$ kJ·mol^{-1}, $\Delta S_p^\ddagger = +54$ J·mol^{-1}·K^{-1}) and ion pairs ($E_p^\ddagger = 21$ kJ·mol^{-1}, $\Delta S_p^\ddagger = -84$·mol^{-1}·K^{-1}) [17]; the latter values are similar to the overall activation parameters for ionic propagation and are quite reasonable. Extrapolation of Kunitake's data to $-80°$ C shows ion pairs being 30 times more reactive than ions [17], which contradicts the available experimental data [213].

2. *Systems Initiated by Stable Carbenium Ions*

The rate constants of nearly all of the elementary reactions in trityl-initiated polymerizations of cyclopentadiene [216], *p*-methoxystyrene [186], vinyl ethers [217], and α-methylstyrene [218] were determined by kinetic measurements, sometimes combined with conductometric measurements. Monomer conversion was followed by either dilatometry, spectroscopy, or calorimetry. Initiation was followed by the decrease in the 410-nm absorption of the trityl carbenium ions (ϵ = 36,000 mol^{-1}·L·cm^{-1}), caused by their reaction with monomer by either direct addition or hydride abstraction. The initiator was assumed not to be consumed in any other reactions. The reaction orders (usually first order in each reagent) and rate constants of initiation were then determined by plotting the rate of initiation versus the initial monomer and initiator concentrations according to Eq. (52).

$$R_i = -d[I]/dt = k_i \cdot [I] \cdot [M] \tag{52}$$

Molecular weights were determined by size exclusion chromatography (SEC).

When sufficiently high molecular weight polymers (DP > 100) are produced, monomer is consumed nearly exclusively (>99%) in propagation, and the rate of polymerization is described by Eq. (48). Because the active species are generated by initiation and consumed by termination, their concentration varies with time according to Eq. 53, assuming that the kinetics is first order.

$$d[\mathrm{P}^*]/dt = R_i - R_t = k_i[\mathrm{M}]\cdot[\mathrm{I}] - k_t\cdot[\mathrm{P}^*] \quad (53)$$

This leads to Eq. (54).

$$d(R_p/[\mathrm{M}])/dt = -d^2\ln[\mathrm{M}]/dt^2 = k_p\cdot(d[\mathrm{P}^*]/dt) = k_p\cdot(R_i - R_t) \quad (54)$$

Initially, the rate of termination can be neglected [Eq. (55)].

$$\{d(R_p/[\mathrm{M}])/dt\}_0 = k_p\cdot R_{io} \quad (55)$$

Combination of the above equations yields Eq. (56), which is integrated to Eq. (57).

$$-d^2\ln[\mathrm{M}]/dt^2 = k_p\cdot(-d[\mathrm{I}]/dt) - k_t\cdot(-d\ln[\mathrm{M}]/dt) \quad (56)$$

$$-d\ln[\mathrm{M}]/dt = k_p\cdot([\mathrm{I}]_0 - [\mathrm{I}]) - k_t\cdot\ln([\mathrm{M}]_0/[\mathrm{M}]) \quad (57)$$

Rate constants k_p and $-k_t$ are then obtained from the slope and intercept, respectively, of a plot of $\{(-d\ln[\mathrm{M}]/dt)/\ln([\mathrm{M}]_0/[\mathrm{M}])\}$ versus $\{([\mathrm{I}]_0 - [\mathrm{I}])/\ln([\mathrm{M}]_0/[\mathrm{M}])\}$, and k_i is determined from Eq. (55) knowing k_p.

In these systems, the degree of polymerization is usually lower than that expected from the ratio of reacted monomer and initiator ($\mathrm{DP}_n < \Delta[\mathrm{M}]/\Delta[\mathrm{I}]$), which indicates that transfer occurs. Therefore, the degree of polymerization is determined by the ratio of the rates of propagation and chain generation by both initiation and transfer. If chain transfer to monomer dominates and k_p is known, Eq. (58) can be used to determine k_{trM} at complete conversion.

$$1/\overline{\mathrm{DP}}_n = \{([\mathrm{I}]_0 - [\mathrm{I}])/[\mathrm{M}]_0\} + k_{\mathrm{trM}}/k_p \quad (58)$$

The validity of this approach has been successfully tested by determination of the content of trityl groups in polymer (equal to expected).

It is difficult to follow the polymerization kinetics if the initiator does not absorb in the UV region. The kinetics of such a system has been derived for α-methylstyrene polymerizations initiated by mixtures of $BuOTiCl_3$ and H_2O [218]. As shown in Eq. 59, reaction between the components of the initiating system is assumed to be the rate-determining step.

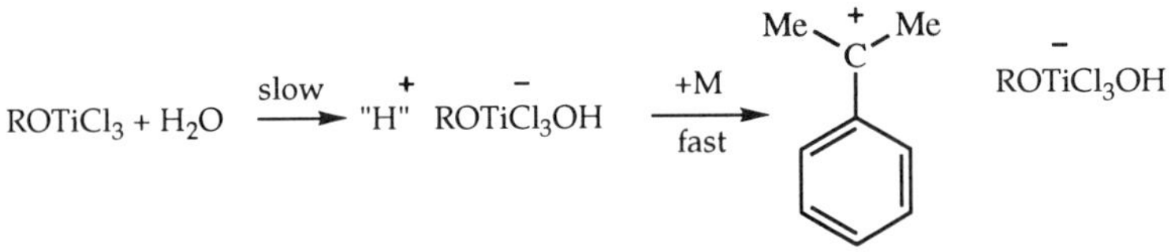

(59)

If termination is unimolecular, $[P^*]$ is obtained by integration of the rate of polymerization from $t = 0$ to t as in Eq. (60).

$$R_p/[M] = k_p \cdot [P^*] = (k_i \cdot k_p/k_t) \cdot [H_2O] \cdot [I]_0 \cdot [1 - \exp(-k_t \cdot t)] \qquad (60)$$

Since chains are formed by both initiation and transfer, the total number of macromolecules ($[N] = [N]_i + [N]_{trM}$) is expressed by Eq. 3.61, where p is the fractional monomer conversion.

$$[N] = \{k_i \cdot [I]_0^2 \cdot t_\infty/(1 + k_i \cdot [I]_0 t_\infty)\} + (k_i \cdot k_{trM}/k_t) \cdot [M]_0 \cdot [I]_0^2 \cdot \int [1 - \exp(-k_t \cdot t)]\,(1 - p)dt \qquad (61)$$

The intercept of a plot of $[N]$ vs. $[M]_0$ yields the number of chains formed by initiation ($[N]_i$), or more precisely, the sum of chains formed by initiation and intramolecular chain transfer to counteranion. The rate constant of initiation k_i is calculated from the intercept, and k_t and k_{trM} are obtained from the slope after solving the integral value graphically.

Sigwalt and co-workers have used this method to determine the rate constants of initiation, propagation, termination, and transfer (Table 14) in polymerizations of cyclopentadiene initiated by $Ph_3C^+SbCl_6^-$ (CH_2Cl_2, 20 to −70° C) [216]; of p-methoxystyrene initiated by $Ph_3C^+SbCl_6^-$ (CH_2Cl_2, 25 to −15° C) [186]; of α-methylstyrene initiated by $BuOTiCl_3 + H_2O$ (CH_2Cl_2, −30 to −70° C) [218]; and vinyl ethers such as IBVE initiated by $Ph_3C^+SbCl_6^-$(CH_2Cl_2, 20 to −70° C) [217]. The rate constants of propagation summarized in Table 14 are smaller than those calculated from both γ-irradiated systems (cf., Section IV.B.3) and stopped-flow measurements. In some cases, the activation energy of propagation is negative. This has been proposed to be due to either exothermic complexation of active centers with monomer before propagation, or exothermic dissociation, provided that ions are much more reactive than ion pairs [217].

These lower k_p values may also be due to overestimating the concentration of growing carbenium ions. For example, although trityl hexachloroantimonate is probably completely ionized, there should also be an equilibrium between the less stable propagating carbenium ions and covalent species formed by collapse of the ion pair [Eq. (62)].

Table 14 Rate Constants in Cationic Polymerizations of Alkenes Initiated by Trityl and Tropylium Salts

Monomer	T (° C)	k_i ($mol^{-1} \cdot L \cdot sec^{-1}$)	k_p ($mol^{-1} \cdot L \cdot sec^{-1}$)	E_p ($kJ \cdot mol^{-1}$)	k_t (sec^{-1})	k_{trM} ($mol^{-1} \cdot L \cdot sec^{-1}$)	Ref.
Cyclopentadiene	−50		$4.5 \cdot 10^1$	−33	$6.8 \cdot 10^{-3}$		[216]
p-MeOSt	+10	0.28	$2.8 \cdot 10^4$	−25	0.08	53	[186]
α-MeSt	−70	17	$2.2 \cdot 10^4$	−29	0.54	30	[218]
MVE	0	0.6	$0.026 \cdot 10^4$	45	0.003	3.5	[217]
EVE	0	2.3	$0.7 \cdot 10^4$	45	0.02	74	[217]
iPVE	0	15	$1.1 \cdot 10^4$	15	2.3	135	[217]
iBVE	0	5.4	$1.5 \cdot 10^4$	23	0.2	180	[217]
p-MeOSt	+10		$>4.8 \cdot 10^3$	21			[219]
MVE	0		$0.014 \cdot 10^4$	58			[220]
EVE	0		$0.15 \cdot 10^4$	45			[220]
cHVE	0		$0.33 \cdot 10^4$	37			[220]
*t*BVE	0		$0.35 \cdot 10^4$	4			[220]
ClEVE	0		$0.02 \cdot 10^4$	29			[220]
iBVE	0		$0.68 \cdot 10^4$	33			[221]
NVC	0		$30 \cdot 10^4$				[222]
NVC	20	130	$10 \cdot 10^4$				[191]
NVC	−70		$0.3 \cdot 10^4$				[223]
NVC	20		$9 \cdot 10^4$				[224]
Styrene	0		$5 \cdot 10^2$				[145]

$$\begin{array}{ll} Ph_3C\text{-}Cl + SbCl_5 \overset{K_1}{\rightleftharpoons} Ph_3C^{+}, SbCl_6^{-} & \\ R\text{-}Cl + SbCl_5 \overset{K_2}{\rightleftharpoons} R^{+}, SbCl_6^{-} & \boxed{K_1 >> K_2} \end{array} \qquad (62)$$

That is, the assumed rate constant of propagation is an apparent rate constant equal to the product of the true propagation rate constant and the equilibrium constant of ionization ($k_p^{app} = K_2 \cdot k_p^{+}$). The ionization equilibrium also accounts for the negative overall activation energy if ionization is sufficiently exothermic to compensate for a relatively low activation energy of carbocationic propagation ($E_a^{app} = E_p + \Delta H_2$) [216,218], [186]. The equilibrium can also be shifted toward inactive species by complexation of the liberated Lewis acid with nucleophiles such as ether groups or double bonds, thereby reducing the concentration of carbenium ions and therefore the rate of polymerization.

The last twelve k_p values reported in Table 14 are lower than those from the Paris group. They are less reliable because they were calculated assuming that initiation of polymerization of vinyl ethers, *N*-vinyl carbazole and *p*-methoxystyrene with trityl or tropylium salts were quantitative and instantaneous, whereas k_i is 10^3–10^5 times smaller than k_p.

3. γ-Radiation

Although the rate constants of carbocationic propagation were previously believed to be $k_p \approx 1\ mol^{-1} \cdot L \cdot sec^{-1}$, alkene polymerizations initiated using γ-irradiation demonstrated that they are actually much higher; i.e., $k_p = 10^5 - 10^8\ mol^{-1} \cdot L \cdot sec^{-1}$. High-energy radiation generates both radicals and ionized species [225], such as radical cations and electrons. Most of these species are annihilated by geminate recombination within less than 10^{-9} sec. However, a small fraction of cations will be sufficiently separated from the electrons to diffuse against Coulombic interactions and evade geminate termination. For example, 100 eV typically generates two to three ions, whereas the initiation yield is only $G_i \approx 0.1$ molecule/100 eV. Thus, less than 5% of the ions generated survive long enough to initiate polymerization. Although the exact mechanism of initiation is not known, mass spectroscopic studies in the gas phase [226], suggest that tertiary cations and allylic radicals are formed by hydrogen atom transfer as shown for isobutene in Eq. (63).

$$CH_2{=}C(CH_3)_2^{\bullet +} + CH_2{=}C(CH_3)_2 \longrightarrow CH_3\text{-}C^+(CH_3)_2 + CH_2{=}C(CH_3)\text{-}CH_2^{\bullet} \tag{63}$$

Alternatively, pulse radiolysis suggests that the radical cations react with monomer to form dimeric radical cations as in Eq. (64).

$$CH_2{=}CH(Ph)^{\bullet +} + CH_2{=}CH(Ph) \longrightarrow {}^{\bullet}CH(Ph)\text{-}CH_2\text{-}CH_2\text{-}{}^{+}CH(Ph) \tag{64}$$

Dimerization is apparently very fast. It is complete within 10 ps [227]. The carbocationic center of the dimeric radical cation reacts with styrene within 20 nsec, which corresponds to a rate constant of carbocationic propagation ($k_p = 4 \cdot 10^6\ mol^{-1} \cdot L \cdot sec^{-1}$) similar to those estimated from conductivity measurements and using carbenium ion scavengers (cf., Table 15).

Carbenium ions apparently are formed ≈100 times less efficiently than radicals, with both radical and carbocationic polymerization operating simultaneously when kinetically possible. Although isobutene does not polymerize radically, styrene readily polymerizes by a radical mechanism. Thus, radical polymerization of styrene dominates in wet systems where the cations are trapped by water to form inactive oxonium ions. Polymerization is ≈100 times faster in super-dry systems, demonstrating that cationic polymerization must dominate, and that the rate constant of cationic polymerization is approximately 10^4 times higher (≈100 × ≈100) than that of the radical polymerization ($k_{p(radical)} = 80\ mol^{-1} \cdot L \cdot sec^{-1}$ at 25° C).

Table 15 Rate Constants of Cationic Propagation Estimated from γ-Radiation Experiments

Monomer	T (° C)	$k_p \cdot 10^{-6}$ ($mol^{-1} \cdot L \cdot sec^{-1}$)	E_p ($kJ \cdot mol^{-1}$)	Comment	Ref.
Styrene	15	400		a	[229]
Styrene	15	3.5		b	[230]
Styrene	25	2.4	9	b	[231]
α-MeSt	0	4.3	~0	b	[230]
p-MeOSt	0	3.0	~0	c	[232]
Isobutene	0	150	9	d	[233]
Isoprene	0	0.003		d	[234]
Cyclopentadiene	−78	600	<9	a	[235]
iBVE	30	0.3	28	b	[230]
iBVE	25	0.12	40	b	[236]
iBVE	0	0.0038	32	c	[237]
iPVE	30	1.3	22	c	[238]
EVE	30	0.006	54	c	[239]

[a] Scavenger method.
[b] Conductivity.
[c] G_i estimated from tables [228].
[d] $G_i = 0.1$ (assumed).

The rate of polymerization in γ-irradiated systems is defined by Eq. (65) in terms of the rate of initiation (R_i), the rate constants of propagation (k_p) and termination (k_{tx}), and the concentrations of monomer ([M]) and terminator ([X]).

$$R_p = \frac{k_p \cdot [M] \cdot R_i}{k_{tx}[X] + (R_i \cdot k_t)^{1/2}} \tag{65}$$

When the concentration of the terminating agent is low, the rate of polymerization is proportional to the square root of the radiation dose ($R_i^{1/2}$), whereas R_p increases linearly with R_i at high concentrations of terminating agent. Thus, the rate of polymerization reduces to Eq. (66) under superdry conditions.

$$R_p = k_p \cdot [M] \cdot (R_i/k_t)^{1/2} \tag{66}$$

The rate constant of propagation is then calculated by determining the rate of monomer consumption using estimated values of the rate of initiation and the rate constant of termination. Rates of monomer consumption, defined as G(-M), vary from 10^2 to 10^7 molecules/100 eV. Rates of initiation are usually determined from conductivity data. These can be correlated with G_i values [228] of various organic molecules, which are approximately $G_i \approx 0.1$ molecule/100 eV. The rate constants of termination are assumed to be approximately $k_t \approx 10^{11}$ mol^{-1}·L·sec^{-1}. Although this may be too high, especially if the viscosity increases during the polymerization, a factor of 100 error in the rate of initiation and the rate constant of termination will result in only a 10-fold error in the rate constant of propagation. Thus, the rate constants of propagation by free ions in systems initiated by γ-irradiation are generally considered correct within 1 or 2 orders of magnitude.

The rate constants of propagation in bulk polymerizations of several alkenes initiated by γ-rays are presented in Table 15. The rate constant of propagation of isobutene is estimated to be 1000 times lower in chlorinated solvents than in bulk [134]. The rate constant of vinyl ether propagation decreases a few times by adding only 1 mol% of methylene chloride [238]. This may be due to either an error in the estimate of G_i, or to specific interactions between growing carbenium ions and solvent molecules; both explanations assume that much less reactive, but still conducting carbenium ions are formed. Nevertheless, recently determined rate constants of propagation of isopropyl and isobutyl vinyl ethers initiated with trityl salts [217] are within a factor of 2 of those calculated from γ-irradiated systems.

The energies of activation of γ-irradiation bulk polymerizations are typically $E_a < 8$ kJ·mol^{-1} for hydrocarbon monomers. This indicates that the energy barrier to electrophilic addition of reactive carbenium ions to alkenes is very low and similar to those of diffusion controlled processes. The entropies of activation range from $\Delta S^\ddagger = -60$ to -100 J·mol^{-1}·K^{-1}, but should be closer to $\Delta S^\ddagger = -125$ J·mol^{-1}·K^{-1} for a simple bimolecular reaction between growing chain end and monomer in which monomer loses three degrees of translational freedom in the transition state. The less negative values of $\Delta S^\ddagger$ may indicate that either solvent or monomer molecules desolvate from the carbenium ion during the propagation step.

The energies of activation of vinyl ether polymerizations are much larger: isobutyl and isopropyl vinyl ether $E_a = 21$ kJ·mol^{-1}; ethyl vinyl ether $E_a = 54$ kJ·mol^{-1}. This indicates that carbenium ions of vinyl ethers are less reactive, probably due to an equilibrium with dormant oxonium ions formed by an intramolecular cyclization [Eq. (67)]. The overall activation energies should also increase to more positive values if formation of the active carbenium ions is endothermic.

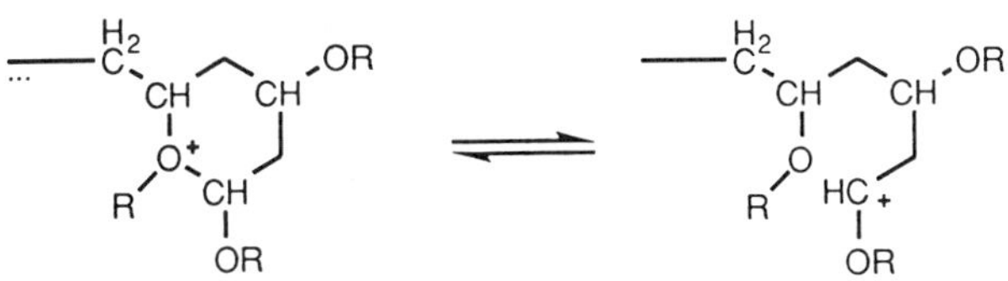

(67)

If both of these ions are in dynamic equilibrium, the overall activation parameters are composites of the enthalpy of equilibrium and the enthalpy of activation of an addition of monomer to carbenium ions.

4. *Conclusions on Carbocationic Rate Constants of Propagation*

As demonstrated in both the model and macromolecular systems described in Chapter 2 and previous sections of this chapter, respectively, electrophilic addition of carbenium ions to alkenes is a fast reaction. The carbenium ion electrophilicity is inversely related to the nucleophilicity of the parent monomers. Because the Hammett ρ values of carbenium ions and alkenes are similar but of opposite sign, the rate constants of electrophilic addition (≡ propagation) should be similar for the various types of monomers. Tables 13 through 15 demonstrate that the rate constants of propagation are indeed similar, with $k_p \approx 10^{5 \pm 1}$ $mol^{-1} \cdot L \cdot sec^{-1}$ for most systems. The slight variations are ascribed to solvent effects which reduce the reactivity of carbenium ions by specific solvation and to imprecise estimates of the yield of carbenium ions generated by initiation.

Based on model systems [240], the rate constants of propagation should be in the range of $k_p \approx 10^{7 \pm 1}$ $mol^{-1} \cdot L \cdot sec^{-1}$ for most alkenes, which is close to the values measured in bulk polymerizations initiated by γ-radiation. However, the estimates based on model reaction were obtained by comparing the rates of electrophilic addition of less reactive carbenium ions and the rates of solvolysis of the corresponding covalent alkyl chlorides. In contrast to the carbenium ions studied in the model systems, more reactive carbenium ions such as those propagating in carbocationic polymerizations may interact very strongly with solvents, thereby reducing their overall reactivity. This strong specific solvation would explain why the energies of activation are higher in solution polymerizations than they are in model reactions and in bulk γ-irradiated polymerizations. Similarly, the lower rate constants of propagation of vinyl ethers may be due to reversible intramolecular oxonium ion formation, which is most favored for vinyl ethers with less bulky methoxy and ethoxy groups. Indeed, methyl and ethyl vinyl ether have the most significantly reduced rate constants and the highest activation energies. (cf., Table 14)

C. Equilibria Between Active Species in Propagation and Their Reactivities

The classic diagram of progressively ionized propagating electrophilic species was presented in Eq. (45). Such species can be considered reactive intermediates rather than transition states if their lifetimes exceed that of a covalent bond vibration ($>10^{-14}$ sec). Therefore, this classic picture does not include species involving the "stretched" covalent bonds recently postulated to operate in some carbocationic systems [105]. Existence of a distinct energy minimum for this ill-defined structure is nebulous unless it is stabilized by a special interaction such as charge transfer. It should instead be considered a transition state between a covalent precursor and its ionized carbenium ion pair. There are only two energy minima in progressive ionization corresponding to sp^3-hybridized covalent species and to sp^2-hybridized carbocationic ion pairs. The next minimum corresponds to free carbenium ion. The proportions, reactivities, and interchangability of these three species must therefore be established to understand the polymerization mechanism well enough to control the structure and properties of the resulting polymers.

1. Ions/Ion Pairs

The relative reactivity of ions and ion pairs are very different in anionic and cationic polymerizations. In anionic systems, the reactivity of ions (k_p^-) is similar to that of solvent-separated ion pairs ($k_p^{\pm s}$), but much higher than that of contact ion pairs ($k_p^{\pm c}$) [12]. For example, the rate constants of propagation of styrene at ambient temperature are $k_p^- \approx 10^5$ $mol^{-1}\cdot L\cdot sec^{-1}$, $k_p^{\pm s} \approx 10^4$ $mol^{-1}\cdot L\cdot sec^{-1}$ and $k_p^{\pm c} \approx 10^1$ $mol^{-1}\cdot L\cdot sec^{-1}$. Although contact ion pairs are still capable of chain growth, ionic aggregates are often dormant as indicated by fractional kinetic orders, especially in nonpolar media [12]. In contrast, there are no data on ionic aggregation and triple ion formation in carbocationic polymerizations and only a few reports of the reactivities of ions and ion pairs in such systems are reliable. Free carbenium ions are apparently only 5 to 20 times more reactive than ion pairs (cf., Section IV.B.1). For example, ions are 20 times more reactive than ion pairs in perchloric acid-initiated polymerizations of styrene at $-80°$ C, but only eight times more reactive at $-60°$ C. These data extrapolate to the two species having similar reactivities at ambient temperature. The interionic distance in contact ion pairs in cationic systems ($\approx$5 Å) corresponds to that of solvent-separated ion pairs in anionic systems. In addition, solvents tend to solvate cations rather than anions. Thus, there is little difference in the reactivities of free ions and ion pairs in cationic systems, whereas they are quite different in anionic systems. Ions and ion pairs also have similar reactivities in cationic ring-opening polymerizations as discussed in Chapter 6.

There is also only limited information available on the dissociation of ion pairs to free carbenium ions, especially from macromolecular systems (cf., Sections II.D and IV.B). The dissociation constants in cationic polymerizations of styrene are approximately $K_D \approx 10^{-6\pm1}$ mol/L in CH_2Cl_2, depending on the temperature and structure of the counteranions

[17,213]. These K_D values are much smaller than those of trityl and benzhydryl salts (10^{-5}–10^{-4} mol/L) (cf., Tables 7 and 16). The K_D values of trityl salts correspond to interionic distances of approximately 5 Å. The smaller ($K_D \approx 10^{-7}$–10^{-6} mol/L) dissociation constants calculated from the common ion effect in styrene polymerizations with triflate and perchlorate anions correspond to interionic distances ≈4 Å. However, perchlorate and triflate anions may not be spherical, and their dipole moments should therefore also be considered in calculating their interionic distances and dissociation constants. As discussed in Section II.D, specific interactions of counteranions with the α-H atoms of the secondary carbenium ions may result in lower dissociation constants [39].

If $K_D \approx 10^{-6}$ mol/L, free ions will dominate when the concentration of carbenium ions is less than 10^{-6} mol/L. However, dissociation is strongly suppressed by adding salts with common ions. For example, addition of 10^{-3} mol/L of salt with common anion decreases the concentration of free ions from approximately 10^{-6}–10^{-9} mol/L. Ion pair dissociation may also be suppressed by the counterions of adventitious oxonium ions formed by alkylation or protonation of impurities such as moisture. Even 10^{-5} mol/L of such oxonium ions reduces the concentration of free ions considerably.

The kinetics of association is fast. For example, the rate constant of association of free carbenium ions and anions is $k_{ass} \approx 10^8$–10^{10} $mol^{-1} \cdot L \cdot sec^{-1}$, with the rate limited by diffusion. This rate constant varies with temperature and viscosity according to Eq. (68).

$$k_{ass} = 8RT/(3000\eta) \tag{68}$$

Although ions and ion pairs have similar reactivities in carbocationic systems, their lifetimes during which monomer units can be added are different. The average lifetime of free carbenium ions is the reciprocal of the frequency of association, which is equal to the product of the rate constant of association and the concentration of counteranions [Eq. (69)].

Table 16 Dissociation Constants of Growing Ion Pairs in Cationic Polymerizations of Styrenes

Monomer	Anion	T (° C)	Solvent	K_D range ($mol \cdot L^{-1}$)	Ref.
Styrene	ClO_4^-	−80	CH_2Cl_2	$4 \cdot 10^{-7}$–$2 \cdot 10^{-6}$	[213]
Styrene	$CF_3SO_3^-$	≈0	$C_2H_4Cl_2$	$2 \cdot 10^{-7}$–$2 \cdot 10^{-6}$	[17]
p-MeOSt	$SbCl_6^-$	+10	CH_2Cl_2	$\approx 3.10^{-5}$	[46]

$$\tau = 1/(k_{ass}\cdot[A^-]) \tag{69}$$

Thus, carbenium ions associate once every 0.1 to 0.001 sec when the concentration of counteranions equals $[A^-] \approx 10^{-8}$–10^{-5} mol/L. Since carbocationic propagation is very fast ($k_p \approx 10^5$ $mol^{-1}\cdot L\cdot sec^{-1}$), long polymer chains with 100 to 10,000 repeating units can be generated during each dissociation event [Eq. (70)].

$$DP = k_p\cdot[M]\cdot\tau \tag{70}$$

However, the chain length produced by free ions is often limited by transfer rather than by association, which results in lower molecular weight polymers. The lifetime of ion pairs may, however, be much shorter as discussed in the next section.

2. *Ions and Covalent Species*

In addition to forming ion pairs, counteranions can react with propagating carbenium ions by either recombination to form a covalent bond, or by abstracting a β-proton to generate an unsaturated end group. Nucleophilic counteranions and additives tend to react by recombination, whereas basic anions favor elimination. Irreversible recombination is not important if it does not occur during the time necessary to complete the polymerization. Unsuccessful polymerizations may be improved by using either longer reaction times or higher initiator concentrations to shift the equilibrium to ions. For example, although styrene conversions are very low in polymerizations initiated by $[I]_0 < 10^{-4}$ mol/L trityl hexachloroantimonate, it goes to complete conversion using $[I]_0 > 10^{-1}$ mol/L [145]. Thus, decreasing the concentration of the initiator 1000 times could reduce the rate 1,000,000 times due to 2:1 equilibria between alkyl halide/$SbCl_5$ and ion pairs [Eq. (71)].

$$\begin{gathered} R - Cl + SbCl_5 \rightleftarrows R^+,SbCl_6^-;\ K \\ [R^+,SbCl_6^-] = K/[R-Cl][SbCl_5] \approx K/[I]_0^2 \end{gathered} \tag{71}$$

In addition, styrene reacts irreversibly with $SbCl_5$ to form 1,2-dichloro-adducts and inactive $SbCl_3$. In fact, all of the Lewis acid is consumed by addition at low concentrations of initiator, resulting in incomplete polymerization [144].

The equilibrium constants of ionization are relatively low in most polymerization systems, resulting in very small proportions of ionic species. The proportion of ionic species may be estimated from the overall polymerization rates by assuming that covalent species are inactive. In polymerization systems with a half-lifetime of monomer in the range of $\tau \approx 20$ min to 3 hr, the concentration of propagating ionic species should be approximately 10^{-8}–10^{-9} mol/L as estimated from Eq. (72) using $k_p{}^i \approx 10^5$ $\text{mol}^{-1}\cdot\text{L}\cdot\text{sec}^{-1}$.

$$\{[C^+] + [C^{\pm}]\} = (\ln 2)/\tau \cdot k_p{}^i \tag{72}$$

The ionic species should be predominantly dissociated at such low concentrations. However, as discussed in the previous section, minute amounts of impurities with common counteranions enhance the proportion of ion pairs.

There are some measurements of the rates of polymerization in systems with reversible formation of covalent species. The equilibrium constants of ionization can be calculated from these kinetic data according to the procedure outlined subsequently in Section IV.D.2.a. The ionization constant depends on the strength of the Lewis acid. For example, the propagating species are almost completely ionized in polymerizations of vinyl ethers with $SbCl_6^-$, BCl_4^-, and $SnCl_5^-$ counteranions, but only partially ionized when the counteranions are I_3^- or Zn_3^-.

The ionization constant also depends on the stabilizing effects of the carbenium ion substituents. Ionization is approximately 100 times less in α-methylstyrene polymerizations than in those of vinyl ethers under comparable conditions. Similarly, the ionization constants in styrene and isobutene polymerizations are $\approx 10^2$ times smaller than in α-methylstyrene polymerizations. Thus, $K_I \approx 10^{-5}$ $\text{mol}^{-1}\cdot\text{L}$ in styrene polymerizations with BCl_4^- counteranions at 20° C in CH_2Cl_2, and $K_I \approx 10^{-2}$ $\text{mol}^{-1}\cdot\text{L}$ in α-methylstyrene polymerizations with BCl_4^- counteranions at −70° C in CH_2Cl_2; $K_I \approx 1$ $\text{mol}^{-1}\cdot\text{L}$ in polymerization of vinyl ethers under similar conditions [241]. In contrast, vinyl ether polymerizations using zinc halides proceed much slower due to relatively low ionization ($K_I \approx 10^{-2}$ $\text{mol}^{-1}\cdot\text{L}$ at −15° C in CH_2Cl_2). Similarly, $K_I \approx 10^{-7}$ $\text{mol}^{-1}\cdot\text{L}$ in styrene polymerizations with $HgCl_3^-$ counteranions at 20° C in CH_2Cl_2 [242]. A large excess of Lewis acid must be used in such systems to complete the polymerization in reasonable time.

Ionization is exothermic and favored at lower temperatures in systems such as trityl, alkoxycarbenium, and benzhydryl derivatives which generate stabilized carbenium ions due to their electron-donating substituents. The exothermicity of ionization of benzhydryl species with BCl_3 in CH_2Cl_2 decreases from $\Delta H = -62$ kJ/mol for di(p-anisyl) to −22 kJ/mol for less stabilized p-tolyl and phenyl derivatives, and to approximately −8 kJ/mol for unsubstituted benzhydryl chloride [193]. These values, based on the solvolysis rates, extrapolate to $\Delta H \approx -12$ kJ/mol for cumyl chloride (HCl adduct of α-methylstyrene) and $\Delta H \approx 0$ kJ/mol for 1-phenylethyl chloride (styrene adduct) and t-butyl chloride (isobutene adduct) [240]. The reported entropies of ionization do not vary significantly and are approximately $\Delta S \approx -125$ J·mol^{-1}·K^{-1} for most systems. These enthalpy and entropy values can be combined to predict the equilibrium constants of ionization of covalent chlorides in polymerizations activated by BCl_3: α-methylstyrene $K_I \approx 10^{-2}$ mol^{-1}·L; styrene and isobutene $K_I \approx 10^{-5}$ mol^{-1}·L. These values are consistent with the experimental data discussed above.

The dynamics and therefore the rate constants of interconversion between active and dormant species determines a polymer's polydispersity. The rate constants of deactivation can therefore be estimated directly from polydispersity. The rate of ionization can also be determined by following the rate of racemization of optically pure compounds. Dynamic NMR can also be used to determine the rate of ionization and has the advantage of not requiring optically active compounds. Rate constants of ionization can be calculated by measuring the rate of conversion of magnetically inequivalent protons in sp^3-hybridized dormant species into equivalent protons via a planar carbenium ion intermediate; the concentration of Lewis acid is adjusted such that the covalent precursor's half-lifetime is approximately $\tau \approx 10^{-2}$ sec. This has been used to demonstrated that 1-isobutoxyethyl chloride ionization using BCl_3 in CH_2Cl_2 at $-70°$ C is rather fast, with $k = 10^4$ mol^{-1}·L·sec^{-1} [11]. Ionization rate constants can also be determined from the rate of ligand exchange using spectroscopic techniques such as dynamic NMR or conventional NMR if ionization is slower ($\tau > 10^2$ sec). For example, ionization of 1-phenylethyl chloride using BBr_3 in CH_2Cl_2 at $-70°$ C ($k = 10^{-2}$ mol^{-1}·L·sec^{-1}) [59] is much slower than that of the vinyl ether derivative [241].

In weakly ionized systems, the rate constants of formation of covalent species are obviously much higher than the rate constants of ionization. The kinetics of these fast reactions are more difficult to determine, but can be studied by generating carbenium ions by flash photolysis or pulse radiolysis and following their decay using scavengers such as halide anions [41,240]. The rate constants of reactions with these scavengers to form covalent species are close to diffusion controlled limits for biomolecular reactions. In contrast, the rates of reaction of less nucleophilic anions such as BCl_4^- or $SbCl_6^-$ with secondary carbenium ions stabilized with α-alkoxy substituents are much slower ($k = 10^4$ mol^{-1}·L·sec^{-1} in CH_2Cl_2 at $-70°$ C) [241].

3. *Equilibria Between Carbenium Ions and Onium Ions*

Carbenium ions react with noncharged nucleophiles to form onium ions rather than covalent species. The thermodynamics and dynamics of exchange between onium and carbenium ions in several relevant systems were summarized in Table 9 in Section II.F.2. This was studied mostly by dynamic NMR and by following decaying carbenium ions generated by flash photolysis and pulse radiolysis in the presence of nucleophiles. The rate of the reverse reaction is correlated with the resulting polymer's polydispersity as described in Section IV.D.3.

The equilibrium constants with nucleophiles such as tertiary amines are so large, that carbenium ions practically do not exist. Thus, tertiary amines and pyridine apparently react with carbenium ions irreversibly and therefore terminate carbocationic polymerizations. Somewhat weaker nucleophiles such as 2,6-dimethylpyridine (lutidine), sulfides, and tris(*p*-chlorophenyl)phosphine are good deactivators in vinyl ether polymerizations because they react reversibly with monomer, thus maintaining a low concentration of carbenium ions without causing elimination. However, the equilibrium constants in styrene and isobutene polymerizations with amines, sulfides, and phosphines are too large to generate a sufficient stationary concentration of carbenium ions to complete polymerization in a reasonable amount of time.

D. Mechanism of Propagation

Carbocationic propagation involves repetitive electrophilic addition of carbenium ions to alkenes. This section discusses the mechanism of propagation by carbenium ions and also potential participation of other species in the chain growth.

1. Mechanism and Stereochemistry of Carbocationic Propagation

Electrophilic addition involves a late transition state in which approximately half of the positive charge is transferred from the attacking carbenium ion to the newly developing carbenium ion center [193]. The α-substituents in the monomer are much more important than the β-substituents at stabilizing the transition state. This results in very high regioselectivity following Markovnikoff's rule and therefore regular head-to-tail polymers. Very few head-to-head errors have been observed in polymers generated by carbocationic polymerizations.

Because the transition state of electrophilic addition is rather open or only very weakly bridged, the stereoselectivity is poor, resulting in primarily atactic polymers. Polystyrene and most poly(vinyl ether)s prepared cationically at ambient temperature are atactic with similar proportions of meso and racemic dyads [243,244]. However, meso addition is slightly preferred with vinyl ethers, and varies from 60 to 70% for most monomers, including isobutyl, neopentyl, *n*-butyl, and ethoxyethyl vinyl ethers [245]. It is higher with benzyl vinyl ether (89%). This tendency to form isotactic polymers increases at lower temperatures; the difference in the activation enthalpies of meso and racemic addition is $\Delta H\ddagger = -7.1$ kJ·mol^{-1} for benzyl, but only -4 kJ·mol^{-1} for isobutyl vinyl ether [245].

Heterogeneous polymerization of isobutyl vinyl ether initiated by $BF_3 \cdot OEt_2$ at $-80°$ C produces a crystalline polymer [246] Poly(isobutyl vinyl ether) (PIBVE) with the highest isotacticity is formed on solid supported catalysts based on sulfates of aluminum, chromium, nickel, and sulfuric acid [247–251]. Stereoregular PIBVE is also reportedly produced using metal oxides such as Cr_2O_3 in nonpolar solvents [252,253]. Modified Ziegler-Natta catalysts such as $VCl_4/AlEt_3$ also lead to crystalline poly(methyl vinyl ether) (m.p. 144° C) [254,255].

Approximately 62% of racemic dyads are formed by cationic polymerization of styrene using Lewis acids as catalysts. This decreases to 56% in the presence of perchlorate salts in CH_2Cl_2, which indicates that the counteranion affects the stereochemistry of addition [256]. Nevertheless, stereocontrol is very poor. The tendency to form racemic dyads increases with more sterically hindered monomers such as α-methylstyrene [257]. The proportion of syndiotactic triads in carbocationic polymerizations of α-methylstyrene is $\approx$80% at $-60°$ C and increases further at lower temperatures. Syndiotacticity is higher in more polar solvents and with smaller counteranions. The preference for syndiotacticity of styrene derivatives compared to the preference for isotacticity of vinyl ethers suggests that the antipenultimate unit of the poly(vinyl ether) backbones may interact with the propagating center [Eq. (67)]. In contrast, the propagating carbenium ion interacts with the penultimate aromatic ring in styrene polymerizations. In fact, these rings react with propagating carbenium ions in a chain transfer reaction to generate indan end groups [Eq. (10)].

Because the microstructure of polymers prepared cationically have only been controlled to a limited extent using various counteranions and nucleophiles, this may be one of the most challenging areas of future research.

2. *Dormant Species and Pseudocationic Propagation*

The majority of propagating chain ends in most cationic polymerizations initiated by protonic acids and/or cocatalyzed by Lewis acids do not exist as carbenium ions, but are instead dormant species. The two major types of dormant species are onium ions and covalent esters or halides. The covalent species are formed by reversible reaction of carbenium ions with nucleophilic anions; onium ions are generated by reaction of carbenium ions with noncharged nucleophiles such as ethers, sulfides, and amines. Because the majority of propagating chain ends exist as dormant species, they are often the only species that can be detected spectroscopically;

the concentrations of carbenium ions are often less than 10^{-6} mol/L. Nevertheless, stopped-flow studies and exchange reactions in model and macromolecular systems indicate that carbenium ions are involved in propagation. Both copolymerization ratios and the minor influence of counteranions and additives on microstructure suggest that the majority, if not the entire polymer chain, is constructed by carbocationic growth.

a. Covalent Species. Covalent species ionize to carbenium ions either spontaneously or by activation with electrophiles such as Lewis or protonic acids. Spontaneous ionization occurs with compounds which are able to form stable carbenium ions, such as trityl and tropylium derivatives, and which contain very good leaving groups such as triflate and perchlorate. For example, trityl chloride ionizes spontaneously in CH_2Cl_2 with an equilibrium constant of ionization $K_I \approx 10^{-4}$ at 20° C; the equilibrium constants of trityl compounds para-substituted with electron-donating alkyl substituents are even higher ($K_I \approx 10^{-2}$). However, only a few currently known propagating chain ends in carbocationic polymerizations ionize spontaneously. The dormant covalent species must therefore be activated by Lewis acids.

1-Phenylethyl chloride is covalent and stable for months at room temperature in CH_2Cl_2; the optical activity of an isolated stereoisomer changes very slowly [47]. Ionization occurs at higher temperatures and in more polar solvents ($k = 10^{-5}$ sec^{-1} in nitromethane at 99° C) and is strongly accelerated by small amounts of Lewis acids, even in nonpolar media [48,50]. Rate constants of ionization (k_i) can therefore be obtained from the rates of racemization (k_α) of optically active compounds ($k_i \approx 2k_\alpha$). These rates of racemization (ionization) correlate well with the kinetics of polymerization and molecular weight distributions, indicating that propagation is carbocationic and not pseudocationic.

In most carbocationic polymerizations, inactive covalent species coexist with ions and ion pairs [Eq. (73)].

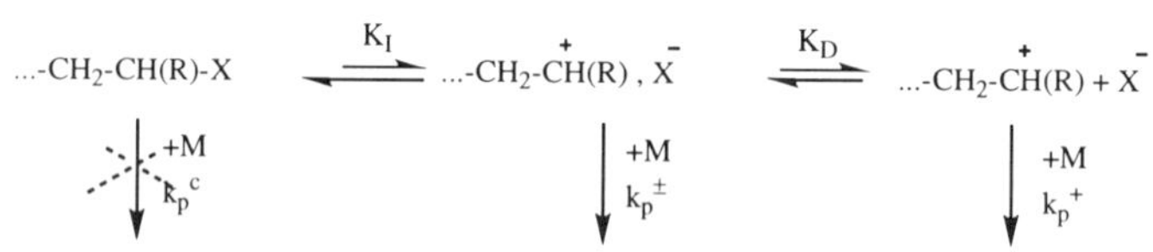

(73)

This ternary system reduces to a binary system involving only ion pairs and covalent species if free ions are suppressed. This can be accomplished in nonpolar solvents and/or in the presence of salts with common counteranions. The rate of monomer consumption is then expressed using either the apparent rate constant and the assumption that initiator converts quantitatively to active species [Eq. (74)], or using the ionic propagation rate constant and assuming that only ions pairs propagate even though covalent species prevail ($K_I \ll 1$) [Eq. (75)].

$$R_p = -d[M]/dt = k_p^{app} \cdot [M] \cdot [I]_0 \tag{74}$$

$$R_p = -d[M]/dt = k_p^{\pm}[M] \cdot [C^{\pm}] \tag{75}$$

$$K_I = [C^{\pm}]/[C^{cov}] \approx [C^{\pm}]/[I]_0 \tag{76}$$

$$K_I = k_p^{app}/k_p^{\pm} \tag{77}$$

Because the reactivities of ions and ion pairs are similar and only weakly affected by the structure of the counteranions, k_p^+ or $k_p^{\pm}$ determined by either stopped-flow studies or γ-radiated systems (cf., Section IV.13) can be used in Eq. (75). The equilibrium constant of ionization can then be estimated from the apparent rate constant of propagation and the rate constant of propagation by carbenium ions [Eq. (77)]. For example, $K_I \approx 10^{-5}$ mol^{-1}·L in styrene polymerizations initiated by R-Cl/$SnCl_4$ [148]. K_I for vinyl ether polymerization catalyzed by Lewis acids can also be estimated by using the available rate constant of ionic propagation ($k_p^{\pm} \approx 10^4$ mol^{-1}·L·sec^{-1} at 0° C) [217]. The kinetic data in Ref. 258 yields $K_I \approx 10^{-3}$ mol^{-1}·L in IBVE polymerizations initiated by HI/I_2 in toluene at 0° C and $K_I \approx 10^{-1}$ mol^{-1}·L initiated by HI/ZnI_2/acetone can be calculated from Eq. (76).

Nevertheless, some research groups challenge the concept of a dynamic equilibrium between dormant covalent species and carbenium ions, and instead insist that covalent species can react *directly* with monomer. They propose that this occurs by a "pseudocationic" mechanism involving a multicenter rearrangements [259], such as those shown in Eq. (78), for the pseudocationic polymerization of styrene initiated by perchloric acid [260].

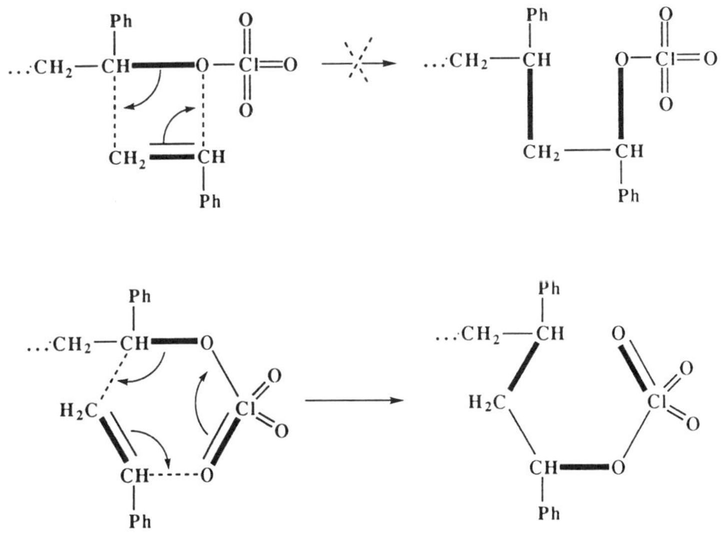

(78)

The four-membered cyclic transition state is not allowed by orbital symmetry theory and parity rules. It requires inversion of configuration at the α-carbon and trans addition to the alkene by a conrotatory process, which is sterically impossible [261,263]. The six-membered transition state is allowed by parity rules, but the relative contributions of this pathway and that by unimolecular ionization depends on their relative rate constants and therefore their free energies of activation. Since the transition state of electrophilic addition to alkenes proceeds with a very late transition state requiring an electrophile with a highly developed charge, covalent species are not sufficiently polarized to react directly with alkenes. Thus, the reaction should occur in two steps rather than by a concerted addition [264].

Nearly all experimental data obtained from model and macromolecular systems support the preliminary ionization mechanism. In addition, the microstructure of polymers obtained in both conventional cationic systems, including those initiated using γ-radiation, and those purportedly prepared in a "pseudocationic" process are very similar. If direct reaction is in fact possible, its contribution should increase with increasing reaction temperature because of the higher activation energy of such a reaction, and with increasing solvent polarity. However, higher temperatures enhance elimination, and more polar media increase ionization. Nevertheless, the concept of a pseudocationic polymerization is attractive because it could lead to polymers with different tacticity and to different reactivity ratios in copolymerizations. Chemoselectivities and regioselectivities in pseudocationic polymerizations may be also very different than those in conventional carbocationic systems.

Alleged improvements in chemoselectivity of propagation by suppressed transfer led to another proposal [105,265] based on covalent propagation involving obscure "stretched or activated, more-covalent-than-ionic" bonds [Eq. (79)].

$$\text{...-C}^{\delta+}\text{ - - -}^{\delta-}\text{XMtX}_n \qquad (79)$$

As discussed previously, such species are more tangibly understood to be intermediates on passing from covalent species to carbenium ions, rather than chemical entities, because their lifetime should be comparable to a bond vibration (Eq. 79). Stretched bond isomerism does not exist unless it is associated with a change in spin number, as in some inorganic compounds [266,267]. Since ionization is accompanied by a change in hybridization from sp^3 to sp^2, it is very unlikely that species with "stretched" covalent bonds could be considered real individual chemical species. Moreover, the chemoselectivity of reactions with alleged "stretched" covalent bonds is not significantly different than that in conventional carbocationic polymerizations. These polymerizations occur with less chain transfer and produce polymers with more narrow molecular weight distributions because the reaction conditions are different rather than because they operate by a different mechanism. In particular, initiation is fast and transfer is hidden by working only at lower molecular weights are attempted.

If carbocationic and pseudocationic mechanisms operated simultaneously but with slow exchange between the two growing centers, a bimodal molecular weight distribution would be produced. Bimodal molecular weight distributions are often observed [16,124,265]. The low molecular weight fraction grows with conversion, whereas the molecular weight of the high molecular weight fraction is usually constant or even decreases with conversion. This suggests that growth of the high molecular weight fraction is limited by transfer reactions. Formation of the high molecular weight fraction is suppressed in nonpolar solvents or by adding salts with common counteranions. This clearly indicates that it is formed by carbocationic propagation of free ions. Since common ions *do not* affect the equilibrium between covalent species and ion pairs, the low molecular weight fraction may be formed by either covalent species or ion pairs. However, dynamic NMR studies of model systems, the use of optically active compounds, as well as ligand exchange clearly indicate that carbenium ions are generated under these conditions. Recent simulations of molecular weight distributions show that bimodal molecular weight distributions are produced when ions and ion pairs have identical reactivities but different lifetimes [268].

b. Onium Ions In addition to covalent species, dormant species can be generated in the form of onium ions. Onium ion formation is especially useful when the equilibrium between covalent species and carbenium ions favors high concentrations of carbenium ions, which result in poorly controlled polymerizations. The concentration of carbenium ions is high when they are stabilized and/or when the counteranions are relatively nonnucleophilic. For example, vinyl ether polymerizations initiated by triflic acid, or activated by either $SnCl_4$ or aluminum halides, are complete within very short times in spite of incomplete initiation, sometimes with strong overheating. In contrast, the polymerization can be controlled and the polymerization rates decreased by adding neutral nucleophiles such as ethers, esters, amines, phosphines, and sulfides.

The added nucleophiles also interact with other electrophilic species, such as Lewis and protonic acids. In contrast to protonic acids which are consumed in the initiation step, Lewis acids act as catalysts and are not consumed during the polymerization. Because they are continuously regenerated, they may form complexes with the added nucleophiles throughout the polymerization. Polymerization of monomers such as styrene and isobutene which generate less stabilized carbenium ions can be controlled with less than an equimolar amount of strong nucleophile versus the Lewis acid. In these systems, the complex is less acidic than the original Lewis acid and also less nucleophilic than the original nucleophile. Because there is no polymerization in the presence of excess nucleophile, the modified Lewis acid is apparently not electrophilic enough to activate covalent bonds; activation proceeds with the remaining original Lewis acid. On the other hand, noncomplexed nucleophile forms onium ions which are too stable to regenerate carbenium ions for propagation. Because the nucleophilicity of the complexed nucleophile is reduced, the equilibrium between onium and carbenium ions is more dynamic.

Excess nucleophile is often needed in polymerization of more nucleophilic monomers. For example, esters, ethers, and amines are used in large excess over aluminum halides and alkylaluminum halides to control polymerization of vinyl ethers [269]. The original Lewis acid is no longer available and covalent species are activated by the Lewis acid/nucleophile complex. Carbenium ions are additionally deactivated by excess nucleophile.

When protonic acids are used as the initiators, nucleophiles do not complex oxyanions and therefore only form onium ions. The kinetic scheme of triflic acid initiated polymerizations of isobutyl vinyl ether in the presence of sulfides is presented in Eq. (80) [58,133]. The rate of polymerization described by Eq. (81) takes into account propagation by both carbenium ($k_p^{\pm}$) and sulfonium ions (k_p^{s}).

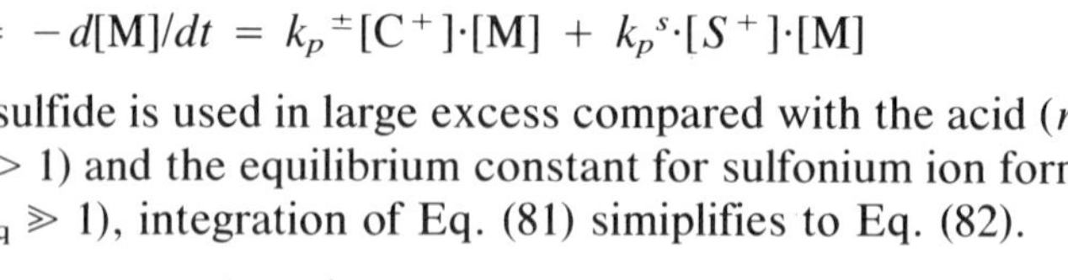

(80)

$$R_p = -d[\mathrm{M}]/dt = k_p^{\pm}[\mathrm{C}^+]\cdot[\mathrm{M}] + k_p^{s}\cdot[S^+]\cdot[\mathrm{M}] \tag{81}$$

Because sulfide is used in large excess compared with the acid ($r = [S]_0/[\mathrm{HOTf}]_0 > 1$) and the equilibrium constant for sulfonium ion formation is large ($K_{eq} \gg 1$), integration of Eq. (81) simiplifies to Eq. (82).

$$-\ln(1 - p) = k_p^{\pm}\cdot t/K_{eq}(r - 1) + k_p^{s}\cdot[\mathrm{HOTf}]_0\cdot t \tag{82}$$

A plot of $-\ln(1 - p)$ versus $1/(r - 1)$ passes through the origin due to an immeasurably small intercept [58]. Thus, no growth occurs via sulfonium ions; sulfonium ions must first regenerate carbenium ions intramolecularly, which only then can react with monomer. Although oxonium ions are considered much more reactive than sulfonium ions, they do not react directly with even the most reactive alkene, *N*-vinyl carbazole [270]. Therefore, direct reaction of less reactive onium ions with less reactive alkenes is very unlikely. (Model reactions involving onium ions are summarized in Tables 9 and 10 in Section II).

Deactivation of growing carbenium ions by reaction with sulfides is evidently very fast. Sulfonium ion formation is exothermic ($\Delta H \approx -40$ kJ/mol) and exoentropic ($\Delta S \approx -74$ J/mol·K) [271]. High equilibrium constants ($K_{eq} \approx 10^4$ mol^{-1}·L) for sulfonium ion formation were calculated from the apparent rate constants of propagation and the rate constants of carbocationic growth. Dynamic NMR experiments of model systems with tetrahydrothiophene indicate that the bimolecular deactivation rate constant is $k_{deact} \approx 10^6$ mol^{-1}·L·sec^{-1} at 0° C ($\Delta H^{\#} = 20$ kJ/mol, $\Delta S^{\#} = -37$ J·mol·K), and that activation is faster than bimolecular exchange ($k_{act} \gg k_e$) [67].

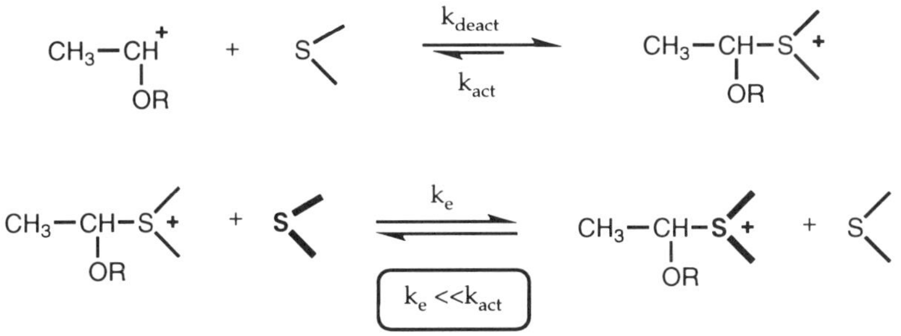

(83)

The equilibrium constants with phosphines are even larger than those with sulfides, with acceptable polymerization rates obtained only when the triaryl phosphine is substituted with three electron-withdrawing groups. The equilibrium constants with tris(*p*-chlorophenyl)phosphine are as high as $K \approx 10^{10}$ mol^{-1}·L [66]. They are even higher with triphenyl phosphine, which can be used as an efficient terminating agent [208].

3. *Effect of Exchange on Polydispersity*

The polydispersity of a given polymer is determined by the relative rates of the various elementary reactions which take place during chain growth. Ideally, transfer and termination are absent, initiation and mixing are fast, and only one type of active species propagates in a homogeneous medium. In this case, the polydispersity is defined by a Poisson distribution [Eq. (84)].

If initiation is slow in an otherwise ideal system, the molecular weight distribution broadens up to a limiting value of $DP_w/DP_n = 1.3$ [272]. More dramatic broadening is caused by transfer and termination. Polydispersity depends on whether transfer and/or termination involves monomer, counterion, polymer, or a transfer agent, and is affected by other variables such as the extent of consumption of the transfer/terminating agent; this is reviewed extensively in Ref. [273].

Probably the most underestimated cause of broadening of molecular weight distributions is slow exchange between active species which either propagate with different rate constants or which have different lifetimes [268]. For example, heterogenous Ziegler Natta polymerizations involve multiple active sites and yield polymers with polydispersities as high as $DP_w/DP_n = 10$. The polydispersity of systems with slow exchange is determined primarily by the chain length and the ratio of the rates of propagation and deactivation from more reactive species to less reactive or dormant species. Polydispersity decreases with increasing chain length and increasing relative rates of deactivation.

The simplest case involves two types of chain ends, with only one species reactive enough to propagate [Eq. (85)]. Propagation occurs with a rate constant k_p, and deactivation to the dormant species occurs with a unimolecular rate constant k_{deact}. If the majority of chain ends are dormant such that their concentration is essentially equal to $[I]_0$, the polydispersity of relatively long chains is described by Eq. (86) at complete conversion [67].

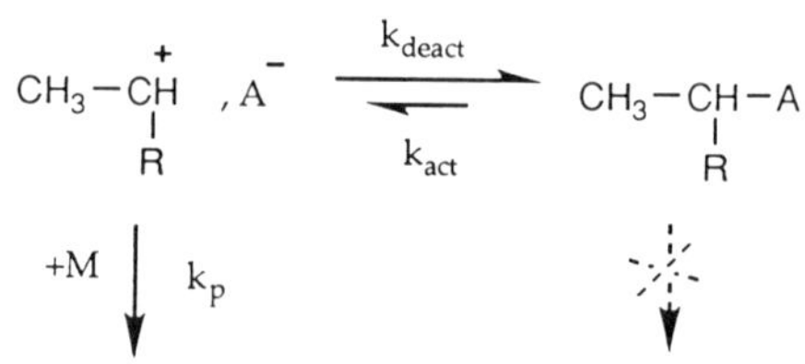

(85)

$$DP_w/DP_n = 1 + [I]_0 \cdot k_p/k_{deact} \tag{86}$$

This equation can be expressed as a function of the length of the total polymer chain (DP_n) and the length of that built during one activation period (λ; initial conditions) [Eq. (87)]

$$DP_w/DP_n = 1 + \lambda/DP_n \tag{87}$$

Polydispersity decreases with decreasing λ/DP_n because more exchanges occur during the growth of each chain. Thus, the probability of growth becomes the same for all chains provided that initiation is completed. This *decrease* in polydispersity with increasing chain length is a distinguishing feature of systems involving relatively slow exchange and is in contrast to conventional systems. That is, the polydispersity of conventional systems dominated by transfer and termination *increases* with increasing chain length.

Although very broad molecular weight distributions will result from slow exchange between one dormant and one reactive propagating species, the distribution will always be monomodal. However, the molecular weight distributions of many carbocationic polymerizations are bimodal. Bimodal molecular distributions are produced in systems with two propagating species with either different reactivities, or with identical reactivity but different lifetimes in their active form [268]. Unfortunately, there is not enough experimental detail on the evolution of M_n and polydispersity as a function of conversion to interpret and explain all of the literature data reported.

In carbocationic polymerizations with bimodal molecular weight distributions, the high molecular weight fraction is usually produced by free carbenium ions whose growth is limited by transfer. If transfer to monomer occurs, the number-average degree of polymerization stays constant because the rates of both transfer and propagation vary directly with monomer concentration. If transfer occurs to counterion, DP_n should decrease with conversion. Although free ions are present at very low concentrations as discussed in Section IV.C.1, they may live long enough to build long chains of up to 1000 repeat units. The low molecular weight fraction is produced by ion pairs whose reactivity may be equal to that of the free ions, but whose lifetime is shorter. For example, ion pairs that convert to covalent species with a unimolecular rate constant $k_{deact} = 10^5$ sec^{-1} will incorporate only one monomer during an average activation period if $k_p = 10^5$ mol^{-1} $L \cdot sec^{-1}$ and [M] = 1 mol/L. This lower molecular weight fraction grows progressively with conversion. The production of a high molecular weight fraction by free ions and a low molecular fraction by ion pairs was demonstrated in perchloric acid-initiated polymerizations of styrene by suppressing formation of the high molecular weight fraction using salts with common counteranions, and by suppressing the low molecular weight fraction using salts with nonnucleophilic SbF_6^- anions which do not convert to covalent species [131].

4. *Methods for Accelerating Exchange*

The previous section demonstrated that the dynamics of exchange between different types of growing species must be fast in order to prepare polymers with narrow molecular weight distributions. As shown by Eq. (86), polydispersity depends on the ratio of the rate constants of ionic propagation and deactivation of the growing species. Thus, only one monomer should react with a given carbenium ion before it converts back to the dormant state to obtain very narrow molecular weight distributions. Although this process formally resembles monomer insertion into a covalent bond, a carbenium ion is actually generated before monomer can react.

Three approaches have been used to decrease the molecular weight distributions. One approach suppresses dissociation of ion pairs to free ions by adding salts with common counteranions; however, this may cause a special salt effect [274]. Addition of a common ion salt shifts the equilibrium between ions and ion pairs toward the latter by mass law. (In spite of speculation to the contrary [275], the common ions can not influence the equilibrium between covalent species and ion pairs.) The kinetics of association is also affected because ion pair formation is a bimolecular reaction whose rate increases with increasing anion concentration. This decreases the lifetime of free ions. In such systems, k_{deact} in Eq. (68) should be replaced with the product of k_{deact} and deactivator $[D] = [A^-]$, in which the deactivator is a counteranion ($DP_w/DP_n = 1 + [I]_0 \cdot k_p/([D] \cdot k_{deact})$).

The common ion effect does not influence the kinetics of collapse of the ion pairs to dormant covalent species since it is a unimolecular reaction. In this case, however, deactivation of the ion pair can be increased by using less stable and more nucleophilic counteranions. The nucleophilicity of both pure halides and complex anions with halide ligands increases in the order F<Cl<Br<I. Therefore, polymerizations of α-methylstyrene with $SnBr_4$ behave better than those with $SnCl_4$. Other examples of polymerizations which are well behaved because the equilibrium is favorable due to nucleophilic counteranions include HI/I_2 initiated polymerizations of vinyl ethers and polymerizations of isobutene and styrene using acetate-based initiators in the presence of BCl_3.

The third approach for increasing the rate of deactivation is to add nucleophiles that form onium ions. As discussed previously, these nucleophiles can also complex with Lewis acids, thereby decreasing the stability of the complex counteranions and accelerating the dynamics of exchange. In contrast to common ion salts, nucleophiles have the advantage that they can convert both dissociated ions and ion pairs to dormant onium ions. In addition, nucleophiles are easier to handle than hygroscopic salts and can be used in greater amounts to shift the equilibria toward dormant species. The choice of deactivator depends on the nucleophilicity of the monomer and reactivity of the propagating carbenium ion. For example, dialkyl sulfides work very efficiently in vinyl ether polymerizations, but terminate styrene polymerizations. To avoid transfer by β-proton elimination, the deactivators also should not be basic.

5. *Solvent and Temperature Effects*

The elementary reactions of carbocationic polymerizations can be separated into three types. Deactivation of carbenium ions with anions and transfer to counteranion are ion-ion reactions, propagation and transfer to monomer are ion-dipole reactions, and ionization is a dipole-dipole reaction [274]. Ion-ion and dipole-dipole reactions with polar transition states experience the strongest solvent effects. Carbocationic propagation is an ion-dipole reaction in which a growing carbenium ion adds electrophilically to an alkene; it should be weakly accelerated in less polar solvents because the charge is more dispersed in the transition state than in the ground state [276]. However, a model addition reaction of bis(*p*-methoxyphenyl)carbenium ions to 2-methyl-1-pentene is two times faster in nitroethane ($\epsilon = 28$) than in methylene chloride ($\epsilon = 9$) at $-30°$ C [193]. However, this is a minor effect which corresponds to only $\Delta\Delta G^{\ddagger} \approx 2$ kJ·mol^{-1}; it may also be influenced by specific solvation, polarizability, etc. [276,277].

However, the apparent rate constants of propagation (calculated assuming that the concentration of growing species equals the initial concentration of the alkyl halide/ester initiator) are often 1000 times larger in CH_2Cl_2 than in nonpolar hydrocarbon solvents. This is not because the carbenium ions unexpectedly add faster to alkenes in more polar solvents, but because the ionization equilibrium has shifted to higher concentrations of carbenium ions. Covalent species with polarized C—*X* bonds ionize faster in more polar solvents because solvation stabilizes the transition state with its partial charge more than the ground state. Because the resulting ionic species are more solvated in more polar solvents, the rate of the reverse reaction (collapse of the ion pairs) decreases significantly as solvent polarity increases. Therefore, solvents with larger dielectric constants enhance the dissociation of ion pairs and increase the concentration of ions. Similar but more significant trends are observed in equilibria between ions and covalent species. Therefore, although propagation by carbocationic species is slightly slower in more polar solvents, the higher concentration of carbenium ions due to increased ionization of the covalent species and retarded ion pair recombination results in larger apparent rate constants of propagation and faster polymerization rates in polar solvents. The dielectric constant also affects the equilibria between carbenium ions and onium ions because the charge is more dispersed in onium ions. The concentration of onium ions should therefore be higher in less polar solvents.

Additional "solvent effects" are caused by decreasing monomer concentration with conversion. For example, the polymerization rate decreases with increasing initial concentration of nonpolar monomers such as styrene. However, there is no variation in the apparent rate constant of propagation with $[M]_0$ in systems in which the initial dielectric constant of the medium is maintained by replacing styrene with toluene or benzene [71].

In addition to the purely physical effect of solvent as a continuum with an average dielectric constant, solvent may be involved in specific interactions to generate onium ions such as chloronium, nitronium, oxonium, and sulfonium ions. Nevertheless, chlorinated solvents are believed to be inert and therefore ideal solvents for cationic polymerizations. This is true for the most stable carbenium ions such as tropylium, trityl, and benzhydryl ions. However, less stabilized carbenium ions such as cumyl, 1-phenylethyl, or *t*-butyl ions tend to interact with any nucleophilic species with a free electron pair and π- or σ-electrons. Thus, in the absence of *n*-donors such as amines or ethers, some carbenium ions will interact with even chloroalkanes to form chloronium cations. For example, halonium [5] and nitronium [278] cations of primary carbenium species have been isolated. Presumably secondary and tertiary carbenium ions can also form chloronium and nitronium ions. Even if chlorinated solvents do not react with carbenium ions to form true onium ions, their specific solvation may be strong enough to reduce the carbenium ions' reactivity. In this case, the addition of a small amount of a chlorinated solvent may cause a drastic decrease in the apparent rate constant of propagation [238]. Once all carbenium ions are converted to less active cations, further additions should have little effect, with the primary effect being due to the change of the dielectric constant and/or polarizability of the reaction medium. Surprisingly, addition of methylene chloride or nitromethane decreases the apparent rate constant of propagation of vinyl ethers much more than addition of diethyl ether [279]. Nevertheless, the rate constants of propagation of trityl salt-initiated polymerizations of isopropylvinyl and isobutylvinyl ether in CH_2Cl_2 [217] are similar to those calculated from polymerizations initiated by γ-radiation (IV.B.1).

More unusual solvent effects have been used to account for anomalous behavior such as the trimodal molecular weight distribution in trityl salt-initiated polymerizations of *p*-methoxystyrene [46]. The authors suggested that the trimodal distribution resulted from propagation by growing species with three distinct types of solvation: solvation by monomer, polymer, and solvent. Because all three fractions of the distribution increase progressively with conversion, the rate of exchange between the three types of solvation would have to be very slow ($\tau \approx 1$ sec). This requires that the lifetimes of the propagating carbenium ions are very long, which is highly unprecedented.

Variations in temperature also have complex effects on carbocationic polymerizations. For example, trityl salt-initiated polymerizations of α-methylstyrene, p-methoxystyrene, and cyclopentadiene apparently have negative enthalpies of activation (cf, IV.B.2) even though all of the elementary reactions should have positive energies of activation. However, these negative enthalpies of activation are based on apparent rather than absolute rate constants for which the concentration of growing carbenium ions was not directly measured. Propagation by at least two different propagating species can result in negative apparent energies of activation if the equilibrium is shifted to the more active species at lower temperature ($k_p^{app} = k_p^{+} \cdot K_I$; $E_a = \Delta H_p^{\#} + \Delta H_I$). The activation energies of bulk cationic propagation by free ions in γ-irradiated systems are close to zero. In contrast, stopped-flow studies indicate that the activation enthalpies ($\Delta H^{\#}$) are approximately 15–30 kJ/mol. These variations may also be affected by specific solvation by chlorinated solvents, which reduces the carbenium ions' reactivity.

6. *Mechanistic Aspects of Cationic Copolymerizations*

The relative reactivities of monomers can be estimated from copolymerization reactivity ratios using the same reference active center. However, because the position of the equilibria between active and dormant species depends on solvent, temperature, activator, and structure of the active species, the reactivity ratios obtained from carbocationic copolymerizations are not very reproducible [280]. In general, it is much more difficult to randomly copolymerize a variety of monomers by an ionic mechanism than by a radical. This is because of the very strong substituent effects on the stability of carbanions and carbenium ions, and therefore on the reactivities of monomers; substituents have little effect on the reactivities of relatively nonpolar propagating radicals and their corresponding monomers. The theoretical fundamentals of random carbocationic copolymerizations are discussed in detail and the available data are critically evaluated in Ref. 280. This review and additional references [281,282] indicate that only a few of the over 600 reactivity ratios reported are reliable.

More reliable estimates of monomer reactivity are available from reactions of model compounds (Chapter 2). For example, the rate constants of addition of the same standard benzhydryl carbenium ion to various substituted styrenes correlate very well to Hammett's σ^{+} parameter to provide $\rho_{(M)^{+}} = 4.9$ [193]. Addition of various p-substituted benzhydryl cations to the same standard alkene yielded $\rho_{(AC)^{+}} = -5.1$ [193]. These results demonstrate that carbocationic polymerizations are extremely sensitive to even small changes in the monomer structure. They also demonstrate that the reactivity of carbenium ions scales nearly perfectly to the inverse of the monomer reactivity, i.e., $\rho_{(AC)^+} = -\rho_{(M)^+}$, and that the rate constants of homopropagation, particularly of styrenes, should be very similar. Indeed, all rate constants of propagation are in the range $k_p \approx 10^{5\pm1}$ mol^{-1}·L·sec^{-1} at ambient temperature [207]. Model reactions (cf., Chapter 2) and more reliable copolymerization studies have established the following general monomer reactivity order: *N*-vinyl carbazole > vinyl ethers > *p*-methoxystyrene > α-methylstyrene > styrene ≈ isobutene > α-olefins.

7. *Living Carbocationic Polymerizations*

One of the most rapidly developing areas of cationic processes is controlled/living carbocationic polymerization which provide well-defined polymers and copolymers. As discussed in Chapter 4, although special mechanisms via nonionic propagating species have been proposed for many of the new controlled living carbocationic systems, the mechanism is quite similar to that of conventional cationic polymerizations [283]. That is, ions have been detected in most of these systems using optically active compounds and/or by various salt, substituent, and solvent effects, and by exchange reactions in polymerizations and model systems. Living carbocationic systems with a slow overall rate of polymerization are controlled by maintaining a low stationary concentration of carbenium ions; the kinetics is usually first order in monomer, initiator and Lewis acid (activator), and negative first order in added nucleophile (deactivator), i.e., $-d[\mathrm{M}]/dt = k[\mathrm{M}][\mathrm{I}][\mathrm{LA}]/[\mathrm{Nu}]$.

The new living systems use a much higher concentration of a well-defined initiator than in conventional systems. This results in lower molecular weight polymers than in conventional systems, which makes transfer less easily detected, and decreases the probability of spontaneous coinitiation with adventitious moisture. In these systems, the degree of polymerization is determined by the ratio of monomer reacted to that of initiator ($\mathrm{DP}_n = \Delta[\mathrm{M}]/[\mathrm{I}]_0$) as in a living system. The molecular weight is defined by $\Delta[\mathrm{M}]/[\mathrm{I}]_0$ rather than limited by transfer. The polydispersities are usually $M_w/M_n < 1.2$ due to fast exchange between active and dormant species; exchange is evident by the *decrease* in polydispersity with conversion. Due to relatively long-lived chain ends, it is possible to functionalize the end groups in many of these systems.

Those systems which are successful at mimicking a living system use fast and efficient initiators in combination with activators and sometimes added deactivator. This results in fast exchange and optimal reaction rates. The components should be nonbasic, thereby reducing the possibility of transfer by β-proton elimination. Although all of these conditions appear to suppress transfer relative to propagation, the inability to detect transfer is mainly due to the low $\Delta[M]/[I]_0$. The highest molecular weight polyisobutene ($M_n > 10^6$) is still produced by classic Lewis acid catalysis at low temperatures, and not by "living" systems. Nevertheless, it may still be possible to carefully choose a polymerization system which truly enhances the selectivity for propagation relative to transfer. We think that it is preferrable to call these systems *controlled* rather than *living* because transfer has been unambiguously detected in most cases. The controlled/living systems are described in detail in Chapter 4.

V. TRANSFER REACTIONS

Transfer reactions generally decrease the polymer molecular weight by generating new chains from the same initiator. Because the concentration of growing species does not change, transfer reactions generally do not affect the polymerization kinetics. Transfer is the most important chain-breaking reaction in cationic polymerizations of alkenes. The two major transfer reactions in carbocationic polymerizations are β-proton elimination to form unsaturated end groups, and Friedel-Crafts alkylation of aromatic rings; both generate protons capable of reinitiation. Occasionally, covalent species transfer a halide, alkoxide, or other group to propagating carbenium ions. Alternatively, termination (degradative transfer) occurs if the transfer reaction produces an unreactive species. Transfer may also be a prerequisite to termination, which occurs subsequent to transfer (cf., Section VI.C). More controlled polymerizations have used inifers [280,284], which act as both initiators and transfer agents. These systems will be discussed in detail later in this chapter and also in Chapter 5.

A. Chemistry of Transfer Reactions

1. *Elimination of β-Protons from Carbenium Ions*

As discussed in Section II, hyperconjugation results in as much as 7–12% of the positive charge being located on each β-hydrogen atom in propagating carbenium ions of isobutene, styrene, and vinyl ether polymerizations. Thus, even weak bases such as monomer, polymer, solvent, counteranion, or an impurity such as water may abstract these β-protons to produce unsaturated end groups; "spontaneous" loss of proton is highly unlikely [Eq. (88)].

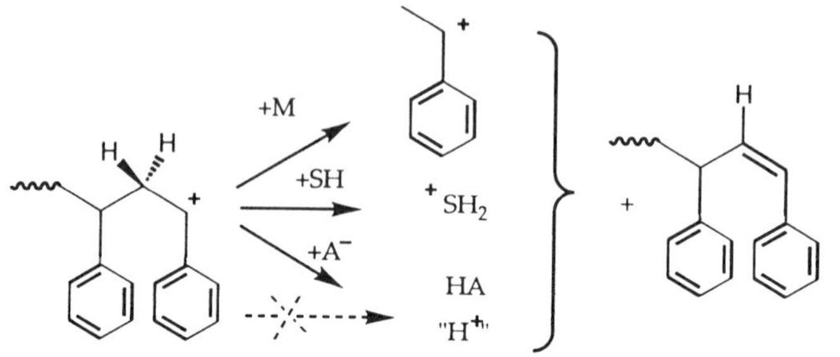

(88)

Nonbasic solvents and counteranions, such as ${}^{-}MtX_{n+1}$ derived from Lewis acids, should therefore be used in carbocationic polymerizations to prevent β-proton abstraction. Hexafluoroantimonate is probably the least basic and least nucleophilic counteranion, whereas the conjugate bases of oxy acids are more basic. Anion basicity, which correlates reciprocally with the acid strength, decreases in the order: $HOClO_3 \approx HOSO_2CF_3 > HOSO_2Cl > HI > HBr > SO_2(OH)_2 > HCl > HO(O)CCH_3$. As discussed in Section B.1, triflate, and perchlorate anions are the least basic and least nucleophilic of the oxy-acid anions.

Proton elimination from propagating secondary carbenium ions of substituted alkenes such as vinyl ethers and styrenes can only generate endo double bonds. However, the tertiary propagating carbenium ions of disubstituted alkenes such as isobutene and α-methylstyrene can eliminate a proton from either the endo or exo position as shown in Eq. (89).

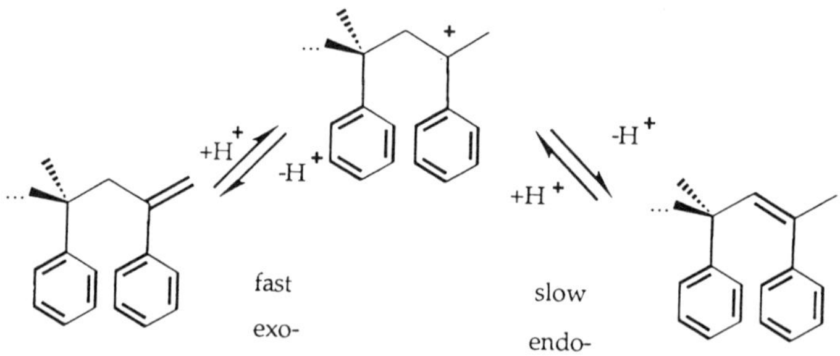

(89)

The exo double bond is formed first in polymerizations of α-methylstyrene, but is later isomerized by protonic acid to the more stable endo isomer [14]. Carbocationic polymerizations initiated by protonic acids with extremely basic counteranions, as in triflic acid-initiated polymerizations of isobutene, produce predominantly the unsaturated dimer [285]. The exo dimer forms first and then isomerizes to the more stable endo isomer [Eq. (90)].

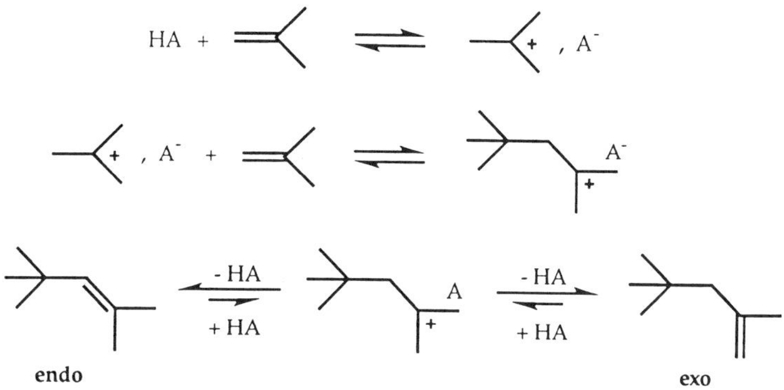

(90)

Although the resulting highly substituted terminal double bonds are usually nonhomopolymerizable, they can sometimes be reprotonated (reversible transfer) [286]. If co- or homopolymerization is possible, branched polymers of higher molecular weight are produced, primarily at high conversions.

In addition, the endo double bonds activate neighboring hydrogen atoms or alkyl groups to hydride and alkyl anion abstraction, respectively, to produce tertiary carbenium ions stabilized by both the double bond and the α-substituent [Eq. (91)].

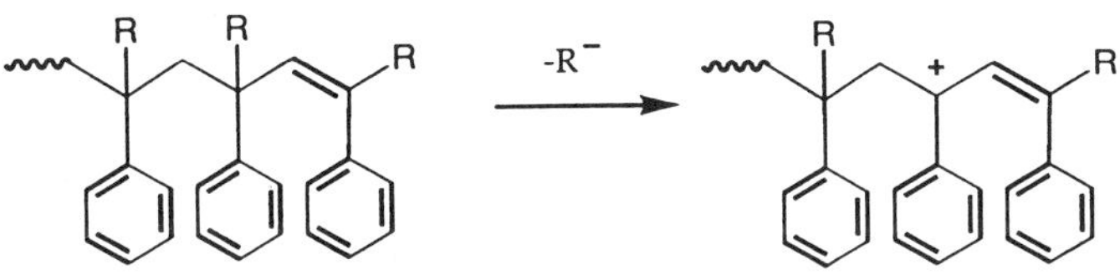

(91)

Because this abstraction produces delocalized and sterically hindered carbenium ions incapable of reacting with monomer, it is a termination reaction (cf., Section VI.A).

2. *Friedel-Crafts Alkylations*

The second major type of transfer reaction by propagating carbenium ions is Friedel-Crafts alkylation of aromatic rings, with concomitant loss of proton. The aromatic ring may be from an aromatic solvent, or from monomer or polymer chain if it is based on a styrene monomer. The most important factors determining the rate of electrophilic aromatic substitution are sterics and the electron-donating ability of substituents on both the aromatic ring and the carbenium ion; solvent, temperature, and Lewis acid are also important [240,287–289]. Substituents such as amino, alkoxy, aryl, vinyl, and alkyl groups which are electron donating by resonance stabilize carbenium ions, and therefore decrease their reactivity and the rate at which they alkylate aromatic rings. These substituents, however, activate aromatic rings to electrophilic substitution and direct the substitution to the *p*- and *o*-positions.

For example, the vinyl group on styrene monomers is activating and ortho-para-directing. This leads to macromonomers of polystyrene with styrene end groups functionalized at the *p*-position [Eq. (92)].

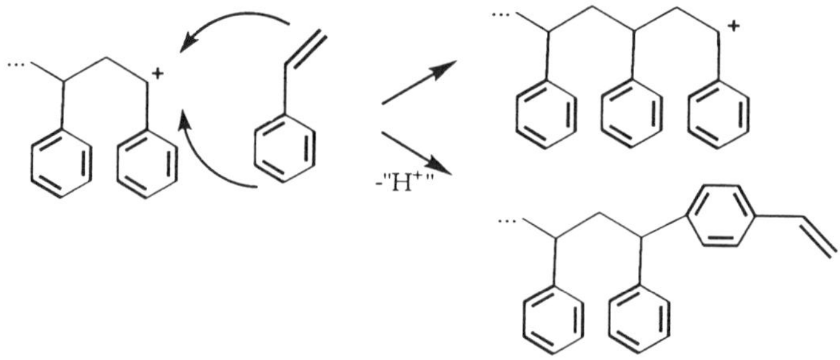

(92)

Because *p*-substituted styrenes are more reactive than unsubstituted styrene, the resulting macromonomers should be able to homopolymerize to produce comb polymers, and/or copolymerize with styrene to produce graft copolymers. Copolymerization should also increase at higher conversions when monomer is depleted.

Formation of macromonomers by bimolecular alkylation of monomer evidently occurs in γ-radiation polymerizations of styrene [290]. That is, higher molecular weight polystyrenes are obtained at low conversion if the *p*-position of the aromatic ring is blocked as in *p*-methylstyrene and *p*-chlorostyrene. In addition, significantly lower molecular weight polymers of *p*-chlorostyrene are produced in benzene than in bulk, whereas benzene does not affect the molecular weight of polystyrene and poly(*p*-methylstyrene). This indicates that propagating carbenium ions destabilized by an aromatic ring substituted with electron-withdrawing Cl react rapidly with benzene; styrene and *p*-methylstyrene are activated relative to benzene, and the molecular weights of their polymers are therefore not affected by dilution with benzene.

Electrophilic aromatic substitution also occurs intramolecularly to generate polymers with indanyl end groups [Eq. (93)]; entropy strongly favors this unimolecular reaction.

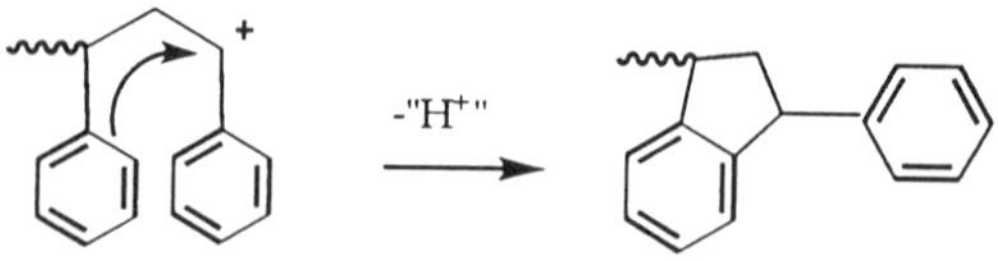

(93)

Indan formation is significant in polymerizations of styrene, α-methylstyrene, and p-alkyl-substituted styrenes, but is less important in p-methoxystyrene polymerizations [14]. Indan formation is not possible in polymerizations of styrene derivatives substituted at both ortho positions and is suppressed in polymerizations of 3,5-disubstituted styrene derivatives due to steric hindrance [105]. Indan formation consumes initiator in attempts at polymerizing isobutene with cumyl chloride and p-dicumyl chloride initiators [280].

Nuyken has optimized indan formation of diisopropenylbenzene to prepare polyindans with both high glass transition temperatures and high thermal stability [Eq. (94)] [291].

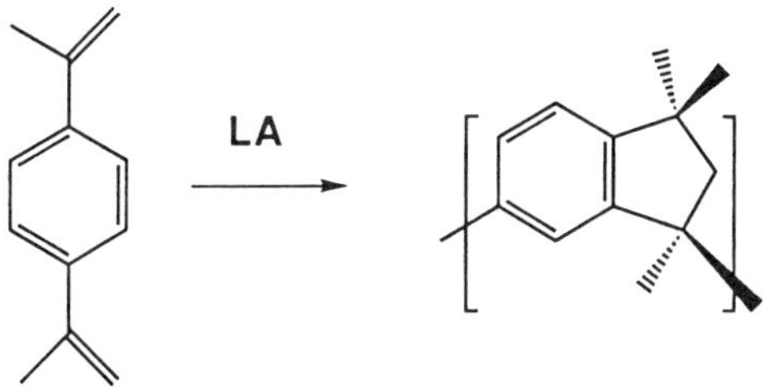

(94)

Indan formation is favored by the relatively high equilibrium monomer concentration of the corresponding α-methylstyrene ($[M]_e \approx 1$ mol/L at ambient temperature). For example, α-methylstyrene itself is dimerized rather than polymerized at $[M]_0 < 1$ mol/L, and the resulting unsaturated dimers cyclize efficiently to provide dimeric indans, especially at elevated temperatures.

3. *Relative Rates of Transfer by Intramolecular Alkylation and Elimination*

Although indan formation is significant in styrene polymerizations, β-proton elimination is much faster than intramolecular alkylation [292]. Unsaturated styrene and α-methylstyrene dimers are prepared quantitatively under high dilution at elevated temperatures without cyclization to indan derivatives [293]. In this case, the carbocations must be quenched before intramolecular cyclization becomes significant at high conversion. However, indan formation competes with depropagation at temperatures above 50° C, which is much too high for unsaturated styrene dimers ($D=$) to homopolymerize. As outlined in Eq. (95), the unsaturated dimers form indans (D^{in}) in the presence of acid.

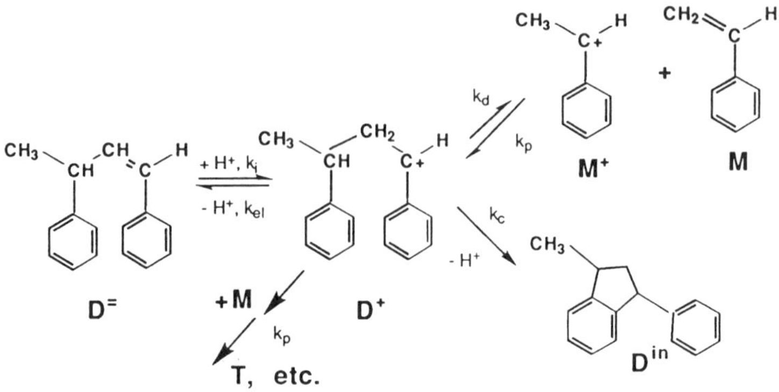

(95)

Surprisingly, trimeric indans (T^{in}) are also formed in nearly comparable amounts [294]. The intermediate trimeric carbenium ions must be formed by addition of dimeric carbenium ions to monomer, which can only be produced by depropagation of dimeric carbenium ions.

At lower temperatures (−70° C) with lower equilibrium monomer concentrations, unsaturated dimers rapidly form tetramers without producing a significant amount of trimers [295]. This indicates that the unsaturated dimer prefers to dimerize, with depropagation and indan formation less probable. 1,2-Hydride shifts may also occur, as indicated by the large variety of dimer stereoisomers formed, as well as by spectral changes in the stopped-flow studies of the dimerization. Eq. (96) outlines the variety of isomeric indan derivatives formed at later stages of the reaction by a combination of hydride shifts (cf., Section V.A.4) and intramolecular cyclizations.

Although transfer generally decreases the average molecular weight obtained from a polymerization, it can also lead to an abrupt increase in molecular weight at high conversion [182,206]. For example, the molecular weight of polystyrenes can increase at high monomer conversion by both intermolecular Friedel-Crafts alkylation and by macromonomer formation followed by copolymerization as shown in Eq. (97).

The selectivity for reaction of carbenium ions with unsaturated oligomers increases not only in the absence of monomer, but also with decreasing temperature [182]. However, unsaturated macromonomers of styrene may dimerize rather than homopolymerize [cf., Eq. (96)]. That is, the molecular weight at complete conversion only doubles, rather than increasing further, because no copolymerization is possible in the absence of monomer. Because molecular weight only doubles [182], dimerization apparently dominates over Friedel-Crafts alkylation. The alkylation may

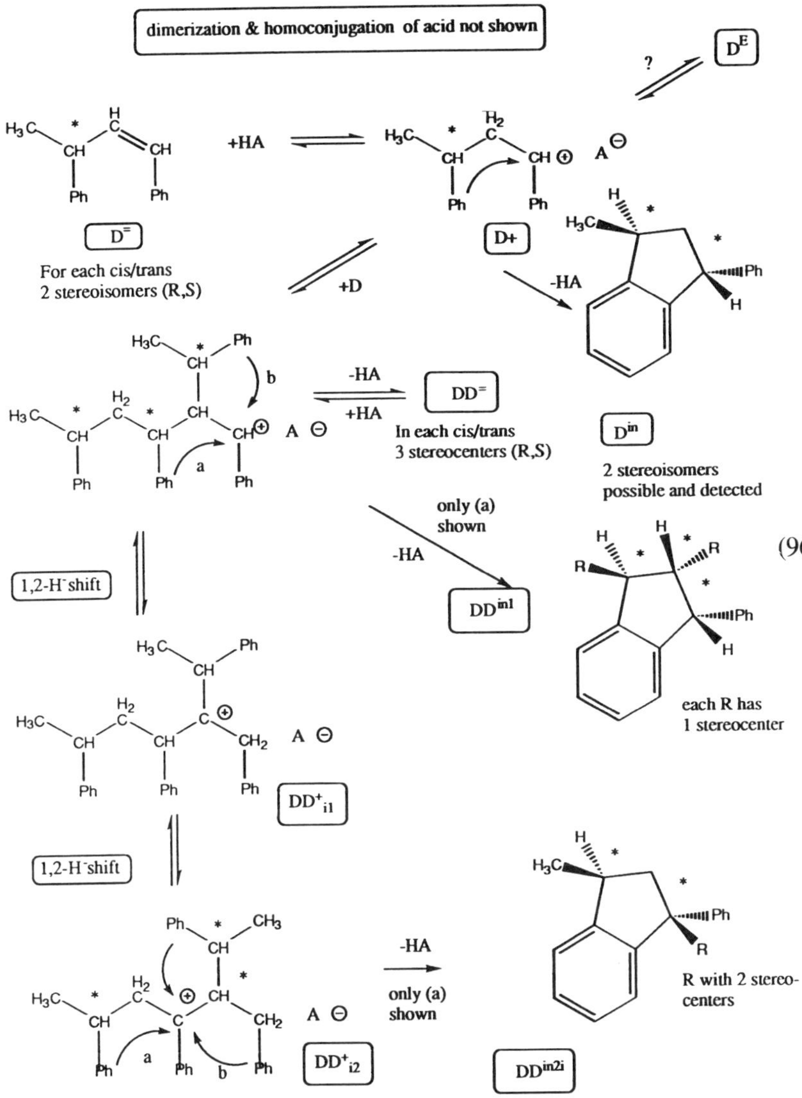

(96)

become the dominant transfer reaction at extremely low temperatures because it has a lower activation energy than elimination [292].

Figure 5 summarizes how temperature affects the individual rate constants of the elementary reactions involved in carbocationic polymerizations of styrene. Although it compiles data from various sources, the

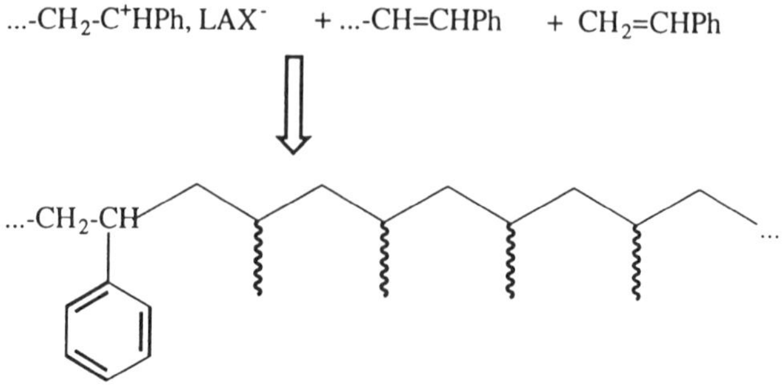

(97)

trends derived from different methods for the same type of reaction are consistent. For example, the rate constants (k_c) of cyclization to idan derivatives were calculated both by low-temperature (−70 to −30° C), stopped-flow measurements and by comparison with the depropagation rate constant (k_d) at 50° C; the latter value is consistent with that calculated from the high-temperature cyclizations of unsaturated dimers discussed above. This plot demonstrates that indan formation should dominate over transfer reactions (k_{trM} and K_{tr}) at temperatures below −70° C. Indeed,

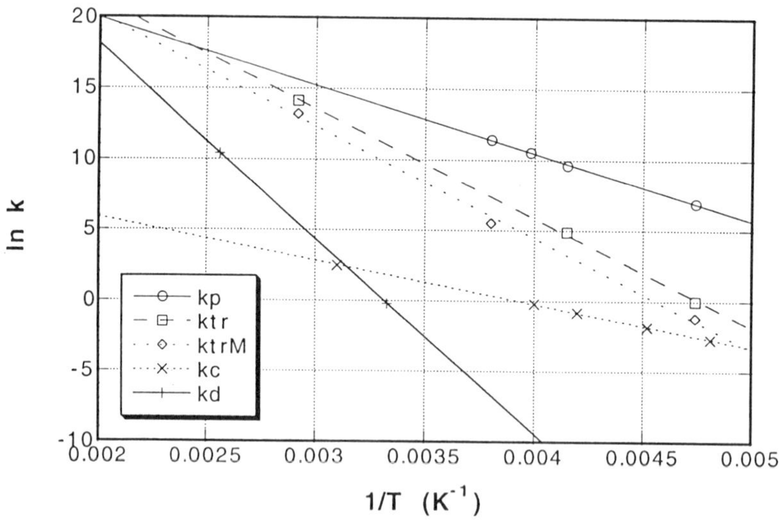

Figure 5 Effect of temperature on rate constants of propagation, depropagation, transfer to monomer, transfer to triflate anion, and indan formation in the carbocationic polymerization of styrene (From Ref. 292).

polymers are successfully formed by intermolecular Friedel-Crafts only at very low temperatures ($< -80°$ C) [296]. Because intramolecular indan formation is irreversible, it also occurs at high temperatures when other reactions such as protonation of double bonds in unsaturated oligomers are reversible [291].

Figure 5 also demonstrates that polymers should not be obtained at temperatures higher than 170° C, because this is the temperature at which the extrapolated rate constant of transfer to monomer (k_{trM}) intersects and then becomes greater than that of propagation (k_p). The extrapolated rate constants of propagation and transfer to (triflate or perchlorate) counteranion (k_{tr}) intersect at 130° C. However, in contrast to propagation and chain transfer to monomer which are both first order in monomer, all other reactions plotted in Figure 5 are zero order in monomer. Thus, the rates of propagation and transfer to monomer vary with monomer concentration, whereas the other reactions are not affected. Therefore, the relative rates of propagation and chain transfer to counteranion depend on the monomer concentration, with "polymers" (nonamers at 130° C) formed in bulk ($[M]_0 \approx 9$ mol/L); only dimers are produced at high dilution, even at lower temperatures [294].

4. *Hydride (and Methide) Transfer*

In addition to the transfer reactions already discussed, propagating carbenium ions also react with nucleophilic hydride and methide anions. This reaction may be bimolecular, or it may occur by an intramolecular hydride shift to form a more stable carbenium ion. The activation energy of hydride transfer is usually higher than that of propagation, and therefore occurs only at elevated temperatures. Nevertheless, hydride transfer is the dominant reaction when α-methylstyrene is initiated by triphenylcarbenium ions. That is, steric hindrance prevents initiation by direct electrophilic addition of the carbenium ion to α-methylstyrene. Instead, it occurs by hydride transfer from monomer to yield triphenylmethane and the primary carbenium ion of α-methylstyrene [cf., Eq. (40)].

Hydride transfer is especially favorable when it generates tertiary carbenium ions which are stabilized by phenyl, alkoxy, and/or vinyl groups. Degradation of poly(p-methoxystyrene) in the presence of acid indicates that hydride transfer occurs in p-methoxystyrene polymerizations [Eq. (98)] [29]. Monomeric and oligomeric carbenium ions derived from α-methyl-p-methoxystyrene were detected by both NMR and UV, which indicates that hydride transfer at the chain end generates a tertiary carbenium ion which subsequently depropagates according to Eq. (98).

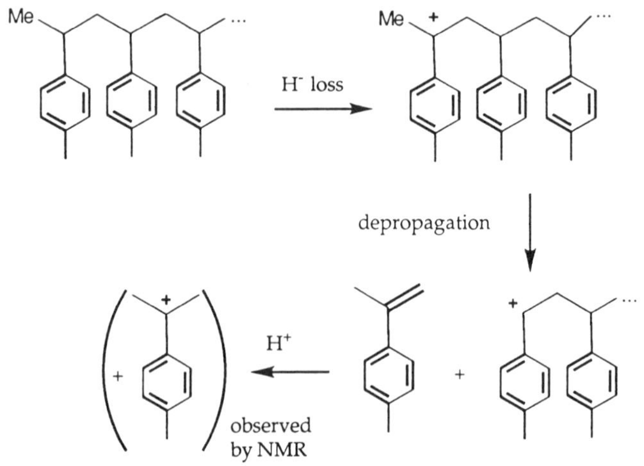

(98)

Similarly, isomerized carbenium ions are formed in polymerizations of α-methylstyrene derivatives by either a 1,3-intramolecular or bimolecular methide anion shift, or by reaction of carbenium ion with an exo-unsaturated oligomer [cf., Eq. (10)] [13].

Because hydride transfer is driven by a decrease in free energy, the carbenium ions generated are usually unreactive. For example, stable indanyl carbenium ions are generated by hydride transfer from indan end groups. Hydride transfer also occurs readily from the —CHPh— group located at an unsaturated chain end [cf., Eq. (10) and Section VI]. However, some hydride transfers generate isomerized carbenium ions which can still propagate. The rate of 1,2-hydride shifts is faster than propagation in carbocationic polymerizations of 3-methyl-1-butene [297], with the isomerized repeat units shown in Eq. (99) formed in preference to the "normal" units.

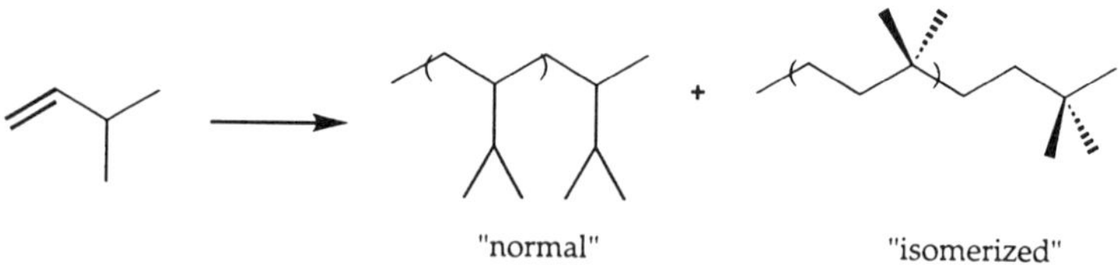

(99)

The isomerized structure dominates even at $-100°$ C, but accounts for only 70% of the repeat units at $-130°$ C. Similar but more complicated structures are formed in 4-methyl-1-butene polymerizations by competing hydride and methide shifts [298]. Other monomers whose propagating carbenium ions isomerize include 5-methyl-1-hexene, 4,4-dimethyl-1-pentene and some terpenes [299].

5. *Transfer of Anions from Covalent Species*

Carbenium ions react with neutral nucleophiles to form onium ions. These onium ions decompose reversibly by releasing either the most weakly bound or the most stable cationic species. As discussed previously, the stability of such species is determined by their structure (tertiary > secondary > primary) and by the stabilizing ability of substituents at the cationic center (α-alkoxy > phenyl > alkyl). However, because protons are only weakly bound to heteroatoms, most proton-containing compounds such as water, alcohols, and acids produce onium ions which release protons. One of the most common transfer reactions in carbocationic polymerizations is reaction with adventitious water. As outlined in Eq. (100), the resulting primary oxonium ion loses a proton and generates an alcohol end group. The alcohol end group can then react further with a propagating carbenium ion to form a secondary oxonium ion, which again releases a proton to generate an ether end group.

$$
\begin{aligned}
&\text{...-}CH_2\overset{+}{C}HPh + H_2O \rightleftharpoons \text{...-}CH_2CHPh\text{-}\overset{+}{O}H_2 \rightleftharpoons \text{...-}CH_2CHPh\text{-}OH + \text{"}H^+\text{"} \\
&\text{...-}CH_2\overset{+}{C}HPh + \text{...-}CH_2CHPh\text{-}OH \rightleftharpoons \text{...-}CH_2CHPh\text{-}\overset{+}{O}H\text{-}CHPh\text{-}CH_2\text{-...} \qquad (100) \\
&\text{...-}CH_2CHPh\text{-}\overset{+}{O}H\text{-}CHPh\text{-}CH_2\text{-...} \rightleftharpoons \text{...-}CH_2CHPh\text{-}O\text{-}CHPh\text{-}CH_2\text{-...} + \text{"}H^+\text{"}
\end{aligned}
$$

In contrast, the onium ion intermediates formed by esters tend to release acyl cations, and those formed by acetals release alkoxycarbenium ions.

To produce well-defined telechelic polymers with both chain ends functionalized, the transfer agent must produce reactive cations. For example, acetal chain transfer agents have recently been used to endcap growing chains with ether end groups; the intermediate oxonium ions release alkoxycarbenium ions which reinitiate a propagating chain functionalized with an ether group. Although these acetals have only been used in vinyl ether polymerizations so far [300], they might be extended to other alkenes such as styrene and isobutene as shown in Eq. (101).

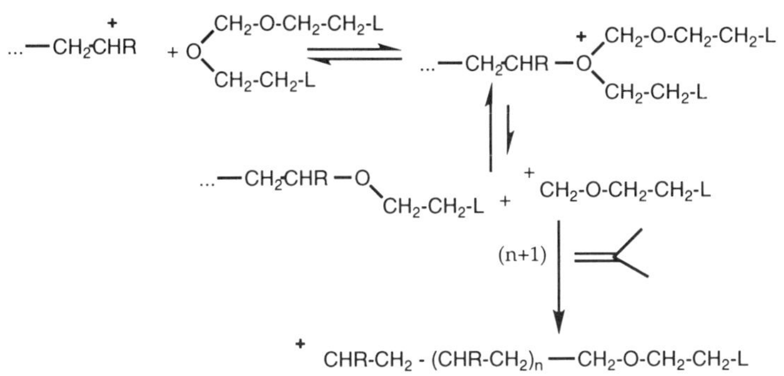

(101)

Either halide or ester end groups L could be used, and then functionalized further as outlined in Eq. (102).

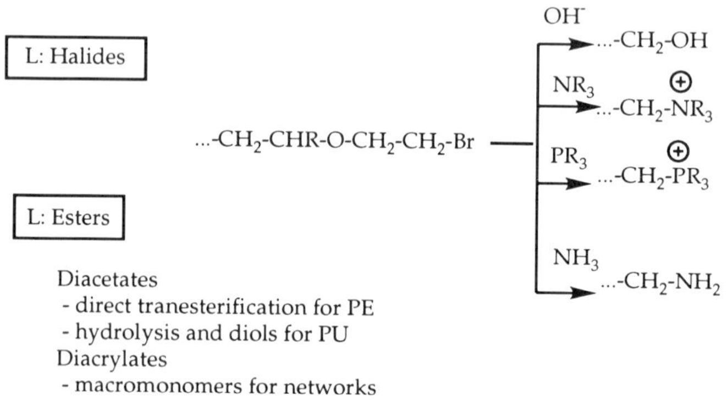

(102)

Silyl group transfer can also be used to functionalize chain ends. For example, allyl silanes, silyl ketene acetals, and silyl enol ethers [301–304] generate polymers with terminal allyl and methacrylate groups [Eq. (103)]. This type of transfer becomes degradative (termination) if reinitiation with silyl halides is not possible.

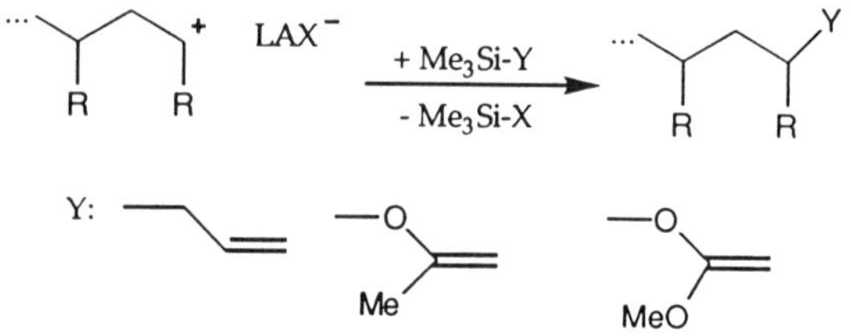

(103)

An interesting extension of this work is to use multifunctional silyl enol ethers to prepare telechelic and star-like poly(vinyl ether)s (cf., Chapter 5) [305].

Since proton reinitiation always produces an inert, nonfunctionalized end group, such transfer agents can not be used to synthesize end-functionalized polymers. In addition, proton reinitiation is often slow, resulting in broader polydispersities. The inifer technique was introduced in 1979 as a method for preparing telechelic polymers and regulating the molecular weight of polyisobutylenes prepared cationically. Inifer refers to compounds which act as both initiators and transfer agents. As shown in Eq. (104), the inifer concept [284] was based on transfer of halide directly from an initiator such as cumyl chloride to propagating carbenium ions, followed by initiation, propagation, and subsequent transfer.

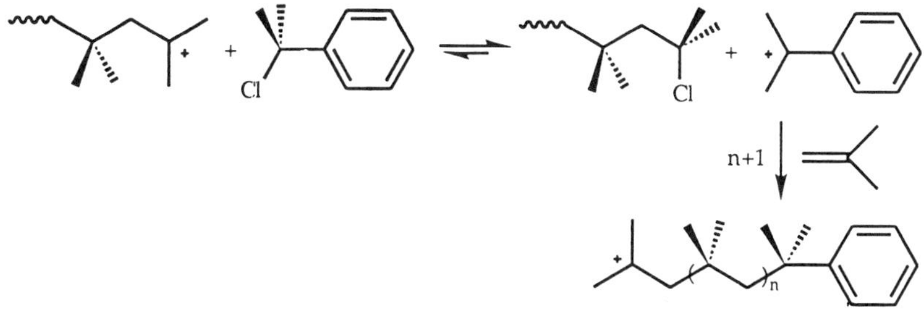

(104)

However, halide is not transferred directly in polymerization of isobutene. Instead, the propagating ion pair first reacts unimolecularly to form a covalent species with regeneration of the Lewis acid [Eq. (105)], which can then activate the inifer [306].

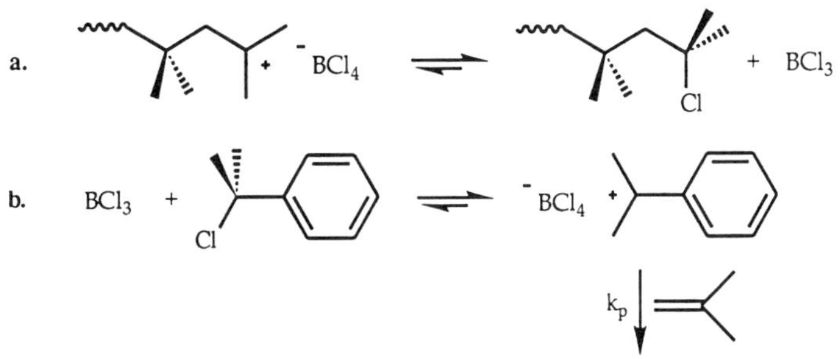

(105)

Nevertheless, bimolecular halide transfer [Eq. (106)] reportedly occurs in polymerization of cyclohexyl vinyl ether [113].

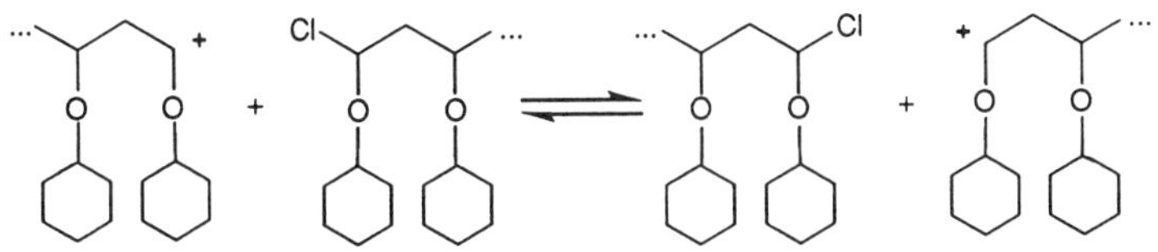

(106)

B. Effect of Transfer on Molecular Weight

The evolution of molecular weight with conversion depends primarily on the rates of initiation and/or termination relative to propagation, and on the effect of monomer concentration on transfer. Transfer reduces molecular weight. Kinetically, transfer can be divided into reactions which are either first or zero order in monomer. First-order reactions occur when monomer is involved in the rate-limiting step of the transfer process, even though the active site may not transfer directly to monomer. For example, transfer is kinetically first order in monomer concentration if reinitiation by protonation of monomer is slow compared to transfer. However, monomer concentration will not affect the rate of transfer if reinitiation is fast. Transfer reactions which are zero order in monomer are also called "spontaneous" transfer reactions. Transfer to counteranion, solvent, and some impurities are typical "spontaneous" transfer reactions. Both types of transfer processes often occur simultaneously and may be differentiated by the approach outlined below.

The number average degree of polymerization is defined as the number of monomer molecules consumed divided by the number of chains formed [Eq. (107)].

$$\mathrm{DP}_n = \Delta[\mathrm{M}]/\Sigma[N] \tag{107}$$

If the polymer has sufficiently high molecular weight, monomer is consumed primarily in propagation, which is first order in both active sites and monomer [Eq. (108)].

$$R_p = k_p[\mathrm{P}^+][\mathrm{M}] \tag{108}$$

The growth of polymer chains is limited by termination and transfer. Therefore, these reactions define the total number of polymer chains in nonliving systems. The rate of termination is usually independent of monomer concentration, and is first order in active species since termination generally occurs by reaction with counteranion [Eq. (109)].

$$R_t = k_t[\mathrm{P}^+] \tag{109}$$

As discussed previously, transfer is either first order in monomer [Eq. (110)],

$$R_{\mathrm{trM}} = k_{\mathrm{trM}}[\mathrm{P}^+][\mathrm{M}] \tag{110}$$

or zero order in monomer [Eq. (111)].

$$R_{\mathrm{tr}} = k_{\mathrm{tr}}[\mathrm{P}^+] \tag{111}$$

The number-average degree of polymerization (DP_n) defined in Eq. (107) is determined by the ratio of the rates of the corresponding reactions [Eq. (112)].

$$\mathrm{DP}_n = \frac{R_p}{R_t + R_{\mathrm{trM}} + R_{\mathrm{tr}}} = \frac{k_p[\mathrm{P}^+][\mathrm{M}]}{k_t[\mathrm{P}^+] + k_{\mathrm{tr}}[\mathrm{P}^+] + k_{\mathrm{trM}}[\mathrm{P}^+][\mathrm{M}]} \tag{112}$$

Eq. (113) is produced by inversion of Eq. (112),

$$\frac{1}{\mathrm{DP}_n} = \frac{k_{\mathrm{trM}}}{k_p} + \frac{k_t + k_{\mathrm{tr}}}{k_p} \cdot \frac{1}{[\mathrm{M}]} \tag{113}$$

which corresponds to the classic "Mayo equations" used in radical polymerizations. Replacement of the ratios k_{tr}/k_p and k_t/k_p by the corresponding transfer and termination constants produces Eq. (114).

$$1/\mathrm{DP}_n = C_{\mathrm{trM}} + (C_{\mathrm{tr}} + C_t)/[\mathrm{M}] \tag{114}$$

If a transfer agent is added, the degree of polymerization decreases further according to Eq. (115), in which $(\mathrm{DP}_n)_0$ is the average degree of polymerization that would be obtained in the absence of the transfer agent.

$$1/\mathrm{DP}_n = 1/(\mathrm{DP}_n)_0 + C_x[X]/[\mathrm{M}] \tag{115}$$

However, this equation must be modified at high conversions [Eq. (116)] and when the change in $[X]$ is small.

$$1/\mathrm{DP}_n = 1/(\mathrm{DP}_n)_0 + C_x[X]_0 \ln ([\mathrm{M}]_0/[\mathrm{M}])/([\mathrm{M}]_0 - [\mathrm{M}]) \tag{116}$$

Transfer to monomer ($C_{\mathrm{trM}} = k_{\mathrm{trM}}/k_p$) and "spontaneous" transfer ($C_{\mathrm{tr}} = k_{\mathrm{tr}}/k_p$) can be differentiated by plotting $1/\mathrm{DP}_n$ versus $1/[\mathrm{M}]_0$ at low conversions using Mayo plots or the integrated equation. In that case the Schultz-Harboth plots should be used [307]. The transfer constant to monomer is estimated from the intercept, whereas the slope presumably estimates k_{tr}/k_p when $k_t \ll k_{\mathrm{tr}}$, or k_t/k_p when k_{tr} is very small. However, these assumptions have rarely been justified by supporting kinetic measurements [200,308,309]. In addition, transfer constants may vary with varying initial monomer concentrations due to changes in the dielectric constant and/or other properties of the reaction medium, leading to difficulties in separating C_{trM} and C_{tr}.

In many carbocationic polymerizations, termination is negligible (cf., Section VI). In this case, the total number of chains equals the total number of chains generated by transfer to monomer $[N_{\mathrm{trM}}]$, by spontaneous transfer $[N_{\mathrm{tr}}]$, and the number of macromolecules $[N]$ still growing. The latter should be equal to the initial initiator concentration $[\mathrm{I}]_0$ if initiation is completed, or to $[\mathrm{I}]_0 - [\mathrm{I}]_t$ at time t. The total number of macromolecules generated by transfer ($\Sigma[N_{\mathrm{tr}}]$) equals the sum of Eqs. (117) and (118), assuming that initiation is rapid ($[N]_0 \approx [\mathrm{I}]_0$). The number of macromolecules formed by transfer which is first order in monomer is proportional to conversion, whereas that formed by transfer which is zero order in [M] is proportional to time.

$$[N_{\mathrm{trM}}] = \int_0^t k_{\mathrm{trM}}[N]_0[\mathrm{M}]\, dt = (k_{\mathrm{trM}}/k_p)([\mathrm{M}]_0 - [\mathrm{M}]_t) \tag{117}$$

$$[N_{\mathrm{tr}}] = \int_0^t k_{\mathrm{tr}}[N]_0 dt = (k_{\mathrm{tr}}/k_p)\ \ln[\mathrm{M}]_0/[\mathrm{M}]_t \qquad (118)$$

Determining the amount of polymer formed by transfer versus that formed by direct initiation can therefore be used to distinguish between the two types of transfer [310]. This has been accomplished by initiating polymerization of various monomers using macromolecular initiators of polyisobutene functionalized with $PhCH_2Cl$ initiating groups along its backbone, which was prepared by copolymerization of isobutene with *p*-chloromethylstyrene [310].

1. *Coefficients of Transfer to Monomer*

Transfer coefficients to monomer ($C_{\mathrm{trM}} = k_{\mathrm{trM}}/k_p$) are tabulated in several reviews [308,309]; these values were determined using Eqs. (113) and (117), and are considered reliable if propagation and transfer to monomer are the same order in monomer concentration. However, many older values may be less reliable because they were determined using either viscosity or osmometry. The disadvantage of calculating coefficients from viscosity measurements is that viscometry determines M_v, which is closer to M_w than M_n; the disadvantage of calculating coefficients from osmometry is that it is less reliable and more sensitive to impurities than size exclusion chromatography (SEC). Monomer, temperature, and solvent also affect the C_{trM} values. If the initiating system affects C_{trM}, transfer may not occur directly from propagating chains to monomer.

The simplest polymerizations to study are those initiated by γ-irradiation because counteranions are not present. In this case, the upper molecular weight limit is regulated by either transfer to monomer or transfer to polymer. The molecular weights of polyisobutenes obtained by γ-radiation-induced polymerizations (SEC $M_n = 330{,}000$ at $-41°$ C, $M_n = 1{,}980{,}000$ at $-70°$ C) are much higher than those initiated by Lewis acids ($AlEtCl_2$, BF_3, $AlCl_3$) in CH_3Cl ($[\mathrm{M}]_0 = 3.1$ mol/L) [311]. However, this comparison between chemically initiated polymerizations and bulk polymerizations initiated by γ-irradiation may not be fair considering the stringent purification procedure of the latter. The very high molecular weights obtained by bulk γ-irradiated polymerizations indicate that transfer to monomer by free ions is very small at $-78°$ C. Nevertheless, the transfer constants in $TiCl_4$-initiated systems are also quite low ($C_{\mathrm{trM}} = 5 \cdot 10^{-4}$ at $-14°$ C in CH_2Cl_2) [309] and, based on similar molecular weights, are comparable to those in cumyl acetate/BCl_3-initiated "living" polymerizations ($-10°$ C) [105].

The coefficients of transfer to styrene are larger than those to isobutene. For example, the molecular weights of polystyrene obtained by bulk

γ-radiation-induced polymerizations (-21 to $33°$ C) are 51,000 and 42,000 (viscosimetric data) [312]. Similar molecular weights were recently obtained under more stringent conditions in CH_2Cl_2 at $-10°$ C (SEC M_n = 39,000) and in benzene at 7° C (SEC M_n = 43,000) [290]. This similarity in the molecular weights regardless of the rates of polymerization ($R_{bulk} > R_{benzene} > R_{CH_2Cl_2}$) indicates that molecular weight is controlled in all three cases by transfer to monomer, with $k_{trM}/k_p \approx 5 \cdot 10^{-4}$–$10^{-3}$.

Although their rates of polymerization are similar, the molecular weights obtained by bulk polymerizations of p-methylstyrene (M_n = $4.6 \cdot 10^5$ at $-10°$ C) and p-chlorostyrene (M_n = 10^6 at $-10°$ C) are much higher than those of styrene [290]. Transfer coefficients in chemicaly initiated polymerizations of p-methoxystyrene at $-15°$ C in CH_2Cl_2 are also small ($C_{trM} = 3 \cdot 10^{-4}$), resulting in higher moleculecular weight polymers ($M_n \approx 4 \cdot 10^5$) [186]. These large differences between styrene and p-substituted styrenes is presumably due to the p-substituents preventing alkylation at that position [cf., Eq. (92)].

Values of k_{trM}/k_p and k_{tr}/k_p were determined by the grafting method mentioned in Section V.B for indene, styrene [310] and α-methylstyrene [313] in (50/50) CH_2Cl_2/methylcyclohexane using $AlEt_2Cl$ as the coinitiator. Only transfer to monomer was detected ($k_{trM}/k_p = 7.5 \cdot 10^{-4}$) for indene polymerizations at $-55°$ C. The transfer coefficient to styrene was approximately $2.4 \cdot 10^{-3}$, although zero order transfer is possible. Transfer coefficients in the α-methylstyrene polymerizations at $-50°$ C were $k_{trM}/k_p = 10^{-4}$ and $k_{tr}/k_p = 4 \cdot 10^{-4}$ mol/L [313]. These values are similar to those obtained in p-methylstyrene and indene polymerizations initiated by cumyl chloride or cumyl methyl ether and catalyzed by $TiCl_4$ at $-40°$ C in CH_2Cl_2 [314]: indene $C_{trM} \approx 6 \cdot 10^{-4}$; p-methylstyrene $C_{trM} \approx 10^{-3}$. That is, chain transfer to monomer is also quite significant in the "living" systems (cf. Chapter 4). The larger transfer constant for p-methylstyrene compared to that determined by γ-radiation experiments indicates that either counteranion influences transfer to monomer or transfer to counteranion occurs.

No comparative data are available for these temperatures in Lewis acid-initiated polymerizations of styrene. However, DP_n increases linearly with polymer yield in $TiCl_4$-initiated polymerizations in CH_2Cl_2 at temperatures from $-90°$ C to $-25°$ C ($DP_n \approx 60$ at $-25°$ C, $DP_n \approx 200$ at $-60°$ C, $[M]_0 \approx 0.25$ mol/L), indicating that transfer to monomer is small ($C_{trM} \approx 4 \cdot 10^{-4}$ and $< 5 \cdot 10^{-5}$) [308].

2. *Coefficients of "Spontaneous" Transfer*

As discussed in Section V.B, the coefficient of "spontaneous" transfer, $C_{tr} = (k_{tr} + k_t)/k_p$, is obtained from the slope of Mayo plots of $1/DP_n$

versus $1/[M]_0$. If $k_t \ll k_{tr}$, k_{tr}/k_p is obtained from the slope, whereas k_t/k_p is obtained when $k_t \gg k_{tr}$. Separate values of k_t and k_{tr} have only been reported in sulfuric acid-initiated polymerizations of styrene in which termination occurs by recombination of counterions to form inactive sulfates [118]. Although termination is not the typical chain-breaking reaction in carbocationic polymerizations, most studies have assumed that $k_t \gg k_{tr}$, and that only transfer to monomer occurs. However, significant variations in molecular weight with variations in the composition of the initiating system indicate that transfer to counteranion is significant in many systems.

Thus, some of the values of k_t/k_p determined from Mayo plots [308] evidently denote k_{tr}/k_p. Most systems have coefficients of "spontaneous" transfer in the range $C_{tr} \approx 10^{-2}$ mol/L at temperatures from 0 to 20° C [308], which are quite similar to monomer transfer constants. However, many of these polymerizations were taken to rather high conversions at which separate values of C_{trM} and C_{tr} determined from simple Mayo plots are incorrect. The unreliability of some of the reported values is also indicated by the unusually rapid decrease in the coefficients of "spontaneous" transfer reported in perchloric acid-initiated polymerizations of styrene with decreasing temperature. That is, C_{tr} reportedly decreases more rapidly than C_{trM} [315], with $\Delta E_a = -24$ kJ/mol for the differences in the free energies of activation of propagation and "spontaneous" transfer, and $\Delta E_a = +5$ kJ/mol for the differences in propagation and transfer to monomer. This unusual positive value would require that the activation energy for transfer to monomer is lower than that for propagation. In contrast, both C_{trM} and C_{tr} decrease rapidly with decreasing temperature in styrene polymerizations initiated by $TiCl_4$ and adventitious moisture. That is, the transfer constant to monomer decreases from $C_{trM} = 4 \cdot 10^{-4}$ at -60° C to $<5 \cdot 10^{-5}$ at -90° C, whereas the "spontaneous" transfer constant decreases from $C_{tr} = 1.3 \cdot 10^{-3}$ mol/L at -60° C to $C_{tr} \approx 6 \cdot 10^{-4}$ mol/L at -90° C [308].

Transfer to counteranion is greater with oxyanions than with anions derived from Lewis acids because the former are more basic. For example, triflates and perchlorates form unsaturated dimers quantitatively at temperatures greater than 60° C in benzene (cf., Section V.A.1); oligomers are generated by Lewis acids under identical conditions [293]. Transfer to counteranion is especially efficient in nonpolar solvents due to its transition state being less polar than its ground state. In contrast, both propagation and transfer to monomer are much less affected by varying solvent polarity because they are ion-dipole reactions. Therefore, the substantial decreases in molecular weight observed in less polar solvents again indicates that they are caused by transfer to counteranion.

3. *Coefficients of Transfer to Transfer Agents*

Carbenium ions react readily with alcohols, ethers, ketones, anhydrides, and various aromatic compounds. Transfer constants to most alcohols are approximately $C_x \approx 1$ in styrene polymerizations [308]. Anhydrides are more reactive, with transfer constants $C_x \approx 10$; ketones and ethers are less reactive ($C_x \approx 0.1$) [120]. In general, the transfer constants to aromatic compounds listed in Table 17 increases as their nucleophilicity increases. This indicates that the rate-determining step in Friedel-Crafts alkylation is formation of arenium ions ("Wheland" intermediates) [Eq. (119)] rather than reinitiation.

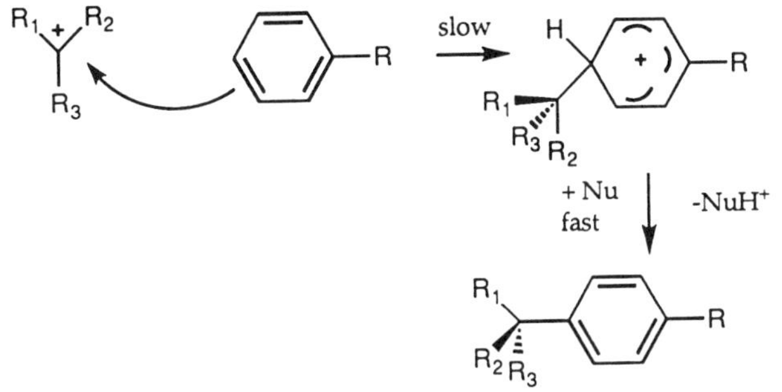

(119)

Some of the transfer constants listed in Table 17 were determined using the simplified Eq. (115) without accounting for monomer conversion and consumption of transfer agent. In addition, the molecular weights used to calculate the transfer constants were usually determined by viscometry and are therefore higher than the number-average molecular weights. Both approximations result in inflated transfer constants. For example, the transfer constant to anisole in styrene polymerizations is slightly lower when SEC is used to determine M_n and the monomer conversion is followed by NMR ($C_x = 0.4$ at $-20°$ C) [317] compared to earlier studies which calculated $C_x = 1.6$ at 0° C [321].

The transfer constants in styrene polymerizations are higher than predicted from model studies using diarylmethylium cations at $-70°$ C (cf., Chapter 2) [289]. In the model studies, (p-MeOPh)PhCH$^+$ adds to styrene approximately 300 times faster than it reacts with anisole, 10^7 times faster than with toluene, and 10^{10} times faster than with benzene. However, if the slightly higher activation enthalpy of reaction with anisole compared to styrene ($\Delta\Delta H\ddagger \approx 12$ kJ/mol) [289,322] is taken into account, extrapolation of their relative rate constants from -70 to 0° C still indicates that anisole is 30 times more reactive than styrene. This is in contrast to the similar reactivities ($C_x \approx 1$ at 0° C) calculated from polymerization studies [317]. That is, the p-methoxydiphenylcarbenium ion is apparently

Table 17 Transfer Coefficients to Various Aromatic Compounds in Cationic Polymerization of Styrene[a]

Compound	k_{tr}/k_p	Ref.
Anisole	1.6	[316]
Anisole[b]	0.4[b]	[317]
Thiophene	1.0	[316]
Anthracene	0.44	[318]
m-Xylene	0.134	[318]
o-Xylene	$8.3 \cdot 10^{-2}$	[318]
Phenanthrene	$4.5 \cdot 10^{-2}$	[318]
Naphtalene	$3.7 \cdot 10^{-2}$	[318]
p-Xylene	$1.05 \cdot 10^{-2}$	[318]
Pentamethylbenzene	$1.0 \cdot 10^{-2}$	[318]
Toluene	$8.8 \cdot 10^{-3}$	[318]
p-Chloroanisole	$8.2 \cdot 10^{-3}$	[319]
Ethylbenzene	$8.0 \cdot 10^{-3}$	[320]
Mesitylene	$7.3 \cdot 10^{-3}$	[318]
p-*t*-Butyltoluene	$6.2 \cdot 10^{-3}$	[319]
	$3.0 \cdot 10^{-3}$	[320]
t-Butylbenzene	$4.5 \cdot 10^{-3}$	[320]
p-Cumene	$4.4 \cdot 10^{-3}$	[320]

[a] Typical reaction conditions: 0°C, $SnCl_4$ as Lewis acid, adventitious moisture as initiator, transfer estimated from the Mayo plots, molecular weights estimated by viscometry.
[b] Reaction run at −15°C, with $[SnCl_4]_o = 0.1$ mol/L in the presence of $[NBu_4Cl]_o = 0.040$ mol/L and initiated by $[1\text{-PhE} + Cl]_o = 0.02$ mol/L, molecular weights from SEC.

less reactive and more selective [240] than the propagating polystyryl cations. Thus, model studies can not always be used to compare quantitatively the reactivities of transfer agents, but do indicate their relative reactivities.

Much less information is available on transfer constants in polymerizations of other alkenes. It appears that the transfer constant to anisole in isobutene polymerizations is smaller than in styrene polymerizations, and much closer to values predicted by model studies $C_x \approx 5 \times 10^{-3}$ [323]. Calculated C_x values are relatively independent of temperature.

C. Transfer in "Living" Carbocationic Polymerizations

Transfer reactions should be absent in living polymerizations but are usually significant in classic carbocationic polymerizations. However, they appear to be suppressed in new "living" cationic polymerizations. This suppression was allegedly ascribed to a new propagation mechanism involving nonionic active sites of stretched covalent bonds. However, as discussed in Section IV.D.2 and also in Chapter 4, propagation proceeds through the same ionic intermediates propagating in conventional systems. In fact, the measured transfer constants to monomer in these new systems are usually at least as high as those in conventional systems (cf., Section V.B.1). There have also been no significant reductions reported in the transfer reactions in "living" systems containing polar additives or added salts with common counterions. For example, the same transfer constant C_{trM} to indene was found in the presence and absence of dimethyl sulfoxide [170]. These additives improve the polymerizations and reduce the polydispersities primarily by reducing the concentration of free ions and therefore the rates of polymerizations. Dissociation to free ions may also be suppressed by decreasing the dielectric constant of the reaction mixture. The "living" systems will be discussed in detail in Chapter 4.

VI. TERMINATION

In contrast to transfer, which limits the growth of only an individual polymer chain, termination stops the kinetic chain. It is unavoidable in radical polymerizations because two growing radicals will always react with each other. However, because neither two growing cations nor two growing anions can react with each other due to electrostatic repulsion, termination can sometimes be avoided in ionic polymerizations if the counterion is sufficiently stable. Although termination does occur in carbocationic polymerizations, transfer is the dominant chain-breaking process. The three general types of termination in carbocationic polymerizations which will be discussed in the next sections are formation of unreactive or "too stable" carbenium ions, irreversible decomposition of counteranions generating inactive covalent species, and reactions with nucleophiles present in the system.

Termination is formally an *irreversible deactivation* of growing species. That is, *reversible termination* is not a real termination process and would be more appropriately labeled *reversible deactivation*. If this reversible deactivation is sufficiently dynamic, the number of growing species remains constant throughout the polymerization and all chains have the same opportunity to grow, resulting in polymers with narrow molecular weight distributions. This will be discussed in detail in Chapter 4.

A. Termination by Formation of Unreactive Carbenium Ions

As discussed in previous sections, the reactivity of carbenium ions depends on both steric and electronic factors. The "normal" growing carbenium ion at the chain end is either a secondary (vinyl ether, styrene) or tertiary (isobutene, α-methylstyrene) cationic center. However, more stable tertiary carbenium ions may be generated in the interior of the chain by hydride or methide transfers [Eq. (120)].

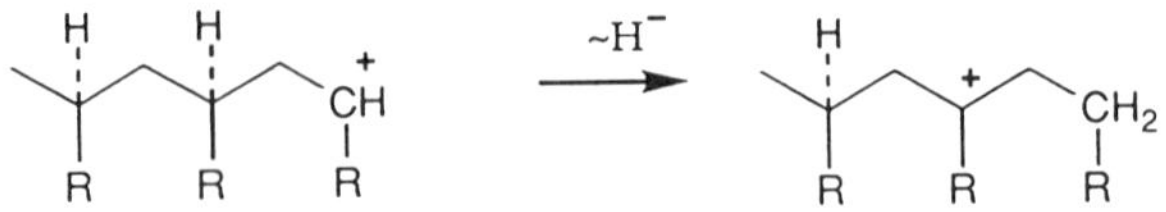

(120)

The resulting cations are much less reactive than growing species due to steric and/or electronic effects. Their generation is therefore a termination reaction.

In many cases, transfer must occur before this termination take place. As outlined below, the requisite transfer reactions include proton transfer reactions and formation of polymer chains with unsaturated or indanyl end groups. The resulting chains then undergo further reactions to generate stable and unreactive carbenium ions, such as indanyl ions, sterically protected tertiary carbenium ions, and highly delocalized protonated polyenes.

Terminal unsaturation favors termination by formation of carbocations which are additionally stabilized by double bonds. This can be illustrated by β-proton elimination and subsequent loss of methide in polymerization of isobutene [Eq. (121)].

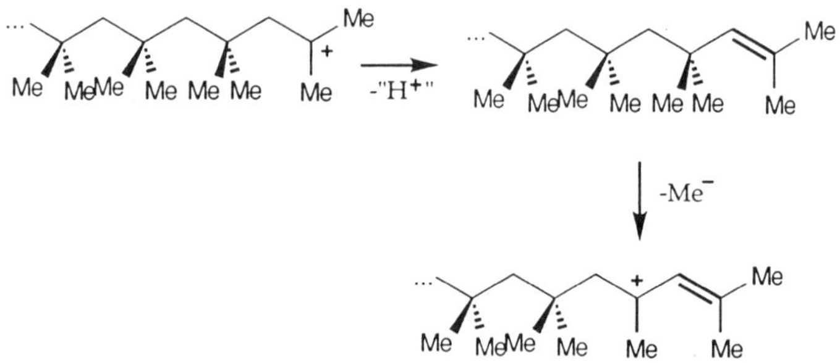

(121)

Similar processes can occur in polymerizations of styrenes and vinyl ethers. Various colors are observed in vinyl ether polymerizations when protonated polyene sequences are generated by loss of several alcohol molecules [Eq. (122)] [324].

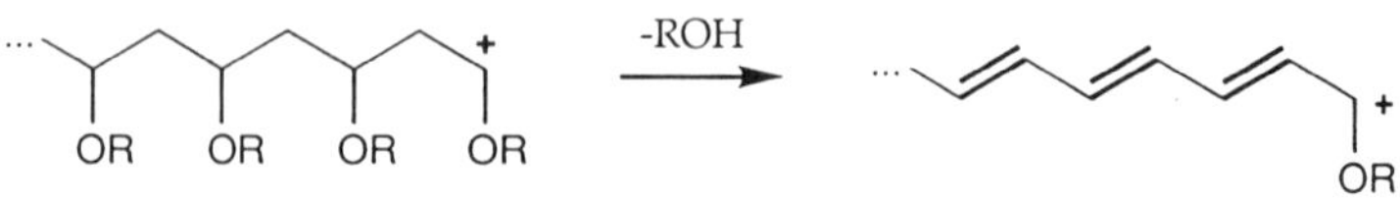

(122)

Absorbances of up to 500 nm indicate that the unsaturated units ($n < 10$) are relatively long. The color depends on the acidity of the medium, with neutralized oligomers being yellow, and acidified oligomers black [Eq. (123)].

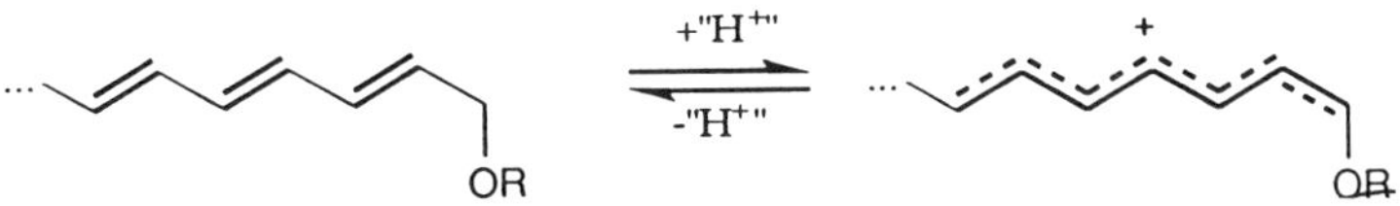

(123)

The formation of stable carbenium ions can be observed visually and/or spectroscopically. For example, styrene and α-methylstyrene polymerizations are generally colorless because the growing carbenium ions absorb at approximately 340 nm (cf., Sections II.B and IV.B.1). However, these systems may turn brown or dark red at longer reaction times due to formation of indanyl carbenium ions ($\lambda \approx 440$ nm) [14,26,325] and other delocalized carbocations similar to those in Eq. (121). The stable cyclic diaryl carbenium ions are generated by hydride transfer from the initially formed indanyl end groups [Eq. (124)] in styrene polymerizations, and by methide transfer in α-methylstyrene polymerizations. The prerequisite for this termination is therefore intramolecular transfer by Friedel-Crafts alkylation; protons liberated in the first stage can then reinitiate polymerization.

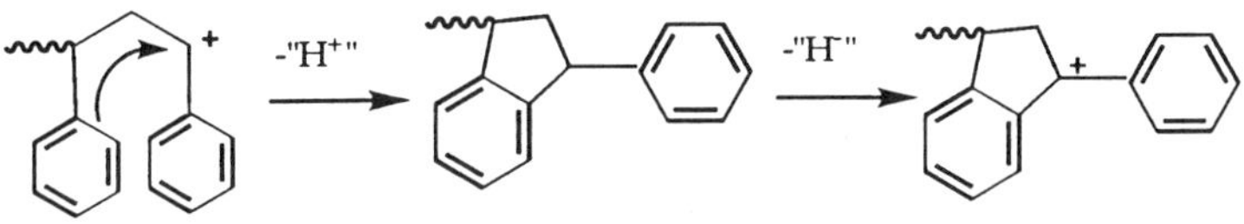

(124)

The red color observed at later stages in polymerization of p-methoxystyrenes is probably due to formation of a charge transfer complex with two adjacent aromatic rings as shown in Eq. (9). The internal tertiary carbocation is sterically not accessible and is unreactive. It may be formed by a 1,3-methide shift or by reaction of unsaturated oligomers with growing species [13]. In the latter case, transfer by elimination occurs before termination.

B. Reaction with Counterions

In contrast to the dynamic deactivation of carbenium ions with counterions based on large polarizable halides [Eq. (125)] (cf., Chapter 4 for the new "living" systems),

$$\text{...-}CH_2CHR\text{-}X + LA \rightleftharpoons \text{...-}CH_2\overset{+}{C}HR,\ L\overset{-}{A}X \qquad (125)$$

exchange with counterions based on fluoride ligands is extremely slow. Both the fluorine-metal and fluorine-carbon bonds are much stronger and more difficult to cleave. Decomposition of the complex anions MtF_n^- is generally irreversible, leading to true termination [Eq. (126)].

$$\text{...-}CH_2C^{+}HPh,\ ^{-}PF_6 \underset{?}{\rightleftharpoons} \text{...-}CH_2CHPh\text{-}F + PF_5 \tag{126}$$

Thus, nearly all propagating chains are in the form of carbenium ions in polymerizations with hexafluorophosphate or hexafluoroantimonate anions, resulting in very fast polymerizations [131]. However, once these ions decompose to alkyl fluorides, the latter can not be reactivated.

Termination also occurs if ligand exchange produces a weaker Lewis acid which is unable to reactivate the covalent adduct. In this case, addition of another aliquot of the stronger acid may reactivate the system. For example, hydroxy or alkoxy groups are exchanged in polymerization systems based on boron trifluoride and adventitious water or alcohol, respectively. Termination occurs infrequently by irreversible cleavage of B—F bond from the BF_3OH^- anion, with irreversible C—F bond formation [Eq. (127)].

$$\text{...-}CH_2CHPh\text{-}OH + BF_3 \rightleftharpoons \text{...-}CH_2C^{+}HPh,\ ^{-}BF_3OH \underset{?}{\rightleftharpoons} \text{...-}CH_2CHPh\text{-}F + BF_2OH \tag{127}$$

The Lewis acid BF_2OH is evidently too weak to reactivate the covalent species, especially covalent species with strong C—F bonds, although analogous reactions occur with the chloride and bromide derivatives. However, excess BCl_3 is required to complete alkyl acetate/BCl_3-initiated polymerizations because ligand exchange leads to the much weaker Lewis acid, BCl_2OAc [104] [Eq. (128)].

$$\begin{gathered}\text{...-}CH_2CHPh\text{-}OAc + BCl_3 \rightleftharpoons \text{...-}CH_2\overset{+}{C}HPh,\ \overset{-}{BCl_3OAc} \underset{?}{\rightleftharpoons} \text{...-}CH_2CHPh\text{-}Cl + BCl_2OAc \\ \text{...-}CH_2CHPh\text{-}Cl + BCl_3 \rightleftharpoons \text{...-}CH_2\overset{+}{C}HPh,\ \overset{-}{BCl_4}\end{gathered} \tag{128}$$

Finally, termination also occurs if and when the Lewis acid is consumed by side reactions. For example, $SbCl_5$ adds rapidly to double bonds and must therefore be used in large excess to complete styrene polymerizations at all but very low temperatures [144]. In this case, $SbCl_3$ is too weak a Lewis acid to reactivate the corresponding alkyl chlorides [Eq. (129)].

$$
\begin{array}{l}
\ldots\text{-}CH_2CHPh\text{-}Cl + SbCl_5 \rightleftharpoons \ldots\text{-}CH_2\overset{+}{C}HPh,\ \overset{-}{S}bCl_6 \\
CH_2{=}CHPh + SbCl_5 \longrightarrow Cl\text{-}CH_2\text{-}CHPh\text{-}Cl + SbCl_3 \\
\ldots\text{-}CH_2CHPh\text{-}Cl + SbCl_3 \not\rightleftharpoons \ldots\text{-}CH_2\overset{+}{C}HPh,\ \overset{-}{S}bCl_4
\end{array}
\qquad (129)
$$

Trityl hexachloroantimonate-initiating systems behave similarly [145]. In contrast, only a small concentration of BCl_3 is required to complete BCl_3-initiated polymerizations in the absence of water because haloboration (Section III.A.3.a.2) is usually slow.

Termination can also occur in Lewis acid-activated systems by transfer reactions producing volatile hydrogen halide which may be lost from the reaction mixture [Eq. (130)].

$$
\ldots\text{-}CH_2\overset{+}{C}HPh,\ \overset{-}{B}Cl_4 \underset{?}{\rightleftharpoons} \ldots\text{-}CH{=}CHPh + BCl_3 + HCl\uparrow \qquad (130)
$$

Such systems are therefore not very efficient, especially at low concentrations.

C. Reactions with Nucleophiles

Carbenium ions react with neutral nucleophiles to produce onium ions. A favorable equilibrium between active carbenium ions and temporarily inactive onium ions can be used to produce well-defined polymers (Chapter 4). However, rather than reacting directly with carbenium ions, nucleophiles may also react with Lewis acids to form strong complexes, thereby reducing their activity and ability to ionize covalent compounds. A third reaction that basic nucleophiles may be involved in is β-proton elimination; this transfer reaction may subsequently result in termination if it involves a strong base [Eq. (131)].

$$
\ldots CH_2\text{-}CHPh^+,\ {}^-LAX + Nu
\begin{cases}
\rightleftharpoons \ldots CH_2\text{-}CHPh\text{-}Nu^+,\ {}^-LAX \\
\rightleftharpoons \ldots CH_2\text{-}CHPh\text{-}X + LA:Nu \\
\rightleftharpoons \ldots CH{=}CHPh + H\text{-}{}^+Nu,\ {}^-LAX
\end{cases}
\qquad (131)
$$

Thus, the extent of termination depends on the equilibrium positions in Eq. (131) and on subsequent reactions. For example, termination occurs when very stable onium ions are generated which do not isomerize back to active carbenium ions during the time of the polymerization or when another moiety (Y or Z) within the nucleophile allows it to release a cationic species such as acyl, alkoxyalkyl, or proton which is less reactive than the original carbenium ion [Eq. (132)]. If reinitiation by this new cationic species is slow, reaction with the nucleophile is a degradative chain transfer reaction; i.e., termination.

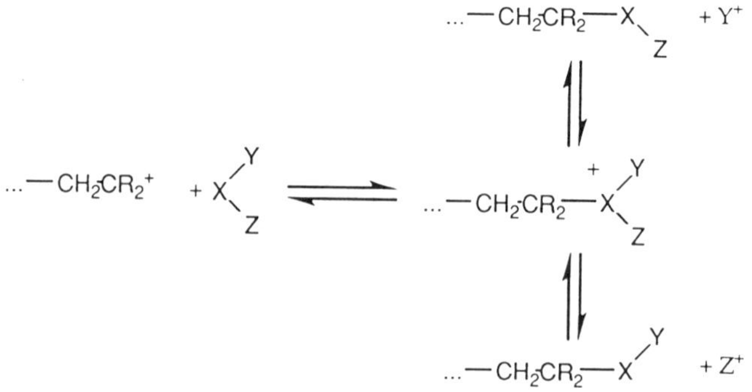

(132)

Water reacts with carbenium ions to produce primary oxonium ions, which usually decompose to alcohols and protonic acids. Alcohol can then react with growing carbenium ions to produce secondary oxonium ions [56], which decompose to a proton and a polymer with an ether end group (cf., Section V.A.3). Such reactions are quite probable in cationic polymerizations because water is the most common impurity in insufficiently purified monomer or solvent. Therefore, cationic polymerizations using a very low initiator concentration ($<10^{-4}$ mol/L) may be unsuccessful due to the concentration of water being considerably higher than that of initiator. However, most new "living" systems use much higher concentrations of initiator ($>10^{-3}$ mol/L), which therefore exceed the concentration of water. That is, the lower sensitivity to water is simply due to its concentration being too low relative to that of the initiator to cause termination, rather than to the altered structure of active sites being less sensitive to water as has been proposed [259]. The sensitivity to water will also depend on the nucleophilicity of the monomer. For example, the reactivity of vinyl ethers is similar to that of water and alcohols [240], although carbenium ions usually react with nucleophiles faster than with most alkenes [41]. Thus, small amounts of water do not affect vinyl ether polymerizations, but have much greater effect on styrene and isobutene polymerizations.

D. Monitoring Termination Reactions

Termination generally reduces the concentration of active species. However, a stationary concentration of active species is established if the rate of initiation is slow and approximately the same as that of termination. This is the case for conventional radical polymerizations, and also for conventional cationic polymerizations initiated by Lewis acids in combination with adventitious H_2O (cf., Section III.A.3). The concentration of growing species also determines how termination affects the polymerization kinetics. If the concentration of carbenium ions is very low, they can be destroyed by impurities. If the concentration of carbenium ions is high relative to that of impurities, the ultimate conversion depends on both the ratio of the rate constants of propagation and termination and on the concentration of initiator. For unimolecular termination, e.g., reactions illustrated by Eqs. (120), (126), and (130), and for propagation which is first order in monomer and in initiator, Eq. (133) can be applied if initiation is fast.

$$\ln ([M]_0/[M]_\infty) = [I]_0 k_p/k_t \tag{133}$$

For example, if $k_p/k_t = 10^3\ \text{mol}^{-1}\cdot\text{L}$, the ultimate conversion will be 1%, 10%, 63%, and 99.9% for $[I]_0 = 10^{-5}$ mol/L, 10^{-4} mol/L, 10^{-3} mol/L, and 10^{-2} mol/L, respectively. Thus, the ratio k_p/k_t can be obtained by determining the monomer conversion as a function of the concentration of initiator, assuming that initiation is quantitative. The same equation can be applied when termination is pseudo-unimolecular, and excess terminating agent, T, reacts slowly with active species. In this case, k_t should be substituted with $k_t\cdot[T]_0$. If transfer is a prerequisite for termination, it may become the rate-determining step, and the actual termination may not be kinetically measurable.

Rate constants of termination can also be determined by directly measuring the concentration of growing species. This is often measured spectroscopically, such as by stopped-flow UV measurements in styrene polymerizations [18]. In systems with reversible deactivation, the concentration of active and dormant growing chains can be determined by first reacting the growing species with nucleophiles and then quantitatively detecting the products spectroscopically. The original papers using this technique assumed that the nucleophiles reacted only with active carbenium ions. However, dormant onium ions and covalent species also react with most of the traps used, and/or are converted to carbenium ions during the time of analysis. Styrene polymerizations initiated by perchlorates were first monitored successfully by UV analysis of naphthyl ether chain ends generated by naphthoxide traps [210]. Both the covalent esters and carbenium ions apparently reacted with the naphthoxide anions according to Eq. (134); only unsaturated chain ends and indanyl-terminated chains do not react.

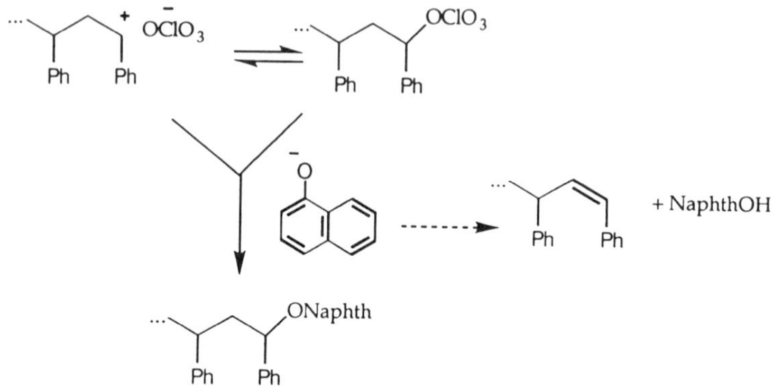

(134)

However, the danger of this approach is that the traps themselves may cause β-proton elimination which lead to undetectably low concentration of active (or temporarily deactivated) chain ends. Malonate traps and ^{1}H NMR analysis of the resulting terminal ester groups have been used successfully to monitor the growing chains in vinyl ether polymerizations [209]. This technique measures the sum of dormant species and growing carbenium ions. Another method first used to monitor the active species in ring-opening polymerizations traps the active and dormant chain ends with phosphines [Eq. (135)].

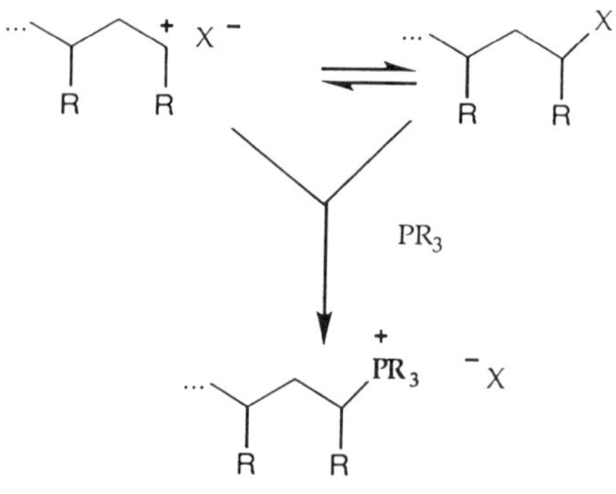

(135)

^{31}P NMR is then used to determine not only the amount of the resulting phosphonium ions, but also their structure [326]. ^{31}P NMR is very sensitive to even the remotest structural changes in the substituents at phosphorus, with chemical shifts spread over a large scale (>500 ppm). This tool is important not only for determining the structure of carbenium ions, but also for determinatining the microstructure, tacticity and comonomer sequence distribution of the corresponding (co)polymers. The overall sensitivity of ^{31}P NMR is also high due to its large abundance and high gyromagnetic ratio. This generates reliable spectra even at concentrations as low as 10^{-3} mol/L, which is close to the concentration of active species in many polymerization systems.

Phosphine trapping and ^{31}P NMR has also been used to monitor polymerizations of vinyl ethers and follow the polymer tacticity [208]. However, less nucleophilic phosphines, such as tris(*p*-chlorophenyl)phosphine, do not terminate the growing chains efficiently, and polymerization continues slowly [66]. Nevertheless, trapping with less nucleophilic phosphines provides information on microstructure and also on kinetics in slow polymerizations.

E. Effects of Medium on Termination

Decreasing temperature usually results in higher polymer yields and higher molecular weight polymers from carbocationic polymerizations of alkenes. This indicates that the activation energies of termination and transfer are higher than those of propagation, and is also compatible with transfer being a prerequisite for some termination reactions.

As discussed previously in section III.D.5, less polar solvents accelerate ion-ion reactions. Therefore, collapse of propagating ion pairs are faster, for example, in hydrocarbon solvents than in chlorinated solvents. Termination reactions of the propagating carbenium ions with impurities or other neutral molecules are ion-dipole reactions, which are also accelerated in less polar solvents, although to a much lower degree. In addition to variations in solvent polarity influencing the relative rates of the elementary reactions in carbocationic polymerizations, the solvent itself can terminate the polymerization if it is too nucleophilic. Solvent can terminate the polymerization by reacting with either growing carbenium ions to form inactive onium ions, or with Lewis acid to form inactive complexes. In addition, solvents that are strong bases will abstract β-protons; this transfer reaction is often followed by termination.

F. Forced Termination

If termination is neglible throughout the polymerization, the propagating chains will retain their activity until complete monomer consumption. These electrophilic chain ends can therefore be functionalized by controlled, forced termination, to generate macromonomers and telechelic oligomers and polymers. Usually anions and other strong nucleophiles are used [327]. In some systems, organosilicon derivatives can be used [328]. These functionalized polymers can then be used to synthesize block, multiblock, graft and star polymers, and copolymers, as well as model networks. Controlled forced termination will be discussed in detail in Chapter 5.

REFERENCES

1. G. A. Olah and P. v. R. Schleyer, *Carbonium Ions*, Wiley, New York, 1970.
2. G. A. Olah, *Friedel-Crafts and Related Reactions* Wiley, New York, 1963.
3. P. Vogel, *Carbocation Chemistry*, Elsevier, Amsterdam, 1985.
4. H. Mayr, *Angew. Chem. 29*:1371 (1990).
5. G. A. Olah, *Halonium Ions*, Wiley, New York, 1975.
6. I. I. Schuster, A. K. Colter and R. J. Curland, *J. Am. Chem. Soc. 90*:4679 (1968).
7. G. Frenkel and D. G. Farnum, *Carbonium Ions* (G. A. Olah and P. v. R. Schleyer, eds.) Wiley, New York, 1968, vol. 1, p. 237.
8. D. P. Kelly and R. J. Spear, *Aust. J. Chem. 31*:1209 (1978).
9. R. N. Young, *Prog NMR Spectroscopy 12*:261 (1979).
10. G. A. Olah, A. L. Berrier, and G. K. S. Prakash, *J. Am. Chem. Soc. 104*: 2373 (1982).
11. M. Teodorescu, C. H. Lin, and K. Matyjaszewski, *Macromol. Chem. Phys., 196*:2149 (1995).
12. M. Szwarc, *Carbanions, Living Polymers and Electron Transfer Processes*, Interscience Publishers, New York, 1968.
13. M. Moreau, K. Matyjaszewski, and P. Sigwalt, *Macromolecules 20*:1456 (1987).
14. K. Matyjaszewski and P. Sigwalt, *Macromolecules 20*:2679 (1987).
15. G. A. Olah, C. U. Pittman, R. Waack, and M. Doran, *J. Am. Chem. Soc. 88*:1488 (1966).
16. D. C. Pepper, *Makromol. Chem. 175*:1077 (1974).
17. T. Kunitake and K. Takarabe, *Macromolecules 12*:1067 (1979).
18. J. P. Vairon, A. Rives, and C. Bunel, *Makromol. Chem., Macromol. Symp. 60*:97 (1992).
19. M. DeSorgo, D. C. Pepper, and M. Szwarc, *Chem. Commun.* 419 (1973).
20. R. A. McClelland, C. Chan, F. Cozens, A. Modro, and S. Steenken, *Angew. Chem. 30*:1337 (1991).
21. K. Takarabe and T. Kunitake, *Makromol. Chem. 182*:1587 (1981).
22. T. Kunitake and K. Takarabe, *J. Polym. Sci., Polym. Symp. 56*:33 (1976).
23. M. Sawamoto and T. Higashimura, *Macromolecules 12*: 581 (1979).
24. T. Higashimura, N. Kanoh, and S. Okamura, *J. Macromol. Sci., Chem. 1*: 109 (1966).
25. D. Teyssie, M. Villesange, and J. P. Vairon, *Polym. Bull. 11*:459 (1984).
26. H. Cheradame, N. A. Hung, and P. Sigwalt, *J. Polym. Sci. Symp. 56*:335 (1976).
27. G. Sauvet, J. P. Vairon, and P. Sigwalt, *J. Polym. Sci., Polym. Symp. 52*: 173 (1975).
28. B. E. Fleischfresser, W. J. Cheng, J. M. Pearson, and M. Szwarc, *J. Am. Chem. Soc. 90*:2172 (1968).
29. M. Moreau, Doctorat d'Etat, University of Paris (1988).
30. J. C. Evans, *Carbonium Ions* (G. A. Olah and P. v. R. Schleyer, eds.), Wiley, New York, 1968, vol. 1, p. 223.

31. G. Olah, *Angew. Chem. 12*:173 (1973).
32. R. G. Ghirardelli, *J. Am. Chem. Soc. 95*:4987 (1973).
33. H. P. J. M. Dekkers and E. C. M. K.-V. Luyt, *Mol. Phys. 31*:1001 (1976).
34. R. Schneider, H. Mayr, and P. H. Plesch, *Ber. Bunsenges Phys. Chem. 91*:1369 (1987).
35. P. M. Bowyer, A. Ledwith, and D. C. Sherrington, *J. Chem. Soc. (B) 1971*: 1511 (1971).
36. N. Kolfoglu and M. Szwarc, *J. Phys. Chem. 72*:2233 (1968).
37. W. Gogolczyk, S. Slomkowski, and S. Penczek, *J. Chem. Soc., Perkin II 1977*:1729 (1977).
38. G. Heublein and D. Bauerfeind, *J. Prakt. Chem. 326*:81 (1984).
39. D. C. Sherrington, *J. Polym. Sci. Symp. 56*:323 (1976).
40. J. T. Denison and J. B. Ramsey, *J. Am. Chem. Soc. 77*:2615 (1955).
41. V. M. DePalma, Y. Wang, and L. M. Dorfman, *J. Am. Chem. Soc. 100*: 5416 (1978).
42. M. Feldman and S. Winstein, *J. Am. Chem. Soc. 83*:3338 (1961).
43. G. Heublein, *J. Macromol. Sci. Chem. A16*:563 (1981).
44. T. Dittmer, S. D. Pask, and O. Nuyken, *Makromol. Chem. 193*:2547 (1992).
45. T. Kunitake and S. Tsugawa, *Macromolecules 8*:709 (1975).
46. G. Sauvet, M. Moreau, and P. Sigwalt, *Makromol. Chem., Macromol. Symp 3*:33 (1986).
47. Y. Pocker, W. A. Miller, F. Naso, and G. Tocchi, *J. Am. Chem. Soc 86*: 5011 (1964).
48. K. Heald and G. Williams, *J. Chem. Soc. 1954*:362 (1954).
49. R. M. Evans and R. S. Satchell, *J. Chem. Soc. B. 1970*:300 (1970).
50. R. S. Satchell, *J. Chem. Soc. 1964*:5464 (1964).
51. D. P. N. Satchell and R. S. Satchell, *Chem. Rev. 69*:251 (1969).
52. Z. Kecki and B. Izdebska, *Rocz. Chem. 40*:1529 (1966).
53. A. Diaz, I. L. Relch, and S. Winstein, *J. Am. Chem. Soc. 92*:7598 (1970).
54. H. L. Goering, R. G. Briody, and G. Sandrock, *J. Am. Chem. Soc. 92*: 7401 (1970).
55. S. Chang and W. J. LeNoble, *J. Am. Chem. Soc. 105*:3708 (1983).
56. K. Matyjaszewski and P. Sigwalt, *Makromol. Chem. 187*:2299 (1986).
57. S. Penczek and R. Szymanski, *Makromol. Chem., Macromol. Symp. 60*: 65 (1992).
58. C. G. Cho, B. A. Feyt, and O. W. Webster, *Macromolecules 23*:1918 (1990).
59. K. Matyjaszewski, *Makromol. Chem., Macromol. Symp. 60*:107 (1992).
60. S. Slomkowski and S. Penczek, *J. Chem. Soc. Perkin II 1974*:1718 (1974).
61. L. M. Dorfman and V. M. DePalma, *Pure Appl. Chem. 51*:123 (1979).
62. J. Bartl, S. Steenken, and H. Mayr, *J. Am. Chem. Soc. 113*:7710 (1991).
63. C. B. Kim and K. T. Leffek, *Can. J. Chem. 53*:3408 (1975).
64. S. Penczek and R. Szymanski, *Polym. J. 12*:617 (1980).
65. M. Bjoroy, B. B. Saunders, S. Esperas, and J. Songstad, *Phosphorus 1976*: 83 (1976).
66. C. H. Lin and K. Matyjaszewski, *Polym. Prep. (Am. Chem. Soc., Div. Polym. Chem.) 35*(1):462 (1994).

67. K. Matyjaszewski and C.-H. Lin, *Makromol. Chem. Macromol. Symp. 47*: 221 (1991).
68. J. P. Kennedy, *Cationic Polymerization of Olefins: A Critical Inventory*, Wiley, New York, 1975.
69. A. F. Johnson and R. N. Young, *J. Polym. Sci., Polym. Symp. 56*:211 (1976).
70. G. Sauvet, and P. Sigwalt, *Comprehensive Polymer Science* (G. Allen and J. C. Bevington, eds.), Pergamon Press, Oxford, 1989, vol. 3, p. 579.
71. A. Gandini and H. Cheradame, *Adv. Polym. Sci. 34/35*:1 (1980).
72. T. Fujinaga and I. Sakamoto, *Electroanal. Chem. 85*:185 (1977).
73. D. Coutagne, Doctorate Thesis, Grenoble University (1973).
74. A. Engelbrecht and B. M. Rode, *Monatsh. Chem. 103*:1315 (1972).
75. M. Bos and E. A. M. F. Dahmen, *Anal. Chim. Acta 63*:185 (1973).
76. I. M. Kolthoff, S. Bruckenstein and M. K. Chantooni, *J. Am. Chem. Soc. 83*:3927 (1961).
77. T. Jasinski, A. A. El-Harakany, F. Halaka and H. Sadek, *Croat. Chem. Acta 51*:1 (1978).
78. F. Bolza and F. E. Treloar, *J. Chem. Eng. Data 17*:197 (1972).
79. B. M. Rode, A. Engelbrecht and J. Schantl, *Z. Phys. Chem. 253*:17 (1973).
80. K. Matyjaszewski, S. Penczek and E. Franta, *Polymer 20*:1184 (1979).
81. K. Matyjaszewski, *Macromolecules 21*:933 (1988).
82. D. M. Coutagne, *J. Am. Chem. Soc. 93*:1518 (1971).
83. Z. I. Grigorovich, Y. I. Malov and V. Y. Rosoloskii, *Bull. Acad. Sci. USSR Chem. 257* (1973).
84. C. P. Brown and A. R. Matheson, *J. Phys. Chem. 58*:1057 (1954).
85. L. W. Reeves and W. G. Schneider, *Trans. Faraday Soc. 54*:314 (1958).
86. L. W. Reeves, *Can. J. Chem. 39*:1711 (1961).
87. F. Thyrion, D. Decroocq, *C. R. Acad. Sci. 260*:2797 (1965).
88. C. P. Brown and A. R. Mathieson, *J. Chem. Soc. 1957*:3625 (1957).
89. T. S. S. R. Murty and K. S. Pitzer, *J. Phys. Chem. 73*:1426 (1969).
90. M. Kirszenbaum, J. Corset and M. L. Josien, *J. Phys. Chem. 75*:1327 (1971).
91. A. C. Pavia, *Rev. Chim. Min. 7*:471 (1970).
92. I. M. Kolthoff and J. M. K. Chantooni, *J. Am. Chem. Soc. 95*:8539 (1973).
93. A. Gandini and S. Prieto, *J. Polym. Sci., Polym. Lett. Ed. 15*:337 (1977).
94. H. Scott, T. P. Konen and M. M. Labes, *J. Polym. Sci., Polym. Lett. B-2*:689 (1964).
95. C. P. Brown and A. R. Mathieson, *J. Chem. Soc. 1958*:3445 (1958).
96. C. P. Brown and A. R. Mathieson, *J. Chem. Soc. 1958*:3507 (1958).
97. A. R. Mathieson, *J. Chem. Soc. 1958*:268 (1958).
98. J. Murphy, L. Roubinek and A. Wassermann, *J. Chem. Soc. 1961*:1964 (1961).
99. A. G. Evans, P. M. S. Jones and J. H. Thomas, *J. Chem. Soc. 1958*:4563 (1958).
100. J. J. Throssell, S. P. Sood, M. Szwarc and V. Stannett, *J. Am. Chem. Soc. 78*:1122 (1956).

101. M. Sawamoto, T. Masuda, T. Higashimura, S. Kobayashi and T. Saegusa, *Makromol. Chem. 178*:389 (1977).
102. W. Obrecht and P. H. Plesch, *Makromol. Chem. 182*:1459 (1981).
103. K. Matyjaszewski and C. H. Lin, *J. Polym. Sci., Part A: Polym. Chem. 26*:3031 (1988).
104. K. Matyjaszewski and C. H. Lin, *J. Polym. Sci., Part A: Polym. Chem. 29*:1439 (1991).
105. J. P. Kennedy and B. Ivan, *Designed Polymers by Carbocationic Macromolecular Engineering. Theory and Practice*, Hanser, Munich, 1992.
106. R. Faust and J. P. Kennedy, *J. Macromol. Sci., Chem. A27*:649 (1990).
107. M. Kamigaito, K. Yamaoka, M. Sawamoto and T. Higashimura, *Macromolecules 25*:6400 (1992).
108. R. C. Fahey, C. A. McPherson and R. A. Smith, *J. Am. Chem. Soc. 96*:4534 (1974).
109. D. J. Pasto, G. R. Meyer and B. Lepeska, *J. Am. Chem. Soc. 96*:1858 (1974).
110. D. J. Pasto and J. F. Gadberry, *J. Am. Chem. Soc. 100*:1469 (1978).
111. J. Pielichowski, *J. Polym. Sci., Polym. Symp. 42*:451 (1973).
112. M. Sawamoto, J. Fujimori and T. Higashimura, *Macromolecules 20*:916 (1987).
113. H. Cramail and A. Deffieux, *Macromolecules 27*:1401 (1994).
114. T. Higashimura, M. Miyamoto and M. Sawamoto, *Macromolecules 18*:611 (1985).
115. Y. Tsuda, *Makromol. Chem. 36*:102 (1960).
116. M. Berthelot, *Bull. Soc. Chim. Fr. 6*:294 (1866).
117. R. E. Burton and D. C. Pepper, *Proc. Roy. Soc. A 263*:58 (1961).
118. M. G. Hayes and D. C. Pepper, *Proc. Roy. Soc. A 263*:63 (1961).
119. A. Albert and D. C. Pepper, *Proc. Roy. Soc. A 263*:75 (1961).
120. D. H. Jenkinson and D. C. Pepper, *Proc. Roy. Soc. A263*:82 (1961).
121. C. Peniche and A. Gandini, *Rev. CENIC (Cuba) 4*:59 (1973).
122. S. Tazuke, T. B. Tjoa and S. Okamura, *J. Polym Sci. A-15*:1911 (1967).
123. T. Masuda, M. Sawamoto, and T. Higashimura, *Makromol. Chem. 177*:2981 (1976).
124. M. Sawamoto and T. Higashimura, *Macromolecules 11*:328 (1978).
125. E. Mathias and P. H. Plesch, *Chem. Ind. (Lond.) 1971*:1043 (1971).
126. D. C. Pepper and P. J. Reilly, *J. Polym. Sci. 58*:291 (1962).
127. A. Gandini and P. H. Plesch, *Eur. Polym. J. 4*:55 (1968).
128. A. Gandini and P. H. Plesch, *Polym. Lett. 3*:127 (1965).
129. M. Chmelir, N. Cardona, and G. V. Schulz, *Makromol. Chem. 178*:169 (1977).
130. M. Chmelir, *Makromol. Chem. 176*:2099 (1975).
131. D. C. Pepper, *J. Polym. Sci., Polym. Symp. 50*:51 (1975).
132. T. Higashimura, Y. Miyoshi, and H. Hasegawa, *J. Appl. Polym. Sci. 28*:241 (1983).
133. C. H. Lin and K. Matyjaszewski, *Polym. Prepr. (Am. Chem. Soc., Div. Polym. Chem.), 31(1)*:599 (1990).

134. P. H. Plesch, *Prog. React. Kinetics 18*:1 (1993).
135. P. Giusti and F. Andruzzi, *J. Polym. Sci. C16*:3797 (1968).
136. L. Bui, H. A. Nguyen and E. Marechal, *Polym. Bull. 17*:157 (1987).
137. L. Balogh, L. Wang, and R. Faust, *Macromolecules, 27*:3453 (1994).
138. D. C. Jordan and A. R. Mathieson, *J. Chem. Soc. 1952*:90 (1952).
139. J. H. Beard, P. H. Plesch, and P. P. Rutherford, *J. Chem. Soc. 1964*:2566 (1964).
140. M. Chmelir, M. Marek, and O. Wichterle, *J. Polym. Sci., Part C 16*:833 (1967).
141. R. Branchu Bourne, H. Cheradame, and P. Sigwalt, *C.R. Hebd. Seances Acad. Sci. 268*:1292 (1969).
142. M. Masure, G. Sauvet, and P. Sigwalt, *J. Polym. Sci., Polym. Chem. Ed. 16*:3065 (1978).
143. I. C. Lewis and L. S. Singer, *J. Chem. Phys. 43*:2712 (1965).
144. W. Bracke, W. J. Cheng, J. M. Pearson and M. Szwarc, *J. Am. Chem. Soc., 91*:203 (1969).
145. A. F. Johnson and D. A. Pearce, *J. Polym. Sci., Polym. Symp. 56*:57 (1976).
146. W. H. Hunter and R. V. Johe, *J. Am. Chem. Soc. 55*:1248 (1933).
147. G. A. Olah, *Makromol. Chem., 175*:1039 (1974).
148. C. H. Lin, J. S. Xiang and K. Matyjaszewski, *Macromolecules 26*:2785 (1993).
149. D. D. Eley, F. L. Isack, and C. H. Rochester, *J. Chem. Soc. 1968*:1651 (1968).
150. T. Higashimura, H. Teranishi, and M. Sawamoto, *Polym. J. 12*:393 (1980).
151. T. Higashimura, Y.-X. Deng, and M. Sawamoto, *Polym. J. 15*:385 (1983).
152. N. A. Ghanem and M. Marek, *Eur. Polym. J. 8*:999 (1972).
153. J. P. Kennedy, S. Y. Huang, and S. C. Feinberg, *J. Polym. Sci., Polym. Chem. Ed. 15*:2869 (1977).
154. D. O. Jordan and F. E. Treloar, *J. Chem. Soc. 1961*:737 (1961).
155. J. P. Kennedy and R. G. Squires, *J. Macromol. Sci. Chem. 1*:995 (1967).
156. C. M. Fontana and G. A. Kidder, *J. Am. Chem. Soc. 70*:3745 (1948).
157. M. Masure, G. Sauvet, and P. Sigwalt, *Polym. Bull. 2*:699 (1980).
158. M. Masure, P. Sigwalt, and G. Sauvet, *Makromol. Chem., Rapid Commun. 4*:269 (1983).
159. J. Maslinska-Solich, M. Chmelir, and M. Marek, *Collect. Czech. Chem. Commun. 34*:2611 (1969).
160. H. Hirata and H. Tani, *Polymer 9*:60 (1968).
161. M. Deville, *Ann. Chim. Phys. 75*:66 (1839).
162. R. O. Colclough and F. S. Dainton, *Trans. Faraday Soc. 1958*:54 (1958).
163. C. G. Overberger, R. J. Ehring, and R. A. Marcus, *J. Am. Chem. Soc. 80*: 2456 (1958).
164. A. G. Evans, and J. Lewis, *J. Chem. Soc. 1957*:2975 (1957).
165. T. Higashimura, M. Kamigaito, M. Kato, T. Hasebe, and M. Sawamoto, *Macromolecules 26*:2670 (1993).
166. K. Matyjaszewski, A. Bon, C. H. Lin, and J. S. Xiang, *Polym. Prep. (Am. Chem. Soc., Div. Polym. Chem.) 34(1)*:487 (1993).

167. P. Lopour, J. Pecka, and M. Marek, *Makromol. Chem. 174*:1 (1973).
168. B. Matyska, L. Petrusova, K. Mach, and P. Svestka, *Collect. Czech. Chem. Commun. 44*:1262 (1979).
169. L. Thomas, A. Polton, M. Tardi, and P. Sigwalt, *Macromolecules 25*:5886 (1992).
170. L. Thomas, M. Tardi, A. Polton, and P. Sigwalt, *Macromolecules 26*:4075 (1993).
171. P. Sigwalt and J. P. Vairon, *Bull. Soc. Chim. 1964*:482 (1964).
172. H. Cheradame and P. Sigwalt, *C. R. Hebd. Seances Acad. Sci. 259*:4273 (1964).
173. H. Cheradame and P. Sigwalt, *Bull. Soc. Chim. Fr. 1970*:843 (1970).
174. H. Cheradame and P. Sigwalt, *Bull. Soc. Chim. Fr. 1970*:849 (1970).
175. R. F. Storey, B. J. Chisholm, and E. Mowbray, *Polym. Prep. (Am. Chem. Soc., Div. Polym. Chem.) 34(1)*:568 (1993).
176. Z. Fodor, M. Gyor, H. C. Wang, and R. Faust, *J. Macromol. Sci., Pure Appl. Chem. A30*:349 (1993).
177. K. Kojima, M. Sawamoto, and T. Higashimura, *Macromolecules 22*:1552 (1989).
178. C. G. Cho and J. McGrath, *J. Macromol. Sci., Chem. A25*:499 (1987).
179. M. Szwarc and M. V. Beylen, *Ionic Polymerization and Living Polymers*, Chapman Hill, New York, 1993.
180. K. Matyjaszewski, *New Polym. Mater. 2*:115 (1990).
181. D. O. Jordan and A. R. Mathieson, *J. Chem. Soc. 1952*:611 (1952).
182. M. Gyor, H. C. Wang, and R. Faust, *J. Macromol. Sci. A29*:639 (1992).
183. C. Schade and H. Mayr, *Makromol. Chem., Rapid Commun. 9*:477 (1988).
184. T. Higashimura, T. Fukushima, and S. Okamura, *J. Macromol. Sci., Chem. 1*:683 (1967).
185. K. Yamamoto and T. Higashimura, *Polymer. 16*:815 (1975).
186. R. Cotrel, G. Sauvet, J. P. Vairon, and P. Sigwalt, *Macromolecules 9*:931 (1976).
187. F. Subira, J. P. Vairon, A. Polton, and P. Sigwalt, *Bull. Soc. Chim. Fr. 1974*:2903 (1974).
188. G. Sauvet, J. P. Varion, and P. Sigwalt, *J. Polym. Sci., Part A-1 7*:985 (1969).
189. F. Subira, J. P. Varion, and P. Sigwalt, *Proc. 28th IUPAC Macromol. Symp.* 151 (1982).
190. F. Subira, G. Sauvet, J. P. Vairon, and P. Sigwalt, *J. Polym. Sci., Polym. Symp. 56*:221 (1976).
191. J. M. Rooney, *J. Polym. Sci., Polym. Symp. 56*:47 (1976).
192. T. Ohmura, M. Sawamoto, and T. Higashimura, *Macromolecules 27*:3714 (1994).
193. H. Mayr, *Angew. Chem. 29*:1371 (1990).
194. H. C. Brown and R. S. Fletcher, *J. Am. Chem. Soc. 71*:1845 (1949).
195. C. H. Lin and K. Matyjaszewski, *J. Polym. Sci., Part A: Polym. Chem. 28*:1771 (1990).

196. M. Sawamoto and T. Higashimura, *Makromol. Chem., Macromol. Symp. 32*:131 (1990).
197. C. G. Cho, B. A. Feit, and O. W. Webster, *Macromolecules 25*:2081 (1992).
198. N. H. Haucort, S. Kashikar, and E. J. Goethals, *Makromol. Chem., Rapid Commun. 14*:489 (1993).
199. J. Hrkach and K. Matyjaszewski, *J. Polym. Sci., Chem. Ed. 33*:285 (1995).
200. F. Yenca, Y. L. Chen, and K. Matyjaszewski, *Polym. Prepr. (Am. Chem. Soc., Div. Polym. Chem.) 28(2)*:222 (1987).
201. K. Matyjaszewski, J. Hrkach, H. K. Kim, and K. Ruehl, *Adv. Chem. Ser. 224*:285 (1987).
202. J. S. Hrkach and K. Matyjaszewski, *Polym. Prep (Am. Chem. Soc., Div. Polym. Chem.) 33(1)*:229 (1992).
203. J. V. Crivello, *Adv. Polym. Sci. 62*:1 (1984).
204. J. L. Dektar and N. P. Hacker, *J. Org. Chem. 56*:1838 (1991).
205. H. Baumann, H. J. Timpe, and H. Z. Bottcher, *Chemistry 23*:102 (1983).
206. D. J. Worsfold, S. Bywater, and P. Black, 6th IUPAC Meeting on Cationic Polymerization Gent, 1983, p. 11.
207. K. Matyjaszewski, *Makromol. Chem., Macromol. Symp. 54*:51 (1991).
208. S. Penczek, P. Kubisa, K. Brzezinska, and M. Basko, *Polym. Prep. (Am. Chem. Soc., Div. Polym. Chem.) 34(1)*:812 (1993).
209. W. O. Choi, M. Sawamoto, and T. Higashimura, *Macromolecules 23*:48 (1990).
210. M. Sawamoto, A. Furukawa, and T. Higashimura, *Macromolecules 16*:583 (1983).
211. M. Villesange, A. Rives, C. Bunel, J. P. Vairon, M. Froeyen, M. V. Beylen, and A. Parsoons, *Makromol. Chem., Macromol. Symp. 47*:271 (1991).
212. A. Gandini and P. H. Plesch, *J. Chem. Soc. 1965*:4826 (1965).
213. J. P. Lorimer and D. C. Pepper, *Proc. Roy. Soc. A351*:551 (1976).
214. S. L. Malhotra, J. Leonard, and P. E. Harvey, *J. Macromol. Sci., Chem. A11*:2199 (1977).
215. D. Souverain, A. Leborgne, G. Sauvet, and P. Sigwalt, *Eur. Pol. J. 16*:855 (1980).
216. G. Sauvet, J. P. Vairon, and P. Sigwalt, *Eur. Polym. J. 10*:501 (1974).
217. F. Subira, J. P. Vairon and P. Sigwalt, *Macromolecules 21*:2339 (1988).
218. M. Villesange, G. Sauvet, J. P. Vairon and P. Sigwalt, *J. Macromol. Sci., Chem. A11*:391 (1977).
219. A. M. Goka and D. C. Sherrington, *Polymer 16*:809 (1975).
220. A. Ledwith, E. Lockett and D. C. Sherrington, *Polymer 16*:31 (1975).
221. C. E. H. Bawn, C. Fitzimmons, A. Ledwith, J. Penfold, D. C. Sherrington and J. A. Weightman, *Polymer 12*:119 (1971).
222. P. M. Bowyer, A. Ledwith and D. C. Sherrington, *Polymer 12*:509 (1971).
223. J. M. Rooney, *Makromol. Chem. 179*:165 (1978).
224. M. Rodriguez and L. M. Leon, *Eur. Polym. J. 19*:585 (1983).
225. V. T. Stannett, J. Silverman and J. L. Garnett, *Comprehensive Polymer Science* (G. Allen and J. C. Bevington, Eds.) Pergamon Press, Oxford, 1989, vol. 4, p. 317.

226. F. W. Lampe, *J. Phys. Chem. 63*:1986 (1959).
227. V. T. Stannett and J. Silverman, *ACS Symp. Ser. 212*:435 (1983).
228. A. O. Allen, NBS Ref. Data System, "Yields of Free Ions Formed in Liquids by Radiation" (1976).
229. K. Ueno, F. Williams, K. Hayashi and S. Okamura, *Trans. Faraday Soc. 63*:1478 (1967).
230. F. Williams, K. Hayashi, K. Ueno, K. Hayashi and S. Okamura, *Trans. Faraday Soc. 63*:1501 (1967).
231. K. Hayashi, K. Hayashi and S. Okamura, *Polym. J. 4*:426 (1973).
232. A. Deffieux, D. R. Squire and V. Stannett, *Polym. Bull 2*:469 (1980).
233. R. B. Taylor and F. Williams, *J. Am. Chem. Soc. 91*:3728 (1969).
234. F. Williams, A. Shinakawa and J. P. Kennedy, *J. Polym. Sci., Polym. Symp 56*:421 (1976).
235. M. A. Bonin, W. R. Bustle and F. Williams, *J. Am. Chem. Soc. 87*:199 (1965).
236. K. Hayashi, K. Hayashi and S. Okamura, *J. Polym. Sci., Chem. Ed. 9*: 2305 (1971).
237. A. M. Goineau, J. Kohler and V. Stannett, *J. Macromol. Sci., Chem. A11*: 99 (1977).
238. A. Deffieux, J. A. Young, W. C. Hsieh, D. R. Squire and V. Stannett, *Polymer 24*:573 (1983).
239. A. Deffieux, W. C. Hsieh, D. R. Squire and V. Stannett, *Polymer 22*:1575 (1981).
240. H. Mayr and M. Patz, *Angew. Chem. 33*:938 (1994).
241. C. H. Lin, M. Teodorescu and K. Matyjaszewski, *Polym. Prepr., Am. Chem. Soc., 35(2)*:561 (1994).
242. M. S. Sambi, *Macromolecules 5*:351 (1970).
243. T. Higashimura, O. Kishiro, K. Matsuzaki and T. Uryu, *J. Polym. Sci., Polym. Chem. 13*:1393 (1975).
244. M. Delfini, F. Conti, M. Paci, A. L. Segre, R. Solario and E. Chiellini, *Macromolecules 16*:1212 (1983).
245. S. Aoshima, Y. Ito and E. Kobayashi, *Polym. J. 25*:1161 (1993).
246. C. E. Schildknecht, A. O. Zoss and F. Grosser, *Ind. Eng. Chem. 41*:2891 (1949).
247. S. A. Mosley, *U.S. Patent 2 549 921* (1951).
248. S. Aoki, K. Nakamura and T. Otsu, *Makromol. Chem. 50*:137 (1968).
249. S. Okamura, T. Higashimura and T. Watanabe, *Makromol. Chem. 50*:137 (1961).
250. J. Lal and J. E. McGrath, *J. Polym. Sci., Part A 2*:3369 (1964).
251. T. Higashimura, T. Watanabe and S. Okamura, *Kobunshi Kagaku 20*:680 (1963).
252. K. Iwasaki, *J. Polym. Sci. 56*:27 (1962).
253. S. Nakano, K. Iwasaki and H. Fukutani, *J. Polym. Sci., Part A 1*:3277 (1963).
254. E. J. Vandenberg, R. F. Heck and D. S. Breslow, *J. Polym. Sci. 41*:519 (1959).

255. F. J. Vandenberg, *J. Polym. Sci., C 1*:207 (1963).
256. T. Higashimura, T. Takeda, M. Sawamoto, K. Matsuzaki and T. Uryu, *J. Polym. Sci., Polym. Chem. 16*:503 (1978).
257. T. Kunitake and C. Aso, *J. Polym. Sci., A-1 8*:665 (1970).
258. M. Sawamoto, C. Okamoto and T. Higashimura, *Macromolecules 20*:2693 (1987).
259. P. H. Plesch, *Makromol. Chem., Macromol. Symp 13/14*:393 (1988).
260. D. J. Dunn, E. Mathias and P. H. Plesch, *Eur. Polym. J. 12*:1 (1976).
261. R. B. Woodward and R. Hoffman, *J. Am. Chem. Soc. 87*:395 (1965).
262. J. Mathieu and A. Rassat, *Tetrahedron 30*:1753 (1974).
263. K. Matyjaszewski, *J. Polym. Sci., Polym. Chem. Ed. 25*:765 (1987).
264. A. Williams, *Chem. Soc. Rev. 23*:93 (1994).
265. I. Majoros, A. Nagy, and J. P. Kennedy, *Adv. Polym. Sci. 112*:1 (1994).
266. G. Parkin and R. Hoffman, *Angew. Chem. 33*:1462 (1994).
267. G. Parkin, *Chem. Rev. 93*:887 (1993).
268. K. Matyjaszewski, R. Szymanski, and M. Teodorescu, *Macromolecules 27*:7565 (1994).
269. T. Higashimura, S. Aoshima, and M. Sawamoto, *Makromol. Chem., Macromol. Symp. 13/14*:457 (1988).
270. G. Turchi, F. Matera, and P. L. Magagnini, *Makromol. Chem. 170*:75 (1973).
271. C. H. Lin, Ph.D. Thesis, Carnegie Mellon University (1991).
272. L. Gold, *J. Chem. Phys. 28*:91 (1958).
273. L. H. Peebles Jr., *Molecular Weight Distributions in Polymers*, Wiley, New York, 1971.
274. K. Matyjaszewski, C. H. Lin, A. Bon, and J. S. Xiang, *Macromol. Symp. 85*:65 (1994).
275. T. Pernecker and J. P. Kennedy, *Polym. Bull. 26*:305 (1991).
276. C. Reichardt, *Solvents and Solvent Effects in Organic Chemistry*, VCH, Weinheim, 1988.
277. E. S. Amis, *Solvent Effects on Chemical Phenomena*, Academic Press, New York, 1973.
278. G. A. Olah and J. R. DeMember, *J. Am. Chem. Soc. 92*:2562 (1970).
279. W. C. Hsieh, A. Deffieux, D. R. Squire, and Stannett, *Polymer 23*:427 (1982).
280. J. P. Kennedy and E. Marechal, *Carbocationic Polymerization*, Wiley Interscience, New York, 1982.
281. J. P. Kennedy, T. Kelen, and F. Tudos, *J. Polym. Sci., Polym. Chem. Ed. 13*:2277 (1975).
282. T. Kelen, F. Tudos, B. Turcsanyi, and J. P. Kennedy, *J. Polym. Sci., Polym. Chem. Ed. 15*:3047 (1977).
283. K. Matyjaszewski and P. Sigwalt, *Polym. Int. 35*:1 (1994).
284. J. P. Kennedy and R. A. Smith, *J. Polym. Sci., Polym. Chem. Ed. 18*:1539 (1980).
285. H. Hasegawa and T. Higashimura, *J. Appl. Polym. Sci. 27*:171 (1982).

286. J. P. Kennedy, T. Kelen, and F. Tudos, *J. Macromol. Sci., Chem. A18*: 1189 (1982).
287. G. Olah, S. Kobayashi, and M. Tashiro, *J. Am. Chem. Soc. 94*:7448 (1972).
288. R. Pock, H. Mayr, M. Rubow, and E. Wilhelm, *J. Am. Chem. Soc. 108*: 7767 (1986).
289. H. Mayr, J. Bartl, and G. Hagen, *Angew. Chem. 31*:1613 (1992).
290. K. B. Wood and V. T. Stannett, *Polym. Bull. 25*:199 (1991).
291. O. Nuyken, G. Maier, D. Yang, and M. B. Leitner, *Makromol. Chem., Macromol. Symp. 60*:57 (1992).
292. K. Matyjaszewski, C. H. Lin, and C. Pugh, *Macromolecules 26*:2649 (1993).
293. T. Higashimura, M. Hiza, and H. Hasegawa, *Macromolecules 12*:217 (1979).
294. M. Sawamoto, and T. Higashimura, *Macromolecules 14*:467 (1981).
295. A. Rives, Ph.D. Thesis, University of Paris, Paris 1994.
296. J. E. Chandler, B. H. Johnson, and R. W. Lenz, *Macromolecules 13*:377 (1980).
297. J. P. Kennedy, L. S. Winkler, G. Wanless, and R. M. Thomas, *J. Polym. Sci. A2*:2093 (1964).
298. J. P. Kennedy and J. E. Johnston, *Adv. Polym. Sci. 19*:57 (1975).
299. C. Corno, G. Ferraris, A. Priola, and S. Cesca, *Macromolecules 12*:404 (1979).
300. L. Peng, N. Haurcourt, and E. J. Goethals, *Macromol. Rapid Commun. 15*:197 (1994).
301. B. Ivan and J. P. Kennedy, *J. Polym. Sci., Polym. Chem. Ed. 28*:89 (1990).
302. A. Verma, A. Nielsen, J. E. McGrath, and J. S. Riffle, *Polym. Bull 23*:563 (1990).
303. H. Fukui, M. Sawamoto, and T. Higashimura, *Macromolecules 26*:7315 (1993).
304. K. Miyashita, M. Kamigaito, M. Sawamoto, and T. Higashimura, *J. Polym. Sci., Polym. Chem. 32*:2531 (1994).
305. H. Fukui, M. Sawamoto, and T. Higashimura, *J. Polym. Sci., Polym. Chem. 32*:2699 (1994).
306. H. Mayr and C. Schade, *Makromol. Chem., Rapid Commun. 9*:483 (1988).
307. G. V. Schulz and G. Harborth, *Makromol. Chem. 1*:104 (1947).
308. P. H. Plesch, *The Chemistry of Cationic Polymerization*, Pergamon Press, Oxford, 1963.
309. P. Sigwalt, *Makromol. Chem., Macromol. Symp. 47*:179 (1991).
310. B. Pary, M. Tardi, A. Polton, and P. Sigwalt, *Eur. Polym. J. 21*:393 (1985).
311. J. P. Kennedy, A. Shinkawa, and F. Williams, *J. Polym. Sci. 9*:1551 (1971).
312. K. Ueno, K. Hayashi, and S. Okamura, *Polymer 7*:434 (1966).
313. M. Tazi, M. Tardi, A. Polton, and P. Sigwalt, *Br. Polym. J. 19*:369 (1987).
314. P. Sigwalt, A. Polton, and M. Tardi, *J. Macromol. Sci., Pure & Appl. Chem. A31*:953 (1994).
315. D. C. Pepper and P. J. Reilly, *Proc. Chem. Soc. 1961*:200 (1961).
316. G. F. Endres and C. G. Overberger, *J. Am. Chem. Soc. 77*:2201 (1955).

317. A. Bon, J. Hartt, C. H. Lin, and K. Matyjaszewski, *Polym. Prepr. (Am. Chem. Soc., Div. Polym. Chem.) 35(1)*:464 (1994).
318. C. G. Overberger and M. G. Newton, *J. Am. Chem. Soc. 82*:3622 (1960).
319. C. G. Overberger and G. F. Endres, *J. Am. Chem. Soc. 75*:6349 (1953).
320. C. G. Overberger, G. F. Endres, and A. Monaci, *J. Am. Chem. Soc. 78*: 1969 (1956).
321. C. G. Overberger and G. F. Endres, *J. Polym. Sci. 16*:283 (1955).
322. H. Mayr, R. Schneider, B. Irrgang, and C. Schade, *J. Am. Chem. Soc. 112*:4454 (1990).
323. J. Penfold and P. H. Plesch, *Proc. Chem. Soc. 1961*:311 (1961).
324. S. Aoshima and T. Higashimura, *Polym. J. 16*:249 (1984).
325. V. Bertoli, and P. H. Plesch, *J. Chem. Soc. B 1968*:1500 (1968).
326. K. Brzezinska, W. Chwialkowska, P. Kubisa, K. Matyjaszewski, and S. Penczek, *Makromol. Chem. 178*:2491 (1977).
327. M. Sawamoto, S. Aoshima, and T. Higashimura, *Makromol. Chem., Macromol. Symp. 13/14*, 513 (1988).
328. H. Fukui, M. Sawamoto, and T. Higashimura, *Macromolecules 27*:1297 (1994).

4

Controlled/Living Carbocationic Polymerization

KRZYSZTOF MATYJASZEWSKI Carnegie Mellon University, Pittsburgh, Pennsylvania

MITSUO SAWAMOTO Kyoto University, Kyoto, Japan

I. INTRODUCTION

Living carbocationic polymerization has been one of the most rapidly developing fields in ionic polymerization in recent years. However, although a large number of systems have been reported which allow preparation of homopolymers and copolymers with predetermined molecular weights, low polydispersities, and desired end functionalities, the understanding of the fundamentals of these reactions has not been adequate. It must be admitted that some time ago both co-authors of this chapter had quite different views on the mechanism of the reactions involved from what is presented in this chapter. However, a new picture emerged with time, with new experimental results, and with ample supporting evidence coming from model studies covered in Chapter 2 and from conventional cationic polymerization and related processes described in Chapter 3. As will be discussed later, these reactions are quite similar to many cationic ring-opening polymerizations (Chapter 6) in which a variety of equilibria between active and dormant species exist. To clarify our views on living carbocationic polymerizations we decided to describe them separately from conventional systems (Chapter 3) and divide the discussion into five sections.

In Section II we discuss fundamentals of living and controlled polymerization and demonstrate that some living systems, i.e., those without irreversible chain-breaking reactions, may provide uncontrolled polymers with very high polydispersities. We also show that it is possible to obtain well-defined polymers even in the presence of chain-breaking reactions when the molecular weight is low enough, and both initiation and exchange reactions are faster than propagation. As will be presented later, in many new living carbocationic polymerizations transfer and termination do exist, and therefore they are not formally living. At the same time they provide controlled polymers and hence should be called controlled polymerizations. Because in many cases the quantitative information on the contribution of chain-breaking reactions is not yet available, we decided to use the term *controlled/living* to describe the still existing uncertainty in these systems.

Section III summarizes the most important features and problems of carbocationic polymerization relevant to understanding new controlled/living systems. The more detailed information is in Chapter 3. The subsequent Section IV is devoted to the general description and methodologies of controlled/living systems and includes a historical background on the development of three operational systems based on nucleophilic counterions, added salts, and nucleophiles. Section V presents the scope of living carbocationic polymerization by reviewing various controlled/living systems for vinyl ethers, isobutylene, styrenes, and other monomers, in which it is emphasized that the structure and concentration of the activating Lewis acids and deactivating salts and nucleophiles have to be carefully selected for each type of monomer. Section VI discusses the chemistry of the controlled/living systems including the structure of active centers, model studies, kinetics, molecular weights, and polydispersities. The final section, Section VII, is devoted to the mechanism of new controlled/living polymerizations systems. It first covers peculiarities of initiation, propagation, transfer, and termination and compares them with conventional systems. We also define the role of each of the components and reaction conditions used in the new systems and propose one unified mechanistic picture of controlled/living systems in which growing carbocations exchange rapidly with dormant species, being either onium ions or covalent species.

II. FUNDAMENTALS OF LIVING POLYMERIZATION

The term *living polymerization* was originally used to describe a chain polymerization in which chain-breaking reactions are absent [1]. In such an ideal system, after initiation is completed, growing or propagating chains would only propagate and would not participate in transfer or termination. Each chain should infinitely retain its ability to react with monomer.

The first living systems based on anionic polymerization of nonpolar monomers, such as styrene and dienes in some hydrocarbon solvents, showed nearly perfect livingness producing very high molecular weight polymers ($M_n \gg 100{,}000$) with low polydispersities ($M_w/M_n < 1.1$), providing that initiation and mixing were fast enough [1–4]. The polymerization resumes with the same rate after addition of a new portion of monomer, and a linear increase of molecular weight with conversion is observed, indicating that the number of active centers is constant (no termination) and that the number of chains is constant (no transfer). In addition, block copolymers are formed by consecutive polymerization of two comonomers. In these systems ions, ion pairs of various structures and reactivities, as well as their aggregates coexist together. Reactivities of ions are sometimes much higher than those of ions pairs ($k_p^-/k_p^\pm \approx 10^5$ in the polymerization of styrene with Li^+ counterion in dioxane at 20° C), and ionic aggregates are much less reactive than ion pairs. Nevertheless, polymers with the degrees of polymerization equal to the ratio of the concentrations of the reacted monomer to that of the active species or the introduced initiator ($DP_n = \Delta[M]/[I]_0$) and with narrow molecular weight distributions have been prepared. This observation indicates that growing species with different reactivities exchange rapidly enough to give the same probability of growth for all chains. It also implies that the temporary decrease of activity (or temporary deactivation) does not interfere with the concept of living polymerization. Therefore, temporary deactivation should not be considered as termination. For the same process, the term "reversible (or temporary) termination" has often been used in the field of cationic polymerization [5], but we suggest using the term "temporary deactivation," instead.

Thus, living polymerization does not allow termination and transfer processes. In real systems, these reactions may take place. However, if transfer and termination cannot be detected under certain reaction conditions using currently available instrumentation, essentially living systems might be achieved.

The contributions of transfer and termination increase with the increasing chain length. It may happen, as discussed in Chapter 3, that for sufficiently short chains no deviation from ideal behavior (straight semilogarithmic kinetic plots, linear increase of M_n with conversion, very narrow polydispersities) can be observed. Such systems could be called living. However, if attempts to prepare higher molecular weight polymers under otherwise identical conditions (initiator, additive, solvent, temperature, etc.) are unsuccessful and if an increase of molecular weights with conversion, polydispersities and polymerization kinetics indicate the presence of chain-breaking reactions, then such a system should not be called living. It might be possible that sooner or later the chain-breaking reactions could be identified in all polymerization systems, even those that are considered today as perfectly living.

It was proposed to determine the rate constants of transfer and termination or their ratios to that of propagation for polymerization systems that allow synthesis of well-defined polymers [6]. This information will be useful for the reproducible synthetic efforts and also for setting limits for the preparation of well-defined high polymers. The direct determination of transfer/termination rate constants may be facilitated by working under "difficult" reaction conditions such as those which usually disallow preparation of well-defined polymers (higher temperatures, lower $[I]_0$, longer chains, etc.), and the obtained results extrapolated to the "usual living" conditions.

Thus, preparation of well-defined polymers by *controlled polymerization* should not be confused with living polymerization. Herein we define a controlled polymerization to be a synthetic method to make polymers with predetermined molecular weights, low polydispersity, controlled functionality, block copolymers, etc. Transfer and termination can happen in controlled polymerization but their contribution is sufficiently reduced by the proper choice of the reaction conditions.

On the other hand, a living polymerization is a chain polymerization without transfer and termination. Living polymerization may lead to well-defined polymers with narrow molecular weight distributions and predetermined molecular weights only if some additional prerequisites are fulfilled:

1. Initiation is faster than propagation.
2. The exchange between species of different reactivities is faster than propagation.
3. The rate of depropagation is substantially lower than propagation.
4. The system is homogeneous and mixing is sufficiently fast.

If these specifications are not met, then a living polymerization may produce polymers with very broad polydispersities and degrees of polymerization much higher than the $\Delta[M]/[I]_0$ ratio. Importantly, these prerequisites must be fulfilled in controlled polymerization as well.

Terms such as *apparently living* or "*living*" polymerization have been used previously to name systems that produced well-defined polymers but in which the presence of transfer or termination was unambiguously proven [7]. The term *controlled* polymerization may describe better the essence of such systems. This discussion is summarized in Scheme 1.

Because in many cases the quantitative information on the contribution of chain-breaking reactions is not yet available, we decided to use the term *controlled/living* polymerization to describe the still existing ambiguity and uncertainty in these systems.

It will be shown below that in some systems without irreversible

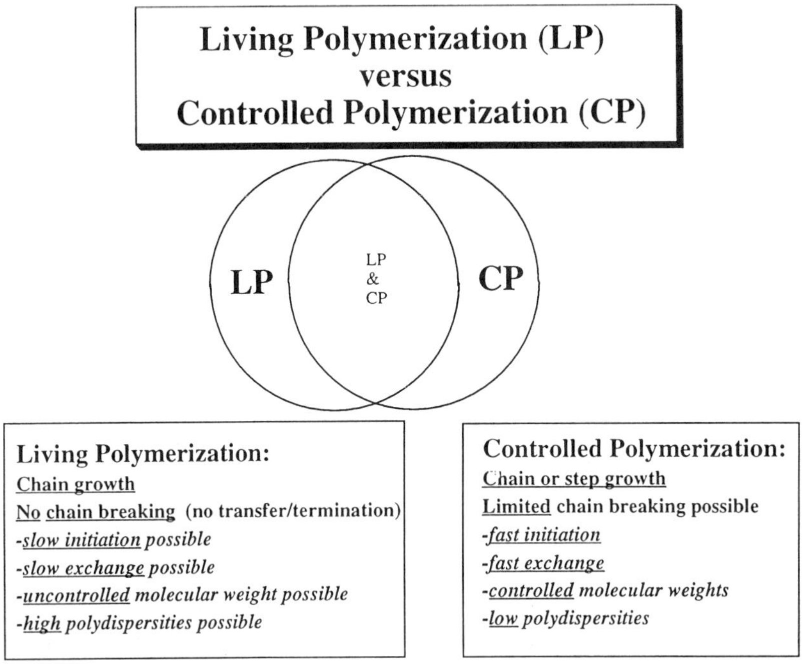

Scheme 1

transfer and termination (i.e., essentially living) broad molecular weight distributions and no increase of molecular weights with conversion may be observed. Therefore, it is instructive to analyze how deviations from ideal systems influence kinetics and molecular weights.

A general kinetic scheme for a typical chain polymerization is shown below (Scheme 2). Initiation [reaction (1)] proceeds by the reaction of initiator (I) with monomer (M) to produce a growing species (P_1^+). This species propagates [reaction (2)]. The reaction of the growing species with monomer may also lead to transfer to monomer and the generation of new chains [reaction (3)]. Transfer may also happen spontaneously (in fact, usually induced by a counterion). If a new species has a reactivity similar to or higher than that of the growing species, then no effect on the kinetics will be observed. Growing species (P^+) may be in equilibrium [reaction (4)] with a species P* of different reactivity. The growing species may also lose reactivity through spontaneous unimolecular termination [reaction (5)] or through bimolecular termination with a terminating agent.

$$\mathrm{I + M \xrightarrow{k_i} P_1^{+}} \quad (1)$$

$$\mathrm{P_n^{+} + M \xrightarrow{k_p} P_{n+1}^{+}} \quad (2)$$

$$\mathrm{P_n^{+} + M \xrightarrow{k_{trM}} P_n^{=} + P_1^{+}} \quad (3)$$

$$\mathrm{P_n^{+} \underset{k_{*/+}}{\overset{k_{+/*}}{\rightleftharpoons}} P_n^{*}} \quad (4)$$

$$\mathrm{P_n^{+} \xrightarrow{k_t} P_n} \quad (5)$$

Scheme 2 Kinetic scheme for chain polymerization.

The simplest system that provides both living and controlled polymerization consists only of reaction (2) and reactions (3) and (5) should be absent. As will be shown later, well-defined polymers can still be formed if the contribution of these reactions is small and if the degrees of polymerization are limited. Living polymerization does allow slow initiation [reaction (1)] which, however, increases the polydispersities and produces polymers of "too high" molecular weights. Multiplicity of growing species with various reactivities and various lifetimes may produce polymers with broad and even polymodal molecular weight distributions. However, a Poisson distribution can be observed if the exchange [reaction (4)] is fast.

In the next sections the quantitative effect of reactions (1), (3), (4), and (5) which can be considered as various deviations from ideal systems, on kinetics, molecular weights, and polydispersities, will be presented. The magnitude of only one variable will be changed each time to demonstrate clearly the effect of slow initiation, termination, transfer, and slow exchange on the polymerization rates and properties of the resulting polymers.

A. Slow Initiation

Figure 1 illustrates the effect of slow initiation on kinetics for a hypothetical system with $[M]_0 = 1$ mol/L, $[I]_0 = 0.01$ mol/L, and where termination is absent and only one type of active species is present. The first-order kinetics in monomer should provide a straight line in a semilogarithmic

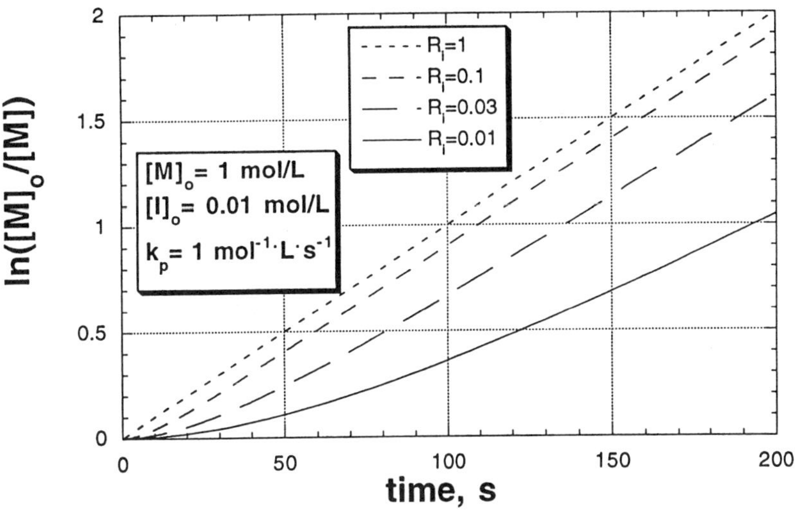

Figure 1 Effect of slow initiation. Dependence of DP_n on conversion for various ratios $R_i = k_i/k_p$.

coordinate for a constant concentration of active sites (instantaneous initiation). The time scale of the polymerization is defined by the product of the concentration of the propagating species and the rate constant of propagation, in which the shape of the plots depends on the ratio of the rate constants of propagation to initiation. For the particular ratio $[M]_0/[I]_0 = 100$, no detectable deviation from the ideal line is found for $R_i = k_i/k_p = 1$. At the ratio $R_i = 0.1$, initiator is nearly completely consumed at approximately 40% monomer conversion. However, at the ratios $R_i = 0.03$ and 0.01, unreacted initiator stays until complete monomer consumption. Thus, continuous accelerations in the semilogarithmic coordinates are observed.

It is much easier to notice the effect of slow initiation by analysis of the evolution of molecular weights with conversion in Figs. 2 and 3. The small increase of the polymerization degree, in relation to the ideal case, disappears for the ratio $R_i = 0.1$ at approximately 40% conversion. However, it is necessary to add subsequent portions of a monomer (conversions > 100%) for ratios $R_i = 0.03$ and $R_i = 0.01$ to asymptotically approach ideal M_n values as shown in Fig. 3. The polydispersities in systems with slow initiation depend on the ratio $[M]_0/[I]_0$ and R_i and are very low for $R_i = 1$ and 10 ($M_w/M_n < 1.02$) but approach $M_w/M_n \approx 1.15$ for the ratio

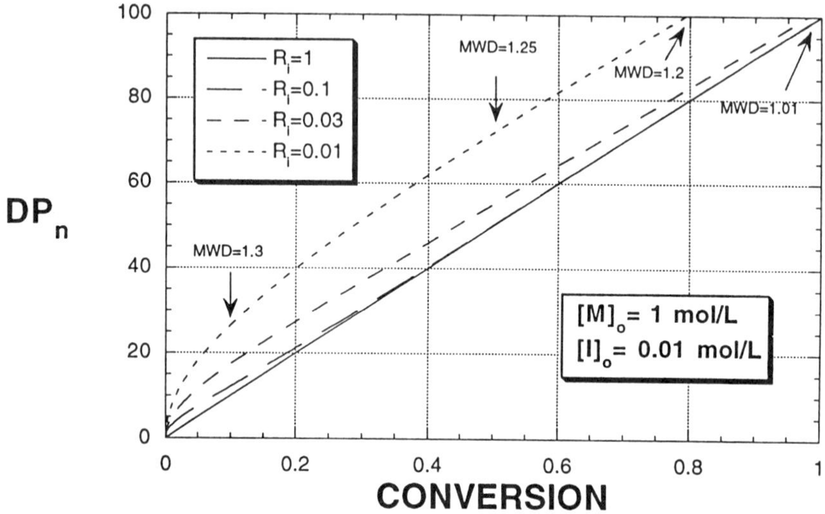

Figure 2 Effect of various ratios $R_i = k_i/k_p$ on kinetics up to 100% conversion for slow initiation.

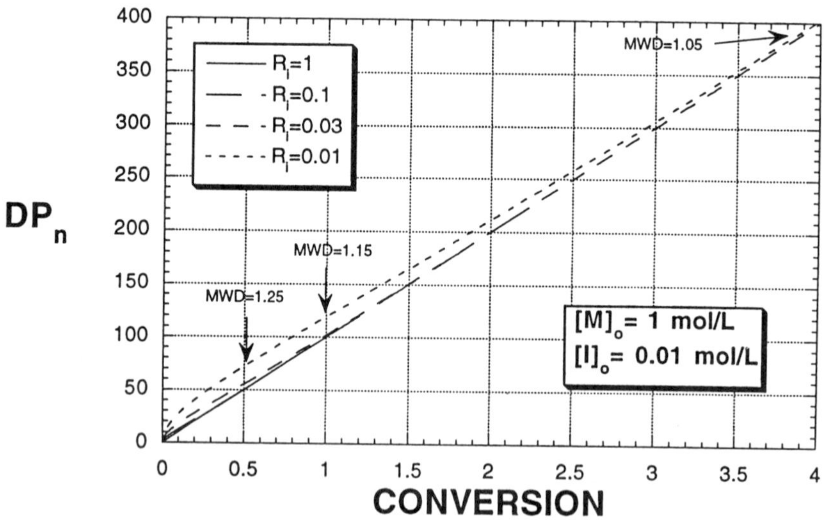

Figure 3 Effect of various ratios $R_i = k_i/k_p$ on kinetics during four consecutive monomer additions for slow initiation.

$R_i = 0.01$. It seems that the highest polydispersity due to slow initiation is $M_w/M_n < 1.35$ [8]. The effect of slow initiation on kinetics, molecular weights and polydispersities has been discussed before in detail for general systems [1,8–10] and for carbocationic polymerization in particular [5].

B. Termination

Termination has no effect on the final number-average molecular weights, because it does not change the total number of chains. However, termination may lead to incomplete polymerization if the initiator concentration is too low. In the case of unimolecular termination, the final monomer conversion ($[M]_\infty$) is set by Eq. (1).

$$\ln([M]_0/[M]_\infty) = [I]_0(k_p/k_t) \tag{1}$$

Thus, termination will have a mostly kinetic effect. Figure 4 shows the semilogarithmic plots for various ratios $R_t = k_t/k_p$, taking $k_p = 1$ $mol^{-1} \cdot L \cdot sec^{-1}$, arbitrary concentrations $[M]_0 = 1$ mol/L, $[I]_0 = 0.001$ mol/L, and assuming instantaneous initiation.

For the ratios $R_t = 10^{-5}$ and 10^{-4} mol/L nearly no deviation from ideal behavior is found. Complete conversions, predicted molecular weights, and polydispersities of less than $M_w/M_n < 1.03$ are calculated. On the other hand, for the ratio $R_t = 10^{-3}$ mol/L, only 63% conversion

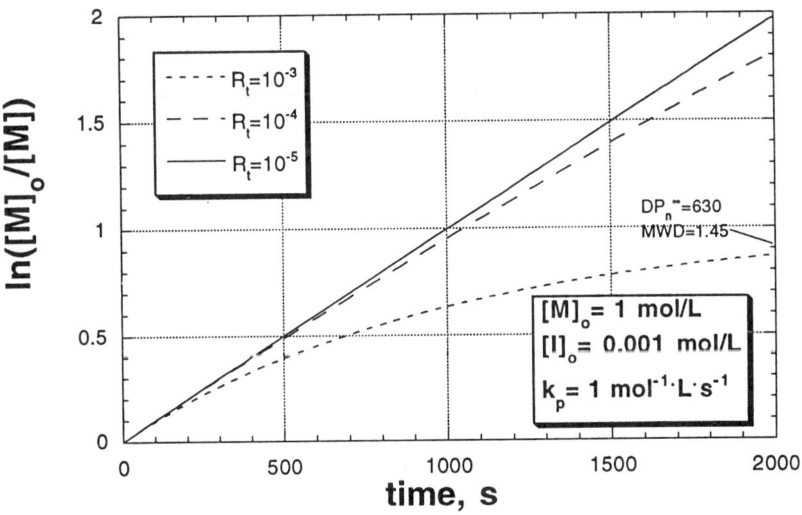

Figure 4 Effect of various ratios $R_t = k_t/k_p$ on kinetics.

is expected at infinite time. Calculations indicate that DP_n = 630 and $M_w/M_n \approx 1.45$ for the final product. The effect of termination on molecular weight distributions has been discussed thoroughly in Refs. 11–13.

C. Transfer

Ideally, transfer has no effect on kinetics but has a pronounced effect on molecular weights and polydispersities [11,14,15].

Figure 5 shows the effect of transfer to monomer on the polymerization degree for various ratios $R_{trM} = k_{trM}/k_p$, taking arbitrary concentrations $[M]_0 = 1$ mol/L, $[I]_0 = 0.01$ mol/L, assuming instantaneous initiation, and assuming the absence of termination. Instead the predicted final $DP_n = 100$, smaller values are computed (DP = 91, 75, and 50) for the ratios $R_{trM} = 10^{-3}$, $3 \cdot 10^{-3}$, and 10^{-2}, respectively. It can be noticed that the deviation becomes smaller at lower DP and increases at the higher DP range for each R_{trM} value.

As shown in Fig. 6, deviation depends not only on the R_{trM} value but also on the ratio of the concentration of monomer to that of initiator, which can be expressed by the parameter $a = (k_{trM}/k_p)\cdot[M]_0/[I]_0$, as shown in Eq. (2):

$$DP/DP^{id} = 1/\{1 + (k_{trM}/k_p)\cdot[M]_0/[I]_0\cdot p\} \tag{2}$$

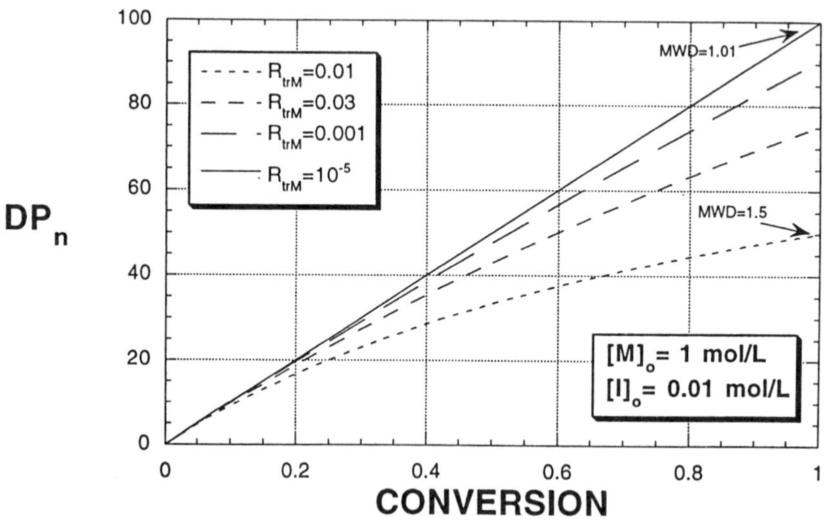

Figure 5 Dependence of DP_n on conversion, p, for various ratios $R_{trM} = k_{trM}/k_p$ for transfer to monomer.

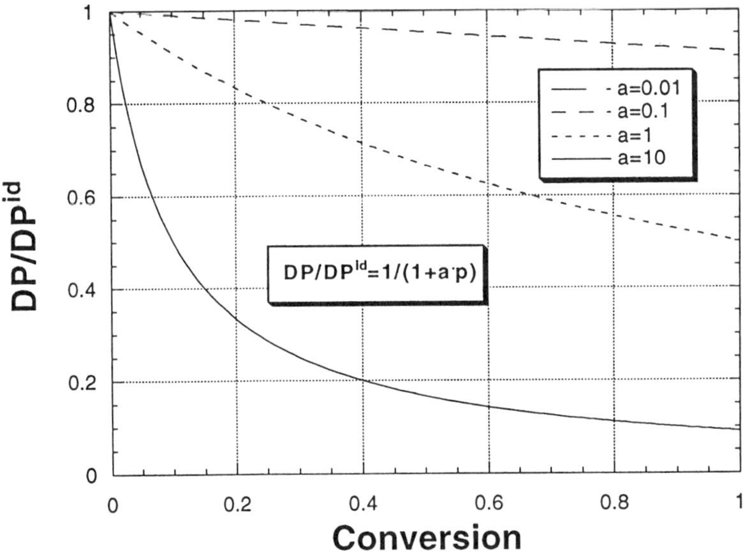

Figure 6 Effect of parameter $a = (k_{trM}/k_p)\cdot(\Delta[M]/[I]_0)$ on deviation from ideal behavior for transfer to monomer (from Ref. 16).

The ratio of the degree of polymerization in the presence of transfer to that predicted for the ideal system without transfer (DP^{id}) decreases with conversion [16]. For the initial conditions $[M]_0 = 1$ mol/L, $[I]_0 = 0.01$ mol/L, and the ratio $R_{trM} = k_{trM}/k_p = 10^{-3}$, a 10% deviation from the ideal behavior is expected (case a = 0.1). A decrease of the initiator concentration to $[I]_0 = 10^{-3}$ mol/L leads to twice lower than ideally predicted values of DP at complete conversion (a = 1), and a decrease to $[I]_0 = 10^{-4}$ mol/L leads to 10 times lower values of DP (a = 10). However, nearly ideal behavior can be reached with $[I]_0 = 0.1$ mol/L (a = 0.01). This shows that it is possible to improve control in polymerization by simple manipulation (increase) of the initiator concentration.

Figure 7 depicts the predicted effect of the parameter a on the polydispersities at complete conversion [15]. With $[M]_0 = 1$ mol/L and $[I]_0 = 0.01$ mol/L, a polydispersity of $M_w/M_n \approx 1.06$ is expected for the ratio $k_{trM}/k_p = 10^{-3}$. A decrease in the concentration of initiator to $[I]_0 = 0.001$ mol/L leads to $M_w/M_n \approx 1.5$.

Figure 8 is similar to Fig. 6, but it illustrates deviations from the ideal behavior in the case of transfer to counterion. Because the rate of transfer

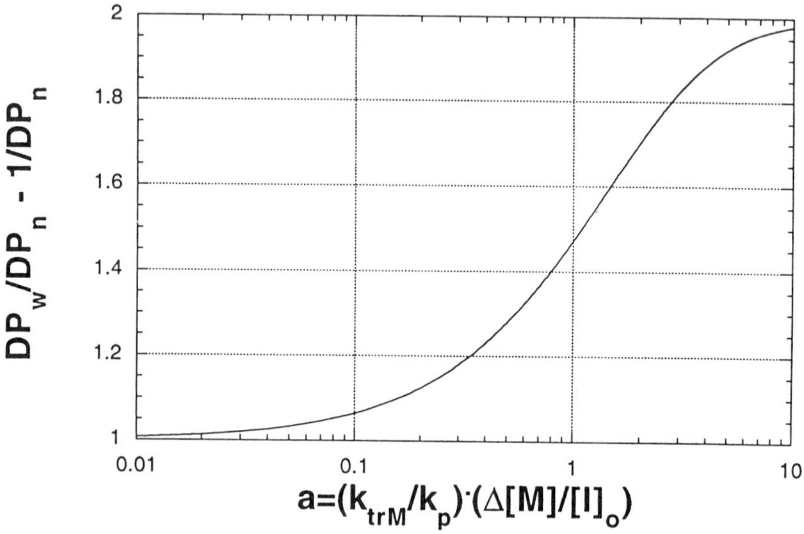

Figure 7 Dependence of MWD on various ratios "a" for transfer to monomer (from Ref. 16).

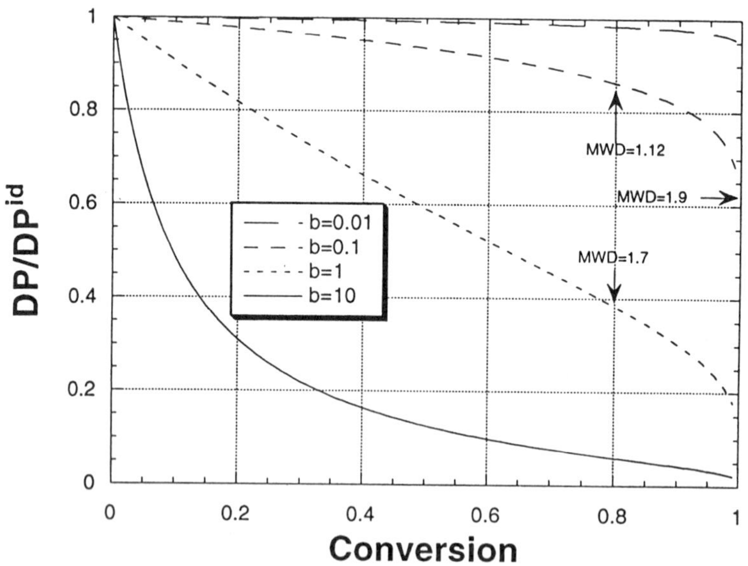

Figure 8 Effect of parameter $b = (k_{tr}/k_p)/[I]_0$ on deviation from ideal behavior for unimolecular transfer (e.g., to counteranion) (from Ref. 16).

to counterion is independent of the monomer concentration, and the propagation rate decreases with conversion, lower molecular weights and a strong increase in polydispersities are expected at high conversion [14,17]. Quite often this effect may not be detected experimentally if the polymer is precipitated rather than analyzed as the entire sample which includes oligomeric products.

The plots in Fig. 8 were calculated for various values of parameters b = $(k_{tr}/k_p)/[I]_0$ using the following equation [18]:

$$DP/DP^{id} = 1/\{1 + \ln[1/(1 - p)]\cdot(k_{tr}/k_p)/[I]_0\} \quad (3)$$

D. Slow Exchange

Exchange between ions and ion pairs of different activities has been analyzed in anionic systems and the small broadening of polydispersities was used for the evaluation of the dynamics of the exchange [2,19]. When exchange becomes slower, as in the case of aggregates of ion pairs, then polydispersities increase and distribution may become bimodal [20].

Slow exchange, in addition to transfer, might be the most important parameter affecting polydispersities in the carbocationic polymerization [21,22]. It is the main reason for polymodal molecular weight distributions. It is intuitively easy to imagine bimodal molecular weight distribution (MWD) when two species of different reactivities do not exchange or exchange slowly in comparison with propagation. As stated in the next section and discussed in Chapter 3, the reactivities of ions and ion pairs are similar while dormant species are inactive. Therefore, is a bimodal MWD possible for two propagating species with the same reactivities? The answer is yes, if their lifetimes are different [23,24].

Figure 9 shows evolution of molecular weights with conversion for a hypothetical system, in which ions and ion pairs have the same reactivities ($k_p^+ = k_p^\pm = 10^5$ mol$^{-1}\cdot$L$\cdot$sec^{-1}), covalent species are inactive, the ionization equilibrium constant is $K_I = 10^{-5}$ mol$^{-1}\cdot$L, and the dissociation constant is $K_D = 10^{-7}$ mol/L. K_I is defined by the ratio of the rate constant of ionization to that of recombination of counterions within the ion pair ($K_I = k_i/k_r$). K_D is defined by the ratio of the rate constants of the dissociation of ion pair and that of the association of free ions ($K_D = k_{dis}/k_{as}$):

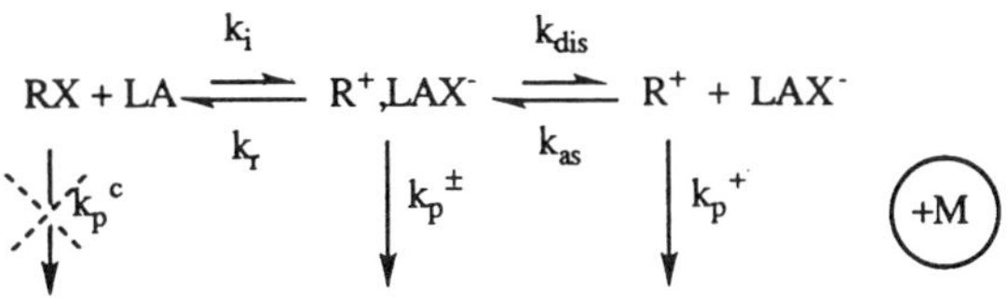

(4)

Initial conditions $[M]_0 = 1$ mol/L, $[I]_0 = 0.01$ mol/L, and $[LA]_0 = 0.1$

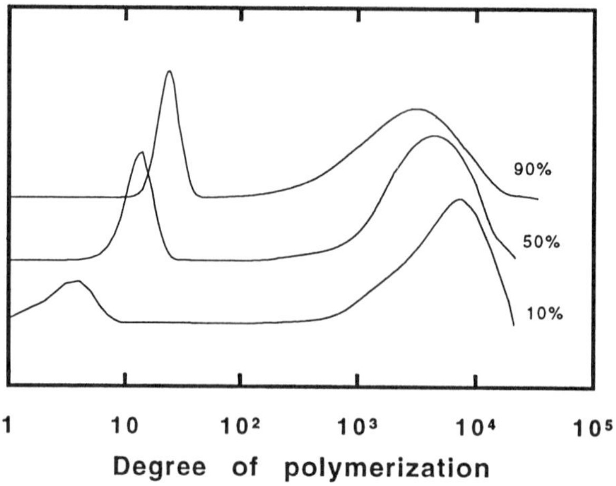

Figure 9 MWD in cationic polymerization as a function of conversion; $[M]_0 = 1\ M$, $[I]_0 = 0.01\ M$, $[LA]_0 = 0.1\ M$; $k_p{}^+ = k_p{}^\pm = 10^5\ M^{-1}\ sec^{-1}$; $K_D = 10^{-7}\ M$; $K_I = 10^{-5}\ M^{-1}$; $k_i = 10^2\ M^{-1}\ sec^{-1}$ (from Ref. 23).

mol/L should lead to the average values of $DP_n = 10$, 50, and 90 at conversions 10%, 50%, and 90%, correspondingly, provided that initiation is quantitative. It can be clearly seen that although ions and ion pairs have the same reactivities, a bimodal MWD is obtained. The low molecular weight peak increases progressively with conversion. The number-average degree of polymerization for the LMW peak increases from 3.2 to 12.2 and to 22.8 with conversion (10%, 50%, and 90%, respectively). The polydispersity of the LMW peak stays fairly narrow ($M_w/M_n = 1.16$, 1.10, and 1.05, respectively).

The HMW peak shifts slightly and its number-average degree of polymerization decreases from 2900 to 2300 and to 1300 with conversion. The polydispersities broaden from $M_w/M_n = 2.2$ to 2.1 to 2.78, respectively. The overall number average degrees of polymerization increase slightly from 11 to 51 to 91, as expected for nearly complete initiation. Overall the polydispersities decrease from $M_w/M_n = 400$ to 70 and to 30, respectively.

There are some differences between the real data and this simulation. We intended in this particular case to show clear separation of both peaks. Closer resemblance to observed systems will be shown later. Experimentally, the molecular weight of the LMW peak increases with conversion as seen in most cases [25,26]. A decrease in the molecular weight of the HMW peak was also reported and it is simply due to monomer consumption and, consequently, the reduction of the ratio of the rates of propagation (first order in monomer) to the deactivation (association of counterions and subsequent recombination), which is zero order in monomer. The degrees of polymerization of the HMW fraction observed experimentally are much lower than shown in Fig. 9, probably due to a transfer process (transfer to monomer or solvent, because there is no counterion next to a free carbocation).

Figure 10 shows the effect of both ionization and dissociation equilibria on the MWD of polymers at 90% conversion with a standard recombination rate constant $k_r = 10^7$ mol^{-1}·L·sec^{-1}. The bottom three traces and the top three traces depict the change in K_D from 10^{-5} to 10^{-6} and to 10^{-7} mol/L for two different ionization equilibrium constants, $K_I = 10^{-5}$ (bottom) and 10^{-3} L/mol (top), respectively.

When ionization is weak, a clear bimodal MWD is observed. The concentration of ion pairs equals $[C^{\pm}] = 10^{-8}$ mol/L, whereas the concentration of free ions changes from $[C^+] = 0.3 \cdot 10^{-7}$, 10^{-7}, and to $3 \cdot 10^{-7}$ mol/L. This means that the proportion of monomer consumed by free ions increases from ≈ 76% to 90% and to 97%, respectively. This is depicted in the relative proportions of the HMW and LMW peaks.

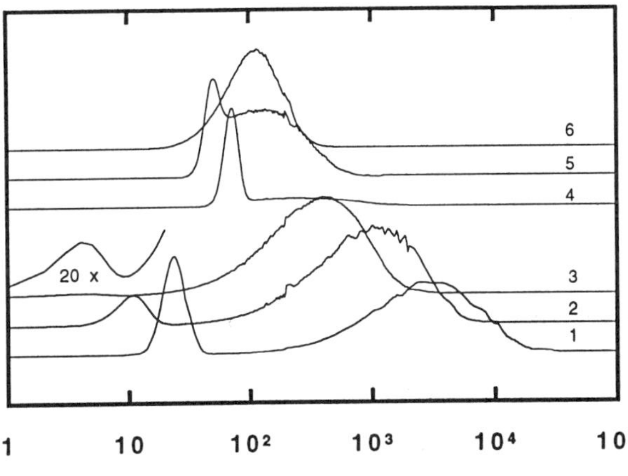

Figure 10 MWD in cationic polymerization as a function of dissociation and ionization equilibrium constants; $[M]_0 = M$; $[I]_0 = 0.01\ M$; $[LA]_0 = 0.1\ M$; $k_p^+ = k_p^{\pm} = 10^5\ M^{-1}$ sec^{-1}; $k_r = 10^7$ sec^{-1}; 1: $K_D = 10^{-7}\ M$, $K_I = 10^{-5}\ M^{-1}$; 2: $K_D = 10^{-6}\ M$, $K_I = 10^{-5}\ M^{-1}$; 3: $K_D = 10^{-5}\ M$, $K_I = 10^{-5}\ M^{-1}$; 4: $K_D = 10^{-7}\ M$, $K_I = 10^{-3}\ M^{-1}$; 5: $K_D = 10^{-6}\ M$, $K_I = 10^{-3}\ M^{-1}$; 6: $K_D = 10^{-5}\ M$, $K_I = 10^{-3}\ M^{-1}$ (from Ref. 23).

For stronger ionization, the concentration of ion pairs equals $[C^{\pm}] = 10^{-6}$ mol/L, and the concentration of free ions varies from $[C^{+}] = 0.3 \cdot 10^{-6}$, 10^{-6}, and to $3 \cdot 10^{-6}$ mol/L. This means that the contribution of free ions to monomer consumption increases from ≈25% to 50% and to 75%, correspondingly. In trace 4, the 25% contribution of free ions can hardly be seen without magnification (broad MWD). In trace 5 nearly equal proportions of both peaks are seen, whereas in trace 6, free ions dominate but the difference between polymerization degrees is so small that peaks do not separate.

It is more difficult to interpret changes in average polymerization degrees of LMW and HMW peaks. Fraction of the LMW peak is determined by the proportion of ion pairs among all carbocations. However, if they exchange very rapidly with covalent species, they can not be distinguished from the dormant species and the average DP_n of the LMW peak is in that case defined by the ratio of the concentrations of reacted monomer to that of covalent species (approximately equal to the concentration of the initiator, provided that initiation is complete). Thus, DP_n of the LMW peak equals approximately:

$$DP_n{}^{L} = \Delta[M]/([I]_0 - [I]) \cdot \{[C^{\pm}]/([C^{+}] + [C^{\pm}])\} \tag{5}$$

The molecular weight of this fraction progressively grows with conversion and the MWD is rather narrow ($M_w/M_n < 1.2$).

The DP_n of the HMW fraction is more difficult to estimate. The DP formed during one activation period depends on the relative rates of propagation (R_p) and deactivation of free ions (the association process, R_{as}).

$$\begin{aligned} DP_n{}^{H} &= R_p/R_{as} \cdot \{[C^{+}]/([C^{+}] + [C^{+}])\} \\ &= \{(k_p \cdot [M])/(k_{as} \cdot [C^{+}])\} \cdot \{[C^{+}]/([C^{+}] + [C^{\pm}])\} \\ &= k_p \cdot [M]/\{k_{as} \cdot ([C^{+}] + [C^{\pm}])\} \end{aligned} \tag{6}$$

It is surprising that DP_n of the HMW peak does not depend on the concentration of ions but rather on the total amount of ionic species. Figure 10 shows this behavior quite clearly. The decrease of $DP_n{}^{H}$ in traces 1, 2 and 3 is due to the increase of the total concentration of carbocations from $4 \cdot 10^{-8}$ to $1.1 \cdot 10^{-7}$ and to $3.3 \cdot 10^{-7}$ mol/L. The values of $DP_n{}^{H}$ in Fig. 9 are in good agreement with those predicted from Eq. (6), assuming one single activation process for this fraction: $DP_n{}^{H} = 1200, 500, 200$. On the other hand, in traces 4, 5 and 6, the values of $DP_n{}^{H}$ are in the range of 100 to 200 and are much higher than those predicted for one single activation process for this population: $DP_n{}^{H} = 80, 20$, and 12, respectively. This indicates that this population must grow with conversion in contrast to the changes shown in Fig. 9. The repetitive activation processes also lead to a more narrow MWD for this fraction which in trace 6 is $M_w/M_n = 1.30$ for both free ions (75%) and ion pairs (25%).

As will be discussed in Section IV.B.3, one of the simple ways to improve control in the polymerization is to suppress the concentration of free ions. This can be accomplished by addition of salts with common ions. Figures 11 and 12 demonstrate the common ion effect. Broad polydispersities in systems without added common ion ($M_w/M_n = 5.3$, 1.74, and 1.3 at 10, 50, and 90% conversion) are reduced to one narrow peak growing steadily with conversion in the presence of added common ion $[A^-] = 10^{-4}\ M^{-1}\cdot sec^{-1}$. The overall polydispersities are $M_w/M_n = 1.12$, 1.02, and 1.01 at 10, 50, and 90% conversion, very close to the ideal Poisson distribution.

The decrease of polydispersities with conversion is in contrast to processes with transfer and usually originates in slow exchange. Slow initiation also leads to a decrease in polydispersities with conversion but the highest polydispersity due to slow initiation is only $M_w/M_n = 1.35$.

The dynamics of exchange is also very important. Figure 13 shows a system nearly identical to the ideal system shown in Fig. 12. The only difference is the dynamics of ionization; the concentrations of monomer, initiator, Lewis acid, and anion are the same, and the equilibrium constants of ionization and dissociation are also the same.

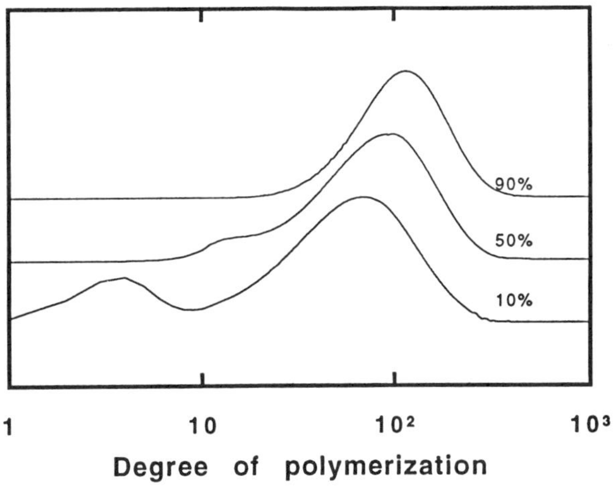

Figure 11 MWD in cationic polymerization as a function of conversion; $[M]_0 = 1\ M$, $[I]_0 = 0.01\ M$, $[LA]_0 = 0.1\ M$; $k_p^+ = k_p^{\pm} = 10^5\ M^{-1}\ sec^{-1}$; $K_D = 10^{-5}\ M$; $K_I = 10^{-3}\ M^{-1}$; $k_i = 10^4\ M^{-1}\ sec^{-1}$; $k_r = 10^7\ sec^{-1}$ (ionization and dissociation equilibria 100 times smaller than shown in Figure 9) (from Ref. 23).

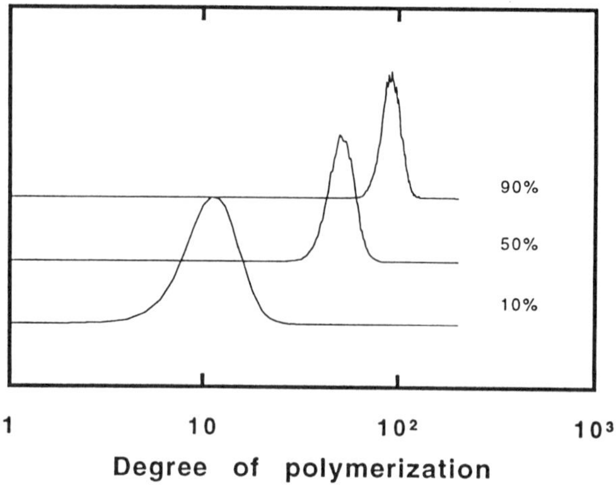

Figure 12 MWD in cationic polymerization as a function of conversion in the presence of common anion; $[M]_0 = 1\ M$, $[I]_0 = 0.01\ M$, $[LA]_0 = 0.1\ M$; $[A^-]_0 = 10^{-4}\ M$; $k_p^+ = k_p^\pm = 10^5\ M^{-1}\ sec^{-1}$; $K_D = 10^{-5}\ M$; $K_I = 10^{-3}\ M^{-1}$; $k_i = 10^4\ M^{-1}\ sec^{-1}$; $k_r = 10^7\ sec^{-1}$ (all conditions except common ion identical to those in Figure 11) (from Ref. 23).

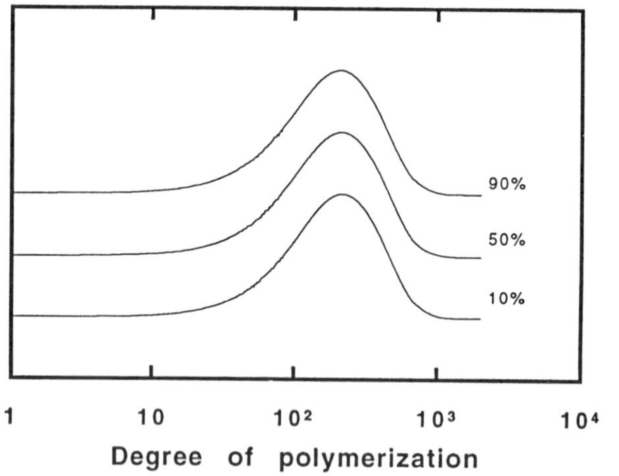

Figure 13 MWD in cationic polymerization as a function of conversion in the presence of common anion; $[M]_0 = 1\ M$, $[I]_0 = 0.01\ M$, $[LA]_0 = 0.1\ M$; $[A^-]_0 = 10^{-4}\ M$; $k_p^+ = k_p^\pm = 10^5\ M^{-1}\ sec^{-1}$; $K_D = 10^{-5}\ M$; $K_I = 10^{-3}\ M^{-1}$; $k_i = 10^0\ M^{-1}\ sec^{-1}$; $k_r = 10^3\ sec^{-1}$ (dynamics of ionization 10^4 times slower than shown in Figure 12) (from Ref. 23).

The surprising result is that the molecular weights are fairly constant and independent of conversion (DP_n = 110, 108, and 104 at 10%, 50%, and 90% conversion) and that the polydispersities are rather broad ($M_w/M_n \approx 2$, independent of conversion). These results indicate clearly that initiation is incomplete even at 90% conversion. The ratio of rate of propagation to that of recombination of counterions within an ion pair is rather high, and at the initial monomer concentration $[M]_0$ = 1 mol/L, ion pairs can propagate on average 100 times before they are converted back to inactive covalent species.

$$DP_n = R_p/R_r = k_p^{\pm}[M][C^{\pm}]/k_r[C^{\pm}] \approx 100 \tag{7}$$

Figures 14 and 15 show the changes of the MWD with conversion and as a function of polymerization degree for systems consisting of only ion pairs and inactive covalent species. This might be the case for systems with common ions when free ions are suppressed. The MWD is unimodal but its breadth depends on the ratio of the rate constants of propagation and recombination of the ion pairs (k_p/k_r).

The polydispersities continuously decrease with conversion and with the increase of the chain length, which is in contrast to systems dominated

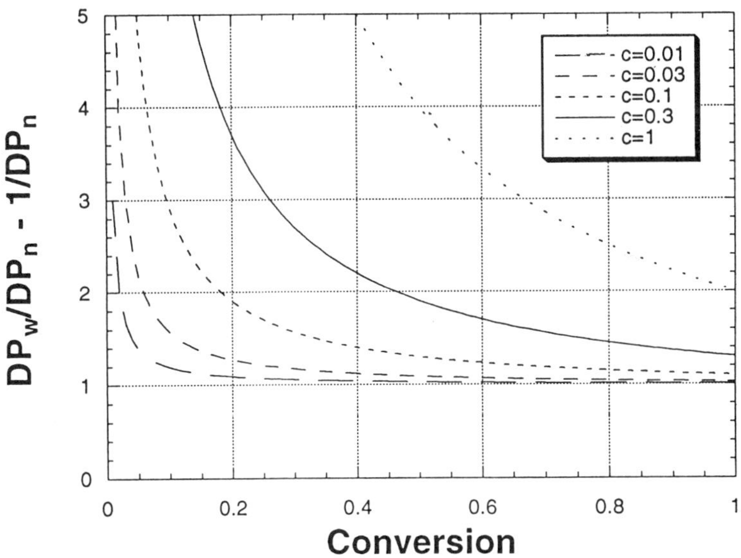

Figure 14 Effect of parameter $c = ([I]_0 - [I])\cdot(k_p/k_r)$ on the decrease in polydispersities with conversion (from Ref. 23).

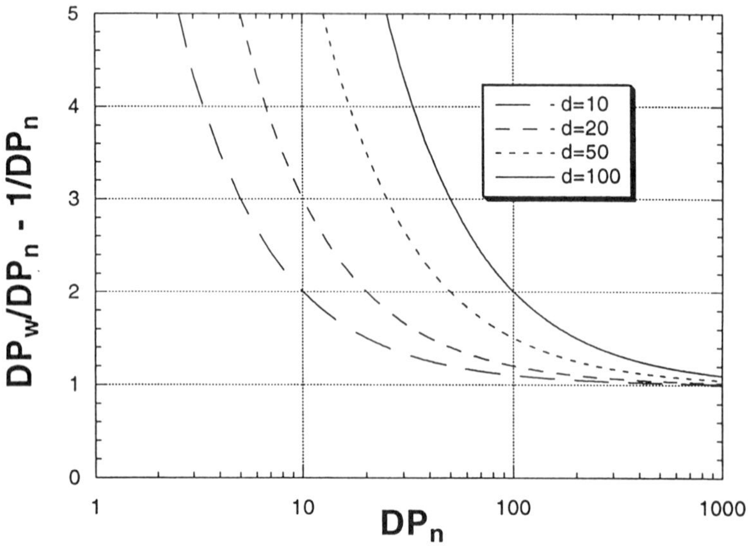

Figure 15 Effect of parameter $d = [M]_0 \cdot (k_p/k_r)$ on the decrease of polydispersities with chain length at complete conversion (from Ref. 23).

	Rate	MW	MWD
slow initiation	slower than ILS (but with acceleration)	initially higher, but approaches ILS	<1.35
termination	slower than ILS (with deceleration)	no effect (limited conversion)	broader than ILS
transfer	no effect	lower than ILS	broader than ILS
slow exchange	no effect	may be higher than ILS	may be very broad, may be polymodal

Scheme 3 Effect of k_i, k_t, k_{tr}, and k_{ex} on kinetics, molecular weights and polydispersities in comparison with an ideal living system (ILS).

by transfer. When exchange is slow, more exchanges may happen during the chain build-up, leading to a more uniform distribution. At lower concentrations of growing species ($[I]_0-[I]$), longer chains with more narrow MWD can be formed. Figure 15 demonstrates that some systems at DP = 10 may have high polydispersities but that they will become nearly ideal at DP = 100. To obtain polymers with lower polydispersities, longer chains have to be formed under slower deactivation. Of course, it may happen that transfer will start to operate at such high molecular weights and the polydispersities, after initial decrease, may start augmenting again.

This discussion of the effects of various imperfections on kinetics, MW, and MWD shows the complex nature of "living" systems. Scheme 3 summarizes these effects.

III. TYPICAL FEATURES OF THE CARBOCATIONIC POLYMERIZATION

Typical features of carbocationic polymerization are related to the structure of the carbenium ions and to the mechanism of the electrophilic addition to alkenes. They have been discussed extensively in Chapters 2 and 3 and will be summarized only briefly below in relation to controlled/living cationic polymerizations. We will focus on four inherent features of carbocationic polymerization which for a long time disabled controlled/living carbocationic polymerization and which have been recently recognized and either controlled or manipulated to successfully prepare well-defined polymers. The four features include transfer of β-protons, high values of propagation rate constants, variety of coexisting electrophilic species and their exchange, and the initiation process.

1. Carbenium ions have only a partial positive charge ($\approx$20%) on the sp^2-hybridized C-center with the formal positive charge [27]. The rest is located on the substituents stabilizing the electron-deficient C atom by resonance, inductive or hyperconjugative effects. Accordingly, partially positively charged β-H atoms (more than 10% positive charge) can be attacked by various nucleophilic and especially by basic components of the reaction mixture leading to transfer reactions. Transfer is the most frequent and most difficult-to-control chain-breaking reaction in the carbocationic polymerization. Usually, the energy of activation of elimination reactions (transfer) is higher than that of the electrophilic addition (propagation). Therefore, the contribution of transfer decreases at lower temperatures and very high molecular weight polymers can be prepared successfully at low temperatures, sometimes even below $-100°$ C. The best control of carbocationic polymerization can be achieved at low tempera-

tures and in the absence of basic components. As discussed in the previous sections, the proportion of chains marked by transfer increases with the chain length. Thus, carbocationic polymerization can be better controlled at lower polymerization degrees, provided that transfer is the only deviation from ideal systems. This approach is very often used in new controlled/living carbocationic polymerizations.

2. Carbenium ions are very strong electrophiles and very rapidly react with alkenes [28]. The electrophilic addition reaction (propagation), most desirable in the carbocationic polymerization, is very fast. Rate constants of propagation are typically in the range $k_p^+ \approx 10^{5\pm1}$ $mol^{-1} \cdot L \cdot sec^{-1}$ at 20° C [29,30]. This means that if only ionic species are present in a system for the targeted synthesis of a high polymer ($M_n \approx$ 100,000, corresponding to $DP_n \approx$ 1,000 for polystyrene; $[M]_0 = 1$ mol/L and $[I]_0 = 10^{-3}$ mol/L), then half of the monomer will be consumed after ≈0.01 sec ($\tau_{1/2} = \ln 2/(k_p^+ \cdot [I]_0)$). In such a short time a lot of heat should be evolved (the polymerization of alkenes is strongly exothermic, ΔH varies from ≈ −10 kcal/mol for isobutene to −20 kcal/mol for styrene), leading to a temperature increase (and therefore more transfer).

It is possible to work at either lower cation concentrations—but this could lead to stronger effects of impurities (adventitious moisture)—or to use more specialized flow reactors. The dynamic equilibration between active and dormant species offers another solution to this problem. In this case, the sensitivity to impurities is low due to the high total number of chains, but the momentary concentration of propagating carbocations is tremendously reduced. This approach is always used in new controlled/living carbocationic polymerizations, as we will discuss in detail in this chapter.

3. As stated above, various active species coexist in carbocationic polymerization: free carbocations and the corresponding ion pairs, as well as some ionic aggregates, onium ions, and covalent species. Model studies show identical rate constants of electrophilic addition to alkenes for benzhydryl cations and the corresponding ion pairs [28]. Polymerization experiments indicate that the reactivities are also similar ($k_p^+/k_p^\pm < 10$ for styrene with $CF_3SO_3^-$ or ClO_4^- in CH_2Cl_2 at −70° C to 20° C) [31,32]. Similar values have been estimated in the polymerization of cyclohexyl vinyl ethers in CH_2Cl_2 at −30° C ($k_p^+/k_p^\pm < 10$) [33]. However, reactivities of carbocations in comparison with onium ions and covalent species are very different, especially in propagation. The electrophilic addition of carbenium ions to alkenes proceeds by a relatively late transition state in which nearly half of positive charge is transferred from the carbocation to a double bond [28]. This transition state may explain why the direct addition of covalent esters or onium ions to alkenes has not yet been

observed. Preliminary ionization or dissociation of carbenium species from onium ion is therefore a prerequisite for electrophilic addition.

The differences in reactivities and the lifetimes of various species may strongly affect the polydispersities of the synthesized polymers and might be the main reason for the often observed polymodalities of molecular weight distributions [23,29]. The synthesis of uniform polymers requires a rate of exchange comparable to propagation. This is especially important for the aforementioned temporary deactivation process. If deactivation is slow then the entire monomer can be consumed by only a small fraction of chains, leading to polymerization degrees much higher than $\Delta[M]/[I]_0$ ratio. As discussed in Sections IV and V, there are a few methods of accelerating the deactivation process. They are based on using Lewis acids which form relatively unstable complex counterions [5,27,34,35], on addition of certain nucleophiles [5,36–38], and salts with counterions capable of facile formation of covalent species [5,25,39]. Additionally, nonpolar solvents may accelerate deactivation in some cases [40,41]. Thus, successful new controlled/living carbocationic polymerizations employ the concept of rapidly equilibrating active and dormant species.

4. Typical initiators for carbocationic polymerization include stable carbocations, strong protonic acids, Lewis acids in combination with either adventitious water or compounds capable of generation of protons/carbocations. Stable carbocations alone, such as trityl cations, are very slow initiators. Sometimes they react with monomer by hydride abstraction rather than directly. Very often initiation is incomplete and some of the trityl salts remain until the end of polymerization. This disables the synthesis of uniform polymers with predetermined molecular weight. Protonic acids generate oxyanions which are quite basic. These anions may also facilitate transfer processes. Acids tend to aggregate and to form homoconjugated anions. Rates of initiation vary with the aggregation state of the acids, resulting in polymers with broad or sometimes even polymodal molecular weight distributions and unpredictable molecular weights. Lewis acids (MtX_n) in combination with adventitious water do not allow control over molecular weights (the amount of water varies unpredictably), initiation is usually slow, and molecular weights are often limited by transfer. An excess of water may terminate polymerization by formation of oxonium ions, and may finally destroy the Lewis acids.

The initiating systems based on Lewis acids with covalent esters and halides offer some advantages. First, the number of chains can be easily controlled by the concentration of the ester or halide used as *an initiator*. Second, the polymerization rate and the proportion of carbocationic species may be easily adjusted by the strength and concentration of the Lewis acid used as *a coinitiator* (*catalyst*). Third, through the correct choice of substituents and leaving groups in the initiator, the ratio of the rates of initiation to that of propagation can be controlled. For systems with a relatively small degree of ionization, it is desirable to use initiators that ionize easier than the dormant macromolecular species. For example, cumyl chloride is used successfully in polymerization of isobutene. Cumyl chloride is also an excellent initiator for polymerization of styrene. A more detailed discussion on the control of the initiation process is presented in Chapter 3 and later sections of this chapter. New controlled/living carbocationic polymerizations typically use a combination of a Lewis acid and a covalent precursor which ionizes easier than dormant macromolecular species. Sometimes it is necessary to use additives such as nucleophiles or salts. If the system is properly designed and molecular weights are not too high, polymers with predetermined polymerization degrees, low polydispersities, and controlled end functionalities, as well as block copolymers, can be prepared [5,42,43].

IV. CONTROLLED/LIVING CARBOCATIONIC POLYMERIZATION: PRINCIPLES AND METHODS

A. Historical Background

1. Preludes

The possibility of controlled/living cationic polymerization of vinyl monomers was reported during the period of 1975–1977 by Higashimura and Kishiro, who polymerized *p*-methoxystyrene (*p*MOS) using iodine [44,45]. They noticed that, in a 1:1 mixture of methylene chloride and carbon tetrachloride, the iodine-initiated poly(*p*MOS) showed a bimodal molecular weight distribution (MWD) where the lower molecular weight peak increased with conversion and the higher polymer fraction remained almost unchanged.

By that time, the Kyoto group had been studying the nature of growing species that give bimodal MWDs of polystyrene and its derivatives, an intriguing but puzzling phenomenon found a few years earlier [32,46–50]. The experiments with *p*MOS were originally intended to determine the scope of such bimodal MWD as a function of the types monomers and initiators. Higashimura concluded that the double-peaked MWD was due to the coexistence of two mutually independent, simultaneously propagating species, then coined either "dissociated" and "nondissociated" species [48,49] or "visible" and "invisible" species [51], which form the higher and lower polymer populations, respectively. It was already known [47,50] that the nondissociated species dominate in nonpolar media and in the presence of a "common ion" salt (such as tetra-*n*-butylammonium perchlorate for acetyl perchlorate initiator) and behave differently than the "dissociated" counterpart, which consists of free carbocations.

Thus, by lowering solvent polarity (CH_2Cl_2/CCl_4 mixtures) or by the addition of a common ion salt ($n$$Bu_4NI$), these researchers obtained poly(*p*MOS) of unimodal and relatively narrow MWDs, whose shape and position corresponded to those of the lower polymer population in the two-peaked distribution. Under these conditions the number-average molecular weights increase with conversion. Upon sequential addition of fresh monomer feeds, and block copolymers of *p*MOS and isobutyl vinyl ether (IBVE) are obtained [52]. Similar results were also obtained for the polymerization of IBVE [53,54].

Nearly at the same time, Johnson and Young reported a similar system for the polymerization of *n*-butyl vinyl ether with iodine, and called the process a "pseudoliving" polymerization [55]. Also, Pepper suggested that the active centers in polystyrene derived from perchloric acid might have a long lifetime at a very low temperature [32,50], and later obtained block copolymers of styrene with *tert*-butylazyridine by a sequential polymerization from the perchlorate-initiated polystyrene end [56].

In the early 1980s, Kennedy and his co-workers reported "quasiliving" polymerizations, which are phenomenologically akin to living polymerizations [57]. These processes involved slow and continuous monomer addition to a stirred initiator solution kept at a relatively low temperature. The monomers used therein included α-methylstyrene, isobutene, styrene, and alkyl vinyl ethers. In most cases, the number-average molecular weights steadily increased with the weight of the added monomer and the formed polymers had relatively narrow MWDs.

All these early examples of "living-like" polymerizations, however, are not well controlled and can not be considered truly living, as evidenced by M_n values not directly proportional to monomer conversion, the rather broad MWDs, and, in some cases, low initiation efficiency. The possible reasons for the difference may be that initiation is relatively slow and inefficient (e.g., with iodine), that the polymerizations are often too fast to control (e.g., due to local heating and resulting chain transfer reactions), and that the exchange reactions among multiple growing ends are not fast enough (see Section IV.B).

2. *Discovery*

In 1982 Higashimura et al. [54] began studies focused on the development of living cationic polymerizations of vinyl monomers. They decided to use IBVE and related alkyl vinyl ethers as monomers because they form the alkoxy-stabilized growing carbocations, along with iodine as the initiator that gives a nucleophilic counteranion, as deduced from earlier results with the *p*MOS/iodine system [44,52]. The initial outcome was rather similar to that with *p*MOS, i.e., steady increases in polymer molecular weights with conversion but not directly proportional, imperfect initiation efficiency, and fairly narrow MWDs.

During the subsequent search for better initiators, the Kyoto group reached hydrogen iodide, which would generate the same nucleophilic iodide counteranion as molecular iodine did. Because they did not observe IBVE polymerization with the protonic acid alone, they combined anhydrous hydrogen iodide with molecular iodine to form a binary initiating system, initially intending to accelerate the polymerization and to generate the triiodide counteranion ($HI + I_2 \rightarrow H^+I_3^-$). The binary system triggered a polymerization much faster than that with iodine alone and eventually led to polymers with surprisingly narrow MWDs (M_w/M_n less than 1.1). The Kyoto group discovered perhaps the first controlled/living cationic polymerization of vinyl monomers [58]. Obviously, the key to the discovery was the binary initiating system (HI/I_2).

With the HI/I_2 initiating system, they also found that the polymerization of IBVE and related alkyl vinyl ethers ($CH_2{=}CHOR$; R = methyl, ethyl, and higher alkyls) in nonpolar solvents led to results that conformed with living polymerizations, as summarized in Figure 16 [58,59]:

a. The number-average molecular weights (M_n; by GPC, VPO, and ^1H NMR) increase in direct proportion to monomer conversion.
b. The M_n is very close to the calculated value assuming that one hydrogen iodide molecule (not iodine) forms one living chain (i.e., $[HI]_0$ = [living polymer]).
c. Upon addition of more monomer to a completely polymerized reaction mixture, polymerization resumes to give a further increase in polymer molecular weights which is also in direct proportion to conversion.
d. The polymers have very narrow, nearly monodisperse or uniform MWDs (M_w/M_n = 1.03–1.10).

The adoption of the two-component system (generally, initiating system) was intended to increase the initiation rate by changing from iodine to hydrogen iodide, whereas keeping the same nucleophilic and iodine-based counteranion (I_3^-), which appeared well suited for controlling the polymerization as well as for obtaining much narrower MWDs of produced polymers. Although the use of anhydrous hydrogen iodide and its combination with iodine in cationic polymerization was already studied by Guisti

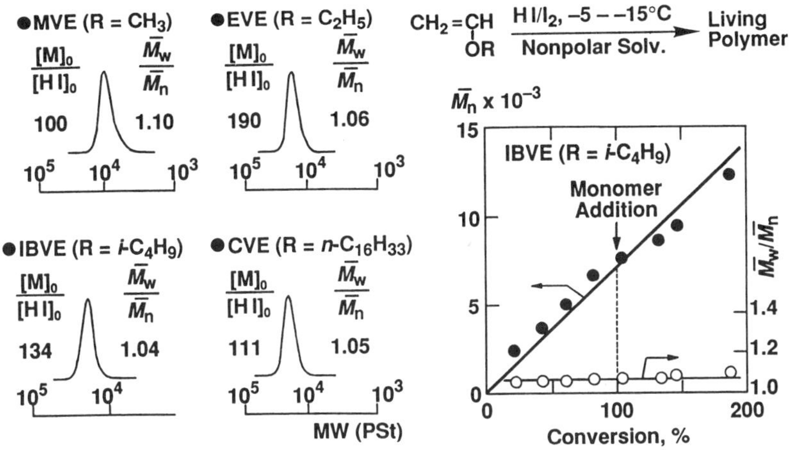

Figure 16 Controlled/living cationic polymerizations of alkyl vinyl ethers with the HI/I_2 initiating systems. (From Ref. 58.)

in the late 1960s [60,61], their research efforts had primarily been directed towards the kinetic analysis of styrene polymerization without regards to M_n, MWD, and the living nature of the resulting polymers. The mechanistic implications of the living cationic processes with the HI/I_2 and related systems will be discussed in Section IV.B.1.a.

In 1986, Kennedy and Faust published a similar controlled/living polymerization of isobutene by the cumyl acetate/boron trichloride [$C_6H_5C(CH_3)_2OC(O)CH_3/BCl_3$] and related binary initiating systems [35,62].

The 1980s and the early 1990s witnessed the discovery of numerous living polymerizations and new initiating systems for a variety of cationically polymerizable vinyl monomers such as vinyl ethers, isobutylene, alkoxystyrenes, styrene, and others [5,36,42,49,63–73]. The discovery has challenged a long-standing notion among the specialists in this field [50] that living cationic polymerizations of vinyl monomers would be inherently impossible due to frequent and unavoidable chain transfer reactions. Interestingly, the subsequent rapid and extensive proliferation of controlled/living cationic polymerizations coincides with the expansion of the scope of controlled/living processes by other mechanisms such as group transfer, ring-opening metathesis, and radical polymerizations [18,67].

Even since the early 1990s, when one of the authors (M.S.) published a comprehensive review of this field [42], in which he tentatively concluded that ". . . the basic principles of living cationic polymerization have been established. . . . ," the scope and understanding of living cationic polymerization have been expanded and deepened considerably, and new methodologies have also been developed. The four-year progress can be seen quite clearly, if one compares the contents of this chapter to the review in 1991 [42] and later [5,7,74].

In light of the knowledge currently available, this chapter aims to discuss the general principles, methodologies, and scope of controlled/living cationic polymerizations and more specific examples for representative classes of monomers. The next chapter will be devoted to the controlled syntheses of new functional polymers using these systems.

B. General Principles and Methodologies

As discussed in the preceding sections of this chapter, the key to living cationic polymerization is to reduce the effect of chain transfer reactions (Scheme 4); because termination is much less important in the cationic polymerization of vinyl monomers. The primary reason for frequent chain transfer reactions of the growing carbocation (1) is the acidity of the β-H atoms, next to the carbocationic center, where a considerable part of the positive charge is localized. Because of their electron deficiency, the protons can readily be abstracted by monomers, the counteranion (B^-), and other basic components of the systems, to induce chain transfer reactions. It is particularly important to note that cationically polymerizable monomers are, by definition, basic or nucleophilic. Namely, they have an electron-rich carbon-carbon double bond that can be effectively poly-

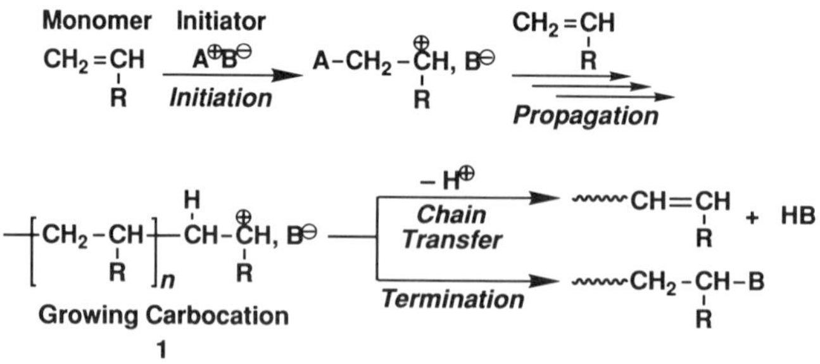

Scheme 4

merized via a carbocationic intermediate, thus chain transfer reactions via β-proton elimination are "built-in" side reactions in cationic vinyl polymerization.

The lifetime of carbocations usually does not exceed a few seconds, unless very low temperature or superacid media are used. This means that lifetime of the growing species is also very short and may not be sufficient to accomplish the desired synthetic task. This is the primary reason why living cationic polymerization has generally been believed to be almost impossible [50]. As discussed in this chapter, we are now increasingly convinced that the solution to this problem is extension of the lifetime of growing chains through the reversible conversion of the carbocations to a dormant state such as covalent species or onium ions, in which they are stable for long times. This type of equilibration can be considered to be a dynamic stabilization of the growing carbocations. This "*stabilization*" and controlled/living cationic polymerizations of vinyl monomers can be achieved, at least operationally, by the following three general methods: (A) with nucleophilic *counteranions*; (B) with *nucleophiles*; (C) with *salts*. There are still a number of points to be clarified in the implications and reaction mechanisms of these approaches, which are currently under extensive study and discussion around the world. However, in the last section of this chapter we will attempt to create a unified picture of controlled/living cationic polymerizations and to explain the role of all constituents of these multicomponent systems. A general and interesting discussion of some of these problems is available in the book by Kennedy and Iván [5]. The following mechanistic and other aspects of living cationic polymerizations by these three methods have been discussed: (a) the definition and critical comparison of living (cationic) polymerizations [7,74–77]; (b) kinetic and operational criteria [6,78–84]; (c) system classification [5,42]; (d) the lifetime of living polymers [75,80,85]; (e) the nature of the (living) growing species in relation to dormant or covalent intermediates [86,87]; (f) the exchange processes among multiple growing species [21,24,27,29,73,88–90]; (g) the effects of additives in living cationic polymerizations, such as nucleophiles [36,64,91–94] and salts [71,72,88] (see also the next sections).

As the preceding discussion indicates, the dynamic "stabilization" of carbocations calls for the use of some nucleophiles, including counteranions of the growing ends, that donate electrons to the carbocation, generate covalent species and onium ions, decrease cationic charges, and decrease the acidity of the neighboring protons. Therefore, method A is reduced to the design of initiating systems that generate suitably nucleophilic counteranions, whereas methods B and C require the search for suitable nucleophilic additives. Let us begin the discussion by considering the general aspects of the three approaches; subsequently, representative examples will be treated more specifically for the respective monomers.

1. *Controlled/Living Polymerizations with Nucleophilic Counteranions*

a. Hydrogen Iodide/Iodine (HI/I_2) Initiating System. According to mechanistic and kinetic studies by the Kyoto group [34], the living polymerization of vinyl ethers with the HI/I_2 initiating system proceeds as shown in Scheme 5:

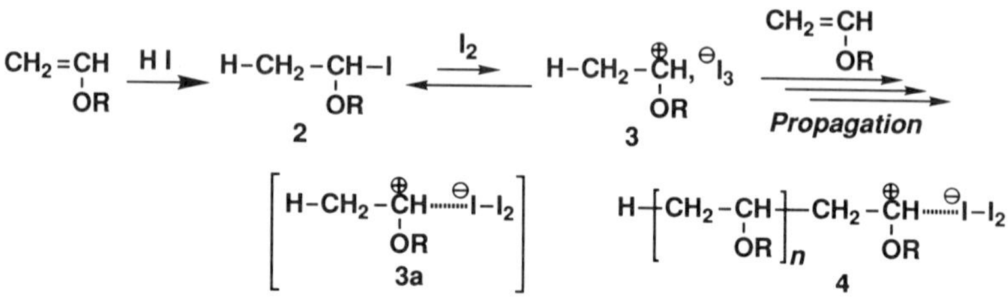

Scheme 5

The first step is a quantitative and selective addition of hydrogen iodide across the double bond of vinyl ether to give the covalent adduct (2), which is too stable to initiate propagation alone. In the absence of I_2 there is no polymerization of most vinyl ethers, even when the monomer is in a large molar excess over hydrogen iodide. The molecular iodine, in turn, activates the carbon-iodine linkage of the adduct to form the carbocation (3). Species (3a) and (41) schematically represent dynamic exchange between covalent species (2) and ion-pair (3). This abbreviated notation will be also used later in this chapter. Successive addition of vinyl ether monomers to this species and its higher homologues leads to the living growing polymer (4). The active site of this polymer is accompanied by a counteranion (I_3^-) that is considered to be suitably nucleophilic to dynamically stabilize the growing carbocation through its nucleophilic interaction. Higashimura's school calls this methodology as "Carbocation Stabilization by Nucleophilic Counteranions" [36,42]. In the earlier papers, such nucleophilic interaction was represented by the dotted lines between a carbocation, a leaving group, and a Lewis acid. However, the real meaning of the "stabilization" is now considered as the dynamic exchange between the cation and the dormant alkyl halide.

In this mechanism, adduct (2) forms a polymer chain with a "dormant" terminal carbon-iodine bond that is essentially identical to (2), whereas molecular iodine acts as a Lewis acid (electron acceptor) that reversibly activates this dormant terminal to enable a reaction with incoming vinyl ethers. The activation herein implies ionization of the carbon-iodine covalent bond. Accordingly, the M_n of the resulting living polymers is simply determined by the initial molar ratio of (2) to monomer, whereas the iodine concentration does not affect polymer molecular weight but accelerates the polymerization. Thus, adduct (2) (or hydrogen iodide) is called "initiator" and iodine "activator" (the latter is also called "coinitiator" or "catalyst," depending on authors [5]). The reaction rate is first order with respect to the initial concentrations of hydrogen iodide and iodine, namely, $-d[\mathrm{M}]/dt = k[\mathrm{M}]^n[\mathrm{HI}]_0[\mathrm{I}_2]_0$, where k is an apparent rate constant and the superscript n is the kinetic order in monomer consumption; n is usually unity but sometimes smaller and even zero (cf., Section VI.B.1). The kinetics of this and other related living polymerizations is the subject of several recent papers [34,78,95–97] and is discussed further in Section VI.B.

Relative to the initiator/activator mechanism shown in Scheme 5, it is interesting to compare vinyl ether polymerizations initiated with the HI/I_2 system and with iodine alone. The former system provides living polymers of controlled molecular weights and very narrow MWD [58], whereas the latter has been known for more than a century but fails to give such controlled polymerizations (cf., Sections IV.A) [49,55]. In the iodine-mediated polymerization, iodine serves as both the initiator and activator; one molecule of iodine first slowly adds across the vinyl ether double bond to give an adduct. The α-carbon-iodine bond is activated by another molecule of iodine [34,95]. Thus, both systems would in fact form the identical growing chain end [$\sim\sim\sim CH_2CH(OR)^+ \cdots\cdots I_3^-$], and the observed distinct difference is, at first glance, rather difficult to understand.

As discussed in Section IV.A.2, the difference indicates that the initiation is much faster with the HI/I_2 initiating system, and the initiation rate difference is, in turn, attributed to the different reactivity of the initial adducts of hydrogen iodide and iodine with vinyl ethers, H—CH_2CH(OR)—I (**I**) and I—CH_2CH(OR)—I (**II**), respectively. ^{1}H NMR analysis of reaction systems [34] shows that the α-carbon-iodine linkage in the hydrogen iodide adduct (**I**) is much more reactive than that in the iodine adduct (**II**). This would be expected from the electron-withdrawing effect of the iodine on the β-methylene carbon in the latter. In retrospect, this subtle difference in electronegativity between hydrogen (in **I**) and iodine (in **II**) on the β-carbon resulted in the marked difference between the HI/I_2- and the iodine-mediated polymerizations [34].

b. Protonic Acid/Lewis Acid (HB/MtX_n) Initiating Systems. The generalization of the initiator/activator mechanism for the HI/I_2-initiation system (Scheme 5) leads to the design of a variety of initiating systems that can also induce controlled/living cationic polymerization of vinyl ethers and related monomers via the ''carbocation stabilization by nucleophilic counteranions'' (Scheme 6) [42,98,99]. In Scheme 6 the activated growing species is shown schematically as **6**, which is a generalized version of species **3a** derived from HI/I_2 in Scheme 5. The carbocation species **6** is considered to be stabilized via the nucleophilic interaction of a counteranion ($BMt\mathbf{X}_n^-$) derived from the initiating system. This is a dynamic process and stabilization occurs via rapid equilibration between dormant species **5** and the corresponding carbocations.

Thus, the *initiators* are not only the iodide-type adduct **2** but also adducts of protonic acids (HB) that carry counteranions (B^-) with suitable nucleophilicity; namely these protonic acids should form adducts **5** quantitatively with vinyl ethers but should not initiate uncontrolled polymerization by themselves. Such protonic acids include hydrogen halides and carboxylic acids.

The *activators* (coinitiators) are not only molecular iodine but also other mild Lewis acids ($Mt\mathbf{X}_n$) that are mostly metal halides such as zinc halides. The suitable metal halides and other Lewis acids should be able to effectively activate the carbon-B linkage of the adducts **5** but they should not be too strongly Lewis acidic to initiate uncontrolled polymerization in the presence of adventitious protogens like water.

Perhaps the most frequently used example is the HI/ZnI_2 system, where the iodine in the HI/I_2 counterpart is now replaced with the mild Lewis acid, zinc iodide [98,99]. A more detailed discussion of the scope and mechanism of the polymerizations by the HB/$Mt\mathbf{X}_n$ systems will be given for respective monomers in the later parts of Section IV. It should be noted here that the ''suitable nucleophilicity'' of the initiator's anion B^- and the ''mild Lewis acidity'' of $Mt\mathbf{X}_n$ strongly depend on the nature and stability of the growing carbocations or the monomers from which they are derived.

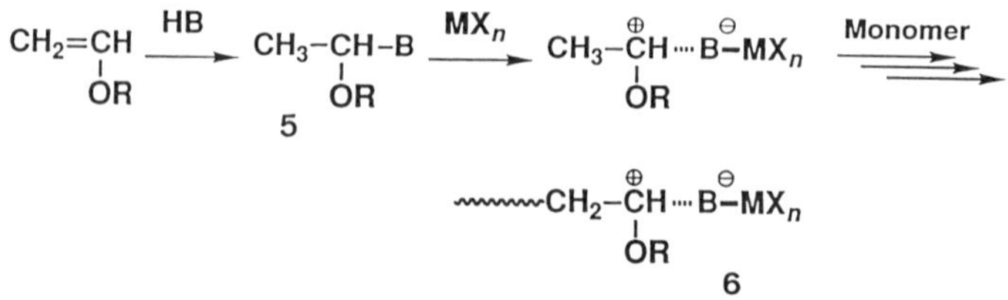

Scheme 6

Binary initiating systems that consist of a protonic acid (or a cationogen) and a Lewis acid (metal halide) have been known for a long time. Typical examples of such classical initiating systems include *tert*-butyl chloride/aluminum trichloride and others. In contrast to the use of mild Lewis acids in the HB/Mt$\mathbf{X}_n$ systems, these conventional systems use almost invariably a very strong Lewis acid ($AlCl_3$, BF_3, etc.) as an activator to obtain polymers in high yield. As a result, although effective in generating carbocations from alkyl halides, they generate carbocations associated by too weakly nucleophilic and too stable counteranions, preventing the dynamic equilibration of carbocations with dormant species. A unique feature of the HB/Mt$\mathbf{X}_n$ initiating systems is that nucleophilicity of the counteranion BMt$\mathbf{X}_n^-$ can be varied by designing combinations of the initiator HB (or the anion B^-) and the activator Mt$\mathbf{X}_n$.

2. *Controlled/Living Polymerizations with Added Nucleophiles*

When adduct **5** (B = Cl or RCOO; R = CH_3, CF_3, etc.) is combined with a metal halide (e.g., $SnCl_4$ and $EtAlCl_2$) that is a stronger Lewis acid than zinc halides, under otherwise the same reaction conditions, vinyl ether polymerizations with such initiating systems are extremely fast and provide uncontrolled polymers with broad MWDs (cf., Figure 17, B and D) [100–105]. This is attributed to the generation of complex counteranions that are much less nucleophilic than I_3^- and ZnI_3^- and that can no longer dynamically stabilize the growing carbocations.

For such systems, it was found [64] that external addition of some weak nucleophiles, such as esters and ethers for vinyl ethers, decelerates the polymerization, narrows the polymer MWDs, and eventually leads to polymers with controlled molecular weights (cf., Figure 17, A, C, and E). In the example shown in Figure 17C, the added nucleophile is tetrahydrofuran. Later studies by Kennedy and associates revealed that a similar methodology is applicable to isobutene [106]. Examples of the externally added nucleophiles are shown in the following sections for respective monomers. Importantly, those nucleophiles that are effective depends on the structure of monomers.

Although the Kyoto group initially called the additives "Lewis bases" [64] and the Akron group "electron donors" [106], herein we call them collectively "nucleophiles," which we believe is a more general term.

The actual role of the added nucleophile is still under discussion [41,88,92] (cf., Sections VI.B.2 and VII.E.4). They may interact with the growing carbocations and stabilize them through the weak solvating interactions [36]. Another possibility is that the added nucleophiles and the growing carbocations form reversibly onium ions that serve as dormant species, as discussed by Penczek [92] and Matyjaszewski [88] (schemati-

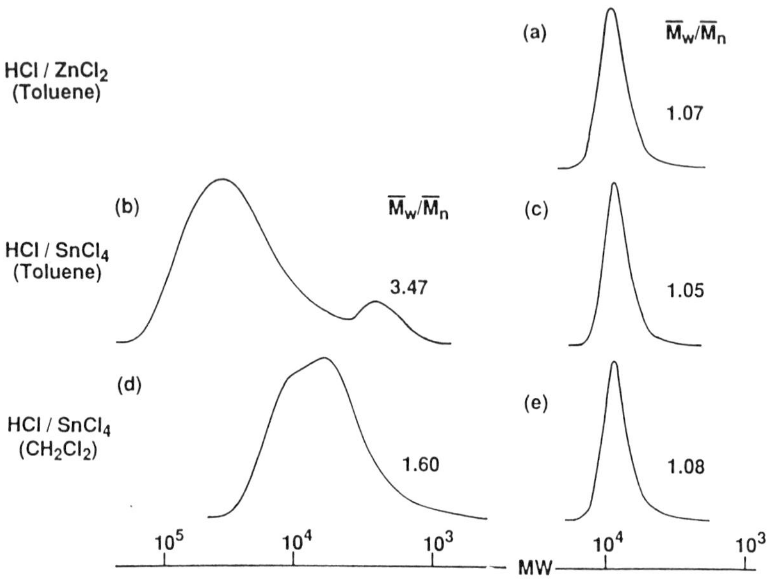

Figure 17 MWDs of poly(isobutyl vinyl ether) to illustrate the three general methods for living cationic polymerizations at $-15°$ C: $[M]_0 = 0.38\ M$; $[HCl]_0 = [ZnCl_2]_0 = [SnCl_4]_0 = 5.0$ mM; conversion = ca.100%. Initiating systems and reaction conditions: (a) $HCl/ZnCl_2$, in toluene without additive; (b) $HCl/SnCl_4$, in toluene without additive; (c) $HCl/SnCl_4$, in toluene with added tetrahydrofuran (100 mM); (d) $HCl/SnCl_4$, in CH_2Cl_2 without additive; (e) $HCl/SnCl_4$, in CH_2Cl_2 with added nBu$_4$NCl (2.0 mM). (From Refs. 73 and 105.)

cally, species **7**, Scheme 7), or alternatively, that they destabilize complex counteranions to convert carbocations to covalent species more rapidly [41].

3. Controlled/Living Polymerizations with Added Salts

The two approaches discussed above are primarily useful in nonpolar solvents (like toluene and n-hexane) where the interactions of carbocations with nucleophiles are strong and favored. In relatively polar solvents like methylene chloride, these methods often fail to give controlled polymerizations, most likely because the interaction is weaker between the growing carbocations and nucleophiles [whether they are "built-in" (counteranions) or "externally added" (esters, etc.)], which facilitates dissociation of the carbocation. The effect of solvent in the latter system, however, is much weaker.

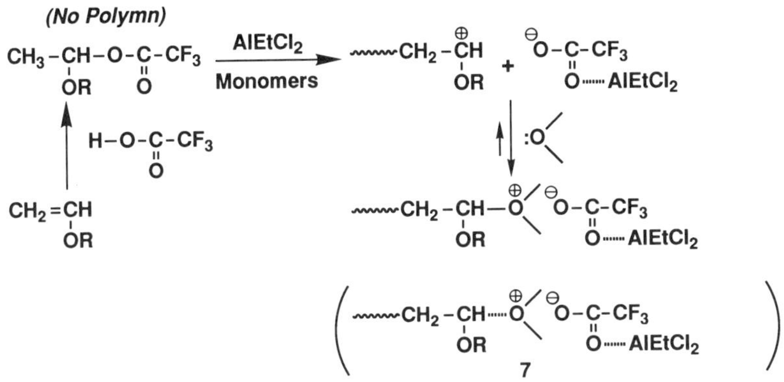

Scheme 7

For example, the polymerization of alkyl vinyl ethers using an HCl/$SnCl_4$ (or adduct **5**/$SnCl_4$) initiating system in methylene chloride is very fast even at −15° C to give polymers with broad and often bimodal MWDs (Figure 17D) [105]. Similar effects of solvent polarity are found in the polymerizations of *p*-alkoxystyrenes [107], styrene [25], and *N*-vinylcarbazole [108].

These apparently nonliving/uncontrolled systems can be converted into controlled/living cationic processes by adding certain salts that carry nucleophilic anions, such as tetra-*n*-butylammonium halides and acetate (nBu_4NX; X = Cl, Br, I, CH_3COO) [25,39,105,107,108]. Upon addition of the salts, the polymerizations are decelerated, give polymers with much narrower MWDs than in the salt-free systems, and fulfill the criteria for a living polymerization as already discussed (cf., Figure 17, D and E).

These "salt effects" are schematically depicted in Scheme 8. As we will discuss later more in detail (Sections VI.B.3 and VII.E.3), mechanistically, salts may act in two different ways. In polar solvents they will suppress the free ions and considerably reduce their lifetime. This often converts bimodal MWD to monomodal MWD and provides controlled polymers. However, in polymerization of vinyl ethers initiated by strong Lewis acids such as $SnCl_4$, where only ion pairs are present after addition of a few percent of salts or in nonpolar toluene, control is still very poor (Fig. 17B). Controlled polymers can be obtained only after addition of a more than equimolar amount of tetra-*n*-butylammonium halides. This implies that the salts change the weakly nucleophilic counterion $SnCl_5^-$ to the more nucleophilic $SnCl_6^{2-}$, which faster converts growing carbocations to covalent species. Another effect of added salts is related to

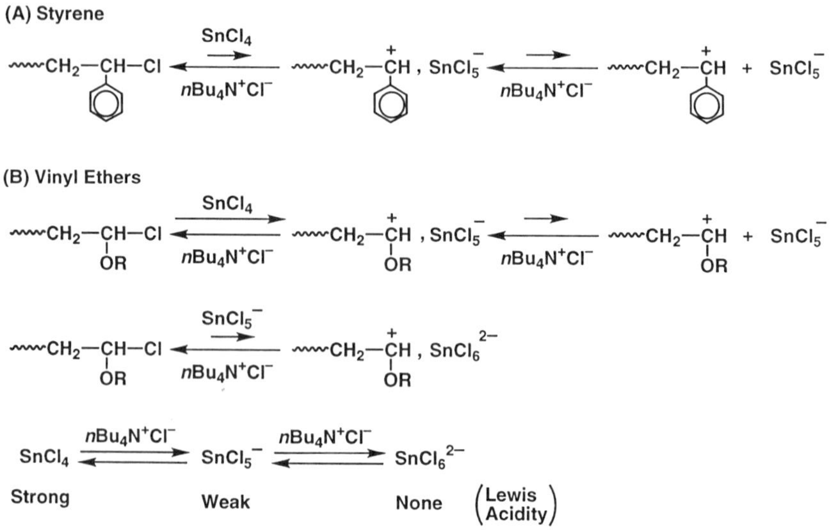

Scheme 8

special salt effects based upon the exchange of counterions. Addition of a salt with a less nucleophilic anion can accelerate the polymerization and provide controlled polymers [109] (see also Section V.A.4), whereas a more nucleophilic anion reduces the polymerization rate [33].

Historically, it would be interesting to note that the early work by Higashimura and Kishiro for the iodine-initiated *p*MOS polymerization (see Section IV.A) [44] also employed an externally added salt ($n\mathrm{Bu_4NI}$) to obtain long-lived polymers in polar solvents. Even before that study, the Kyoto group [47,48], and Pepper [50], observed marked effects of added salts ($n\mathrm{Bu_4NClO_4}$, etc.) on the so-called "bimodal MWD" of polystyrene formed with acetyl perchlorate [47,110], perchloric acid [50], and related oxygen-containing protonic acids [26,111]. These effects have been attributed to the suppression of ionic dissociation of the growing species by the salts via "common-ion" or "mass law" effects. In retrospect, the discovery of controlled/living cationic polymerizations is a logical consequence stemming from the accumulated knowledge of cationic polymerizations.

4. *Evolution of Mechanistic Views on Controlled/Living Carbocationic Polymerization*

The controversies, attention, and active discussions of the mechanisms of new controlled/living systems in the 1990s probably originate in an early statement: ". . . it seems unlikely that any cationic polymerizations (of alkenes) will display living characteristics in their full perfection, i.e., satisfy the molecular weight as well as rate criteria," ascribed to frequent and unavoidable chain-breaking reactions [50]. However, as demonstrated in the earlier parts of this section, polymerizations of vinyl ethers, isobutene, and styrenes can be controlled in a way typical for living systems, in which straight kinetic plots in semilogarithmic coordinates as well as linear evolution of molecular weights with conversion are observed [5,36,42]. Probably because of this contradiction, new mechanisms have been proposed [58,62]. They could have been supported additionally by relatively slow polymerization rates and apparent rate constants of propagation thousands or even millions time lower than the rate constants reported for carbocationic propagation.

Mechanisms based on the insertion or multicenter rearrangement (so called pseudocationic polymerization) had been offered previously for styrene polymerization initiated by perchloric acid [112,113]. However, later studies showed unambiguously that a vast majority, if not all, of monomer was consumed in a typical carbocationic growth [31,32]. Propagation on "stretched or activated, more-covalent-than-ionic" bonds have been proposed to occur in new controlled/living systems [5]. However, such species generated on passing from covalent species to carbocations are not considered as individual chemical entities, because their lifetime is comparable to that of a bond vibration. Therefore, they should not be considered as growing species. Instead, according to physical organic chemistry (cf., Chapter 2), carbocations can dynamically equilibrate with either onium ions or covalent species which per se do not add to alkenes [28]. These dormant species must first convert to carbocations which only then can react with vinyl monomers [27].

Thus, although the conventional carbocationic mechanism of propagation had endorsement from various model studies and additional support from similar stereo-, regio-, and chemoselectivities, it was not immediately accepted. Indeed, many phenomena, such as bimodal molecular weight distributions, decrease or increase of polydispersities with conversion, unusual effects of added nucleophiles, strange salt effects, and others could not be easily explained. For example, how could a bimodal molecular weight distribution lincarly be obtained if all monomer is consumed in ionic propagation and if both ions and ion pairs have similar reactivities? Or, why could the molecular weights increase with conversion but polydispersities are very high ($M_w/M_n > 5$)? Both observations were very different from anionic systems which were always used as standards for living systems [1].

The new controlled/living carbocationic polymerizations were, in some sense, quite similar to new controlled/living anionic systems developed for acrylates, such as group transfer polymerization (GTP) [114] and lithium salt-mediated systems [115]. In both systems, the typical ionic polymerizations were very fast and difficult to control. However, the use of special initiating systems with labile ligands, such as esters and halides for cationic systems and silyl groups for methacrylates, allowed the preparation of polymers with degrees of polymerization (although usually limited to relatively low values, $DP_n \approx 200$) predetermined by the ratio of concentrations of reacted monomer to the introduced initiator. In GTP, as well as in cationic systems, the rates were controlled by the concentration of added catalysts which, depending on the system, could be at very low but also at high concentrations.

The key to understanding both of these reactions, but especially controlled/living cationic polymerization, was to recognize the importance of the dynamics of exchange reactions and their effect on polydispersities [20–22,88,116]. The appropriate selection of Lewis acid and ligands, the structure, and the concentration of nucleophiles and salts all reduce the lifetimes of the growing carbocations and transform them rapidly to the dormant state as well as provide well-defined polymers. If only a few monomer molecules can be added to a growing cation before it converts into a dormant form, the probability of the growth for all chains becomes similar resulting in narrow molecular weight distribution. Additional support for the exchange reactions and the presence of intermediate carbocations came from ligand exchange experiments and dynamic NMR [41,105,117]. It was also demonstrated by computer simulations (cf., Section II) that a bimodal MWD is possible when ions and ion pairs have identical reactivities but different lifetimes, as well as that slow exchange may initially provide polymers with very high polydispersities which should decrease with conversion, in contrast to systems dominated by transfer [23].

One of the most controversial issues is the "living nature" of new controlled carbocationic systems and the apparent suppression of transfer. It has been demonstrated that extension of molecular weight range in controlled/living systems to that typically used in conventional systems may be accompanied by a loss of control [7,76]. Thus, in these cases there is no significant improvement in chemoselectivities, and polymers with the highest, albeit not predetermined, molecular weights are obtained with free carbocations such as those generated by γ-radiation or strong Lewis acids/adventitious moisture at low temperatures (cf., Chapter 3).

It must be stressed that MWD of the resulting polymers can not be considered as the only criterion of the "living nature." In many systems, the decrease of polydispersities achieved by the appropriate initiation system originates in faster exchange between carbocations and dormant species and not in a reduction of transfer and termination, i.e., an improved "living nature."

The mechanistic details and roles of all constituents of the multicomponent initiating systems for new controlled/living carbocationic polymerization are also discussed in Section VI. At this stage it suffices to say that in both the new systems and conventional carbocationic polymerization, monomer is consumed by the repetitive electrophilic addition of growing carbocations whether or not in dynamic equilibrium with either covalent species or onium ions.

V. SCOPE OF CONTROLLED/LIVING CARBOCATIONIC POLYMERIZATION: MONOMERS AND INITIATING SYSTEMS

Since its discovery for vinyl ethers and isobutene in the 1980s, the scope of controlled/living cationic polymerization has been expanded rapidly in terms of monomers and initiating systems. Figure 18 shows a partial list of representative monomers for which controlled/living cationic polymerizations are available. They cover virtually all classes of cationically polymerizable vinyl compounds, such as vinyl ethers, isobutene, styrene and its derivatives, and *N*-vinylcarbazole. A rough estimate indicates that the total number of monomers for controlled/living cationic polymerization

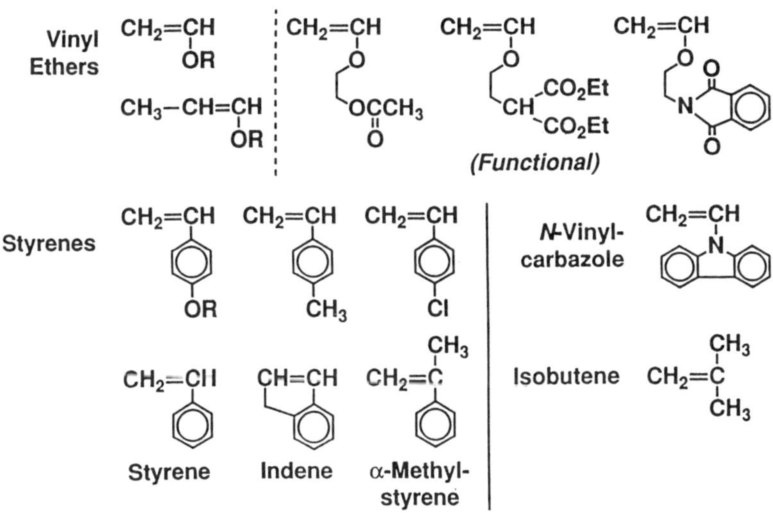

Figure 18 Vinyl monomers for which controlled/living cationic polymerizations are feasible at the end of 1994.

Table 1 Controlled/Living Carbocationic Polymerization: The Status Quo for Monomers and Methods[a]

Monomer	Type of initiating systems[b]		
	Counteranion	Added nucleophile	Added salt
Vinyl ethers	Yes	Yes	Yes
Isobutylene	Yes	Yes	Yes
Styrenes	Yes	Yes(?)	Yes
N-Vinylcarbazole	Yes	(?)	Yes

[a] Yes: controlled/living cationic polymerization is feasible; (?): controlled/living polymerizations based on this method are not yet reported.
[b] See Section IV.B for the three general methods for initiator design.

may reach 100 by the end of 1994. As Table 1 shows, the three general approaches discussed above (Section IV.B) are almost invariably applicable to all these monomers in designing initiating systems, and a number of new initiating systems have been developed recently.

Following the general description of the principles and methodologies of controlled/living cationic polymerizations in the preceding section, we herein take an overview of which class of vinyl monomers provide controlled/living cationic polymerizations and which initiating systems are developed for a particular class of monomers. Thus, the next sections will discuss specific examples of initiating systems for each class of monomers. Earlier compilations of monomers and initiating systems for controlled/living cationic polymerizations can be found in recent reviews [5,42]. By discussing the numerous examples, this section is also intended to attest to the generality of the three methodologies and, additionally, to illustrate how initiating systems for a particular monomer should be designed in relation to the structure and stability of the growing carbocations derived therefrom.

Throughout this book, and particularly in this section, the term "initiating system" refers to a combination of an initiator (protogen or carbocation source) and an activator (or a coinitiator). The activator is a Lewis acid that assists in the formation of a carbocation from the initiator and triggers propagation (cf., Scheme 6). For example, in the combination of hydrogen iodide and zinc iodide, the former is the initiator, the latter is the activator, and the pair is called the "HI/ZnI_2 initiating system."

A. Vinyl Ethers

Since the discovery of the first controlled/living cationic polymerization of isobutyl vinyl ether [IBVE; $CH_2{=}CH\{OCH_2CH(CH_3)_2\}$] with the HI/

I_2 initiating system [58], vinyl ether polymerizations have served as the basis for developing a number of initiating systems and for elucidating the general principles of controlled cationic polymerization. This is not too surprising because alkyl vinyl ethers are among the most reactive vinyl monomers in cationic polymerization and, more importantly, the pendant alkoxyl groups provide the growing vinyl ether carbocation with a high stability. Therefore, vinyl ether polymerizations are highly suited for studying and designing living cationic polymerizations and the initiating systems for them.

A number of initiating systems are now available for the controlled/living polymerizations of vinyl ethers. Figure 19 summarizes the representative examples; some MWD data were already given in Figure 17 along with a general discussion. The partial list in Figure 19 demonstrates that all three approaches are viable for vinyl ethers.

1. *Initiating Systems with Nucleophilic Counteranions*

The first generation of initiating systems for vinyl ether controlled/living polymerizations consists of a protonic acid (HB, initiator) with a Lewis acid (MtX_n, activator), [63,69,70]. The reported examples of HB and MtX_n

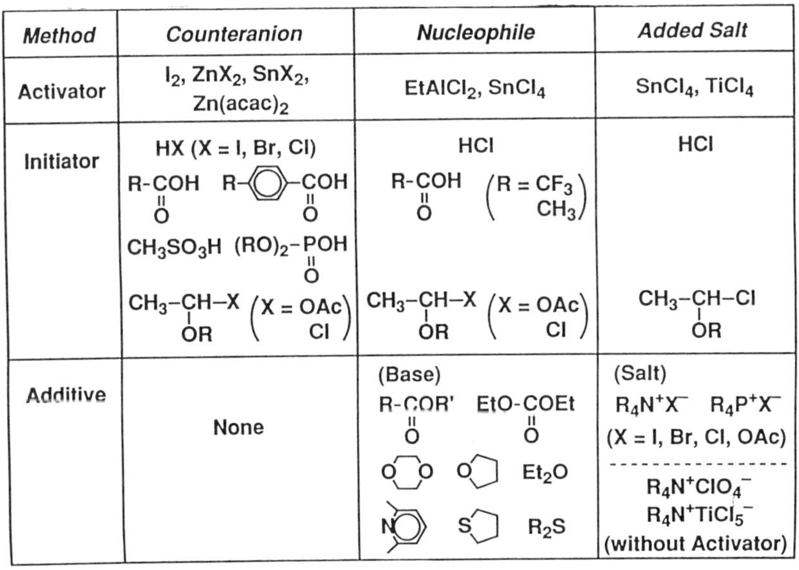

Figure 19 Typical initiating systems for vinyl ethers classified by the three general methods. See Section V.A and Table 2 for references.

that fulfill the above-discussed criteria (Section IV.B.1.b) are summarized in Fig. 20 [72], and can be classified as follows:

HB (protonic acids)
- Hydrogen halides (HX) [58,59,98,99,105,118–124]
- Acetic acids (RCOOH) [104,125]
- Benzoic acids ($R\text{-}C_6H_4\text{-}COOH$; R = NO_2, etc.) [126]
- Phosphoric and phosphinic acids ($R_2P(O)OH$) [127,128]

$Mt\mathbf{X}_n$ (metal halides or weak Lewis acids)
- Iodine [58,59,118,119]
- Tin(II) halides ($Sn\mathbf{X}_2$) [99]; tin tetrabromide ($SnBr_4$) [105]
- Zinc halides ($Zn\mathbf{X}_2$) [90,98,99,104,122–124,126–128]
- Zinc acetates ($Zn(OCOR)_2$) [125]
- Metal acetylacetonates ($Mt(acac)_n$; Mt = Al, Zn, etc.) [120]

Scheme 9 illustrates the pathway for the controlled/living polymerization by a typical HB/$Mt\mathbf{X}_n$ system, $CF_3COOH/ZnCl_2$ [104]. The key feature of this reaction is the interaction of $ZnCl_2$ with the carbonyl group of the acid-IBVE adduct (**10**). This interaction activates the dormant ester linkage to enable propagation with a vinyl ether monomer. The resulting activated form of the growing species (**11**) carries a complex counteranion,

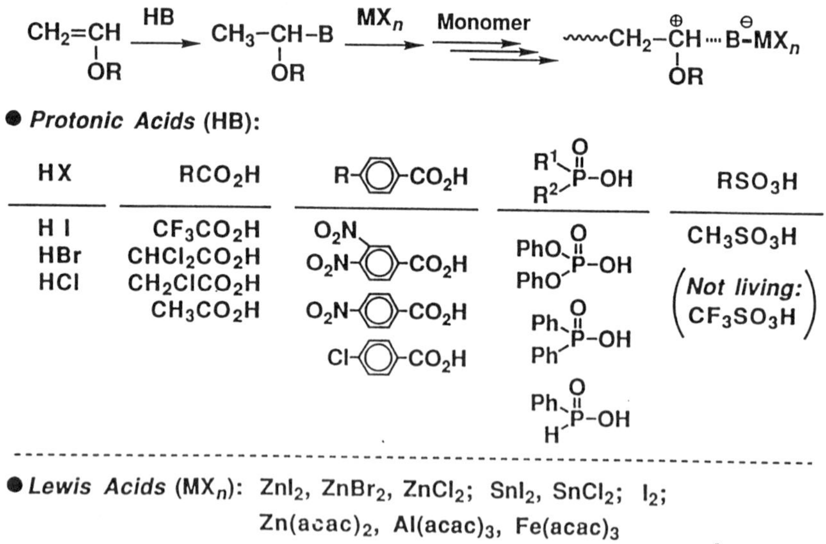

Figure 20 Protonic acid/Lewis acid (HB/$Mt\mathbf{X}_n$) initiating systems for living cationic polymerizations of vinyl ethers. See Section V.A.1 for references.

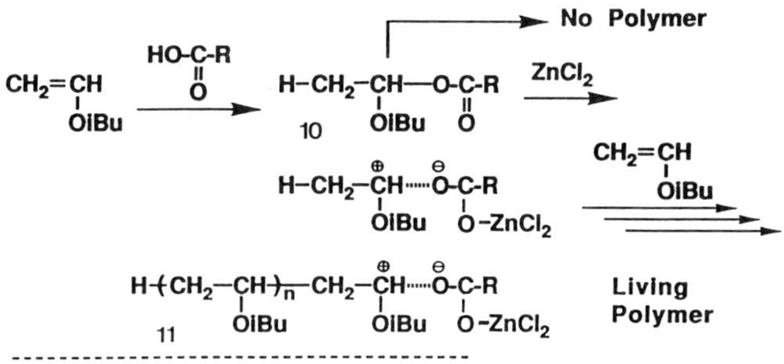

● R: CF_3, CCl_3, $CHCl_2$, CH_2Cl, CH_3

Scheme 9

$[CF_3C(O)O\text{-}ZnCl_2]^-$, that is suitably nucleophilic to stabilize the growing carbocationic center by reversible formation of the dormant species.

A systematic search for suitable protonic acids (with $ZnCl_2$ as an activator) [34,70,126–128] revealed that there is a certain range of nucleophilicity of the acid's counter anions or the acidity (pK_a) of the parent acids. For example, strong acids (e.g., CF_3SO_3H) with too weakly nucleophilic counteranions (e.g., $CF_3SO_3^-$), induce rapid and uncontrolled polymerizations. For relatively weak acids like acetic [104] and benzoic [126] acids and their derivatives, it is needed to introduce an electron-withdrawing group in the anion's part to decrease its nucleophilicity or increase the acidity of the initiator. Thus, the overall polymerization rate of IBVE with $RCO_2H/ZnCl_2$ systems [104] increases with the electron-withdrawing power of the substituent R: $CH_3 < CH_2Cl < CHCl_2 < CCl_3 < CF_3$. Along with the rate changes, the MWD of the polymers narrows in the same order, indicating that the ionization of the RCO_2H-IBVE adduct (e.g., **10**, Scheme 9) and the interconversion between the dormant and the activated or carbocationic species (Scheme 6) are faster with less nucleophilic carboxylate anions carrying a more electron-withdrawing substituent.

As Lewis acid activators, weak metal halides such as zinc halides are effective [98,99]; too-strong Lewis acids, such as $SnCl_4$ and $EtAlCl_2$, induce uncontrolled polymerization [104]. In some cases, the latter metal halides also initiate polymerization in the presence of a trace of water, which is unavoidable under usual experimental conditions (under dry nitrogen or argon atmosphere, via the syringe technique). Thus, zinc halides [98,99] are used most frequently, but tin(II) halides [99] and some acetylacetonate complexes [120] can also be effective.

For easy handling and a better control of its initial concentration, the protonic acid HB is often employed in the form of its preformed adduct **5** [$CH_3CH(O\textit{i}Bu)$-B; Scheme 3] with a vinyl ether [90,104,122,123]. With trifluoroacetic acid, for instance, such an adduct is the ester **10** in Scheme 6. These vinyl ether adducts can be synthesized cleanly and quantitatively by electrophilic addition of HB to a vinyl ether (e.g., via addition of a solution of trifluoroacetic acid or bubbling dry hydrogen chloride through a solution of IBVE in *n*-hexane at 0° C), and in most cases the products can be distilled and stored for a month or longer at low temperatures [104,122,129]. A wide variety of mono- and bifunctional hydrogen halide-vinyl ether adducts (haloethers) can be synthesized and used for living cationic polymerizations [34,118,122,123]. Recently, Deffieux applied the classical haloether formation from aldehydes and alcohols [e.g., $CH_3CH{=}O + HOR + H\mathbf{X} \rightarrow CH_3CH(OR){-}\mathbf{X}$] for the synthesis of various HCl-adducts of vinyl ethers [124].

Ability to use the preformed vinyl ether adducts (**5**) as initiators implies that one can separate the initiation process from the subsequent propagation step and ensure the initiating species [$CH_3CH(OR)^+$] to be virtually identical to the growing poly(vinyl ether) carbocation [$\sim\sim\sim CH_2CH(OR)^+$].

With strong protonic acids such as HI and CF_3CO_2H, the addition to an alkyl vinyl ether is very fast and quantitative, and thus quantitative initiation efficiency and narrow or nearly monodisperse MWDs, can be achieved even without using preformed adducts. However, the addition processes with weaker acids, like acetic acid, are much slower, and the advance preparation of these adducts is recommended.

Among various combinations of HB and $Mt\mathbf{X}_n$ (Fig. 20), perhaps hydrogen chloride, trifluoroacetic acid, and their vinyl ether adducts **5** [Scheme 3; B = Cl and $OCOCF_3$ (**10**; Scheme 5)] are the most convenient initiators, that can be combined with zinc halides as activators, because of the easier handling and the commercial availability of the starting materials. Hydrogen iodide adducts with vinyl ethers (**5**, B = I, Scheme 3) are more reactive than the chloride counterparts (**5**, B = Cl) and useful in most cases, but the preparation and handling of anhydrous hydrogen iodide is rather cumbersome and requires experience.

The $HB/Mt\mathbf{X}_n$-initiated polymerizations of vinyl ethers are typically carried out in nonpolar media such as toluene and *n*-hexane (depending on the solubility of the products) at temperatures below 0° C. In some cases, however, polar solvents (e.g., methylene chloride) may be used at appropriate initiator/activator mole ratios [119]; and, specifically with the HI/ZnI_2 system, controlled/living polymerization is feasible even at room temperature (+25° C) [98,99].

2. Initiating Systems with Added Nucleophiles

The HB/MtX_n initiating systems with stronger Lewis acids than zinc halides induce very rapid or almost instantaneous polymerizations of alkyl vinyl ethers and are not suited for controlled/living cationic polymerizations (Section IV.B.2). These initiating systems include:

Initiators
- H_2O (added or adventitious) [100–102,130,131]
- CH_3CO_2H, CF_3CO_2H [101,103]
- $CH_3CH(OiBu)$—B (B = $OCOCF_3$ [101,103,104], $OCOCH_3$ [101,103, 132,133], Cl [105,131])
- CF_3SO_3H (without activators) [37,38,134,135]
- $(C_6H_5)_3C^+SbCl_6^-$ [135]

Activators
- $EtAlCl_2$ [100–104,130–132,136–140]
- $Et_{1.5}AlCl_{1.5}$, Et_2AlCl; $AlCl_3$ [131]; $SnCl_4$ [105]

For these initiating systems, externally added nucleophiles are necessary to induce controlled/living cationic polymerizations of vinyl ethers [36,64]. Table 2A lists nucleophiles (Lewis bases) that are effective for such purposes and includes esters (carboxylates and carbonates) [100,101,130–133], ethers (linear and cyclic) [102–104,137–140], methylpyridines [140], and phosphines [21,141]. CF_3SO_3H-initiated polymerizations, sulfides are also effective [37,38,134,135].

Table 2 also indicates that the nucleophiles effective for vinyl ethers are relatively mild, when compared with those for isobutene (cf., Section V.B.2). In fact, stronger bases lead to inhibition or severe retardation of polymerization [36,64] ketones aldehydes, amides, acid anhydrides, dimethyl sulfoxide (retardation); alcohols, aliphatic amines, pyridine (inhibition). The choice of nucleophiles is determined by their Lewis basicity (as measured by pK_b, etc. [64,103]), and this factor determines the effective concentrations of the nucleophiles. For example, the required amounts of esters and ethers decrease in the order of increasing basicity (i.e., a stronger base is more effective and therefore less is needed) [101,103]: tetrahydrofuran $<$ 1,4-dioxane $\sim$ ethyl acetate $<$ diethyl ether. On the other hand, for amines not only basicity but also steric factors play an important role [142]; thus, unsubstituted pyridine is an inhibitor, while 2,5-dimethylpyridine is an effective nucleophile for controlled/living polymerization, although the latter is more Lewis basic.

Recently Aoshima and Kobayashi reported an interesting series of studies on the effects of the structure of added carboxylate esters. For example, with methyl carboxylates $R^1CO_2CH_3$, the rate of polymerization of IBVE by $CH_3CH(OiBu)OCOCH_3/EtAlCl_2$ increases with increasing

Table 2 Nucleophiles for Controlled/Living Carbocationic Polymerization

Nucleophile	Examples		References
A. Vinyl ethers			
Esters	$CH_3CO_2C_2H_5$		[100,101,130,131]
	$C_6H_5CO_2C_2H_5$		[101]
	$(p)\mathbf{X}C_6H_4CO_2C_2H_5$ ($\mathbf{X}$ = CH_3O, CH_3, Cl)		[132]
	$C_2H_5OCO_2C_2H_5$		[101]
	$Cl_nCH_{3-n}CO_2CH_3$ (n = 1–3)		[133]
	RCO_2CH_3 (R = CH_3, C_2H_5, i-C_3H_7, t-C_4H_9)		[132,133]
Ethers	1,4-Dioxane[a]		[102–104, 137–140]
	$C_2H_5OC_2H_5$		[102,103]
	Tetrahydrofuran		[102,103]
Sulfides	R_2S (R = CH_3, C_2H_5, n-C_3H_7, etc.)		[37,134]
	Tetrahydrothiophene		[38,135]
Pyridines	2,6-Dimethylpyridine[b]		[142]
	2,4,6-Trimethylpyridine		[142]
Phosphines	Triarylphosphine		[28,38,141]
B. Isobutene			
Sulfoxide	$(CH_3)_2S{=}O$ (DMSO)		[106,143–150]
Amides	$CH_3CON(CH_3)_2$ (DMA)		[106,149–155]
	$HCON(CH_3)_2$ (DMF)		[156]
	1-Methyl-2-pyrrolidinone[d]		[157,158]
Amines	$(C_2H_5)_3N$		[152,156,159]
	1-Methylpyrrolidine[e]		[152]
Pyridines	Pyridine		[156,160,161]
	2,6-Di(t-butyl)pyridine (DTBP)[c]		[150,162]
Ester	$CH_3CO_2C_2H_5$		[150,154]
C. Styrenes			
Sulfoxide	$(CH_3)_2S{=}O$ (DMSO)	(*indene*)	[163,164]
Amide	$CH_3CON(CH_3)_2$ (DMA)	(*styrene*)	[165,166]
		(*p-t-butylstyrene*)	[167]
		(*p-chlorostyrene*)	[168–170]
		(*indene*)	[171,172]
Amine	$(C_2H_5)_3N$	(*p-methylstyrene*)	[172,173]

a
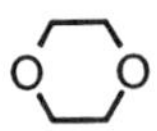

b
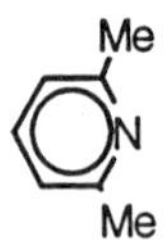

c
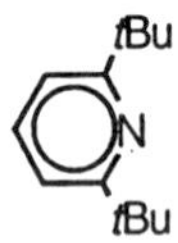

d
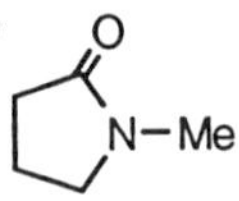

e
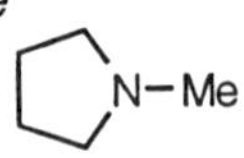

bulkiness of the ester's substituent R^1($CH_3 < C_2H_5 < i$-$C_3H_7 < t$-C_4H_9) [132] and with increasing electron-withdrawing power of R^1 ($CH_3 < CH_2Cl < CHCl_2 < CCl_3$) [133]. According to these authors, the effects may be accounted for by changes in the interaction between the added esters and the growing carbocation; for example, the bulkier the ester's R^1, the weaker the interaction and thereby the faster the polymerization. In addition, esters can be selected according to the reactivity of vinyl ethers; i.e., a more weakly basic ester is used for a less reactive monomer (e.g., $CCl_3CO_2CH_3$ for 2-chloroethyl vinyl ether [133]).

An important advantage of the use of such added nucleophiles is that it allows controlled/living cationic polymerization of alkyl vinyl ethers to proceed at +50 to +70° C [101,103], relatively high temperatures at which conventional cationic polymerizations fail to produce polymers but result in ill-defined oligomers only, due to frequent chain transfer and other side reactions. Recently, initiators with functionalized pendant groups [137] and multifunctional initiators [138–140] have been developed for the living cationic polymerizations with added nucleophiles.

3. *Initiating Systems with Added Salts*

Vinyl ether polymerizations with the $HCl/SnCl_4$ initiating system in methylene chloride are extremely rapid even at −40° C and cannot provide controlled polymers [71,72,105]. The uncontrolled nature of the reaction is attributed to the high polarity of the solvent and to the high Lewis acidity of the tin chloride as the activator, both of which promote the ionization of the growing end.

Under these reaction conditions, the use of ammonium and related onium salts with nucleophilic anions has been found effective at converting the $HCl/SnCl_4$-initiated, uncontrolled polymerizations into controlled/living processes [105]. Similar results are reported for $TiCl_4$-based polymerizations [174,175]. Effective salts include tetraalkylammonium and phosphonium salts: $R_4N^+\mathbf{Y}^-$ and $R_4P^+\mathbf{Y}^-$ ($\mathbf{Y}$ = I, Br, Cl, CH_3COO; R = CH_3, C_2H_5, n-C_4H_9, etc.). As added nucleophiles do in nonpolar solvents, the added salts retard the polymerization, narrow the MWD of the polymers, and render their M_n values directly proportional to conversion and close to the calculated values (one living chain per initiator molecule).

In regards to the structure of these salts, two points deserve noting:

1. The anions $\mathbf{Y}^-$ play a primary role in achieving controlled/living polymerizations, and they should be nucleophilic (less nucleophilic anions such as ClO_4 and CF_3SO_3 are totally ineffective).
2. The anions $\mathbf{Y}^-$ are not necessarily the common ions of the counteran-

ions of the initiator and the growing end (e.g., for the $HCl/SnCl_4$ system, $\mathbf{Y}^-$ can be a halide other than chloride).

These salts suppress the free ionic (dissociated) growing species and accelerate the conversion of carbocations to covalent species (see Sections IV.B.3 and IV.B.3). The salt-mediated living polymerizations of vinyl ethers in polar media also provide useful systems in which polar monomers and their polymers that can be insoluble in nonpolar solvents can be polymerized successfully.

4. *Lewis Acid-Free Initiating Systems with Added Salts*

In a way different from what has been discussed, ammonium salts and related additives have also been used by Kröner and Nuyken to achieve controlled/living vinyl ether polymerizations [109,129,176,177]. They found that the inert adduct of hydrogen iodide with an alkyl vinyl ether [$CH_3CH(OR)I$; R = isobutyl, etc.], can initiate a controlled polymerization in methylene chloride in the absence of a metal halide activator but in the presence of a suitable ammonium salt. The most useful salt is apparently tetra-*n*-butylammonium perchlorate, $n\text{Bu}_4\text{N}^+\text{ClO}_4^-$. Under similar conditions, complexes of lithium and potassium perchlorates with crown ethers (12,-crown-4 and 18-crown-6, respectively) may also be used to activate the adduct [109]. Similarly, $n\text{Bu}_4\text{N}^+\text{TiCl}_5^-$ is reported to be effective in polymerization with the HCl-IBVE adduct [178].

Recently, Cramail and Deffieux [179] reported that cyclohexyl vinyl ether can be polymerized using vinyl ether adducts of hydrogen halide (initiators; **5**, R = Cl, Br), without any added Lewis acid activator or salt. Such Lewis acid-free polymerizations, also known for *N*-vinylcarbazole [108], are attributed to the high reactivity of the monomers. In these processes, the addition of $n\text{Bu}_4\text{N}^+\text{ClO}_4^-$ accelerates the reaction but fails to give living polymers; in contrast, the addition of $n\text{Bu}_4\text{N}^+\text{I}^-$ leads to controlled/living polymerizations, again demonstrating the importance of nucleophilic anions. These results suggest that the counteranion exchanges with the halide growing end [90,122,123,177].

Perchlorate salts may undergo a reversible counteranion exchange with the terminal halogen in the growing (dormant) end; such an anion exchange has been observed in some cases [21,90,177] and would be considered equivalent to activation by metal halides. Alternatively, it was suggested that the perchlorate salts may act as polar compounds facilitating the dissociation of the iodide adduct as do Lewis acid activators in nonpolar solvents [129].

5. *Pendant-Functionalized Vinyl Ethers*

Almost all of the initiating systems discussed in this section can be applied to the living cationic polymerizations of vinyl ethers that carry a variety of functional pendant groups, in a general form [42,43,65,66,180]:

$$\begin{array}{l} CH_2{=}CH \\ \quad\;\; | \\ \quad\;\; OCH_2CH_2{-}\mathbf{X} \end{array} \qquad \mathbf{X} = \begin{array}{l} OCOCH_3;\ CH(COOC_2H_5)_2; \\ N(COO^tC_4H_9)_2;\ Cl; \\ (OCH_2CH_2)_mOR,\ \text{etc.} \end{array}$$

In particular, HI/ZnI_2, $HCl/ZnCl_2$, and CF_3SO_3H/Me_2S (the sulfide as a nucleophilic additive) are among the most frequently used initiating systems.

The scope of the living cationic polymerizations and synthetic applications of these functionalized monomers will be treated in the next chapter on polymer synthesis (see Chapter 5, Section III.B). One should note that the feasibility of living processes for these polar monomers further attests to the formation of controlled and "stabilized" growing species. Conventional nonliving polymerizations, esters, ethers, and other nucleophiles are known to function as chain transfer agents and sometimes as terminators. In addition, the absence of other acid-catalyzed side reactions of the polar substituents, often sensitive to hydrolysis, acidolysis, etc., demonstrates that these polymerization systems are free from free protons that could arise either from incomplete initiation (via addition of protonic acids to monomer) or from chain transfer reactions (β-proton elimination from the growing end).

6. *Propenyl Ethers and Unsaturated Cyclic Ethers*

Propenyl ethers ($CH_3{-}CH{=}CH{-}OR$; R = ethyl, isobutyl, etc.; *cis*- and *trans*-isomers) and 3,4-dihydrofuran are linear and cyclic α,β-unsaturated ethers, that can be regarded as β-substituted vinyl ether derivatives. For these monomers a few controlled/living cationic polymerizations have been reported. The HI/I_2 system is generally effective for both linear and cyclic monomers [181,182,183], whereas a recent study by Nuyken indicates that the IBVE-HI adduct coupled with $n\mathrm{Bu_4NClO_4}$ is suited for 3,4-dihydrofuran (see Section V.A.4) [184]. A variety of mono- and bifunctional propenyl ethers can readily be prepared by the ruthenium complex-catalyzed isomerization of corresponding allyl ethers [185].

B. Isobutene

In 1986 Faust and Kennedy reported the first example of controlled/living cationic polymerization of isobutene, which was initiated by a cumyl acetate/BCl_3 initiating system [35]. Since then the Akron group extensively contributed to the development of a number of initiating systems for the controlled polymerization of this industrially important hydrocarbon monomer, as summarized in the book by Kennedy and Iván [5]. As with vinyl ethers (Figure 19, Section V.A), these initiating systems can be classified in the same framework (the three methods for carbocation "stabilization") as proposed in Section IV.B and shown in Fig. 21. A more detailed structural list is also available [42]. Table 3 gives a comprehensive reference list of these initiating systems, in terms of the methods (A–C), activators (BCl_3 vs. $TiCl_4$), initiator types (counteranions or leaving groups), and their functionality (*F*; mono- to tetrafunctional).

The initiating systems for these controlled/living isobutene polymerizations are invariably binary, consisting of an initiator (cationogen) and an activator (coinitiator or Lewis acid). Although the activators for vinyl ethers are mostly halides of zinc, tin(II), tin(IV), and aluminum (e.g., ZnX_2, $SnCl_4$, and $EtAlCl_2$), those for isobutylene are almost always chlorides of boron and titanium (BCl_3 and $TiCl_4$). Although the early series of the initiating systems were all based upon BCl_3, the publications in the last few years from Kennedy's group indicate that $TiCl_4$ is the activator of choice, probably because of its higher activity and lower price.

Method	*Counteranion*	*Nucleophile*	*Added Salt*
Activator	BCl_3	BCl_3, $TiCl_4$	BCl_3, $TiCl_4$
Initiator	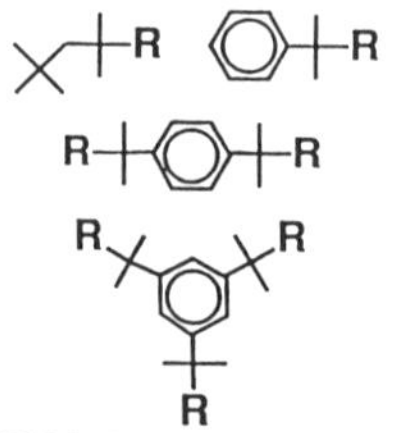 R = OAc, OMe, OH	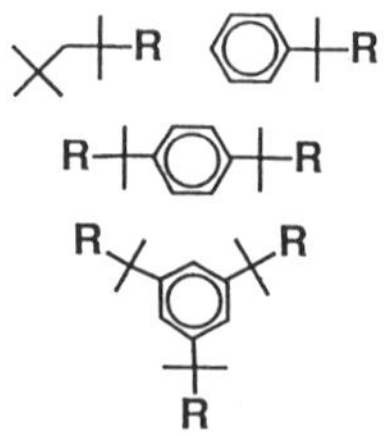 R = OAc, OMe, OH, Cl	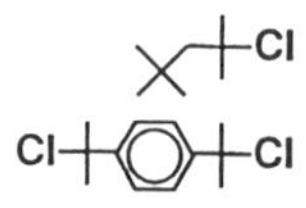
Additive	None	(Base) $Me_2S{=}O$, EtOAc, Et_3N, Me_2NCHO, Me_2NAc, 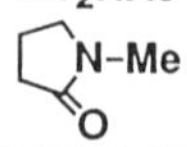	(Salt) $n Bu_4N^+Cl^-$, $n Bu_4N^+ I^-$, KCl / crown

Figure 21 Typical initiating systems for isobutene classified by the three general methods. See Section V.B and Tables 2 and 3 for references.

Table 3 Initiating Systems for Controlled/Living Polymerization of Isobutene: A Reference List[a]

Method[b]	Activator (MtX_n)	F[c]	Ester (tROCOR)	Initiators ether ($tROCH_3$)	Alcohol (tROH)	Chloride (tRCl)
A. Counteranion	BCl_3	1	[34,186,187]	[188]	[189]	–
		2	[190–195]	[196–198]	[189,199]	–
		3	[200][d]	[201]	[199]	–
	$TiCl_4$	1	–	[106]	–	–
		2	–	[106,202]	–	–
B. Nucleophile	BCl_3	1	[91]	–	[158]	[106,150]
		2	[143,157]	–	[106,152,153, 159]	[106,151]
		3	[144]	[106,144–148]	–	[154]
	$TiCl_4$	1	[203]	–	–	[106,149,151, 155,156]
		2	[162]	[162]	–	[160–162]
		3	–	–	–	[160]
C. Added salt	BCl_3	2	–	–	–	[204]
	$TiCl_4$	1	–	–	–	[205–207]

[a] See Figure 6 for the structure of representative initiators.
[b] The three methods for carbocation stabilization; see Section IV.B.
[c] Functionality of initiators: 1, monofunctional; 2, bifunctional; 3, trifunctional.
[d] F = tetrafunctional.

1. *Initiating Systems with Nucleophilic Counteranions*

For isobutene, this group refers almost exclusively to the BCl_3-based initiating systems without external additives. As listed in Table 3.A, combinations of BCl_3 with tertiary esters, ethers (methoxides), or alcohols induce controlled/living polymerization of isobutene in CH_3Cl or other solvents at temperatures below $-30°$ C. Scheme 10 illustrates the proposed pathway for the polymerization initiated with the cumyl acetate (**12**)/BCl_3 system [35]:

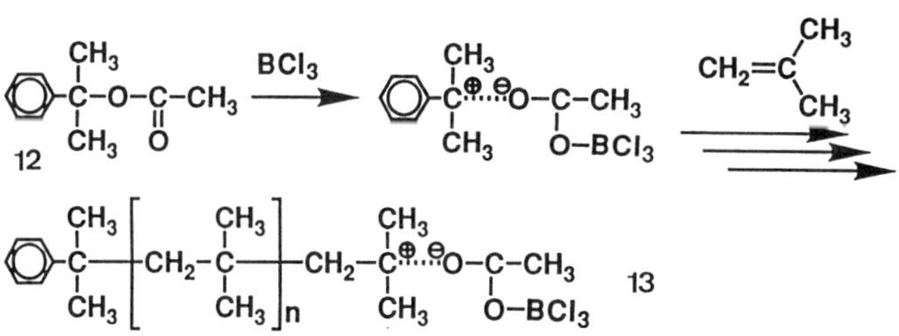

Scheme 10

A comparison between Schemes 9 and 10 reveals that this pathway (Scheme 10) is very similar, at least formally, to that for the vinyl ether polymerization with the RCO_2H/ZnI_2 or $CH_3CH(O\textit{i}Bu)OCOR/ZnI_2$ system. Namely, the initial process is the interaction of the Lewis acid activator with the initiator's ester carbonyl to activate (ionize) the ester linkage, and the growing species is believed to carry a complex counteranion, $[CH_3C(O)O\text{-}BCl_3]^-$, as shown in **13**.

Figure 21 and Table 3.A show typical features of these initiating systems: (1) The activator is either BCl_3 or $TiCl_4$. (2) The initiators (cationogens) are invariably tertiary compounds, among which 2,4,4-trimethylpentyl (TMP), 2-phenyl-2-propyl (cumyl), and their multifunctional derivatives are the most convenient. Evidently, the use of these tertiary cationogens is to generate initiating tertiary carbocations that are similar or almost identical with the growing cation to facilitate the initiation processes.

Although the reaction pathway shown in Scheme 10 is simple and reasonable, there are some ambiguities that should be clarified. For example, the polymer chain ends obtained with all of the BCl_3-based systems are tertiary chlorides [~~~$C(CH_3)_2$-Cl] [35], regardless of the initial counterion of the initiators (esters, ethers, and alcohols) and of the type of terminating agents (mostly methanol). This suggests that whatever the initiator is, the growing end soon undergoes a halogen exchange with BCl_3 (free or interacting with the ester group) to form tertiary chloride chain ends [117,151,208].

2. *Initiating Systems with Added Nucleophiles*

Following the first generation based on BCl_3 (see above), the second generation of initiating systems was developed by the Akron group, that use externally added nucleophiles ("electron donors" according to the original authors). Table 3.B gives a list of initiating systems that can be used in combination with added nucleophiles, and Table 2.B shows the nucleophilic additives thus far used for isobutene. There are several features of these initiating systems relative to those in Table 3.A (with nucleophilic counteranions without additives):

1. The activators are BCl_3 and $TiCl_4$.
2. The initiators may be tertiary chlorides, tertiary esters (acetates), ethers (methoxides), and alcohols.
3. The solvents are almost invariably CH_3Cl for the BCl_3-based systems and $CH_3Cl/n\text{-}C_6H_{14}$ [4/6 (v/v)] mixture for the $TiCl_4$-based systems; i.e., the latter require less polar polymerization media.
4. The nucleophiles include amides (and lactams), esters, aliphatic and

aromatic amines, and sulfoxides (Table 2.B), among which *N*,*N*-dimethylacetamide (DMA) and dimethyl sulfoxide (DMSO) are most frequently used.

5. These nucleophiles are much stronger than those used for vinyl ethers (Section V.A.4; Table 2.A), and their concentrations are usually lower than that of the Lewis acid activator, in sharp contrast to the cases for vinyl ethers where the nucleophiles are usually in a large excess over the Lewis acid (the concentrations depend upon the Lewis acidity, however). It is reported, for example, that the concentration of ethyl acetate should be below the $TiCl_4$ concentration, otherwise no polymerization occurs [91].

A few papers have examined the range of Lewis bases that are effective for isobutene controlled/living polymerization [91,149,156]. One proposal is that effective nucleophiles (electron donors) should have relatively high donor numbers (DN > 26) [149]. Another screening shows that triethylamine is exceedingly effective for isobutene [156]. Pratap et al., also reported recently the use of cyclic amides (lactams; 1-methyl-2-pyrrolidone) [158] and cyclic amines [152] for the dicumyl acetate/BCl_3 initiating system.

As in the corresponding vinyl ether polymerizations (Section V.A.2), the roles of the added nucleophiles are not fully clarified and are under some controversy (see, for example, Refs. 41 and 151). Perhaps the most serious question is "would the nucleophilic additives indeed stabilize the growing carbocation as originally postulated?" although there are some effects of these additives such as the better control of polymer molecular weights and the narrowing of MWDs. For example, Sigwalt [163,164] showed that at low temperatures (below $-80°$ C) the polymerization of indene with the cumyl chloride/$TiCl_4$ initiating system is controlled/living even in the absence of DMSO and other nucleophiles. In a slightly different context, Faust examined the effects of DMSO as well as 2,6-di-*tert*-butylpyridine (a proton trap) on isobutene polymerization and pointed out the absence of "cation stabilization" by DMSO [162].

Another point of discussion is related to the fact that, irrespective of the structure of the initiators (esters, ethers, etc.), the ω-end group of the polyisobutene obtained in these living polymerizations with $TiCl_4$ and BCl_3 is a tertiary chloride, due to rapid halogen exchange at the growing end. If this process occurs, for example, with the tertiary ester-type initiator [$\sim\sim\sim C{-}OC(O)CH_3 + TiCl_4 \rightarrow \sim\sim\sim C{-}Cl + TiCl_3(OC(O)CH_3)$], the resulting mixed metal halide would also act as a weak nucleophile. The possibility of the formation of such an "in situ electron donor" was suggested [151,203].

The $TiCl_4$-based initiating systems with appropriate external nucleophiles provide perhaps the best systems for controlling molecular weights and MWDs of polyisobutene. In particular, the cumyl chloride or cumyl methyl ether/$TiCl_4$ system in conjunction with *N*,*N*-dimethylacetamide (DMA) gives polyisobutene with high molecular weights ($M_n > 10^5$) and very narrow MWDs ($M_w/M_n \leq 1.1$) at a temperature of −80° C [106].

3. Initiating Systems with Added Salts

For isobutene, as with vinyl ethers, it is possible to use onium salts as additives to achieve controlled/living cationic polymerization at −80° C (Table 3C), although the method has been used less frequently than the two discussed above. The initiators are alkyl chlorides (2,2,4-tetramethylpentyl and dicumyl) coupled with either $TiCl_4$ or BCl_3; the salt is nBu$_4$NCl or nBu$_4$NI. An interesting extension of these findings is to use the complex of potassium chloride with 18-crown-6 as a substitute for nBu$_4$NCl [204]. Interestingly, the solvent used is more polar than those used for other methods, i.e., CH_2Cl_2 alone or 4:6 v/v mixtures of n-C_6H_{14} with CH_3Cl or CH_2Cl_2.

C. Styrene and Its Derivatives

The development of controlled/living cationic polymerizations of styrene and its derivatives has been behind those for vinyl ethers and isobutene, despite the fact that the first evidence suggestive of the formation of long-lived polymers in cationic polymerization had been obtained with *p*-methoxystyrene [44] (Section IV.A). This is because the growing cations derived from styrenic monomers are, in general, less stable than those from vinyl ethers, and thus cationic polymerizations of styrene and its derivatives have been considered more difficult to control. Another reason is that styrene can be polymerized to well-defined living polymers by anionic polymerizations, although they often require stringent vacuum-line technique. However, since the late 1980s and particularly in the early 1990s [42], when the first-phase research to develop controlled/living cationic polymerizations was almost complete for vinyl ethers and isobutene, intensive efforts have been directed toward styrene derivatives. As a consequence, a large number of controlled/living polymerization systems are now available for these monomers, as summarized in Figure 22 in terms of the structure of monomers and methodologies.

Again, the three methodologies for carbocation "stabilization" (Section IV.B) are applicable to styrene and its derivatives. Generally, the three methods also call for the rather specific use of the following solvents: (A, counteranions) toluene and similar nonpolar solvents; (B, added nu-

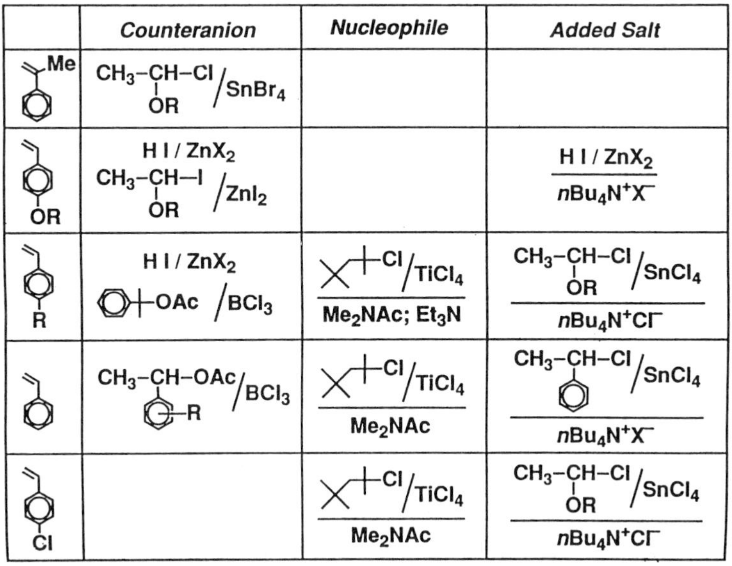

Figure 22 Typical initiating systems for styrene derivatives classified by the three general methods. See Section V.C and Table 2 for references.

cleophiles) $CH_3Cl/n\text{-}C_6H_{14}$ (4:6 v/v); (C, added salts) CH_2Cl_2. The following parts describe some representative examples for each class of styrenic monomers. Table 2C) shows examples of added nuleophiles that have been used for styrenes; the list is much shorter than that for vinyl ethers and isobutylene.

1. *p-Alkoxystyrenes*

When compared with vinyl ethers, *p*-alkoxystyrenes might be considered as analogs with a conjugating phenyl ring is inserted between the vinyl ether's alkoxy and vinyl groups. Because of such structural similarity and electron donation by the alkoxy group, *p*-alkoxystyrenes are the most reactive among styrene derivatives in cationic polymerization and form relatively stable growing carbocations, although they are clearly less reactive than alkyl vinyl ethers. Therefore, their cationic polymerizations are similar to those of vinyl ethers, and controlled/living cationic polymerizations of *p*-methoxystyrene [107,209–211] and *p-tert*-butoxystyrene (a protected form of *p*-hydroxystyrene or *p*-vinylphenol) [212,213] are also feasible under similar conditions; namely, with the HI/ZnI_2 initiating system in toluene (salt free) or in methylene chloride with added $n\mathrm{Bu_4NI}$ salt [107]. In the latter polar solvent, the MWD of the polymers is bimodal without salts, as will be discussed for styrene (Section V.C.2). The HI adducts of vinyl ethers [$CH_3CH(OCH_2CH_2X)$-I; X = $OCOCH_3$ and other functional groups] can also be used with zinc iodide as initiating systems [210]; they are specific for introducing α-end functional groups into poly(p-alkoxystyrenes) [211].

These controlled/living polymerizations operate even at +25° C [209]; the only difference from the vinyl ether systems is that styrenic monomers require higher concentrations of zinc activators to ensure sufficiently fast polymerizations. Under these conditions, the polymers have controlled molecular weights (one living chain per initiator), relatively high molecular weights (DP_n up to 1000), and very narrow MWDs ($M_w/M_n < 1.1$).

2. Styrene

In contrast to p-alkoxystyrenes, styrene lacks an electron-donating, carbocation-stabilizing substituent, and thus it is much less reactive and forms a much less stable growing carbocation. It has therefore been believed that controlled/living cationic polymerization of styrene would be very difficult.

In 1988 Faust and Kennedy reported that controlled/living styrene polymerization would be possible with a combination of 1-(p-methylphenyl)ethyl acetate [$CH_3C_6H_4CH(CH_3)$-$OCOCH_3$; the adduct of acetic acid with p-methylstyrene] and BCl_3 in CH_3Cl solvent below −30° C [214]. Similar systems were also reported by Matyjaszewski and Lin [27,117,208]. Although the M_n of the polymers increases linearly with conversion, the controlled/living nature of these polymerizations is rather obscured by very broad MWDs where the M_w/M_n ratios sometimes exceed 6.

Two years later, the Kyoto group developed another polymerization system for styrene, where polystyrene with controlled molecular weights and narrow MWDs (M_w/M_n = 1.1–1.2) can be produced [25,215,216]. Typically, the initiating system consists of 1-phenylethyl chloride [$C_6H_5CH(CH_3)Cl$ (PhEtCl); formally, the adduct of hydrogen chloride with styrene] and $SnCl_4$, as initiator and activator, respectively, in conjunction with added $n\mathrm{Bu_4NCl}$ salt. In CH_2Cl_2 solvent below −15° C, the M_n is directly proportional to conversion and close to the calculated value based on the assumption that one living chain is formed per initiator (PhEtCl), and increases further upon addition of more styrene at the end of the first-stage polymerization. Thus, this is a controlled/living cationic process based on added salts.

A later study by the same group showed that, as the initiator, not only PhEtCl but also its bromide counterpart (1-phenylethyl bromide) can be used; that, as the added salt, not only the chloride but also the bromide and the iodide ($n\mathrm{Bu_4NY}$; Y = Br, I) can be used (see Figure 23) [25]. As in isobutene polymerization, the polymer's ω-end (tail group) is a chloride [~~~$CH_2CH(C_6H_5)$-Cl], irrespective of the type of initiator (chloride or bromide) and terminating agent (mostly methanol). Some kinetic and mechanistic aspects of the PhEtCl-based systems were studied by Lin et al. [39].

The design of the PhEtCl/$SnCl_4$/$n\mathrm{Bu_4NCl}$ and related initiating systems are understood as follows [25,39]: (1) the use of PhEtCl as the initiator ensured a smooth and quantitative initiation via the styryl-type initial cation virtually identical to the growing polymer cation; (2) the use of $SnCl_4$ as a strongly Lewis acidic activator ensured a sufficient polymerization rate by promoting the ionization of the carbon-chlorine terminal bond; (3) the use of CH_2Cl_2 as a relatively polar solvent served the same purpose as item 2; and (4) the use of $n\mathrm{Bu_4NCl}$ as added salt suppressed the ionic dissociation of the growing carbocation.

In methylene chloride solvent without added salt, the polymerization is not controlled and generates polystyrene with bimodal MWDs (e.g., Figure 23a), which are very similar to those found in nonliving cationic

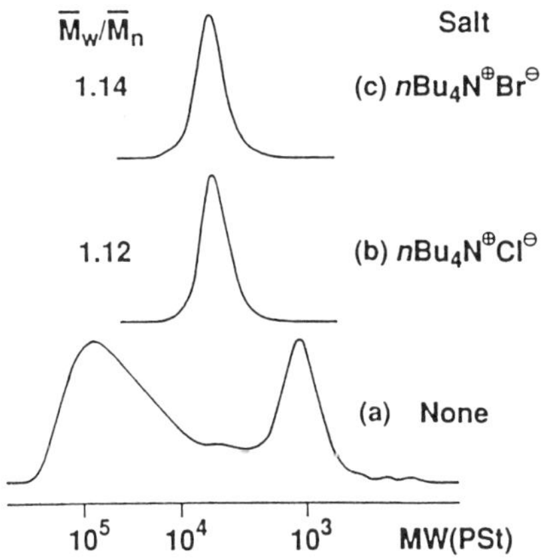

Figure 23 MWDs of polystyrene obtained with the 1-phenylethyl chloride/$SnCl_4$ initiating system in CH_2Cl_2 solvent at −15° C: $[\mathrm{styrene}]_0$ = 1.0 *M*; $[\mathrm{1\text{-}phenylethyl\ chloride}]_0$ = 20 m*M*; $[SnCl_4]_0$ = 100 m*M*; $[\mathrm{salt}]_0$ = 40 m*M*; conversion >90%. Additives: (a) none (salt-free); (b) $n\mathrm{Bu_4NCl}$; (c) $n\mathrm{Bu_4NBr}$. (From Ref. 25.)

polymerizations using perchlorates and related oxo acids [217,218]. As already discussed (Sections II.D and IV.A.1), the bimodal MWDs show the existence of two growing species that propagate simultaneously and exchange slowly one with another. Upon addition of the salt ($n\text{Bu}_4\text{NY}$), the higher MW polymer peak of the bimodal MWD is suppressed or eliminated completely to give a unimodal and narrow MWD. Very similar suppression of the dissociated species is observed upon decreasing the solvent polarity from CH_2Cl_2 to its mixtures with carbon tetrachloride or similar less polar media [25,219]. Thus, the nondissociated growing species, that is responsible for the formation of the lower polymers in the bimodal MWD, behaves as the living growing species, and the added ammonium salt suppresses the ionic dissociation of the polystyrene growing end.

As an extension of these findings, functionalized initiating systems have been developed. Thus, a series of adducts of hydrogen chloride with pendant functionalized vinyl ethers [$CH_3CH(OCH_2CH_2X)Cl$; X = $OCOC(CH_3){=}CH_2$, etc.] also serve as initiators for controlled/living styrene polymerization when they are coupled with $SnCl_4$ (activator) and $n\text{Bu}_4\text{NCl}$ (added salt) in methylene chloride solvent [220]. The resulting polymers carry the functional group X in the α-end derived from the initiator. An important mechanistic consideration is that the vinyl ether type carbocations derived from these initiators successfully initiate styrene polymerization. This has been used for the synthesis of vinyl ether-styrene block copolymers by sequential cationic polymerization (see Chapter 5).

More recently, Kennedy reported another initiating system that controls styrene polymerization with an added nucleophile: 2,2,4-trimethylpentyl chloride (TMP-Cl)/$TiCl_4$ with *N,N*-dimethylacetamide (DMA) in CH_3Cl/methylcyclohexane (4:6 v/v) mixture at $-80°$ C [165]. The use of another additive, 2,6-di-*tert*-butylpyridine (proton trap), is described as beneficial. The molecular weight and MWD are controlled in this system, but the role of the added DMA is still ambiguous [166]. This system with the aliphatic *tert*-chloride was designed to extend to the synthesis of isobutene-styrene block copolymers via sequential cationic polymerization (Chapter 5).

3. *p-Alkylstyrenes and Related Derivatives*

For this class of styrenic monomers, controlled/living cationic polymerizations have been reported for *p*-methylstyrene, *p*-*tert*-butylstyrene, and 2,4,6-trimethylstyrene (Figure 22). Structurally, these monomers lie between styrene and *p*-alkoxystyrenes, and the moderately electron-donating alkyl substituents permit the use of initiating systems employed for both extremes.

The first reported controlled/living polymerization of *p*-methylstyrene with acetyl perchlorate in the presence of $n\mathrm{Bu_4NClO_4}$ [221] was based on the added salt method. Later, it was reported that cumyl acetate/BCl_3 and related initiating systems induce controlled/living polymerizations of *p*-methylstyrene [117,222] and 1,3,5-trimethylstyrene [223]. The $HI/ZnCl_2$ system, suited for vinyl ethers and *p*-alkoxystyrenes, can also be used for *p*-methylstyrene [224], but the lower reactivity of the monomer requires a much higher concentration of the zinc activator (ca. 100 m*M* for 10 m*M* HI) to obtain a sufficient polymerization rate. These systems function without added nucleophilic additives and can be classified under the counteranion method.

Alternatively, additives may be used to effect the controlled/living polymerization of *p*-alkylstyrenes. For instance, $SnCl_4$-based initiating systems for styrene, such as vinyl ether-HCl adduct/$SnCl_4$ with an added salt ($n\mathrm{Bu_4NCl}$), can also be used for *p*-methylstyrene [220]. The $TiCl_4$-based system with TMP-Cl also works when the added nucleophile is triethylamine [173] or DMA [167], but in these cases initiation is slow and incomplete. Thus, for alkylstyrenes all three methodologies for carbocation "stabilization" are applicable.

4. *p-Chlorostyrene*

In contrast to *p*-alkylstyrenes, this chloro derivative has reduced reactivity due to the electron-withdrawing *p*-substituent. Despite this, the initiating systems effective for styrene and *p*-methylstyrene can also be used for the controlled/living polymerization of *p*-chlorostyrene, such as TMP-Cl/$TiCl_4$ (with added DMA) [168–170] and PhEtCl/$SnCl_4$ (with added $n\mathrm{Bu_4NCl}$) [225]. Interestingly, the latter system permits the controlled/living process to proceed at room temperature.

5. *α-Methylstyrene*

Relative to living cationic polymerization, the structure of α-methylstyrene is both advantageous and disadvantageous. Because of the additional methyl group on the α-carbon, the growing carbocation is tertiary and should be thermodynamically more stable, but it would also be prone to undergo β-proton elimination (chain transfer) due to the increase in the number of abstractable protons. Another important aspect of this monomer is its low ceiling temperature that requires low temperatures for polymerization.

Because of these complexities, a controlled/living cationic polymerization of α-methylstyrene has not been developed until recently. Higashimura et al. found that combinations of vinyl ether-HCl adduct and tin tetrabromide [$CH_3CH(OCH_2CH_2X)Cl/SnBr_4$; X = Cl, $OCOC(CH_3){=}CH_2$, etc.] were the best initiating systems (Figure 7) [40,226]. Similar initiating systems using BCl_3 have also been reported recently [227]. For example, in CH_2Cl_2 solvent at $-78°$ C, the initiating systems with X = Cl induce a controlled/living polymerization of α-methylstyrene where polymers with controlled molecular weights and narrow MWDs can be obtained up to DP_n = 1000 [40]. The important aspects of initiators for α-methylstyrene polymerization are (1) to use the vinyl ether adduct to induce quantitative and rapid initiation and (2) to use the mildly Lewis acidic tin bromide to obtain a suitable ionization state of the growing species (without any additives). For example, the replacement of tin bromide with its chloride counterpart (a stronger Lewis acid) or of the initiator with cumyl chloride (the HCl-adduct of α-methylstyrene) fails to give well-controlled polymerizations.

On the other hand, Matyjaszewski et al. reported that cumyl chloride/BCl_3 coupled with $n Bu_4NCl$ gives a controlled polymerization [227]. A more recent report by Tsunogae and Kennedy [228] indicated that the use of triethylamine as an additive enables the controlled/living polymerization of α-methylstyrene, where the initiating system consists of TMP chloride and $TiCl_4$ in CH_3Cl/n-hexane mixtures at $-80°$ C. Therein the initiation is apparently slower than the propagation, rendering the initiation efficiency below 100%. These authors propose that the amine acts both as a nucleophile (electron donor) and as a proton trap.

Recently, Soares studied α-methylstyrene polymerization with iodine in liquid sulfur dioxide [229,230], which is known as a unique solvent [231]. Polymers are formed in a 1:1 mixture of liquid SO_2 and methylene chloride, the latter being needed to keep the solution homogeneous. The molecular weight control is, however, not as good as in the other systems, accompanied by broad MWDs.

6. *Indene*

The controlled/living cationic polymerization of indene, which can be regarded as a cyclic analog of β-methylstyrene, has been examined partly because of the high glass transition temperature of the polymer and partly because of the expected absence of chain transfer via indan formation. Sigwalt and co-workers studied the polymerizations of indene using cumyl methyl ether [163] or cumyl chloride [164] at $-40°$ C in conjunction with $TiCl_3(OBu)$ (without additive) or with $TiCl_4$ and dimethyl sulfoxide additive [232]. Kennedy employed a TMP chloride/$TiCl_4$ system with DMA as an additive at $-80°$ C, where the polymerization apparently involves a slow initiation [171,172]. Both groups observed a linear increase in polymer molecular weight with polymer yield, although the MWDs of the products were rather broad and the initiation efficiency was not always quantitative.

Sigwalt proposed, based on calculated transfer constants, that for indene and related isobutene polymerizations, the use of nucleophilic additives does not stabilize the growing carbocation but adjusts the relative rates of initiation and propagation to achieve efficient initiation and narrow MWDs of the polymers [163,164,232].

D. *N*-Vinylcarbazole

With a large conjugating substituent and electron-donating nitrogen, this monomer is among the most reactive in cationic polymerization and forms the most stable carbocation. Because of these favorable factors, *N*-vinylcarbazole can be polymerized effectively using hydrogen iodide without iodine or any metal halide activator to form controlled/living polymers [108], in sharp contrast to vinyl ethers and styrene derivatives, which almost always need activators to be polymerized. This difference shows that the terminal carbon-iodine bond strength depends on the parent monomer structure, and the highly labile species derived from *N*-vinylcarbazole are reactive enough to grow even in the absence of Lewis acid.

The HI-mediated controlled/living polymerizations may be carried out in CCl_4 or CH_2Cl_2/CCl_4 mixtures [108]. In earlier studies by the Kyoto group, long-lived polymers have also been obtained with iodine as the initiator in CH_2Cl_2 and in the presence of added $n$$Bu_4NI$ salt [233]. No polymerization systems for *N*-vinylcarbazole that use added nucleophiles have been yet reported.

E. Special Initiating Systems

1. *Multifunctional Initiators*

Controlled/living cationic polymerizations of vinyl ethers, isobutene, and styrene derivatives can be initiated with multifunctional initiators or initiating systems. Figure 24 shows representative examples of such multifunctional initiators. In general, these initiators carry two to four initiating sites [such as —$CH_2CH(OR)$**X** in **19**] that have originally been developed for monofunctional versions [such as $CH_3CH(OR)$**X**]. These multifunctional systems are important for the synthesis of telechelics, block copoly-

(A) Isobutene and Styrenes

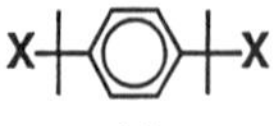

X = OC(=O)CH$_3$, OCH$_3$, OH, Cl

14

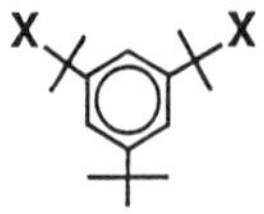

15

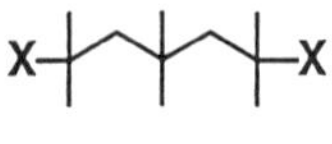

18

16

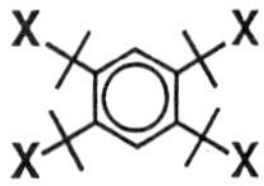

17

(B) Vinyl Ethers and Styrenes

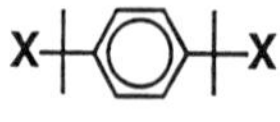

X = Cl, Br, I, OC(=O)CF$_3$

14

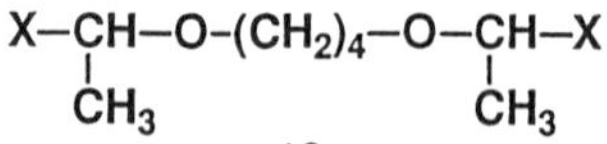

19

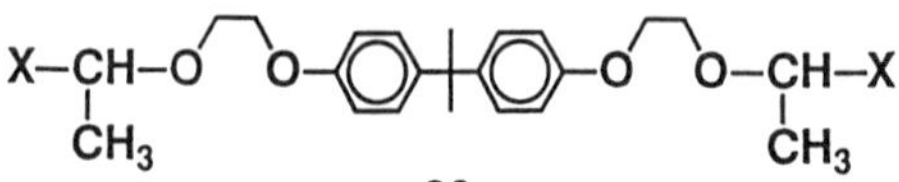

20

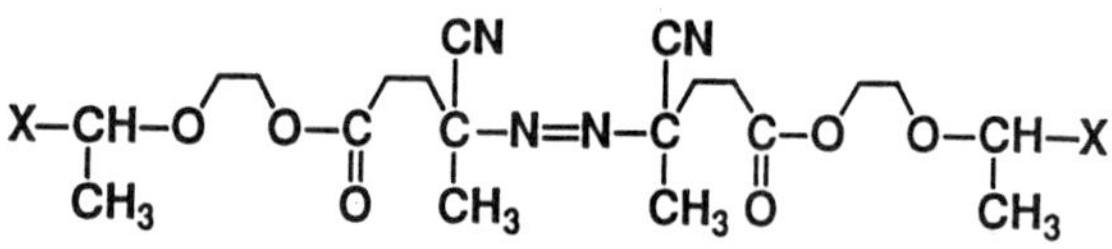

21

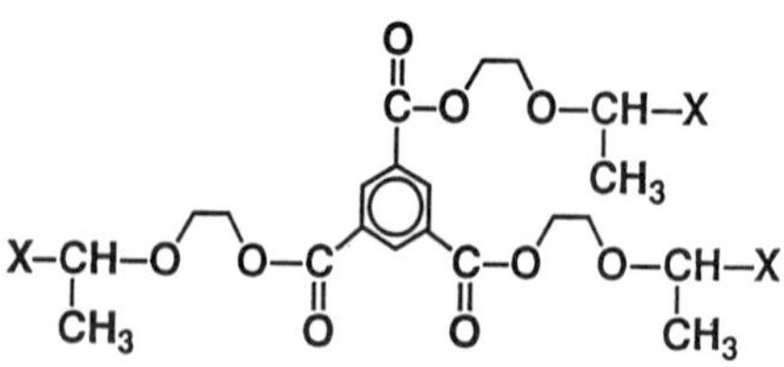

22

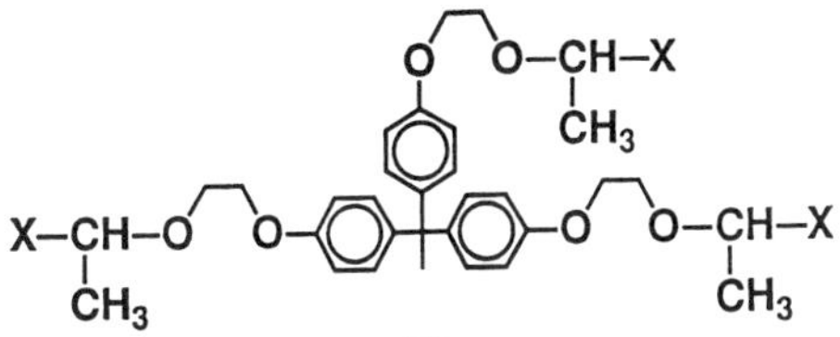

23

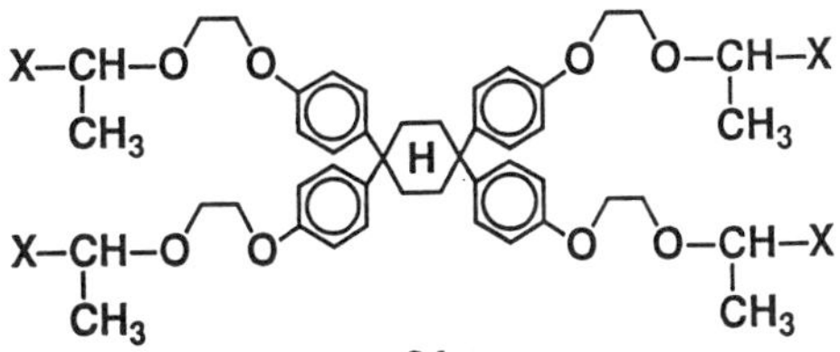

24

Figure 24 Typical multifunctional initiators for controlled/living cationic polymerizations. See also Table 3 for references with isobutene.

mers, and multiarmed or star polymers, as will be discussed in detail in the next chapter. Herein we will consider the design of multifunctional initiating systems.

As with their monofunctional counterparts, the multifunctional initiators are combined with Lewis acid (metal halide) activators to induce controlled/living cationic polymerizations. When necessary, bases or salts are added. That means that all of the three methodologies for the design of initiating systems have been applied to these multifunctional versions. Figure 24 shows that these initiators may be classified into two groups, where the initiating sites are cumyl [—(aryl)—$C(CH_3)_2$**X**] (**14–17**) or a vinyl ether-adduct [—$CH_2CH(OR)$**X**] (**19–24**).

The cumyl-type initiators (**14–17**, Figure 24A) are primarily used for isobutene (Section V.B), but are also applicable to styrene derivatives (see references included in Table 3). Other than those shown in Figure 24A, bifunctional initiators with aliphatic backbones (spacers), such as **18**, are also available for isobutene (see Figure 21). The leaving groups **X** (or counteranions) are typically chloride or methoxide, which can be combined with BCl_3 or $TiCl_4$ activators. Although highly effective and versatile, these initiating systems sometimes require special conditions to prevent deactivation processes via intramolecular Friedel-Crafts reactions to form indan rings, which reduce the effective functionality of the initiating systems. Examples of such conditions include lowering the polymerization temperature and solvent polarity [234] or to introduce a protective group into the pertinent position of the initiator [235].

The vinyl ether-adducts (**19–24**) (Figure 24B) are designed for the polymerization of vinyl ethers and *p*-alkoxystyrenes but they may be used also for styrene [220] and α-methylstyrene [226]. With **X** = CF_3CO_2, for example, controlled/living polymerizations of vinyl ethers are induced with $C_2H_5AlCl_2$ as the activator in the presence of 1,4-dioxane as an added nucleophile [140]. Experiments with **22**, where the three initiating sites are connected to an aromatic core via hydrolyzable ester linkages, showed the attachment of three arm chains with controlled and uniform lengths; after the polymerization, the ester linkages of the initiator residue (core) in the polymers are hydrolyzed to give "arm" polymer chains that have MWDs as narrow as that of the parent polymer [139]. Initiators with **X** = halogen (hydrogen halide adducts), **14** and **19–23**, coupled with iodine or zinc halides, can initiate controlled/living polymerizations of vinyl ethers [118,122,176] and *p*-alkoxystyrenes [210]. Thus, the leaving groups (counteranions) in these multifunctional initiators and the corresponding methodologies for system design should be selected carefully according to the structure of the monomers employed.

In the development of the tetrafunctional initiator **24**, the spatial shapes of initiator molecules turned out to be crucial for obtaining well-defined initiators [140]. As shown in Scheme 11, **24** is prepared from the corresponding tetrafunctional phenol via a reaction with 2-chloroethyl vinyl ether to attach vinyl ether moieties, followed by addition of trifluoroacetic acid or hydrogen iodide. In this acid addition, the four vinyl ether groups should be well separated spatially. If the vinyl ether groups are located too close to each other, the treatment with the acid leads to intramolecular cyclization and other side reactions.

2. *Trimethylsilyl Halides as Initiators*

In organic chemistry, the trimethylsilyl group is known as an equivalent of proton, and is sometimes called a "bulky proton" [236]. It is therefore expected that trimethylsilyl esters (Me_3SiY) would initiate cationic polymerization, as do HY (protonic acids), but this process turned out to be not so straightforward. For example, Hall attempted to use trimethyl and triisopropyl trifluoromethanesulfonate (triflate) ($R_3SiOSO_2CF_3$; R = Me, *i*Pr) for isobutyl vinyl ether, *N*-vinylcarbazole, *p*-methoxystyrene, α-methylstyrene, and styrene polymerizations [237]. Polymerizations occurred at −78° C in methylene chloride to give polymers of high molecular weights (10^4–10^5) in 80% yield or above. However, later studies [134,238,239] showed that the polymerizations are not initiated directly

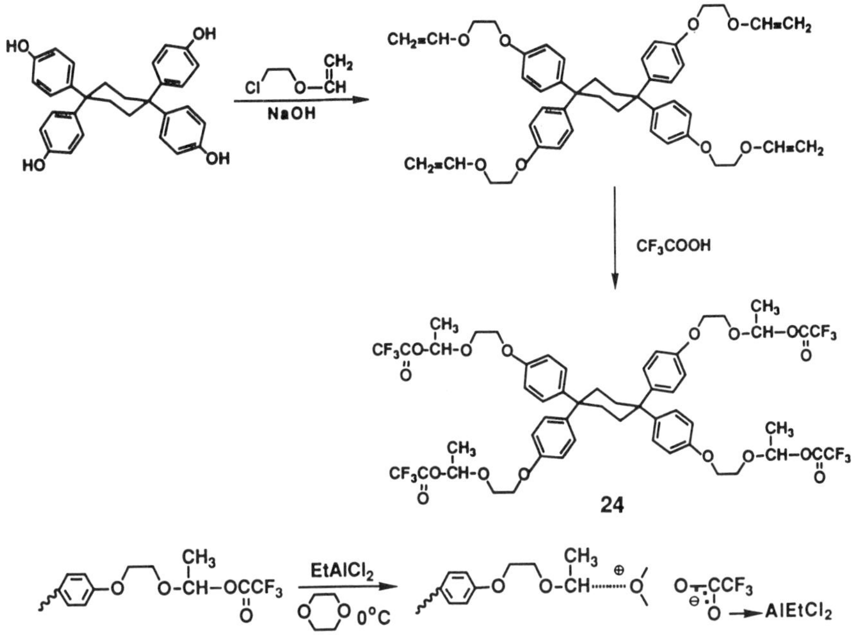

Scheme 11

through the addition of the trimethylsilyl group to the monomers but rather by CF_3SO_3H generated in situ via hydrolysis of the silyl ester. Similar results were obtained with other Me_3SiY [Y = $OP(O)(OC_6H_5)_2$ and OSO_2Me] [238].

Similarly, trimethylsilyl iodide (Me_3SiI) would be equivalent to hydrogen iodide in the HI/I_2 and HI/ZnI_2 initiating systems for living cationic polymerizations. The Kyoto group examined the polymerization of isobutyl vinyl ether (IBVE) with a series of $Me_3Si\mathbf{X}/Zn\mathbf{X}_2$ (**X** = I, Br, Cl) systems [240–243]. In contrast to the trifluoromethanesulfonate counterpart, Me_3SiI does not hydrolyze under these conditions. Controlled/living cationic polymerizations are in fact possible with these initiating systems, but the initiation reaction is not the direct addition of the silyl compound to monomer. Instead, acetone and other carbonyl compounds (such as benzaldehyde) are needed for efficient and quantitative initiation to occur from $Me_3Si\mathbf{X}$; the initial process of the polymerization is illustrated in Scheme 12 A for the Me_3SiI/ZnI_2/benzaldehyde system [241,243]:

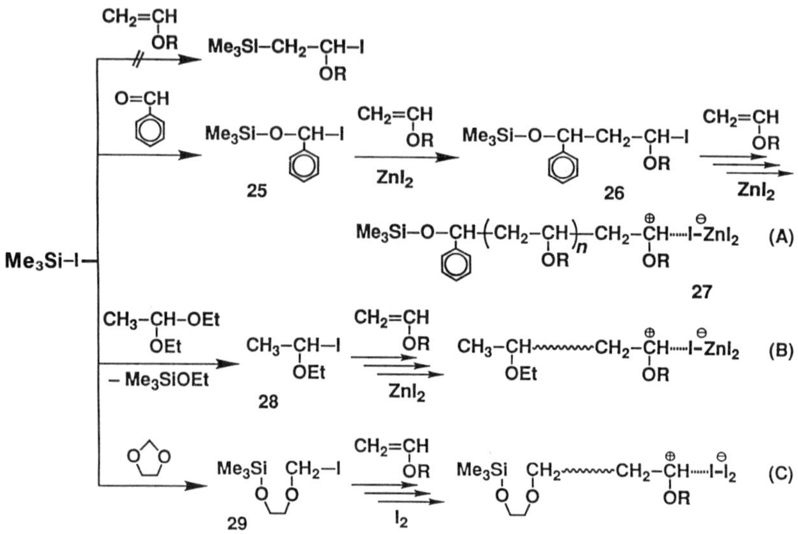

Scheme 12

Due to silicon's higher affinity toward oxygen than carbon, the silyl iodide does not add directly across the vinyl ether double bond, but first reacts with the carbonyl compound to form a silyloxyalkyl iodide (**25**). In the presence of zinc iodide or iodine, this adduct adds to a vinyl ether, and the resulting vinyl ether-iodide adduct (**26**) initiates ZnI_2-mediated propagation (via **27**).

As an extension of the $HOSO_2CF_3/SMe_2$ system (SMe_2 as an added nucleophile) [37], it was reported that $Me_3SiOSO_2CF_3$ initiates controlled/living cationic polymerization of IBVE in the presence of acetone (for initiation) and SMe_2 [134,244].

The carbonyl compounds used in these initiating systems include acetone, benzaldehyde, and its ring-substituted derivatives with varying functionalities [134,242]; acetophenone is less effective. Consistent with the proposed pathway, the polymerization affords poly(IBVE) with a silyloxyl group at the α-end (polymer **27**; Scheme 12A) [242]. Hydrolysis of the terminal end group gives alcohol functionalities (secondary from aldehydes, tertiary from ketones), thus providing a method for the synthesis of hydroxy-capped polymers that is complementary to those with the acetate functionality (giving primary alcohols only) derived from $CH_3CH(OCH_2CH_2OCOCH_3)$—**X** (**X** = halogen, etc.) and related functionalized initiators (see Chapter 5).

The affinity of the trimethylsilyl group toward oxygen was also used to generate the HI-vinyl ether adduct or its analogs from linear and cyclic acetals [245] (Scheme 12, B and C). In conjunction with zinc iodide, these iodoethers (**28** and **29**) can initiate controlled/living cationic polymerizations of vinyl ethers (ethyl, 2-chloroethyl, and isobutyl), as does the HI/ZnI_2 system. When the starting acetal has a polymeric substituent, e.g., $H[CH_2CH(O\textit{i}Bu)]_n$—$CH_2CH(O\textit{i}Bu)$—OMe, the products are vinyl ether block copolymers [246]. Such acetal-capped poly(vinyl ethers) are readily obtained by quenching the corresponding living polymers [e.g., ~~~$CH_2CH(O\textit{i}Bu)^+,ZnI_3^-$] with excess methanol.

VI. CHEMISTRY OF CONTROLLED/LIVING CARBOCATIONIC POLYMERIZATION

In previous sections we discussed the typical methodologies and practical aspects of the controlled/living carbocationic polymerizations of vinyl ethers, isobutene, styrene, and other monomers. It is possible to select optimal conditions such as the structure and concentrations of initiators, Lewis acids, additives (nucleophiles and salts), solvent, and temperature for each class of monomers in order to control and to prepare well-defined polymers. However, further progress requires a better understanding of the mechanisms of the involved reactions.

The enhanced control achieved in many new systems could be explained by either entirely new mechanisms of polymerization or by the careful selection of polymerization conditions or both. To establish a mechanism of a polymerization, it is necessary to determine the structure of chain carriers and the polymerization products as well as analyze kinetics. Because the active species are present at very low concentrations, model studies help in the studies of their nature.

In this section we will review some details concerning the chemistry of these systems with a special emphasis on the structure of the active species, kinetics, as well as on the molecular weights and the evolution of molecular weight distributions with conversion. In the last Section VII, we will discuss the mechanistic features of new controlled carbocationic polymerizations.

A. Active Species and Model Studies

1. *UV-Visible Spectroscopy*

The active centers in the polymerizations of vinyl ethers absorb weakly in the range ($\lambda_{max} < 250$ nm) precluding their direct observation by UV-visible spectroscopy. In polymerizations of styrene and its derivatives, carbenium ions have been observed as transient species at 330 to 380 nm by UV-visible stopped-flow methods [31,32,247,248]. These systems are usually poorly controlled, and the monomers are consumed within a fraction of a second. Typical half-lifetimes of controlled/living cationic polymerizations are in the range of 1 min to 3 hr. Assuming that an average value of the rate constant of propagation is $k_p = 10^5\ \mathrm{mol^{-1}{\cdot}L{\cdot}sec^{-1}}$, the concentration of propagating carbocations should be in the range $[C^+] \approx 10^{-7}$ to 10^{-9} mol/L, which is below the direct detection limit by UV spectroscopy. Therefore, it is desirable to conduct stopped-flow UV measurements on systems closely resembling controlled/living systems but at increased concentrations of Lewis acids and initiators. It is expected that the growing carbenium ions will be observed under such conditions in the near future.

2. *Trapping Study*

Trapping agents, such as malonate anions, naphthoxides, and phosphines have been used to determine the concentration of chain carriers in controlled/living and other carbocationic systems [85,249,250]. These strong nucleophiles react with all sufficiently electrophilic species, including not only carbocations but also onium ions and covalent esters. Thus, the discussed measurements can provide only the total concentration of active and dormant end groups. In principle, the kinetics of formation of the product in the trapping experiments could resolve more and less active species but only if they are present at comparable concentrations. As discussed before, carbocations are present in ppm quantities in comparison with dormant species.

Trapping experiments with malonate anions in the controlled/living polymerization of vinyl ethers initiated by mixtures of hydrogen iodide (HI) and Lewis acids revealed that the total concentration of the growing species (the sum of dormant and active) stays nearly constant and equal to that of the introduced initiator when monomer is present in the reaction mixture. However, after complete monomer consumption, the concentration of the growing species decays relatively rapidly [251,252].

This unusual behavior can be explained in two ways. The first one is related to the stabilization effect of the monomer. This may involve special solvation such as π-complexation by a monomer's double bond. The second explanation is based on transfer to counterion [253]. In the latter case, if the contribution of transfer by β-H^+ elimination is relatively small, the released protons will continuously reinitiate new chains, and the total number of growing chains, detectable by malonate anion, stays the same [Eq. (8); LA: Lewis acid]:

$$\text{...-}CH_2\text{-}CH(OR)\text{-}X + LA \underset{\text{v. fast}}{\overset{\text{fast}}{\rightleftharpoons}} \text{...-}CH_2\text{-}C^{+}H(OR), LAX^{-} \underset{?}{\overset{\text{slow}}{\rightleftharpoons}} \text{...-}CH{=}CH(OR) + \text{"}H^{+}\text{"}\ LAX^{-} \quad (+M) \tag{8}$$

However, once all monomer is consumed, the acid formed by elimination accumulates, and the total number of growing species decreases. It is possible that both phenomena contribute to the dependence shown in Figure 25.

3. *NMR Spectroscopy*

Because the concentration of carbocations in a real polymerization is very low, model NMR studies have been used to obtain a deeper insight into the nature of the growing species. These experiments are restricted to sufficiently stable carbocations, such as those derived from vinyl ethers. Styrene derivatives are not stable enough and participate in Friedel-Crafts alkylation. For example, derivatives of α-methylstyrene easily deprotonate, dimerize and then form intramolecularly indan derivatives.

The methine protons in secondary esters and halides that resemble active species in the polymerization of styrene and vinyl esters absorb in the range of 5 to 6 ppm, usually as quartets coupled to the neighboring methyl group, whereas those in the corresponding carbenium ions absorb at much lower field, 9–10 ppm [105,123,254,255]:

$$\underset{\approx 5\text{ to }6\text{ ppm}}{CH_3\text{-}CH(R)\text{-}X + MtX_n} \overset{\text{fast}}{\rightleftharpoons} \underset{\approx 9\text{ to }10\text{ ppm}}{CH_3\text{-}C^{+}H(R), MtX_{n+1}^{-}} \tag{9}$$

The overall changes in the chemical shift of the methine protons can be directly correlated with the amounts of ions if ionization is significant (>1%). However, even minute amounts of intermediate cations can be detected in some systems by dynamic NMR. Because ionization leads to the formation of a planar carbenium ion, the chirality at the carbon atom is lost. In the particular case of the isobutyl vinyl ether derivatives, the isobutoxy group has a built-in probe ($CH_2\text{-}CHMe_2$) separated from the chiral center by oxygen atom. The methylene protons on the ether group are magnetically nonequivalent due to the presence of four different substituents at the electrophilic carbon center. They become equivalent only

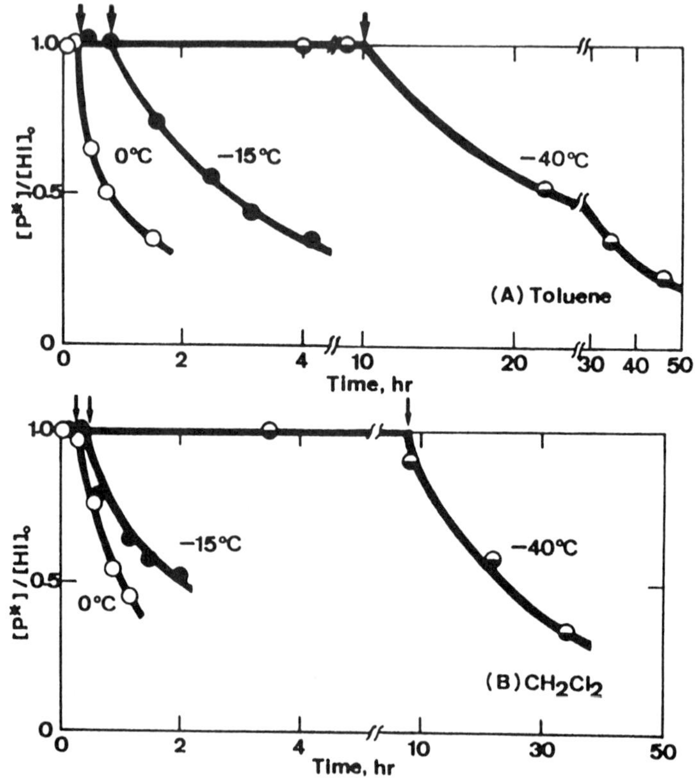

Figure 25 The total concentration of living end ([P*]/$[HI]_0$; P*: active + dormant species) as a function of time in the polymerization of isobutyl vinyl ether with the HI/I_2 initiating system in toluene (A) and CH_2Cl_2 (B) at temperatures 0 to −40° C: $[M]_0$ = 0.38 *M*; $[HI]_0$ = 10 m*M*; $[I_2]_0$ = 5.0 m*M* (in toluene) or 0.20 m*M* (in CH_2Cl_2). [P*] is determined by quenching the reaction with the sodium salt of ethyl malonate followed by ^{1}H NMR end group analysis of the product. The vertical arrows indicate the time for 100% conversion at each temperature. (From Ref. 85.)

after ionization of the covalent species, i.e., after the formation of the flat sp^2-hybridized carbocation:

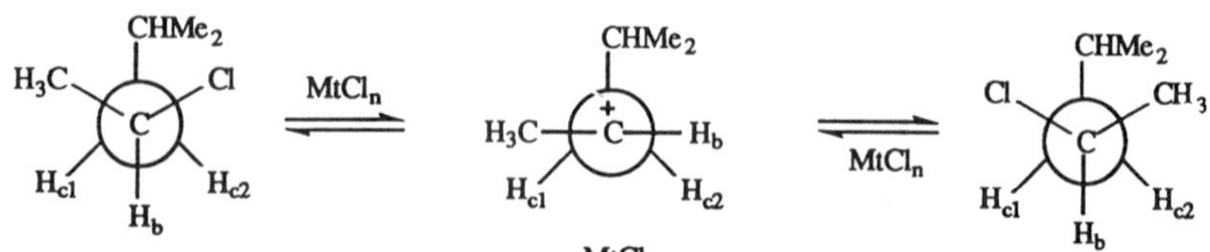

(10)

The following changes in the NMR spectra occur upon ionization: (1) Averaging of the methylene protons (H_{c1} and H_{c2}) of the isobutyl group due to the ionization process. This process is sensitive to the rate of ionization and can be detected even if very small amounts of the carbocations are present. (2) Shifting of the methine proton from the covalent region (≈5.5 ppm) to the carbocationic region [≈10.2 ppm, Eq. (9)]. This process allows one to determine the equilibrium constants of ionization and is useful only if the degree of ionization is sufficiently large (>5%).

The averaging of the chemical shifts of the substituents in the isobutyl group can be used successfully to assess the activity of various Lewis acids and to confirm the presence of carbenium ions in the studied systems [89,105,256].

Figure 26 presents typical ^1H NMR spectra of covalent esters under various conditions.

The increase of the ionization power or the concentration of Lewis acids leads not only to the averaging of the diastereotopic protons H_{c1} and H_{c2} but also shifts the entire spectrum downfield due to the exchange of covalent species with carbenium ions. When ionization is strong, the shifts are significant due to the high proportion of carbocations. Under such conditions polymerization is very fast and difficult to control and therefore "not living." Well-defined polymers can be prepared in the presence of $ZnCl_2$ and $SnBr_4$, when the amount of ions is very low and polymerization is slow enough. Nevertheless, even in these systems averaging of the H_{c1} and H_{c2} protons (Figure 26E) and considerable broadening (Figure 26F) are clearly noted.

Figure 27 shows the changes in the ^1H NMR spectra of these adducts when ionization becomes substantial.

A linear increase in the average chemical shifts with the concentration of the Lewis acid is anticipated until an equimolar amount of the Lewis acid is added if the equilibrium constant of ionization is large enough (cf., broken line in Figure 28). It seems that some Lewis acids such as $SnCl_4$ can ionize more than one covalent species. Half of an equivalent of $SnCl_4$ (Figure 27C) shifts the methine proton signals to a position indicating ≈60% ionization. This suggests that not only $SnCl_4$ but also $SnCl_5^-$ anion can ionize the covalent species at −78° C.

On the other hand, BCl_3, a Lewis acid of a strength comparable with $SnCl_4$, can coordinate only one ligand and form BCl_4^- anion. However, this Lewis acid may also coordinate to the nucleophilic ether moiety to form a complex. This complexation and the potential formation of oxonium ions by the reaction of the ether with carbocations is less important with an excess of BCl_3 over the HCl-IBVE adduct [IBVCl; $CH_3CH(OR)$-Cl] and allows the estimate of the equilibrium constant, *K*, for ionization

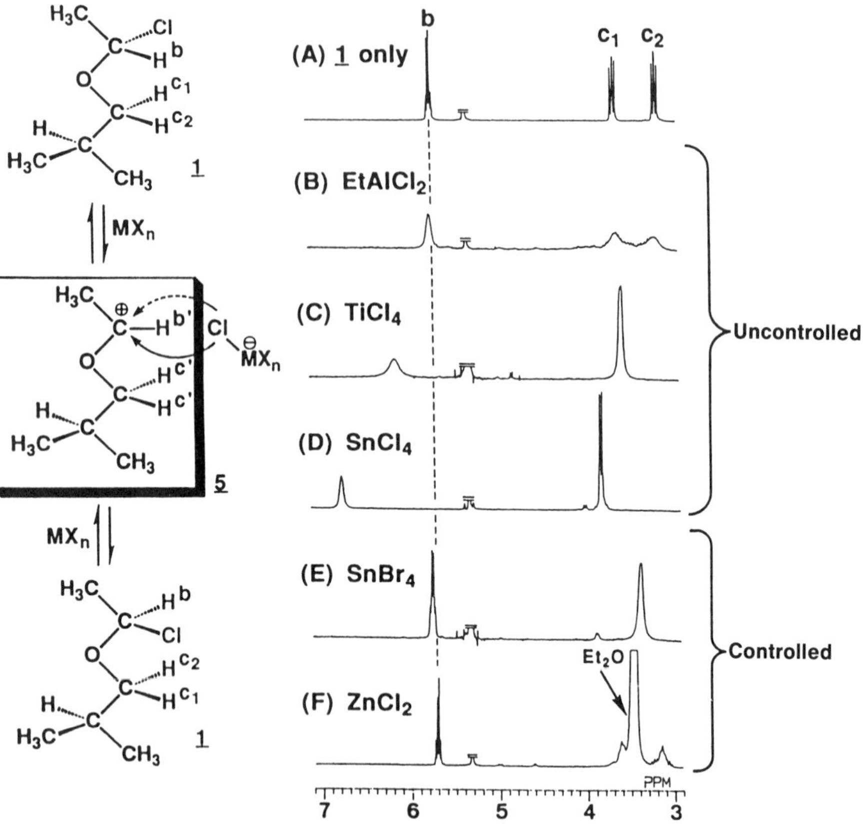

Figure 26 ^{1}H NMR spectra (270 MHz) of the HCl-adduct of isobutyl vinyl ether (A) and its mixtures (B–F) with various Lewis acids (**MX**$_n$) in CD_2Cl_2/n-hexane (9:1 v/v) at −78° C: [adduct]$_0$ = 100 mM; [**MX**$_n$]$_0$ = 20 mM. **MX**$_n$: $EtAlCl_2$ (B); $TiCl_4$ (C); $SnCl_4$ (D); $SnBr_4$ (E); $ZnCl_2$ (F). Only for $ZnCl_2$ the solvent is CD_2Cl_2/n-hexane/Et_2O (8:1:1 v/v). (From Ref. 105.)

[256]. At relatively low concentrations of the Lewis acid, the values of K vary indicating some contribution of side reactions. A typical dependence of the chemical shifts of methine protons on the $[BCl_3]_0/[IBVCl]_0$ ratio at −53° C is shown in Figure 28. The broken line depicts the theoretical behavior for an infinitely large equilibrium constant; the experimental data indicate a moderate value of the equilibrium constant.

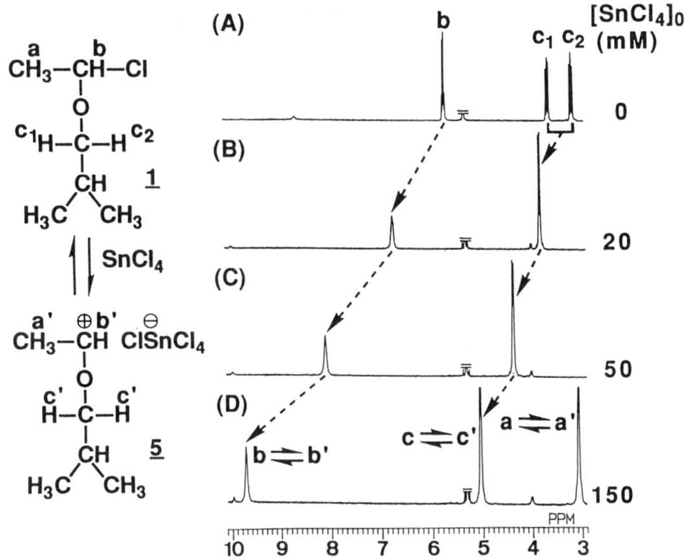

Figure 27 ^{1}H NMR spectra (270 MHz) of the HCl-adduct of isobutyl vinyl ether (A) and its mixtures (B–D) with $SnCl_4$ in CD_2Cl_2/*n*-hexane (9:1 v/v) at −78° C at varying $SnCl_4$ concentrations: $[\text{adduct}]_0$ = 100 m*M*. $[SnCl_4]_0$: 0 (A), 20 (B); 50 (C); 150 (D) m*M*. (From Ref. 105.)

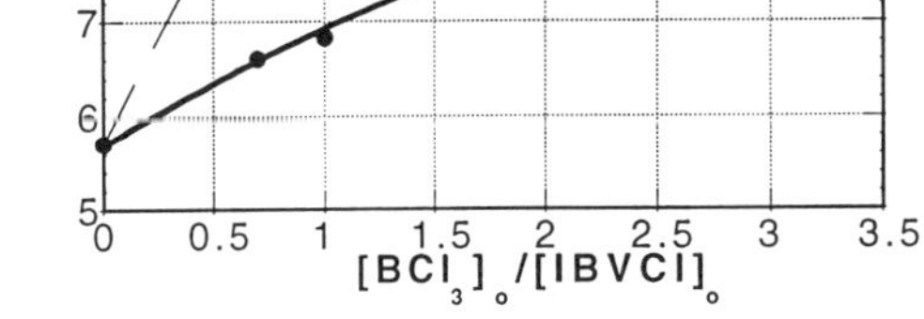

Figure 28 The change of the chemical shift of the methine protons as a function of $[BCl_3]_0/[IBVCl]_0$ ratio (IBVCl: HCl-adduct of isobutyl vinyl ether). T = −53° C. Broken line is for an infinitely large equilibrium constant. (From Ref. 256.)

The equilibrium constant is calculated from the following equation:

$$K = \frac{\alpha}{[\mathrm{IBVCl}]_0(1 - \alpha)([\mathrm{BCl_3}]_0/[\mathrm{IBVCl}]_0 - \alpha)} \tag{11}$$

where: $\alpha = \Delta\delta/\Delta\delta^{\max}$; $\Delta\delta$ = the difference between the chemical shifts of the methine protons at the intermediate stage of ionization and in IBVCl; $\Delta\delta^{\max}$ = the difference between the chemical shifts of the methine protons in the carbocation and in IBVCl, where the plateau value (ca. 10.2 ppm) is assumed to represent the chemical shift of the completely ionized carbocation.

Ionization is exothermic, and the equilibrium constant between covalent species and carbenium ions decreases substantially with temperature:

$$CH_3\text{–}CH(OR)\text{–}Cl + BCl_3 \underset{CH_2Cl_2}{\overset{K}{\rightleftharpoons}} CH_3\text{–}C^+H(OR),BCl_4^-$$

$$\begin{array}{lll} T = -78°\mathrm{C} & & K = 40\ M^{-1} \\ -25°\mathrm{C} & & 1\ M^{-1} \\ +25°\mathrm{C} & & 0.1\ M^{-1} \end{array} \tag{12}$$

Not only the chemical shifts but also the width of the signal for the methine protons depends on temperature. The width (S) at half-height of this signal can be used to estimate the lifetime (τ) of the carbocations and of the covalent species, according to Eq. (13).

$$1/\tau = \frac{4\pi\alpha(1 - \alpha)^2(\Delta\delta^{\max})^2}{S - S_0} \tag{13}$$

where S_0 is the half-height width of the methine signal at very fast exchange and α is the proportion of carbocations. The estimated values of the rate constants (k_r) for the recombination of the counterions in the ion pairs as well as the rate constants for the ionization (k_I) of IBVCl in the presence of BCl_3 are plotted in Figure 29, as a function of temperature.

It seems that the rate constants for the deactivation of the carbocations (k_r) are higher than the reported for propagation rate constants, k_p [257]. This indicates that polymerization of IBVE in the presence of a small amount of BCl_3 could have led to well-defined polymers. Unfortunately, BCl_3 adds to alkenes via a haloboration and this concept could not be verified [256,258].

The addition of salts may lead to two phenomena: suppression of free ions and modification of Lewis acids. Tetrabutylammonium chloride and Lewis acids ($Mt\mathbf{X}_n$) form salts with complex anions $NR_4{}^+Mt\mathbf{X}_{n+1}{}^-$.

Thus, addition of large amounts of salts effectively reduces the concentration of Lewis acids. As shown in Figure 30, the signals move upfield

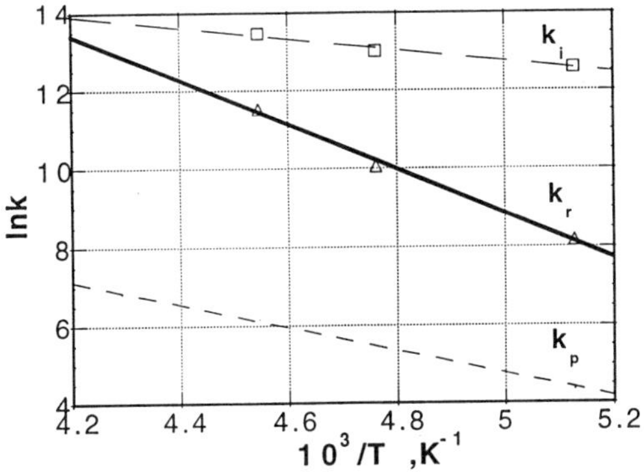

Figure 29 The dependence of rate constants on temperature. $[BCl_3]_0/[IBVCl]_0$ = 2:1 (IBVCl: HCl-adduct of isobutyl vinyl ether). (From Ref. 256.)

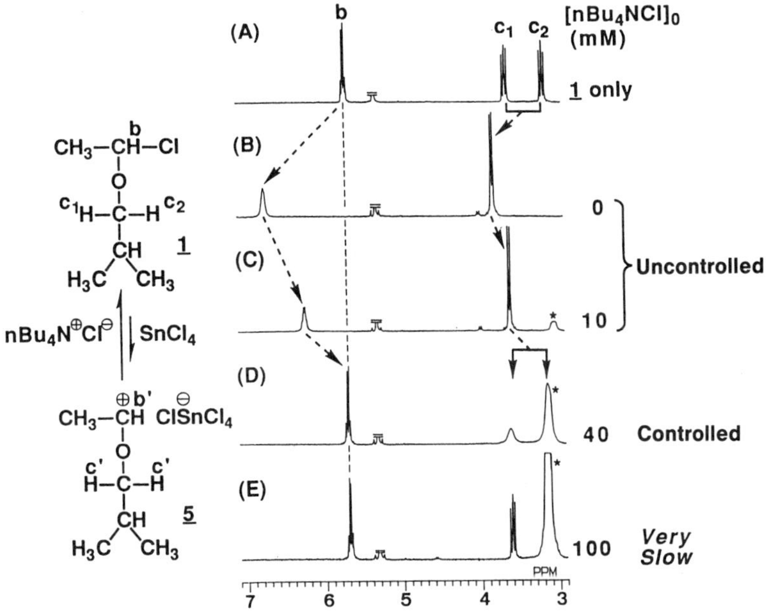

Figure 30 ^1H NMR spectra (270 MHz) of the HCl-adduct of isobutyl vinyl ether (A) and its mixtures (B–E) with $SnCl_4$ and nBu_4NCl in CD_2Cl_2/*n*-hexane (9:1 v/v) at −78° C at varying nBu_4NCl concentrations: $[adduct]_0$ = 100 m*M*; $[SnCl_4]_0$ = 20 m*M*. $[nBu_4NCl]_0$: 0 (B); 10 (C); 40 (D); 100 (E) m*M*. The signals marked with an asterisk: $(CH_3CH_2CH_2CH_2^*)_4NCl$. (From Ref. 105.)

and, depending on the structure of Lewis acid, ionization may or may not be entirely suppressed [105]. However, with two equivalents of salt (Figure 30D), ionization can be detected by the broadening of the H_c signals. Addition of only 5 equivalents of salt reduces the rate of this process below the NMR time scale and is accompanied by impractically slow polymerization (Figure 30E).

Thus, if the complex anions have sufficient electrophilicity ($SnCl_5^-$), ionization may still occur, resulting in a controlled/living polymerization.

Other methods for the deactivation of carbenium ions are based on the use of nucleophiles; for example, sulfides act as deactivators in the polymerization of IBVE initiated by triflic acid [37,38]. In this case, the sulfides react with the carbenium ions to form sulfonium ions. Dynamic NMR measurements were used to estimate the rate of the exchange process and demonstrated that exchange occurs predominantly by dissociation of the sulfonium ions to carbenium ions rather than bimolecularly via exchange of the sulfonium ion with the sulfide [21]. The rate constant of activation was $k_{act} = 60\ sec^{-1}$ and the rate constant of deactivation is $k_{deact} = 10^6\ mol^{-1}\cdot L\cdot sec^{-1}$ at $-30°$ C in CH_2Cl_2.

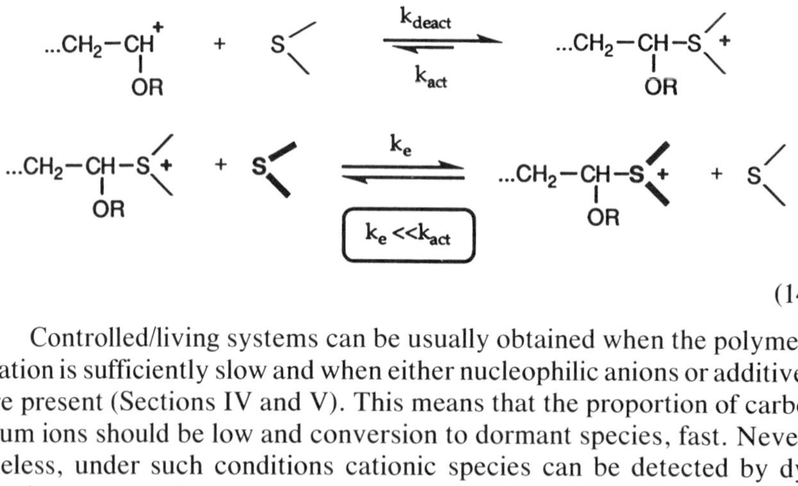

(14)

Controlled/living systems can be usually obtained when the polymerization is sufficiently slow and when either nucleophilic anions or additives are present (Sections IV and V). This means that the proportion of carbenium ions should be low and conversion to dormant species, fast. Nevertheless, under such conditions cationic species can be detected by dynamic NMR, by ligand exchange, salt, and solvent effects, and by other methods discussed in Chapters 2, 3, and in this section. Under typical controlled/living conditions, dormant species such as onium ions and covalent esters predominate. It is possible that the active species are strongly solvated by monomer and by some additives. These interactions may lead to a stabilization of the carbocations. However, in the most general case, this stabilization has a dynamic sense and can be described by the reversible exchange between carbocations and dormant species.

B. Kinetics

In physical organic chemistry, kinetics has been used successfully to distinguish between S_N1 and S_N2 mechanisms. When ion formation is the rate-determining step, the reaction rate does not depend on the concentration of other reagents such as solvent or monomer. Second-order kinetics does not necessarily mean that a direct bimolecular reaction (e.g., S_N2) takes place between the dormant species, D, and monomer, M. If the covalent precursor, D, is in dynamic equilibrium with the carbenium ion, C, and only the latter reacts with M to give the product P, then the overall kinetics depends on the ratio of the rate constants k_{-1} and k_2:

$$D \underset{k_{-1}}{\overset{k_1}{\rightleftharpoons}} C \qquad C + M \xrightarrow{k_2} P \tag{15}$$

$$-d[M]/dt = k_2[C][M] \tag{16}$$

Assuming a stationary state for [C]:

$$d[C]/dt = k_1[D] - k_{-1}[C] - k_2[C][M] \approx 0 \tag{17}$$

$$[C] = k_1[D]/(k_{-1} + k_2[M]) \tag{18}$$

$$-d[M]/dt = k_2k_1[D]\,[M]/(k_{-1} + k_2[M]) \tag{19}$$

Thus, if $k_{-1} \gg k_2[M]$, then

$$-d[M]/dt = k_2k_1[D]\,[M]/k_{-1} = k^{app}[D]\,[M] \tag{20}$$

If, however, $k_{-1} \ll k_2[M]$, then

$$-d[M]/dt = k_2k_1[D]\,[M]/k_2[M] = k_1[D] \tag{21}$$

In the former case [Eq. (20)], the rate does depend on [M]. Thus, kinetics alone may erroneously indicate that the covalent species D reacts directly with M (monomer), although it first ionizes to C. The latter case [Eq. (21)] results in zero-order kinetics in monomer. However, this may not happen in the polymerization process. If $k_{-1} \ll k_2[M]$, then once the ions C are generated, they can react many times with monomer before deactivation. Thus, the kinetics may be first order with respect to monomer and resemble a system with slow initiation and fast propagation.

1. *Systems with Lewis Acids*

Kinetic studies have been reported only for a few controlled/living carbocationic systems initiated with alkyl halide/Lewis acid combinations (cf.,

Section IV.B.1). The reactions are usually first order in monomer, in alkyl halide, and in Lewis acid [39,98,150,259].

$$R_p = -d[\mathrm{M}]/dt = k\cdot[\mathrm{M}]\cdot[\mathrm{R}X]_0\cdot[\mathrm{LA}]^n \qquad (22)$$

In some cases, a higher order in Lewis acids (e.g., $n = 2$ for $TiCl_4$) was reported and ascribed to the participation of the corresponding dimers of the Lewis acids, i.e., $(TiCl_4)_2$ [161,166]. Indeed, as discussed in Chapter 2, $TiCl_4$ dimerizes easily and the dimer is a much stronger Lewis acid than monomeric $TiCl_4$. In the earlier studies first-order kinetics in Lewis acid was reported [5,260], probably because the proportion of dimers depends on the concentration of Lewis acid. However, second order kinetics would also be observed if a more stable $Ti_2Cl_9^-$ anion is formed. Eq. (22) conforms to the system in which covalent species (RX) are ionized reversibly by Lewis acids to form carbocations (C^+) capable of propagation.

The kinetic studies in the presence of added salts support the presence of free ions and ion pairs as propagating species [33,39]. In the presence of salts with common ions or in nonpolar solvents, the proportion of free ions can be neglected and polymerization can be described by Scheme 13.

If the initiator RX has the same or higher ionizing ability as the dormant species P_nX, and if the reactivity of the initiating cation, R^+, is similar to that of the growing carbenium ions, P_n^+, then only the last two equations in Scheme 13 should be considered. This results in the following kinetic expression:

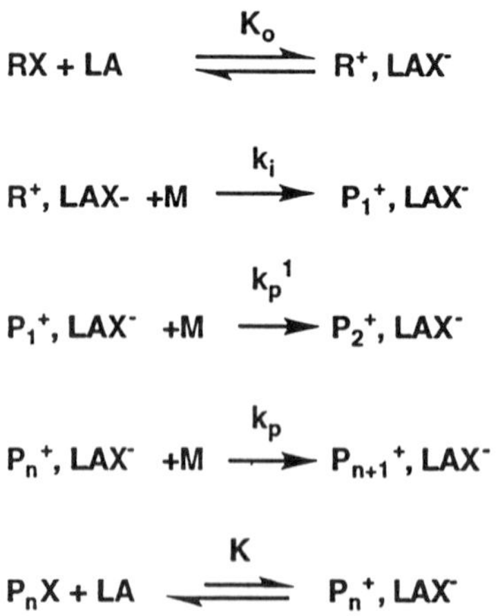

Scheme 13

$$R_p = -d[\mathrm{M}]/dt = k_p \cdot [\mathrm{M}] \cdot [\mathrm{P}^+, \mathrm{LA}\mathbf{X}^-] = k_p \cdot K \cdot [\mathrm{M}] \cdot [\mathrm{R}\mathbf{X}]_0 \cdot [\mathrm{LA}] \quad (23)$$

A comparison of Eq. (22) with Eq. (23) reveals that the apparent rate constant, k, equals the product $k_p \cdot K$. Thus, the rate of polymerization is affected by the concentrations of monomer, initiator, and Lewis acid, as well as by the rate and equilibrium constants. Although the ionic propagation rate constant is not very sensitive to the nature of the counterion, solvent and temperature, the equilibrium constant K usually depends strongly on the temperature, solvent, Lewis acid, and the leaving group **X**.

If ionization is strongly exothermic, polymerization may be faster at lower temperatures due to the increased concentration of carbenium ions. Temperature affects kinetics by the activation enthalpy of propagation ($\Delta H_p^{\ddagger}$) and by the enthalpy of the ionization equilibrium (ΔH). The apparent activation enthalpy is therefore a sum of both components and may become negative, if $\Delta H > \Delta H_p^{\ddagger}$:

$$\Delta H_p^{\ddagger\mathrm{app}} = \Delta H + \Delta H_p^{\ddagger} \quad (24)$$

Indeed, overall negative activation energies of propagation have been observed in some carbocationic polymerizations [261,262]. In most controlled/living systems, however, weak Lewis acids ($SnCl_4$, BCl_3, $ZnCl_2$, etc.) are used, and the overall activation energies are positive, because the activation energy of propagation is higher than the enthalpy of ionization. With strong Lewis acids, more exothermic ionization results in overall negative activation energies.

Rates of controlled/living carbocationic polymerizations increase with polarity of the medium. The proportion of ions increases strongly with the solvent's dielectric constant. Because solvents rate constants of propagation are not strongly affected by solvent [28,41], the observed changes in the apparent rate constant of propagation can be ascribed to the changes in the ionization equilibrium constant. An empirical Kosower's Z parameter sometimes gives better correlation than dielectric constant for reactions involving carbocations [28,263]. Parameters a and b proposed by Swain may also be used for predicting solvent effects in various reactions [264].

Zero order kinetics in monomer was reported [95,97] for the polymerization of vinyl ethers initiated by HI/I_2 in hexane and was ascribed to the formation of a complex between iodine (Lewis acid) and monomer. It has been proposed that monomer reversibly forms a complex with the growing chain (. . .–$CH_2CH(OR)I,I_2$) which then slowly inserts monomer in the rate-determining step [74]. Another possibility is that the ionization of the dormant species the rate-determining step where the cation subsequently undergoes the rapid reaction with monomer and then soon collapses back to the covalent species [265]. Apparently, with stronger Lewis acids and in more polar solvents such as methylene chloride, "normal" first-order kinetics in monomer is observed [96,98].

2. *Systems with Nucleophiles*

Nucleophiles affect the polymerization in two different ways (cf., Section VII.E.4). They form complexes with Lewis acids reducing their strength, and/or they also form onium ions with the growing carbocations. In both cases they reduce the polymerization rates.

When protonic acids are used as initiators, the resulting anions do not interact with nucleophiles. Therefore, kinetic analysis of these systems is simpler than for polymerizations carried out in the presence of Lewis acids. A typical example is the polymerization of vinyl ethers initiated by triflic acid in the presence of dialkyl sulfides [37,38].

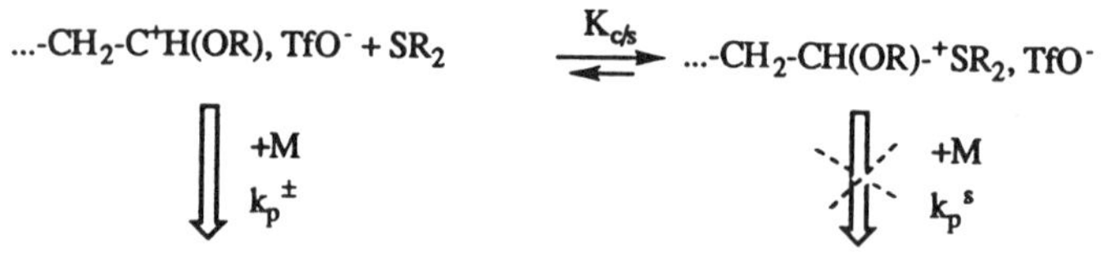

(25)

The reactions are conducted at relatively high acid concentrations ($[HOTf]_0 > 10^{-3}$ mol/L), and the ionic species are mostly in the form of ion pairs. Monomer can be consumed in the reaction with carbenium ions ($k_p^{\pm}$) and with sulfonium ions (k_p^{s}).

$$R_p = -d[\mathrm{M}]/dt = k_p^{\pm}[\mathrm{C}^{\pm}]\cdot[\mathrm{M}] + k_p^{s}\cdot[\mathrm{S}^{\pm}]\cdot[\mathrm{M}] \quad (26)$$

Because the sulfide is used in a large molar excess with respect to the acid ($r = [\mathrm{S}]_0/[\mathrm{HOTf}]_0 > 1$), and the equilibrium constant for sulfonium ion formation is very large ($K_{c/s} \gg 1$), integration of Eq. (26) leads to Eq. (27).

$$-[\ln(1 - p)]/t = k_p^{\pm}/K_{c/s}(r - 1) + k_p^{s}[\mathrm{HOTf}]_0 \quad (27)$$

A plot of $-[\ln(1 - p)]/t$ versus $1/(r - 1)$ passes through the origin, indicating that no growth occurs via sulfonium ions [37]. The sulfonium ions must first convert into carbenium ions which only then can react with monomer. It has been reported [21,38] that propagation is first order in acid and negative first order in sulfide.

$$-d\ln[\mathrm{M}]/dt = k_p^{\mathrm{app}}\cdot[\mathrm{HOTf}]_0\cdot[\mathrm{S}]_0^{-1} \quad (28)$$

If the initiator (acid, HOTf) quantitatively forms the growing species, the equilibrium constant is large, and the sulfide is used in large excess, then:

$$K_{c/s} = [\mathrm{S}^{\pm}]/([\mathrm{C}^{\pm}][\mathrm{S}]) \approx [\mathrm{HOTf}]_0/([\mathrm{C}^{\pm}][\mathrm{S}]_0) \quad (29)$$

The equilibrium constants for sulfonium ion formation were calculated from the apparent rate constants of propagation and the rate constants of carbocationic propagation [257], using Eq. (31) obtained by the combination of Eq. (30) with Eqs. (28) and (29).

$$-d\ln[\mathrm{M}]/dt = k_p^{\pm}[\mathrm{C}^{\pm}] \tag{30}$$

$$K_{c/s} = k_p^{\pm}/k_p^{\mathrm{app}} \tag{31}$$

The ratio of the ionic propagation rate constant ($k_p^{\pm} = 1.1 \cdot 10^3$ $\mathrm{mol}^{-1} \cdot \mathrm{L} \cdot \mathrm{sec}^{-1}$ $-35°$ C) [257] to the experimentally observed apparent propagation rate constant ($k_p^{\mathrm{app}} = 2.8 \cdot 10^{-2}$ $\mathrm{mol}^{-1} \cdot \mathrm{L} \cdot \mathrm{sec}^{-1}$) provides the equilibrium constant $K_{c/s} = 4 \cdot 10^4$ mol/L at $-35°$ C in CH_2Cl_2 with tetrahydrothiophene (THT) as a nucleophile [38]. Measurements at variable temperatures indicate that the formation of the sulfonium species is exothermic, $\Delta H_{c/s} = -40$ kJ/mol. In this calculation it was assumed that the equilibrium between triflate esters and the corresponding carbocations was strongly shifted to the carbenium side. The primary triflates (CH_3O-CH_2OTf) were already partially ionized ($K_{\mathrm{eq}} > 10^{-4}$) [266] and the secondary species should be ionized much more strongly. Moreover, the apparent rate constants of propagation in polymerization of IBVE, initiated with trityl hexachloroantimonate in the presence of THT, are nearly the same as in the systems initiated with triflic acid, confirming a high degree of ionization [135].

Phosphonium salts, formed in the presence of phosphines, are more stable than sulfonium salts. With triphenylphosphine no polymerization of IBVE by triflic acid was observed [250]. Polymerization proceeds slowly only with the most weakly nucleophilic phosphines such as tris(*p*-chlorophenyl)phosphine. The equilibrium constant is much higher than for sulfides, $K_{c/p} = 10^{10}$ mol/L [141]. In that case, Eq. (29) should be modified to Eq. (33), because only a small excess of phosphine can be used:

$$[\mathrm{P}] = [\mathrm{P}]_0 - [\mathrm{HOTf}]_0 \tag{32}$$

$$d\ln[\mathrm{M}]/dt = k_p^{\mathrm{app}}\,[\mathrm{HOTf}]_0/([\mathrm{P}]_0 - [\mathrm{HOTf}]_0) = k_p^{\mathrm{app}}\,[\mathrm{C}^+]\,K_{c/p} \tag{33}$$

Kinetics of polymerizations coinitiated by Lewis acids are more complicated (cf., Section IV.B.2), because nucleophiles complex not only with the growing carbocations but also with Lewis acids. In the polymerization of IBVE coinitiated with $AlCl_3$ along with ethyl acetate, a negative first order is reported in respect to large amount of the ester [267]. This can be explained by a weak and reversible deactivation of the growing cation by the nucleophile. Ethyl acetate complexes with $AlCl_3$, reducing its

Lewis acidity but Lewis acid still preserves its ability to ionize the corresponding dormant species.

A different behavior is found when strong nucleophiles are added to a polymerization of isobutene coinitiated with Lewis acids. When the concentration of nucleophile reaches that of Lewis acid, no polymerization is observed [5,91,268]. Quite often a precipitate, identified as a complex between Lewis acid and the nucleophile, is detected. This indicates that the complexed Lewis acid is no longer capable of ionization of the covalent species. A fractional negative order (−0.3) was reported in the polymerization of isobutene initiated by alkyl chloride/$TiCl_4$ with added pyridines [268]. It is possible that a small amount of pyridine complexes with the Lewis acid, reduces its concentration, and thereby additionally shifts the equilibria from the more reactive dimeric to the less reactive monomeric $TiCl_4$. Nevertheless, a small amount of nucleophile apparently has a beneficial effect, because polymers with lower polydispersities are formed. The plausible explanation of the role of nucleophiles in these systems will be offered in Section VII.E.4.

3. *Systems with Salts*

Salts usually have a dual effect on cationic polymerization (cf., Section VII.E.3). Tetrabutylammonium chloride coordinates strongly with Lewis acids, e.g., with BCl_3 or $SnCl_4$ to form tetrabutylammonium tetrachloroborate and pentachlorostannate, respectively. In the latter case, an excess of salt may lead to the formation of hexachlorostannate dianions. Thus, salts scavenge Lewis acid and reduce the polymerization rate in a way similar to nucleophiles. Recent analysis of model reactions by ^{119}Sn NMR spectroscopy demonstrate a transformation of $SnCl_4$ in the presence of the added salt:

$$\begin{aligned} SnCl_4 + R_4N^+, Cl^- &\rightleftharpoons R_4N^+, SnCl_5^- \\ R_4N^+, SnCl_5^- + R_4N^+, Cl^- &\rightleftharpoons (R_4N^+)_2, SnCl_6^{2-} \end{aligned} \tag{34}$$

Salts also affect the equilibria between free ions and ion pairs. The most obvious action is the second salt effect (common ion effect), namely suppression of free ions:

$$\text{...-}CH_2C^+HR, SnCl_5^- \underset{\boxed{+\ R_4N^+ + SnCl_5^-}}{\overset{K_D}{\rightleftharpoons}} \text{...-}CH_2C^+HR + SnCl_5^- \tag{35}$$

Thus, after addition of a few percent of salt (relative to Lewis acid), a few fold drop in the polymerization rate is observed as shown in Figure 31.

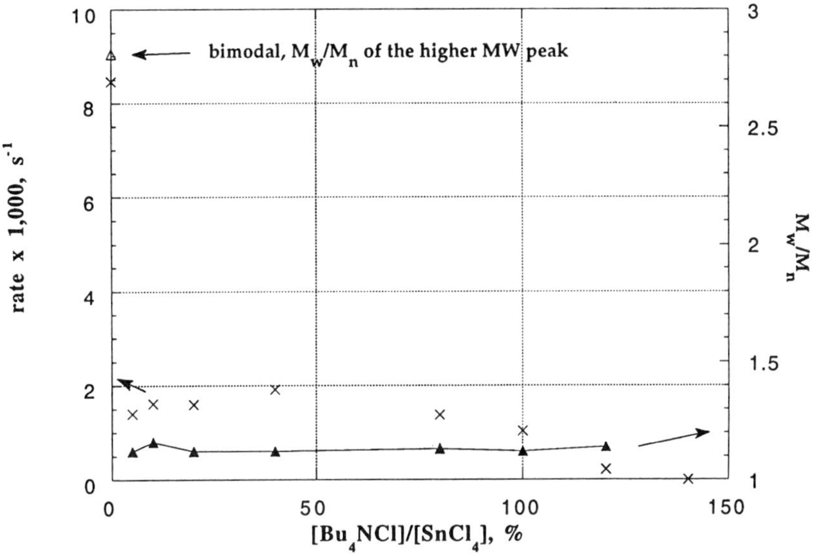

Figure 31 Effect of added salt on the polymerization rate and polydispersity in CD_2Cl_2 at 20° C ($[\text{styrene}]_0 = 1\ M$, $[\text{1-phenylethyl chloride}]_0 = 0.02\ M$, $[SnCl_4]_0 = 0.1\ M$, $[Bu_4NCl]_0 = 0.005$–$0.14\ M$). (From Ref. 39.)

The initial decrease of the apparent rate constants is due to the suppression of free ions in the system. Typically a 2- to 5-fold rate reduction is observed which corresponds to 50 to 80% population of free ions in the system (assuming similar reactivities of ions and ion pairs). The addition of larger amounts of salt has two effects. The first one is the entrapment of Lewis acid and reduction of the polymerization rate. The second effect, which increases the concentration of carbocations, is related to the formation of aggregates of anions (e.g., ionic triplets) as found for $SnCl_4$ [39,269].

$$\left.\begin{matrix} \ldots\text{-}CH_2C^+HR,\ SnCl_5^- + R_4N^+,\ Cl^- \\ R_4N^+,\ SnCl_5^- + \ldots\text{-}CH_2CHR\text{-}Cl \end{matrix}\right\} \rightleftharpoons \ldots\text{-}CH_2C^+HR,\ SnCl_6^{--},\ R_4N^+, \qquad (36)$$

The involvement of the carbenium ion pairs in the aggregates should increase the overall concentration of carbocations and may lead to the apparent increase (or rather, the lack of decrease) of the overall polymerization rate with the addition of the salt. In that case, the polymerization rate stays fairly constant until an equimolar amount of salt is added, and then drops rather dramatically.

A similar effect was reported in polymerization of cyclohexyl vinyl ether (CHVE) initiated by HI alone [33]. The small addition (1%) of tetrabutylammonium halides led to a strong initial decrease in rate due to the suppression of free ions. Subsequent additions of tetrabutylammonium iodide had no effect on kinetics; however, addition of tetrabutylammonium bromide and chloride resulted in an additional rate decrease. This can be explained by the special salt effect and exchange of anions. The corresponding HBr and HCl adducts with CHVE do not initiate polymerization. Kinetic analysis allowed the estimate of the relative reactivities of ions and ion pairs ($k_p{}^+/k_p{}^\pm < 10$) and dissociation constants ($K_D \approx 10^{-5}$ mol/L) in CH_2Cl_2 at $-30°$ C.

Salts with less nucleophilic ions accelerate the polymerization of vinyl ethers initiated by HI adducts alone. This can be again ascribed to a special salt effect and exchange of counterions [176].

C. Molecular Weights and Molecular Weight Distribution

Most controlled/living cationic systems provide well-defined polymers of relatively low molecular weight, typically in the range $M_n = 5{,}000$ to 20,000. The degrees of polymerization are equal to the ratio of concentrations of the reacted monomer to that of the introduced initiator ($DP_n = \Delta[M]/[I]_0$), and the polydispersities remain relatively low, $M_w/M_n < 1.2$. The addition of new portions of monomer (sometimes called incremental monomer addition, IMA) [5] leads to the expected increase of molecular weights, and the addition of another comonomer results in block copolymers, whereas the addition of terminating reagents provides end-functionalized macromolecules (cf. Chapter 5).

All of the observations above indicate the presence of living systems; however, attempts to extend these well-defined systems above a limit of $M_n \approx 100{,}000$ have been unsuccessful, except at very low temperatures ($< -70°$ C) [270]. Thus, polymers with predetermined polymerization degrees, low polydispersities, and with desired end groups can be obtained only for a sufficiently low molecular weight range. This indicates that contribution of transfer increases with temperature and with chain length [cf. Eq. (2) in Section II.C]. In the presence of transfer and termination polydispersities increase with the chain length and with conversion.

There are, however, cases in which relatively broad or even polymodal molecular weight distributions decrease at higher molecular weights [271]. The contribution of chain-breaking reactions in these systems is smaller and the main reason for higher polydispersities is slow exchange between species of either different reactivities or different lifetimes (cf., Section II.D). The number of exchange processes between dormant and active species increases with molecular weight, and therefore more chains have the same probability of growth, leading to the reduction of polydispersities. Thus, when polydispersities decrease significantly with conversion, then slow exchange is responsible for the deviations from the ideal systems [7]. Slow initiation will give the same effect but the highest polydispersity due to slow initiation is $M_w/M_n \approx 1.35$. In systems with slow initiation, molecular weights higher than predicted are observed. In systems with slow exchange, molecular weights might be either higher or equal to those predicted for ideal or living behavior.

D. Reasons for the Deviations from Ideal Behavior

1. Molecular Weights

Molecular weights lower than predicted by the ideal law ($DP_n = \Delta[M]/[I]_0$) indicate the presence of transfer and/or additional initiation with adventitious impurities. There are two reasons for higher than expected molecular weights. The first one is incomplete initiation. This can be due either to slow initiation or to the partial deactivation of the initiator by impurities. In the first case the molecular weights will increase initially much faster than expected and then asymptotically reach those predicted by the ideal low as shown in Figures 2 and 3. Apparent rate constants (slopes in semilogarithmic plots) continuously increase in that case, until they reach those for the system with complete initiation. When the initiator is partially deactivated, the molecular weights increase linearly with conversion, but they are higher than for the ideal system.

The second possibility for higher than expected molecular weights is formation of branched polymers. They can be formed by either grafting-through or grafting-onto processes. Grafting-through occurs when unsaturated chain ends, resulting from transfer of β-protons, subsequently react with growing chains as macromonomers [cf. Eq. (97) in Chapter 3]. In this case the molecular weights increase at relatively high conversion, as reported [166].

Grafting-onto occurs when growing chains react with the polymer backbone; this is possible in polymerization of styrenes:

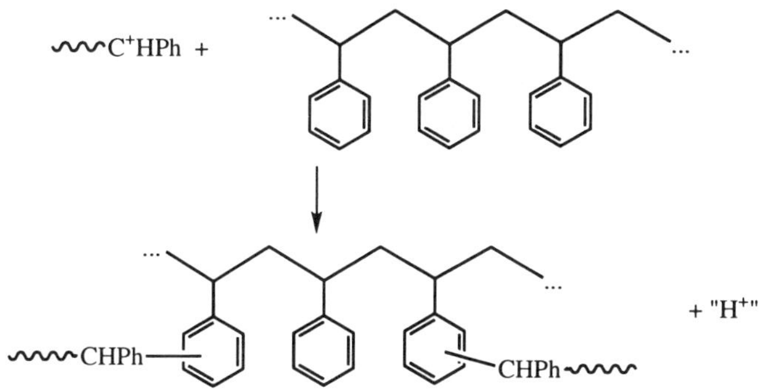

(37)

It must be stressed that the observation of an increase in molecular weight with conversion should not be used as the only criterion of "livingness." For example, termination alone will not affect a linear increase of molecular weights with conversion, because the number of chains will remain constant. However, termination leads to a rate reduction and additionally to a broadening of the MWD with conversion. Therefore, the combination of a linear increase in molecular weight with conversion and successful sequential monomer addition experiments, should be used to demonstrate the occurrence of controlled/living polymerizations.

2. Polydispersity

When a Poisson distribution ($M_w/M_n \approx 1$) is observed, the system must behave as a living one. Incomplete initiation by partial deactivation of the initiator is the only possible side reaction. Polydispersities higher than described by Poisson distribution originate from various side reactions, and only analysis of the MWD together with the evolution of molecular weights with conversion and with kinetics may help to identify the reason for the loss of control.

Slow initiation alone leads to the maximal polydispersity of $M_w/M_n \approx 1.35$. This value decreases significantly when initiation is completed before all monomer is consumed. Slow and imperfect mixing can be another reason for higher polydispersities, as discussed by Szwarc [1].

Transfer and termination lead to the substantial increase of polydispersities with conversion. Transfer to monomer being the only side reaction provides polymers with polydispersities lower than $M_w/M_n \approx 2$. In that case, the rates of propagation and transfer are affected in the same way by the monomer concentration. In the case of transfer to counterion, solvent, or a transfer agent, the rate of propagation decreases with conversion faster than that of transfer. This leads to a significant drop in molecular weights at high conversions and, accordingly, to a strong increase in polydispersities.

Slow exchange may lead to extremely broad polydispersities and often to polymodal MWDs. It is recommended to study the evolution of molecular weights with conversion and especially the proportion and position of various peaks in the MWD by size-exclusion chromatography (SEC). Use of scavengers is helpful in the identification of the origin of peaks in SEC (MWD) traces. For example, salts with common ions suppress free ions and reduce the intensity of peaks formed by free ions. Hindered pyridines trap protons and reduce peaks resulting from protonic initiation, especially in systems with adventitious moisture. Apparently, stability of complex anions MtX_{n+1}^- and MtX_nOH^- can be different and slow exchange may lead to polymodal MWD.

This brief discussion was to emphasize that a complete analysis of new controlled/living cationic systems can not be accomplished by one method and should be studied by a combination of spectroscopic, kinetic, molecular weight, and MWD analysis, as well as by model studies.

VII. MECHANISM OF CONTROLLED/LIVING CARBOCATIONIC POLYMERIZATION

In the previous sections (II–V) of this chapter, we discussed the fundamentals of controlled and living polymerizations, summarized the typical features of carbocationic polymerization, and reviewed the history, phenomenology, and general methodologies of controlled carbocationic systems. In the preceding section (VI), we described the chemistry of these systems in terms of the structure of the active centers, kinetics, molecular weights, and molecular weight distribution of polymers formed. In this section we propose a comprehensive mechanistic picture for controlled/living carbocationic polymerizations with active and dormant species being in a dynamic equilibrium.

The enhanced control can be ascribed to the dynamic "stabilization" of growing cations. The "stabilization" can be understood as an extension of the lifetime of the growing species which only for a small fraction of time are in the carbocationic form and during the vast majority of time they are in the dormant form. The nature and concentrations of various activators and deactivators define the equilibrium position and dynamics of the exchange processes. There are three general approaches to controlled systems (Section IV.B):

1. Use of relatively weak Lewis acids which only partially, but rapidly and reversibly ionize covalent species
2. Use of nucleophiles which lead to the formation of onium ions and the formation of complexes with Lewis acids
3. Use of salts which suppress the free ions and may additionally modify the nature of Lewis acids

A number of variations of these systems are possible. For example, salts with either common anions or other counterions can be added (Section V.A.3); nucleophiles may be a part of the initiating system or could be added externally (Section V.A.2). In a recent monograph on carbocationic living systems [5], seven different cases were discussed as separate approaches without showing their common mechanistic features. In contrast, we would like to propose one unifying picture that fits most, if not all, of the controlled/living carbocationic systems.

A complete mechanistic picture of a polymerization should include structures of the active species participating in all elementary reactions (initiation, propagation, transfer, and termination), a mechanism (in the organic chemist's sense) of all of these reactions with a special emphasis on propagation which is responsible for the construction of nearly the entire macromolecule (except end groups), and an explanation of various structural effects in the monomer, active centers, additives, medium, etc., which affect rates, molecular weights, and MWDs.

Before going into details, let us make a brief statement that propagation in new controlled/living carbocationic systems has nearly the same mechanism as in the conventional systems discussed in Chapter 3, which consists of the electrophilic addition of carbenium ions to alkenes. The main difference is that carbenium ions are in dynamic equilibria with dormant species (covalent esters and onium ions). The correct choice of structures and concentrations of activators and nucleophilic additives as well as those of initiator allows for the preparation of polymers with predetermined molecular weights, low polydispersities, and controlled end functionality, including block copolymers (see Chapter 5).

The design of new controlled systems must take into account four typical features of carbocationic systems discussed in Section III.

Limitation of molecular weights to a level when transfer reactions (β-H^+ elimination) can be neglected; thus, typically $M_n < 20{,}000$, but this limit can be higher at lower temperatures or in some selected systems.

Decrease of the polymerization rate by shifting the equilibrium from active to dormant species; thus, typically $[I]_0 \approx 10^{-3}$ to 10^{-2} mol/L, but $[C^+] \approx 10^{-8}$ to 10^{-6} mol/L.

Sufficiently rapid exchange among species of different reactivities and different lifetimes; thus, the rate of deactivation (return to the dormant species from the cationic counterpart) should be comparable with that of propagation. This can be accomplished by using weak Lewis acids, nucleophiles, salts, as well as nonpolar media.

Well-defined initiators with an activation ability similar to the macromolecular dormant species.

In the following discussions, we will sometimes employ the term "deactivation" in the sense that the active growing carbocation is converted into the corresponding dormant or covalent species (e.g., $\sim\sim\sim C^+, SnCl_5^- \rightarrow \sim\sim\sim C\text{—}Cl + SnCl_4$). Accordingly, added nucleophiles and salts are sometimes referred to as "deactivators."

A. Initiation

Controlled polymerization requires that the initiation rate is at least comparable to that of propagation. Initiation in controlled/living carbocationic systems is usually carried out using models of growing species in their dormant state (e.g., the adducts of a monomer with protonic acids). This enables a similar set of equilibria to be established between carbocations and dormant species for initiation and for propagation. For example, 1-phenylethyl halides have similar reactivity as the macromolecular dormant species in styrene polymerizations, and 1-alkoxyethyl derivatives are as reactive as the macromolecular species in the polymerization of vinyl ethers [Eq. (38)]:

Ionization Ability

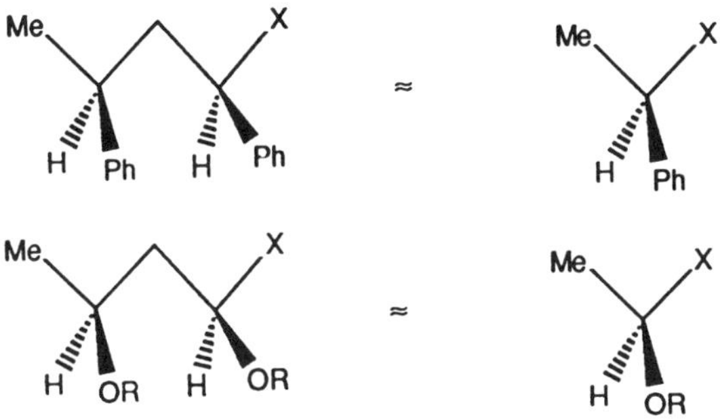

(38)

Sometimes, an apparently ideal model of the growing species such as t-butyl halide for isobutene polymerization and cumyl halide for α-methylstyrene polymerization may not be sufficiently reactive. In both cases the ionization ability and initiation efficiency for the monomeric species is much lower than that for the dimeric/macromolecular species:

Ionization Ability

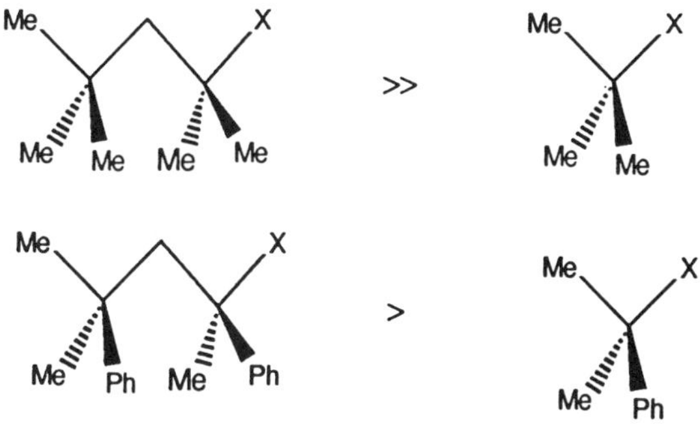

(39)

This is due to B-strain [272] rather than to any additional inductive effects from the alkyl chain as compared to a methyl group.

To assure a sufficient initiation rate, it is recommended to use a compound that ionizes more readily than the macromolecular covalent species. This is the case for 1-alkoxyhaloethanes used as initiators for α-methylstyrene and cumyl halides for isobutene polymerizations [5,40]. The latter system may, however, lead to intramolecular cyclization. Blocking either the ortho or meta position in the aromatic ring improves the efficiency of initiation with cumyl derivatives [5].

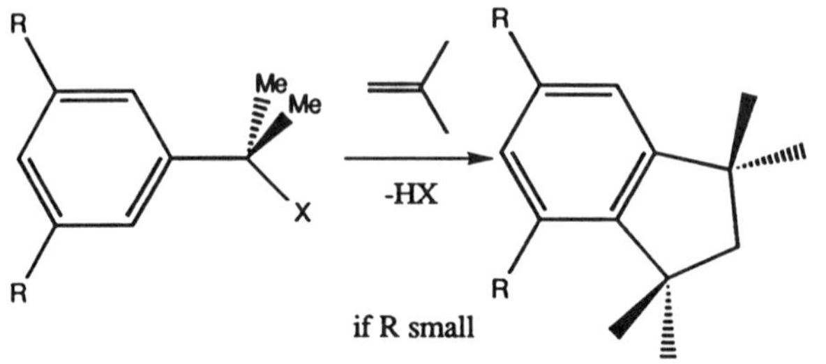

(40)

Initiation efficiency can also be improved by using an initiator with a leaving group better than the one in the growing species. This is the case for alkyl acetates used as initiators with boron halides as Lewis acids as activators [35,117]. The initiating acetates are ionized stronger and faster than macromolecular alkyl halides ($k_i^0 > k_i^n$), resulting in an enhancement of the relative initiation rate:

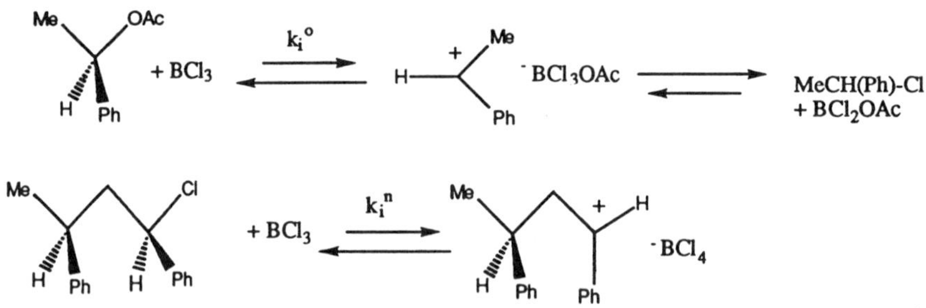

(41)

An excess of BCl_3 is necessary in these polymerizations because BCl_2OAc is a Lewis acid too weak to complete polymerization within a reasonable time. A variation of this approach employs alkyl bromides as initiators and tin tetrachloride as Lewis acid [273].

Another method for enhancing the efficiency of initiation is to deactivate for growing species to a larger degree than the initiator. The use of sulfides as deactivators in the polymerization of vinyl ethers initiated by triflic acid and by trityl hexachloroantimonate are examples [37,38,135].

Sulfides are strong nucleophiles but weak bases and, therefore, they deactivate carbocations more strongly than protonic acids. This reduces the relative rate of propagation and increases the ratio of the rate of initiation to that of propagation. For example, when equimolar amounts of isobutyl vinyl ether and triflic acid are mixed, even at −78° C, a polymer and the unreacted acid are observed. Under similar conditions in the presence of an excess sulfide (R_2S), a monomeric sulfonium salt is the only product:

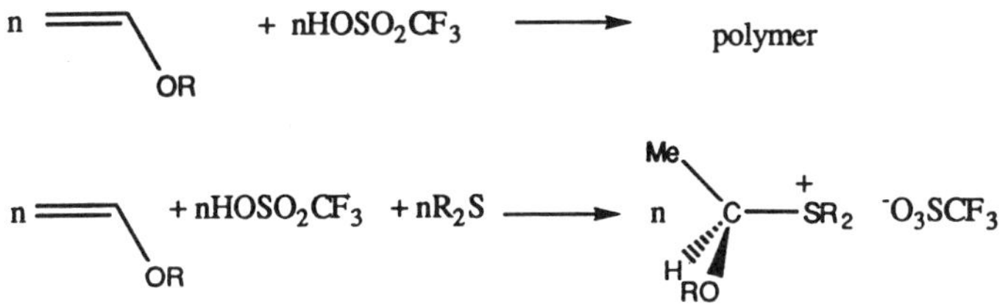

(42)

A similar effect was observed for trityl derivatives. Initiation of vinyl ether polymerization with trityl salts is very slow and often incomplete [257]. This precludes preparation of well-defined polymers with predetermined molecular weights and narrow MWDs. However, polymerization of vinyl ethers initiated by trityl salts in the presence of tetrahydrothiophene leads to controlled polymers [135]. The equilibrium constant for the formation of sulfonium ions is much smaller for trityl salts than for the growing species ($K_i \ll K_p$), which increases the ratio of the apparent initiation to the propagation rate constants a thousand times (Scheme 14):

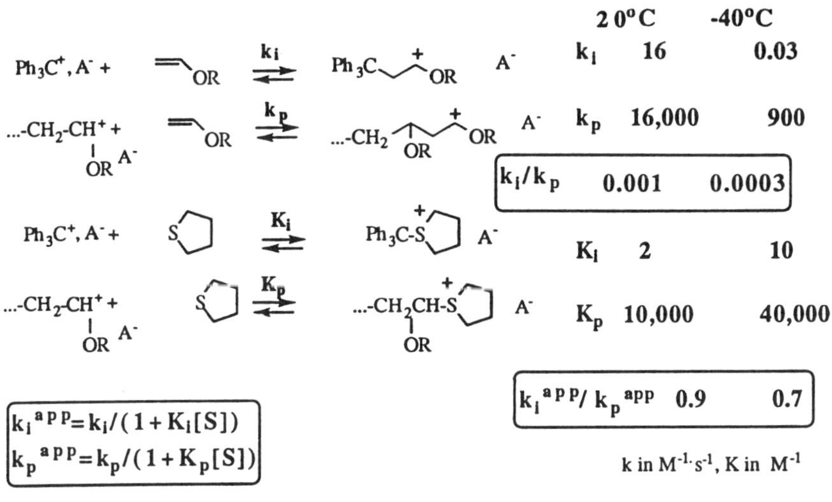

Scheme 14

In summary, initiation in the new controlled/living carbocationic polymerizations is performed using well-defined compounds (initiators) that have similar or higher ability of ionization than the macromolecular dormant species. This can be accomplished by using either more cation-stabilizing substituents in alkyl moiety, better leaving groups, or stronger deactivators for the propagating than the initiating species. Very often the adducts of protonic acids with alkenes are used as initiators (Section V.A.1). The addition of protonic acids to alkenes is usually slow and should be carried out as a separate reaction, and only then should the preformed adduct be used as an initiator. If the addition is carried out in situ, it is recommended to use nucleophiles as deactivators.

B. Propagation

Propagation is the most important elementary reaction in which a macromolecular chain is formed. Control in new carbocationic polymerizations, in which well-defined polymers are prepared, might be explained by new mechanisms of propagation and new types of active centers involved. However, as discussed briefly in Section IV.B.4, we believe that only two types of species with different degrees of ionization are involved: sp^3-hybridized dormant species and sp^2-hybridized carbenium ions [Eq. (43)]:

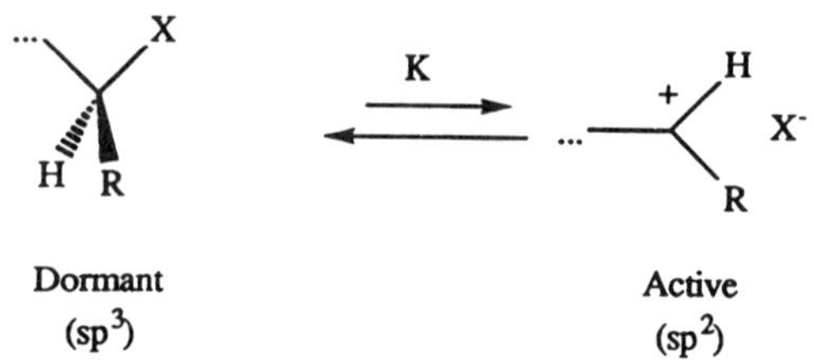

(43)

Covalent species are dormant species that do not react directly with alkenes. The ionization process may be spontaneous or facilitated by activators, most often by Lewis acids. In the case of onium ions, the leaving group **X** is not charged but otherwise Eq. (43) remains the same.

Any species formed on passing from sp^3 covalent species to sp^2 carbocations should not be considered as chemical entities, because their lifetime should be comparable to a bond vibration. Stretched bond isomerism does not exist unless it is associated with a change in spin number, as in some inorganic compounds [274]. Therefore, the often-adopted symbolism [Eq. (44)] should be understood not as an intermediate which can propagate but as a schematic description of dynamically exchanging carbocation and covalent species:

$$\ldots\text{—}C^{\delta+}\text{- - -}^{\delta-}\mathbf{X}\text{Mt}\mathbf{X}_n \tag{44}$$

Carbenium ions can exist as free ions, ion pairs, and, at high concentrations, as aggregated structures:

$$1/n(\ldots\text{-}C^+, \mathbf{X}^-)_n \underset{K_A}{\rightleftharpoons} \ldots\text{-}C^+, \mathbf{X}^- \overset{K_D}{\rightleftharpoons} \ldots\text{-}C^+ + \mathbf{X}^- \tag{45}$$

The precise reactivities of various types of carbenium ions (free, paired, and aggregates) are not known, but it seems that the differences in reactivities are not very large. This may be due to solvation effects, which are similar for all types of carbocations, and also due to the large size of the counterions which interact weakly with the cations. A positive charge in carbocations is located partially on the sp^2-hybridized C atom and is widely delocalized over the β-protons and substituents, especially in the case of aromatic and alkoxy α-substituents.

Propagation proceeds by the electrophilic addition of carbenium ions to double bonds with the regeneration of carbocations. The transition state is relatively late, and it was estimated that approximately half of the charge is transferred into the developing carbocation (Chapter 2). This may explain the fact that dormant species (covalent esters and onium ions) do not react directly with alkenes. The charge on the α-C atoms in the dormant species is not sufficient for the formation of the transition state. A multicenter rearrangement process is additionally disfavored by entropy. In contrast, a two-step process in which carbocations are formed and then very rapidly add to alkenes is free of this difficulty.

It has been proposed [62] that the polymerization of isobutene initiated by alkyl acetates/BCl_3 may occur via a pseudocationic pathway [112], in which the acetate end group is continuously activated by Lewis acids to provide an "insertion" of the monomer without the formation of carbocations:

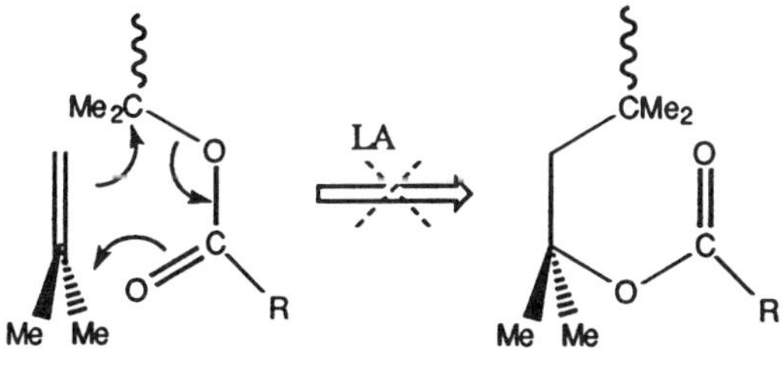

(46)

As discussed in Section V.B.2, it has been shown for model and polymeric systems that ligands exchange rapidly and that the corresponding macromolecular chlorides are formed in addition to BCl_2OAc [cf., Eq. (41)]. Moreover, the growing species in new controlled/living cationic polymerizations and in old conventional carbocationic systems have similar chemo-, regio-, and stereoselectivities. The first observation is related to the contribution of chain-breaking reactions and will be discussed in the next section. The second observation means that in both systems Markovnikov addition and perfect head-to-tail structure almost invariably prevail:

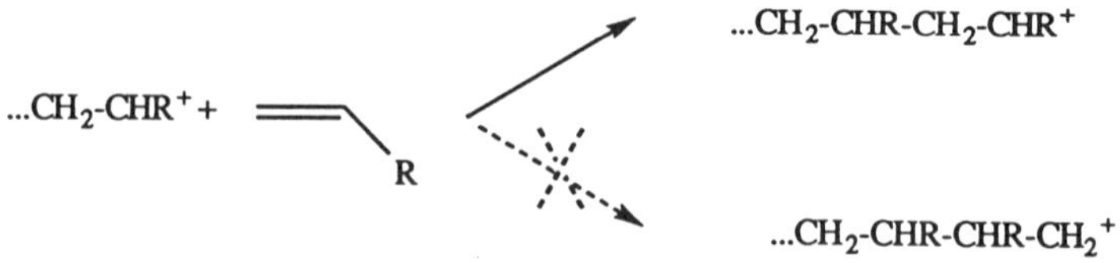

(47)

The last observation means that, in both cases, the control of tacticity is similar and rather poor.

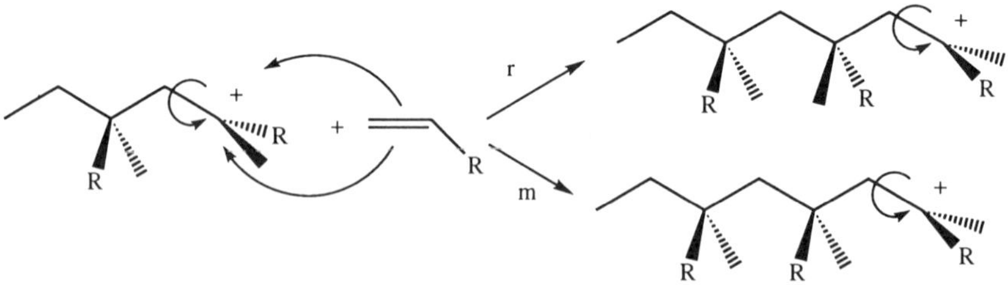

(48)

In carbocations, rotation around . . .CH_2–CHR^+ bond is quite facile. One conformer with the lower steric hindrance might be preferred if the R groups are sufficiently bulky. Monomer may attack the sp^2-hybridized cation from both sides of the nodal plane. Backside attack gives isotactic (*m*) enchainment and frontside attack leads to syndiotactic polymer(s). Stereoselectivity in controlled/living and conventional polymerization is similar, which supports a similar mechanism of propagation. Although only a small degree of control over the microstructure of polymers prepared by carbocationic mechanisms has been achieved so far by changing counterions and nucleophiles, the tacticity control is one of the most challenging areas of research in the future.

C. Transfer and Termination

Transfer and termination should be absent in a living polymerization. As discussed in Chapter 3, termination is rare in carbocationic polymerization. However, transfer is very difficult to avoid, unless at sufficiently low temperatures. There are many systems in which transfer can not be detected at low polymerization degrees but it can be observed clearly for higher molecular weight polymers. These systems have been named as *living* or *apparently living*, because the ratio of rate constants of transfer to propagation may be very similar for "living" and "nonliving" systems [7,76]. Broad HWDs and unpredictable molecular weights observed for the latter category may originate from slow initiation and from slow exchange but not necessarily from transfer and termination.

It is useful to re-examine Chapter 3, Section VI describing conventional carbocationic polymerizations. Termination reactions are not very common and involve either the formation of "too stable" carbocations, unreactive alkyl halides, usually alkyl fluorides, or unreactive onium ions via reaction with very strong nucleophiles (impurities or intentionally added compounds):

$$\ldots CH_2\text{-}CHR\text{-}CH_2\text{-}CHR^+,\ ^-MtX_{n+1} \nearrow \ldots CH_2\text{-}C^+R\text{-}CH{=}CHR,\ ^-MtX_{n+1}\ (+H_2) \quad (A)$$

$$\ldots CH_2\text{-}CHR\text{-}CH_2\text{-}CHR^+,\ ^-MtX_{n+1} \longrightarrow \ldots CH_2\text{-}CHR\text{-}CH_2\text{-}CHR\text{-}X + MtX_n \quad (B)$$

$$\ldots CH_2\text{-}CHR\text{-}CH_2\text{-}CHR^+,\ ^-MtX_{n+1} \xrightarrow{+Nu} \ldots CH_2\text{-}CHR\text{-}CH_2\text{-}CHR\text{-}Nu^+,\ ^-MtX_{n+1} \quad (C)$$

(49)

In controlled/living systems reactions B and C can be avoided or converted into reversible ones, if ligands such as fluorides are not used, if the concentration of moisture is very low in comparison with initiator and if weakly basic/nucleophilic components (additives, counteranions) are used. Contribution of reaction A is reduced at low temperatures but can not be eliminated completely.

Transfer processes can be caused by monomer, counterion, and other components of the reaction mixture (additives, solvents, impurities). The latter reactions are sometimes called spontaneous because they are zero order in monomer. However, the spontaneous elimination of β-protons is very unlikely, and proton elimination must be assisted by some basic reagent. The ratio of the rate constants of β-proton elimination to that of electrophilic addition depends on several factors. The relative rate of transfer decreases with temperature, and therefore polymers with higher molecular weights are formed at sufficiently low temperatures. The effect of solvent and counterion is not yet sufficiently understood.

Elimination is favored in the presence of basic components. It may also be favored in less polar solvents if the anion is the proton abstractor. The effect of ion-pairing and nucleophiles on the relative rate of elimination has not been studied in detail. It is possible that the presence of either a counterion or a nucleophile in the closest neighborhood of the carbocation may polarize the electronic cloud around the cation and may shift the positive charge toward either the sp^2-hybridized carbon atom or the β-H atoms. In the former case, an increase of "livingness" is expected, in the latter case the "livingness" will be reduced. Steric effects may also contribute to the relative rates of transfer and propagation. It has been recently reported that transfer to monomer in the polymerization of IBVE was slightly reduced ($\approx$3 times) in the presence of ethyl acetate [267]. This is the first quantitative information on a reduced transfer coefficient. Nevertheless, the 3-fold change in relative rate constants correspond to a very small difference in activation free energies ($\Delta G^{\ddagger} \approx 2$ kJ/mol) between the two reactions and is too small to be discussed in terms of physical organic chemistry.

D. Copolymerization

As discussed in Chapters 2 and 3, the reactivities of monomers are very strongly affected by substituents and increase in the order: isobutene $\approx$ styrene $\ll$ α-methylstyrene $\ll$ vinyl ethers. This trend is in agreement with the stabilization effect of the electron-donating group at the newly developed carbocation. The sign $\ll$ indicates approximately 10 to 10^2 difference in reactivity. Reactivities of carbocations follow precisely the opposite direction, and the differences are also approximately 10 to 10^2 times. Because propagation is the reaction between a carbocation and monomer both effects cancel one another and give similar rate constants for carbocationic propagation for most monomers ($k_p^+ = 10^{5\pm1}$ $\text{mol}^{-1}\cdot\text{L}\cdot\text{sec}^{-1}$ at 0° C).

The overall polymerization rates and the apparent rate constants of propagation ($k_p^{\text{app}} = R_p/[\text{M}][\text{I}]_0$) for the same initiating system are, however, very different for each class of monomers. For example, the same initiating system, that will polymerize α-methylstyrene (αMeSt) in 1 h, will complete polymerization of vinyl ether within less than 1 min but would require a few days to polymerize styrene and isobutene under otherwise identical conditions. This trend is due to the equilibria between dormant and active species. In this case the apparent rate constant of propagation is the product of the rate constant of propagation (weakly depending on monomer structure) and the ionization constant ($k_p^{\text{app}} \approx k_p^+ \cdot K_I$). This equilibrium constant is much higher for more stable cations derived from vinyl ethers than from αMeSt, than styrene or isobutene.

These equilibria also strongly affect copolymerization. Monomer reactivity ratios in controlled/living systems should be identical to those in conventional cationic copolymerizations, if the comonomers react exclusively with carbocationic species. The equilibrium between active and dormant species should affect the overall kinetics of copolymerization strongly, but it should not affect the copolymer composition and monomer reactivity ratios. The differences in reactivity ratios might reflect either the contribution of different species, the effect of ion pairing on reactivities, or specific solvation by monomer. One comonomer may preferentially assist a dormant species in ionization and react with it as a component of the first solvation shell rather than as a molecule arriving from the "outside." This would indicate that the microscopic environment of the active centers is significantly different from the bulk environment. Such behavior was proposed for the anionic copolymerization of dienes with styrene [275].

Model studies discussed in previous chapters show that the reactivity of cations and alkenes are very strongly affected by inductive and resonance effects in the substituents. Correlation of the rate constants of addition of benzhydryl cation to various styrenes with Hammett σ^+ parameters shows $\rho^+ = -5.0$ [28]. The addition of various substituted benzhydryl cations to a standard alkene (2-methyl-2-pentene) gave also good correlation and $\rho^+ = 5.1$ [28]. The large ρ value signals difficult copolymerizations between alkenes, even of similar structures. Thus, in contrast to radical copolymerization which easily provides random copolymers, cationic systems have a tendency to form either mixtures of two homopolymers or block copolymer (if the cross-over reaction is possible).

There are very few reports on random carbocationic copolymerization in controlled/living systems. *p*-Methoxystyrene was copolymerized with isobutyl vinyl ether using the HI/I_2 initiating system [276]. The molecular weights increased linearly with conversion and a random copolymer was formed. Measurement of the individual polymerization rates of the two comonomers during copolymerization indicated that the vinyl ether was about 10 times more reactive, but the precise values of the reactivity ratios were not reported. In another report it was found that 2-chloroethyl vinyl ether copolymerizes faster than αMeSt in the system activated by $SnBr_4$ [226]. However, the rate of αMeSt homopolymerization was higher than that of the vinyl ether under otherwise identical conditions. This apparent discrepancy was explained by the partial deactivation of the Lewis acid by ether bonds which did not occur in the homopolymerization of αMeSt. Thus, results from random copolymerizations are more meaningful for the determination of relative reactivities.

Differences in reactivity ratios must be taken into account in the synthesis of block copolymers. Synthetic aspects of block copolymerization are discussed in detail in Chapter 5. Ideally, cross-propagation should occur from the more reactive growing species to the more reactive monomer. Here, by more reactive growing species we mean not the more reactive cation but the species that will have a higher apparent rate constant of addition to a standard monomer ($k^{app} \approx k^+ \cdot K_I$). Thus, although the styryl cation is more reactive than the alkoxy-substituted cation, it is present at a relatively much lower concentration, and the apparent propagation rate constant in the polymerization of styrene is smaller than that in the polymerization of vinyl ethers under similar conditions. Because ionization equilibrium constants (K_I) are more sensitive to the structure of substituents than the intrinsic reactivities of the carbenium ions (k^+), the apparent rate constant (k^{app}) of addition of the covalent precursors to a standard alkene ($k^{app} \approx k^+ \cdot K_I$) increases with the reactivity of monomers until approximately half of the species are ionized (cf., Chapter 2). Then, the apparent rate constants start to follow the reactivity trends of pure carbocations. In all controlled/living systems the ionization degree is very low and therefore, under standard conditions, polymerization is faster for more reactive monomers.

Thus, how should block copolymers between styrene and a vinyl ether be prepared? Starting with styrene or with a vinyl ether? In the former system, the propagating styryl cation is intrinsically more reactive but present at much lower concentration. A rough estimate of the ratio of cation reactivities is $\approx 10^3$ but the ratio of carbocations concentrations is $\approx 10^{-5}$. Thus, the ratio of apparent rate constants of addition is 10^{-2}. Macromolecular species derived from styrene should add to a standard alkene one hundred times slower than those derived from vinyl ethers. Thus, one cross-over reaction St* → VE* will be accompanied by ≈ 100 homopropagation steps VE* → VE*. Therefore, in addition to a small amount of block copolymer, a mixture of two homopolymers will be formed. Blocking efficiency should be very low, accordingly.

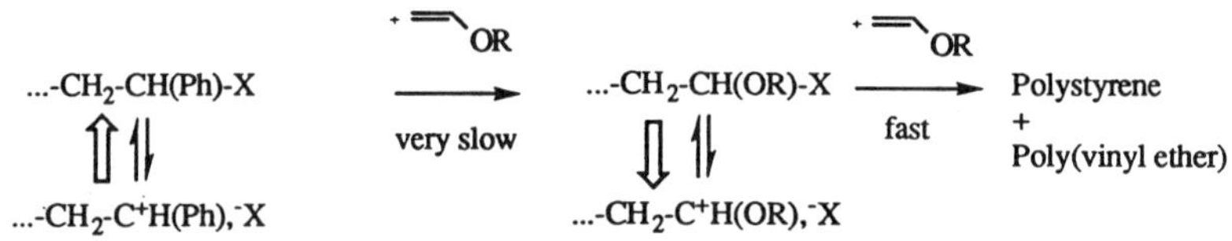

(50)

A hypothetical solution to this problem would be to use a system ionized strongly enough to assure nearly complete ionization of the growing species derived from styrene. Unfortunately, such a system would have an extremely short lifetime (<<1 sec), and successful blocking could not be accomplished due to difficulties with mixing and heat transfer. Recently, another solution has been proposed by adding a strongly ionizable but nonpolymerizable alkene such as 1,1-diphenylethylene [277,278].

Another approach is to use the vinyl ether first and then add styrene, thereby increasing the ionizing power of the system [279]. However, poly(vinyl ether) acts as a nucleophile and deactivates Lewis acid. Therefore it is important to use higher concentrations of Lewis acid relative to that of etheral functions.

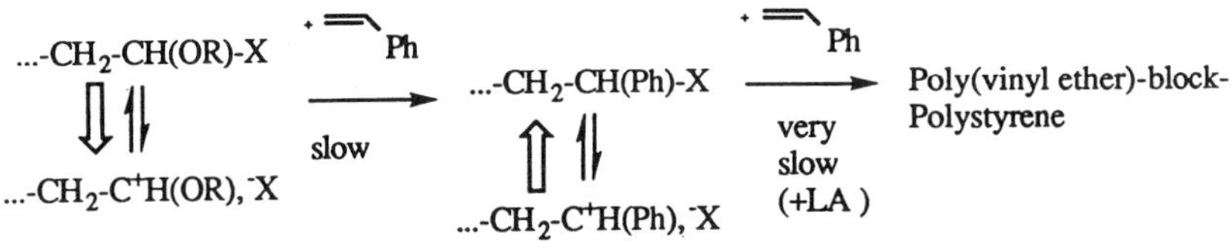

(51)

Indeed, blocking efficiency was much higher in the block copolymerization starting from isobutyl vinyl ether and then adding less reactive *p*-methoxystyrene as a second monomer rather than starting from *p*-methoxystyrene and subsequently adding isobutyl vinyl ether [280]. In the latter case, in addition to a block copolymer, a homopoly(*p*-methoxystyrene) was also observed. Similarly, the synthesis of block copolymers starting from living polyisobutene and then adding isobutyl vinyl ether as a second monomer has low crossover efficiencies (35%), resulting in a mixture of homopolyisobutene and a block copolymer [207,281]. Poor results were also reported when αMeSt was added to living polyisobutene. However, the blocking efficiency was higher and the polydispersities lower when isobutene was added to the growing poly(αMeSt) [282]. In a similar manner polystyrene was successfully block copolymerized with living poly(*p*-methylstyrene) [283].

Block copolymerization between monomers of similar reactivities such as isobutene and various styrenes (styrene, *p*-chloro-, *p*-methyl-, and *p*-*t*-butylstyrenes, and indene) [284–288], as well as αMeSt and 2-chloroethyl vinyl ether [226], involves fewer limitations and is more successful. A detailed description of the copolymerization of α-methylstyrene with 2-chloroethylvilnyl ether has been reported [226]. Because the reactivities of both monomers are similar, AB and BA block copolymers were prepared. However, the enhanced formation of indan derivatives was observed when vinyl ether was used as the first block.

It is important to control the time at which the second monomer is added to avoid formation of a "tapered" middle block between the two homoblocks [161]. If the second monomer is added too early, then "tapered" structures are formed. If the second monomer is added too late, then some chains may terminate and, in addition to block copolymer, homopolymer may be formed.

A-B-A triblock copolymers containing stiff polystyrene end blocks and a flexible polyisobutene central block are very interesting thermoplastic elastomers with higher thermal and oxidative stability than those using polyisoprene and polybutadiene as central soft blocks [161,289].

E. Role of Major Components of Controlled/Living Systems

Controlled/living carbocationic polymerizations occur usually in multicomponent systems with carefully selected initiators, activators, additives (nucleophiles or deactivators), and modifiers. These polymerizations occur in nonbasic solvents, and usually at sufficiently low temperatures, as discussed earlier in this chapter. The roles of each of the components of these systems [290] are discussed below.

1. Covalent Initiators (Alkyl Esters and Halides, RX)

Covalent initiators (R**X**) should rapidly form growing species in the presence of Lewis acids (activators). The structure of R often mimics the structure of the growing species, e.g., 1-phenylethyl derivative for polystyrene or 2,4,4-trimethylpentyl derivative for polyisobutene [117]. Sometimes the covalent precursors ionize easier than the growing species; for example, cumyl derivatives in the polymerization of isobutene and alkoxy derivatives in the polymerization of α-methylstyrene [5,40]. If, due to steric, strain or electronic effects, the covalent precursor is less reactive than a growing chain, like *t*-butyl halide in the polymerization of isobutene, then the leaving group **X** should be more nucleofugic than the leaving group in the growing chain. This is the case for acetate derivatives in the polymerization of styrene and isobutene [117].

2. Lewis Acids

Covalent species (R**X**) react reversibly with Lewis acids (Mt$\mathbf{Y}_n$) to form carbocations with complex counteranions:

$$\mathrm{R{-}}\mathbf{X} + \mathrm{Mt}\mathbf{Y}_n \underset{}{\stackrel{K}{\rightleftharpoons}} \mathrm{R}^+,\ {}^-\mathrm{Mt}\mathbf{XY}_n \qquad (52)$$

The position of the equilibrium affects the overall polymerization rate, but the dynamics of exchange influences the polydispersity of the resulting polymers [cf., Section VI.B.1)]. The strength of a Lewis acid and its concentration should be adjusted for each particular monomer. Thus, styrene and isobutene require relatively strong Lewis acids such as $SnCl_4$, $TiCl_4$, or BCl_3, whereas vinyl ethers can be polymerized using iodine or zinc halides (Sections IV and V). Lewis acids should not contain ligands **Y** that form strong C—**Y** bonds, such as fluorine. Ligands **Y** and **X** should be very labile. The strength of Lewis acids may be adjusted by using nucleophilic additives. For example, esters can coordinate to Lewis acids, such as $AlCl_3$, and form strong complexes [291]. Lewis acids should be strong enough to complete polymerization in a reasonable time and should form sufficiently nucleophilic anions to ensure adequate dynamics of exchange leading to polymers with low polydispersities.

3. *Salts*

Salts induce three types of effects: an increase in ionic strength, the suppression of free ions, and the special salt effect including anion/ligand exchange (see also Section VI.B.3). The common ion effect is most often observed in ionic polymerizations. Free ions are usually present at low concentrations, and their lifetime is long enough for the growth of high polymer with a molecular weight that is often limited by transfer. It is enough to add very small amount of salts (10^{-3} mol/L) to shift the equilibrium entirely to the side of ion pairs [71]. Thus, polymers with monomodal distribution usually are formed in the presence of salts [39,205,215].

In some systems it is necessary to add a large amount of salts to obtain polymers with low polydispersities. This happens when salts participate in ligand/anion exchange (special salt effect) and when they enhance ionization of covalent compounds through the increase of ionic strength. The special salt effect may either reduce or enhance ionization. Strong rate increases observed in the polymerization of isobutyl vinyl ether initiated by an alkyl iodide in the presence of tetrabutylammonium perchlorate or triflate can be explained by the special salt effect [109]. The reduction in polymerization rate of cyclohexyl vinyl ether initiated by its HI adduct in the presence of ammonium bromide and chloride can be also ascribed to the special salt effect [33]. The breadth of MWD depends on the relative rate of conversion of ion pairs to covalent species and is affected by the structure of the counterions.

4. *Nucleophiles (Donors)*

Nucleophiles (or electron donors) may react with cationic species in three different ways (see Section VI.B.2; see also Table 2, Section V.A.2 for examples). They can reversibly form onium ions, complexes and covalent species, and if they are basic enough, they may eliminate β-protons [cf., Chapter 3, Eq. (131)]. In the first reaction, nucleophiles are used to control the polymerization rate because onium ions are inactive dormant species. They can be added at concentrations higher than salts, and can considerably reduce the lifetimes of both unpaired cations and ion pairs by converting them into dormant onium species. In the second reaction, nucleophiles can also affect the polymerization rates by coordination to Lewis acids and reducing their strength. Both reactions are beneficial for controlled polymerization. The third reaction, favored by strongly basic species, should be avoided.

Some nucleophiles may complex with growing species stronger than with an initiator and may increase the ratio of the rate of initiation to propagation. This is the case for sulfides in polymerization of IBVE initiated by triflic acid and by trityl salts [37,38,135]. The assistance of the nucleophile was proposed by Penczek to accelerate the transformation of covalent species to carbocations and reduce polydispersities [92].

Complexes of nucleophiles with Lewis acids have ambivalent reactivities. The nucleophilicity of the complex is weaker than that of the pristine nucleophile, and the electrophilicity is weaker than that of the original Lewis acid. The complexes lead to different equilibrium constants between carbocations and dormant species and to different dynamics of exchange in comparison with systems with either nucleophiles or Lewis acids alone. For example, the addition of less than an equimolar amount of strong nucleophiles such as amides or sulfoxides to the polymerization of isobutene initiated by $RCl/TiCl_4$ system decreases the polydispersities. Noncomplexed $TiCl_4$ is responsible for ionization, because it is a strong Lewis acid, but $DMSO{\cdot}TiCl_4$ can act as a weak nucleophile. DMSO alone is a too powerful nucleophile (deactivator) and inhibits styrene polymerization initiated by triflic acid. The complex has lower nucleophilicity but is still sufficient for the reversible deactivation of growing carbocations. The complex is not electrophilic enough in comparison with the "naked" Lewis acid to ionize covalent species, and therefore the system can be described by the highlighted upper part of Scheme 15.

Another possible explanation for the action of strong nucleophiles in these systems involves the rapid exchange of ligands between LAX^- and LA : Nu, leading to a faster collapse of the ion pairs to the covalent species.

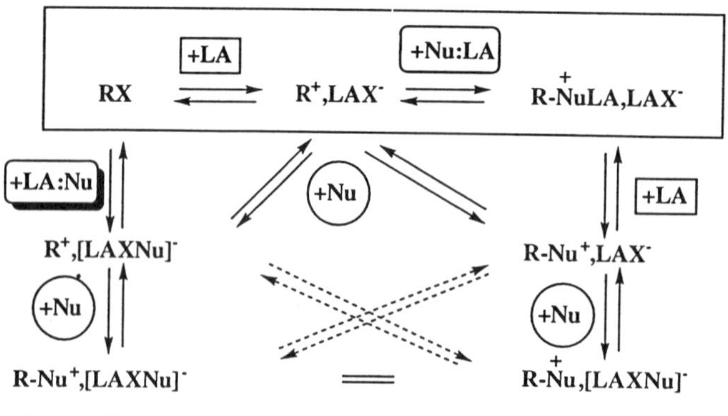

Scheme 15

In the polymerizations of more reactive monomers such as vinyl ethers, an excess of nucleophile is often used (Section V.A.2) [64,102,131]. This indicates that the complex between Lewis acid and the nucleophile has a sufficient reactivity to ionize covalent species. Excess nucleophile acts as an additional deactivator to convert carbocations to onium ions. In that case, the concentration of "naked" Lewis acid may be too low to efficiently assist the ionization and the reaction can be described by the highlighted left part of Scheme 16.

5. Solvents

In controlled/living polymerizations, solvents influence the ionization equilibria and the rates of exchange of growing species through their polarity (estimated from their dielectric constants), but they may also interact in a more specific way with the intermediates by the formation of onium ions. Solvents of low dielectric constants are generally recommended, because ion pairs are less dissociated therein. However, initiation may also be slower and incomplete. On the other hand, the rate of the unimolecular transfer, a reaction between an anion and a β-proton of the growing end (ion–ion reaction), is accelerated in less polar solvents in comparison with propagation, which is a reaction between an ion and a dipole. Thus, in less polar solvents, lower polydispersity due to faster exchange could be observed, but lower molecular weights and higher polydispersity due to stronger transfer could also be observed.

Solvents must be purified and dried in appropriate ways. Water may act as an adventitious initiator, disabling the control of molecular weights [271]. Additional complications may take place when the complex anions

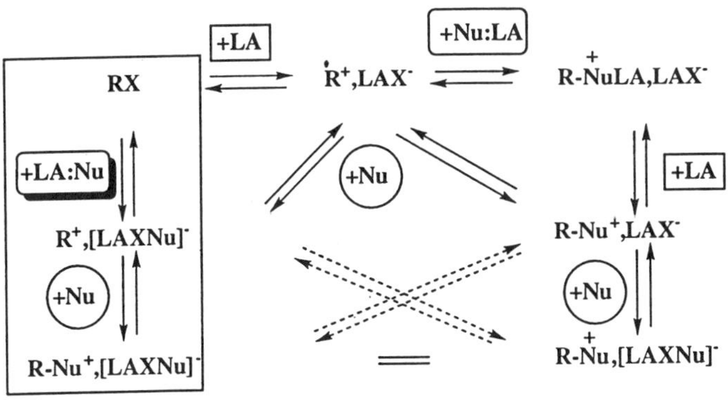

Scheme 16

with OH^- ligand have different stabilities and lifetimes than those with ligands provided by the initiator and the Lewis acid. Water may also act as a terminator or at least as a transfer reagent. The relative reactivities of water and alkenes towards carbocation are similar (water is only 10 times more reactive than vinyl ethers toward benzhydryl cations [292]). Thus, if the water concentration is maintained at a level below 10 mol % of the initiator, well-defined polymers can still be prepared, because $[M]_0/[H_2O]_0 \approx 1000$.

6. *Temperature*

β-Proton elimination (transfer) has a higher activation energy than electrophilic addition (initiation and propagation). Thus, higher molecular weight polymers are formed at lower temperatures, due to reduced transfer. Temperature affects all elementary reactions, and it might happen that at lower temperatures, polymers with broader MWDs can be formed. This can be the case for systems with slow initiation and slow exchange. In the conventional cationic polymerization of styrene, Friedel-Crafts alkylation (e.g., indan formation) has a lower activation energy than propagation, and at very low temperatures it may become the dominating transfer reaction. In such a case, the decrease of temperature should initially increase molecular weights but then may reduce them when Friedel-Crafts alkylation will become faster than elimination [81].

7. *Other Components*

Hindered pyridines can act in three different ways [150,293–295]. The first is to trap protonic impurities and prevent adventitious initiation by water. This improves the control of molecular weights. The second role is similar to that of salts with common anions. The pyridinium salts formed in the system are accompanied by complex anions which may scavenge free ions in a similar manner as tetrabutylammonium salts. Hindered pyridines may also act as nucleophiles (or donors) and interact with some Lewis acids. These interactions will be directed toward the aromatic ring rather than the nitrogen atom which is protected by bulky *tert*-butyl groups in ortho position [293].

8. *Monomer Concentration*

An increase in the initial monomer concentration, $[M]_0$, should provide higher molecular weights if transfer to counterion dominates. However, a high concentration of monomer may affect the dielectric constant of the reaction mixture in the way discussed in Section VII.E.5. When polymerization is carried out at very low monomer concentration, then the rate of propagation is reduced in comparison to the rate of exchange between various active sites. This might lead to narrower MWDs, as observed in the so-called quasi-living polymerization [57]. A similar approach is often used in new controlled anionic polymerizations of acrylates, such as group transfer polymerization [114] and lithium salts-mediated systems [20,115,296].

F. Overview

In concluding this chapter, we summarize the general requirements for achieving controlled/living carbocationic polymerizations. One of the advantages of these systems is the possibility of preparation of polymers with predetermined molecular weights and narrow molecular weight distributions. The low contribution of chain-breaking reactions is a necessary but not sufficient requirement for the preparation of well-defined polymers. In addition, all of the chains should start growing simultaneously and have a similar probability of growth.

Relative to these factors, as seen in the preceding discussions, there are four basic requirements for the synthesis of well-defined polymers:

1. Limitation of molecular weights to levels where chain-breaking reactions, if any, can be neglected (e.g., $Dp_n < 0.1\ k_p/k_{trM}$)
2. Rate of polymerization should be adjusted to a range convenient for synthetic manipulations (half-lifetimes in monomer consumption $\approx$ minutes to hours)
3. Initiation is fast enough to control molecular weights and end groups ($R_i \geq R_p$)
4. There is rapid exchange between various types of active species ($R_{exch} \geq R_p$).

All components of controlled systems must be selected with respect to their low basicity in order to decrease the possibility of β-proton elimination from growing carbocations.

The four aforementioned requirements can be achieved in the following ways.

1. After the proper initiator is selected, it should be used at concentrations that yield polymers with desired molecular weights in a range not marked by transfer/termination. This range strongly depends on temperature and on other components of the system.

2. The equilibrium position between carbocations and dormant species (covalent or onium) should be adjusted to provide convenient rates (i.e., an appropriately low concentration of the growing carbocation at a given time). For a particular monomer, the equilibrium position (and overall rate) depends on the nature and concentrations of the leaving group **X** in the initiator **RX**, of the activator (Lewis acid), and of the deactivator

(added nucleophile or salt), as well as on the reaction medium (solvent and temperature).

3. The initiators should have an ability for ionization similar to that of the dormant end groups and should provide carbocations of reactivity similar to the growing carbocations. Trityl cation, although fully ionized, is not reactive enough in the polymerization of most alkenes. Additional manipulation is possible by varying the structure of the leaving group in the initiator.

4. One of the most underestimated problems in the cationic polymerization of alkenes is related to the multiplicity of active sites and their slow interconversions [21,22]. Monomodal and monodisperse polymers can be prepared only if exchange is fast enough [23]. The dynamics of exchange between active and dormant species can be enhanced by more nucleophilic *and* more nucleofugic counteranions and additives. Namely, complex anions derived from Lewis acids should contain the appropriate ligands to assure rapid exchange between carbocations and dormant species. Nucleophiles may decrease the strength of Lewis acids and accelerate exchange via the formation of not only onium ions but also covalent species. The lifetime of free carbocations can be reduced by addition of various salts (for example, common anions). These three methodologies have been described for a variety of monomers in this chapter (Sections III–V), and all of the initiating systems for controlled/living polymerizations are designed to fulfill the above criteria.

Thus, various components of controlled/living systems reduce the overall polymerization rate and accelerate the exchange between active and dormant species. Usually they do not decrease the ratio of the rate constants of transfer to propagation. New "living" and conventional carbocationic systems behave mechanistically in the same way, and it is not necessary to postulate new types of active species and new modes of propagation with "stretched" covalent bonds or "stabilized" carbocations. The apparent "stabilization" of the growing carbocations should be considered as a dynamic phenomenon reducing the momentary concentration of carbocations but not affecting their structure and reactivity. The preparation of well-defined polymers in new controlled/living systems originates not in a new mechanism but in the proper selection of several components of the polymerization system which provide rapid initiation, slow overall polymerization, and the dynamic equilibration of active and dormant species (covalent and onium ions). The development of controlled/living carbocationic polymerizations, mechanistically similar to conventional systems, has shed new light not only on the synthetic possibilities of carbocationic polymerization but also on the understanding of the thermodynamics and kinetics of equilibria between carbocations and dormant species.

REFERENCES

1. M. Szwarc, *Carbanions, Living Polymers and Electron Transfer Processes*, Wiley, New York, 1968.
2. G. V. Schultz, *Chem. Technol.* 220 (1973).
3. S. Bywater, *Adv. Polym. Sci. 4*: 66 (1965).
4. M. Morton, *Anionic Polymerization: Principles and Practice*, Academic Press, New York, 1983.
5. J. P. Kennedy and B. Iván, *Designed Polymers by Carbocationic Macromolecular Engineering: Theory and Practice* Hanser, München, 1992.
6. K. Matyjaszewski, *Macromolecules 26*: 1787 (1993).
7. K. Matyjaszewski and P. Sigwalt, *Polym. Int. 35*: 1 (1994).
8. L. Gold, *J. Chem. Phys. 28*: 91 (1958).
9. M. Litt, *J. Polym. Sci. 58*: 429 (1962).
10. V. S. Nanda and R. K. Jain, *Trans. Farad. Soc. 60*: 949 (1964).
11. J. L. H. Peebles, *Molecular Weight Distributions in Polymers*, Wiley, New York, 1971.
12. R. Chiang and J. J. Hermans, *J. Polym. Sci. A1, 4*: 2843 (1966).
13. D. Yan and C. Yuan, *J. Macromol. Sci. Chem. A23*: 781 (1986).
14. P. Sigwalt, *Makromol. Chem., Macromol. Symp. 47*: 179 (1991).
15. C. Yuan and D. Yan, *Makromol. Chem. 187*: 2629 (1986).
16. K. Matyjaszewski, *J. Phys. Org. Chem. 8*: 197 (1995).
17. C. Yuan and D. Yan, *Makromol. Chem. 188*: 341 (1987).
18. D. Greszta, D. Mardare and K. Matyjaszewski, *Macromolecules 27*: 638 (1994).
19. R. V. Figini, *Makromol. Chem. 107*: 170 (1967).
20. D. Kunkel, A. H. E. Mueller, M. Janata and L. Lochman, *Makromol. Chem. Macromol. Symp. 60*: 315 (1992).
21. K. Matyjaszewski and C.-H. Lin, *Makromol. Chem., Macromol. Symp. 47*: 221 (1991).
22. J. E. Puskas, G. Kaszas and M. Litt, *Macromolecules 24*: 5278 (1991).
23. K. Matyjaszewski, R. Szymanski and M. Teodorescu, *Macromolecules 27*: 7565 (1994).
24. K. Matyjaszewski, *J. Polym. Sci. Part A Polym. Chem. 31*: 995 (1993).
25. T. Higashimura, Y. Ishihama and M. Sawamoto, *Macromolecules 26*: 744 (1993).
26. M. Sawamoto, M. Masuda and T. Higashimura, *Makromol. Chem. 177*: 2995 (1976).
27. K. Matyjaszewski, *Makromol. Chem., Macromol. Symp. 13/14*: 433 (1988).
28. H. Mayr, *Angew. Chem. Int. Ed. Engl. 29*: 1371 (1990).
29. K. Matyjaszewski, *Makromol. Chem., Macromol. Symp. 54/55*: 51 (1992).
30. P. H. Plesch, *Prog. React. Kinet. 18*: 1 (1993).
31. T. Kunitake and K. Takarabe, *Macromolecules 12*: 1061 (1979).
32. D. C. Pepper, *Makromol. Chem. 175*: 1077 (1974).
33. H. Cramail and A. Deffieux, *Macromolecules 27*: 1401 (1994).
34. T. Higashimura, M. Miayamoto and M. Sawamoto, *Macromolecules 18*: 611 (1985).

35. R. Faust and J. P. Kennedy, *Polym. Bull. 15*: 317 (1986).
36. T. Higashimura, S. Aoshima and M. Sawamoto, *Makromol. Chem., Macromol. Symp. 13/14*: 457 (1988).
37. C. G. Cho, B. A. Feit and O. W. Webster, *Macromolecules 23*: 1918 (1990).
38. C.-H. Lin and K. Matyjaszewski, *Polym. Prep. Div. Polym. Chem. Am. Chem. Soc. 31(1)*: 599 (1990).
39. C.-H. Lin, J. S. Xiang and K. Matyjaszewski, *Macromolecules 26*: 2785 (1993).
40. T. Higashimura, M. Kato, T. Hasebe, M. Kamigaito and M. Sawamoto, *Macromolecules 26*: 2670 (1993).
41. K. Matyjaszewski, C. H. Lin, A. Bon and J. S. Xiang, *Macromol. Symp. 85*: 65 (1994).
42. M. Sawamoto, *Prog. Polym. Sci. 16*: 111 (1991).
43. M. Sawamoto, S. Aoshima and T. Higashimura, *Makromol. Chem., Macromol. Symp. 13/14*: 513 (1988).
44. T. Higashimura and O. Kishiro, *Polym. J. 9*: 87 (1977).
45. T. Higashimura and O. Kishiro, *Polym. Prep. Jpn. 24(2)*: 106 (1975).
46. T. Masuda and T. Higashimura, *J. Polym. Sci. Part B 9*: 783 (1971).
47. T. Higashimura and O. Kishiro, *J. Polym. Sci. Polym. Chem. Ed. 12*: 967 (1974).
48. T. Higashimura, *J. Polym. Sci. Polym. Symp. 56*: 71 (1976).
49. T. Higashimura and M. Sawamoto, *Adv. Polym. Sci. 62*: 49 (1984).
50. D. C. Pepper, *J. Polym. Sci. Polym. Symp. 50*: 51 (1975).
51. M. Sawamoto and T. Higashimura, *Macromolecules 11*: 501 (1978).
52. T. Higashimura, M. Mitsuhashi, and M. Sawamoto, *Macromolecules 12*: 178 (1979).
53. T. Ohtori, Y. Hirokawa, and T. Higashimura, *Polym. J. 11*: 471 (1979).
54. T. Higashimura, M. Miyamoto and M. Sawamoto, *Polym. Prep. Jpn. 31(2)*: 200 (1982).
55. A. F. Johnson and R. N. Young, *J. Polym. Sci. Polym. Symp. 56*: 211 (1976).
56. P. K. Bossaer, E. J. Goethals, P. J. Hackett and D. C. Pepper, *Eur. Polym. J. 13*: 489 (1979).
57. J. P. Kennedy, ed., *J. Macromol. Sci. Chem. A18*: 1185–1382 (1982/83).
58. M. Miyamoto, M. Sawamoto and T. Higashimura, *Macromolecules 17*: 265 (1984).
59. M. Miyamoto, M. Sawamoto, and T. Higashimura, *Macromolecules 17*: 2228 (1984).
60. P. Giusti, G. Puce and F. Andruzzi, *Makromol. Chem. 98*: 170 (1966).
61. P. Giusti and F. Andruzzi, *J. Polym. Sci. C 16*: 3797 (1968).
62. R. Faust and J. P., Kennedy, *J. Polym. Sci. Polym. Chem. Ed. 25*: 1847 (1987).
63. M. Sawamoto and T. Higashimura, *Makromol. Chem., Macromol. Symp. 3*: 83 (1986).
64. T. Higashimura, M. Sawamoto, S. Aoshima, Y. Kishimoto, and E. Takeu-

chi, *Frontiers of Macromolecular Science* (T. Saegusa, T. Higashimura, and A. Abe, eds.), Blackwell, Oxford, 1989, p. 67.
65. T. Higashimura and M. Sawamoto, *Comprehensive Polymer Science* (G. C. Eastmond, A. Ledwith, S. Russo, and P. Sigwalt, eds.), Pergamon Press, Oxford, 1989, vol. 3, Chap. 42.
66. M. Sawamoto and T. Higashimura, *Encyclopedia of Polymer Science and Engineering*, John Wiley, New York, 1989, Suppl. Vol., 2nd Ed, p. 399.
67. O. W. Webster, *Science 251*: 887 (1991).
68. M. Sawamoto, *Trends Polym. Sci. 1*: 111 (1993).
69. M. Sawamoto and T. Higashimura, *Makromol. Chem., Macromol. Symp. 47*: 67 (1991).
70. M. Sawamoto and T. Higashimura, *Makromol. Chem., Macromol. Symp. 54/55*: 41 (1992).
71. M. Sawamoto and T. Higashimura, *Makromol. Chem., Macromol. Symp. 60*: 47 (1992).
72. M. Sawamoto and T. Higashimura, *Makromol. Chem., Macromol. Symp. 67*: 299 (1993).
73. M. Sawamoto, *Makromol. Chem., Macromol. Symp. 85*: 33 (1994).
74. M. Szwarc and M. V. Beylen, *Ionic Polymerization and Living Polymers* Chapman & Hall, New York, 1993.
75. M. Szwarc, *Makromol. Chem. Rapid Commun. 13*: 141 (1992).
76. P. Sigwalt, *Makromol. Chem., Macromol. Symp. 47*: 179 (1991).
77. P. Sigwalt, *J. Macromol. Sci. Chem. A28*: 949 (1991).
78. P. H. Plesch, *Polym. Bull. 19*: 145 (1988).
79. K. Matyjaszewski, *J. Polym. Sci. Polym. Chem. Ed. 31*: 995 (1993).
80. K. Matyjaszewski and C. Pugh, *Makromol. Chem., Macromol. Symp. 67*: 67 (1993).
81. K. Matyjaszewski, C.-H. Lin and C. Pugh, *Macromolecules 26*: 2649 (1993).
82. E. J. Goethals, *J. Polym. Sci. Polym. Symp. 56*: 271 (1976).
83. P. Rempp, *Makromol. Chem., Macromol. Symp. 67*: 1 (1993).
84. S. Penczek, P. Kubisa, and R. Szymanski, *Makromol. Chem., Rapid Commun. 12*: 77 (1991).
85. W. O. Choi, M. Sawamoto and T. Higashimura, *Macromolecules 23*: 48 (1990).
86. K. Matyjaszewski, *J. Polym. Sci. Part A Polym. Chem. 25*: 765 (1987).
87. P. H. Plesch, *Makromol. Chem., Macromol. Symp. 60*: 11 (1992).
88. K. Matyjaszewski, *Makromol. Chem., Macromol. Symp. 60*: 107 (1992).
89. Y. H. Kim, *Macromolecules 24*: 2122 (1991).
90. M. Kamigaito, M. Sawamoto and T. Higashimura, *Macromolecules 25*: 2587 (1992).
91. R. Faust, B. Iván and J. P. Kennedy, *J. Macromol. Sci. Chem. A28*: 1 (1991).
92. S. Penczek, *Makromol. Chem., Rapid Commun. 13*: 147 (1992).
93. S. Penczek, *Makromol. Chem., Macromol. Symp. 60*: 65 (1992).
94. M. Szwarc, *Makromol. Chem., Macromol. Symp. 67*: 83 (1993).

95. C. G. Cho and J. E. McGrath, *J. Macromol. Sci. Chem. A25*: 499 (1988).
96. M. Sawamoto and T. Higashimura, *Makromol. Chem., Macromol. Symp. 32*: 131 (1990).
97. V. Héroguez, A. Deffieux, and M. Fontanille, *Polym. Bull. 18*: 287 (1987).
98. K. Kojima, M. Sawamoto, and T. Higashimura, *Macromolecules 22*: 1552 (1989).
99. M. Sawamoto, C. Okamoto, and T. Higashimura, *Macromolecules 20*: 2693 (1987).
100. S. Aoshima and T. Higashimura, *Polym. Bull. 15*: 417 (1986).
101. S. Aoshima and T. Higashimura, *Macromolecules 22*: 1009 (1989).
102. T. Higashimura, Y. Kishimoto, and S. Aoshima, *Polym. Bull. 18*: 111 (1987).
103. Y. Kishimoto, S. Aoshima, and T. Higashimura, *Macromolecules 22*: 3877 (1989).
104. M. Kamigaito, M. Sawamoto, and T. Higashimura, *Macromolecules 24*: 3988 (1991).
105. M. Kamigaito, Y. Maeda, M. Sawamoto, and T. Higashimura, *Macromolecules 26*: 1643 (1993).
106. G. Kaszas, J. E. Puskas, and J. P. Kennedy, *Polym. Bull. 20*: 413 (1988).
107. K. Kojima, M. Sawamoto, and T. Higashimura, *Macromolecules 23*: 948 (1990).
108. M. Sawamoto, J. Fujimori, and T. Higashimura, *Macromolecules 20*: 916 (1987).
109. O. Nuyken, H. Kröner, and S. Aechtner, *Makromol. Chem., Macromol. Symp. 32*: 181 (1990).
110. T. Higashimura, T. Takeda, O. Kishiro, and M. Sawamoto, *J. Polym. Sci. Polym. Chem. Ed. 16*: 503 (1978).
111. M. Sawamoto, T. Masuda, T. Higashimura, S. Kobayashi, and T. Saegusa, *Makromol. Chem. 178*: 1497 (1977).
112. P. H. Plesch, *Makromol. Chem., Macromol. Symp. 13/14*: 393 (1988).
113. A. Gandini and P. H. Plesch, *J. Chem. Soc.* 4826 (1965).
114. O. W. Webster, *Makromol. Chem., Macromol. Symp. 60*: 287 (1992).
115. R. Fayt, R. Forte, R. Jacobs, R. Jerome, T. Ouhadi, P. Teyssie, and S. K. Varshney, *Macromolecules 20*: 1442 (1987).
116. K. Matyjaszewski, Proceedings Sympol-90 Paris, France, 1990, p. 63.
117. K. Matyjaszewski and C.-H. Lin, *J. Polym. Sci. Part A Polym. Chem. 29*: 1439 (1991).
118. M. Miyamoto, M. Sawamoto, and T. Higashimura, *Macromolecules 18*: 123 (1985).
119. T. Enoki, M. Sawamoto, and T. Higashimura, *J. Polym. Sci. Polym. Chem. Ed. 24*: 2261 (1986).
120. M. Sawamoto, K. Kojima, and T. Higashimura, *Makromol. Chem. 192*: 2479 (1991).
121. V. Héroguez, A. Deffieux, and M. Fontanille, *Makromol. Chem., Macromol. Symp. 32*: 199 (1990).
122. M. Schappacher and A. Deffieux, *Macromolecules 24*: 2140 (1991).
123. M. Schappacher and A. Deffieux, *Macromolecules 24*: 4221 (1991).

124. H. Cramail and A. Deffieux, *Makromol. Chem. 193*: 2793 (1992).
125. Y. H. Kim and T. Heitz, *Makromol. Chem. Rapid Commun. 11*: 525 (1990).
126. M. Kamigaito, K. Yamaoka, M. Sawamoto, and T. Higashimura, *Macromolecules 25*: 6400 (1992).
127. M. Sawamoto, M. Kamigaito, and T. Higashimura, *Polym. Bull. 20*: 407 (1988).
128. M. Kamigaito, M. Sawamoto, and T. Higashimura, *J. Polym. Sci. Polym. Chem. 31*: 2987 (1993).
129. O. Nuyken and H. Kröner, *Makromol. Chem. 191*: 1 (1990).
130. G. Wu, X. Han, and B. Cheng, *Polym. Bull. 22*: 341 (1989).
131. J.-M. Leleu, M. Tardi, A. Polton, and P. Sigwalt, *Makromol. Chem., Macromol. Symp. 47*: 253 (1991).
132. S. Aoshima, K. Shachi, E. Kobayashi, and T. Higashimura, *Makromol. Chem. 192*: 1749 (1991).
133. S. Aoshima, K. Shachi, and E. Kobayashi, *Makromol. Chem. 192*: 1759 (1991).
134. C. G. Cho, B. A. Feit, and O. W. Webster, *Macromolecules 25*: 2081 (1992).
135. C.-H. Lin and K. Matyjaszewski, *Polym. Prep. Div. Polym. Chem. Am. Chem. Soc. 34(2)*: 398 (1993).
136. S. Aoshima, Y. Ito, and E. Kobayashi, *Polym. J. 25*: 1161 (1993).
137. H. Shohi, M. Sawamoto, and T. Higashimura, *Polym. Bull. 21*: 357 (1989).
138. T. Yoshida, M. Sawamoto, and T. Higashimura, *Makromol. Chem. 192*: 2317 (1991).
139. H. Shohi, M. Sawamoto, and T. Higashimura, *Macromolecules 24*: 4926 (1991).
140. M. Sawamoto, H. Shohi, H. Sawamoto, and T. Higashimura, *J. Macromol. Sci. Pure Appl. Chem. A31*: 1609 (1994).
141. C. H. Lin and K. Matyjaszewski, *Polym. Prep. Div. Polym. Chem. Am. Chem. Soc. 35(1)*: 462 (1994).
142. T. Higashimura, S. Okamoto, Y. Kishimoto, and S. Aoshima, *Polym. J. 21*: 725 (1989).
143. M. Zsuga and J. P. Kennedy, *Polym. Bull. 21*: 5 (1989).
144. M. Zsuga, T. Kelen, and J. Borbély, *Polym. Bull. 26*: 417 (1991).
145. M. Zsuga, L. Balogh, T. Kelen, and J. Borbély, *Polym. Bull 23*: 335 (1990).
146. L. Balogh, L. Fábian, I. Majoros, and T. Kelen, *Polym. Bull. 23*: 75 (1990).
147. M. Zsuga, T. Kelen, L. Balogh, and I. Majoros, *Polym. Bull. 29*: 127 (1992).
148. T. Kelen, M. Zsuga, L. Balogh, I. Majoros, and G. Deak, *Makromol. Chem., Macromol. Symp. 67*: 325 (1993).
149. G. Kaszas, J. E. Puskas, J. P. Kennedy, and C. C. Chen, *J. Macromol. Sci. Chem. A26*: 1099 (1989).
150. L. Balogh and R. Faust, *Polym. Bull. 28*: 367 (1992).
151. G. Kaszas, J. E. Puskas, C. C. Chen, and J. P. Kennedy, *Macromolecules 23*: 3909 (1990).
152. G. Pratap, S. A. Mustafa, and J. P. Heller, *J. Polym. Sci. Part A Polym. Chem. 31*: 2387 (1993).

153. C. C. Chen, J. Si, and J. P. Kennedy, *J. Macromol. Sci. Pure Appl. Chem. A29*: 669 (1992).
154. R. F. Storey, and Y. Lee, *Polym. Prep. Div. Polym. Chem. Am. Chem. Soc. 30(2)*: 162 (1989).
155. J. Si and J. P. Kennedy, *J. Macromol. Sci. Pure Appl. Chem. A30*: 863 (1993).
156. J. P. Kennedy and A. Hayashi, *J. Macromol. Sci. Chem. A28*: 197 (1991).
157. G. Pratap and J. P. Heller, *J. Polym. Sci. Part A Polym. Chem. 30*: 163 (1992).
158. G. Pratap, S. A. Mustafa, W. K. Hollis, and J. P. Heller, *J. Appl. Polym. Sci. 44*: 1069 (1992).
159. G. Pratap, S. A. Mustafa, and J. P. Heller, *J. Appl. Polym. Sci. 46*: 559 (1992).
160. R. F. Storey and Y. Lee, *J. Macromol. Sci. Pure Appl. Chem. A29*: 1017 (1992).
161. R. F. Storey and B. J. Chisholm, *Macromolecules 26*: 6727 (1993).
162. M. Gyor, H.-C. Wang, and R. Faust, *J. Macromol. Sci. Pure Appl. Chem. A29*: 639 (1992).
163. L. Thomas, A. Polton, M. Tardi, and P. Sigwalt, *Macromolecules 25*: 5886 (1992).
164. L. Thomas, M. Tardi, A. Polton, and P. Sigwalt, *Macromolecules 26*: 4075 (1993).
165. G. Kaszas, J. E. Puskas, J. P. Kennedy, and W. G. Hager, *J. Polym. Sci. Part A Polym. Chem. 29*: 421 (1991).
166. Z. Fodor, M. Gyor, H.-C. Wang, and R. Faust, *J. Macromol. Sci., Pure Appl. Chem. A30*: 349 (1993).
167. J. P. Kennedy, N. Meguriya, and B. Keszler, *Macromolecules 24*: 6572 (1991).
168. J. P. Kennedy and J. Kurian, *Macromolecules 23*: 2880 (1990).
169. J. P. Kennedy and J. Kurian, *Polym. Bull 23*: 259 (1990).
170. J. P. Kennedy and J. Kurian, *J. Polym. Sci. Part A Polym. Chem. 28*: 3725 (1990).
171. J. P. Kennedy, S. Mishra and B. Keszler, *Macromolecules 26*: 424 (1993).
172. Y. Tsunogae, I. Majolos, and J. P. Kennedy, *J. Macromol. Sci. Pure Appl. Chem. A30*: 253 (1993).
173. Y. Tsunogae and J. P. Kennedy, *Polym. Bull. 27*: 631 (1992).
174. T. Pernecker, J. P. Kennedy, and B. Iván, *Macromolecules 25*: 1642 (1992).
175. T. Pernecker and J. P. Kennedy, *Polym. Bull. 29*: 27 (1992).
176. O. Nuyken, H. Kröner, and S. Aechtner, *Makromol. Chem. Rapid Commun. 9*: 671 (1988).
177. H. Cramail, A. Deffieux, and O. Nuyken, *Makromol. Chem., Rapid Commun. 14*: 17 (1993).
178. A. V. Lubnin and J. P. Kennedy, *Polym. Bull. 29*: 9 (1992).
179. H. Cramail and A. Deffieux, *Macromol. Chem. Phys. 195*: 217 (1994).
180. T. Higashimura, S. Aoshima, and M. Sawamoto, *Makromol. Chem., Macromol. Symp. 33*: 99 (1986).

181. T. Higashimura, A. Tanizaki, and M. Sawamoto, *J. Polym. Sci. Polym. Chem. Ed. 22*: 3173 (1984).
182. M. Sawamoto, K. Ebara, A. Tanizaki, and T. Higashimura, *J. Polym. Sci. Polym. Chem. Ed. 24*: 2919 (1986).
183. Y. Ogawa, M. Sawamoto, and T. Higashimura, *Polym. J. 16*: 415 (1984).
184. O. Nuyken and S. Aechtner, *Polym. Bull. 28*: 117 (1992).
185. J. V. Crivello and D. A. Conlon, *J. Polym. Sci. Polym. Chem. Ed. 22*: 2105 (1984).
186. R. Faust and J. P. Kennedy, *J. Macromol. Sci. Chem. A27*: 649 (1990).
187. Á. F. Fehérvári, R. Faust and J. P. Kennedy, *J. Macromol. Sci. Chem. A27*: 1571 (1990).
188. M. K. Mishra and J. P. Kennedy, *J. Macromol. Sci. Chem. A24*: 933 (1987).
189. M. K. Mishra, C. C. Chen and J. P. Kennedy, *Polym. Bull. 22*: 455 (1989).
190. R. Faust, A. Nagy, and J. P. Kennedy, *J. Macromol. Sci. Chem. A24*: 595 (1987).
191. A. Nagy, R. Faust, and J. P. Kennedy, *Polym. Bull. 15*: 411 (1986).
192. M. Zsuga, J. P. Kennedy, and T. Kelen, *Polym. Bull. 19*: 427 (1988).
193. M. Zsuga, J. P. Kennedy and T. Kelen, *J. Macromol. Sci. Chem. A26*: 1305 (1989).
194. M. Zsuga, R. Faust and J. P. Kennedy, *Polym. Bull. 21*: 273 (1989).
195. R. Faust, M. Zsuga and J. P. Kennedy, *Polym. Bull. 21*: 125 (1989).
196. M. K. Mishira and J. P. Kennedy, *Polym. Bull. 17*: 7 (1987).
197. B. Wang, M. K. Mishra and J. P. Kennedy, *Polym. Bull. 17*: 205 (1987).
198. B. Wang, M. K. Mishra and J. P. Kennedy, *Polym. Bull. 17*: 213 (1987).
199. C. C. Chen, G. Kaszas, J. E. Puskas and J. P. Kennedy, *Polym. Bull. 22*: 463 (1989).
200. H. K. Huang, M. Zsuga and J. P. Kennedy, *Polym. Bull. 19*: 43 (1988).
201. M. K. Mishra, B. Wang and J. P. Kennedy, *Polym. Bull. 17*: 307 (1987).
202. G. Kaszás, J. Puskás and J. P. Kennedy, *Polym. Bull. 18*: 123 (1987).
203. G. Kaszás, J. Puskás and J. P. Kennedy, *Makromol. Chem., Macromol. Symp. 13/14*: 473 (1988).
204. G. Deák, M. Zsuga and T. Kelen, *Polym. Bull. 29*: 239 (1992).
205. T. Pernecker and J. P. Kennedy, *Polym. Bull. 26*: 305 (1991).
206. T. Pernecker and J. P. Kennedy, *Polym. Bull. 29*: 15 (1992).
207. T. Pernecker, T. Kelen and J. P. Kennedy, *J. Macromol. Sci., Pure Appl. Chem. A30*: 399 (1993).
208. C. H. Lin and K. Matyjaszewski, *Polym. Prep. Div. Chem. Am. Chem. Soc. 28(1)*: 176 (1987).
209. T. Higashimura, K. Kojima and M. Sawamoto, *Polym. Bull. 19*: 7 (1988).
210. H. Shohi, M. Sawamoto and T. Higashimura, *Macromolecules 25*: 53 (1992).
211. H. Shohi, M. Sawamoto and T. Higashimura, *Makromol. Chem. 193*: 2027 (1992).
212. H. Shohi, M. Sawamoto and T. Higashimura, *Makromol. Chem. 193*: 1783 (1992).

213. T. Higashimura, K. Kojima and M. Sawamoto, *Makromol. Chem. Suppl. 15*: 127 (1989).
214. R. Faust and J. P. Kennedy, *Polym. Bull. 19*: 21 (1988).
215. Y. Ishihama, M. Sawamoto and T. Higashimura, *Polym. Bull. 23*: 361 (1990).
216. Y. Ishihama, M. Sawamoto and T. Higashimura, *Polym. Bull. 24*: 201 (1990).
217. M. Sawamoto, T. Masuda and T. Higashimura, *J. Polym. Sci. Polym. Chem. Ed. 16*: 2675 (1978).
218. M. Sawamoto, T. Masuda and T. Higashimura, *Makromol. Chem. 178*: 1497 (1977).
219. O. Kwon, Y. Kim, S. Kwon, B. Choi and S. Choi, *Makromol. Chem. 194*: 3007 (1993).
220. K. Miyashita, M. Kamigaito, M. Sawamoto and T. Higashimura, *Macromolecules 27*: 1093 (1994).
221. A. Tanizaki, M. Sawamoto and T. Higashimura, *J. Polym. Sci. Polym. Chem. Ed. 24*: 87 (1986).
222. R. Faust and J. P. Kennedy, *Polym. Bull. 19*: 29 (1988).
223. R. Faust and J. P. Kennedy, *Polym. Bull. 19*: 35 (1988).
224. K. Kojima, M. Sawamoto and T. Higashimura, *J. Polym. Sci. Part A Polym. Chem. 28*: 3007 (1990).
225. Y. Eika, S. Kanaoka, M. Sawamoto and T. Higashimura, *Polym. Prep. Jpn. 42*: E81 (1993).
226. M. Sawamoto, T. Hasebe, M. Kamigaito and T. Higashimura, *J. Macromol. Sci. Pure Appl. Chem. A31*: 937 (1994).
227. K. Matyjaszewski, A. Bon, C.-H. Lin, and J. S. Xiang, *Polym. Prepr Div. Polym. Chem. Am. Chem. Soc. 34(1)*: 487 (1993).
228. Y. Tsunogae and J. P. Kennedy, *J. Polym. Sci. Polym. Chem. Ed. 32*: 403 (1994).
229. B. G. Soares, A. D. Silva and A. S. Gomes, *Polym. Bull. 29*: 253 (1992).
230. A. D. Silva, A. S. Gomes and B. G. Soares, *Polym. Bull. 30*: 133 (1993).
231. R. Asami and N. Tokura, *J. Polym. Sci. 42*: 553 (1960).
232. L. Thomas, A. Polton, M. Tardi and P. Sigwalt, *Macromolecules 28*: 2105 (1995).
233. T. Higashimura, H. Teranishi, and M. Sawamoto, *Polym. J. 12*: 393 (1980).
234. V. S. C. Chang and J. P. Kennedy, *Polym. Bull. 4*: 513 (1981).
235. R. Santos, A. Fehérvári, and J. P. Kennedy, *J. Polym. Sci. Polym. Chem. Ed. 22*: 2685 (1984).
236. J. R. Hwu and J. M. Wetzel, *J. Org. Chem. 50*: 3946 (1985).
237. M. S. Gong and J. H. K. Hall, *Macromolecules 19*: 3011 (1986).
238. H. K. Hall Jr., A. B. Padias, M. Atsumi, and T. F. Way, *Macromolecules 23*: 678 (1990).
239. C.-H. Lin and K. Matyjaszewski, *J. Polym. Sci. Part A Polym. Chem. 28*: 1771 (1990).
240. M. Sawamoto, M. Kamigaito, K. Kojima, and T. Higashimura, *Polym. Bull. 19*: 359 (1988).

241. M. Kamigaito, M. Sawamoto, and T. Higashimura, *Macromolecules 23*: 4896 (1990).
242. M. Kamigaito, M. Sawamoto, and T. Higashimura, *Makromol. Chem. 194*: 727 (1993).
243. M. Kamigaito, M. Sawamoto, and T. Higashimura, *J. Polym. Sci. Polym. Chem. 29*: 1909 (1991).
244. J. Hrkach and K. Matyjaszewski, *J. Polym. Sci. Polym. Chem. Ed. 33*: 285 (1995).
245. D. V. Meirvenne, N. Haucourt, and E. J. Goethals, *Polym. Bull. 23*: 185 (1990).
246. N. H. Haucourt, E. J. Goethals, M. Schappacher, and A. Deffieux, *Makromol. Chem. Rapid Commun. 13*: 329 (1992).
247. M. Villesange, A. Rives, C. Bunel, J. P. Vairon, M. Froeyen, M. V. Beylen, and A. Persoons, *Makromol. Chem., Macromol. Symp. 47*: 271 (1991).
248. J.-P. Vairon, A. Rives, and C. Bunel, *Makromol. Chem., Macromol. Symp. 60*: 97 (1992).
249. M. Sawamoto, A. Furukawa, and T. Higashimura, *Macromolecules 16*: 518 (1983).
250. S. Penczek, P. Kubisa, K. Brzezinska, and M. Basko, *Polym. Prep. Div. Chem. Am. Chem. Soc. 34(1)*: 812 (1993).
251. W. O. Choi, M. Sawamoto, and T. Higashimura, *J. Polym. Sci. Chem. Ed. 28*: 2923 (1990).
252. W. O. Choi, M. Sawamoto, and T. Higashimura, *J. Polym. Sci. Chem. Ed. 28*: 2937 (1990).
253. B. Ivan and A. H. E. Mueller, Presented at the 35th IUPAC Meeting Akron, 1994, pp. 85.
254. K. Matyjaszewski, *Macromolecules 21*: 933 (1988).
255. D. P. Kelly and R. J. Spear, *Austr. J. Chem. 31*: 1209 (1978).
256. K. Matyjaszewski, M. Teodorescu, and C. H. Lin, *Macromol. Chem. Phys., 196*: 2149 (1995).
257. F. Subira, J. P. Vairon, and P. Sigwalt, *Macromolecules 21*: 2339 (1988).
258. L. Balogh, L. Wang, and R. Faust, *Macromolecules 27*: 3453 (1994).
259. R. Faust and J. P. Kennedy, *J. Macromol. Sci. Chem. A27*: 649 (1990).
260. H. Cheradame and P. Sigwalt, *Bull. Soc. Chim. Fr. 1970*: 849 (1970).
261. R. Cotrel, G. Sauvet, J. P. Vairon, and P. Sigwalt, *Macromolecules 9*: 931 (1976).
262. M. Villesange, G. Sauvet, J. P. Vairon, and P. Sigwalt, *J. Macromol. Sci. Chem. A11*: 391 (1977).
263. C. Reichardt, *Solvent Effects in Organic Chemistry*, Verlag Chemie, Wein heim, 1979.
264. C. G. Swain, M. S. Swain, A. L. Powell, and S. Alunni, *J. Am. Chem. Soc. 105*: 502 (1983).
265. K. Matyjaszewski, *New Polym. Mater. 2*: 115 (1990).
266. S. Penczek and R. Szymanski, *Makromol. Chem. Macromol. Symp. 60*: 65 (1992).
267. J. M. Lelou, V. Bernardo, A. Polton, M. Tardi, and P. Sigwalt, *Polym. Int. 37*: 219 (1995).

268. R. F. Storey, B. J. Chisholm, and L. B. Brister, *Macromolecular 28*: 4055 (1995).
269. R. Szymanski and K. Matyjaszewski, *Macromol. Theory Simul. 4*: 335 (1995).
270. B. Ivan and J. P. Kennedy, *Macromolecules 23*: 2880 (1990).
271. I. Majoros, A. Nagy, and J. P. Kennedy, *Adv. Polym. Sci. 112*: 1 (1994).
272. H. C. Brown and R. S. Fletcher, *J. Am. Chem. Soc. 71*: 1845 (1949).
273. K. Matyjaszewski, *Macromol. Rapid Commun. 16*: 219 (1995).
274. G. Parkin and R. Hoffman, *Angew. Chem. Int. Ed. Engl. 33*: 1462 (1994).
275. A. A. Korotkov, *Angew. Chem. Int. Ed. Engl. 70*: 85 (1958).
276. M. Sawamoto, T. Ohtoyo, T. Higashimura, K. H. Guhrs, and G. Heublein, *Polym. J. 17*: 929 (1985).
277. D. Li and R. Faust, *Macromolecules 28*: 4893 (1995).
278. Z. Fodor and R. Faust, *J. Macromol. Sci., Pure Appl. Chem. A32*: 575 (1995).
279. T. Ohmura, M. Sawamoto, and T. Higashimura, *Macromolecules 27*: 3714 (1994).
280. K. Kojima, M. Sawamoto, and T. Higashimura, *Macromolecules 24*: 2658 (1991).
281. A. V. Lubnin and J. P. Kennedy, *J. Polym. Sci. Polym. Chem. Ed. 31*: 2825 (1993).
282. Y. Tsunogae and J. P. Kennedy, *J. Polym. Sci. Polym. Chem. 32*: 41 (1994).
283. J. Oh, S. Kang, O. Kwon, and S. Choi, *Macromolecules 28*: 3015 (1995).
284. J. P. Kennedy, *Makromol. Chem., Macromol. Symp. 51*: 169 (1991).
285. J. P. Kennedy, N. Meguriya, and B. Keszler, *Macromolecules 24*: 6572 (1991).
286. Y. Tsunogae and J. P. Kennedy, *J. Macromol. Sci., Pure Appl. Chem. A30*: 269 (1993).
287. J. P. Kennedy, S. Midha, and Y. Tsunogae, *Macromolecules 26*: 429 (1993).
288. J. E. Puskas, G. Kaszas, J. P. Kennedy, and W. G. Hager, *J. Polym. Sci., Polym. Chem. 30*: 41 (1992).
289. J. E. Puskas, G. Kaszas, J. P. Kennedy, and W. G. Hager, *J. Polym. Sci. Polym. Chem. 29*: 427 (1991).
290. K. Matyjaszewski, *Polym. Mat. Sci. Eng. 72*: 234 (1995).
291. D. P. N. Satchell and R. S. Satchell, *Chem. Rev. 69*: 251 (1969).
292. H. Mayr and M. Patz, *Angew. Chem. Int. Ed. Eng. 33*: 938 (1994).
293. M. Masure, P. Sigwalt and G. Sauvet, *Makromol. Chem., Rapid Commun. 4*: 269 (1983).
294. J. P. Kennedy and R. T. Chu, *J. Macromol. Sci. Chem. A18*: 3 (1982).
295. J. M. Moulis, J. Collomb, A. Gandini, and H. Cheradame, *Polym. Bull. 3*: 197 (1980).
296. J. S. Wang, P. Bayard, R. Jerome, S. K. Varshney, and P. Teyssie, *Macromolecules 27*: 4890 (1994).

5

Controlled Polymer Synthesis by Cationic Polymerization

MITSUO SAWAMOTO Kyoto University, Kyoto, Japan

I. INTRODUCTION

As discussed in Chapter 4, it is now possible to design and carry out controlled cationic polymerizations for almost all classes of cationically polymerizable vinyl/alkene monomers using a wide variety of initiating systems. In these modern systems, in sharp contrast to the conventional cationic polymerizations, the molecular weights and the molecular weight distributions of the polymers, as required for living polymerizations, can be precisely controlled, which demonstrates the suppression of chain transfer, termination, and other undesirable side-reactions that have been believed extremely difficult to eliminate in the conventional acid-catalyzed or carbocationic polymerizations. As already seen, there is still a controversy about whether some of cationic polymerization is "living" or "controlled." Throughout this chapter, for simplicity, however, we use the term "living" for the polymerizations that might be considered as "controlled" in the sense that has already been discussed in the preceding chapters. The development of such a new generation of cationic polymerizations in the 1980s and 1990s, whatever its mechanistic implications might be (see Chapter 4), has in turn opened a new vistas of polymer syntheses via cationic processes in which the precise control of polymer architectures is feasible [1–3].

When one looks at a polymer chain, one will soon realize that there are a number of structural factors that should or may be controlled in syntheses (Fig. 1). These factors may include: (1) molecular weight (chain length); (2) molecular weight distribution (MWD); (3) end group; (4) pendant group; (5) sequence (the arrangement of constitutional repeat units

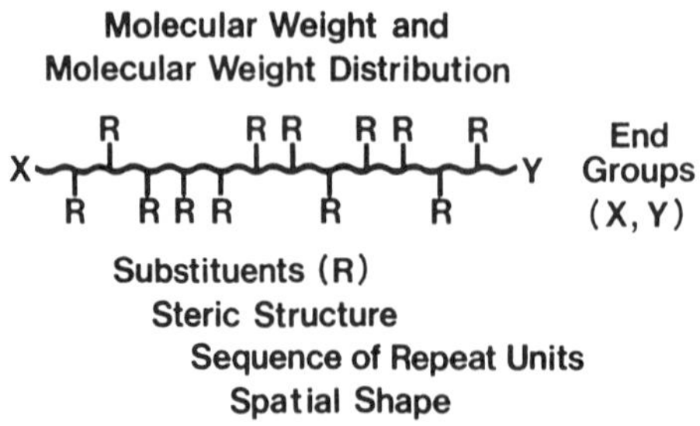

Figure 1 Structural factors to be controlled in precision polymer synthesis.

and segments along a main chain); (6) steric structure; and (7) three-dimensional or spatial shape (as described as spherical, cylindrical, multi-armed, star-shaped, rod-like, graft, etc.). Among these, the most fundamental are the molecular weight and MWD, and we have already seen in Chapter 4 that, in cationic polymerization, the control of both structural factors is now feasible through living (or at least controlled) polymerizations. Namely, the control of these can be paraphrased as to develop living cationic polymerizations that are specifically suited for particular monomers.

In "controlled polymer synthesis," in addition, it is particularly important that the "control" herein implies not only the simple regulation of molecular weights, MWD, and other structural factors but also the precise introduction of "functional groups" into specific positions of polymers with well-defined architectures. Namely, the control of one or more of these structural factors, then, leads to a variety of polymers of synthetic interest, as some of them illustrated schematically in Fig. 2:

a. Polymers with pendant functional groups
b. Block copolymers
c. Polymers with terminal functional polymers
d. Polymers and oligomers with regulated sequences of repeat units
e. Star-shaped or multi-armed polymers
f. Graft polymers
g. Macrocyclic polymers
h. Amphiphilic polymers

Polymers e–g may also be called collectively as "polymers with controlled spatial shapes"; amphiphilic polymers (h) may include block, star-shaped, and graft polymers covered in classes b, e, and f. Comparison of Figs. 1 and 2 also tells us that, unlike the anionic and coordination (Zieglar-Natta) counterparts, cationic polymerization still fails to provide general methods to control the steric structures of polymers, although the first indication

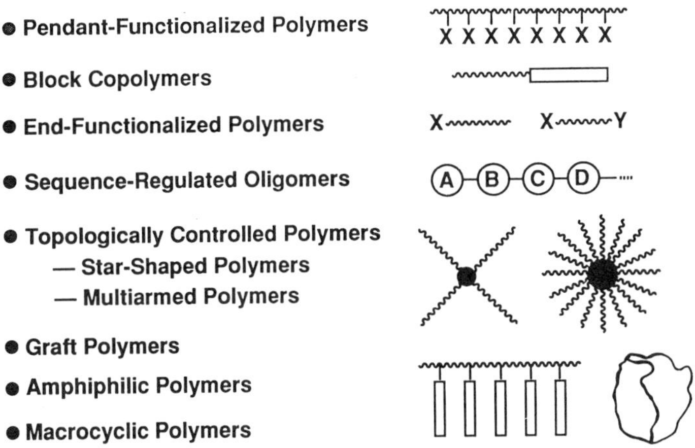

Figure 2 Precision polymer synthesis: polymer structures.

of the formation of stereoregular polymers was in fact found in the cationic polymerization of alkyl vinyl ethers almost half a century ago [4–7]. Another area of difficulty lies in the control of sequences of constitutional repeat units along the main chain, as abundantly found in genes, proteins, and other biologically synthesized polymers (see Section V).

This chapter presents an updated overview of the current status of the controlled polymer syntheses via the modern generation of cationic polymerizations that are mostly "living" or "controlled": what and how one can design and eventually synthesize novel polymers with well-defined structures and functionalities. Thus, the following sections are devoted to each of these classes of polymers (Fig. 2), with emphasis on the general methodologies and specific examples. The last section (Section VII) briefly covers the experimental procedures in living cationic polymerization and related polymer synthesis.

II. POLYMERS WITH PENDANT FUNCTIONAL GROUPS

A. General Methodology and Scope

The simplest examples of so-called functional or functionalized polymers are perhaps those with pendant functional groups. The introduction of pendant functional groups to synthetic polymers, in general, may be achieved by two methods: (1) polymerization of monomers with pendant functional groups and (2) chemical transformation of the pendant groups in preformed polymers. Both of them involve advantages and disadvantages, and most of the currently available polymers of this type are produced by the second method via polymer reactions, such as chelating and ion-exchange polystyrenes and other resins with ionic pendant substituents. Herein, on the other hand, let us consider the first method that is specifically based on living cationic polymerizations, because it is suited for the synthesis of "pendant-functionalized" polymers with controlled molecular weights and narrow MWDs.

In the conventional cationic polymerizations, such as those with boron trifluoride etherate (with adventitious water) or protonic acids (perchloric acid, etc.), polar functional groups such as esters often induce chain transfer and termination reactions (see Chapter 3). It has therefore been difficult to polymerize vinyl monomers with polar pendant functionalities in a controlled manner, although there are examples of cationic polymerizations of such monomers [8], e.g., vinyl ethers with pendant cinnamate esters [9,10]. Despite this pessimistic expectation, the initiating systems originally developed for the living cationic polymerizations of alkyl vinyl ethers and styrene derivatives (Chapter 4, Section V) proved to successfully polymerize a variety of "pendant-functionalized monomers" without undesirable side reactions to form polymers with controlled molecular weights and narrow MWDs. The tolerance toward polar functional groups of these living processes, in turn, provides the most important basis of the controlled polymer synthesis to be discussed in this chapter.

In some cases where the intended functional group may adversely affect living cationic polymerization, it should be suitably "protected" (e.g., a hydroxyl group into a carboxylate) to prevent side reactions to be induced, as in the anionic counterparts that have recently been developed for ring-substituted styrenes by Nakahama and Hirao [11,12]. By definition, the "protecting" groups in monomers should be such that they do not cause side reaction during the living cationic polymerization but can readily be transformed ("deprotected") into the originally intended form by a polymer reaction which does not deteriorate the other parts of the parent polymers. Scheme 1 illustrates typical examples of these

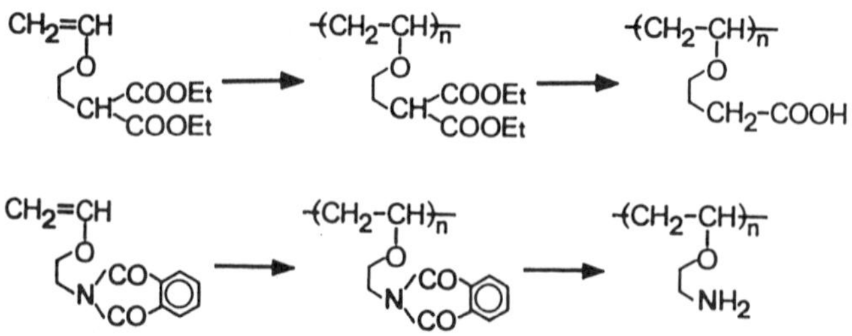

Scheme 1 Typical examples of the synthesis of pendant-functionalized poly(vinyl ethers) [23,29].

syntheses, consisting of living cationic polymerizations of pendant-functionalized vinyl ethers and subsequent deprotections of the product polymers (see below for details).
In such syntheses, therefore, one deals with a hybrid process of methods 1 and 2, discussed above.

B. Pendant-Functionalized Polymers by Living Cationic Polymerization

1. *Vinyl Ethers with Polar Functional Groups*

Since the development of living cationic polymerizations of alkyl vinyl ethers (Chapter 4, Sections IV and V.A), considerable efforts have been made to synthesize and polymerize vinyl ether derivatives carrying polar functional substituents, and thereby it is now possible to obtain a variety of pendant-functionalized poly(vinyl ethers) of controlled molecular weights and narrow MWDs [1,2,13]. Figure 3 lists typical examples of vinyl ethers carrying various pendant functionalities for which living cationic polymerizations are available [14–35]. These monomers are synthesized most conveniently from 2-chloroethyl vinyl ether, now commercially available,

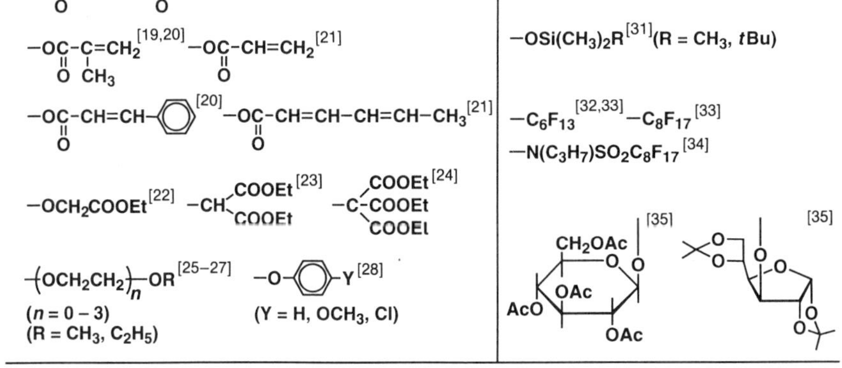

Figure 3 Vinyl ethers with pendant functional groups for which living cationic polymerization is feasible. The numerals in brackets attached to each entry indicate reference numbers.

where the pendant alkyl chloride provides an excellent reaction site for pendant transformations via nucleophilic substitution. The two methylene units, inherited from the precursor, serve as a spacer that effectively insulates the vinyl ether moiety from the electronic and steric effects of the functional substituents; some of them, such as ester and imide groups, are electron-withdrawing and, when attached directly to the vinyl group, will adversely affect the reactivity of the monomers.

In Fig. 3 one finds a variety of functional groups that may be subdivided as follows: haloalkyl groups [14–16], saturated esters (acetate [17], and benzoate [18,19]); unsaturated esters (methacrylate [19,20], acrylate [21], cinnamate [20], and others [21]); malonate and related esters [22–24]; ethers and oligo(oxyethylenes) [25–28]; imides [29,30]; alkylsilyloxy groups [31]; perfluoroalkyls [32–34]; and protected carbohydrate groups [35]. Vinyl ethers with mesogenic (liquid-crystal forming) substituents will be treated separately in Section II.B.2.

Living cationic polymerizations of almost all of these monomers can be effected with such initiating systems as hydrogen iodide/iodine (HI/I_2) [36], hydrogen halide/zinc halide (HX/ZnX_2; X = halogen) [37], and $CH_3CH(O\textit{i}Bu)OCOCH_3/EtAlCl_2$/1,4-dioxane [38,39]. Readers may refer to Refs. [14–35] given in Fig. 3 in terms of specific initiating systems and reaction conditions for respective monomers (see also Chapter 4, Section V.A). It may suffice herein to point out that, despite their polar functional groups, the overall polymerization patterns of these vinyl ethers are basically similar to those of the alkyl counterparts [36,40] and lead to polymers with controlled molecular weights and narrow MWDs. During the polymerizations, no particular side reactions of the pendant groups have been observed. In some cases, however, the pendant functions affect the overall polymerization rates and the reactivity of the monomers, although they are usually separated from the vinyl groups via methylene units [41]. For instance, the polymerization of 2-acetoxyethyl vinyl ether (a monomer with an ester group) is clearly slower than those of ethyl and isobutyl vinyl ethers under similar conditions [17,41]. In such a case, due care should be taken to control polymerization rate and other factors, like the concentration of initiators and the activators, according to the monomer's reactivity.

The pendant functions listed in Figure 3 are often useful and of synthetic interest per se. For example, methacrylate and acrylate esters are polymerizable (cross-linking sites) [19–21]; the cinnamate is photoresponsive (for the photo-induced dimerization of its unsaturated groups) [20]; oligo(oxyethylene) [25–27] and carbohydrate groups [35] give hydrophilic and water-soluble polymers, whereas perfluoroalkyl moieties [32–34] enhance hydrophobicity. Thus, poly(vinyl ethers) with cinnamate functions have been used for the early generation of high-resolution photoresist materials [9,10]. In aqueous media, poly(vinyl ethers) with the oxyethylene pendant groups also undergo selective transport of metal cations through liquid membranes [27] and thermosensitive solubility changes (cloud points) [42].

On the other hand, these pendant substituents also serve as "protecting groups" for functions that, if present as they are, would deteriorate the living cationic polymerization, namely:

(Protected)	→	(Deprotected)	[Ref]
—OCOR	→	—OH	[17–19]
$—OSiR_3$	→	—OH	[31]
$—CH(CO_2R)_2$	→	$—CH_2CO_2H$	[23]
$—N(COR)_2$	→	$—NH_2$	[29,30]

Among these, in particular, the acetate [17] and the silyloxyl [31] derivatives are often used as the protecting groups for the hydroxyl (alcohol) function. For example, polymers of 2-acetoxyethyl vinyl ether are readily transformed into a polyalcohol, poly(2-hydroxyethyl vinyl ether), by alkaline hydrolysis [17]. Due to the polar pendant functions, the polymers are of course hydrophilic and often water-soluble, and serve as hydrophilic segments in so-called "amphiphilic" polymers, as will be discussed later (Sections III.D and VI.B.5). Other important protecting groups include the malonate [23] and the imides [29,30], which lead to polymeric carboxylic acids and amines, respectively (Scheme 1).

Another important role of the pendant-functionalized vinyl ethers is that they can be precursors of initiators for living cationic polymerization of other vinyl ether and styrene derivatives, from which polymers with terminal functional groups can be prepared (see Section IV).

2. *Vinyl Ethers with Mesogenic Groups*

In response to the growing interest in side-chain liquid crystalline polymers, vinyl ethers carrying a variety of mesogenic substituents have been prepared and cationically polymerized [15,43–63]. Figure 4 shows examples of these vinyl ether monomers; propenyl ethers ($CH_3CH{=}CH{-}OR$) with similar mesogenic substituents (R) are also available [46]. In the syntheses of these monomers, the ether exchange reactions with a palladium complex are often used, in addition to the nucleophilic substitution of the 2-chloroethyl vinyl ether. As in the corresponding methacrylates, the mesogens therein are most often biphenyl and similarly rigid aromatic moieties with polar substituents, where the cyano group is most preferred [47–56]. For the "insulation" (decoupling) between the main chain and the pendant mesogens, the aromatic part (mesogen) is connected to the vinyl ether group through a flexible methylene or oxyethylene spacer with varying lengths (see Fig. 4).

Most of these monomers can be polymerized into narrowly distributed polymers with relatively low degrees of polymerization (10–50; $M_n <$ 6000) by such initiating systems as HI/I_2 [36], HI/ZnI_2 [37], and $CF_3SO_3H/S(CH_3)_2$ [64], among which the latter two are most frequently used. The polymerizations are usually carried out in dichloromethane solvent where these monomers and their polymers are soluble.

Polymers of the mesogenic vinyl and propenyl ethers mostly assume thermotropic liquid crystalline phases where smectic forms are frequently

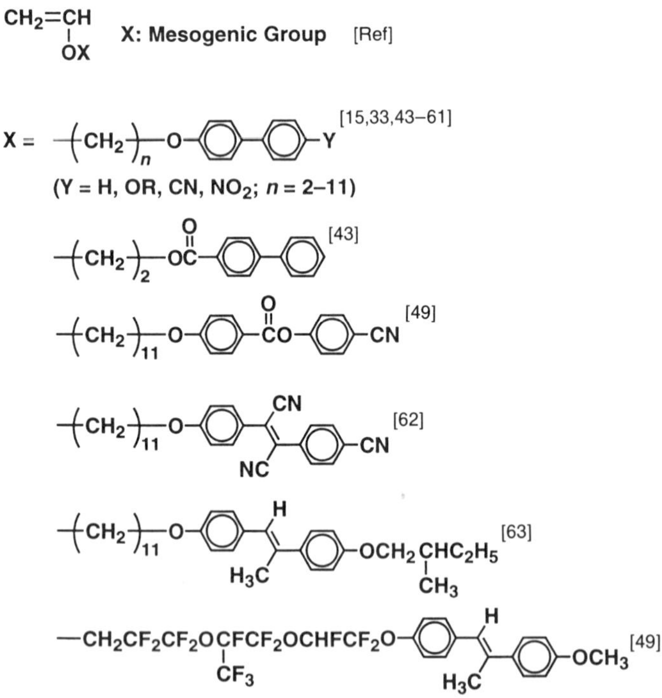

Figure 4 Vinyl ethers with pendant mesogenic groups for side-chain liquid-crystalline polymers. The numerals in brackets attached to each entry indicate reference numbers.

observed; hysteresis upon repeated heating-cooling cycles are also noted frequently. Taking advantage of the controlled nature of the polymerizations, Sagane and Lenz [43,44] and Percec et al. [46–63] systematically studied the mesogenic phase formation, phase diagrams, and related aspects as a function of polymers' degree of polymerization, spacer length, copolymer composition, pendant mesogen structures, and other factors.

For example, a series of exhaustive and systematic study in Percec's group use $CH_2{=}CH{-}O(CH_2)_n{-}O\text{-}Ar{-}CN$ (Ar = *p*-biphenyl; $n = 2$–11) to understand the effects of the spacer length (n) in homopolymerizations [47,49,54,56], random copolymerizations [51–53,57,58], and block copolymerization [59]. Among these, the homopolymerizations of the monomers with $n = 6$ and 8 are carried out with CF_3SO_3H in the presence of dimethyl sulfide in methylene chloride at 0° C to prepare polymers with the degrees of polymerization (DP) ranging from 2 to 31 [49]. With $n =$ 6, the monomer shows a nematic phase, but the polymers give liquid crystalline phases that depend on DP: (DP) 3–7, nematic + smectic s_A; 8–14, smectic s_A; > 23, smectic s_A + another undefined smectic phase. These phase transitions are all enantiotropic, whereas the monomer's transition is monotropic. In the copolymerization of the monomers with $n = 2$ and 8 [51], the liquid crystalline phases depend on the comonomer compositions, from smectic s_C ($n = 2$) through nematic to smectic s_A ($n = 8$). Another series with $CH_2{=}CH{-}O(CH_2)n{-}OR$ (R = mesogen; see Fig. 4) [50,62,63] was studied to clarify the relation between the mesogen structures and homopolymerization.

3. *Styrene Derivatives with Polar Functional Groups*

The aromatic rings in styrene derivatives serve as the points to which functional substituents can be attached. The simplest form would be *p*-chlorostyrene, for which living cationic polymerization is possible (see Chapter 4, Section V.C.4) [65–67].

Due to the potential usefulness as photoresist materials for the dry etching lithography [68], more extensive studies have been directed to the protected forms of *p*-hydroxystyrene (*p*-vinylphenol) and α-methyl-*p*-hydroxystyrene, where the protecting groups include methoxy [69,70], *t*-butoxy [71,72], acetoxy [73–75], and *t*-butoxycarbonyloxy [73,75]. It should be noted herein that the alkoxy groups are electron donating and suited for living cationic polymerization, whereas the latter two ester functions are moderately electron withdrawing or carbocation destabilizing. Among these monomers, therefore, living cationic polymerization has been achieved only for *p*-methoxy- [69,70] and *p*-*t*-butoxystyrenes [71] with the HI/I_2 initiating system. The polymerization is feasible even at room temperature to give polymers with relatively high, controlled molecular weights ($>10^5$ in some cases).

Another class of pendant-functionalized styrene derivatives carries a glycidyl ether (epoxy) group at the para-position in styrene [76] or α-methylstyrene [77]; the p-epoxy function is attached to the aromatic ring via an ether oxygen to increase the monomer reactivity and the stability of the resultant growing carbocations. With the HI/I_2 and HI/ZnI_2 initiating systems, these vinyl/epoxy-type, "hetero-bifunctional" monomers, i.e., monomers with cationically polymerizable different functions, undergo living cationic polymerizations selectively via the vinyl groups alone with the oxirane functions remaining intact, thereby giving completely soluble polymers. In contrast, the conventional cationic polymerizations with BF_3-etherate, $EtAlCl_2$, etc. result in cross-linked insoluble products under similar reaction conditions.

Despite these examples, one can readily notice that the living cationic polymerization of pendant-functionalized styrene derivatives has been studied much less extensively than those of vinyl ethers, and further progress is anticipated accordingly.

III. BLOCK COPOLYMERS

A. General Methodology and Scope

The synthesis of block copolymers is obviously one of the polymer synthesis areas where living polymerization is most effective and convenient [78,79]. In general, block copolymers may be synthesized either (A) via sequential living polymerization or (B) via reactions of end-functionalized polymers that include living polymers; both of the two methods are also available in cationic polymerizations:

A. Sequential living cationic polymerization (*: living end)

$$m\mathrm{A} \longrightarrow -(\mathrm{A})_m-* \xrightarrow{n\mathrm{B}} -(\mathrm{A})_m\!-\!-(\mathrm{B})_n-$$

B. Reactions of end-functionalized polymers

B-1. Polymerization from macroinitiator (X: functional group)

$$-(\mathrm{A})_m-X + n\mathrm{B} \longrightarrow -(\mathrm{A})_m\!-\!-(\mathrm{B})_n-$$

B-2. Polymer coupling (X, Y: functional group)

$$-(\mathrm{A})_m-X + Y-(\mathrm{B})_n- \longrightarrow -(\mathrm{A})_m\!-\!-(\mathrm{B})_n-$$

1. *Sequential Living Cationic Polymerization*

The sequential living polymerization of two monomers (method A) is perhaps the most straightforward method to synthesize block copolymers, and sometimes the successful formation of block copolymers by this way is taken as evidence for the occurrence of living polymerization. Namely, monomer A is polymerized into a living cationic polymer, from the living end of which another monomer B is polymerized again in a living fashion into an AB-block copolymer. Thus, the synthesis can be carried out simply in a one pot, but the propagation mechanisms for the two monomers should accordingly be the same, which sometimes limits the range of monomer (segment) combinations. Mechanistically, it is also essential that a common initiating system applicable to the two comonomers be found, along with reaction conditions that are optimized for the respective monomers to ensure their living propagation and, perhaps more important, the perfect cross-propagation from poly(A) to monomer B (see Section III.B).

2. *Reactions of End-Functionalized Polymers*

The reactions of end-functionalized polymers (method B) may be divided into two subclasses, the living polymerization from macroinitiators (B-1) and the polymer coupling reactions (B-2). The former (B-1) uses end-functionalized polymers with a terminal function (X) that can initiate the living polymerization of a second monomer. The latter (B-2) involves two end-functionalized polymers, where the terminal functions, X and Y, react with each other to combine them into a block copolymer; X and/or Y may also be living ends. A notable characteristic of the polymer reaction methods (B), relative to method (A), is that one can select propagation mechanisms most suited for each monomer (e.g., cationic for monomer A and anionic for monomer B), so that the range of monomer pair or segmental combinations is, in principle, much wider and freer in selection than in the sequential propagation. In this regard, the method enables one to transform polymerization mechanism on going from monomer A to monomer B. However, the synthesis as a whole is multistep, more time-consuming, and cumbersome, at least involving the preparation of end-functionalized polymers, their separation and purification, and the subsequent living polymerization or polymer coupling reaction (see Section III.C).

B. Block Copolymers by Sequential Living Cationic Polymerization

1. *Vinyl Ethers*

A large variety of AB- and ABA-type block copolymers have been prepared from vinyl ethers via sequential living cationic polymerizations. For example, as shown in Scheme 2 (A), isobutyl vinyl ether (IBVE) is first polymerized with the HI/ZnI_2 or $HCl/ZnCl_2$ initiating system, and from the resulting living polymer, the second monomer, 2-acetoxyethyl vinyl

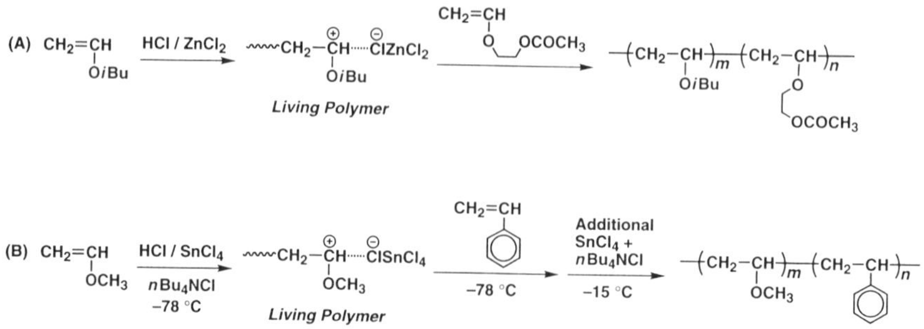

Scheme 2 Examples of the synthesis of AB-block copolymers by sequential living cationic polymerization [80,87].

ether (AcOVE), is further polymerized to form an AB-block copolymer [80]. Figure 5 summarizes other typical examples of these vinyl ether-vinyl ether AB-block copolymers, where the monomer combinations may be selected from alkyl vinyl ethers [36,81] and their pendant-functionalized versions [31,42,80,82–85]. The alkyl substituents cover a fairly wide range of carbon numbers, from methyl to hexadecyl (cetyl). The pendant-functionalized vinyl ethers may be those listed in Figs. 3 and 4, among which AcOVE is the most frequently employed [31,82,83]. In particular, the sequential block copolymerizations of these polar monomers with alkyl derivatives often lead to amphiphilic block copolymers (Section III.D).

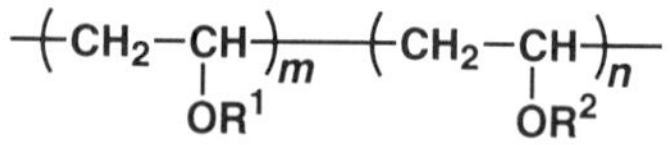

R^1 (A)	R^2 (B)	[Ref]
$nC_{16}H_{33}$	CH_3	[36,81]
	C_2H_5	[36]
CH_2CH_2OH	iC_4H_9	[31,82,83]
	nC_4H_9, iC_8H_{17}, $nC_{16}H_{33}$	[80]
$CH_2CH_2CH_2COOH$	nC_4H_9, iC_8H_{17}, $nC_{16}H_{33}$	[84]
$CH_2CH_2NH_2$	nC_4H_9, $nC_{16}H_{33}$	[85]
$CH_2C_6H_5$ (→ H)	iC_4H_9	[42]

Figure 5 Block copolymers of vinyl ethers: typical examples.

As seen in Scheme 2 (A), the most of the syntheses have been carried out with the HI/I_2 and HX/ZnX_2 (X = halogen) initiating systems, because these systems can effectively polymerize a large variety of vinyl ethers, including those with pendant functions, into well-defined living polymers [1]. In this way, the sequential living cationic polymerizations of two vinyl ethers are mostly "reversible"; i.e., both A → B and B → A polymerization sequences are operable. This is in sharp contrast to the block copolymerization of a vinyl ether with a styrene derivative or isobutylene (see below), where such "reversibility" often fails to work.

In the reversible combinations, however, adjustment of reaction conditions is sometimes needed to establish the cross-propagation from A to B or vice versa. The most important reason for this is the marked difference in reactivity among alkyl and polar vinyl ethers, and the optimum reaction conditions for a pair of vinyl ethers often differ from each other. For example, AcOVE is much less reactive than ethyl or isobutyl vinyl ethers [41], and thus, when AcOVE is polymerized from the preformed living poly(IBVE), an additional amount of ZnX_2 is added to accelerate the second-stage polymerization. Varying polymerization temperature according to the monomer reactivity is another factor to be considered in sequential polymerizations.

In an extreme case the reactivity difference between the two monomers is positively utilized to obtain block copolymers. For example, Goethals recently polymerized IBVE with CF_3SO_3H in the presence of thietane (a cyclic thioether) [86]. Because the thioether is much less reactive than vinyl ethers, it cannot polymerize and serves as a nucleophilic additive in the first-phase vinyl ether polymerization [64], but once IBVE has been completely polymerized, the cyclic monomer now polymerizes from the living end to form block polymers.

2. *Styrene Derivatives*

As the range of styrene derivatives for living cationic polymerization expands (Chapter 4, Section V.C), a variety of block copolymers with styrenic segments have been synthesized. Most of the reported examples involve combinations of styrene derivatives with vinyl ethers or isobutene. Some examples of styrene derivative-vinyl ether block copolymers are listed in Fig. 6 [16,87–89]. Monomers that can form similar block copolymers with isobutylene are listed in Fig. 7 (Section III.B.3).

In general, styrene and its substituted derivatives are less reactive than vinyl ethers in cationic polymerization, although the reactivity depends considerably on the nature of the substituents. This in turn requires some care in synthesizing block copolymers of styrene derivatives by sequential living cationic polymerization. For example, styrene-methyl

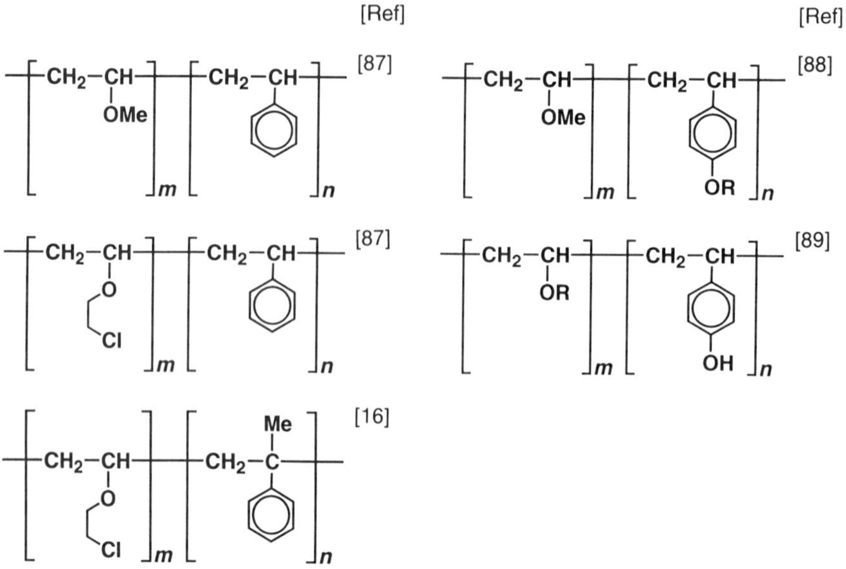

Figure 6 Block copolymers of styrene derivatives: typical examples.

vinyl ether AB-block copolymers may be synthesized as shown in Scheme 2 (B) [87]. For both monomers, living cationic polymerizations are feasible with the $HCl/SnCl_4$ system in the presence of a salt, $n Bu_4NCl$, in methylene chloride solvent. Thus, the first process is the living cationic polymerization of methyl vinyl ether (MVE) at −78° C. Into the resulting living poly(MVE), styrene is added, but no polymerization occurs under the conditions where MVE smoothly polymerizes. Therefore, additional amounts of $SnCl_4$ and the salt are added, temperature is also raised to −15° C, and thereby the second-stage polymerization of styrene eventually takes place to give the target block copolymers. The styrene-MVE block copolymers are of interest, partly because blends of their respective homopolymers are well known to undergo unique phase-separation processes.

In block copolymerizations with vinyl ethers, as seen in this example, a styrene derivative should be polymerized after the first-stage polymerization of a vinyl ether component, and an additional dose of a Lewis acid (activator) is usually needed to accelerate the second-phase polymerization. The reverse polymerization sequence (from a styrene derivative to a vinyl ether) often results in a mixture of block copolymers and homopolymer(s). This is because the reaction conditions for the first stage are too severe (too high reagent concentrations, etc.) for a vinyl ether that its subsequent polymerization is often too fast and difficult to control; namely, the cross-propagation is probably slower than propagation. Examples of such cases may also be found for the combinations of vinyl ethers with *p*-alkoxystyrenes [88] and *p*-alkylstyrenes [89].

3. *Isobutene*

Considerable efforts have been directed, primarily in Kennedy's group [3], to synthesize a series of block copolymers of isobutene with isoprene [90,91], styrene derivatives [92–104], and vinyl ethers [105–107]. Figure 7 lists the monomers that have been used for the block copolymerizations with isobutene. The reported examples include not only AB- but also ABA- and triarmed block copolymers, depending on the functionality of the initiators (see Chapter 4, Section V.B, Table 3). Obviously, the copolymers with styrene derivatives, particularly ABA versions, are mostly intended to combine the rubbery polyisobutene-centered segments with glassy styrenic side segments in attempts to prepare novel thermoplastic elastomers. These styrene monomers are styrene, *p*-methylstyrene, *p*-chlorostyrene, α-methylstyrene, and indene.

The sequential living polymerizations for these syntheses begun with isobutene, initiated by the *tert*-alkyl chloride or methoxide/$TiCl_4$ initiating systems with a nucleophilic additive such as dimethylacetoamide (DMA)

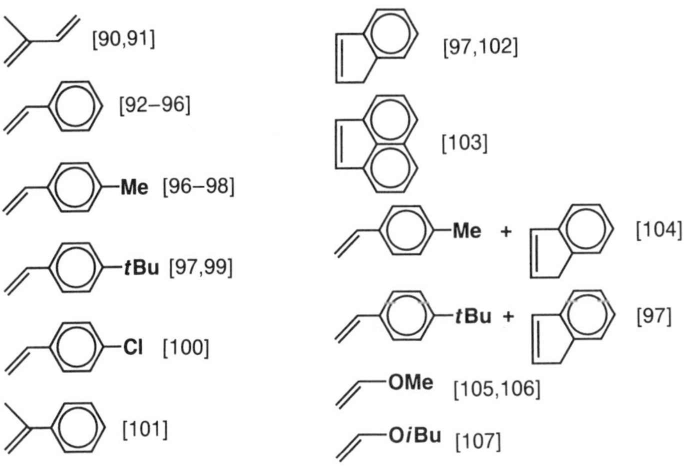

Figure 7 Block copolymers of isobutene: a list of comonomers.

at a low temperature ($-80°$ C) [92,100]. The most favored system utilizes a bifunctional initiator, *p*-dicumyl methyl ether [1,4-bis(2-methoxypropyl)benzene], in methyl chloride/methylcyclohexane (4:6 v/v) mixture as solvent to give ABA block copolymers [92–104]. For the block copolymerizations with vinyl ethers, the *tert*-alkyl chloride/$TiCl_4$ system is combined with an added salt in a slightly more polar solvent, methyl chloride/methylcyclohexane (6:4 v/v) [105–107]. In the latter systems with vinyl ethers, it is also pointed out, as also discussed above for styrenes, that further addition of $TiCl_4$ is necessary to effect the second-phase vinyl ether polymerizations, which is attributed to the complexation of $TiCl_4$ with the vinyl ether monomers [105,106].

C. Block Copolymers by Reactions of End-Functionalized Polymers

In addition to the sequential living polymerizations, block copolymers may be synthesized by various reactions of end-functionalized polymers. The method may be subdivided into two classes, as discussed in Section III.A. Most of the examples via living cationic polymerizations are based on vinyl ethers, and Fig. 8 lists the second monomers used in these vinyl ether-based block copolymers. Similar syntheses based on the inifer methods are discussed elsewhere [1,3].

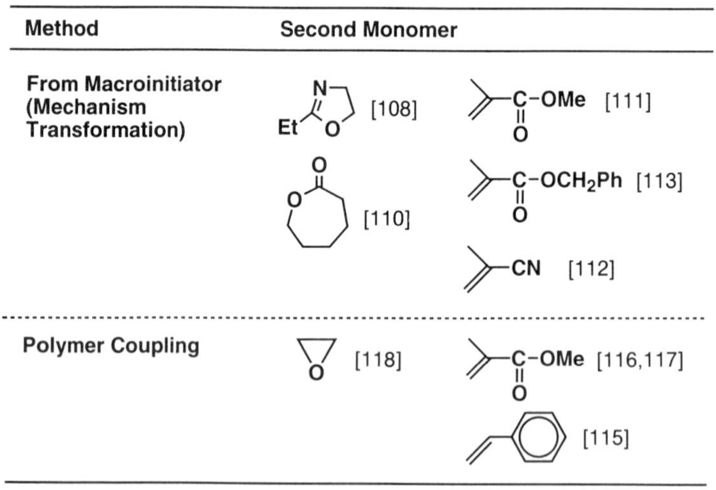

Figure 8 Monomers for the block copolymer synthesis by transformation of mechanisms.

1. Polymerization from Macroinitiators

As will be discussed in Section IV, polymers with a variety of terminal functions can be synthesized by living cationic polymerization, and some of the end functions may be useful for initiating another polymerization to give block copolymers. Mechanistically, therefore, this method involves transformation of a living end into another with a completely different nature.

a. Cationic → Onium Poly(IBVE) with a primary chloride terminal [Cl—$CH_2CH_2OCH(CH_3)$-poly(IBVE); from 2-chloroethyl vinyl ether] [108] has been used to initiate ring-opening cationic polymerization of 2-ethyl-2-oxazoline. Before the oxazoline polymerization, the terminal chloride is converted into the more reactive iodide counterpart by treating the polymer with sodium iodide. The living cationic polymerizations of isobutene and *p*-chlorostyrene with tertiary alkyl chlorides, although quenched with methanol, usually give chlorine-capped polymers. These polymeric tertiary or benzylic chlorides (from isobutene and *p*-chlorostyrene, respectively) can be combined with silver salts to initiate ring-opening polymerization of tetrahydrofuran [65,109].

b. Cationic → Anionic In the presence of a stannous catalyst, poly(IBVE) with a hydroxyl terminal initiates anionic ring-opening polymerization of ϵ-caprolactone to form poly(vinyl ether)-polyester block copolymers [110]. In another example, the chloride terminal of poly(isobutene) is transformed into a benzyl anion to initiate anionic polymerization of methyl methacrylate [111].

c. Cationic → Radical Methacrylonitrile [112] and methyl methacrylate [113] are radically polymerized from poly(vinyl ethers) with terminal azo-initiating sites prepared by living cationic polymerization. Obviously, the second-stage radical polymerization is not living, so that the product may be a mixture of the intended block polymers and the respective homopolymers.

d. Anionic → Cationic As a reversed case for polymerization b, polybutadiene diol, a commercially available telechelic polymer, is converted into an macroinitiator with two haloether terminals that initiates vinyl ether polymerization in the presence of zinc halides [114].

2. Polymer Coupling Reactions

The simplest examples of this class are the quenching living cationic polymers with living anionic or nucleophilic polymers. Namely, living poly(vinyl ethers) derived from the HI/ZnI_2 system are allowed to react with living anionic polystyrene with the lithium counterion [115], poly(methyl methacrylate) with a silyl ketene acetal terminal by group transfer polymerization [116,117], and polyethylene glycol with hydroxyl terminals [118].

In a reversed way, cationically prepared end-functional polymers are used to quench other living polymers. For example, living anionic polystyrene may be terminated by polyisobutenes with silylchloride terminals [119,120] or epoxide ends [121,122] and by poly(vinyl ethers) with acetal terminals [123]. The former case is reported to give H-shaped, tetraarmed block copolymers.

D. Amphiphilic Block Copolymers

Block copolymers that consist of hydrophilic and hydrophobic segments are typical "amphiphilic" polymers, a variety of which have been synthesized by living cationic polymerization. Figure 9 schematically illustrates the structures of some of these amphiphilic polymers thus far obtained; though the examples therein are based on poly(vinyl ether) segments, any other appropriate segments may be incorporated. As seen in the illustrations, macromolecular amphiphiles are not necessarily linear AB- and ABA-type block copolymers but may be graft, multiarmed, and network polymers, where the basic components are amphiphilic block copolymers.

In addition to their traditional applications as surfactants, dispersants, etc., amphiphilic polymers have recently been attracting active interest in terms of their behavior at liquid-liquid, solid-liquid, and other interfaces (micellization, segmental conformation, etc.), along with their biocompatibility. In this section, amphiphilic block copolymers alone are briefly discussed. Graft and multiarmed polymers with amphiphilic arms will be treated later in this chapter (Section VI).

Most typically, amphiphilic block copolymers may be synthesized by sequential living cationic polymerization (cf., Section III.B). For example, 2-acetoxyethyl vinyl ether (AcOVE) and isobutyl or other alkyl vinyl ethers are sequentially polymerized with HI/I_2 or $HCl/ZnCl_2$ into block copolymers [Scheme 2(A)], and the pendant acetate esters in the poly (AcOVE) segments are converted into hydroxyl groups by alkaline hydrolysis to give amphiphilic block copolymers of 2-hydroxyethyl vinyl ether and alkyl vinyl ether (see formula **1** in Fig. 9) [80]. Similarly, amphiphilic poly(vinyl ethers) with amino [85] and carboxyl [84] groups are synthesized (**2** and **3**, Fig. 9). Depending on their pendant functionalities, these polymeric amphiphiles may be nonionic (**1–3**) or ionic (**4** and **5**), and they invariably function as efficient surface active agents that reduce the surface tension of water from 78 to 30 N/m (dyn/cm) or below [80,84]. More recently, the micelle structures of these block polymers are being determined by small-angle x-ray and neutron scattering [124].

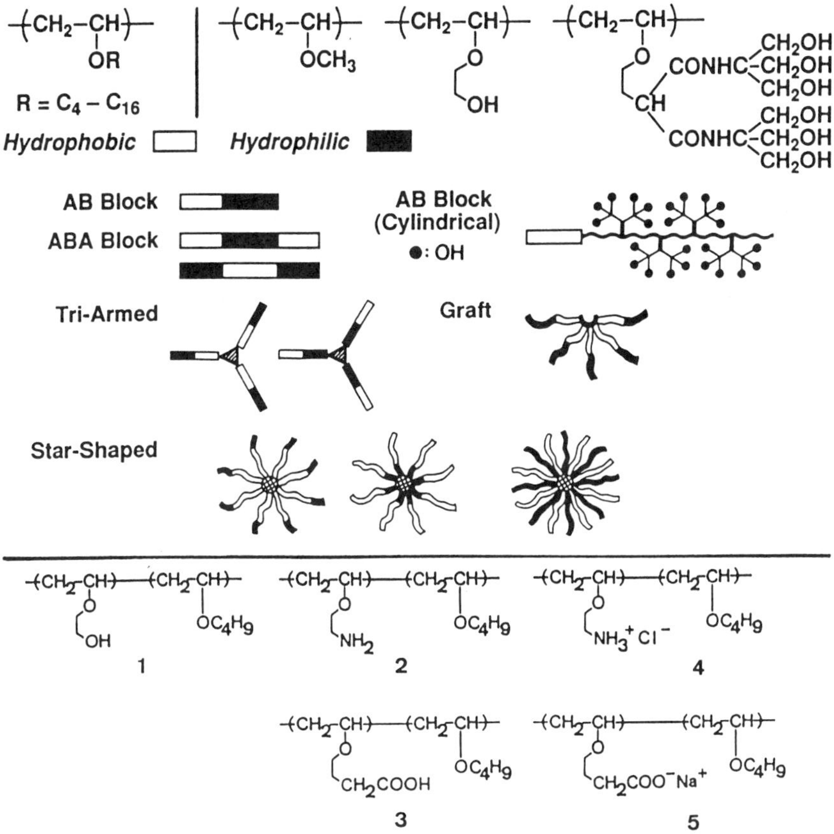

Figure 9 Schematic illustrations of amphiphilic block copolymers of varying spatial shapes and thypical examples of amphiphilic AB-block copolymers of vinyl ethers (**1–5**; [80–85]).

Some of block copolymers with styrene derivatives are also amphiphilic, such as methyl vinyl ether–*p*-alkoxystyrene [88] and alkyl vinyl ether–*p*-hydroxystyrene (from *p*-*t*-butoxystyrene) [89] (Fig. 6).

Another example of amphiphilic block copolymers is illustrated in Scheme 3 [125]. The first step is the sequential living polymerization of the malonate-bearing vinyl ether (**6**) and 1-hexadecyl vinyl ether. The geminal esters in the malonate in the product block copolymer (**7**) are then amidated with a primary amine [$H_2NC(CH_2OH)_3$] carrying three hydroxyl groups. The product is thus an amphiphilic polymer (**8**) whose hydrophilic segment consists of regularly branched repeat units, each with as many

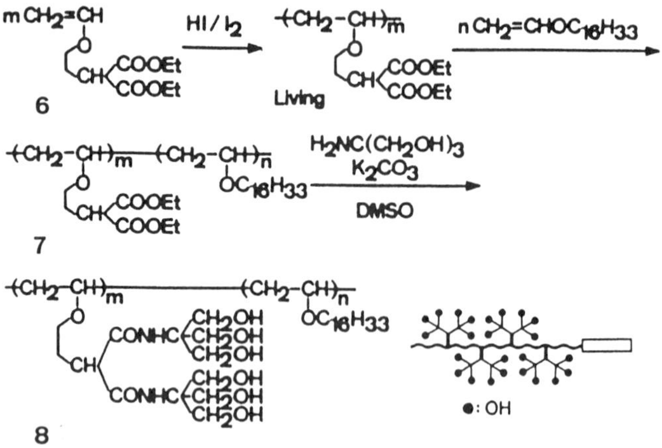

Scheme 3 Synthesis of amphiphilic block copolymers of a "cylindrical" shape where the pendant substituent carries six hydroxyl groups attached to a regularly branched spacer [125].

as six hydroxyl groups. Inspection of molecular models indicates that the hydrophilic segments with the bulky pendant groups may assume a cylindrical form, the surface of which is covered by a number of hydroxyl groups. Transmission electron microscopic analysis of their cast films also revealed interesting phase separation phenomena. These topologically unique polymers are relevant to so-called dendrimers [126] and related multibranched polymers with amphiphilic nature (cf., Section VI).

IV. POLYMERS WITH TERMINAL FUNCTIONAL GROUPS

A. General Methodology and Scope

Along with block copolymers, polymers with terminal functions, or "end-functionalized" polymers, are another typical class of well-designed polymers that living polymerizations can provide. On the basis of the absence of chain transfer and termination, when coupled with the quantitative and selective initiation from a well-defined initiator, living polymerizations offer two basic methods to prepare end-functionalized polymers, as Scheme 4 illustrates for cationic processes:

A. Functional Initiator Method:
   Initiation from functionalized initiators;
B. Functional Terminator Method:
   Termination with functionalized terminators.

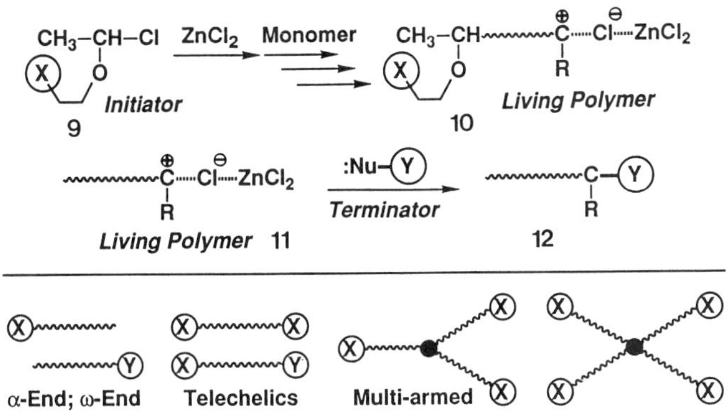

Scheme 4 Synthesis of end-functionalized polymers by living cationic polymerization: Methodologies and typical architectures.

where *X* and *Y* are functional groups for terminal functions, which often are protected to prevent side reactions during the living polymerizations.

1. *Functional Initiator Method*

The functional initiator method requires specially designed initiators that carry some functional group (*X*) and are able to initiate living polymerizations quantitatively. In cationic polymerization, such functionalized initiators (e.g., **9**, Scheme 4) may be obtained from vinyl ethers with a pendant function (Fig. 3) by the quantitative electrophilic addition of hydrogen halides [15,16,127,131]. In conjunction with an appropriate activator (coinitiator) such as $ZnCl_2$, initiator **9** and its analogs initiate living cationic polymerizations of vinyl ethers, styrene and its derivatives, and others to form polymers (**10**) where the first repeat unit (the α-end or head group) is derived from the initiator and thus carries a terminal function *X*. Similar initiation processes might also be feasible for isobutene.

These functional initiators should fulfill the following criteria: (1) By itself or in conjunction with an appropriate activator (Lewis acid; e.g., $ZnCl_2$), the initiator should be able to initiate, quantitatively, living cationic polymerization of a vinyl monomer; (2) the functional group *X* in the initiator (e.g., **9**) should not induce undesirable side reactions that deteriorate living cationic processes; thus they often take protected forms accordingly. Other criteria for the functional initiators are basically the same as for initiators to be used for living cationic polymerizations; namely, the counteranions and the activators, along with reaction conditions, should be carefully designed according to the structure of monomers.

2. *Functional Terminator Method*

The functional terminator method in cationic polymerization calls for nucleophiles (:Nu-*Y*) with functional groups (*Y*). As required, the nucleophilic part of the terminator, but not the function *Y*, should combine with the living end (e.g., **11**, Scheme 4) and thereby attach the function *Y* to the last repeat unit (the ω-end or tail group) of polymer **12**. The terminators should meet the following criteria:

a. The quenching reaction occurs exclusively with the growing end, not with the β-proton, which is another electron-deficient site that the added nucleophile might attack. Recall that addition of nucleophiles to conventional cationic polymerization systems often leads to the β-proton elimination, leaving less useful olefin terminals.

b. The quenching reaction with the living end should form a stable covalent linkage that withstands the subsequent chemical treatment of the polymer (such as transformation of the terminal functions). This criterion, of course, must be considered in accordance with the monomer structure.

c. The quenching reaction should be fast and efficient enough to avoid possible spontaneous decay that the living end might undergo [128], specifically in the virtual absence of monomer where, by definition, the end-capping reaction is carried out most frequently.

As expected, combination of the two methods, i.e., quenching the α-functionalized living polymer **11** with the terminator :Nu-*Y*, leads to "telechelic" or α,ω-bifunctional polymers (see Section IV.B.4). These methods are also applicable in the living polymerization with multifunctional initiating systems or with multifunctional terminating agents to provide "multiarmed" and end-functionalized polymers (see Section VI.C). The end groups also include polymerizable groups such as methacrylate, and the macromonomer synthesis will be treated separately in Section IV.B.5.

3. *"Inifer" Method*

Before the development of living cationic polymerization in the 1980s, Kennedy and his co-workers devised another way to synthesize end-functionalized polymers, which uses special reagents called "inifer," or initiator-chain transfer agents [129]. The method is primarily for the synthesis of polyisobutene with a tertiary chlorine terminal, which is, however, a synthon for a variety of other functional groups. These developments have been reviewed extensively [1,3,130] and fall outside the scope of this chapter.

Rather interestingly, most of the end-functionalized polyisobutenes have been prepared by the inifer method, in sharp contrast to that those for vinyl ether and styrenic counterparts come from living polymerizations (Scheme 4), as discussed in the following sections.

B. End-Functionalized Polymers by Living Cationic Polymerization

1. *Vinyl Ethers*

The feasibility of the two general methods for end-functionalization in cationic polymerization (Scheme 4) has been examined and confirmed with these monomers (for example, see Ref. 131). Thus, a wide variety of end-functionalized poly(vinyl ethers) are available, as summarized in Table 1 [30,47,81,82,108,116,117,131–147]. Inspection of this table indicates that almost all of important end groups are covered, including hydroxyl, carboxyl, amino, and aldehyde.

For the functional initiator method, initiators [**9**; $CH_3CHZ(OCH_2CH_2—X)$; Z = Cl, I; X = functional group] are prepared in advance by the electrophilic addition of dry hydrogen iodide or chloride with a pendant-functionalized vinyl ether [15,16,127,131]. Mostly the addition reactions are selective and quantitative. For the initiator synthesis, almost all of the functionalized monomers listed in Figure 3 (Section II.B.1) may be used, and thus the functional groups include hydroxyl, amino, carboxylic acid, etc., which are usually protected (e.g., as acetate for hydroxyl group). As the functionalized initiators, the adducts are used in conjunction with iodine, zinc halides, and other metal halides as activators. Usually, the hydrogen chloride adduct and zinc chloride are preferred to the iodide counterparts, due to their easy handling and commercial availability. The initiating systems with ethylaluminum dichloride and an added nucleophile (1,4-dioxane, etc.) may also be used, which can induce living cationic polymerizations and end-functionaliztion well above room temperature [30,145].

Besides the initiation with the vinyl ether adducts, trimethylsilyl halides in conjunction with oxolane [135] or a carbonyl compound [136–141] also provide an interesting method of end-functionalization. As discussed in Chapter 4, Section V.E.2 (also Figure 9 therein), the α-end group is $(CH_3)_3SiO—$, derived from the silyl compound, to be converted into the hydroxyl group [140,141]. Depending on the structure of the carbonyl compounds, it is either secondary (from aldehyde) [136–139] or tertiary (from ketone) [137,138,140,141], both of which are difficult to obtain from the vinyl ether adducts (note that the adduct of AcOVE leads to a primary alcohol [30,31]).

The terminators for vinyl ether polymerizations include the sodium salt of ethyl malonate [sodiomalonic ester; $Na^{+}\,{}^{-}CH(CO_2C_2H_5)_2$] [131],

Table 1 End-Functionalized Polymers of Vinyl Ethers[a]

Monofunctional Polymers

α-End Group (X)	[Ref]	ω-End Group (Y)	[Ref]
$Z—CH_2CH_2OCH(CH_3)—$		$—CH_2COOH$	[131]
Z: $HOOCCH_2—$	[30,131]	$—C(CH_3)_2COOCH_3$	[116,117, 142]
$HOOCCH_2O—$	[132]	$—CH_2COR^1$	[142]
$HO—$	[30,133]	$—CH_2CHO$	[117,133]
$H_2N—$	[30,133]	$—CH_2CH_2OH$	[117,133]
$Cl—$	[108]	$—NH\textit{n}Bu$	[81]
$CH_3CH(OCOCH_3)—$	[134]	$—NH(CH_2)_4NH_2$	[81]
$HOCH_2CH_2OCH_2—$	[135]	$—NH(C_6H_4—Z)$	[143]
$HOCH(R^1)—$	[136–139]	$—CH_2CH(C_6H_4CH_3)NH\textit{n}Bu$	[144]
$HOC(R^1)(R^2)—$	[137,138, 140,141]		

Telechelic Polymers

α-End Group (X)	ω-End Group (Y)	[Ref]
$HOOCCH_2CH_2CH_2OCH(CH_3)—$	$—CH_2COOH$	[131,145]
$HOCH_2CH_2OCH(CH_3)—$	$—CH_2CH_2OH$	[133]
$H_2NCH_2CH_2OCH(CH_3)—$	$—C(COCH_3)_2(CH_2)_3NH_2$	[146]
$HOOCCH_2—$	$—CH_2COOH$	[82]
$H(O)CCH_2—$	$—CH_2CHO$	[47]
$HOCH_2CH_2—$	$—CH_2CH_2OH$	[47]
$HOCH_2CH_2OCH(CH_3)—$	$—CH(CH_3)OCH_2CH_2OH$	[147]
$HOCH_2CH_2OCH(CH_3)—$	$—CH_2COOH$	[145]
$H_2NCH_2CH_2OCH(CH_3)—$	$—CH_2COOH$	[145]
$HOOCCH_2CH_2CH_2OCH(CH_3)—$	$—C_6H_5$	[145]
$CH_3COOCH_2CH_2OCH(CH_3)—$	$—CH_2CHO$	[133]

[a] $X—\left(—CH_2—\underset{\textstyle OR}{CH}—\right)_n—Y$ (R = Me, Et, *n*Bu, *i*Bu, CH_2CH_2Cl)

silyl enol ethers [116,117,142], alkyl amines [81], ring-substituted anilines [143], water [117,133], and alcohols [40]. Among these, the malonate and silyl enol ethers would be the best, because their termination reactions lead to stable carbon-carbon bonds. On the other hand, amines and alcohols are highly effective but need some care, because they form acid-sensitive aminoether [$—CH_2CH(OR)NR'_2$] and acetal [$—CH_2CH(OR)(OR')$] groups upon termination with the vinyl ether living ends. The quenching with water takes this unfavorable situation into advantage, where the resulting labile hemiacetal terminal [—$CH_2CH(OR)(OH)$] is immediately converted by acid-catalyzed dealcoholation into an aldehyde (—$CH_2CH{=}O$) [117,133].

2. *Styrene Derivatives*

Rather unexpectedly, the hydrogen halide-vinyl ether adducts (**9**) turn out to be effective in initiating living cationic polymerizations of styrene and its derivatives, although the vinyl ether-type initiating carbocations differ considerably from those derived from styrenic monomers. Table 2 lists these examples [16,148–151] obtained by the functional initiator method. The monomers include styrene [151], α-methylstyrene [16,148], *p*-methylstyrene [151], *p*-chlorostyrene [67], *p*-methoxystyrene [149], and *p-tert*-butoxystyrene [150]. According to the reactivity of the monomers, appropriate activators (Lewis acids) should be used, as already discussed in Chapter 4, Section V.C. Thus, unlike those for vinyl ethers, the activators are mostly strongly Lewis acidic; for example, tin tetrachloride is used for styrene in the presence of a salt [151,152]. Despite such operational differences, the basic features of the synthesis are very similar to those for vinyl ethers, and the polymers carry the α-end groups arising from the vinyl ether adduct (initiator) where the terminal functions cover hydroxyl, carboxyl, and amino groups (cf., Scheme 4).

Table 2 End-Functionalized Polymers of Styrene Derivatives

	Monomer [Ref][b]				
α-End Group (X)[a]	*p*MOS	*p*BOS	*p*MS	St	αMS
—CH_2CO_2H	[149]	[150]			
—NH_2	[149]		[151]	[151]	[16]
—OH	[149]	[150]	[151]	[151]	[148]
—Cl			[151]	[151]	[148]

[a]
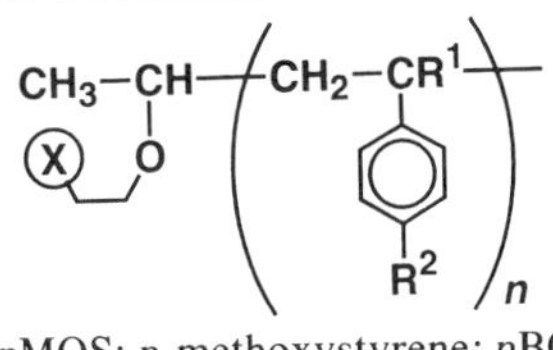

R^1 = H, CH_3
R^2 = H, CH_3, OCH_3, O*t*Bu

[b] *p*MOS: *p*-methoxystyrene; *p*BOS: *p*-*t*-butoxystyrene; *p*MS: *p*-methylstyrene; St: styrene; αMS: α-methylstyrene.

Importantly, the selection of functional terminators for styrene derivatives needs some care, due to considerable difference in the structure and reactivity of the living end from the styrene monomers relative to those from vinyl ethers. Thus, for *p*-alkoxystyrenes the malonate anion is totally ineffective, whereas functionalized alcohol (HOR—X; X = functional group) are probably of the best choice, because they form stable ether linkages [—CH_2CHAr—OR—X; Ar = C_6H_4—$OCH_3(p)$, etc.] with the living ends. Such alcoholic terminators include 2-hydroxyethyl acetate and 2-hydroxyethyl methacrylate [$HOCH_2CH_2$—$OCOR^1$; R^1 = CH_3, $C(CH_3)$=CH_2] [149,150]. However, these alcohols are too weakly nucleophilic to terminate the living end of styrene, giving instead chloride terminal [—$CH_2CH(C_6H_5)$—Cl] derived from the initiating systems [153,154]. The terminal benzylic chloride would be an interesting function. Although not highly effective, selected organisilicon compounds such as truimethylsilyl esters [$(CH_3)_3SiOCOR^2$; R^2 = CH_3, $C(CH_3)$=CH_2] may also be used to terminate the growing ends from styrene and *p*-methylstyrene [16,155].

3. *Isobutene*

As pointed out already, rather few end-functionalized polyisobutenes have been obtained from isobutene via living cationic polymerization, whereas abundantly via the inifer method followed by various chemical reactions to convert the resulting tertiary chloride terminal (Section IV.A.3) [3].

Recently, a functional initiator method has been reported for isobutene, where the initiator is $CH_3OC(O)$—Ar—$C(CH_3)_2$—Cl (Ar = *t*-BuC_6H_3) [156]. In the presence of $TiCl_4$ (activator) and *N*,*N*-dimethylacetamide (as an added nucleophile), the cumyl-type moiety of the initiator initiates living cationic polymerization; the acetate moiety serves as the protected carboxylic acid.

In sharp contrast to the vinyl ether systems, attempts to attach terminal functions via the terminator methods usually result in the tertiary chloride group [—$CH_2C(CH_3)_2$—Cl] derived from the initiating system [157]. In this respect the situation is similar to the polymerization of styrene with the chloride-based initiating systems (see Section IV.B.2) [152–155]. These facts show that the growing ends of both monomers interact strongly with the choride anion. An exception is probably the termination with allyltrimethylsilane [CH_2=$CHCH_2Si(CH_3)_3$] to give an allylic ω-end [—$CH_2C(CH_3)_2$—CH_2CH=CH_2] [158].

4. *Telechelic Polymers*

Telechelic or α,ω-bifunctional polymers carry functional groups at both terminals. By varying combinations of initiation and termination reactions, telechelic polymers may be prepared in living polymerizations via three ways, as summarized in Scheme 5:

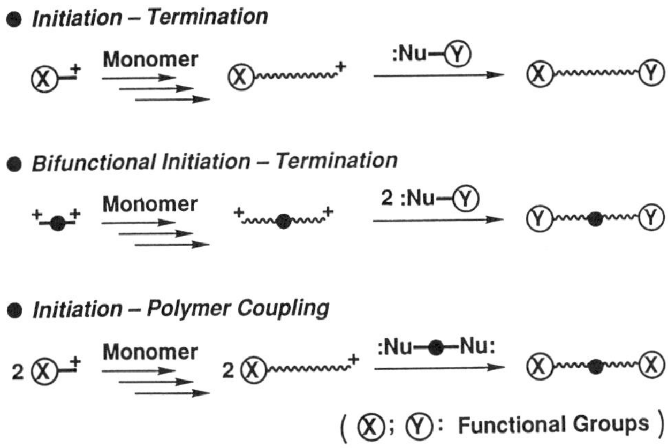

Scheme 5 Synthesis of telechelic polymers: methodologies for homo- and heterotelechelic polymers.

Namely,

	(Initiation)	/	(Termination)	[Ref]
A.	Monofunctional	/	Monofunctional	[131,145,146]
B.	Bifunctional	/	Monofunctional	[82,158]
C.	Monofunctional	/	Bifunctional	[147]

All these methods have been used successfully in living cationic polymerizations. In telechelic polymers ($X\sim\sim\sim\sim\sim Y$; Scheme 5), the two terminal functions (X and Y) may be the same as, or different from each other, and in the former case the polymers may be called "homotelechelic" (or symmetrically telechelic; $X = Y$), whereas in the latter "heterotelechelic" (or asymmetrically telechelic; $X \neq Y$). By definition, methods B and C give exclusively homotelechelic polymers, whereas method A can give either homo- or heterotelechelic polymers, depending on the combinations of the initiator and the terminator.

For example, Scheme 6 gives an example of the synthesis of heterotelechelic poly(vinyl ether) by method A, where α-end-functional living polymers, derived from functional initiators, are terminated with the malonate

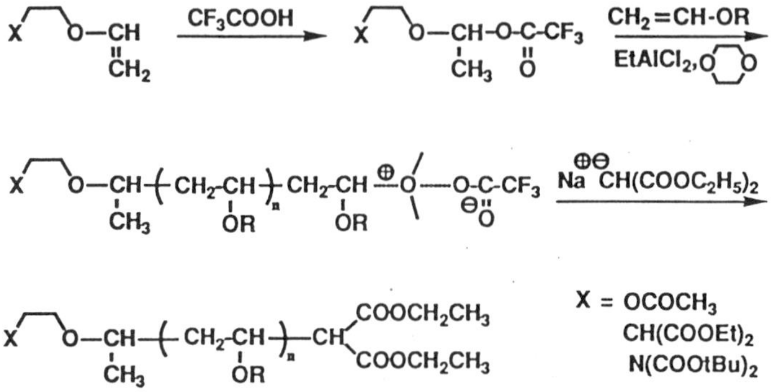

Scheme 6 Synthesis of heterotelechelic poly(vinyl ethers) [145].

anion (for the carboxyl ω-end group) [145]. Similarly, a series of telechelic polymers have been prepared from vinyl ethers (Table 1) [47,82,131, 133,145–147] as well as p-methoxystyrene [149]. As seen, examples are rather ample for vinyl ethers.

The synthesis, starting from a bifunctional initiator followed by quenching the double-headed living ends, gives homotelechelic polymers (method B). Carboxylate-capped telechelic poly(isobutyl vinyl ether) has been obtained in this way [82], where the adduct of a bifunctional vinyl ether with trifluoroacetic acid is the initiator, and the quencher is the malonate anion. For method C, a bifunctional trimethylsilyl enol ether, $CH_2{=}C[OSi(CH_3)_3]C_6H_4OCH_2CH_2OC_6H_4[(CH_3)_3SiO]C{=}CH_2$, is a useful terminator (chain coupler) for vinyl ethers [142,147] and α-methylstyrene [159] (see also Section VI.B.4).

5. *Macromonomers*

A class of end-functionalized polymers with polymerizable terminal groups are generally called "macromonomers." By both functional initiator and terminator methods, a variety of macromonomers have been synthesized in living cationic polymerization of vinyl ethers, styrenes, and isobutene, as summarized in Table 3 [16,31,147,149–151,155,158–171]. Some of these macromonomers are used in the synthesis of graft polymers (Section VI.C).

The polymerizable groups include methacryloyl, styrenic, epoxide, vinyl ether, and others, among which the methacryloyl-capped macromonomers are most widely available from vinyl ethers, styrene, and its derivatives. For example, the α-end methacryloyl group can be introduced by the functional initiator method, with the hydrogen halide-adduct of 2-

Table 3 Macromonomers by living cationic polymerization

Method	Polymerizable End Group	Monomer [Ref][a]
Initiation (α-End)	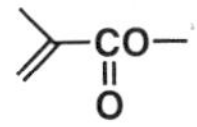	VE [31,160,161], *p*MOS [151], St [151], αMS [16]
	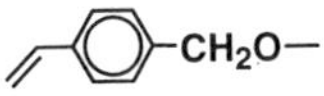	VE [147,162]
	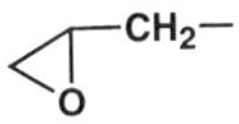	VE [161]
Termination (ω-End)	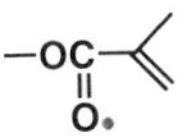	VE [163], *p*MOS [149], *p*BOS [150], St [155], IB [171]
	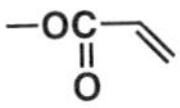	*p*MOS [149], *p*BOS [150]
	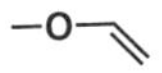	VE [164,165], IB [170]
	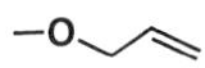	VE [163]
		IB [158], St [155], αMS [16]

[a] VE: vinyl ethers; *p*MS: *p*-methylstyrene; St: styrene; αMS: α-methylstyrene; *p*MOS: *p*-methoxystyrene; *p*BOS: *p*-*t*-butoxystyrene; IB: isobutene.

(methacryloyl)ethyl vinyl ether as initiator [$CH_3CH(X)OCH_2$-CH_2O-$COC(CH_3){=}CH_2$], into polymers of vinyl ethers [31,160,161] and styrene derivatives [16,151], whereas the quenching with methacrylate-containing alcohol [149,150,163], amine [163], and silane [155] affords the corresponding ω-ends. The vinyl ether terminal (ω-end) may be attached with a malonate-type anion, $—C(CO_2C_2H_5)_2CH_2CH_2OCH{=}CH_2$, as the quencher derived from the corresponding vinyl ether [164].

Apart from the syntheses by living cationic polymerizations, a variety of polyisobutene macromonomers have been prepared by converting the chloride terminal derived from the inifer method (Section IV.A.3) [166–170] and more recently from living cationic polymerization [3,157,171]. The polymerizable terminals include methacryloyl [166,167,171], styryl [168], acrylonitrile [169], and vinyl ether [170]. Bi- [166–168] and trifunctional [171] isobutylene macromonomers are also available; for example, the trifunctional version has recently been employed in toughening poly(methyl methacrylate) via network formation [171].

V. SEQUENCE-REGULATED OLIGOMERS AND POLYMERS

A. General Methodology and Scope

Among the structural factors that should be controlled in polymer syntheses (Fig. 1, Section I), perhaps the least exploited is the sequence of constitutional repeat units along a polymer main chain. We have already discussed the syntheses of block copolymers, where two or more homopolymer segments are connected, such as ···AAAAA-BBBBB···, which is among the most primitive examples of sequence control in synthetic polymers.

Herein let us consider more advanced sequences, typically ····A-B-C-D-E···, where each constitutional repeat unit (A, B, C,···) originates from a different monomer. Such precise control of repeat unit sequences can be found abundantly in the biological synthesis of genes and polypeptides [172–176], but that in addition polymerizations is currently far beyond our reach. Although some kind of template processes may be useful to achieve the sequence control, there has not been yet any conclusive success. If one adopts a way of sequence control similar to those in vivo, one must consider at least the following: (a) how to input information about the predetermined sequence into the polymerization reaction system; (b) according to this information, how to recognize one particular monomer among many in the reaction system; and (c) how to polymerize the monomer thus recognized into polymers. Thus, currently no general methodology has been established in the nonbiological synthesis of "sequence-regulated" polymers.

B. Sequence Regulation by Living Cationic Polymerization

Apart from the template synthesis in vivo, there has been an attempt to control the repeat unit sequence, though rather primitively, via living cationic polymerization [177–179]. The work is based on the assumption that, if the living end is stable enough to maintain its activity for a sufficiently long time, repeated sequential additions of different monomers will permit some control in repeat unit sequences; Scheme 7 illustrates an example in vinyl ether oligomerization [178]. Although primitive thus far, the approach shown in Scheme 7 implies that, among the three conditions (a)–(c) listed above, the sequential addition of different monomers will "manually" fulfill the first two, and the remaining point (c) will be whether the living end survives long enough and reacts rapidly enough to prevent homopropagation.

Thus, the reaction starts with the addition of hydrogen iodide to *n*-butyl vinyl ether, which is considered the first repeat unit. To the resulting

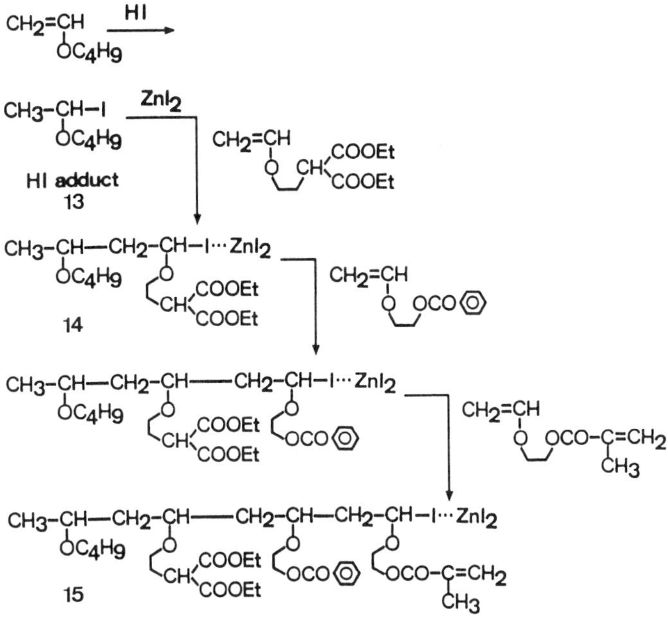

Scheme 7 Synthesis of sequence-regulated tetramers of vinyl ethers where each repeat unit carries a different pendant group [178].

adduct (**13**) is added a second monomer, equimolar to the first monomer; zinc iodide is further added to trigger the second-step addition (propagation) to give an AB-type heterodimer (**14**). Continuing additions of third and fourth vinyl ether monomers, again each equimolar to the active site, thereby give a tetramer (**15**) in which four different repeat units are connected in the predetermined sequence.

Obviously, although a new monomer is fed equimolar to the growing end at each step, it is imperative to prevent homopropagation (into a sequence such as A-B-B-··· instead of the intended A-B-C-···). Monitoring the product distribution along with the multistep additions in fact reveals that the further the synthesis proceeds, the more the amount of such undesired products. Further examination also indicated that sequence selectivity is higher for sequences where the reactivity of monomers progressively decreases [177].

Despite these problems, it has at least been shown that some control of repeat unit sequences is possible by living cationic polymerization to give up to tetramers (A-B-C-D; e.g., **15**, Scheme 7). In these "sequence-regulated'' oligomers such as **15**, importantly, varying pendant functional groups are placed in the predetermined order along the main chain. Multiple coupling reactions of these oligomers might give higher polymers with repetitive but regulated sequences [179], as often observed in biologically obtained polypeptides [180].

VI. POLYMERS WITH UNIQUE SPATIAL SHAPES

A. General Methodology and Scope

In addition to the repeat unit sequence, another area of current interest in polymer structural control (Fig. 1) may be the spatial or three-dimensional shapes of macromolecules. In fact, the recent development of star [181–184] and graft [185] polymers, as well as starburst dendrimers [126], arborols [186,187], and related multibranched or multiarmed polymers of unique and controlled topology, has been eliciting active interest among polymer scientists. In this section, let us consider the following macromolecules of unique topology for which living cationic polymerizations offers convenient synthetic methods that differ from the stepwise syntheses (polycondensation and polyaddition) [126,186,187].

1. *Star-shaped* or *multiarmed* polymers
2. *Graft* polymers
3. *Macrocyclic* polymers.

B. Star-Shaped Polymers (Multiarmed Polymers)

1. General Methodology

The synthesis of star-shaped or multiarmed polymers may be achieved at least by three methods via living cationic polymerization, as illustrated in Scheme 8:

A. Multifunctional Initiator Method:
   Living polymerization from a multifunctional initiating system
B. Multifunctional Terminator Method:
   Coupling of living polymers with a multifunctional terminator
C. Polymer Linking Method:
   Linking of living polymers with a bifunctional vinyl monomer.

A. Multifunctional Initiator Method For example, living polymerization may be initiated with a multifunctional initiating system to grow multiple polymer chains from a central initiator core to give multiarmed or star polymers [Scheme 8(A)]. The key to this approach is of course the development of multifunctional initiators, which has already been

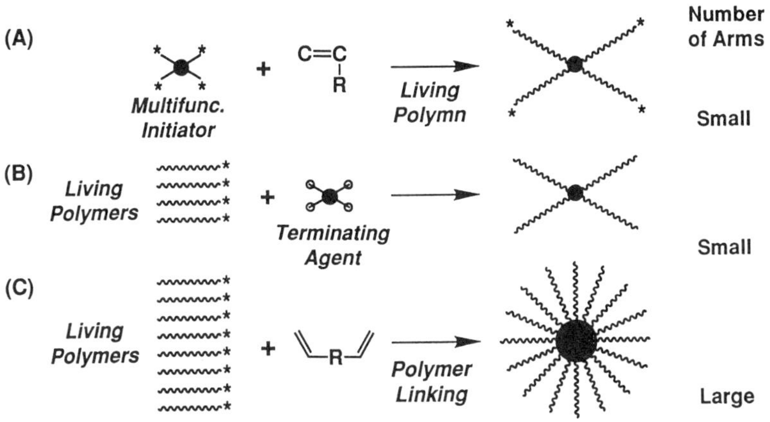

Scheme 8 Synthesis of multiarmed polymers by living cationic polymerization: three typical methodologies.

achieved to some extent, as summarized in Chapter 4, Section V.E.1. These multiple initiating systems may involve living ends similar to their monofunctional counterparts, but an additional consideration and design would be needed to ensure the nearly simultaneous initiation from the more than two initiating sites located in a relatively small, single molecule. Obviously, in these multifunctional initiators, steric crowding and intramolecular side reactions among initiating sites should be minimized by molecular design [e.g., 188,189].

B. Multifunctional Terminator Method Alternatively, monofunctional, linear living polymers may be coupled with a multifunctional terminating agent into star polymers [Scheme 8(B)]. This method highly depends on the development of multifunctional terminators, which should not only fulfill the criteria for monofunctional terminators (Section IV.A.2) but also work under the equimolar conditions with the living ends ([quenching site] = [living end]); otherwise, for example, an incomplete reaction of a tetrafunctional terminator might give a triarmed polymer.

In the multifunctional initiation (A) and the multifunctional termination (B), living polymers grow outward from the initiator core and inward into the terminator core, respectively, but both processes lead to similar polymers. If they operate properly, these methods give multiarmed polymers that carry arms in a precisely controlled or predetermined number per molecule (= the functionality number of the initiator or terminator).

On the other hand, it is probably difficult to implant a large number (>10) of initiating or terminating sites into a relatively small molecule.

C. Polymer Linking Method The third method, sometimes called the "polymer linking" or "microgel" method [181–184], adopts a different approach from these two [Scheme 8(C)]. Namely, when monofunctional, linear living polymers are treated with a small amount of a bifunctional vinyl monomer (usually less than 10 equivalents to the living end), the linear chains are connected together onto a microgel formed from the bifunctional monomer (see also Scheme 11, Section VI.B.4 below). Because of the low content of the bifunctional monomer, the gellation or cross-linking does not spread over the entire polymerization system under proper reaction conditions, and thereby the resulting polymers are completely soluble in reaction media and thoroughly characterizable.

The microgel method (C) is advantageous over the multifunctional initiation (A) and termination (B) methods in that the product polymers may have a large number of arm chains, often exceeding 50 or sometimes 100. Because the microgellation process is statistical, however, there is by definition a distribution in the arm number in a particular polymer sample. In these regards, the polymer-linking method (C) is complementary with the multifunctional initiation (A) and the multifunctional termination (B).

Common for these three approaches is that living cationic polymerization permits the introduction of a variety of interesting functional groups at specific positions (outer arm layer, arm ends, central core, etc.) of the multiarmed polymer architectures. The combinations of these functional groups and the unique molecular topology would lead to physical properties and functions that cannot be found in the conventional linear counterparts.

2. *Multifunctional Initiator Method*

As summarized in Chapter 4, Section V.E.1, a variety of multifunctional initiators are currently available for the living cationic polymerizations of vinyl ethers [83,188,189], alkoxystyrenes [149,190], and isobutene [191–201], and up to tetraarmed polymers with controlled arm lengths are prepared by the use of these initiating systems. Scheme 9 exemplifies such a synthesis for vinyl ethers [188]. The details for the design of these initiating systems are found in Chapter 4.

The key to this initiation is the trifunctional vinyl ether (**16**) from which the corresponding trifunctional initiator (**17**) is prepared by the addition of 3 equivalents of trifluoroacetic acid. Note that the three initiating sites in **17** are well separated spatially from each other by the rigid and planar triphenylmethyl core to avoid intramolecular side reactions be-

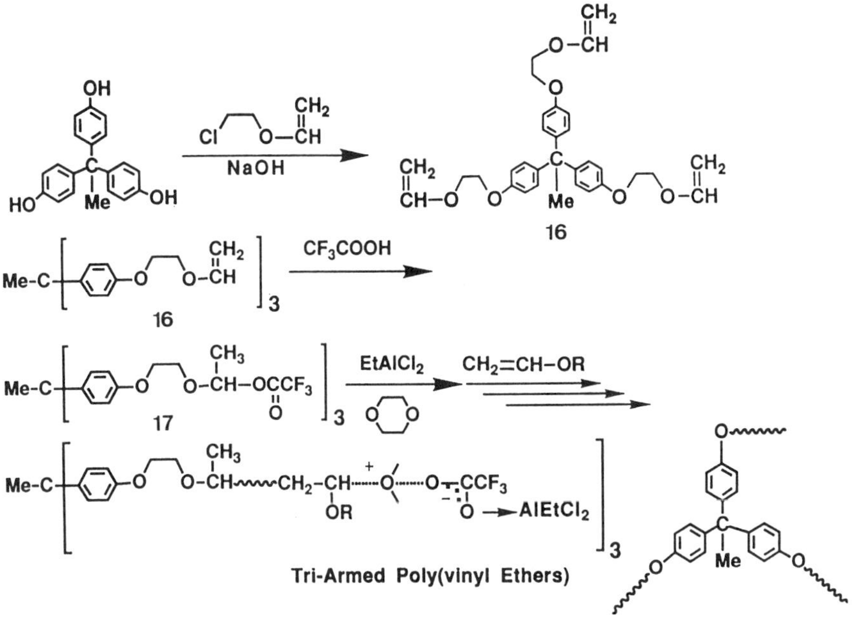

Scheme 9 Synthesis of triarmed poly(vinyl ethers) with a trifunctional initiator [83].

tween a preformed initiating site (trifluoroacetate) and an unreacted vinyl ether during the initiator synthesis.

With the use of the same trifunctional initiating system as in Scheme 9, amphiphilic triarmed block copolymers of vinyl ethers are synthesized [83]. The monomer pair consists of isobutyl vinyl ether (IBVE) for hydrophobic segments and 2-hydroxyethyl vinyl ether (HOVE) for hydrophilic segments; 2-acetoxyethyl vinyl ether (AcOVE) is used as the protected form of HOVE during the polymerization. Thus, IBVE and AcOVE are sequentially polymerized from the trifunctional initiator to form triarmed block copolymers, and the subsequent hydrolysis of the AcOVE parts lead to the target IBVE-HOVE triarmed block copolymers. By reversing the polymerization sequence of the two monomers, it is possible to place the HOVE segment either at the inside or the outside parts of the three arms (cf., Fig. 9).

The multifunctional living polymers thus obtained may be quenched with a monofunctional terminator (Section IV.A.2) to attach a terminal functional group at each of the arm chains. Such end-functionalized, tri- or tetraarmed polymers have been obtained from vinyl ethers [145] and *p*-alkoxystyrenes [149,189,190].

3. *Multifunctional Terminator Method*

When compared with the multifunctional initiators, the corresponding terminators are less available in cationic polymerization [202]. The situation is in sharp contrast to anionic living polymerization, where a variety of multifunctional terminators are developed (e.g., $Cl_2MeSiCH_2CH_2SiMeCl_2$) [203,204]. However, a series of multifunctional silyl enol ethers were recently found to be effective in multiple termination of living cationic polymers of vinyl ethers [142,147,205,206] and α-methylstyrene [159,207] (Scheme 10).

A systematic search of terminating functions [142] indicates that an electron-rich silyl enol ether [$CH_2{=}C(OSiMe_3){-}C_6H_4{-}OCH_2{-}$] is highly suited for rapid and quantitative quenching of the living end derived from vinyl ethers as well as α-methylstyrene. The termination process involves the attack of the growing cation onto the enolate double bond, followed by the release of trimethylsilyl chloride. In this regard, as the good leaving group, the trimethylsilyl moiety serves as an equivalent with the chloride anion in living anionic polymerization.

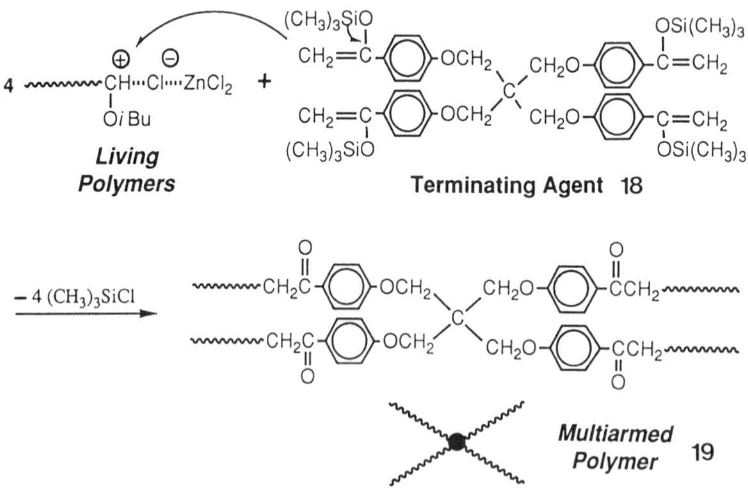

Scheme 10 Synthesis of tetraarmed polymers with a tetrafunctional terminating agent (silyl enol ether) [142,205].

On the basis of these studies, a tetrafunctional silyl enol ether (**18**) carrying four such enolate functions has been prepared; the corresponding trifunctional compound is also available [142]. These silyl enolates effectively couple living poly(vinyl ethers) with the chloride counteranion to form tri- and tetraarmed polymers (e.g., **19**, Scheme 10) [205]. Similar chemistry also operates with living cationic poly(α-methylstyrene) but specifically needs the additional use of an amine to accelerate the release of the trimethylsilyl groups [159,207].

The silyl enolate coupling processes with **18** are further applicable to the synthesis of end-functionalized polymers [147,159] and amphiphilic block copolymers [206,207] with the tetraarmed architectures. The former involves the coupling of four living chains of IBVE or α-methylstyrene carrying an α-end-functional group (by the functional initiator method; Section IV.A.1). The tetraarmed amphiphilic block copolymers are obtained by coupling four chains of AB-block living copolymers of IBVE and AcOVE, followed by alkaline hydrolysis of the latter segments into poly(HOVE) [206]. The resulting polymers are virtually identical with those obtained by the multifunctional initiation (Section VI.B.2), and it is similarly possible to alter the arrangement of the hydrophilic and the hydrophobic segments within the arms by reversing the polymerization sequence of the two comonomers. The use of α-methylstyrene in place of IBVE leads to more rigid arms than those in the vinyl ether versions [207]. Comparison of solubility and conformational changes in different solvents in fact shows such interesting effects of arm rigidity on polymer properties.

4. *Polymer Linking Method*

As illustrated in Scheme 11, treatment of linear living chains with a small amount of bifunctional monomer (**20** as a vinyl ether) gives multiarmed polymers [181–184]. The initial process is considered to be its block copolymerization from the living end via one of the two vinyl groups. The resulting polymer (**21**) thus carries a short segment derived from **20** with a few unreacted pendant vinyl groups, and it then undergoes inter- and intramolecular reactions of the pendant vinyl groups with the newly generated living ends to form a microgel core to which a relatively large number of the linear chains are connected radially. Thus, a star-shaped, multiarmed polymer (**22**) results, which is completely soluble in polymerization solvents.

Such a synthesis is feasible in living cationic polymerization of vinyl ethers [208]. For bifunctional vinyl ethers (**20**), a series of compounds are examined (Scheme 11), among which **20a** turned out the best, probably due to the appropriate rigidity and length of the spacer. By adjusting reaction

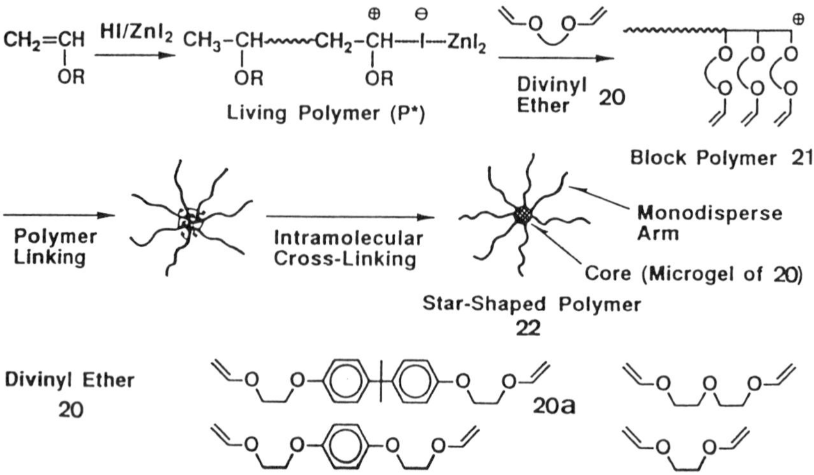

Scheme 11 Synthesis of multiarmed polymers by the polymer-linking reaction with a bifunctional vinyl ether [208].

conditions, such as the mole ratio of **20a** to the living end and the concentration of the latter, the number of arms per polymer can be controlled empirically from 5 to 60. Importantly, the process is also feasible with pendant-functionalized vinyl ethers.

By the use of the polymer-linking method with **20a**, a variety of star-shaped poly(vinyl ethers) have been synthesized (Scheme 12) [208–212]. A focus of these syntheses is to introduce polar functional groups, such as hydroxyl and carboxyl, into the multiarmed architectures. These functionalized star polymers include star block (**23a**,**23b**) [209,210], heteroarm (**24**) [211], and core-functionalized (**25**) [212] star polymers. Scheme 12 also shows the route for the amphiphilic star block polymers (**23b**) where each arm consists of an AB-block copolymer of IBVE and HOVE [209] or a vinyl ether with a pendant carboxyl group [210]. Thus, this is an expanded version of triarmed and tetraarmed amphiphilic block copolymers obtained by the multifunctional initiation (Section VI.B.2) and the multifunctional termination (Section VI.B.3). Note that, as in the previously discussed cases, the hydrophilic arm segments may be placed either the inner or the outer layers of the arms.

Similarly, the microgel-mediated linking of living poly(isobutene) chains with divinylbenzene gives star-shaped rubbery polymers [213]. The living isobutene polymerization is initiated with a tertiary chloride/$TiCl_4$

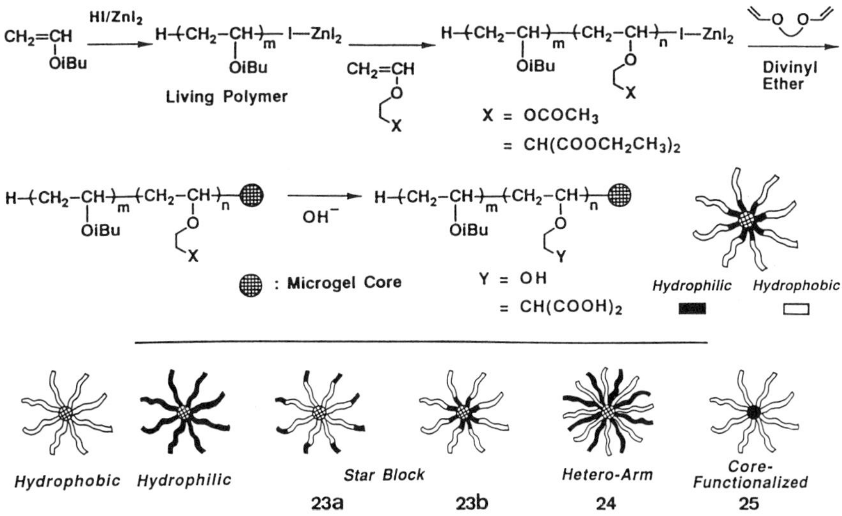

Scheme 12 Synthesis of multiarmed amphiphilic polymers by the polymer-linking reaction [209,210].

system at −40° C in the presence of triethylamine (as an added nucleophile), and the resultant living polymer solution was treated with a mixture of divinylbenzene and ethylvinylbenzene (both *m*- plus *p*-isomers). Detailed analysis of the reaction as well as the products showed the formation of multiarmed star polyisobutylene with a number-average arm number of 56 and $M_w = 113 \times 10^5$.

5. *Amphiphilic Polymers and Host-Guest Interaction*

As seen in the preceding discussion, the multiarmed polymers obtained by various methods are designed to be amphiphilic, where the arm chains are often AB-block copolymers consisting of hydrophilic and hydrophobic segments (see Fig. 9 and Scheme 12). In addition to traditional amphiphilic linear block copolymers (Section III.D), these multiarmed polymeric amphiphiles with unique multiarmed or star-shaped architectures and the controlled placement of hydrophobic and hydrophilic moieties, are of particular interest in their behavior at interfaces etc. [214]. Recently, amphiphilic networks of biological interest have been prepared from polyisobutene, as reviewed by Kennedy and colleagues elsewhere [3,215–218].

For example, amphiphilic star block copolymers (Scheme 12) may be characterized by its high accumulation of hydrophilic alcohol groups in the arm segments, along with its nearly spherical molecular shape [219]. In fact, this unique topology and dense accumulation of functional groups induce specific host-guest interaction with small organic molecules (Fig. 10) [220]. The star block copolymers (**23a**) with hydroxyl arm-pendant groups, as hosts, interact specifically with benzoic acid, 2-(acetoxy)benzoic acid, and 3-(methoxycarbonyl)phenol, as guests, through the hydrogen bonding between the host's hydrophilic arm segments and the guest's carboxyl or phenolic hydroxyl groups. Blocking the latter functions by esterification etc. inhibits such host-guest interaction. The host polymers also recognizes the positional isomers, such as 2- and 3-(methoxycarbonyl)phenols; the former is a poor guest due to the blocking of its phenolic group via intramolecular hydrogen bonding with the vicinal carbonyl.

Similar host-guest interactions are found not only with the amphiphilic star block copolymers [210,220] but also with the corresponding hetero-arm [211] and core-functionalized [212] versions. Overall, these star-shaped polymers induce the interaction more efficiently than their linear counterparts [220].

C. Graft Polymers

Graft polymers may be prepared by a variety of methods [221], including:

A. "Grafting-from" Method:
   Initiation of new polymer chains from the in-chain/pendant active sites of other polymers;

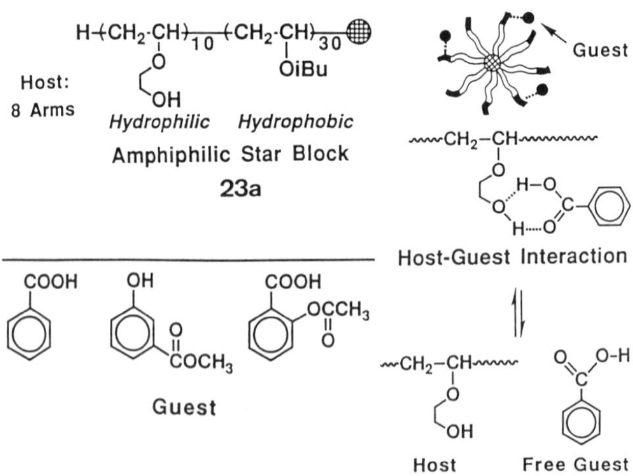

Figure 10 Host-guest interaction between multi-armed amphiphilic poly(vinyl ether) and small organic molecules [220].

B. "Grafting-onto" Method:
Reaction of growing polymers with the in-chain/pendant functional groups of other polymers;

C. "Macromonomer" Method:
"Polymerization" of polymers with polymerizable terminal groups.

Some recent syntheses employ the first method (A), where, for example, living cationic polymerizations of isobutene [222], (*t*-butyl)dimethylsilyl vinyl ether [223,224], and 2-methyloxazoline [225] are initiated from appropriate pendant functional groups.

The macromonomer method (C) has also been adopted in cationic polymerization. For instance, amphiphilic graft polymers of vinyl ethers are synthesized by the cationic polymerization of a vinyl ether-capped macromonomer (**26**) with a block copolymer chain consisting of IBVE and AcOVE segments, followed by alkaline hydrolysis of the latter part into the HOVE units [165]. This graft polymer also undergoes a host-guest interaction similar to those with amphiphilic star block copolymers [220].

$$H{-}[CH_2CH(OR^1)]_m{-}[CH_2CH(OR^2)]_n{-}C(CO_2C_2H_5)_2CH_2CH_2OCH{=}CH_2$$

26 $[R^1 = CH_2CH_2OCOCH_3;\ R^2 = CH_2CH(CH_3)_2]$

D. Macrocyclic Polymers

Noting the clear difference in reactivity between the vinyl ether and the styrenic double bonds, Deffieux and his co-workers synthesized macrocyclic polymers via living cationic polymerization [123,162,226]. Thus, the French group started the synthesis from an asymmetrically hexafunctional monomer, with three vinyl ether and three *p*-alkoxystyrenic groups (**27**, Fig. 11) [162]. Addition of three equivalents of hydrogen iodide selectively

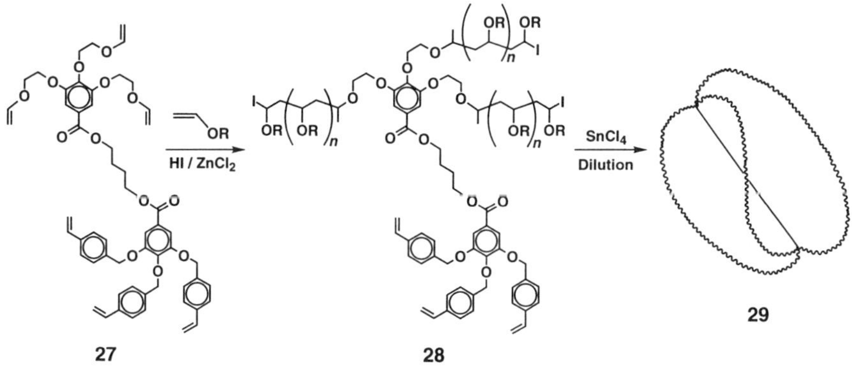

27 28 29

Figure 11 Synthesis of macrocyclic poly(vinyl ethers) starting from a asymmetrically hexafunctional monomer [226].

across the former unsaturated groups gives a trifunctional adduct from which, in conjunction with zinc chloride, 2-chloroethyl vinyl ether is polymerized in a living fashion. The resulting triarmed polymer (**28**) carries three iodoether terminals and three unreacted styrenic moieties. Although, due to the mild Lewis acidity of the zinc halide, the latter three do not react at all during and after the polymerization, the subsequent addition of tin tetrachloride, a stronger Lewis acid, to the as-prepared polymer solution in turn activates the iodoether terminals into vinyl ether-derived carbocationic species that are reactive enough to attack the central styrene double bonds to complete cyclization under highly diluted conditions. The product is an interesting tricyclic poly(vinyl ether) (**29**) in which three nearly uniform polymer chains are connected at two trifunctional junctions [226]. A similar strategy also leads to a monocyclic version [162]. These are good examples of controlled polymer syntheses based on careful consideration of the initiating system design in living cationic polymerization, as already discussed in Chapter 4, Sections V.A and V.C.

VII. EXPERIMENTAL PROCEDURES

Most of the cationic polymerizations and the polymer syntheses thereby, discussed in Chapters 4 and 5, may be carried out conveniently under the dry and inert gas atmosphere (nitrogen or argon) by the so-called syringe technique. Except for highly elaborated kinetic experiments, stringent high-vacuum technique is not required to maintain the control of the polymerizations and polymer architectures.

The following delineates some typical examples of experiments for living cationic polymerizations and related polymer syntheses. They are for vinyl ethers, but the procedures for other monomers are basically the same. On the other hand, polymerization experiments with isobutene are usually carried out in a nitrogen-filled dry box to which are attached a large cryogenic cooling bath, stirrer, and gas inlets to supply the gaseous monomer, dry nitrogen, and methyl chloride (often employed as solvent).

A. General Remarks

1. Polymerization Procedures

All monomers and solvents should be distilled at least twice over calcium hydride. For vinyl ether monomers, see Section VII.A.2. Immediately after the distillation, the monomers are sealed in brown glass vials under dry nitrogen and stored in a refrigerator. The final cut of the distilled solvent is collected over 4A molecular sieves in a glass flask equipped with a three-way stopcock and stored in a desiccator; rubber septa may be used in place of three-way stopcocks but not highly recommended. All polymerizations are carried out under dry nitrogen or argon in glass tubes (Schlenk tubes) or round-bottomed flasks. These glasswares are dried overnight in an oven (>110° C), cooled in a desiccator, equipped with a three-way stopcock, and purged with nitrogen while the walls are heated (baked) with a heat gun immediately before use. Reagents are transferred to the glass tubes via dry syringes through the three-way stopcock against a dry nitrogen flow.

2. *Alkyl Vinyl Ether Monomers*

Alkyl vinyl ethers (CH_2═CH—OR; R = ethyl or higher alkyl) are washed successively with 10% aqueous sodium hydroxide solution and deionized water and distilled at least twice over calcium hydride. The final cut is distributed into small brown ampoules under dry nitrogen and sealed immediately before being stored in a refrigerator. The following show some physical properties needed for experiments:

CH_2═CH—OR			
R (substituent)	C_2H_5	$CH_2CH(CH_3)_2$	$CH_2CH_2OCOCH_3$
Abbreviation	EVE	IBVE	AcOVE
Molecular weight	72.11	100.2	130.2
Density (g/mL)	0.754	0.7638	0.9903
b.p. (uncorrected)	33° C	83° C	75° C/35 Torr

3. *Synthesis of 2-Acetoxyethyl Vinyl Ether*

This monomer is prepared by the substitution reaction of 2-chloroethyl vinyl ether (CEVE) with sodium acetate in the presence of a phase-transfer catalyst (tetra-*n*-butylammonium iodide) [17]. All reagents are of commercial sources and employed without further purification except for vacuum drying when necessary.

In a 500-mL, three-necked, round-bottomed flask equipped with a reflux condenser, a Teflon paddle stirrer, and a drying tube (calcium chloride), and a thermometer are placed CEVE (240 mL, 2.4 mol), sodium acetate (82 g, 1.0 mol), and tetra-*n*-butylammonium iodide (ca. 2 g). The mixture is stirred for 8 hr at 80–90° C in a water bath and cooled to room temperature. The resulting sodium chloride is filtered off and extracted with 200 mL diethyl ether. The ether extract is combined with the reaction mixture, and the ether and unreacted CEVE are removed by evaporation.

The oily crude product is distilled twice over calcium hydride under reduced pressure (75° C/35 Torr) to give AcOVE as a colorless oil. The first distillation should be carried out carefully, where the pressure should be slowly reduced before heating the flask to allow the remaining traces of the ether and CEVE to boil first. The distillate after the double distillation is of gas-chromatographic purity better than 99.5%; the final isolated yield is approximately 50% from the sodium acetate charge (the reaction is quantitative, and the loss is due to the first cuts in the distillation). The distilled AcOVE is immediately sealed in brown ampoules under dry nitrogen and stored in a refrigerator.

4. Preparation of Anhydrous Hydrogen Iodide Solution

The set-up consists of two 200-mL, three-necked, round-bottomed, baked flasks (A and B) that are connected with each other via a drying tube packed with phosphorous pentoxide (2.5 cm i.d., 20 cm long). Flask A (reactor) is equipped with a nitrogen inlet, 50-mL pressure-equilibrating addition funnel, and gas outlet connected to the drying tube with a Tygon tubing [poly(vinyl chloride)]. Flask B (cold trap) is equipped with a three-way stopcock and two gas inlets, one of which is connected to the drying tube and the other to a gas-washing bottle containing a dilute sodium hydroxide solution. The three-way cock is as a pressure-releasing safety valve.

In flask A is placed approximately 50 g of phosphorous pentoxide and to flask B is added 100 mL *n*-hexane via dry syringe under dry nitrogen flow. Flask A is immersed in an ice-brine bath, while flask B is cooled to −78° C in a dry ice-methanol bath, and the entire set-up is flushed with dry nitrogen.

Via the addition funnel on flask A, 25 mL of commercial 57% hydroiodic acid (aqueous solution of hydrogen iodide) is carefully dropped onto the phosphorous pentoxide with vigorous manual shaking, at such a rate as to maintain the temperature just around 0° C and the pressure slightly higher than the atmospheric pressure. Immediately upon the addition of the acid solution, a very vigorous exothermic reaction takes place, and the evolving hydrogen iodide gas is introduced into flask B where it readily dissolves in the cooled *n*-hexane. It should be reminded that, during the gas dissolution, the pressure inside flask B must be kept slightly above the atmospheric pressure.

When the addition is completed, the three outlets of flask B are closed, the vessel is detached from the assembly, while kept immersed in the cooling bath at −78° C. The hydrogen iodide solution thus prepared is then transferred into nitrogen-filled small brown ampoules via dry and cold syringe under dry nitrogen, immediately sealed, and kept in a deep freezer or a Dewer filled with crashed dry ice. These ampoules must be sealed as quickly as possible while the solution is cool, otherwise evolving hydrogen iodide gas renders sealing impossible. Care must be taken to avoid moisture contamination during sealing. The hydrogen iodide solutions are stored and handled at −78° C in the dark to avoid partial decomposition into iodine, which gives a cherry-red solution.

The concentration of the hydrogen iodide solution (usually 0.4–1.0 *M*) is determined by extracting the acid with deionized water and subsequently titrating the aqueous phase with 0.02 *N* standard sodium hydroxide solution with the aid of a pH meter.

B. Living Cationic Polymerization of Ethyl Vinyl Ether with the HI/ZnI_2 System

In a 100-mL round-bottomed flask are mixed toluene (60.0 mL), ethyl vinyl ether (EVE) (3.1 mL), and carbon tetrachloride (1.9 mL; for GC). A 12-mL portion of this monomer solution is transferred to a 20-mL glass tube and cooled to −40° C in a methanol bath. Solutions of anhydrous hydrogen iodide (1.0 mL; 150 m*M* in *n*-hexane; kept at −78° C; see Section VII.D.2 below) and ZnI_2 (2.0 mL; 1.5 m*M* in diethyl ether; cooled to −15° C) are added in this order, and the mixture is thoroughly mixed, where $[EVE]_0$ = 0.40 *M*; $[HI]_0$ = 10 m*M*; $[ZnI_2]_0$ = 0.20 m*M*.

After 50 min, the reaction is terminated with prechilled methanol (5 mL) containing a small amount of ammonia; EVE conversion = 93% by GC. The reaction mixture is diluted with toluene (20 mL), washed by 10 w/v% sodium thiosulfate solution (10 mL × 3) and then with deionized water (10 mL × 2), evaporated dryness under reduced pressure (ca. 40° C; 40 Torr), and vacuum dried overnight to give poly(EVE): M_n = 3910, M_w/M_n = 1.05 by size-exclusion chromatography (SEC) in chloroform at 30° C with a polystyrene calibration.

It is empirically known that the SEC molecular weight for poly(EVE) is always larger than the calculated value from 72.11 × (%conv/100) × $[EVE]_0/[HI]_0$; in this example, M_n(calculated) = 2680. However, the M_n value based on ^1H NMR end groups analysis [from the peak intensity ratio of the methanol-derived terminal acetal -C*H*(OEt)OMe or the HI-derived head methyl CH_3CH(OEt)- to the pendant ethyl of poly(EVE)] is in good agreement with the calculated value.

C. Living Polymerizations of Isobutyl Vinyl Ether with $HCl/ZnCl_2$ and $CF_3CO_2H/ZnCl_2$ Systems

Table 4 shows typical reaction conditions and results of the living cationic polymerization of isobutyl vinyl ether (IBVE) with the $HCl/ZnCl_2$ [127] and the $CF_3CO_2H/ZnCl_2$ [227] initiating systems. The former system is now among the most convenient for the synthesis of poly(alkyl vinyl

Table 4 Living Cationic Polymerization of IBVE with HB/$ZnCl_2$ Initiating Systems

Initiating System	HI/$ZnCl_2$	HCl/$ZnCl_2$	CF_3CO_2H/$ZnCl_2$
Solvent	Toluene	Toluene	Toluene
Temperature (° C)	−40	0	0
$[IBVE]_0$ (M)	0.38	0.38	0.38
$[HB]_0$ (mM)	5.0	5.0	5.0
$[ZnCl_2]_0$ (mM)	1.0	1.0	2.0
Reaction time	30 min	30 min	10 hr
Conversion	98%	100%	98%
M_n (GPC)[a]	7700	7000	8000
M_w/M_n (GPC)	1.05	1.07	1.06

[a] M_n (calculated) = 7600 for 100% conversion.

ethers) with controlled molecular weights and very narrow MWDs. For the HCl/$ZnCl_2$ system (5.0/1.0 mM), each run is 5.0 mL in total volume; the experiments for the CF_3CO_2H/$ZnCl_2$ counterpart (5.0/2.0 mM) are the same except that the $ZnCl_2$ concentration is doubled to accelerate the slower reaction. The reagents, HCl, CF_3CO_2H, and $ZnCl_2$, can be obtained commercially, for example, from Aldrich:

Reagent	Aldrich Catalog No.	State/Purity
HCl	29483-7	1.0 M solution in diethyl ether; water < 0.005%
CF_3CO_2H	29953-7	99+%, in sealed ampoules (1 mL each)
$ZnCl_2$	27683-9	1.0 M solution in diethyl ether
	22999-7	99.999%, solid

As shown in Table 4, typically, a stock monomer solution (for 4 to 5 runs) is prepared in a 50-mL round-bottomed flask by mixing toluene (21.0 mL), IBVE (1.50 mL), and carbon tetrachloride (1.50 mL). To 10–20 mL glass tubes are distributed 4.0-mL portions of the monomer solution, and they are cooled to −78° C in a methanol bath. Separately in 50-mL round-bottomed flasks, solutions of HCl (50 mM) and $ZnCl_2$ (10 mM) are prepared by diluting their commercial solutions (1.0 M in diethyl ether) with toluene; the HCl solution should be prepared and kept at −78° C. These diluted solutions are then added, first HCl and then $ZnCl_2$, to the monomer solution at −78° C with vigorous manual mixing, and the temperature is raised to 0° C in an ice bath to initiate living cationic polymerization. The reaction reaches quantitative IBVE conversion in 30 min and is then quenched with prechilled methanol (2.0 mL) containing a small amount of ammonia. The work-up of the polymer is the same as for the HI/ZnI_2-initiated polymerization (see Section VII.B), but the treatment with sodium thiosulfate may be omitted (thus, washing is with water only).

Similarly, living IBVE polymerization can be initiated by sequential addition of CF_3CO_2H and $ZnCl_2$ solutions in this order to the same monomer solution as for $HCl/ZnCl_2$; see Table 4.

D. Block Copolymerization of 2-Acetoxyethyl and Isobutyl Vinyl Ethers with HI/ZnI_2 [177,209]

1. *Sequential Living Polymerization*

Into a glass tube is added toluene (3.6 mL), 2-acetoxyethyl vinyl ether (AcOVE; 0.20 mL; see Section VII.A.3), and bromobenzene (0.20 mL; the internal standard for GC), and the mixture is cooled to −15° C in a methanol bath. In separate glass tubes, solutions of hydrogen iodide (100 m*M* in *n*-hexane) and ZnI_2 (20 m*M* in diethyl ether) are prepared; the hydrogen iodide solution must be prepared and kept afterward at −78° C until used. Aliquots (0.50 mL each) of these solutions are then added, sequentially in this order, to the monomer solution at −15° C, and the mixture is thoroughly mixed. These procedures lead to a polymerization system where total volume = 5.0 mL; $[\mathrm{AcOVE}]_0$ = 0.30 *M*; $[\mathrm{HI}]_0$ = 10 m*M*; $[\mathrm{ZnI_2}]_0$ = 2.0 m*M*.

The AcOVE polymerization reaches approximately 100% conversion after 50 min, at which moment 1.0 mL of an IBVE solution is added with the mixture kept at −15° C. The IBVE solution has been prepared beforehand by mixing IBVE (0.66 mL), carbon tetrachloride (0.64 mL; the internal standard for GC), and toluene (8.7 mL); thus, the AcOVE/IBVE mole ratio = 30:10 in the reaction mixture. After an additional 35 min, where IBVE conversion reaches about 100%, the second stage polymerization is terminated at −15° C with 2.0 mL of prechilled methanol containing about 2 vol% aqueous ammonia.

The quenched reaction mixture is diluted with 20 mL of toluene and washed with 10 w/v% aqueous sodium thiosulfate solution (20 mL × 3) and then with deionized water (20 mL × 2), evaporated to dryness (30° C; ca. 40 Torr), and vacuum dried overnight to give the AcOVE-IBVE block copolymer in almost quantitative yield; the AcOVE/IBVE mole ratio in the polymer (^{1}H NMR) = 30:9, being close to the initial feed ratio of the two monomers.

2. Hydrolysis into Amphiphilic Block Copolymers

The recovered block copolymer is further purified by freeze-drying from 1,4-dioxane. In a 100-mL round-bottomed flask, the polymer (0.29 g) is dissolved in acetone (15 mL), and 2 *N* aqueous sodium hydroxide (4.4 mL; 5 equivalents to the ester groups in the polymer) is added. The mixture is magnetically stirred at room temperature for 3 hr, during which period the initially colorless and transparent solution becomes yellowish and heterogeneous. The acetone is evaporated, and the residue is dissolved in water (15 mL). The transparent yellowish solution is then stirred magnetically for an additional 2 days, condensed by evaporation into 2–3 mL, dialyzed (Spectrapor 7; cut-off MW = 1000; for 2 days), evaporated to dryness, and vacuum dried overnight to give 2-hydroxyethyl vinyl ether-IBVE block polymer.

REFERENCES

1. M. Sawamoto, *Prog. Polym. Sci. 16*: 111 (1991).
2. M. Sawamoto and T. Higashimura, *Encyclopedia of Polymer Science and Engineering*, 2nd ed., Suppl. Vol., John Wiley, New York, 1989, p. 399.
3. J. P. Kennedy and B. Iván, *Designed Polymers by Carbocationic Macromolecular Engineering: Theory and Practice*, Hanser, München, 1992.
4. C. E. Schildknecht, A. O. Zoss, and C. McKinley, *Ind. Eng. Chem. 39*: 180 (1947).
5. C. E. Schildknecht, S. T. Gross, H. R. Davidson, J. M. Lambert, and A. O. Zoss, *Ind. Eng. Chem. 40*: 2104 (1948).
6. G. Natta, I. Bassi, and P. Corradini, *Makromol. Chem. 18/19*: 455 (1956).
7. C. E. Schildknecht, *Vinyl and Related Polymers*, Wiley, New York, 1952, p. 153.
8. M. Sawamoto, *Prog. Polym. Sci. 16*: 152–157 (1991).
9. M. Kato, T. Ichijo, K. Ishii, and M. Hasegawa, *J. Polym. Sci. A1*, *9*: 2109 (1971).
10. M. Kato and H. Nakane, *Photogr. Sci. Eng. 23*: 207 (1979).
11. S. Nakahama and A. Hirao, *Prog. Polym. Sci. 15*: 299 (1990).
12. A. Hirao and S. Nakahama, *Trends Polym. Sci. 2*: 267 (1994).
13. T. Higashimura and M. Sawamoto, in *Comprehensive Polymer Science* (G. C. Eastmond, A. Ledwith, S. Russo, and P. Sigwalt, eds.), Pergamon, Oxford, Vol. 3, Chapter 42 (1989).
14. T. Higashimura, Y.-M. Law, and M. Sawamoto, *Polym. J. 16*: 401 (1984).
15. V. Héroguez, A. Deffieux, and M. Fontanille, *Makromol. Chem. Macromol. Symp. 32*: 199 (1990).
16. M. Sawamoto, T. Hasebe, M. Kamigaito, and T. Higashimura, *J. Macromol. Sci., Pure Appl. Polym. Chem. A31*: 937 (1994).

17. S. Aoshima, T. Nakamura, N. Uesugi, M. Sawamoto, and T. Higashimura, *Macromolecules 18*: 2097 (1985).
18. T. Higashimura, S. Aoshima, and M. Sawamoto, *Makromol. Chem., Macromol. Symp. 3*: 99 (1986).
19. S. Aoshima and T. Higashimura, *Polym. Bull. 15*: 417 (1986).
20. S. Aoshima, O. Hasegawa, and T. Higashimura, *Polym. Bull. 13*: 229 (1985).
21. S. Aoshima, O. Hasegawa, and T. Higashimura, *Polym. Bull. 14*: 417 (1985).
22. E. Takeuchi, T. Hashimoto, M. Sawamoto, and T. Higashimura, *J. Polym. Sci., Polym. Chem. Ed. 27*: 3303 (1989).
23. T. Higashimura, T. Enoki, and M. Sawamoto, *Polym. J. 19*: 515 (1987).
24. M. Minoda, M. Sawamoto, and T. Higashimura, *Polym. Bull. 17*: 107 (1987).
25. S. Aoshima, H. Oda, and E. Kobayashi, *J. Polym. Sci. Part A Polym. Chem. 30*: 2407 (1992).
26. T. Nakamura, S. Aoshima, and T. Higashimura, *Polym. Bull. 14*: 515 (1985).
27. T. Higashimura, T. Nakamura, and S. Aoshima, *Polym. Bull., 17*: 389 (1987).
28. W. O. Choi, M. Sawamoto, and T. Higashimura, *Polym. J., 19*, 889 (1987).
29. T. Hashimoto, H. Ibuki, M. Sawamoto, and T. Higashimura, *J. Polym. Sci., Polym. Chem. Ed. 26*: 3361 (1988).
30. H. Shohi, M. Sawamoto, and T. Higashimura, *Polym. Bull. 21*: 357 (1989).
31. T. Higashimura, K. Ebara, and S. Aoshima, *J. Polym. Sci. Polym. Chem. Ed. 27*: 2937 (1989).
32. C. Vandooren, R. Jérôme, and Ph. Teyssié, *Polym. Bull. 32*: 387 (1994).
33. J. Höpkin, M. Möller, M. Lee, and V. Percec, *Makromol. Chem. 193*: 275 (1992).
34. W. O. Choi, M. Sawamoto, and T. Higashimura, *Polym. J. 20*: 201 (1988).
35. K. Yamaoka, M. Minoda, and T. Miyamoto, *Polym. Prep. Jpn. Engl. Ed. 41*: E787 (1992).
36. M. Miyamoto, M. Sawamoto, and T. Higashimura, *Macromolecules 17*: 265, 2227 (1984).
37. M. Sawamoto, C. Okamoto, and T. Higashimura, *Macromolecules 20*: 2693 (1987).
38. S. Aoshima and T. Higashimura, *Macromolecules 22*: 1009 (1989).
39. Y. Kishimoto, S. Aoshima, and T. Higashimura, *Macromolecules 22*: 3877 (1989).
40. T. Higashimura, M. Miyamoto, and M. Sawamoto, *Macromolecules 18*: 611 (1985).
41. M. Sawamoto and T. Higashimura, *Makromol. Chem. Macromol. Symp. 32*: 131 (1990).
42. S. Aoshima, S. Iwasa, and E. Kobayashi, *Polym. J. 26*: 912 (1994).
43. T. Sagane and R. W. Lenz, *Polym. J. 20*: 923 (1988).
44. T. Sagane and R. W. Lenz, *Macromolecules 22*: 3763 (1989); *Polymer 30*: 2269 (1989).
45. V. Héroguez, M. Schappacher, E. Papon, and A. Deffieux, *Polym. Bull. 25*: 307 (1991).

46. J. E. Rodriguez-Parada and V. Percec, *J. Polym. Sci., Polym. Chem. Ed. 24*, 1363 (1986).
47. V. Percec, M. Lee, and H. Johnson, *J. Polym. Sci. Part A Polym. Chem. 29*: 327 (1991).
48. V. Percec and M. Lee, *Macromolecules 24*: 1017 (1991).
49. H. Johnson, V. Percec, and A. Hult, *Polym. Bull. 25*: 115 (1991).
50. V. Percec and M. Lee, *Polymer 32*: 2862 (1991).
51. V. Percec and M. Lee, *Polym. Bull. 25*: 123 (1991).
52. V. Percec and M. Lee, *Polym. Bull. 25*: 131 (1991).
53. V. Percec and M. Lee, *J. Macromol. Sci., Chem. A28*: 651 (1991).
54. V. Percec, M. Lee, and C. Ackerman, *Polymer 33*: 703 (1992).
55. V. Percec and M. Lee, *Macromolecules 24*: 2780 (1991).
56. V. Percec and M. Lee, *Macromolecules 24*: 4963 (1991).
57. V. Percec and Q. Zheng, *Polym. Bull. 29*: 485, 493 (1992).
58. V. Percec and Q. Zheng, *Polym. Bull. 29*: 501 (1992).
59. V. Percec and M. Lee, *J. Macromol. Sci., Pure Appl. Chem. A29*: 723 (1992).
60. V. Percec, M. Lee, P. Rinaldi, and V. E. Litman, *J. Polym. Sci. Part A, Polym. Chem. 30*: 1213 (1992).
61. V. Percec, M. Lee, and D. Tomazos, *Polym. Bull. 28*: 9 (1992).
62. V. Percec, A. S. Gomes, and M. Lee, *J. Polym. Sci. Part A Polym. Chem. 29*: 1615 (1991).
63. V. Percec, C.-S. Wang, and M. Lee, *Polym. Bull. 26*: 15 (1991).
64. C. G. Cho, B. A. Feit, and O. W. Webster, *Macromolecules, 23*, 1918 (1990).
65. J. P. Kennedy and J. Kurian, *Polym. Bull. 23*: 259 (1990).
66. J. P. Kennedy and J. Kurian, *Macromolecules 23*: 3736 (1990).
67. Y. Eika, S. Kanaoka, M. Sawamoto, and T. Higashimura, *Polym. Prep. Jpn. Engl. Ed. 42*: E81 (1993); *Macromolecules*, in press.
68. C. G. Willson and M. J. Bowden, *CHEMTECH 19*: 182 (1989).
69. T. Higashimura, K. Kojima, and M. Sawamoto, *Polym. Bull. 19*: 7 (1988).
70. K. Kojima, M. Sawamoto, and T. Higashimura, *Macromolecules 23*: 948 (1990).
71. T. Higashimura, K. Kojima, and M. Sawamoto, *Makromol. Chem., Suppl. 15*: 127 (1989).
72. D. A. Conclon, J. V. Crivello, J. L. Lee, and M. J. O'Brien, *Macromolecules 22*: 509 (1989).
73. J. M. J. Fréchet, E. Eicheler, H. Ito, and C. G. Willson, *Polymer 24*: 995 (1983).
74. R. F. Storey and D. C. Hoffman, *Polym. Prep. Div. Polym. Chem., Am. Chem. Soc. 30*(2): 121 (1989).
75. H. Ito and C. G. Willson, *Macromolecules 16*: 510 (1983).
76. T. Hashimoto, M. Sawamoto, and T. Higashimura, *J. Polym. Sci. Polym. Chem. Ed. 25*: 2827 (1987).
77. T. Hashimoto, M. Sawamoto, T. Higashimura, and N. Saito, *J. Polym. Sci. Polym. Chem. Ed. 25*: 1073 (1987).

78. M. Szwarc, *Carbanions, Living Polymers and Electron Transfer Processes*, Interscience, New York, 1968.
79. O. W. Webster, *Science 251*: 887 (1991).
80. M. Minoda, M. Sawamoto, and T. Higashimura, *Macromolecules 20*: 2045 (1987).
81. M. Miyamoto, M. Sawamoto, and T. Higashimura, *Macromolecules 18*: 123 (1985).
82. T. Yoshida, M. Sawamoto, and T. Higashimura, *Makromol. Chem. 192*: 2317 (1991).
83. H. Shohi, M. Sawamoto, and T. Higashimura, *Polym. Bull. 25*: 529 (1991).
84. M. Minoda, M. Sawamoto, and T. Higashimura, *Macromolecules 23*: 1897 (1990).
85. S. Kanaoka, M. Minoda, M. Sawamoto, and T. Higashimura, *J. Polym. Sci. Polym. Chem. Ed. 28*: 1127 (1990).
86. N. H. Haucourt, L. Peng, and E. J. Goethals, *Macromolecules 27*: 1329 (1994).
87. T. Ohmura, M. Sawamoto, and T. Higashimura, *Macromolecules 27*: 3714 (1994).
88. K. Kojima, M. Sawamoto, and T. Higashimura, *Polym. Bull. 23*: 149 (1990).
89. K. Kojima, M. Sawamoto, and T. Higashimura, *Macromolecules 24*: 2658 (1991).
90. G. Kaszás, J. E. Puskás, and J. P. Kennedy, *J. Appl. Polym. Sci. 39*: 119 (1990).
91. J. E. Puskás, G. Kaszás, and J. P. Kennedy, *J. Macromol. Sci. Chem. A28*: 65 (1991).
92. G. Kaszás, J. E. Puskás, J. P. Kennedy, and W. G. Hager, *J. Polym. Sci. Part A Polym. Chem. 29*: 427 (1991).
93. R. F. Storey, B. J. Chisholm, and K. R. Choate, *J. Macromol. Sci. Pure Appl. Chem. A31*: 969 (1994).
94. R. F. Storey and B. J. Chisholm, *Macromolecules 26*: 6727 (1993).
95. K. Koshimura and H. Sato, *Polym. Bull. 29*: 705 (1992).
96. H. Everland, J. Kops, A. Nielsen, and B. Iván, *Polym. Bull. 31*: 159 (1993).
97. J. E. Puskás, G. Kaszás, J. P. Kennedy, and W. G. Hager, *J. Polym. Sci. Part A Polym. Chem. 30*: 41 (1992).
98. Y. Tsunogae and J. P. Kennedy, *Polym. Bull. 27*: 631 (1992).
99. J. P. Kennedy, N. Meguriya, and B. Kaszler, *Macromolecules 24*: 6572 (1991).
100. J. P. Kennedy and J. Kurian, *J. Polym. Sci. Part A Polym. Chem. 28*: 3725 (1990).
101. Y. Tsunogae and J. P. Kennedy, *J. Polym. Sci. Part A Polym. Chem. 32*: 403 (1994).
102. J. P. Kennedy, S. Midha, and Y. Tsunogae, *Macromolecules 26*: 429 (1993).
103. Zs. Fodor and J. P. Kennedy, *Polym. Bull. 29*: 697 (1992).
104. Y. Tsunogae and J. P. Kennedy, *J. Macromol. Sci. Pure Appl. Chem. A30*: 269 (1993).
105. T. Pernecker, J. P. Kennedy, and B. Iván, *Macromolecules 25*: 1642 (1992).

106. T. Pernecker and J. P. Kennedy, *Polym. Bull. 29*: 27 (1992).
107. A. V. Lubnin and J. P. Kennedy, *J. Polym. Sci. Part A Polym. Chem. 31*: 2825 (1993).
108. Q. Liu, M. Konas, R. M. Davis, and J. S. Riffle, *J. Polym. Sci. Part A Polym. Chem. 31*: 1709 (1993).
109. A. Gadkari and J. P. Kennedy, *J. Appl. Polym. Sci. Appl. Polym. Symp. 44*: 19 (1989).
110. A. Verma, M. Glagola, A. Prasad, H. Marand, and J. S. Riffle, *Makromol. Chem. Macromol. Symp. 54/55*: 95 (1992).
111. J. P. Kennedy, J. L. Price, and K. Koshimura, *Macromolecules 24*: 6567 (1991).
112. O. Nuyken, H. Kröner, and S. Aechtner, *Makromol. Chem. Rapid Commun. 9*: 671 (1988).
113. O. Nuyken, H. Kröner, and S. Aechtner, *Polym. Bull. 24*: 513 (1990).
114. H. Cramail and A. Deffieux, *Makromol. Chem. 193*: 2793 (1992).
115. S. Creutz, C. Vandooren, R. Jérôme, and Ph. Teyssié, *Polym. Bull. 33*: 21 (1994).
116. A. Verma, A. Nielsen, J. E. McGrath, and J. S. Riffle, *Polym. Bull. 23*: 563 (1990).
117. A. Verma, A. Nielsen, J. M. Bronk, J. E. McGrath, and J. S. Riffle, *Makromol. Chem. Macromol. Symp. 47*: 239 (1991).
118. T. Loontjens, F. Derks, and E. Kleuskens, *Polym. Bull. 27*: 519 (1992).
119. T. R. Fang and J. P. Kennedy, *Polym. Bull. 10*: 82 (1983).
120. T. R. Fang and J. P. Kennedy, *Polym. Bull. 10*: 90 (1983).
121. J. P. Kennedy, V. S. C. Chang, and W. P. Francik, *J. Polym. Sci. Polym. Chem. Ed. 20*: 2809 (1982).
122. G. Heublein, H.-H. Rottmayer, D. Stadermann, and J. Vogel, *Acta Polym. 38*: 559 (1987).
123. A. Deffieux, M. Schappacher, and L. Rique-Lurbet, *Polymer 35*: 4562 (1994).
124. H. Yamaoka, H. Matsuoka, K. Sumaru, S. Harada, M. Imai, and G. D. Wignall, *Physica B. 213/214*: 700 (1995).
125. M. Minoda, M. Sawamoto, and T. Higashimura, *Macromolecules 25*: 2796 (1992).
126. D. A. Tomalia, A. M. Naylor, and W. A. Goddard III, *Angew. Chem. Int. Ed. Engl. 29*, 138 (1990); *Angew. Chem. 102*: 119 (1990).
127. M. Kamigaito, Y. Maeda, M. Sawamoto, and T. Higashimura, *Macromolecules 26*: 1643 (1993).
128. W. O. Choi, M. Sawamoto, and T. Higashimura, *Macromolecules 23*: 48 (1990).
129. J. P. Kennedy and R. A. Smith, *J. Polym. Sci. Polym. Chem. Ed. 18*: 1523 (1980).
130. J. P. Kennedy, *J. Appl. Polym. Sci. Appl. Polym. Symp. 39*: 21 (1984); *J. Macromol. Sci. Chem. A21*: 929 (1984); *Makromol. Chem. Suppl. 7*: 171 (1984).
131. M. Sawamoto, T. Enoki, and T. Higashimura, *Macromolecules 20*: 1 (1987).

132. T. Hashimoto, E. Takeuchi, M. Sawamoto, and T. Higashimura, *J. Polym. Sci. Part A Polym. Chem. 28*: 1137 (1990).
133. T. Hashimoto, S. Iwao, and T. Kodaira, *Makromol. Chem. 194*: 2323 (1993).
134. S. Chakrapani, R. Jérôme, and Ph. Teyssié, *Macromolecules 23*: 3026 (1990).
135. D. V. Meirvenne, N. Haucourt, and E. J. Goethals, *Polym. Bull. 23*: 185 (1990).
136. M. Kamigaito, M. Sawamoto, and T. Higashimura, *J. Polym. Sci. Part A Polym. Chem. 29*: 1909 (1991).
137. M. Kamigaito, M. Sawamoto, and T. Higashimura, *Makromol. Chem. 194*: 727 (1993).
138. C. G. Cho, B. A. Feit, and O. W. Webster, *Macromolecules 25*: 2081 (1992).
139. N. H. Haucourt, E. J. Goethals, M. Schappacher, and A. Deffieux, *Makromol. Chem. Rapid Commun. 13*: 329 (1992).
140. M. Sawamoto, M. Kamigaito, K. Kojima, and T. Higashimura, *Polym. Bull. 19*: 359 (1988).
141. M. Kamigaito, M. Sawamoto, and T. Higashimura, *Macromolecules 23*: 4896 (1990).
142. H. Fukui, M. Sawamoto, and T. Higashimura, *Macromolecules 26*: 7315 (1993).
143. M. Sawamoto, T. Enoki, and T. Higashimura, *Polym. Bull. 18*: 117 (1987).
144. C. G. Cho and J. E. McGrath, *Polym. Prepr. Div. Polym. Chem. Am. Chem. Soc. 28*(2): 356 (1987).
145. H. Shohi, M. Sawamoto, and T. Higashimura, *Macromolecules 25*: 58 (1992).
146. M. Sawamoto, S. Aoshima, and T. Higashimura, *Makromol. Chem. Macromol. Symp. 13/14*: 513 (1988).
147. H. Fukui, M. Sawamoto, and T. Higashimura, *J. Polym. Sci. Part A Polym. Chem. 32*: 2699 (1994).
148. T. Higashimura, M. Kamigaito, M. Kato, T. Hasebe, and M. Sawamoto, *Macromolecules 26*: 2670 (1993).
149. H. Shohi, M. Sawamoto, and T. Higashimura, *Macromolecules 25*: 53 (1992).
150. H. Shohi, M. Sawamoto, and T. Higashimura, *Makromol. Chem. 193*: 1783 (1992).
151. K. Miyashita, M. Kamigaito, M. Sawamoto, and T. Higashimura, *Macromolecules 27*: 1093 (1994).
152. Y. Ishihama, M. Sawamoto, and T. Higashimura, *Polym. Bull. 24*: 201 (1990).
153. R. Faust and J. P. Kennedy, *Polym. Bull. 19*: 21 (1988).
154. T. Higashimura, Y. Ishihama, and M. Sawamoto, *Macromolecules 26*: 744 (1993).
155. K. Miyashita, M. Kamigaito, M. Sawamoto, and T. Higashimura, *J. Polym. Sci. Part A Polym. Chem. 32*: 2531 (1994).
156. J. Si and J. P. Kennedy, *J. Macromol. Sci. Pure Appl. Chem. A30*: 863 (1993).

157. R. Faust and J. P. Kennedy, *Polym. Bull. 15*: 317 (1986); *J. Polym. Sci. Polym. Chem. Ed. 25*: 1847 (1987).
158. B. Iván and J. P. Kennedy, *J. Polym. Sci. Part A Polym. Chem. 28*: 89 (1990).
159. H. Fukui, M. Sawamoto, and T. Higashimura, *Macromolecules*, in press.
160. S. Aoshima, K. Ebara, and T. Higashimura, *Polym. Bull. 14*: 425 (1985).
161. K. Ebara, M. Kitaoka, S. Aoshima, and T. Higashimura, *Polym. Prep. Jpn. 35*: 1312 (1986).
162. M. Schappacher and A. Deffieux, *Makromol. Chem. Rapid Commun. 12*: 447 (1991).
163. E. J. Goethals, N. H. Haucourt, A. M. Verheyen, and J. Habimana, *Makromol. Chem. Rapid Commun. 11*: 623 (1990).
164. M. Sawamoto, T. Enoki, and T. Higashimura, *Polym. Bull. 16*: 117 (1986).
165. S. Kanaoka, M. Sueoka, M. Sawamoto, and T. Higashimura, *J. Polym. Sci. Part A Polym. Chem. 31*: 2513 (1993).
166. J. P. Kennedy and M. Hiza, *J. Polym. Sci. Polym. Chem. Ed. 21*: 1033 (1983); *Polym. Bull. 10*: 161 (1983).
167. T.-P. Liao and J. P. Kennedy, *Polym. Bull. 6*: 135 (1981).
168. B. Kaszler, V. S. C. Chang, and J. P. Kennedy, *J. Macromol. Sci. Chem. A21*: 307 (1984).
169. J. P. Kennedy, S. Midha, and A. Gadkari, *J. Macromol. Sci. Chem. A28*: 209 (1991).
170. S. Nemes, T. Pernecker, and J. P. Kennedy, *Polym. Bull. 25*: 633 (1991).
171. J. P. Kennedy and G. C. Richard, *Macromolecules 26*: 567 (1993).
172. K. P. McGrath, D. A. Tirrell, M. Kawai, T. L. Mason, and M. J. Fournier, *J. Biotechnol. Prog. 6*: 188 (1990).
173. D. A. Tirrell, M. J. Fournier, and T. L. Mason, *Curr. Opin. Struc. Biol. 1*: 638 (1991).
174. H. S. Creel, M. J. Fournier, T. L. Mason, and D. A. Tirrell, *Macromolecules 24*: 1213 (1991).
175. K. P. McGrath, M. J. Fournier, T. L. Mason, and D. A. Tirrell, *J. Am. Chem. Soc. 114*: 727 (1992).
176. M. T. Krejchi, E. D. T. Atkins, A. J. Waddon, M. J. Fournier, T. L. Mason, and D. A. Tirrell, *Science 265*: 1427 (1994).
177. M. Minoda, M. Sawamoto, and T. Higashimura, *Macromolecules 23*: 4889 (1990).
178. M. Minoda, M. Sawamoto, and T. Higashimura, *Polym. Bull. 23*: 133 (1990).
179. M. Minoda, M. Sawamoto, and T. Higashimura, *J. Polym. Sci. Part A Polym. Chem. 31*: 2789 (1993).
180. J. H. Waite, *CHEMTECH 17*: 692 (1987).
181. J. G. Zilliox, P. Rempp, and J. Parrod, *J. Polym. Sci. Part C 22*: 145 (1968).
182. D. J. Worsfold, J. G. Zilliox, and P. Rempp, *Can. J. Chem. 47*: 3377 (1969).
183. B. J. Bauer and L. J. Fetters, *Rubber Chem. Technol. 51*: 406 (1978).
184. S. Bywater, *Adv. Polym. Sci. 30*: 90 (1979).
185. P. Rempp and E. Franta, *Adv. Polym. Sci. 58*: 1 (1984).

186. G. R. Newkome, Z.-q. Yao, G. R. Baker, and V. K. Gupta, *J. Org. Chem. 50*: 2003 (1985).
187. G. R. Newkome, Z.-q. Yao, G. R. Baker, V. K. Gupta, P. S. Russo, and M. J. Saunders, *J. Am. Chem. Soc. 108*: 849 (1986).
188. H. Shohi, M. Sawamoto, and T. Higashimura, *Macromolecules 24*: 4926 (1991).
189. M. Sawamoto, H. Shohi, H. Sawamoto, and T. Higashimura, *J. Macromol. Sci. Pure Appl. Chem. A31*: 1609 (1994).
190. H. Shohi, M. Sawamoto, and T. Higashimura, *Makromol. Chem. 193*: 2027 (1992).
191. R. F. Storey and Y. Lee, *Polym. Prepr. Div. Polym. Chem. Am. Chem. Soc. 30*(2): 162 (1989).
192. R. F. Storey and Y. Lee, *J. Macromol. Sci. Pure Appl. Chem. A29*: 1017 (1992).
193. M. K. Mishra, B. Wang, and J. P. Kennedy, *Polym. Bull. 17*: 307 (1987).
194. G. Kaszás, J. E. Puskás, and J. P. Kennedy, *Polym. Bull. 20*: 413 (1988).
195. M. Zsuga, T. Kelen, and J. Borbély, *Polym. Bull. 26*: 417 (1991).
196. M. Zsuga, L. Balogh, T. Kelen, and J. Borbély, *Polym. Bull. 23*: 335 (1990).
197. L. Balogh, L. Fábian, I. Majoros, and T. Kelen, *Polym. Bull. 23*: 75 (1990).
198. M. Zsuga, T. Kelen, L. Balogh, and I. Majoros, *Polym. Bull. 29*: 127 (1992).
199. T. Kelen, M. Zsuga, L. Balogh, I. Majoros, and G. Deak, *Makromol. Chem. Macromol. Symp. 67*: 325 (1993).
200. H. K. Huang, M. Zsuga, and J. P. Kennedy, *Polym. Bull. 19*: 43 (1988).
201. C. C. Chen, G. Kaszás, J. E. Puskás, and J. P. Kennedy, *Polym. Bull. 22*: 463 (1989).
202. H. Fukui, M. Sawamoto, and T. Higashimura, *J. Polym. Sci. Part A Polym. Chem. 31*: 1531 (1993).
203. M. Morton, *Anionic Polymerization: Principles and Practice*, Academic Press, London, 1983.
204. M. J. Stewart, *New Methods of Polymer Synthesis* (J. R. Ebdon, ed.) Blackie, London, Chap. 4, 1991.
205. H. Fukui, M. Sawamoto, and T. Higashimura, *Macromolecules 27*: 1297 (1994).
206. H. Fukui, M. Sawamoto, and T. Higashimura, *Macromolecules 28*: 3756 (1995).
207. H. Fukui, M. Sawamoto, and T. Higashimura, *Macromolecules*, in press.
208. S. Kanaoka, M. Sawamoto, and T. Higashimura, *Macromolecules 24*: 2309 (1991).
209. S. Kanaoka, M. Sawamoto, and T. Higashimura, *Macromolecules 24*: 5741 (1991).
210. S. Kanaoka, M. Sawamoto, and T. Higashimura, *Makromol. Chem. 194*: 2305 (1993).
211. S. Kanaoka, T. Omura, M. Sawamoto, and T. Higashimura, *Macromolecules 25*: 6407 (1992).
212. S. Kanaoka, M. Sawamoto, and T. Higashimura, *Macromolecules 26*: 254 (1993).

213. T. M. Marsalkó, I. Majoros, and J. P. Kennedy, *Polym. Bull. 31*: 665 (1993).
214. A. Halperin, M. Tirrell, and T. P. Lodge, *Adv. Polym. Sci. 100*: 31 (1992).
215. J. P. Kennedy, *Trends Polym. Sci. 1*: 381 (1993); *CHEMTECH 24*(2): 24 (1994); *Macromol. Symp. 85*: 79 (1994).
216. A. Gadkari, J. P. Kennedy, M. M. Kory, and D. L. Ely, *Polym. Bull. 22*: 25 (1989).
217. J. E. Puskás, G. Kaszás, C. C. Chen, and J. P. Kennedy, *Polym. Bull. 20*: 253 (1988).
218. D. Chen, J. P. Kennedy, and A. L. Allen, *J. Macromol. Sci. Chem. A25*: 389 (1988).
219. S. Kanaoka, M. Sawamoto, T. Higashimura, J. Won, C. Pan, T. P. Lodge, M. Fujisawa, D. M. Hedstrand, and D. A. Tomalia, *J. Polym. Sci. Part B Polym. Phys. 33*: 527 (1995).
220. S. Kanaoka, M. Sawamoto, and T. Higashimura, *Macromolecules 25*: 6414 (1992).
221. J. P. Kennedy, ed., *J. Appl. Polym. Sci. Appl. Polym. Symp. 30*: 1–192 (1977).
222. Y. Jiang and J. M. J. Fréchet, *Polym. Prep. Div. Polym. Chem. Am. Chem. Soc. 30*(1): 127 (1989).
223. D. Y. Sogah and O. W. Webster, *Macromolecules 19*: 1775 (1986).
224. D. Y. Sogah and O. W. Webster, *Recent Advances in Mechanistic and Synthetic Aspects of Polymerization* (M. Fontanile and A. Guyot, eds.), D. Reidel, Dordrecht, p. 61, 1987.
225. G. Sinai-Zindge, A. Virma, Q. Liu, A. Brink, J. Bronk, D. Allison, A. Goforth, N. Patal, H. Marand, J. E. McGrath, and J. S. Riffle, *Polym. Prep. Div. Polym. Chem. Am. Chem. Soc. 31*(1): 63 (1990).
226. M. Schappacher and A. Deffieux, *Macromolecules 25*: 6744 (1992).
227. M. Kamigaito, M. Sawamoto, and T. Higashimura, *Macromolecules 24*: 3988 (1991).

6

Cationic Polymerization of Heterocyclics

PRZEMYSŁAW KUBISA Center of Molecular and Macromolecular Studies, Polish Academy of Sciences, Lodz, Poland

I. INTRODUCTION

This chapter deals with the cationic polymerization of heterocyclic monomers, i.e., cyclic compounds containing one or more (identical or different) heteroatoms within the ring. Polymerization proceeds by a ring-opening reaction. Traditionally, this subject is discussed separately from the cationic polymerization of alkenyl monomers (vinyl polymerization) proceeding by opening of a double bond. This is justified, because both processes have their special features, making them distinctly different.

On the other hand, however, these two areas of cationic polymerization are not completely separated fields. In spite of the differences, both processes proceed on electron-deficient active species: cations or species with a partial positive charge. Thus, propagation in both cases involves attack of the nucleophile (double bond or heteroatom) on electrophilic active centers. Several basic principles will therefore hold for both vinyl and ring-opening cationic polymerization.

With accumulating knowledge on both processes, the borderline between them becomes more diffuse and, indeed, if one looks at the active species in polymerization of vinyl ethers (typical vinyl polymerization) and 1,3-dioxolane (typical ring-opening polymerization), there is not much difference between their structures.

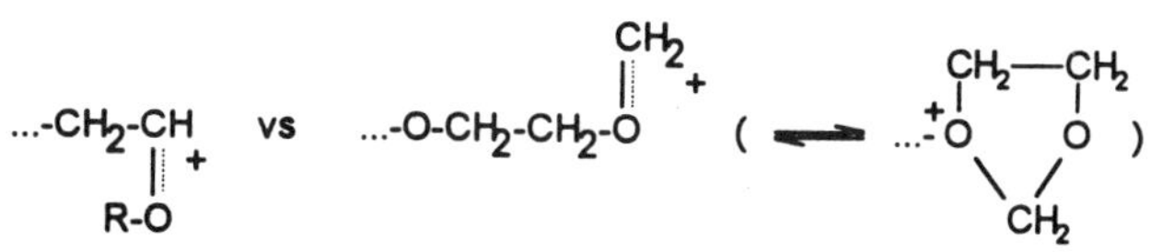

(1)

(In this and the following schemes, the counterions are omitted; the nature of the counterion is indicated, however, wherever it is important.)

Therefore, remembering all the differences, one should look for the similarities and analogies to prepare the ground for the uniform treatment of all polymerization processes proceeding by cationic mechanism [1].

Let us formulate the main differences between vinyl and ring-opening cationic polymerization. First of all, polymerization of vinyl monomers leads to polymers having all-carbon chains (although various heteroatoms may be incorporated in the side groups). Polymerization of heterocyclic monomers gives polymers containing heteroatom(s) within the main chain. Because there are several possible combinations within the cyclic monomer molecule, ring-opening polymerization allows the preparation of polymers with various sequences of carbon atoms and heteroatoms within the main chain. The typical example is the polyether series $[(CH_2)_nO]_m$, where polymers with $n = 2, 3, 4, 6$ can be easily obtained.

The mechanistic consequence of this difference is the different structure of active species. Cationation of a vinyl monomer gives carbenium ion, whereas cationation of heterocyclic monomer results in onium ion:

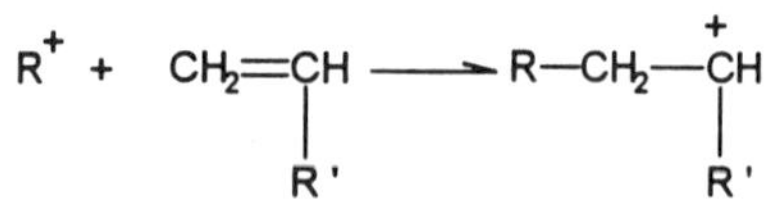

(2)

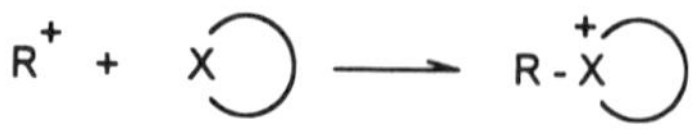

(3)

X = heteroatom (O, N, S, P, etc.)

Although the real situation may be more complex (at least in some systems) (cf., Section II.B.6.b), the vinyl polymerization is essentially carried out on carbenium active species whereas ring-opening polymerization is carried out on onium active species.

Due to the much lower reactivity (thus much higher stability) of onium ions, as compared with carbenium ions, various side reactions (i.e., leading to transfer and termination) are much easier to avoid in ring-opening polymerization and consequently conditions of living process can be approached in several systems in cationic ring-opening polymerization.

The second major difference stems from the fact that the nucleophilic site of vinyl monomer, i.e., double bond, is consumed in the propagation step, whereas the nucleophilic site of the heterocyclic monomer, i.e., heteroatom, is still present in the formed chain. Thus, the saturated all-carbon chain becomes a neutral component of the polymerizing system, while the heteroatom-containing chain may still participate in the reaction:

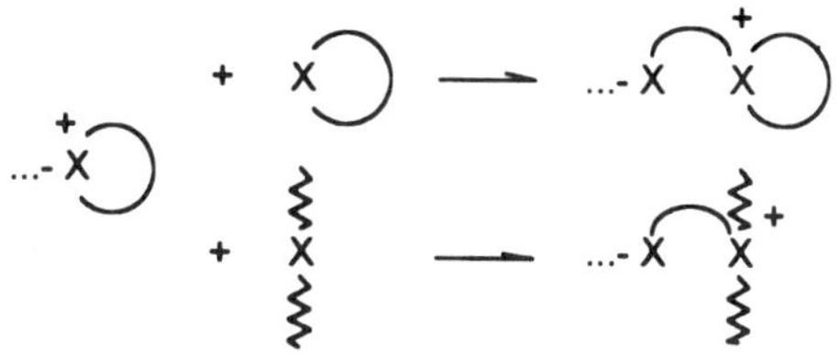

(4)

Thus, chain transfer to polymer is a general phenomenon in cationic ring-opening polymerization. Consequences will be discussed in more detail in Section II.C.

Finally, the polymerization of vinyl monomers usually proceeds as a practically irreversible reaction. Exceptions are heavily substituted monomers like α-methylstyrene for which propagation is clearly reversible. In the ring-opening polymerization, the driving force for polymerization comes from ring strain, thus it varies greatly for different monomers.

Highly strained 3- and 4-membered rings polymerize practically irreversibly but polymerization of 5-, 6-, 7-, and higher member rings, important from both a basic and practical point of view, is highly reversible. Thus, in these systems, reversibility of propagation step introduces an additional factor which should be taken into account in kinetic and mechanistic studies and which has practical consequences discussed in more detail in Section II.B.1.

There are several books and book chapters dealing with cationic ring-opening polymerization [2–6].

A. Monomers

Heterocyclic compounds, which polymerize by cationic mechanism, contain one or more heteroaroms within a ring. Depending on the nature of heteroatoms and their arrangement, these monomers belong to the different classes of organic compounds.

Typical heterocyclic monomers, which polymerize by cationic mechanism, are listed below:

Cyclic ethers: 

Cyclic sulfides: 

Cyclic amines: 

Cyclic formals: 

Lactones: 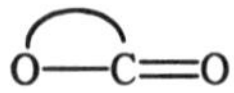

Cyclic oxaza compounds: 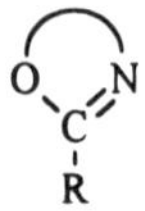

Lactams: 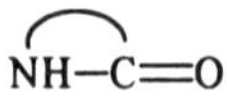

Cyclic phosphates, phosphites: 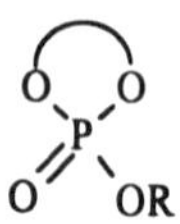 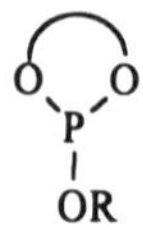

Siloxanes: 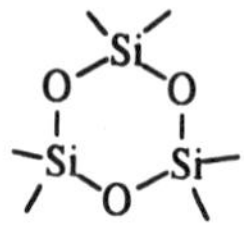

Phosphazenes: 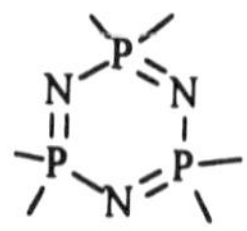

Bicyclic and spiro compounds: ethers, acetals, esters, oxalactams.

The thermodynamic polymerizability of cyclic monomers depends on the ring strain, thus 6-membered rings containing one heteroatom do not polymerize, whereas polymerization of 5-membered rings as well as 7- and higher member rings is an reversible process (cf., Section II.B.1). The situation may differ, however, for heterocyclic monomers containing more than one heteroatom, thus several 6-membered rings, e.g., 1,3,5-trioxane, glycolide, valerolactone, hexamethyltrisiloxane, and phosphazenes, undergo polymerization.

II. ELEMENTARY REACTIONS IN THE CATIONIC RING-OPENING POLYMERIZATION

A. Initiation

Initiation is the reaction in which active species are generated through interaction of initiator and monomer molecules. In cationic ring-opening polymerization this process does not necessarily have to proceed in one step. Comparison of the polymerization of vinyl and heterocyclic monomers initiated by protonic acid illustrates the difference.

$$HA + CH_2{=}CHR \longrightarrow CH_3{-}\overset{+}{C}HR\ ,\ A^- \qquad (5)$$

(6)

Protonation of vinyl monomers directly produces species with a structure similar to the structure of propagating species:

$$CH_3{-}\overset{+}{C}HR,\ A^- \quad \text{vs.} \quad \ldots{-}CH_2{-}\overset{+}{C}HR,\ A^- \qquad (7)$$

while in the case of heterocyclic monomers, for example, cyclic ethers, protonation produces secondary oxonium ions whereas propagation proceeds on tertiary oxonium ions:

(8)

The reactivity of both species may be quite different. Because the protonation, at least when it involves proton exchange, is fast, the rate-determining step in initiation may be the reaction of protonated monomer with the next monomer molecule to form tertiary oxonium ion.

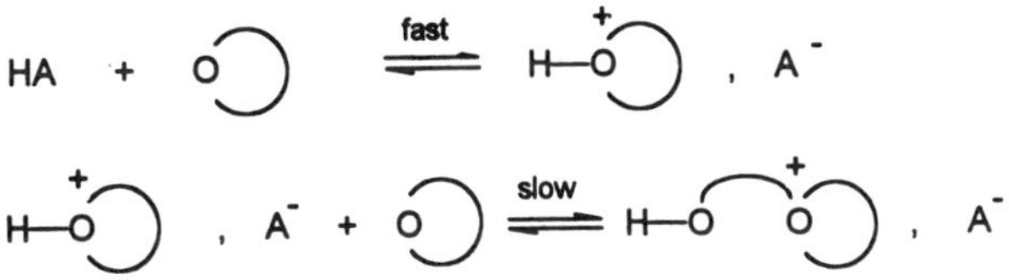

(9)

(10)

Formally similar scheme may also operate for other initiating systems. Thus, for polymerization of cyclic acetals initiated with triphenylmethylium (trityl) salts, the cationation of the monomer is a fast, reversible reaction [7]. The next reaction, however, due to the steric hindrance, is very slow and consequently active species are formed by a parallel path involving hydride transfer (cf., Section II.A.2) [8,9].

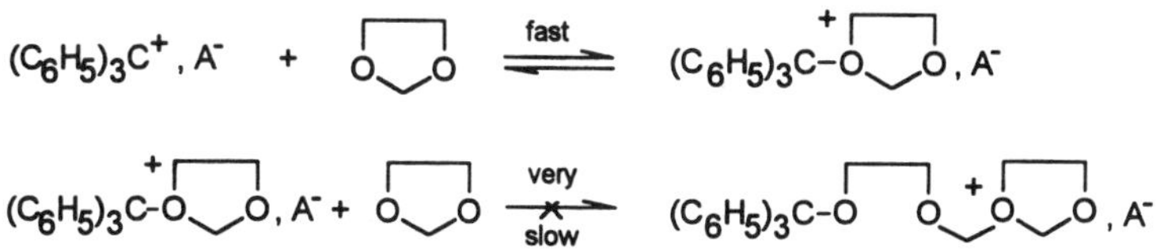

(11)

(12)

Thus, it must be remembered, that in the polymerization of heterocycles, species originally formed by interaction of initiator with monomer, may differ significantly (both in structure and reactivity) from the propagating active species. Initiation may involve the sequence of at least two reactions and the second one may be the slow, thus rate-determining step in initiation.

1. Initiation with Protonic Acids

Protonic acids as initiators of the cationic polymerization of heterocyclic monomers, should fulfil two requirements.

1. They should be the strong acids, i.e., the acid-base equilibrium should be shifted to the right-hand side:

(13)

2. They should provide the weakly nucleophilic anions to avoid the collapse of ionic active species (termination).

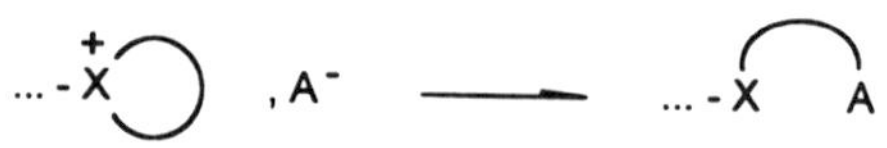

(14)

Although the pK_a or H_0 values for several acids are known [10,11], the definition of strong acid is somewhat arbitrary, because the position of equilibrium (13) depends on the basicity of heterocyclic monomer. Because this basicity varies from rather low (e.g., cyclic acetals) to rather high (e.g., cyclic amines) no universal rule describing the behavior of particular protonic acids in ring-opening polymerization exists.

Several strong protonic acids are commercially available. Trifluoromethanesulfonic (triflic) acid, fluorosulfonic acid, and perchloric acid may be obtained and stored in a pure state. The first two can be conveniently purified by distillation (b.p. 162° C and 165° C, respectively) [12], perchloric acid is less frequently used due to its oxidative properties and difficulties in handling (explosive). Complex acids HPF_6 ($HF + PF_5$) and $HSbF_6$ ($HF + SbF_5$) are available as complexes with ethers. Acids of $H^+BF_3OH^-$ type are often the real initiators of polymerization initiated with Lewis acids (e.g., BF_3) if water is not rigorously excluded from the system.

More recently, heteropolyacids like 12-molybdophosphoric acid ($H_3PMo_{12}O_{40}$) have been found to be the very efficient initiators of cationic ring-opening polymerization [13]. They are easily available, inexpensive initiators with an acid strength comparable to perchloric acid; they are easily purified and conveniently handled in a pure state [14].

Some other acids, like HJ or HCl, are still the strong acids, providing relatively strongly nucleophilic anions. Thus, these acids protonate the heterocyclic monomers, but, except for highly nucleophilic monomers, reaction with counterion leads to 1:1 addition product. (Actually, HCl or HBr is used as titrating agent for a quantitative determination of oxirane rings [15]).

In the polymerization of strongly nucleophilic cyclic monomers, however (oxazolines, cyclic amines), monomer may compete successfully with Br^- or I^- counterion (cf., Section II.B.6.c.). For these monomers initiation with HBr, for example, may lead to high molecular weight polymers.

2. *Initiation with Stable Organic Salts*

The most commonly used initiators belonging to this group may be divided into few categories [2–6]:

Carbenium salts:	Triphenylmethylium [$(C_6H_5)_3C^+$, A^-]
	Tropylium ($C_7H_7^+$, A^-) and related
Oxocarbenium salts:	Methoxymethylium ($CH_3OCH_2^+$, A^-)
	Dioxolenium (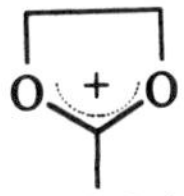 , A^-)
Carboxonium salts:	Benzoylium ($C_6H_5CO^+$, A^-)
	Acetylium (CH_3CO^+, A^-)
Oxonium salts:	Triethyloxonium [$(C_2H_5)_3O^+$, A^-] and related.

where A^- = stable complex counterions, e.g., BF_4^-, $SbCl_6^-$, PF_6^-, AsF_6^-, SbF_6^-,

Other stable organic salts are used less frequently; for example, application of dithiadiazolium salts as mono- or multifunctional initiators of polymerization of tetrahydrofuran has been described [16].

Stability of triphenylmethylium (trityl) salts is due to the resonance stabilization and the lack of β-protons which could undergo elimination. Several trityl salts are commercially available, they may also be easily prepared by direct mixing of components or by the silver salt method, e.g. [17,18]:

$$(C_6H_5)_3CF + SbF_5 \rightarrow (C_6H_5)_3C^+, SbF_6^- \tag{15}$$

$$(C_6H_5)_3CCl + AgSbF_6 \rightarrow (C_6H_5)_3C^+, SbF_6^- + AgCl \tag{16}$$

Oxocarbenium and carboxonium ions have an oxygen atom in the α-position to the carbon atom formally bearing the positive charge. Oxocarbenium salts are usually generated in situ [19], although some dioxolenium salts are stable and may be isolated and stored in pure form [20]. The oxocarbenium salts are prepared either by direct mixing of components in neutral (e.g., Freon 112) solvent [21]:

$$RCOF + SbF_5 \rightharpoonup RCO^+, SbF_6^- \qquad (17)$$

or by silver salt method:

$$RCOCl + AgSbF_6 \rightharpoonup RCO^+, SbF_6^- + AgCl \qquad (18)$$

In oxonium salts (alkylated ethers), the positive charge is located formally on the oxygen atom. It has to be remembered, however, that this is only the formal notation. In fact, as follows from CNDO-2 calculations, the charge is located mainly on carbon and hydrogen atoms [22].

Triethyloxonium tetrafluoroborate was first prepared by Meerwein from epichlorohydrin and $BF_3 \cdot Et_2O$ complex. This is still the most convenient method for preparation of triethyloxonium tetrafluoroborates and hexachloroantimonates [23].

Direct reaction:

$$RF + R_2O \cdot BF_3 \rightharpoonup R_3O^+, BF_4^- \qquad (19)$$

is very slow, and have little preparative value [24]. Other methods involve reaction of secondary salts (protonated ethers) with diazoalkanes [25] or reactions of oxycarbenium salts with ethers [26]:

$$RCO^+, SbF_6^- + 2R_2'O \rightharpoonup R_3'O^+, SbF_6^- + RCOOR' \qquad (20)$$

Several triethyloxonium salts are commercially available.

In the presence of monomers, all discussed initiators form the corresponding onium ions by alkylation or acylation, as shown schematically below, for the polymerization of cyclic ethers:

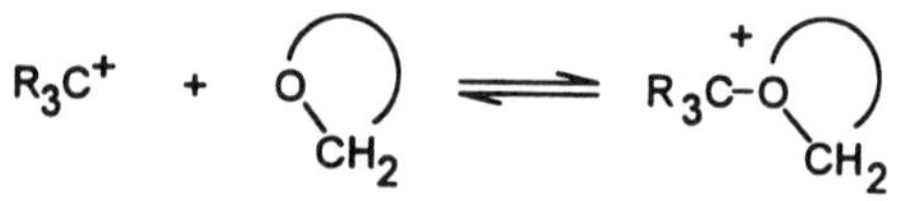

(21)

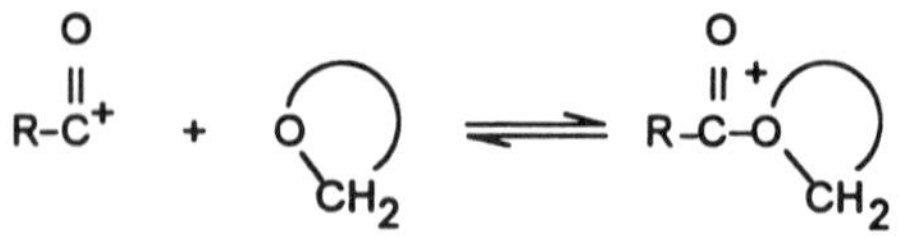

(22)

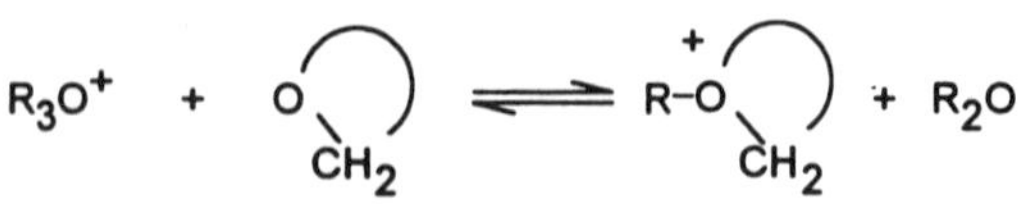

(23)

For oxocarbenium and trialkyloxonium salts this is typically a true initiation. An attack of monomer on α-carbon follows, leading to formation of active species and shifting Equilibria (21–23) practically completely to the right-hand side which leads to the incorporation of initiator moiety as the head groups. This is not always the case with stable carbenium salts. In the polymerization of cyclic acetals initiated with triphenylmethylium salts, oxonium ion is formed quickly. The attack of the next monomer molecule on α-carbon is very slow, however, and the alternative reaction pathway operates, involving hydride transfer to carbenium ion remaining in equilibrium, as shown below for the polymerization of 1,3-dioxolane [7–9]:

(24)

(25)

(26)

Because Reaction (24) is reversible, eventually all the carbenium salt is consumed in irreversible Reaction (25). Similar behavior was observed for polymerization of tetrahydrofuran initiated with trityl salts. In this case, however, hydride transfer from tetrahydrofuran molecule is followed by proton expulsion to form 2,3-dihydrofuran, which complicates the initiation mechanism [27,28]:

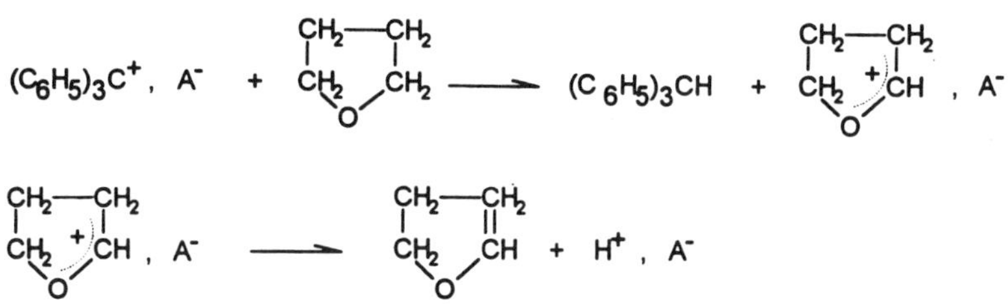

(27)

The studies of initiation of tetrahydrofuran polymerization with differently substituted carbenium salts shed some light on the reasons of low reactivity of oxonium ion formed by direct alkylation of monomer. With triphenylmethylium cation, reaction proceeded exclusively by hydride transfer; the increasing proportion of initiation by direct addition was observed with decreasing the number of aromatic substituents (diphenyl methyl, phenyl dimethyl) [29].

Thus, apparently the less substituted carbenium ion gives a more reactive oxonium ion. This may be related either to decreasing steric hindrance for an attack of the monomer or to increasing charge density of the α-carbon atom.

3. *Initiation with Covalent Compounds*

Covalent compounds, which are strong alkylating or acylating agents may initiate the cationic polymerization of heterocycles. Again, as in the case of acids, classification of these agents is relative; for example, alkyl bromide will efficiently alkylate strongly nucleophilic monomers like cyclic amines or oxazolines and thus initiate their polymerization whereas it will be uneffective in the polymerization of cyclic ethers.

Esters of very strong protonic acids (trifluoromethanesulfonic, fluorosulfonic, perchloric), however, are sufficiently strong alkylating agents to initiate the polymerization of even weakly nucleophilic monomers (cyclic acetals, ethers) [2–6]. Also their anhydrides (e.g., triflic anhydride) are efficient initiators. This last compound is especially interesting, because in the polymerization of cyclic ethers it leads to macromolecules with two identical growing chain ends (difunctional initiator) [30]:

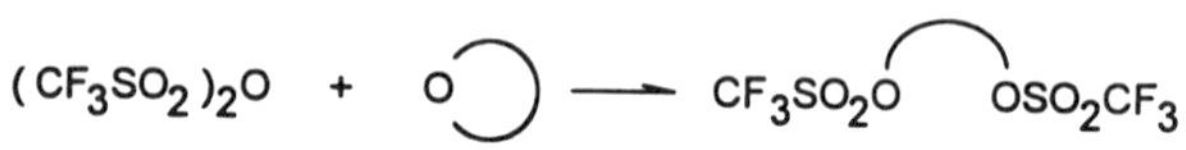

(28)

(For the discussion of the reactivity of covalent active species, see Section II.B.6.c.)

It has been reported that boron, aluminum, or gallium tristriflates are effective initiators of polymerization of tetrahydrofuran [31].

4. *Initiation with Lewis Acids*

A number of different Lewis acids has been used in the past to initiate cationic polymerization of heterocyclic monomers. The strongest Lewis fluorine acids, as measured in the solution, on the pH scale by electrochemical titration are [32]:

$$SbF_5 > AsF_5 > TaF_5 > BF_3 > NbF_3 > PF_5 \qquad (29)$$

although this order may depend on the method of determination. Lewis acids containing other halides may be stronger (depending on definition), thus for boron series the following order is quoted [32]:

$$BJ_3 > BBr_3 > BCl_3 > BF_3 \quad (30)$$

For ionization of alkyl halides, the fluorine-containing Lewis acids show exceptional ability, far exceeding those of $AlCl_3$ and other conventional Lewis acids. They also show a high tendency to form stable complex anions of MtF_{n+1}^- or MtF_nOH^- type. Thus, the following values of stabilization energy for formation of complex anions are reported [33]:

$$X^- + BF_3 \rightleftharpoons BF_3X^- \quad (31)$$

where $X = F^-$ 75 kcal/mol; $X = OH^-$ 79 kcal/mol; $X = Cl^-$ 25 kcal/mol; $X = Br^-$ 11 kcal/mol.

The stability of fluorine-containing complex anions is apparently related to steric reasons (the smallest size of F atom) and to the highest electronegativity of fluorine.

From fluorine-containing Lewis acids, BF_3 is most often used as initiator of the cationic ring-opening polymerization. This is at least partly due to the fact that BF_3 forms a stable, well-defined complex with ethers (e.g., $BF_3 \cdot Et_2O$, m.p. = −59° C, b.p. = 125° C) [34]. Complexes of BF_3 with ethers or alcohols are the commercial products, whereas gaseous BF_3, if needed, may be conveniently obtained by thermal decomposition (>100° C) of relatively less stable complex with anisole.

PF_5 (gaseous in a pure state) gives a stable complex with tetrahydrofuran, whereas no ether complexes of AsF_5 (gas) or SbF_5 (liquid, b.p. = 150° C) are known [35]. SbF_5, being the strong Lewis acid, is also the strong fluorinating and oxidating agent. Thus, the most commonly used Lewis acid type initiator is BF_3 and, to a much lesser extent, PF_5, usually in the form of complexes with ethers.

In discussing the mode of initiation, one has to be very careful not to generalize results, because behavior of various Lewis acids in similar systems may be quite different. Thus, complexes of BF_3 with ethers are stable, whereas BCl_3 reacts with ethers [35]:

$$BCl_3 \cdot O(C_2H_5)_2 \rightarrow C_2H_5Cl + C_2H_5OBCl_2 \quad (32)$$

On the other hand, complexes of PF_5 with ethers undergo disproportionation [36]:

$$3\ R_2O \cdot PF_5 \rightarrow 2\ R_3O^+PF_6^- + F_3P{=}O \quad (33)$$

In the case of BF_3 there were several claims that this Lewis acid alone or in form of a complex with ether may generate the cationic species spontaneously, e.g.:

$$R_2O \cdot BF_3 \rightleftharpoons R^+BF_3OR^- \quad (34)$$

or

$$R_2O\cdot BF_3 \underset{}{\overset{+BF_3}{\rightleftharpoons}} R_2OBF_2^{+}, BF_4^{-} \quad (35)$$

There is, however, no evidence supporting this view; on the contrary, the electron diffraction studies show that the ether molecule is unchanged in the complex and the only change is an increase of B—F bond length from 1.30 Å in BF_3 to 1.43 Å in the complex (the distance B—O is equal to 1.50 Å) [37]. Thus it seems, that in the majority of cases, the presence of coinitiator is necessary. The most common coinitiator is water. Taking into account the high tendency of BF_3 to form complex anions and the high acidity of the resulting acid, it appears that typically initiation proceeds by the following scheme:

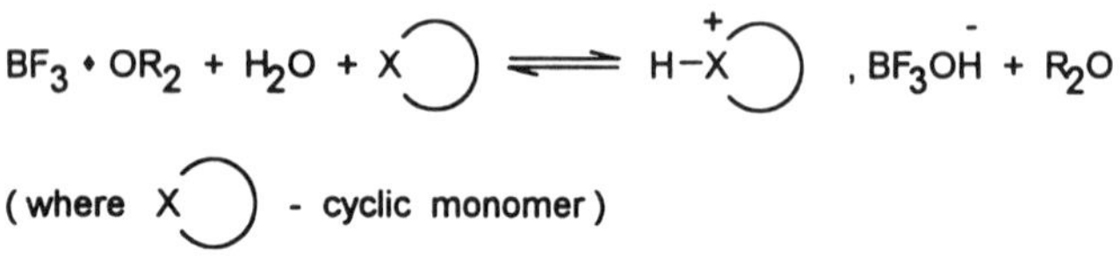

(36)

Indeed, at least in one case (polymerization of 1,3,5-trioxane), it has been shown that BF_3 does not initiate the polymerization of rigorously purified monomer, although polymerization proceeds in less thoroughly dried systems [38]. Also in the presence of proton trap, BF_3 does not initiate the polymerization of 1,3,5-trioxane [39].

Different mechanisms, however, may operate for other Lewis acids. Thus, the following mechanism was proposed for polymerization of tetrahydrofuran initiated with PF_5 on the basis of ^{31}P NMR studies [40].

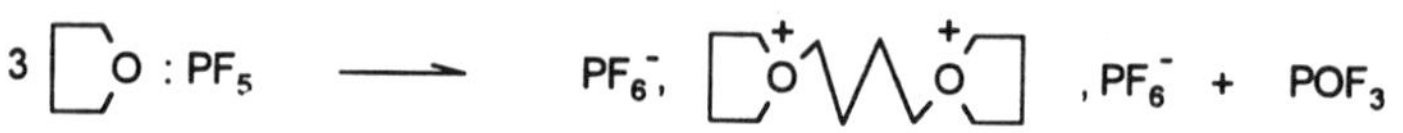

(37)

In general however, it seems that the most frequently used Lewis acid, BF_3, requires the presence of a coinitiator. Because in such a case the concentration of the true initiator is not known, this initiator is not well suited for any quantitative studies, although it is still used in synthetic work.

5. *Initiators Containing Silicon Atom*

There are several reports in recent literature on the application of silicon-containing compounds as the initiators of cationic ring-opening polymerization. This apparently is related to the attempts to prepare block copolymers containing polysiloxane or polysilane segments. $(CH_3)_3SiCl\cdot AgClO_4$ system was used to initiate cationic polymerization of tetrahydrofuran

[41]. Bis(chlorodimethylsilyl)benzene·$AgPF_6$ system was shown to act as a bifunctional initiator of substituted oxirane polymerization [42]. Trimethylsilyl iodide and triflate were used also as initiators of the cationic polymerization of oxazolines [43]. In this system, however, in contrast to typical initiation mechanism of oxazoline polymerization, *O*-silylation leads to initiation, because of the unfavorable charge distribution in *N*-silylated species:

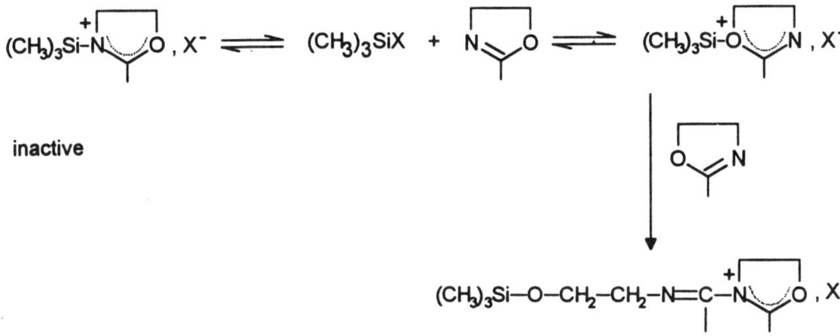

(38)

Another novel class of silicon-based initiators has been described [44]. Compounds containing Si—H bond in the presence of platinum catalysts (PtI_2, H_2PtBr_6, $PtCl_2(C_6H_5CN)_2$) are effective initiators of cationic polymerization of cyclic ethers.

The following mechanism of polymerization was proposed: platinum catalyst is reduced by Si-H compound to platinum colloid, which activates another molecule of Si-H compound and facilitates a nucleophilic attack of oxygen atom in the monomer on silicon atom (e.g., for substituted oxirane):

(39)

6. *Photoinitiation*

For some applications, for example curing of epoxy resins, the "in situ" generation of cationic initiators is required. This may be achieved by photochemical decomposition of suitable promotors to radicals, which in the presence of diazonium, sulfonium, or halonium salts (not able to initiate the polymerization by itself) are converted into corresponding cationic initiators [45–48].

$$R_2 \xrightarrow{h\vartheta} 2R^{\cdot}$$

$$R^{\cdot} + R_2'I^{+}, A^{-} \longrightarrow R^{+}, A^{-} + R'I + R'^{\cdot} \qquad (40)$$

7. Radiation-Induced Polymerization

Polymerization of some cationically polymerizable monomers like cyclohexene oxide [49], 3,3-bis(chloromethyl)oxetane [50], or 1,3,5-trioxane [51] may be initiated by irradiation with γ-rays both in liquid or solid state.

The ions or radical ions are formed by electron transfer or the heterolytic cleavage of the ring. Although potentially this is an attractive route of converting monomer into pure polymer, the results are highly irreproducible and the polymerizations are difficult to control [52].

B. Propagation

1. Reversibility of Propagation

Thermodynamics of ring-opening polymerization has been treated in several monographs [53–55].

The conversion of monomer into polymer is possible when the change of free energy ΔG is negative:

$$n\,\mathrm{M} \rightleftharpoons [\mathrm{M}]_n; \qquad \Delta G < 0 \qquad (41)$$

ΔG is related to changes of enthalpy ΔH and entropy ΔS accompanying the process

$$\Delta G = \Delta H - T\Delta S \qquad (42)$$

Typically, for polymerization, ΔH is negative and so is ΔS (consequently, $-T\Delta S$ is positive). Exceptions from this rule are known, however, and ring-opening polymerization of some monomers may be entropy driven. Typically, however, the driving force for polymerization is the passage from higher energy to lower energy form (for cyclic monomers release of the ring strain), whereas the entropy factor acts in the opposite direction. The enthalpy factor is temperature independent (except for correction for nonideality of the system), whereas entropy factor ($T\Delta S$) increases with increasing a temperature. If ΔH value is relatively high (polymerization of monomers containing unsubstituted or monosubstituted double carbon-carbon bond or highly strained ring), negative ΔH value will always exceed positive $-T\Delta S$ value (for the practically available temperature range, i.e., below decomposition point). If however ΔH value is lower (di- or higher substituted double carbon-carbon bond or moderately strained rings), at certain temperature within the available range, $-T\Delta S$ value becomes equal to ΔH value. At this temperature ΔG is equal to zero and reaches a positive value with further increase of temperature. This temperature, at and above which polymerization is thermodynamically prohibited, is called ceiling temperature.

Treating the propagation as a reversible reaction:

$$M_n^* + M \underset{k_d}{\overset{k_p}{\rightleftharpoons}} M_{n+1}^* \tag{43}$$

where: M_n^*, M_{n+1}^* active species; M = monomer; k_p = rate constant of propagation; k_d = rate constant of depropagation, remembering that for sufficiently large n, $[M_n^*] = [M_{n+1}^*]$ and expressing the equilibrium constant of propagation K as:

$$K = \frac{k_p}{k_d} \tag{44}$$

for equilibrium conditions one finds:

$$\begin{aligned} k_p[M_n^*][M]_e &= k_d[M_{n+1}^*] \\ k_p/k_d &= K = 1/[M]_e \end{aligned} \tag{45}$$

where $[M]_e$ -equilibrium mononomer concentration. Remembering that:

$$\Delta G^0 = -RT\ln K \tag{46}$$

and, that in equilibrium.

$$\Delta G = 0 \qquad \text{thus } \Delta H = T_c\,\Delta S \qquad \text{thus } T_c = \Delta H/\Delta S \tag{47}$$

one finally arrives at equation:

$$T_c = \Delta H^0/(\Delta S^0 + R\ln[M]_e) \tag{48}$$

This equation describes the relation between two parameters characterizing the reversible polymerization: ceiling temperature and equilibrium monomer concentration. For any temperature there is a certain value of $[M]_e$ [Eq. (48)]. If the starting concentration of monomer is lower than this value, i.e., $[M_0] < [M]_e$, the polymerization will not proceed. If $[M]_0$ is higher than $[M]_e$, polymerization will proceed until concentration of monomer reaches $[M]_e$ (i.e., consumption of monomer will be equal to $[M]_0 - [M]_e$).

In Eq. (48) both ΔH^0 and ΔS^0 are negative whereas $R\ln[M]_e$ for $[M]_e > 1$ is positive. Thus at higher temperatures the equilibrium monomer concentration is higher. Temperature at which $[M]_e$ is becoming equal to the highest attainable $[M]_0$ (i.e., in bulk) is thus the ceiling temperature for bulk polymerization, the value that is usually cited in the literature. Thermodynamically more informative is the ceiling temperature for standard 1 *M* solutions, because then $\ln[M]_e = 0$ and T_c is simply equal to $\Delta H^0/\Delta S^0$.

Before discussing in more detail the factors influencing the enthalpy and entropy of polymerization of heterocyclic monomers, it is worth reviewing some practical consequences of the reversibility of polymerization:

a. The polymerization stops at the moment when monomer concentration reaches $[M]_e$ value at given temperature. Consequently, the highest possible yield of polymerization is equal to $([M]_0 - [M]_e)/[M]_0 \cdot 100\%$. As $[M]_e$ decreases with decreasing temperature, the lower the temperature, the higher the yield of polymerization allowed by thermodynamics.

b. After monomer concentration reaches $[M]_e$ value, the system is in dynamic equilibrium (providing, that at least some active species are still present). If the temperature is raised or the reaction mixture is diluted before termination, depolymerization may occur.

c. In the course of polymerization, both propagation and depropagation occur.

$$M_n^* + M \underset{k_d}{\overset{k_p}{\rightleftharpoons}} M_{n+1}^* \tag{49}$$

Thus, kinetics of monomer consumption is given by equation:

$$-\frac{d[M]}{dt} = k_p[M_n^*][M] - k_d[M_{n+1}^*] \tag{50}$$

because $k_d/k_p = [M]_e$, i.e., $k_d = k_p[M]_e$, then

$$-\frac{d[M]}{dt} = k_p[M^*]([M] - [M]_e) \tag{51}$$

Thus, the knowledge of $[M]_e$ is indispensable for determination of rate constant of propagation.

The other consequence of the reversibility of polymerization is the broadening of the molecular weight distribution. Even if initiation is fast and all the chains grow simultaneously, M_w/M_n is equal to 2 under equilibrium conditions. Polymers with narrower MWD, however, may be isolated at early stages of polymerization (at low conversion) when the contribution of depropagation is not very significant, thus the preparation of polytetrahydrofuran with $M_w/M_n \sim 1.1$ at $<10\%$ conversion has been described [56].

2. *Thermodynamics of Reversible Polymerization in Real Systems*

In the preceding discussion, the basic relationships describing the process of reversible polymerization in general were briefly outlined. For any real process, the physical state of substrate, monomer, and product, polymer, has to be considered. This is denoted by subscripts, e.g., ΔG_{gg}, ΔG_{ss}, ΔG_{lc}, $\Delta G_{lc'}$ where first letter denotes the physical state of monomer and the second of the polymer. Thus ΔG_{gg} corresponds to (hypothetical) conversion of gaseous monomer into gaseous polymer, ΔG_{ss} corresponds to conversion of dissolved monomer into dissolved polymer, ΔG_{lc} corresponds to the conversion of liquid (bulk) monomer into condensed (amorphous) polymer, whereas $\Delta G_{lc'}$ corresponds to the conversion of the liquid monomer into crystalline polymer. The subscripts should always be specified because the numerical values of thermodynamic parameters depend on the physical state of monomer and polymer. This is especially evident for the lc′ case, i.e., conversion of liquid monomer into crystalline polymer. This process involves not only the chemical reaction but also phase transition (crystallization), and the enthalpy and entropy of crystallization add to the parameters of chemical reaction. It may happen that the contribution of phase transition is dominating and the process, which would be highly reversible in solution, proceeds almost irreversibly due to the precipitation of crystalline polymer. This is the case of cationic polymerization of 1,3,5-trioxane [3,53].

In addition to the physical state of reactants, it should be remembered that the ideal behavior is encountered only in the gaseous state. As the polymerization processes involve liquid (solution or bulk) and/or solid (condensed or crystalline) states, the interactions between monomer and monomer, monomer and solvent, or monomer and polymer may introduce sometimes significant deviations from the equations derived for ideal systems. The quantitative treatment of thermodynamics of nonideal reversible polymerizations is given in Ref. 54.

To illustrate the magnitude of the problem, it may be remembered that in the cationic polymerization of 5-membered cyclic ether, tetrahydrofuran (THF), the $[M]_e$ value changes from ~3.1 mol·L^{-1} in bulk to ~4.0 mol·L^{-1} in CCl_4 solution ($[M]_0 \sim 6$ mol·L^{-1}) and to ~5.7 mol·L^{-1} in CH_3NO_2 solution ($[M]_0 \sim 6$ mol·L^{-1}) at 25° C [57].

a. Factors Affecting the Entropy of Polymerization The entropy of polymerization of heterocyclic monomers is much less dependent on the structure of monomer than the enthalpy of polymerization. Thus, it is convenient to discuss the magnitude of entropy factor, first, because it shows what is the barrier which should be counterbalanced by enthalpy factor to make polymerization thermodynamically feasible.

Analysis of the changes of entropy upon passage from monomer to polymer unit leads to the conclusion, that the loss of rotational entropy of monomer is nearly balanced by the gain in internal rotation and vibrational entropy in polymer unit [58,59]. Thus, the overall entropy of polymerization is governed mainly by the loss of translational entropy of monomer.

It is quite difficult to analyze the data obtained in real systems, because the scattering of the results given by different authors is significant; moreover, in many cases the nonideality of the system has not been considered. The analysis of the collected experimental data [53] indicates, however, that the upper limit for ΔS values is not very far from this predicted on the assumption that the magnitude of the entropy of polymerization is governed mainly by the loss of translational entropy of monomer.

Taking the value $\Delta S \sim -120$ J/mol·K as the typical value for the polymerizations at ambient temperatures (300° C), one can calculate the contribution of entropy term to the free energy of polymerization as close to: $T\Delta S = (300°\text{K})(-120\ \text{J/mol·K}) = -36$ kJ/mol.

b. Factors Affecting the Enthalpy of Polymerization The enthalpy of polymerization of heterocyclic monomers is related to the ring strain. The ring strain depends on nonbonded repulsions, bond-angle distortions and bond-length deformations (stretches or compressions) [60,61].

The large strain of small (3,4-membered) rings is caused by excessive bond-angle distortions. In larger rings (8–11-membered), transannular interactions between substituent atoms (including hydrogens), pointing "into" the ring are the main source of the ring strain. The lowest strain is observed for 5–7-membered ring (with minimum at 6), due to the fitting of the valency angles and eliminating the nonbonded interactions (substituents pointing "out of" the ring) (Table 1).

The 3-membered cyclopropane ring is flat; the 4-membered (and higher) rings are puckered to minimize the strain. Bending of the ring usually diminishes bond angle distortions (as for 6-membered ring in chair conformation), but it may lead to increasing nonbonding interaction. Thus, the preferred conformation is the one in which the sum of all sources of strain is minimized.

Heteroatoms in small rings affect bond angles, bond lengths, and bond strengths through a combination of factors including their intrinsic hybridization, magnitude of covalent radii, angle-bonding constants, nonbonded interactions, and long-range electronic effects.

The extent of these effects depends on the heteroatom. Some bonds between carbon and heteroatom have more angular flexibility than others; —Si—O—Si— bond angle for example vary from ~125° to ~142.5° without incurring appreciable angular strain [61]. In general, the bonds with

Table 1 Strain Energies (kJ/mol) of Cyclic Compounds

Ring size					
3	115	117	77.8	96.1	
4	109		79.0		
5	25.5	28.0	4.1		30.5
6	0.4		9.2		12.1
7	25.5				
8	40.5				
9	52.5				

Source: Ref. 62.

more ionic character and participation of *d*-orbitals are more flexible while nonpolar sp-type bonding results in low angular flexibility.

In Table 2, the ring strains, bond lengths, and bond angles for some 3- and 4-membered heterocyclics are compared with these of corresponding cycloalkanes [63].

The data of Table 2 indicate that replacement of one or two CH_2 groups by O, NH, CO-NH, or CO-O groups does not change significantly the geometry of the ring and the resulting ring strain. For sulfur-containing heterocycles the C—*X*—C bond angle is considerably lower, and the ring strain is also lower.

The pattern observed for 3- and 4-membered rings may not always hold for larger rings which adopt minimum strain conformation as a result of bending or twisting the ring. The presence of —C(O)—O— or —N—C(O)— group, in which the resonance stabilization requires planar structure, may introduce additional strain to 5-, 6- and 7-membered rings.

It should be remembered that although enthalpy of polymerization is governed mainly by the magnitude of the ring strain, in fact it is related to the difference in enthalpies of monomer and polymer unit. If the release of the ring strain is compensated by nonbonded interactions appearing in

Table 2 Ring Strains, Bond Lengths, and Bond Angles of Some 3- and 4-Membered Heterocyclic Compounds

3 - membered 

	Cyclopropane $X:CH_2$	Oxirane X:O	Thiirane X:S	Aziridine X:N-
Ring strain kJ/mol	115	114	83	113
Bond lengths A°				
X-C	1.51	1.436	1.819	1.488
C-C	1.51	1.472	1.492	1.483
C-H		1.082	1.078	
Bond angles,°				
C-X-C	60	61.4	48.4	59.6
X-C-C	60	59.3	65.8	60.2
H-C-H		116.57	116.0	

polymer unit, the enthalpy of polymerization may be lowered considerably.

As indicated earlier, nonbonding interactions between substituents (including hydrogen atoms) are not important for medium size (5–7-membered) rings. They may however become important in a polymer unit as indicated below [53]:

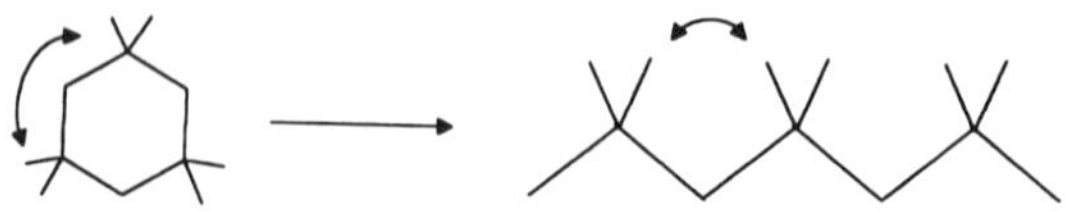

(52)

Table 2 (Continued)

4 - membered 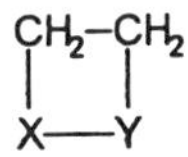

	Oxetane	Thietane	Oxetanone	Azetidinone
	X:O,Y:CH_2	X:S,Y=CH_2	X:O,Y:CO	X:NH,Y:CO
Bond lengths A°				
X-Y			1.45	1.39
Y-C			1.53	1.49
C-C	1.54	1.55	1.53	1.51
C-X	1.46	1.85	1.45	1.38
Bond angles,°				
C-X-Y	94.5	78	89	95
X-Y-C			94	92
Y-C-C	88.5	97	83	96
C-C-X			94	87

Source: Ref. 63.

For unsubstituted repeating units (i.e., hydrogen atoms as substituents) these interactions are relatively weak; however, they may become important when larger substituents are present. This is illustrated by the data of Table 3 where the calculated enthalpies of (hypothetical) polymerization of cyclopentanes are compared [64]. Thus, in general, substitution, decreasing the enthalpy of polymerization, decreases the thermodynamic polymerizability of heterocyclic compounds.

c. The Free Energy of Polymerization In the previous paragraph it was shown that the entropy factor barrier, which should be overcome in order to make polymerization possible, is close to ~ −36 kJ/mol. This is easily achieved for polymerization of 3- and 4-membered rings, where the

Table 3 Calculated Enthalpies of Polymerization of Cyclopentanes

R_1	R_2	ΔH (kJ/mol)
H-	H-	-21.9
CH_3-	H-	-17.2
CH_3-	CH_3-*	-13.4

Source: Ref. 64.
* Substituents at the same carbon atom.

enthalpy of polymerization is usually above − 100 kJ/mol. The minimum of the dependence-relating ring strain with the size of the ring (cf., Table 1) occurs at $n = 6$. These rings, with the exception of lactones and lactams (for the reasons discussed earlier) and some cycles composed exclusively of heteroatoms like siloxanes or phosphazenes, will not polymerize ($\Delta H \sim 0$). Five- and seven-membered rings constitute a transient group; polymerization is usually possible, although it is highly reversible. Larger rings (8–12-membered) polymerize more easily, although such systems are seldom studied.

The enthalpy of polymerization of 3- and 4-membered rings is so much higher than the entropy factor that substitution does not significantly reduce their polymerizability. Disubstituted oxiranes (e.g., isobutylene oxide) or oxetanes (3,3-dimethyloxetane) still polymerize practically irreversibly. Substitution may prohibit polymerization of 5-membered monomers, however.

3. *Structure of Active Species*

The nucleophilic site of heterocyclic monomer is a heteroatom. Reaction of heterocycle with a cation (e.g., initiation) leads to onium ion.

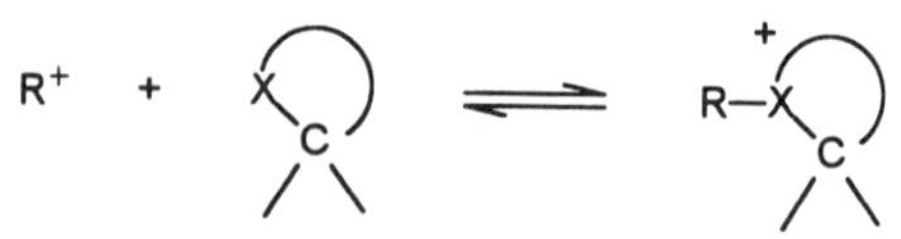

(53)

(Let us assume, for the sake of simplicity, that there is only one heteratom in the molecule.)

The question arises of whether the onium ion (Scheme 53) is the only form in which product of reaction may exist. Neglecting for the time being the other components of the system, one can envisage two possible forms:

cyclic (thus strained, at least for strained heterocycles) onium ion and open-chain carbenium ion, resulting from unimolecular ring-opening:

(54)

Before discussing the real situation existing in polymerization, let us discuss the factors affecting position of Equilibrium (54). The unimolecular opening of the ring relieves the ring strain; the magnitude of this effect was discussed in Section II.B.2.b. Although the strain of onium ion ring may be different than the strain of parent cyclic molecule due to different hybridization of heteroatom and thus different bond angles and bond lengths, the difference is probably not very significant.

On the other hand, unimolecular opening of the ring according to Eq. (54) transforms more stable onium ion (lower energy) into less stable (higher energy) carbenium ion. In most cases in cationic ring-opening polymerization, the heteroatom on which the charge is located is oxygen, sulfur, or nitrogen; thus we deal with oxonium, sulfonium, or ammonium ions. The magnitude of the energy difference between onium and carbenium ions may be estimated from data available for simple model reactions:

where X = O, S or N - (55)

Some related data concerning the gas-phase reactions are shown in Table 4 [65,66]. Data shown in Table 4 indicate that, in the case of primary carbonium ions (e.g., $C_2H_5^+$ cation), the enthalpy of conversion of carbenium into oxonium or ammonium ion (>220 kJ/mol) by far exceeds the strain of even the most strained rings (<115 kJ/mol).

Increasing stabilization of carbocation, however, may change this situation considerably. For *t*-butyl cation, the enthalpy of carbenium-oxonium ion conversion is already lower than the ring strain of the 3-membered ring. Thus, for the isobutylene oxide case, the presence of carbenium ion in equilibrium with oxonium ion can not be excluded a priori:

(56)

A similar situation exists for other types of highly stabilized carbenium ions, e.g., having oxygen atom in α-position to carbon atom as in

Table 4 Enthalpy of Carbenium-Onium Ion Equilibria in the Gas Phase

Reaction					ΔH (kJ/mol)
CH_3^+	+	OH_2	⇌	$CH_3-\overset{+}{O}H_2$	−276
$C_2H_5^+$	+	OH_2	⇌	$C_2H_5-\overset{+}{O}H_2$	−155
$s-C_3H_7^+$	+	OH_2	⇌	$s-C_3H_7-\overset{+}{O}H_2$	−95
$t-C_4H_9^+$	+	OH_2	⇌	$t-C_4H_9-\overset{+}{O}H_2$	−47
CH_3CO^+	+	OH_2	⇌	$CH_3CO-\overset{+}{O}H_2$	−79
$C_2H_5CO^+$	+	OH_2	⇌	$C_2H_5CO-\overset{+}{O}H_2$	−73
$C_2H_5^+$	+	C_2H_5OH	⇌	$C_2H_5-\overset{+}{O}(H)C_2H_5$	−226
$C_2H_5^+$	+	NH_3	⇌	$C_2H_5-\overset{+}{N}H_3$	−275

Source: Ref. 65, 66.

acylium or alkoxymethylium ions:

(57)

As shown in Table 4, the enthalpies of corresponding interconversions are within the range of strains of highly strained rings; thus, also in these cases, the carbenium ion participation may be not insignificant.

In the preceding discussion, only the enthalpy factor was considered. The position of equilibrium is affected also by an entropy of reaction. For the model reactions, the entropy is negative, decreasing the net free energy and thus the equilibrium constant. This is mainly due to the loss of translational entropy resulting from association. Contribution of entropy factor is relatively smaller, however (corresponding to contribution of 0.5 unit to lnK value). It should still be smaller for the opening of the cyclic onium ion because this process does not involve significant changes in translational entropy (monomolecular reaction).

There is no quantitative information on the enthalpies of Reaction (55) in solution for unstabilized carbocations. The only available information concerns highly stabilized carbocations. It may be expected that the enthalpy of carbenium-onium ion interconversion in solution will be lower than in the gas phase, because more electrophilic carbocations will be solvated more strongly than onium ions. The magnitude of this effect is illustrated by the data of Table 5.

The preceding discussion shows that in the case of heterocyclic monomers with a general formula:

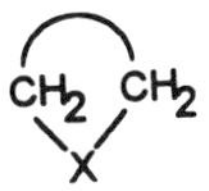

(58)

the concentration of carbenium active species is negligible. This comprises

Table 5 Enthalpy of Carbenium-Oxonium Ion Equilibria in solution, CH_2Cl_2, 25°C, AsF_6^-, SbF_6^-, PF_6^- Counterions

Reaction				ΔH kJ/mol
$(C_6H_5)_3C^+$	+	H_2O	⇌ $(C_6H_5)_3C\text{-}\overset{+}{O}H_2$	-38.9
$(C_6H_5)_3C^+$	+	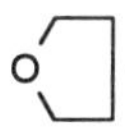	⇌ 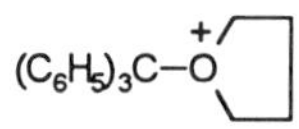	-58.5
$(C_6H_5)_3C^+$	+	$O(C_2H_5)_2$	⇌ $(C_6H_5)_3C\text{-}\overset{+}{O}(C_2H_5)_2$	-71.1
$(C_6H_5)_3C^+$	+		⇌ 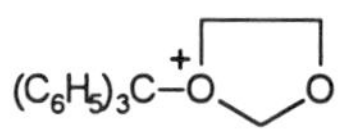	-30.5
$(C_6H_5)_3C^+$	+	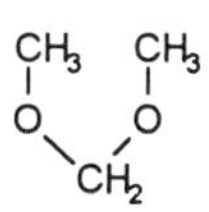	⇌ 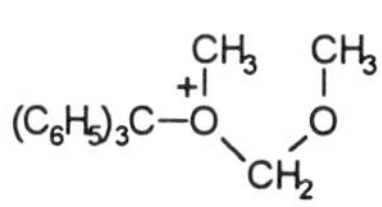	-34.7

Source: Ref. 7.

such monomers as unsubstituted cyclic ethers, cyclic sulfides, and cyclic imines. Indeed, in several systems belonging to this group, active species have been directly observed by ^{1}H or ^{13}C NMR and the only species detected were onium ions [2–6].

On the other hand, when highly stabilized carbenium ion results from unimolecular ring opening, the carbenium active species may be present in the system in detectable concentrations. Indeed, in the polymerization of cyclic acetals —O$\cdots$CH$_2{}^+$ active species exist in equilibrium with tertiary oxonium ion active species (cf., Section II.B.6.b.).

It should be stressed at this point, that the preceding discussion concerned only the question of in what form active species exist in the system. The prevailing presence of one type of species does not necessarily mean that these species are mainly participating in propagation. It may happen that much less abudant, but much more reactive species contribute mainly to propagation. This problem is addressed in more detail in Section II.B.6.b.

4. *Reactions of Active Species*

In the previous paragraph we discussed the structure of isolated cationic active species. In the polymerizing system these active species face at least three different nucleophiles: monomer, polymer repeating unit, and counterion.

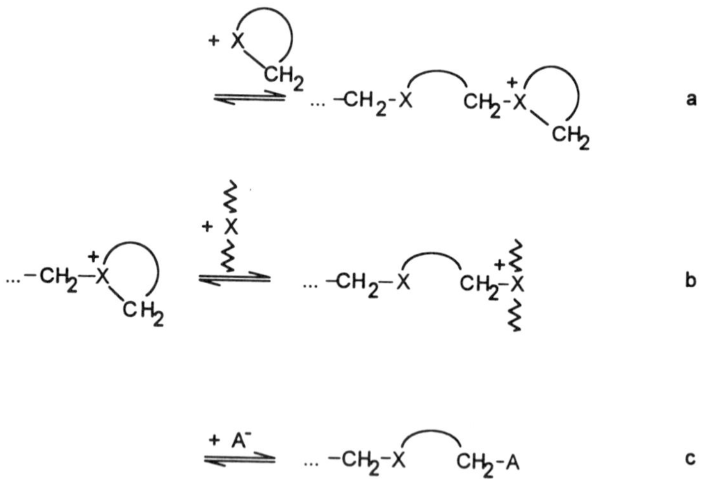

(59)

(For the present discussion, we assume that the active species is onium ions and the situation is essentially analogous for carbenium active species.)

All three reactions are nucleophilic substitutions, proceeding according to S_N2 mechanism. Thus, all three are competing and their relative contributions depend on the nucleophilicity of the monomer, polymer unit, and counterion.

Reaction a is a propagation reaction:

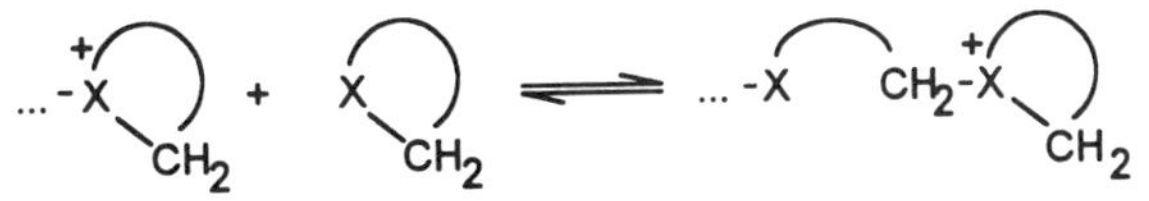

(60)

Depending on the size of the ring it may proceed as reversible (5- and higher member rings) or practically irreversible (3,4-membered rings) reaction.

Reaction b is a chain transfer to polymer; it may involve heteroatom of the foreign or its own macromolecule. In the later case it leads to formation of macrocycle:

(61)

or

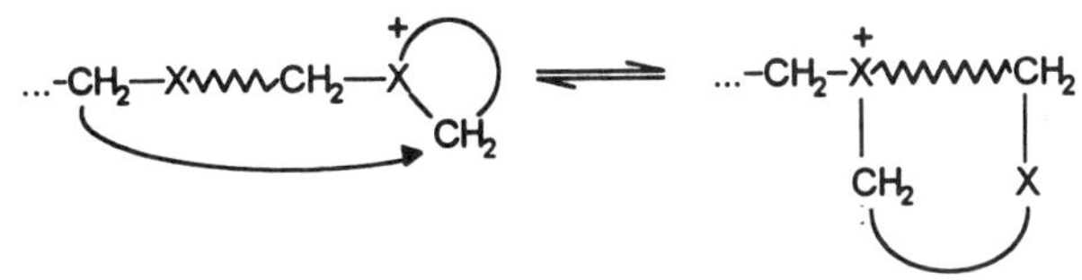

(62)

Again, these reactions, depending on the structure of monomer, may be reversible or irreversible. If resulting branched or macrocyclic onium ions are not reactive, i.e., they can not re-form the original active species either by intramolecular cyclization or by reaction with the next monomer molecule, then these reactions lead to termination (such a situation exists in the polymerization of cyclic sulfides or amines; cf., Section III.D.E.).

If, however, Reaction (61) is reversible and/or branched onium ions may react with next monomer molecule:

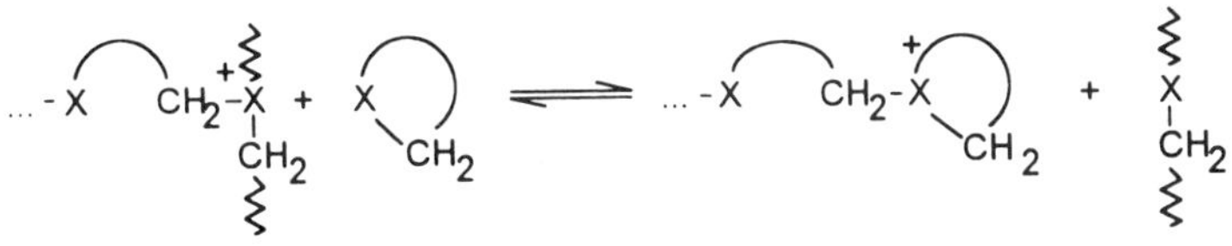

(63)

then branched or macrocyclic onium ions should be considered just another form of active species. Reaction (63) plays important role in the polymerization of cyclic acetals.

The third competing reaction, i.e., reaction with counterion, also may proceed as a irreversible or reversible reaction:

(64)

Irreversible reaction terminates the propagation. If Reaction (64) is reversible or covalent form may react with next monomer molecule:

(65)

then covalent structure should be considered still another form of active species. Eq. (59), and its consequences exemplified by Eqs. (60)–(65) illustrates the most characteristic feature of ring-opening polymerization, as compared with vinyl cationic polymerization. Because the nucleophilic site of the monomer, i.e., heteroatom, is preserved in polymer repeating unit, polymer is not a neutral component of the system. It may participate in polymerization according to Eq. (63) and in some systems (e.g., polymerization of cyclic acetals), when the nucleophilicity of polymer repeating unit is higher than that of monomer, the whole polymerization process may be dominated by reactions involving polymer chain.

The second characteristic feature of ring-opening polymerization stems from the fact that monomers (and polymer repeating units) are relatively strong nucleophiles. Thus, even if ionic species collapse to covalent ones by recombination with counterion, nucleophilic monomer or polymer unit may displace counterion, reforming the ionic species. Whether these reactions are possible or not depends on the relative nucleophilicity of monomer and counterion. Strongly nucleophilic nitrogen-containing heterocycles may displace even relatively strongly nucleophilic bromide:

(66)

Thus, bromide-terminated macromolecules are considered covalent active species in polymerization of cyclic imines or oxazolines (cf., Section III.E, F). The same termini would be inactive, however, in the presence of much weaker nucleophiles, e.g., cyclic ethers. Recombination of active species with Br^- anion would thus result in termination.

(67)

On the other hand, counterions of much lower nucleophilicity like perchlorate ($OClO_3^-$) or triflate ($OSO_2CF_3^-$) are displaced by oxygen-containing bases:

(68)

Thus, corresponding macroesters are the covalent active species in the polymerization of cyclic ethers.

The preceeding discussion shows that, not for all systems, propagation in the cationic ring-opening polymerization can be represented by a single equation. Still, for a number of systems, it is justified to represent a propagation reaction as simple bimolecular S_N2-type reaction.

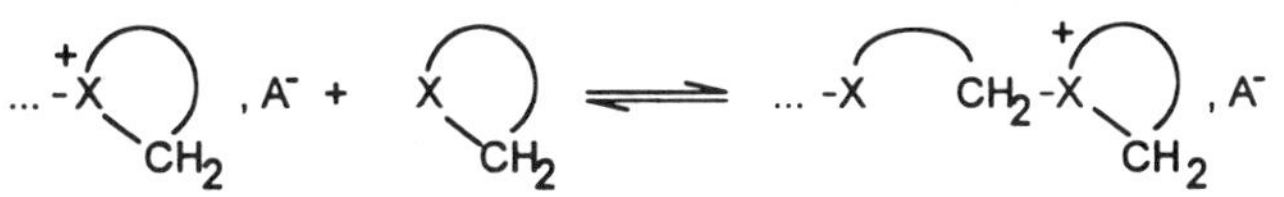
(69)

Polymerization of 5-membered cyclic ether, tetrahydrofuran with SbF_6^- counterion, is the best example of such a system.

Let us discuss at first those cases in which the kinetics of propagation process was analyzed in terms of Eq. (69).

5. *Rate Constants of Propagation*

To determine the values of propagation rate constants it is necessary to know the concentration of active species. The following methods of determination of active species concentration in the cationic ring-opening polymerizations have been used.

a. Direct Determination by Spectroscopic Methods Because onium ions do not absorb in UV, only NMR methods have been used [67]. The disadvantage of the method is its low sensitivity and inapplicability in the case of signals overlapping, although these problems may sometimes be overcome by using modern NMR techniques.

b. Determination after Trapping of Active Species Two approaches have been developed: phenoxide end-capping:

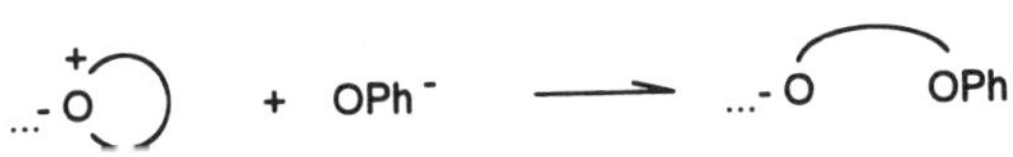
(70)

with the following UV determination of phenoxide end groups after isolation and purification of polymer [68]; and phosphine ion trapping:

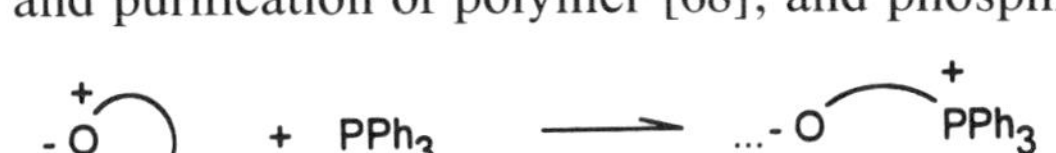
(71)

with the following ^{31}P NMR determination of concentration of phosphonium ion using known excess of phosphine as internal standard. This method does not require isolation and purification of polymer and provides additional information on the structure of active species [69,70].

c. Indirect Methods These are methods in which concentration of active species was assumed to be equal to starting concentration of initiator after clarifying the initiation mechanism and/or estimated from DP_n of the polymer formed.

The collection of rate constants of propagation of different heterocyclic monomers are given in Ref. 2.

6. *Kinetics of the Propagation in the Systems in Which Two or More Types of Ionic Active Species Coexist*

a. Ions vs. Ion Pairs Ionic active species may exist in form of free ions or ion pairs. The dissociation constants have been measured for model systems and in real polymerizing systems [71–74]. The K_D values do not depend strongly on the structure of macrocation-counterion pair, some typical values are given in Table 6.

The numerical values indicate, that in polar solvents (CH_3NO_2, $C_6H_5NO_2$) the fractions of free ions are relatively high, thus their reactivity could be determined for several systems and compared with the reactivity of ion pairs, according to the methodology developed by Szwarc [75].

For a number of systems, it was shown that the reactivities of free and paired active species in cationic ring-opening polymerizations are essentially the same. This is explained by stronger solvation of free ions (predominantly by monomer), leveling off the difference between free and paired cationic species. Thus, the rate constants of propagation on ionic species will be discussed in the following sections without making distinction between free and paired ions. The detailed discussion on the reactivities of ionic active species in cationic ring-opening polymerizations is presented in Ref. 2.

b. Oxonium vs. Carbenium Ions As discussed in Section II.B.3, in the system in which unimolecular ring-opening would lead to stabilized carbocations, carbenium active species may exist in equilibrium with onium ions and participate in propagation. Carbenium ions are much more reactive toward nucleophiles than onium ions. Factors that increase the stability of carbenium ions, at the same time decrease their reactivity. Thus, these carbenium ions, which may coexist with onium ions in significant concentrations, should have considerably reduced reactivity in comparison with primary carbenium ions.

Table 6 Dissociation Constants of Ion Pairs

Cation	Counterion	Solvent	t,°C	K_{diss} mol·l^{-1}	Ref.
$(C_2H_5)_3O^+$	BF_4^-	CH_2Cl_2	0	$4.4 \cdot 10^{-6}$	[71]
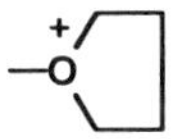	BF_4^-	CH_2Cl_2	0	$5.4 \cdot 10^{-6}$	[71]
$(C_2H_5)_3S^+$	BF_4^-	$C_6H_5NO_2$	20	$1.3 \cdot 10^{-2}$	[71]
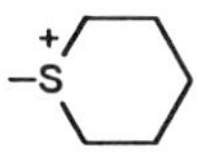	BF_4^-	$C_6H_5NO_2$	20	$1.7 \cdot 10^{-2}$	[71]
$(C_6H_5)_3C^+$	SbF_6^-	CH_2Cl_2	25	$1.6 \cdot 10^{-4}$	[71]
$(C_6H_5)_3C^+$	ClO_4^-	$C_2H_4Cl_2$	21.5	$1.3 \cdot 10^{-3}$	[71]
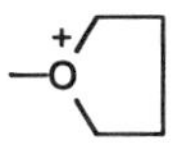	SbF_6^-	CH_2Cl_2	25	$1.5 \cdot 10^{-5}$	[72]
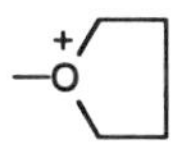	SbF_6^-	CH_3NO_2	25	$2.0 \cdot 10^{-3}$	[73]
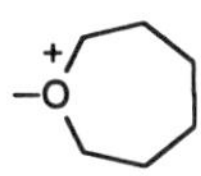	SbF_6^-	CH_2Cl_2	25	$2.8 \cdot 10^{-5}$	[74]
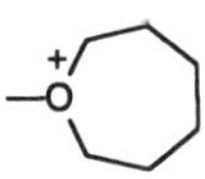	SbF_6^-	CH_3NO_2	25	$1.6 \cdot 10^{-3}$	[74]

The system for which the reactivity of carbenium and onium active species have been quantitatively compared is the polymerization of cyclic acetals [76]:

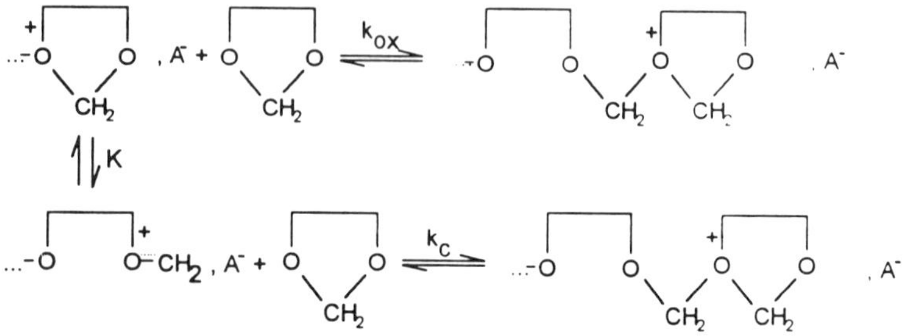

(72)

(73)

By dynamic ^{1}H NMR studies of model reaction:

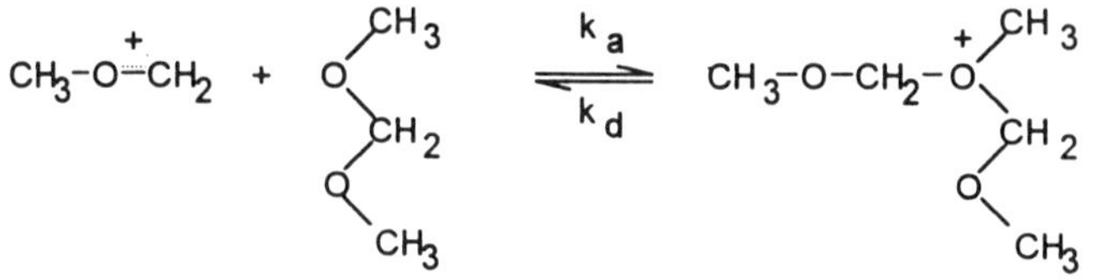

(74)

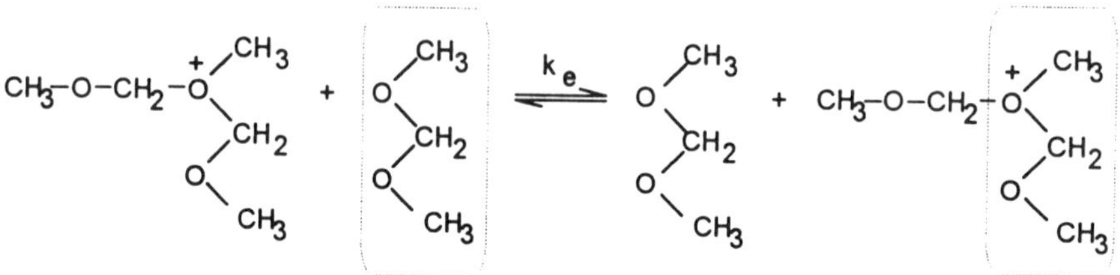

(75)

the following values of the rate constants have been obtained ($-70°$ C, SO_2 solution) $k_a = 2 \cdot 10^6$ mol^{-1}·L·sec^{-1} [corresponding to propagation on carboxonium ions (k_c)] $k_e = 1.9 \cdot 10^4$ mol^{-1}·L·sec^{-1} [corresponding to propagation on oxonium ions (k_{ox})].

At the same time, the equilibrium constants have been determined as equal to $K = k_a/k_d = 3 \cdot 10^3$ mol^{-1}·L [76]. From these sets of data it follows that at the conditions of measurements ([M] = 1 mol·L^{-1}, [M*] = 10^{-3} mol·L^{-1}), carbenium active species would constitute 0.03% of all active species, but would account for ~3% of propagation (in terms of monomer consumed).

c. Ionic vs. Covalent Species As already mentioned, in some systems in cationic ring-opening polymerization, reaction between the ionic active species (onium ions) and counterions is reversible, giving covalent species, capable of reacting with monomer. The, best studied system, in which both ionic and covalent species participate in propagation, is the cationic polymerization of tetrahydrofuran with $CF_3SO_3^-$ counterion [1–6,77,78]:

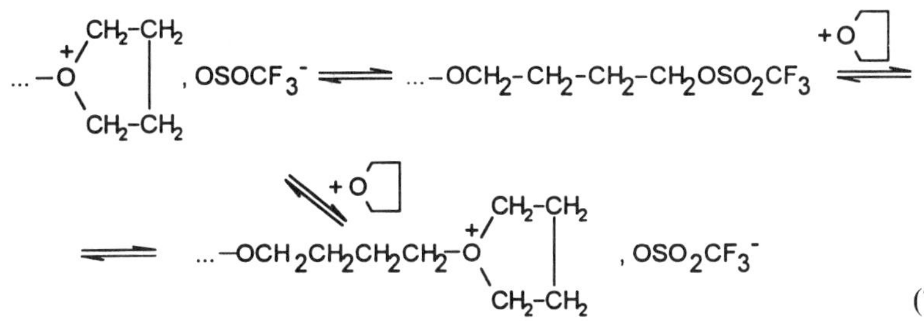

(76)

The other system, in which similar scheme operates, is the polymerization of oxazolines with halide counterion [79]:

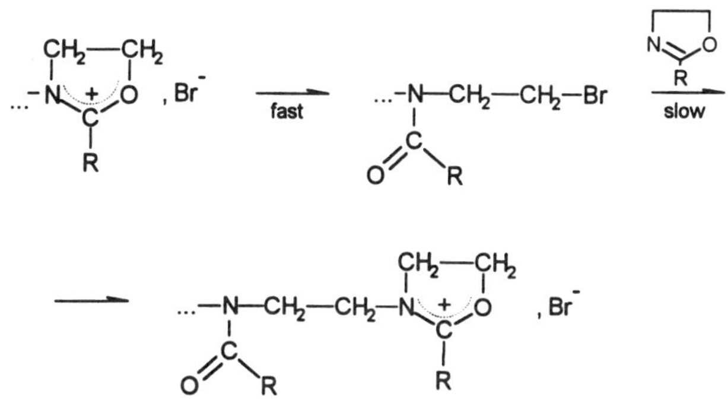

(77)

The simplified Eq. (76) indicates the complexity of the former system. Covalent species may be converted into ionic species by intramolecular cyclization (reverse to collapse of the ion pair) or by intermolecular reaction with a monomer molecule (propagation).

The depropagation of ionic species may give either ionic species (an attack of oxygen atom of the terminal unit on the exocyclic α-carbon) or covalent species (an attack of counterion on exocyclic carbon). Direct conversion of covalently terminated chain into covalently terminated chain with one more unit is not possible, as well as the corresponding backward reaction. The detailed analysis of the complete reaction scheme is given in a recently published review [6].

The situation is relatively simpler in the polymerization of oxazolines [Eq. (77)], where reactions of both ionic and covalent species are practically irreversible. Thus, the simple sequence of reactions operates: one or more propagation steps on ionic species followed by the collapse to the covalent species and one covalent propagation step re-forming ionic species.

In both systems, described by Eqs. (76) and (77), ionic and covalent species were observed directly by ^{1}H NMR and their concentrations determined by integrating the spectra. This, coupled with kinetic measurements, allowed the determination of rate constants of propagation on the covalent species and their comparison with rate constants of propagation or ionic species [57,80–82].

The data of Table 7 show that the reactivity of covalent species is generally much lower than that of ionic species (with exception of 7-membered cyclic ether; this has been attributed to the steric factors), but, as expected, the difference in reactivity decreases with increasing the polarity of the solvent.

C. Chain Transfer to Polymer

As discussed in Section II.B.4, in some systems (cyclic amines, cyclic sulfides), formation of branched onium ion by reaction of active species with polymer unit leads to termination. In such systems, these events

Table 7 Rate Constants of Propagation on Ionic and Covalent Active Species

Monomer [Ref.]	Counterion	Solvent	Temp. (°C)	Type of Propagating Species	k_p ($mol^{-1} \cdot L \cdot sec^{-1}$)
Tetrahydrofuran [57]	$CF_3SO_3^-$	CCl_4	25	Ionic	$4 \cdot 10^{-2}$
				Covalent	$6 \cdot 10^{-5}$
		CH_2Cl_2	25	Ionic	$3.1 \cdot 10^{-2}$
				Covalent	$1.7 \cdot 10^{-4}$
		CH_3NO_2	25	Ionic	$2.4 \cdot 10^{-2}$
				Covalent	$5 \cdot 10^{-4}$
Oxepane [80]	$CF_3SO_3^-$	CH_3NO_2	25	Ionic	$1.3 \cdot 10^{-4}$
				Covalent	$4 \cdot 10^{-4}$
Oxozoline [82]	$CH_3C_6H_4O^-$	CD_3CN	40	Ionic	$1.9 \cdot 10^{-3}$
	J^-			Covalent	$1.8 \cdot 10^{-5}$
2-Methyloxazoline [82]	Br^-	CD_3CN	40	Ionic	$1.2 \cdot 10^{-4}$
	Cl^-			Covalent	$3 \cdot 10^{-6}$

do not have the further history, the branched ion becomes the "dead" component of the reaction mixture (this will be discussed further in Section II.D).

In other systems (cyclic ethers, acetals, esters, amides, siloxanes), the branched ions participate in further reactions, sometimes (cyclic acetals, siloxanes) with the rates not much different than the rates of reaction of cyclic onium ion active species.

Let us discuss the full kinetic and synthetic consequences. First of all it should be stressed that, in contrast to typical chain transfer, the reaction described as chain transfer to polymer does not lead to increase of the number of macromolecules and thus to decreasing the molecular weight. The growing macromolecule reacts with a chain of another (possibly growing) macromolecules; whatever happens next, the number of macromolecules does not change:

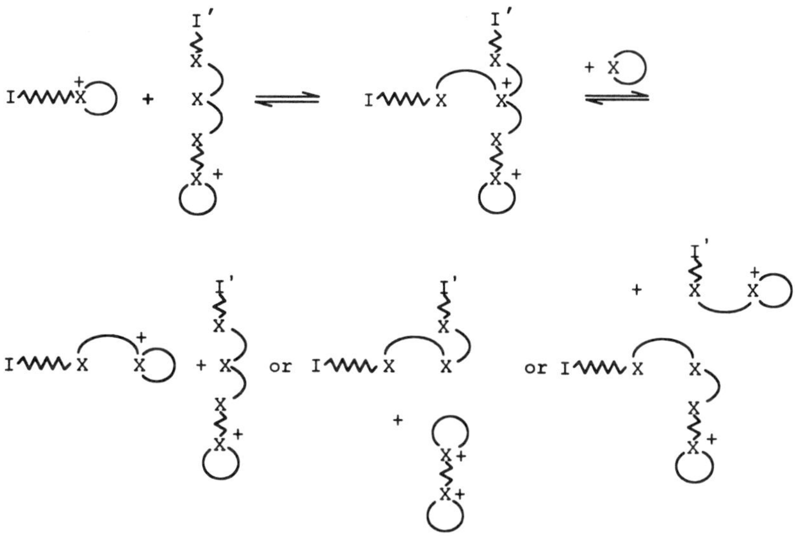

(78)

Thus, chain transfer to polymer does not influence the number average DP_n, it may however alter the molecular weight distribution. If the reversible chain transfer to polymer described in Eq. (78) occurred frequently, it would lead to statistical distribution, i.e., $M_w/M_n = 2$. The other consequence is that if the two originally present chains are different, the repetition of reaction sequence will lead to segmental exchange (so called "scrambling"). Both effects are clearly detectable, for example, in the cationic polymerization of cyclic acetals as it will be discussed in Section III.B.

Chain transfer to polymer may proceed also as a intramolecular reaction:

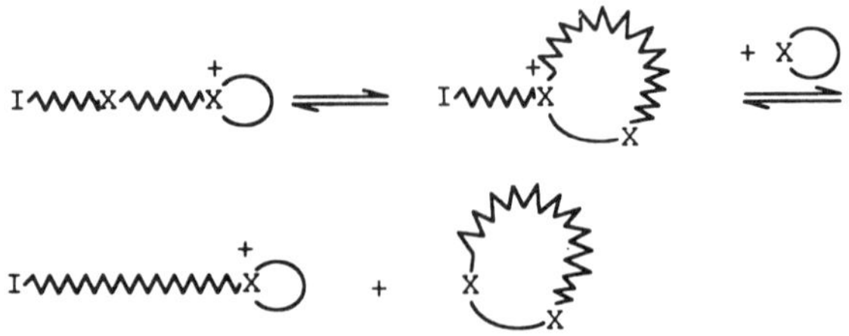

(79)

providing a route for formation of cyclic oligomers.

It is not always remembered that, from a purely thermodynamic point of view, conversion of monomer into polymer should lead to the equilibrium mixture of linear and cyclic polymer. This problem was analyzed quantitatively by Kuhn [83] and Flory [84].

Formation of cyclic macromolecules is observed in polycondensation, where macromolecules contain reactive groups both at the ends and within the chain, e.g., for polycondensation of hydroxyacids:

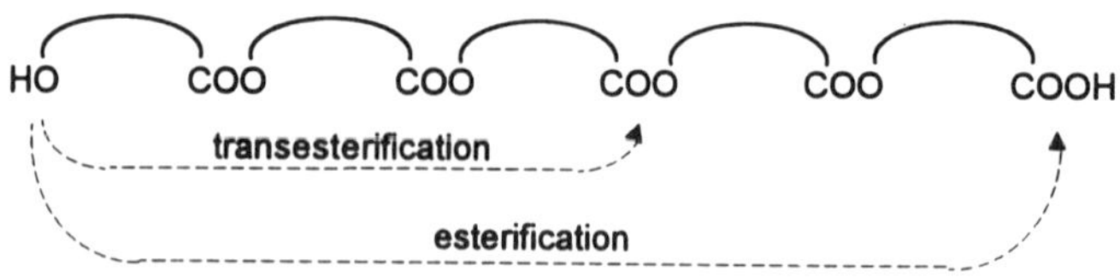

(80)

In vinyl polymerization, there is no mechanism available to bring the system to the thermodynamic equilibrium, thus only the kinetic product, linear polymer, is formed.

In cationic ring-opening polymerization the situation is similar to the one observed for polycondensation; heteroatoms along the chain may participate in reaction with terminal active species:

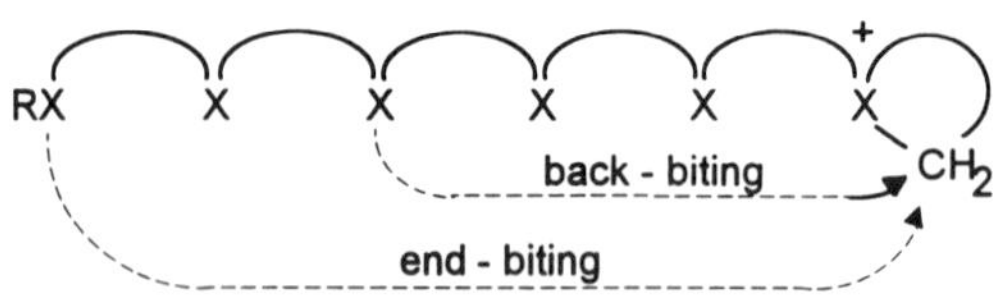

(81)

This reaction, called back-biting, provides a route for cyclization.

1. *Cyclization in Cationic Ring-Opening Polymerization*

The approach to the equilibria between linear and cyclic macromolecules is based on theoretical considerations of Kuhn [83] and Flory [84]. The linear macromolecule is treated as flexible thread, which can form a cycle (loop) between any two points along the chain:

(82)

For such an idealized system the concentration of a given cyclic macromolecule in equilibrium can be expressed as Eq. [85]:

$$[\overline{M_n}] = A \cdot n^{-5/2} \tag{83}$$

where n-polymerization degree (the number of monomer units within a cycle).

The system in cationic ring-opening polymerization, which most closely approaches this ideal situation, is the polymerization of cyclic siloxanes [86–88]. The chain, composed of —O—Si— units is highly flexible and provides a reaction sites for cyclization. It is also well suited for experimental studies, because concentrations of cyclic oligomers in equilibrium, due to their volatility, can be determined with accuracy by gas-liquid chromatography. The slope of the experimentally derived plot is indeed very close to the theoretically predicted value of -2.5.

The other system for which agreement between experimentally observed and theoretically predicted distribution of macrocyclics was observed is the cationic polymerization of 1,3-dioxolane [89].

Typically, the deviation from the theoretically predicted line is seen for smaller rings (up to ~25 atoms in the ring). It is a simple consequence of the fact that the theory relates the equilibrium constant exclusively to entropy changes, neglecting the enthalpy factor (this is equivalent to the assumption that all the rings are strainless). Thus, it applies well for larger strainless rings, while giving too-high equilibrium concentrations of smaller rings, when enthalpy factor can not be neglected.

Thermodynamic theory gives a relatively simple picture, concentration of macrocyclics is governed by a simple equation (with exception of medium-size rings), and consequently the distribution of macrocyclics can be rationalized. In the majority of systems, however, this simple treatment does not apply. First of all, the theory deals with systems being in true thermodynamic equilibrium. This means that not only monomer-polymer equilibrium had been reached, but also equilibrium between linear polymer and cyclic polymers. If propagation (establishing the former equilibrium) is fast as compared with back-biting (establishing the later one), monomer can be converted into purely linear polymer and only afterward, if active species are still present, the cyclic oligomers are formed slowly. This has been observed for cationic polymerization of THF [90].

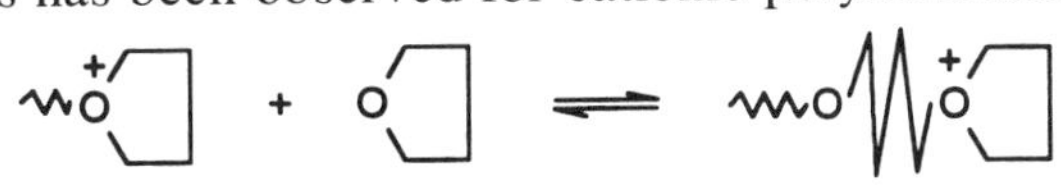

(84)

relatively fast

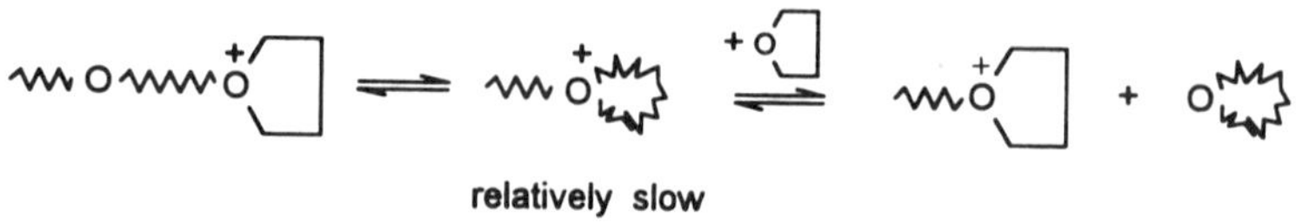

(85)

This factor alone should not affect the distribution of macrocyclics, but the equilibrium conditions would be reached after an unrealistically long time.

The other reason of deviation from theory stems from inapplicability of the theory to medium-size rings. The theory considers only the conformational entropy of macromolecules. For the polymerization of smaller rings, however, the entropy factor is usually unfavorable (cf., Section II.B.2.a) because of the loss of translational entropy. Thus, if medium size, strainless ring may be formed by back-biting, zero enthalpy change and negative entropy change will prevent it from polymerization. The formation of these particular oligomers will thus be irreversible and oligomers will accumulate in the system.

Cationic polymerization of ethylene oxide exemplifies this behavior [91]. Due to the lack of strain of 6-membered 1,4-dioxane ring, this cyclic dimer of ethylene oxide is formed readily by back-biting. Its polymerization would involve $\Delta H \sim 0$ and negative ΔS thus it is thermodynamically prohibited (equilibrium concentration of 1,4-dioxane is higher, than the highest attainable concentration, i.e., concentration in bulk).

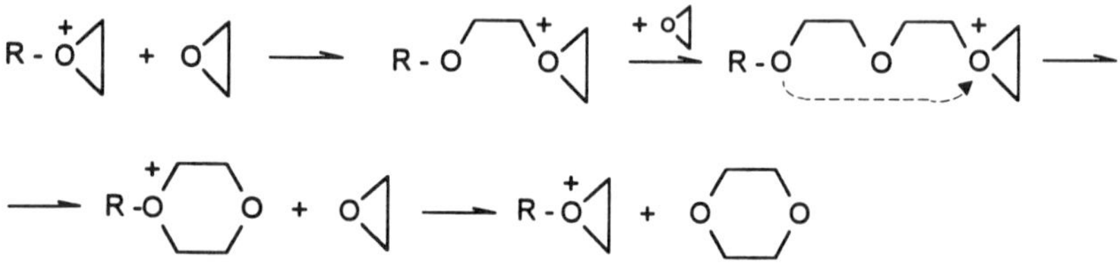

(86)

Cyclic dimer is accumulating in the system, and cationic polymerization of ethylene oxide may give 1,4-dioxane almost exclusively [92]. Depending on the structure of monomer (the sequence of atoms in the ring, the number and nature of substituents), cyclic oligomers with the different ring sizes may have the least favorable ΔG value for polymerization, and these particular oligomers will be accumulating in the system, constituting the main portion of a cyclic fraction.

For the irreversible formation of macrocyclics, the rate of their formation will also be affected by such factors as the nature of solvent (influencing the conformation of the chain in solution) or the nature of counterion (some conformations leading to closure of particular ring may be sterically unfavorable). Thus, in many real systems, the observed distribution of cyclic oligomers can not be rationalized easily. The systems for which quantitative information on the nature and concentration of cyclic oligomers formed are available are listed in Ref. 93.

For a synthetic polymer chemist the important question is whether the cyclization processes in cationic ring-opening polymerization can be controlled. If the preparation of linear polymer is attempted, then cyclic oligomers are undesirable side products. This is especially important in synthesis of telechelic polymers containing reactive end groups, because macrocycles would be unreactive admixtures. On the other hand, cyclic polymers, if prepared selectively, could be a valuable materials.

As already indicated, the overall concentration and distribution of macrocyclics may be to some extent altered by changing the reaction conditions (temperature, solvent, counterion). These effects, however, are not significant and difficult to predict.

Let us discuss briefly the other possibilities of controlling the distribution of linear and cyclic polymers in cationic ring-opening polymerization. In cationic ring-opening polymerization, formation of linear polymer and cyclization may be described by the scheme:

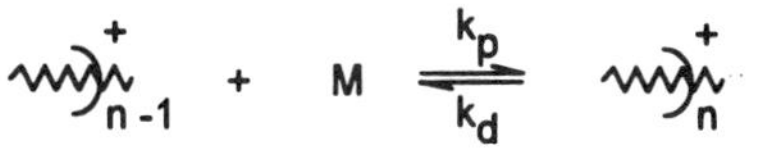

(87)

(88)

Thus, independently of thermodynamically controlled (equilibrium) distribution, exclusively linear polymer can be obtained if $k_p >>> k_c$. In such a system, the consumed monomer is converted practically quantitatively into linear chains, which, if the mechanism is available, i.e., if active species are living, will slowly undergo cyclization until thermodynamically controlled distribution is attained. If, however, polymerization is terminated before significant cyclization occurs, the purely linear polymer will be the product.

In some systems, for typical polymerization conditions, indeed $k_p \gg k_c$ and purely linear polymer is formed "by itself." This is, for example, a case of cationic polymerization of THF [90]. After reaching monomer-polymer equilibrium, concentration of cyclic oligomers is undetectably low. Their concentration, however, increases if the living system is kept for a prolonged period of time (cf., Section III.A.3).

In a number of important systems, however, the ratio of k_p/k_c is not favorable and cyclic oligomers are formed simultaneously with formation of linear macromolecules. The ratio of rate of propagation to the rate of cyclization may be altered in a desirable direction by changing the reaction mechanism. There are well-evidenced examples of kinetic enhancement in macrocycles. Thus, in cationic polymerization of cyclic acetals, protonic acid-initiated polymerization gives higher fractions of macrocyclics than polymerization initiated with alkylating agents [94–96]. This is due to the fact that in later systems only back-biting operates, whereas in the former ones, the presence of highly nucleophilic hydroxyl end group introduces an additional pathway for cyclization, namely end-to-end closure (end biting):

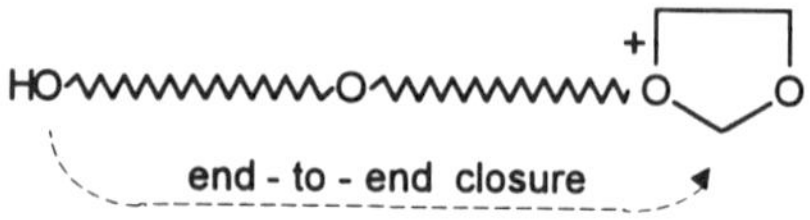

(89)

Thus, by introducing an end group more reactive toward the active species than groups along the chain, one can kinetically enhance the cyclization.

The opposite effect, i.e., decreasing of cyclic fraction content, would however be more desirable from the synthetic point of view. This calls for modifying the system in such a way, which would increase the ratio of rate of propagation to the rate of cyclization (or in more general terms, rate of reaction of active species with chain units). This may be achieved by introducing the substituents which would provide significant steric hindrance for the reaction of chain unit with active species. It is indeed observed for several groups of cyclic monomers that, with increasing number and size of substituents, the contribution of chain transfer to polymer is limited. This is especially pronounced and well documented in the cationic polymerization of oxiranes, oxetanes, thietanes, aziridines, or azetidines (cf., Sections III.A, III.D, III.E).

This is, however, still not the solution of the problem of how to decrease the content of cyclic fraction in the cationic ring-opening polymerization of a particular monomer. As the nature (and thus the reactivity) of the chain unit is fixed in such a case, the only variable left is the nature of active species (growing chain end). This approach has been applied in devising the process for cationic polymerization of oxiranes proceeding by activated monomer (AM) mechanism. In this process, described in more detail in Section III.A.1, propagation involves neutral growing chain end (hydroxyl group) and protonated (activated) monomer molecule:

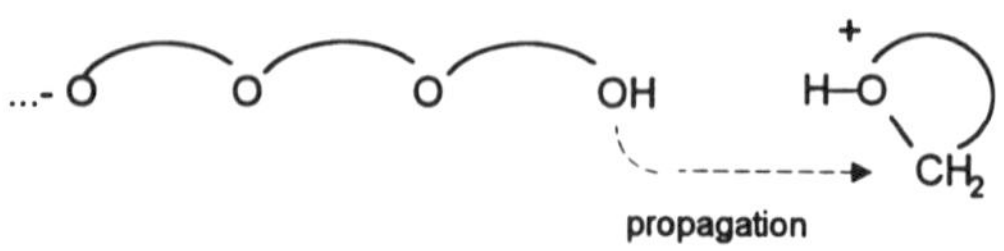

(90)

Growing chain end does not carry a positive charge and back-biting and thus cyclization is effectively eliminated in the systems in which AM mechanism operates.

D. Transfer and Termination

As discussed in Section II.C, in any system in cationic ring-opening polymerization, a reaction of active species with polymer repeating unit may lead to chain transfer to polymer or termination (if the resulting branched or cyclic onium ions are not active), whereas recombination with counterion leads to termination in the case of irreversible reaction. The later reaction may be avoided by the proper choice of counterion. As the onium ions are generally inherently stable there is no other termination reaction, provided that impurities that may act as terminating agents are eliminated.

Thus, in general, if impurities are eliminated, the proper counterion is used and chain transfer to polymer is reversible, cationic polymerization of heterocycles should proceed as process without termination. The absence of termination is the basic requirement for a living system. There is no unequivocal agreement concerning the conditions that should be met by the system in order to be called living. The original Szwarc definition involved several requirements, including narrow molecular weight distribution [97]. In some ring-opening polymerizations, even if the initiation is fast and termination and/or transfer are absent, narrow molecular weight distribution cannot be attained due to the reversibility of propagation.

1. *Terminationless Polymerization*

As discussed already, termination of ring-opening polymerization may proceed by: (a) irreversible recombination with counterion and (b) irreversible chain transfer to polymer. Other sources of termination are also possible, depending on the system: (c) reaction with other components of the system, solvent or impurities and (d) different reactions of more reactive species existing in equilibrium with stable onium species.

Let us discuss in more detail all these factors. First of all, it should be stressed that onium ions (e.g., trialkyloxonium, tetraalkylammonium, trialkylsulfonium) are inherently stable; different onium salts may be isolated and stored in a pure state, some of them are commercially available. This is in contrast to cationic vinyl polymerization, where carbenium ion species are not inherently stable due to the possibility of proton expulsion:

$$...\!-\!CH_2\!-\!\overset{+}{C}HR \rightarrow ...\!-\!CH\!=\!CHR + H^+ \qquad (91)$$

No such process is possible (at least at ambient temperatures) for onium salts. However, if (as discussed in Section II.B.3) onium salt exists in equilibrium with carbenium species, reaction analogous to Eq. (91) should be considered as a possibility, e.g.:

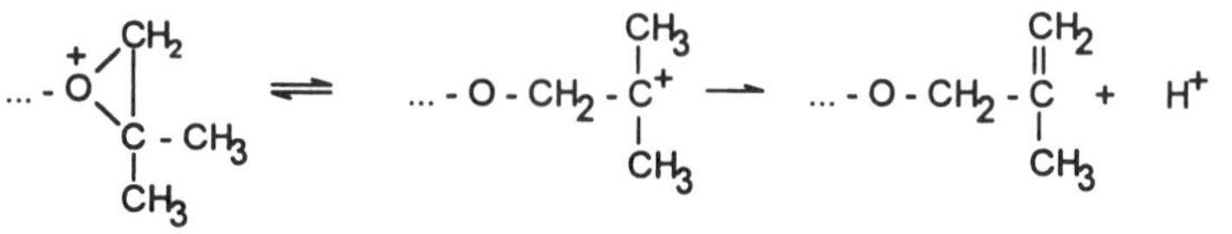

(92)

2. *Termination by Irreversible Recombination with Counterion*

Irreversible collapse of ionic species via recombination with counterion is the most trivial potential source of termination in cationic ring-opening polymerization. The relations discussed in Section II.B.6.c may also be applied to the present discussion. The termination will occur if counterion is nucleophilic enough (as compared with monomer) to compete successfully with monomer in S_N2 nucleophilic substitution and monomer is not nucleophilic enough to displace the counterion from the covalent termini:

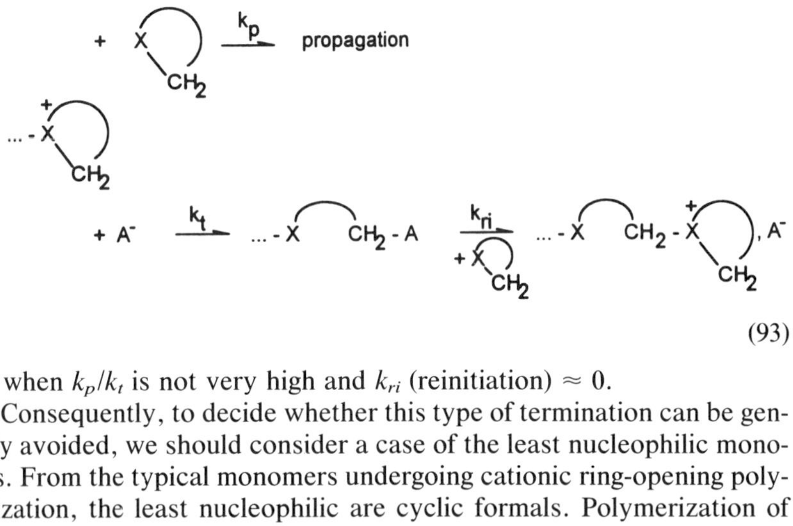

(93)

i.e., when k_p/k_t is not very high and k_{ri} (reinitiation) ≈ 0.

Consequently, to decide whether this type of termination can be generally avoided, we should consider a case of the least nucleophilic monomers. From the typical monomers undergoing cationic ring-opening polymerization, the least nucleophilic are cyclic formals. Polymerization of this group (e.g., 1,3-dioxolane) has been studied in detail. In the polymerization of 1,3-dioxolane it has been shown conclusively that if most stable counterions like SbF_6^- or AsF_6^- are used, there is no detectable termination due to interaction with counterion. Thus, in the polymerization initiated with $C_6H_5CO^+SbF_6^-$ and terminated with $P(C_6H_5)_3$, the numbers of head and tail end groups were equal even in polymers with M_n as high as $1.3 \cdot 10^5$ [96].

The kinetic evidences, however, were obtained that concentration of active species in 1,3-dioxolane polymerization decreases with time in the presence of BF_4^- counterion [98]. As discussed in Section II.A.4, BF_4^-, although considered stable anion, releases nucleophilic F^- anion more readily than SbF_6^- or AsF_6^-.

Thus, even in the least favorable case of polymerization of cyclic acetals, the counterions that do not interact with growing species may be selected, which indicates that, in general, the proper choice of counterion eliminates the possibility of termination by recombination with counterion.

3. *Termination by Irreversible Chain Transfer to Polymer*

In many systems, the polymer repeating unit differs from monomer only by the lack of ring strain containing the same nucleophilic group. In comparing the reactivity of cyclic onium ion toward either monomer or polymer repeating unit, the ring strain is irrelevant, because the ring of reacting monomer is not opened at this stage of reaction:

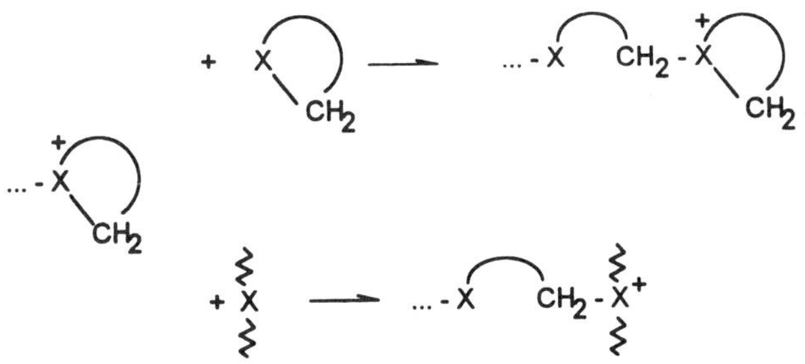

(94)

Thus, the relative reactivities of monomer and polymer repeating units are governed by relative nucleophilicities of both entities and steric factors (if significant).

There is no general rule relating the nucleophilic reactivity of cyclic monomer and linear polymer repeating unit, it depends on the nature of heteroatom and the size of the ring which affects the electronic structure of heteroatom. It is a common practice to estimate the order of nucleophilicities on the basis of basicities. Although it is only partly justified, this procedure enables semiquantitative comparisons of known pK_a values whereas no universal scale of nucleophilicity exists. Some typical values of pK_a for cyclic compounds and their linear analogs are given in Table 8 [99,100].

Data of Table 8 indicate that in some systems (e.g., cyclic acetals) polymer repeating unit is a stronger base (nucleophile) than monomer. Whether or not reaction involving repeating units results in termination depends on the ability of the branched onium ion to re-form the cyclic

Table 8 pK_a Values of Some Cyclic and Linear Organic Bases[a]

Cyclic	Linear
 −2.1; 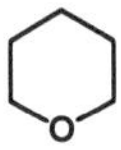 −2.8	$C_2H_5-O-C_2H_5$ −3.6; $C_4H_9-O-C_4H_9$ −5.4
 −2.0; 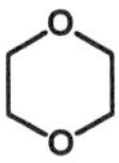 −3.2; 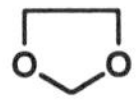 −6.5	$CH_3OCH_2CH_2OCH_3$ −3.3
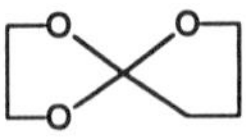 −7.3	
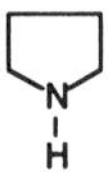 11.3; 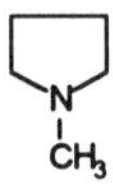 10.5; 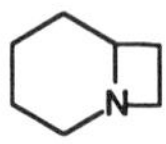 11	$C_2H_5-NH-C_2H_5$ 10.6; $C_2H_5-N(C_2H_5)-C_2H_5$ 9.8
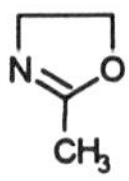 3.4; 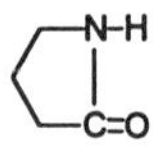 0	$CH_3-CH_2-CH_2-C(=O)-N(CH_3)_2$ −0.6
	polyamide : gly - gly −3.1
 −4.4	$C_2H_5-S-C_2H_5$ −5.4

[a] $pK_a = H_o + \log \frac{[BH^+]}{[B]}$.

Source: Ref. 99.

onium ion. This may occur either as bimolecular reaction with a monomer molecule:

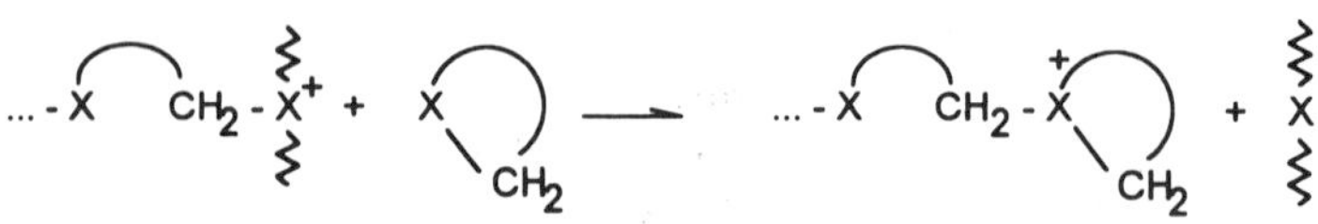 (95)

or by unimolecular cyclization, reaction reverse to formation of branched ion:

(96)

The factors, governing the reactivities of branched onium ions are of complex nature and will not be discussed here. It has been shown, however, that in the polymerization of cyclic sulfides and cyclic amines the formation of branched onium ions of the type:

(97)

is an irreversible process resulting in termination.

Polymerization of 3-membered cyclic amines-aziridines is a good example to illustrate the influence of steric factor on the relative rates of propagation and termination.

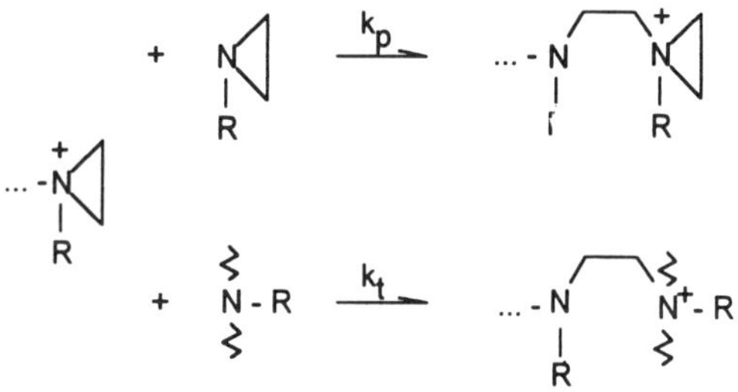

(98)

The ratio k_p/k_t has been measured for differently substituted aziridines [100] (Table 9).

The results show that the presence of bulky substituent on a polymer chain may effectively inhibit the termination proceeding by this mechanism. The results presented at this point may be summarized as follows: chain transfer to polymer is a general feature of cationic ring-opening polymerization although for different systems the contribution of this reaction may vary; only in some systems this process results in termination (These systems involve, e.g., cyclic amines (3- and 4-membered) and cyclic sulfides (3- and 4-membered); and the contribution of the reaction is reduced for substituted chains.

Table 9 Values of k_p/k_t Ratios for the Polymerization of *N*-Substituted Aziridines

Substituent	k_p/k_t ($mol^{-1} \cdot L$)
C_2H_5-	6
$C_6H_5CH_2CH_2$-	14
S-C_3H_7-	21
$NCCH_2CH_2$-	82
$C_6H_5CH_2$-	85
t-C_4H_9	1200

Source: Ref. 100.

4. *Termination by Reactions with Other Components of the System: Solvents or Impurities*

Solvents used typically for conducting cationic ring-opening polymerization involve chlorinated hydrocarbons (CH_2Cl_2, $CHCl_3$, $C_2H_4Cl_2$, CCl_4), aliphatic or aromatic hydrocarbons (e.g., C_6H_{14}, cyclo-C_6H_{12}, C_6H_6, $C_6H_5CH_3$) or nitro compounds (CH_3NO_2, $C_6H_5NO_2$). There is no information on the interaction of properly purified solvents with onium ions leading to termination and these solvents are considered inert. Acetonitrile, which has been sometimes used, should be avoided, because it is already too nucleophilic ($pK_a = -10$) [99].

Even in thoroughly purified systems, impurities can not be completely eliminated. The most common impurity, water, is detrimental for cationic ring-opening polymerization, although it acts rather as a chain transfer agent and not a terminating agent, because usually the proton formed can initiate the new chain:

(99)

The problem of water in ionic ring-opening polymerization is much less critical than in vinyl polymerization, because, if present, water in the former system has to compete with relatively nucleophilic monomer present in large excess. Thus, certain polymerizations (e.g., polymerization of cyclic amine-conidine) can be conducted even in alcohols as solvents.

When applying the chlorinated solvents, it should be remembered that, upon storage on light, they may release HCl. Cl^- anion, due to its relatively high nucleophilicity, will react irreversibly with growing species at least in polymerization of weakly nucleophilic monomers (e.g., cyclic ethers).

5. *Termination by Reactions of More Reactive Species Existing in Equilibrium with Stable Onium Species*

As already discussed, in the systems, in which unimolecular ring-opening of cyclic onium ion leads to highly stabilized carbocationic species, a concentration of the latter species in equilibrium with onium ions may be significant. This is, for example, the case of cationic polymerization of cyclic acetals, where carboxonium ions exist in equilibrium with their oxonium counterpart:

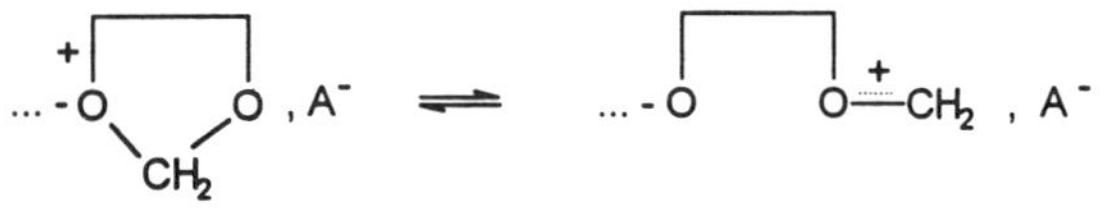

(100)

The carboxonium ions may participate in chain transfer through hydride abstraction reaction:

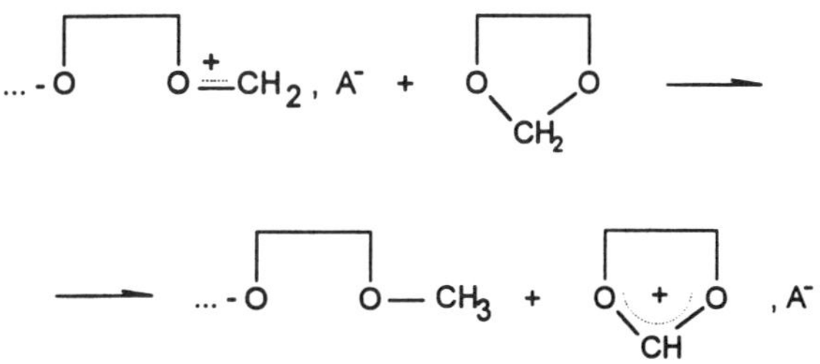

(101)

This type of transfer is of some importance in the cationic polymerization of 1,3,5-trioxane, where it has been shown that some macromolecules are terminated with $—OCH_3$ groups [101].

III. CATIONIC RING-OPENING POLYMERIZATION OF DIFFERENT GROUPS OF CYCLIC MONOMERS

Different heterocyclic monomers containing heteroatoms such as oxygen, sulfur, nitrogen, phosphorus, and silicon within a ring can be polymerized by cationic mechanism. These monomers differ by the nature of the chemical bond which is opened in the propagation step and by the ring size. Thus, it is not always possible to give the general description of the mechanism of polymerization, even for uniform group of monomers, because the course of the polymerization may depend strongly on the ring size.

There are several reviews in which the cationic polymerization of different groups of cyclic monomers is discussed and which provide complete list of references [2–6]. Thus, the present discussion will only highlight the most characteristic features of the polymerization of various groups of cyclic monomers. The importance of cationic ring-opening polymerization stems from the fact that variety of polymers containing different sequences of atoms along the chain can be obtained by this method.

The discussion of the characteristic features of polymerization of different groups of monomers will focus on these factors, which are governing the possibility of preparing purely linear polymers with well-defined structure, including control over the molecular weights and the nature of the end groups.

This problem is stressed here, because, as discussed in Section II.C.1, intramolecular chain transfer to polymer, which in principle may occur in any cationic ring-opening polymerization, leads to formation of cyclic oligomers in addition to linear polymers. The presence of cyclic fraction in high molecular weight polymers usually has a disadvantageous influence on the performance of the polymers, still in several commercially useful polymers, e.g., condensation polymers (polyesters, polyamides) a certain amount of cyclic oligomers can be tolerated. Also the commercial polymer made by cationic ring-opening polymerization, polyoxymethylene (polyformaldehyde, polyacetal), contains certain proportion of cyclic fraction (cf., Section III.B.2).

If, however, the polymerization process is aimed at the preparation of functional polymers (telechelics, macromonomers), then the problem of cyclization becomes much more important, because cyclic oligomers do not contain functional end groups.

As already mentioned, cationic ring-opening polymerization offers several possibilities of preparing useful high molecular weight polymers, which may find applications as such. The main interest, however, is in preparation of reactive polymers, which are used in synthesis of block or graft copolymers.

In the following sections, the cationic polymerization of various groups of cyclic monomers will be briefly reviewed. The systems, in which controlled synthesis of well-defined polymers is possible, will be discussed in more detail and, wherever necessary, some comments on the reaction mechanism will be presented.

A. Cyclic Ethers

Cyclic ethers of various ring size polymerize by cationic mechanism, including oxiranes (3-membered), oxetanes (4-membered), oxolanes (5-membered), oxepanes (7-membered), and larger rings. Six-membered cyclic ethers do not polymerize because of the thermodynamic restrictions. Oxiranes may be also polymerized by anionic mechanism, whereas for other groups the cationic polymerization is the only mechanism of polymerization.

Due to the high ring strain of 3- and 4-membered rings, oxiranes and oxetanes polymerize practically irreversibly. Thermodynamic polymerizability of these groups of monomers is not significantly affected by substitution, thus polymerization of even heavily substituted rings is still practically irreversible. Polymerization of 5- and 7-membered cyclic ethers is reversible, due to relatively low ring strain (relatively low ΔH of polymerization).

1. Oxiranes

Cationic polymerization of the unsubstituted oxirane (ethylene oxide) leads to the mixture of relatively low molecular weight ($M_n < 10^3$) linear polymer and up to >90% of cyclic oligomers, predominantly cyclic dimer (1,4-dioxane) [91,102]. The same behavior was observed for polymerization of substituted oxiranes, propylene oxide [103], epichlorohydrin [104], and other oxiranes having one or more substituents in the ring, although the distribution of cyclic fraction varied, depending on the structure of monomer.

Thus, cationic polymerization of oxiranes is of little synthetic value, if the preparation of linear polymers is attempted. The high tendency for cyclization may be employed, however, for preparation of macrocyclic polyethers (crown ethers). Polymerization of ethylene oxide in the presence of suitable cations (e.g., Na^+, K^+, Rb^+, Cs^+) leads to crown ethers of a given ring size in relatively high yields, due to the template effect [105]. Thus, with Rb^+ or Cs^+ cations, cyclic fraction contained exclusively 18-crown-6.

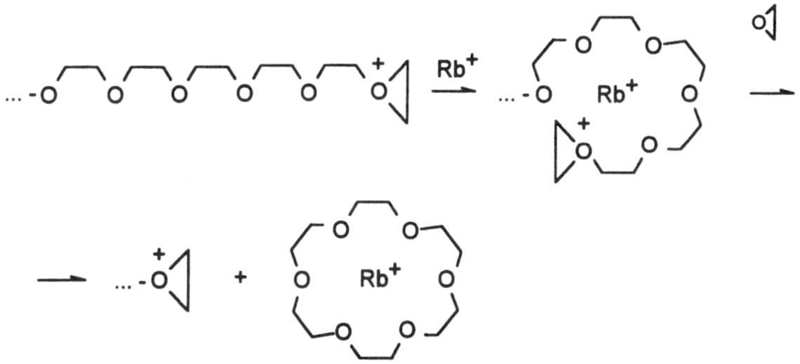

(102)

Cyclization by back-biting cannot be avoided if active species of propagation are tertiary oxonium ions located at the growing chain end, i.e., if propagation proceeds by Active Chain End (ACE) mechanism. It has been found, however, that in the presence of hydroxyl group containing compounds, the other mechanism of propagation competes with ACE mechanism [106]. This may be illustrated by the following scheme:

(103)

(H^+ denotes proton exchanging quickly between all bases present in the system.)

(104)

(105)

In this sequence of reactions, it is the monomer that forms oxonium ion [thus Activated Monomer (AM) mechanism] and the growing chain end is neutral. As shown in the series of papers [107–115], if the conditions are created, when AM mechanism predominates, by keeping the low instantaneous ratio of [monomer]/[HO-] (slow addition of monomer to reaction mixture), back-biting is effectively eliminated. Linear polymers, free of cyclic fraction are obtained under these conditions. The mechanism and kinetics of AM polymerization of oxiranes is discussed in detail in recent monograph [6].

Although there are some limitations on the molecular weights of the linear polymers which may be obtained by this method, AM polymerization offers an attractive, synthetic route for preparation of functional, medium molecular weight polymers by cationic polymerization of oxiranes. As the process involves the extension of the chain of hydroxyl group containing compound used to initiate the polymerization (initiator), the method is especially well suited for preparation of oligodiols (low molecular weight diols as initiators) and macromonomers (for example, hydroxyethyl acrylate as initiator):

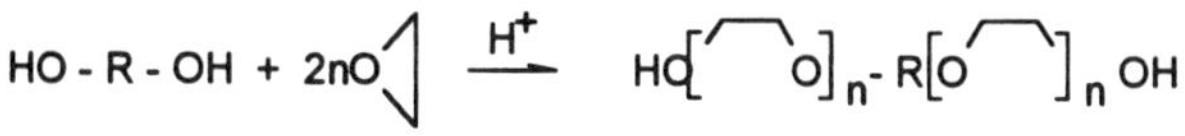
(106)

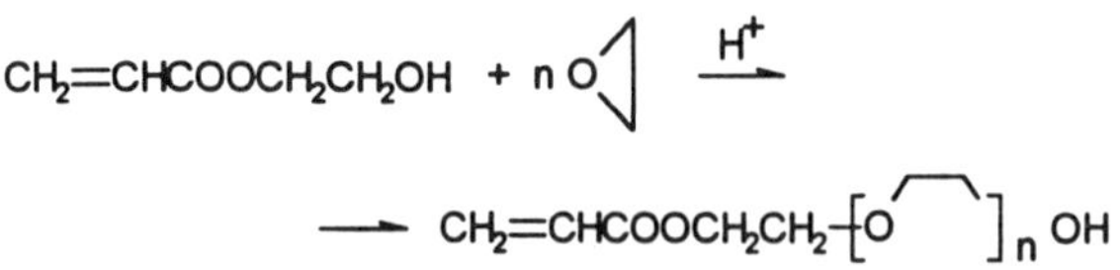
(107)

Preparation of epichlorohydrin oligodiols [114] and epichlorohydrin or propylene oxide macromonomers [115] has been described. AM polymerization of oxiranes containing nitromethyl [116] or *N*-vinyl carbazoyl substituents [117] has also been studied. Photoinitiated AM polymerization of 2,3-epoxypropylphenylether has been reported [118].

2. Oxetanes

Cationic polymerization of 4-membered cyclic ethers, oxetanes, provides the characteristic example of the influence of the number and size of substituents on the extent of the chain transfer to polymer in cationic ring-opening polymerization. Due to the high strain of 4-membered ring, polymerization of mono- and disubstituted oxetanes is still practically irreversible.

In the cationic polymerization of unsubstituted oxetane, formation up to 50% of cyclic tetramer has been observed, in addition to linear polymer for polymerization conducted at 100° C, initiated with BF_3 [119]. The content of cyclic fraction decreased, however, with decreasing temperature.

In the polymerization of 3,3-dimethyloxetane initiated with $(C_2H_5)_3O^+BF_4^-$ at 20° C, up to 20% of cyclic fraction was observed with cyclic tetramer as a main component [120].

By introducing still larger (but at the same time electron-withdrawing) chloromethyl substituents (3,3-bis(chloromethyl)oxirane BCMO), the content of cyclic fraction was further considerably reduced. Only ~2% of cyclic trimer was isolated from the product of polymerization of BCMO initiated with $(C_2H_5)_3Al\text{-}H_2O$ system at temperature as high as 180° C [121].

Cyclization results from intramolecular chain transfer to polymer, being also indicative of the extent of intermolecular chain transfer leading to formation of branched (nonreactive in the case of oxetanes) tertiary oxonium ion (cf., Section II.D.3).

Thus, in the polymerization of unsubstituted oxetane, polymers with limited molecular weights (up to $M_n \sim 35{,}000$) and relatively broad molecular weight distributions (M_w/M_n between 1.5 and 4.2) were obtained in the polymerization initiated with $(C_2H_5)_3O^+PF_6^-$ at −20° C [122], whereas polymerization of substituted oxetanes led to polymers with much higher molecular weights and narrower molecular weight distributions. In the polymerization of 3-methyl-3-chloromethyloxetane polymers with M_n up to 760,000 and $M_w/M_n < 1.3$ were obtained with $R_3Al\text{-}H_2O$ initiating system at 20° C [123]. Polymerization of BCMO, even at 70° C ($R_3Al\text{-}H_2O$ as initiator) gave polymers with M_n up to 600,000 [124].

Although the two possible modes of chain transfer to polymer (i.e., intra- and intermolecular reactions) does not have to be directly interrelated (intramolecular reaction for example may be prohibited due to the stiffness of chain), having the same origin, they should in many systems proceed simultaneously. Thus, if one of these processes is detected in any system, this is a strong indication that the other reaction also occurs.

Cationic polymerization of 3,3-bis(chloromethyl)oxetane (BCMO) was in the past employed in the commercial process for making chlorinated polyether-Penton®, polymer having very good chemical resistance toward aggressive media (e.g., conc. H_2SO_4 up to 120° C). Relatively high price, due to the difficult monomer synthesis, was the reason, that Penton lost its markets to newer products and the production was stopped in the mid sixties.

3. *Oxolanes*

Ring strain of 5-membered cyclic ethers is relatively low, thus even the polymerization of unsubstituted monomer, tetrahydrofuran, is highly reversible. The substitution further decreases the thermodynamic polymerizability, thus 2-methyltetrahydrofuran does not polymerize [125], while 3-substituted monomer undergoes polymerization, however ceiling temperature is low (4° C) [126]. Thus, the only monomer of this group which has been studied in great detail is tetrahydrofuran.

Cationic polymerization of tetrahydrofuran is one of the few systems in cationic ring polymerization in which chain transfer to polymer may be practically avoided. The reasons for that are of purely kinetic nature.

In the polymerization of tetrahydrofuran (THF) initiated with $HOSO_2CF_3$ in CD_3NO_2 solvent at 35° C (molar ratio $[THF]/[CD_3NO_2]/[CF_3SO_3H] = 20.8/24.6/1$), the equilibrium monomer concentration (i.e., the ultimate conversion of monomer to polymer) is reached in ~2 hr. At this stage of polymerization, concentration of cyclic oligomers is still very low [90].

When the reaction mixture is kept for a prolonged period of time without terminating, the formation of cyclic oligomers can be detected by gas chromatography coupled with mass spectrometry and their concentration increases with time, as shown in Table 10 [90]. The results show that

Table 10 Relative Concentrations of Cyclic Oligomers in $THF/CF_3SO_3H/CD_3NO_2$ System

Time (hr)	Peak Areas (gc)			
	Trimer	Tetramer	Pentamer	Hexamer
0.5	0.29	1.88	0.47	0.16
1	1.06	4.41	1.65	1.53
2	1.76	6.60	2.84	2.75
3	2.52	8.88	4.38	4.12
4	3.01	9.83	5.45	5.40
7	3.23	11.22	7.12	6.95
24	4.13	12.43	10.90	9.83

Source: Ref. 90.

after the monomer-polymer equilibrium has been reached, there is a slow, continuous build-up of a cyclic oligomer concentration.

No absolute concentrations were measured; it was however estimated that cyclic tetramer constituted ~0.4% mol of the final products. Thus, the overall concentration of the cyclic fraction would correspond to few mol% (this may be still not the equilibrium distribution).

The important point is that cyclic oligomers are not formed concurrently with linear polymer, chain transfer to polymer is slow as compared to propagation and its contribution is becoming noticeable only after keeping the reaction mixture unterminated for a prolonged period of time. Thus, within a time scale required to complete the polymerization, the contribution of chain transfer to polymer can be neglected.

This characteristic feature of cationic polymerization of THF allows the important synthetic application of this process for preparation of oligodiols used in polyurethane technology and in manufacturing of block copolymers with polyesters and polyamides (cf., Section IV.A). On the other hand, the cationic polymerization of THF not affected by contribution of chain transfer to polymer is a suitable model system for studying the mechanism and kinetics of cationic ring-opening polymerization.

Mechanistic and synthetic aspects of cationic polymerization of THF has been reviewed in several monographs and book chapters [2–5,127]. More recently, the detailed analysis of kinetics and mechanism of this process was presented in Ref. 6. The main features of cationic polymerization of THF and the problems which have been solved by studying this system may be summarized as follows: (1) The quantitative and fast initiation is possible with several initiators including oxonium and oxycarbenium salts; (2) The only ionic active species are tertiary oxonium ions; (3) In the presence of suitable counterions like SbF_6^- the tertiary oxonium ions are long living; (4) Kinetics of propagation has been studied in the living systems. By combining the measurements of reaction rates and conductivities, the rate constants of propagation on ions and ion pairs were measured and it was found that $k_p^+ = k_p^\pm$. (5) In the systems, in which ionic species may be reversibly converted into covalent esters (e.g., with trifluoromethanesulfonic or fluorosulfonic counterion), both types of species were observed directly by ^{1}H NMR, their concentration measured and corresponding equilibrium constants and rate constant determined. As a result, cationic polymerization of THF is the best understood system in cationic ring-opening polymerization, for which the complete kinetic description can be presented as shown in Eq. (108) [57]. Rate constants of all reactions involved in Eq. (108) were measured, enabling quantitative determination of contribution of ionic and covalent species to the propagation.

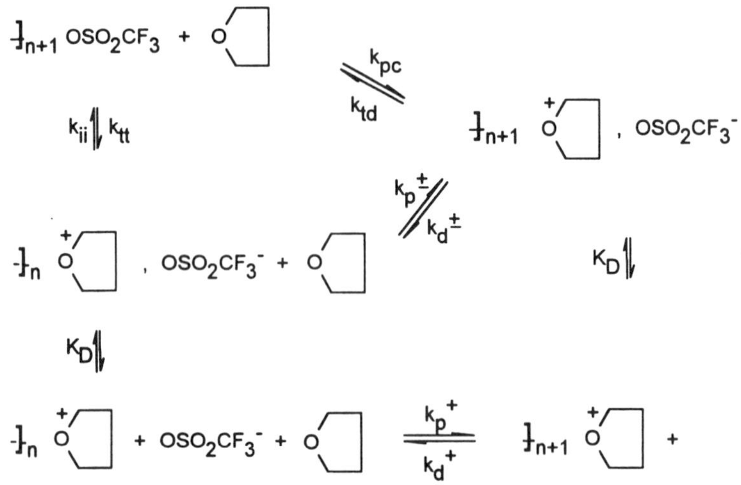

(108)

The values of rate constants were determined in different solvents (CCl_4, CH_2Cl_2, CH_3NO_2) [57]. For 7 mol·L^{-1} solution of THF in CH_3NO_2 at 25° C, for example, the following values were obtained

$$k_{ii} = 12 \cdot 10^{-2} \text{ sec}^{-1}$$
$$k_{tt} = 0.3 \cdot 10^{-2} \text{ sec}^{-1}$$
$$k_{pc} = 5 \cdot 10^{-4} \text{ mol}^{-1} \cdot \text{L} \cdot \text{sec}^{-1}$$
$$k_{td} = 1.10^{-4} \text{ sec}^{-1}$$
$$k_p^{\pm} = k_p^{+} = 2.4 \cdot 10^{-2} \text{ mol}^{-1} \cdot \text{L} \cdot \text{sec}^{-1}$$
$$K_D = 2 \cdot 10^{-3} \text{ mol} \cdot \text{L}^{-1} \text{ } (SbF_6^{-} \text{ counterion})$$

Cationic polymerization of THF fulfills all the requirements of living polymerization. With several initiators, the initiation is relatively fast and quantitative, and propagation proceeds without transfer or termination. The dependence $DP_n = ([M]_0 - [M]_e)/[I]_0$ holds up to high polymerization degrees. The only limitation is that, due to the reversibility of propagation, the molecular weight distribution is broadened and reaches the value of $M_w/M_n = 2$ in equilibrium. Polymers with narrower MWD were obtained by terminating the polymerization at lower conversion [56].

The quantitative character of initiation and the living character of active species permits the preparation of functional polymers (telechelic oligomers) as well as block copolymers (cf., Section IV.A.).

B. Cyclic Acetals

Cyclic acetals are the monomers containing acetal bond —O—C—O— as a part of the ring. The smallest possible ring is 5-membered 1,3-dioxolane (ethylene glycol formal). Cyclic acetals, having medium-size rings, polymerize reversibly, with exception of 6-membered 1,3-dioxane, which does not polymerize ($\Delta H \sim 0$). Polymerization of unsubstituted rings was studied mostly because even monosubstituted cyclic acetals have low ceiling temperatures.

$(CH_2)_2-O-CH_2-O$ Ethylene glycol formal-1,3-dioxolane

$(CH_2)_4-O-CH_2-O$ Butylene glycol formal-1,3-dioxepane

$CH_2CH_2OCH_2CH_2OCH_2O$ Diethylene glycol formal-1,3,6-trioxocane

$CH_2CH_2OCH_2CH_2OCH_2CH_2OCH_2O$ Triethylene glycol formal-1,3,6,9-tetraoxaundecane

It is interesting to note that ring-opening polymerization of these monomers leads to polymers being perfectly regular (alternating) copolymers, composed of oxyethylene (EO) (or oxybutylene (BO) and oxymethylene (MO) units. The monomers listed above give the following structures of the chain, respectively:

$$[\text{EO-MO}]_n, \quad [\text{BO-MO}]_n, \quad [\text{EO-EO-MO}], \quad [\text{EO-EO-EO-MO}]_n$$

Another monomer that belongs to this group is cyclic trimer of formaldehyde, 1,3,5-trioxane. Cationic polymerization of 1,3,5-trioxane leading to polyoxymethylene (polyformaldehyde, polyacetal) is one of the few industrially important processes in cationic ring-opening polymerization.

Cyclic acetals polymerize exclusively by cationic mechanism; it should be noted, however, that polyoxymethylene may be obtained also by anionic polymerization of formaldehyde. Both processes are used in industry.

In contrast to the previously discussed case of THF polymerization, where chain transfer to polymer is slow as compared to propagation, in the polymerization of cyclic acetals, chain transfer to polymer is fast as compared to propagation and the polymerization is dominated by reactions involving polymer chains. Polymerization of the two best studied monomers of this group, 1,3-dioxolane and 1,3,5-trioxane, shows certain specific features. Thus both systems will be discussed separately in the following sections, with special emphasis on the consequences of the chain transfer to polymer.

1. 1,3-Dioxolane

Cyclic acetals contain —O—CH_2—O— group within a cycle. This group readily participates in hydride transfer reaction because resulting carboxonium ion is stabilized by the electron pairs of the two neighboring oxygen atoms [20].

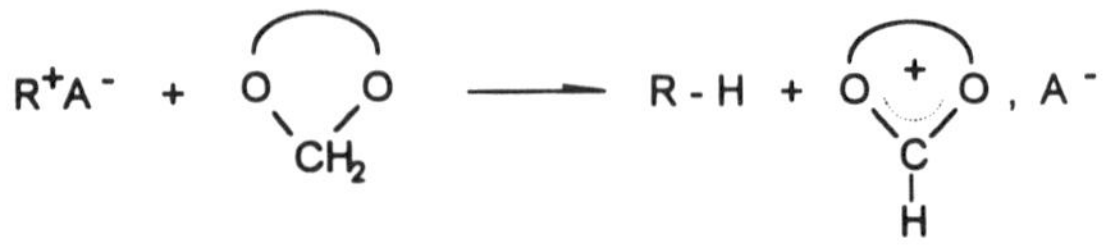

(109)

Thus, if compounds capable of acting as hydride ion acceptors are used as initiators, initiation may proceed through intermediate dioxolenium ion formation (cf., Section II.A.2.) [8,9,128], e.g.:

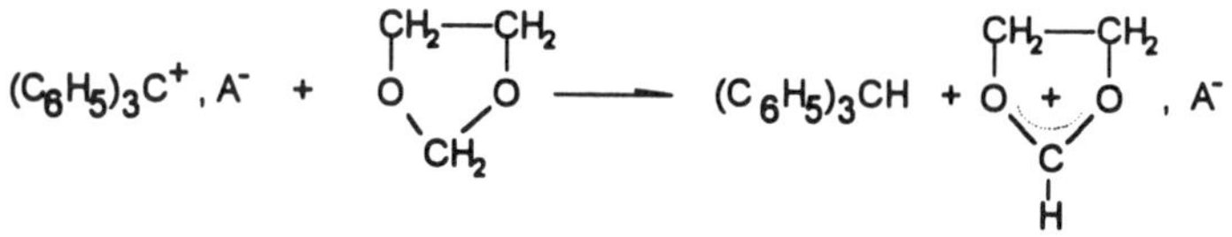

(110)

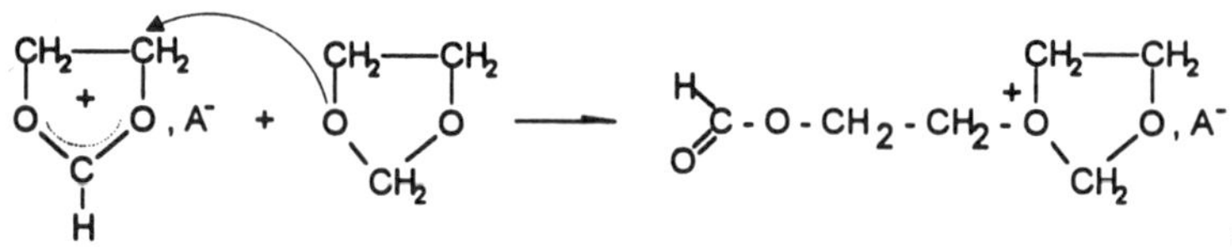

(111)

Initiators such as triethyloxonium salts, oxycarbenium salts (e.g., $C_6H_5CO^+A^-$), or dioxolenium salts initiate the polymerization by direct addition.

The other consequence of the presence of —O—CH_2—O— group in the molecule is the possibility of unimolecular opening of tertiary oxonium ion active species to form corresponding carboxonium ion.

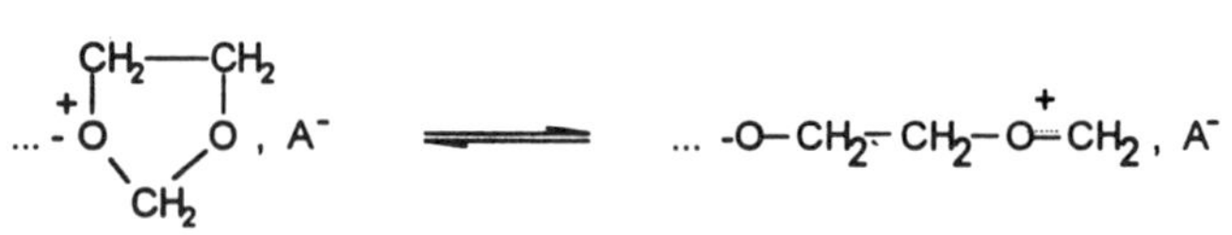

(112)

There was a long-lasting controversy concerning the participation of both isomeric species in the propagation.

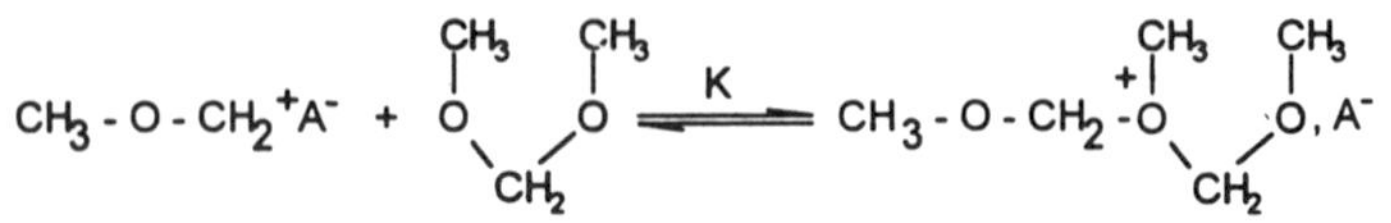

(113)

By studying the suitable model system (113) it was eventually shown that carboxonium ions do exist in equilibrium [76]. In SO_2 solvent at $-70°$ C, the equilibrium constant $K = 3 \cdot 10^3$ mol^{-1}·L, thus at the concentration of active species at the level of 10^{-3} mol·L^{-1} and concentration of acetal equal to 3 mol·L^{-1}, fraction of carboxonium ion would be ~0.03 mol% of all species.

The rate constants of the reaction modeling the propagation on carboxonium and oxonium active species were found equal (at the conditions given above) to:

$$k_{\text{carbox}} = 2 \cdot 10^6 \text{ mol}^{-1} \cdot \text{L} \cdot \text{sec}^{-1} \qquad k_{\text{ox}} = 1.9 \cdot 10^4 \text{ mol}^{-1} \cdot \text{L} \cdot \text{sec}^{-1} \text{ [76]}$$

Thus, the contribution of carboxonium active species to propagation is small, but not negligible.

Both oxonium and carboxonium active species are relatively strong electrophiles (stronger than in the case of tetrahydrofuran). Thus, to avoid the termination by interaction with counterion, the stable counterions of low nucleophilicity are required. It has been shown that only the most stable complex anions: SbF_6^- and AsF_6^- provide the living active species, whereas BF_4^-, $SbCl_6^-$, and even PF_6^- anions cause termination [98].

If however the suitable initiator is used, e.g., $C_6H_5CO^+SbF_6^-$, the initiation is relatively fast and termination or transfer reactions are absent. The degree of polymerization in such a system is equal to $([M]_0 - [M]_e)/[I]_0$ ratio up to high M_n value (polymers with $M_n \sim 100{,}000$ have been prepared) [129].

The most characteristic feature of the cationic polymerization of cyclic acetals, however, is an excessive participation of the polymer chain in the polymerization processes. This is exemplified by the results of attempted synthesis of block copolymer containing segments of poly(1,3-dioxolane, DXL) and poly(1,3-dioxepane, DXP) [130].

When 1,3-dioxolane was added to the solution of living (nonterminated) poly(1,3-dioxepane) or vice versa, further polymerization ensued and the increase of molecular weight indicated that polymerization of added monomer proceeded exclusively on living active species of the former monomer. The isolated copolymer was analyzed by ^{13}C NMR spectroscopy and it was found that, instead of a block copolymer, the copolymer with nearly statistical distribution of DXL and DXP units was formed practically from the beginning of the process. This is a clear indication that chain transfer to polymer leads to branched oxonium ions, which participate in further reactions with a rate comparable to the rate of propagation.

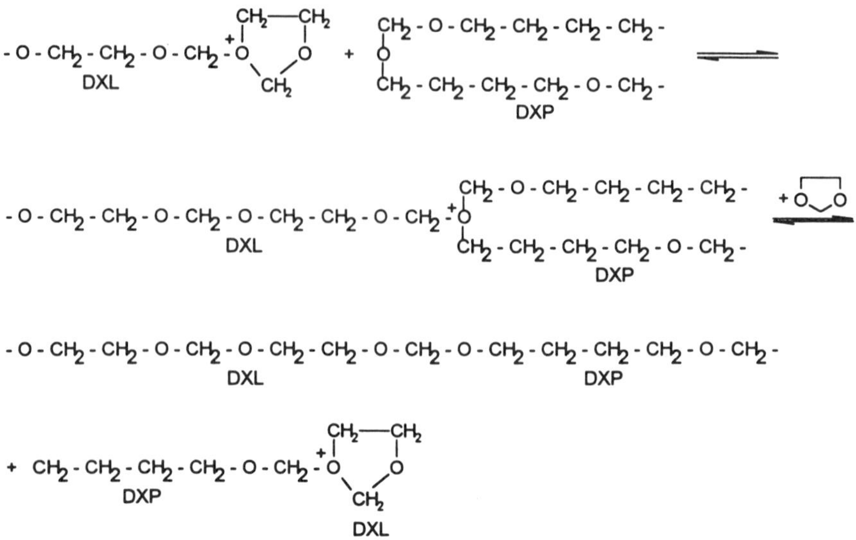

(114)

Several repetitions of the reaction sequence shown in Eq. (114) would lead to complete randomization (scrambling) and the fact that this is indeed observed in the course of polymerization indicates that the rate of this process is comparable to the rate of chain growth.

Intramolecular chain transfer to polymer (back-biting) should lead to the formation of cyclic oligomers:

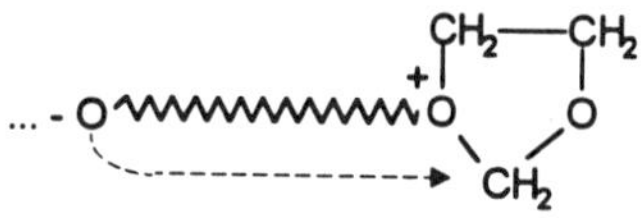

(115)

Indeed, it was shown that cyclic oligomers are always formed in the cationic polymerization of cyclic acetals and their distribution agrees well with the thermodynamic distribution calculated on the basis of Jacobson-Stockmayer theory (with exception of small rings with $n = 2$–4) [89].

The fact that thermodynamic distribution of ring sizes is attained in the course of polymerization provides another indication that involved processes are relatively fast.

Only low molecular weight cyclic oligomers (up to heptamer) were detected and identified by gas-liquid chromatography. Assuming that the dependence [Eq. (83)] holds also for larger rings, one can easily calculate the overall equilibrium concentration of cyclics with n from 2 to infinity. Such calculation gives a value of 0.85 mol·L^{-1} of cyclics [131]. Thus, for bulk polymerization at room temperature, at complete conversion, i.e., at $[M]_0 - [M]_c \sim 10$ mol·L^{-1}, the content of cyclic fraction in high molecular weight polymers may be reduced to <10% mol.

In the polymerization of 1,3-dioxolane cyclization may also lead to formation of isomerized 7-membered ring [132]:

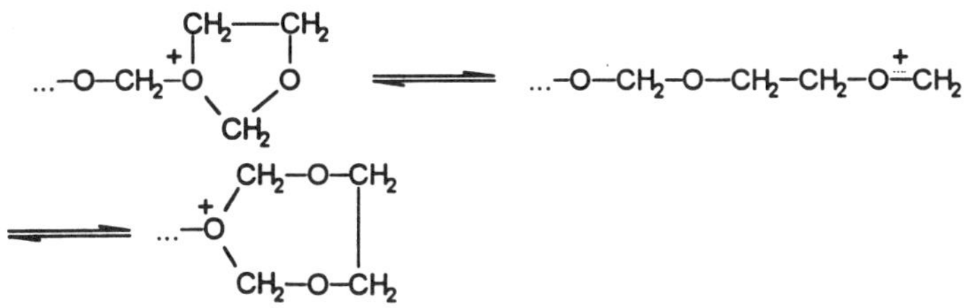

(116)

The presence of isomerized active species has been confirmed by their direct observation (^{1}H, ^{13}C NMR) [132].

In the polymerization of 1,3-dioxolane initiated with protonic acid, another reaction pathway for cyclization is becoming possible [133]. Initiation with protonic acids introduces HO- head group, considerably more nucleophilic than acetal oxygens along the chain. Thus, end-to-end closure reaction (end-biting) proceeds extensively, especially at the early stages of reaction, when the distance between both ends of the macromolecule is still relatively short:

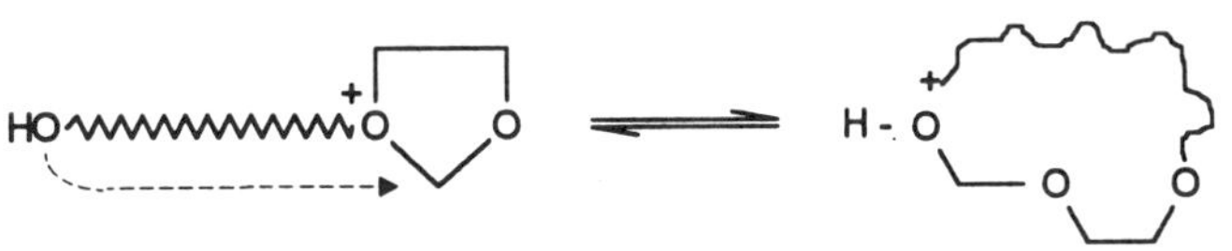

(117)

On the basis of high content of cyclic structures (secondary oxonium ions), the ring expansion mechanism was originally proposed for protonic acid-initiated polymerization of 1,3-dioxolane [134]. It was shown conclusively, however, that cyclic oligomers formed at the early stages of the polymerization in concentrations higher than these calculated on the basis of Jacobson-Stockmayer theory are, in the further course of polymerization, consumed and their concentration, after passing through maximum, approaches the equilibrium concentration [135]. On the other hand, using phosphine capping method:

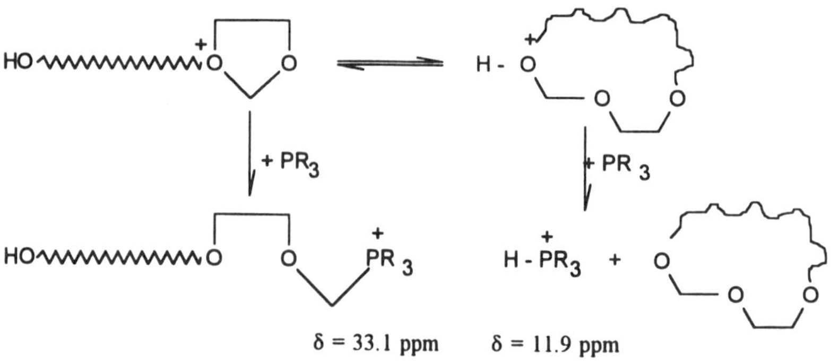

(118)

(for R = n-C_4H_9 in ^{31}P NMR, SbF_6^- counterion, CH_2Cl_2, 25° C) it was shown, that secondary oxonium ions exist in equilibrium with tertiary ones, and the position of the equilibrium is shifted toward tertiary oxonium ions with increasing the chain lengths in agreement with end-to-end cyclization scheme [136].

The preceding discussion shows that in the cationic polymerization of cyclic acetals chain transfer to polymer can not be avoided. If however polymerization is carried out at high initial monomer concentration (preferentially in bulk) the content of cyclic fraction may be limited to a few percent. As the cyclic fraction is composed mainly of medium-size rings, the high molecular weight polymer may be separated from cyclic fraction by fractionation.

The intermolecular chain transfer to polymer (scrambling), however, is detrimental for preparation of functional polymers, leading to disproportionation of monofunctional macromolecules, as shown schematically below:

(119)

If R = R′ (bifunctional polymers), reaction (119) does not affect the functionality but leads to the broadening of the molecular weight distribution, which is occurring anyway, due to the reversibility of propagation. Thus, several bifunctional polymers of 1,3-dioxolane were prepared and used, for example, to form the networks containing degradable and hydrolyzable polyacetal blocks (cf., Section IV.B). Reaction (119), however, may effectively prohibit the preparation of monofunctional polymers, e.g., macromonomers. Indeed, two recent attempts to prepare macromonomers by cationic polymerization of cyclic acetals led to nearly statistical distribution of products having two, one, and none of the functional groups [137,138].

2. 1,3,5-Trioxane

Cationic polymerization of 1,3,5-trioxane provides one of a few examples of industrial application of cationic ring-opening polymerization. Polymerization leads to polyoxymethylene (polyformaldehyde, polyacetal), important engineering plastic. Polyformaldehyde may also be obtained by anionic polymerization of formaldehyde and this process is also used in industry.

Polymers prepared by anionic polymerization of formaldehyde contain hydroxyl end groups. These hemiacetal end groups are thermally unstable and their decomposition initiates degradation of polymer to monomeric formaldehyde (paraformaldehyde, used in a laboratory practice to generate gaseous formaldehyde is medium molecular weight, hydroxyl-terminated polyformaldehyde). Thus, to stabilize the polymer, the hydroxyl end groups are converted into ester groups.

The process based on cationic polymerization of 1,3,5-trioxane employs a different principle for stabilization of polymer. Trioxane is copolymerized with a few percent of 1,3-dioxolane (or ethylene oxide). The sequence of $—OCH_2—$ units is then separated from time to time by $—OCH_2CH_2—$ units. The product of copolymerization is subsequently heated to eliminate the terminal units (unstable fraction). Depropagation proceeds until the stable $—CH_2CH_2OH$ group is reached:

$$—OCH_2CH_2O[CH_2O]_nCH_2OH \rightarrow —OCH_2CH_2OH + (n + 1)CH_2O \quad (120)$$

The whole concept relies on the random distribution of comonomer units along the chain.

It has been shown, however, that in the copolymerization much more reactive 1,3-dioxolane polymerizes first and is practically completely consumed at still low conversion of 1,3,5-trioxane [139]. Thus, by polymerization alone, it would not be possible to achieve the random distribution of 1,3-dioxolane units along the chain. Both analysis of the microstructure of the chain [140], as well as thermal behavior of the copolymer [141], indicate, however, that nearly random distribution of comonomer units is indeed attained by scrambling process similar to that described earlier for sequential polymerization of 1,3-dioxolane and 1,3-dioxepane.

It has to be remembered, however, that this process may be complemented by the reversibility of the propagation step. In the copolymerization of 1,3,5-trioxane with 1,3-dioxolane (or ethylene oxide), the complex set of equilibria is established, involving comonomers but also formaldehyde, higher cyclic oligomers of formaldehyde and cyclic oligomers composed of oxymethylene and oxyethylene units [142]. Cationic copolymerization of 1,3,5-trioxane provides an interesting example of the process in which chain transfer to polymer, usually undesirable, leads to the required random copolymer, which would be difficult to prepare in the absence of scrambling.

It is striking that the time period needed to achieve high conversion to polymer is sufficient to obtain nearly complete randomization in spite of the fact that polymer, being insoluble in the monomer (and typical organic solvents), precipitates out of the reaction mixture and propagation as well as scrambling proceed in solid phase.

Detection and identification of cyclic oligomers in 1,3,5-trioxane polymerization are difficult because of the insolubility of polymers. Using however the high-temperature (150° C) gel permeation chromatography (GPC) in DMF solution, the low molecular weight tail on GPC curves was observed and attributed to the presence of cyclic oligomers [143]. Recently, the formation of cyclic oligomers in cationic polymerization of 1,3,5-trioxane has been studied in more detail [144]. Homopolymers of trioxane prepared with different cationic initiators were subjected to alkaline hydrolysis. The polyoxymethylenes terminated with unstable hemiacetal groups would decompose quantitatively to monomer under these conditions. It was found, however, that certain a fraction of polymer was resistant to hydrolysis and the amount of this stable fraction coincided with the fraction of low molecular weight material appearing as a separate peak on the GPC curve (polymer was dissolved in 1,1,1,3,3,3-hexafluoro-2-propanol for GPC determination). The GPC curves are shown in Fig. 1 [144]. The quantitative information is collected in Table 11 [144].

Formation of alkaline stable fraction was observed a long time ago and it was assumed, but not proven, that it is composed of cyclic macromolecules [145]. The alkaline stable fraction could be either cyclic or terminated at both ends with $—OCH_3$ groups. $CH_3O—$ end groups are indeed formed in this system, due to hydride transfer reaction to carboxonium ion:

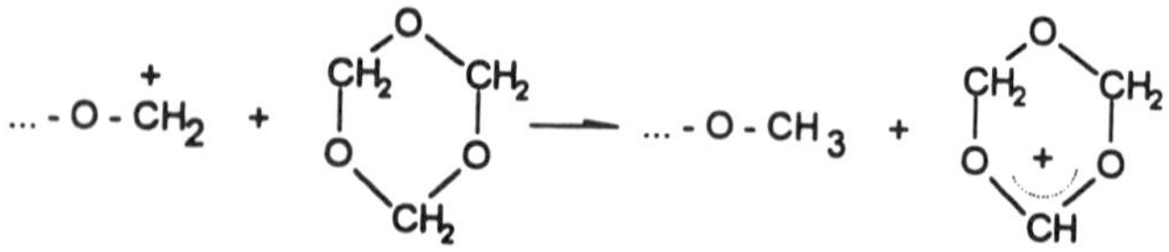

(121)

Transacetalization may lead subsequently to formation of certain fraction of macromolecules terminated at both ends with $—OCH_3$ groups. Both chemical analysis (using the Zeissel method) and careful ^1H NMR studies have shown, however, that $CH_3O—$ groups are absent in alkaline

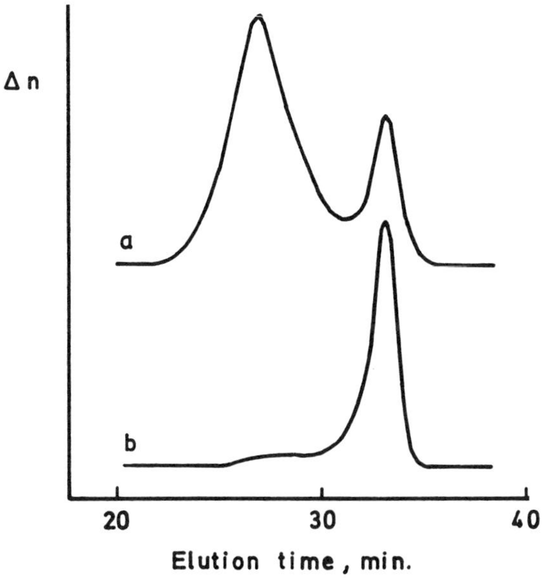

Figure 1 GPC curves of polyoxymethylene: (a) obtained by bulk polymerization of 1,3,5-trioxane at 70° C with $BF_3 \cdot OBu_2$ as initiator, $M_w = 55{,}600$; (b) its alkaline-stable fraction.

stable fraction [144]. Other end groups, which could possibly be formed (e.g., ester groups), would not be stable under the conditions of hydrolysis, thus the only possibility left is that alkaline stable fraction is indeed cyclic [144]. It is striking, that not only the molecular weight of cyclic fraction is almost independent on the molecular weight of linear polymer

Table 11 Polymerization Conditions, Amounts of Alkaline-Stable Fraction, and Low Molecular Weight Fraction in the Cationic Polymerization of 1,3,5-Trioxane

Catlyst ($\cdot 10^{-6}$ mol)	Time (min)	Conversion (%)	M_w	Alkaline-Stable Fraction,%	Fraction Low Molecular GPC Peak,%
$BF_3 \cdot OBu_2$ (9.6)	3	78	56,600	13	17
CF_3SO_3H (0.15)	7	50	70,500	15	21
$(C_2H_5)_3O^+PF_6^-$ (8.1)	13	65	58,100	13	25

Source: Ref. 144.

(M_n of cyclics ~1500–1700), but the molecular weight distribution is quite narrow (M_w/M_n ~ 1.1–1.25). This is in obvious disagreement with random back-biting mechanism, which should lead to much broader distribution of ring sizes.

Cationic polymerization of 1,3,5-trioxane, however, is different from other typical ring-opening polymerizations, because polymer is insoluble in its own monomer or in any typical organic solvent (it may be dissolved however in DMF at ~150° C and perfluorinated alcohols at room temperature). Thus, at any polymerization system, already at the early stage of polymerization, polymer precipitates out of solution and further reaction involves active species in crystalline phase.

The bimodal distributions, like the one shown in Fig. 1, were obtained for bulk, suspension, and solution polymerization (monomer in solution) but not from the solid-state, radiation-induced polymerization [144] (Fig. 2).

It is known that in a solid state polymerization occurs along the certain axis of the monomer crystal and extended chain crystals are formed [146]. In the polymerization from liquid phase, on the other hand, lamellar crystals are formed. This indicates that back-biting reaction proceeds not at random but in a specific manner, forced by the nature of crystals. The model of growth of lamellar crystals of polyoxymethylene is known [147]. According to this model, the subsequent layers are formed on the surface of the crystal by growth of folded chain as shown schematically below:

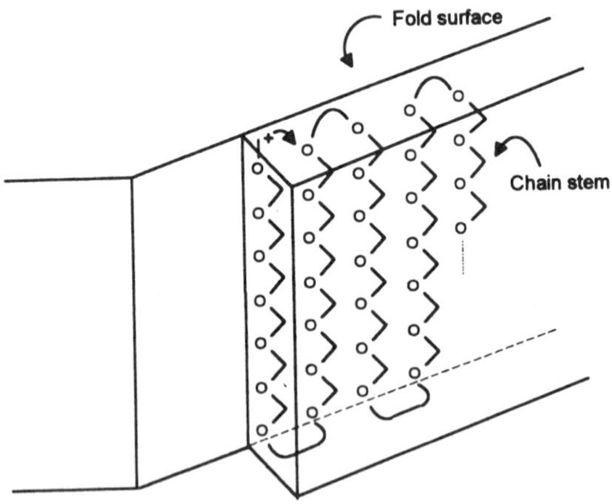

(122)

From the point of view of the mobility of active sites and the reactivity of acetal oxygen, the probability of back-biting reaction is the highest at the surface of the crystal, i.e., at the chain folds.

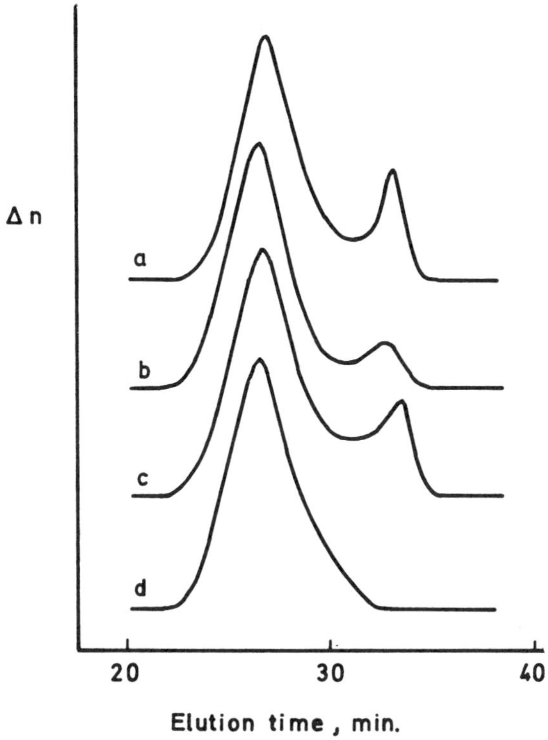

Figure 2 GPC curves of polyoxymethylene obtained by various polymerizations: (a) bulk polymerization at 70° C initiated by $BF_3 \cdot OBu_2$, $M_w = 55{,}600$; (b) suspension polymerization at 70° C initiated by $BF_3 \cdot OBu_2$, $M_w = 64{,}200$; (c) solution polymerization at 32° C initiated by CF_3SO_3H, $M_w = 68{,}600$; (d) γ-ray-induced solid-state polymerization at 30° C, $M_w = 98{,}000$.

For such mechanism the size of the ring of cyclic polymer formed would be closely related to the thickness of lamellar crystals. The thickness of crystals has been measured by SAXS method and good linear dependence between the molecular weight of cyclic polymer ($M_n \sim$ 1300–2100) and the long period of the crystal ($\sim$7–9 nm) was observed. The quantitative relation between M_n of cyclic fraction and parameters of crystalline phase can be expressed as follows:

$$M_n(\text{cycles}) = (2d_c/c)\cdot 9M \tag{123}$$

where c is the crystal lattice spacing along the helical axis ($c = 1.73$ nm),

d_c is lamella thickness, and M is the molecular mass of repeating unit ($M = 30$) [144].

The conformations and crystalline structure of isolated fractions were analyzed by infrared and raman spectroscopy and convincing evidence was obtained, confirming macrocyclic structure.

These developments in cationic polymerization of 1,3,5-trioxane are discussed in more detail, because in this system the problems related to the mechanism of cyclization are now well understood. Cyclic oligomers were identified, isolated, their molecular weight distribution was determined, and the plausible explanation for observed distribution was given. From the synthetic point of view, the cationic polymerization of 1,3,5-trioxane offers the possibility of preparing macrocyclic polymers with relatively narrow molecular weight distribution and predictable (within discussed limits) molecular weights. The cyclic polymers can be prepared easily in relatively large quantities and conveniently separated from linear polymer by alkaline hydrolysis of the latter.

C. Bicyclic Ethers and Acetals: Anhydrosugars

The anhydrosugars are monosaccharide derivatives, which in addition to the 5-membered furanose ring or 6-membered pyranose ring, contain another oxygen-containing ring formed through elimination of water from any two hydroxyl groups of sugar. Thus, anhydrosugars containing additional 3-, 4-, or 5-membered rings may be obtained (only the skeleton shown in the scheme below):

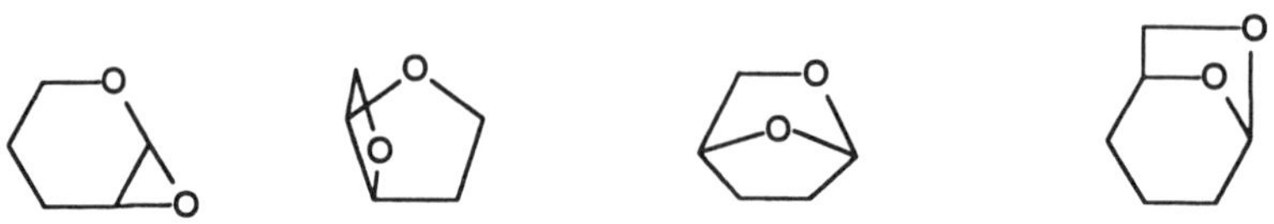

I — 3-membered — 1,2-anhydrosugar
II — 4-membered — 3,5-anhydrosugar
III — 5-membered — 1,4-anhydrosugar
IV — 5-membered — 1,6-anhydrosugar

(124)

Both unsubstituted bicyclic monomers and related anhydrosugars polymerize by cationic mechanism [148]. In sugar derivatives, the remaining free hydroxyl groups have to be blocked before polymerization by typical methods known in carbohydrate chemistry.

Cationic polymerization of anhydrosugars give (after deblocking) polysaccharides with reasonably high molecular weights (DP_n in the range of 10^2–10^3) and in some systems, e.g., in the polymerization of 2,3,4-tris-

O-benzyl-α-D-glucopyran [structure IV in Eq. (124)] initiated with PF_5, the measured DP_n values are close to calculated as $[M]_0/[I]_0$ ratio up to $DP_n \sim 100$ [149].

In the polymerization of anhydrosugars, however, it is desirable to prepare the polymers having the structure identical with that occurring in natural polysaccharidess. This can not always be achieved.

Polymerization of bicyclic acetal (1,4-anhydrosugar) composed of fused 5- and 6-membered rings may involve opening of both rings, because a 6-membered ring exists in enforced unfavorable boat conformation [150–152]:

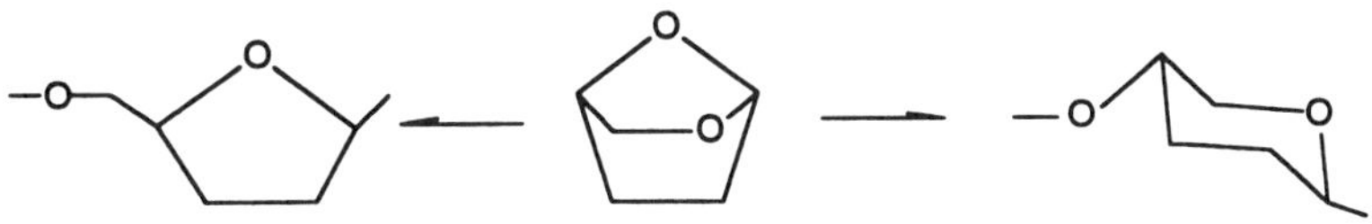

(125)

The lack of specificity is thus related to similar ring strain of both fused rings. 1,6-Anhydrosugars, in contrast, polymerize exclusively with opening of 5-membered ring because the 6-membered ring, being in more favorable conformation, is essentially strainless [153].

The other source of inhomogenity, related to the mechanism of polymerization is the possibility of racemization on carbon atom, which is attacked by the oxygen atom of incoming monomer. In most natural polysaccharides, configurations on the carbon atoms next to oxygen are identical throughout the macromolecule.

(126)

In the cationic polymerization of 1,2-anhydrosugars, however, racemization is clearly detected indicating unimolecular opening of 3-membered ring [154]:

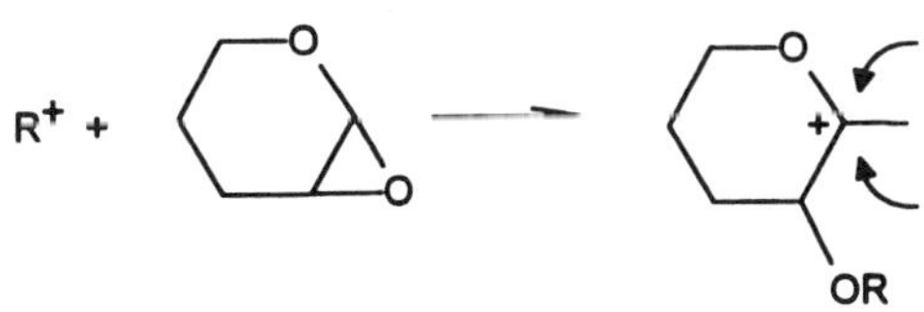

(127)

The unimolecular opening of the ring is apparently related to the high strain of 3-membered ring and to the stabilization of carbocationic center by oxygen atom in α-position.

In the polymerization of 1,6-anhydrosugars due to the lack of strained ring, propagation proceeds as S_N2 reaction with inversion of configuration, thus typically, sterically uniform polysaccharides are obtained [153].

D. Cyclic Sulfides

The cyclic sulfides that can be polymerized by cationic mechanism include 3-membered rings, thiiranes, and 4-membered rings, thietanes. Five-membered rings do not undergo polymerization, although the 6-membered sulfur analog of 1,3,5-trioxane, namely 1,3,5-trithiane, has been reported to polymerize [155]. Polymerization of thiiranes and thietanes is practically irreversible.

1. *Thiiranes*

There is a certain analogy between cationic polymerization of thiiranes and its oxygen-containing analogs, oxiranes. Unsubstituted thiirane (ethylene sulfide) generally gives very low molecular weight products, while mono- and disubstituted thiiranes can be converted to medium molecular weight polymers.

These systems are of little synthetic value, however, because extensive chain transfer to polymer leads to cyclization and formation of branched sulfonium ions even for mono- and disubstituted thiiranes [156].

Kinetic studies revealed that active species of methylthiirane polymerization were rapidly deactivated by chain transfer to polymer [157].

The polymeric sulfonium ions (both cyclic and branched) are not completely unreactive, although the reaction re-forming active species is very slow:

(128)

The main components of cyclic fraction in thiirane polymerization are cyclic dimer and cyclic tetramer [158].

The additional feature of thiiranes polymerization is the possibility of formation of isomerized products containing sulfur-sulfur bond, with expulsion of an olefin as shown for dimethylthiirane polymerization.

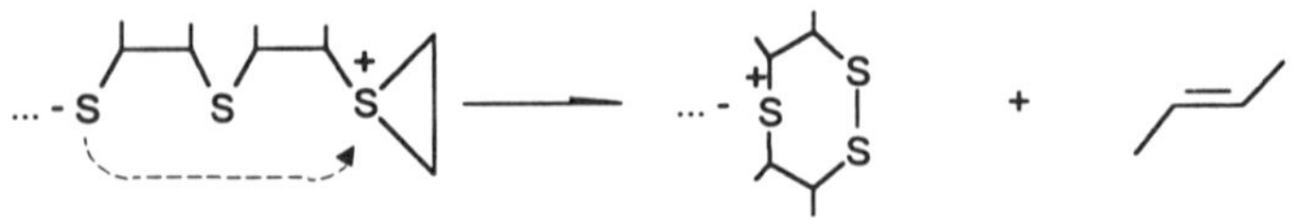

(129)

This is due to the unique ability of sulfur to attack sulfonium ion with formation of disulfide bond.

Because the chain transfer to polymer is fast as compared with reformation of active species of propagation [Eq. (128)] and there is a reaction pathway, which due to the formation of isomerized products is irreversible [reaction (129)], continuous degradation of the already formed polythiirane chains occurs if the reaction system is kept unterminated [159]. Also isolated polymers, treated with cationic initiators degrade to low molecular weight, predominantly cyclic oligomers. Consequently, cationic polymerization of thiiranes is very strongly affected by chain transfer to polymer processes.

2. Thietanes

As in the case of thiiranes, also in the cationic polymerization of 4-membered cyclic sulfides, thietanes, chain transfer to polymer effectively competes with propagation. Intramolecular chain transfer, leading to formation of branched structures is well documented in these systems, because branched ions has been observed directly by ^{1}H NMR [160]:

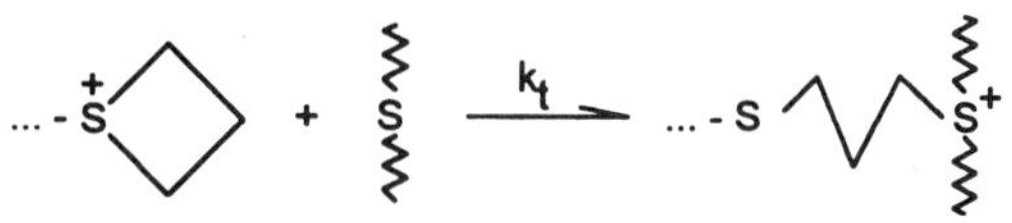

(130)

The branched sulfonium ions are not reactive, thus Reaction (130) is an irreversible termination. The measured ratios of rate constants k_p/k_t [(rate constant of propagation to rate constant of termination according to Eq. (130)] reflect the general phenomenon that by increasing the number and size of the substituents the contribution of chain transfer to polymer may be considerably reduced [161] (Table 12).

Thus, in the polymerization of unsubstituted thietane, polymerization stops at limited conversion (~20% in CH_2Cl_2 at 20° C) while for 2,2-diethylthietane nearly quantitative conversion (96% in CH_2Cl_2 at 20° C) to polymer can be obtained.

Table 12 Values of k_p/k_t for the Polymerization of Different Thietanes

Substituent	k_p/k_t
None	1.1
2-Methyl	2.4
3,3-Dimethyl	28
3,3-Diethyl	450

3. 1,3,5-Trithiane

The cationic polymerization of trithiane is assumed to proceed by similar mechanism on the polymerization of trioxane. Polymer is insoluble in common organic solvents and, like unstabilized polyoxymethylene, is thermally unstable [155].

E. Cyclic Amines

Cationically polymerizable monomers of this group include 3-membered rings (aziridines) and 4-membered rings (azetitidines). Larger rings undergo polymerization only if additional strain is introduced to the ring, like in the case of (1,4-diazabicyclo[2.2.1]octane) containing two fused rings.

1. Aziridines

Cationic polymerization of unsubstituted aziridine leads to branched poly(ethylene imine). This is attributed to the reaction of protonated aziridine formed either directly (protonic acids as initiators) or through proton transfer from alkylated aziridine (alkylating agents as initiators), with NH groups along the chain [162,163]:

(131)

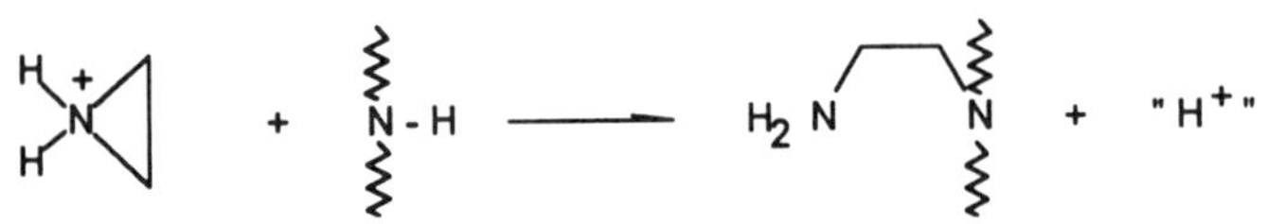

(132)

The mechanism of the polymerization is rather complex and resembles very much the activated monomer polymerization of oxiranes described in Section III.A.1. In the AM polymerization of ethylene oxide, reaction of protonated monomer with chain oxygen leads to tertiary oxonium ion, which undergoes further reactions. In the polymerization of aziridine, the analogous reaction leads to protonated tertiary amines, which, by proton expulsion, gives tertiary amino group as a stable branching point. Typically, as determined by ^{13}C NMR, polymer with $M_n \sim$ 20,000 obtained by cationic polymerization of aziridine, contains ~25% of primary amino groups, 50% of secondary amino groups, and 25% of tertiary amino groups [164].

Purely linear poly(ethylene imine) may be obtained, however, by cationic polymerization of oxazolines (discussed in the next section), followed by the hydrolysis of the resulting polymeric amide.

The formation of covalent branching points is avoided in the cationic polymerization of *N*-substituted aziridines. Chain transfer to polymer occurs, however, leading to formation of branched (or cyclic) quaternary ammonium ions, unreactive in further propagation [165].

(133)

As in the case of other systems discussed earlier, the extent of this reaction may be reduced considerably by introducing the large substituents (cf., Table 9 in Section II.D.3).

Consequently, for polymerization of *N*-substituted aziridines with small substituents like *N*-methylaziridine, limited conversions to relatively low molecular weight polymers were observed [165,166]. On the other hand, in the polymerization of *N*-*t*-butylaziridine chain transfer to polymer is practically eliminated [167]. No other transfer or termination reaction involving quaternary ammonium active species with, e.g., BF_4^- counterion has been detected:

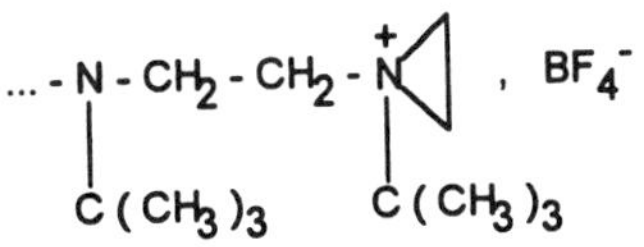

(134)

For this highly nucleophilic monomer the initiation proceeds relatively quickly, even with typical initiators like triethyloxonium salts and polymerization is practically irreversible. Of all the groups of monomers discussed to this point, the cationic polymerization of *t*-butylaziridine approaches most closely the conditions of living polymerization, allowing the preparation of purely linear polymers having predictable molecular weights and narrow molecular weight distribution. The living character of *t*-butylaziridine polymerization has been used in the synthesis of telechelic polymers [168], macromonomers [169], block copolymers [170], and polymer networks [171] (cf., Section IV.C). It should be noted that polymerization of *N*-*t*-butylaziridine proceeds as a living process only when initiated with alkylating agents, e.g., triethyloxonium salts. Initiation with protonic acids leads to formation of secondary amine head group. This leads to coupling of the linear chains [172]:

(135)

and/or cyclization, if reaction between head group and growing chain end proceeds as an intramolecular reaction [173]:

(136)

2. *Azetidines*

Cationic polymerization of unsubstituted 4-membered cyclic amine proceeds in a way similar to that of aziridine, giving branched polymer containing, according to ^{1}H NMR, 20% of primary, 60% of secondary, and 20% of tertiary amino groups [174]. Purely linear polymer can be obtained by polymerization of 5,6-dihydro-4H-1,3-oxazine, followed by hydrolysis [175]:

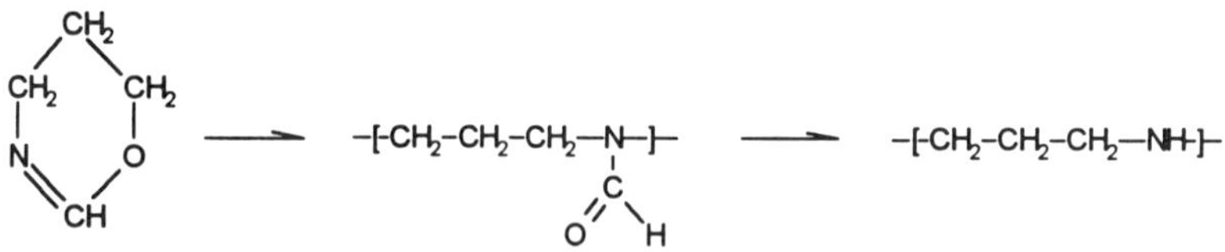

(137)

Substitution of proton on the nitrogen with alkyl group, as in the case of aziridines, simplifies the mechanism of polymerization. In the polymerization of 1-methylazetidine, termination due to the chain transfer to polymers is still detectable [176]. Further substitution at carbon atoms reduces the extent of chain transfer to polymer and in the cationic polymerization of 1,3,3-trimethylazetidine the concentration of active species, quaternary azetidinium ions:

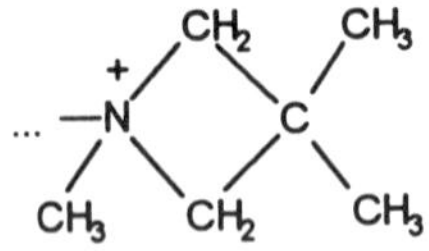

(138)

measured directly by ^{1}H NMR spectroscopy, remains constant even a long time after reaching the complete conversion of monomer [177].

Also the polymerization of bicyclic monomer containing azetidine ring, 1-azabicyclo[4.2.0]octane (conidine), proceeds without appreciable transfer and/or termination [178].

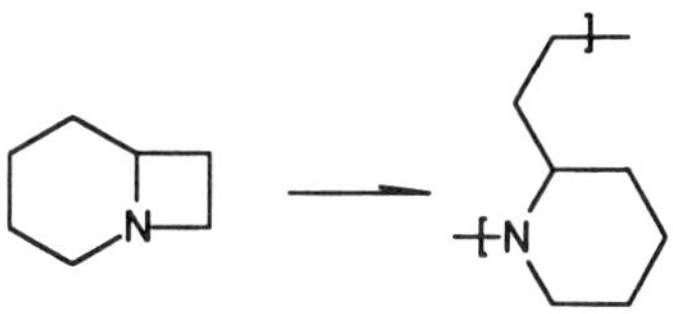

(139)

The synthetic application of azetedine polymerization is limited, however, due to the difficulties with monomer synthesis (the closure of 4-membered ring is much more difficult than the closure of 3-, 5-, or higher member rings).

F. Cyclic Iminoesters: Oxazolines

The lowest member of this group of monomers is 5-membered ring, 1,3-oxazoline. Polymerization of both substituted and unsubstituted oxazolines receives much attention. The characteristic feature of the polymerization of this group of monomers is that isomerization occurs during the propagation step, thus the structure of the repeating unit is not directly related to the structure of the ring as in other ring-opening polymerizations:

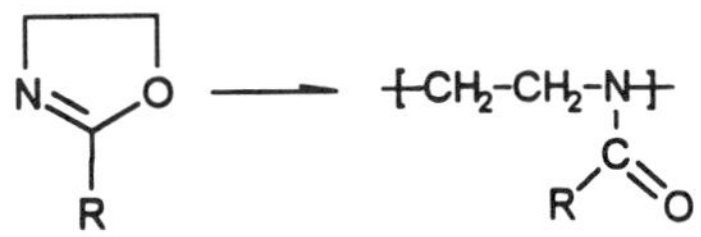

(140)

Six-membered rings can also be polymerized by cationic mechanism and it is believed that the driving force for polymerization of this essentially unstrained monomer comes from conversion of imino ether group into an amide group, which has more resonance stabilization [175]:

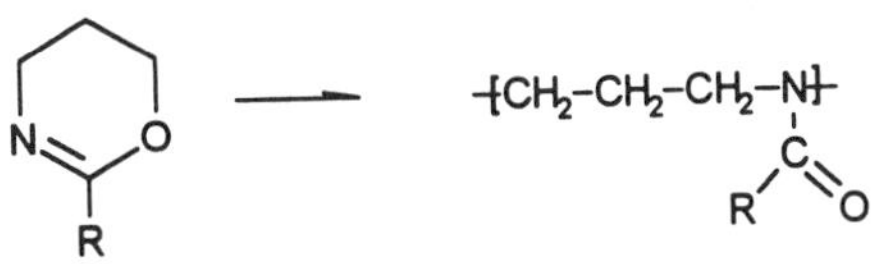

(141)

1. *1,3-Oxazolines*

Polymerization of oxazolines is initiated by typical cationic initiators. Alkylation occurs on the more nucleophilic nitrogen atom (cf., however, Section II.A.5). Propagation may be visualized by the scheme:

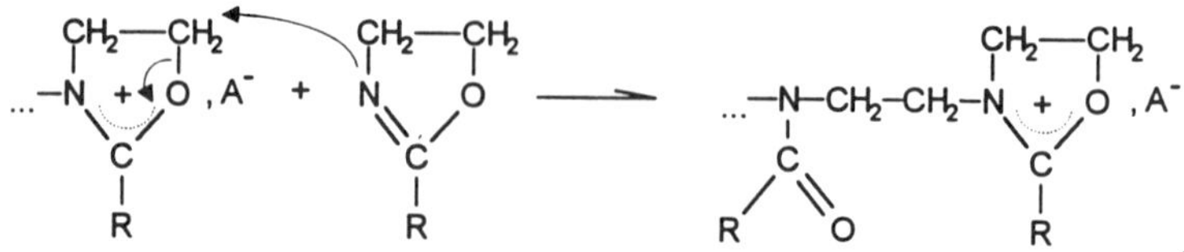

(142)

Polymerization of oxazolines is the rare example of cationic ring-opening polymerization in which chain transfer to polymer is of no importance for the whole group of monomers. This is because the imino group of the monomer is converted upon propagation into an amide group having much lower nucleophilicity. Thus, the cationic polymerization of oxazolines is believed to be of living type, although at least one side reaction affecting the living character has been identified in the polymerization of oxazolines having —CH_2— group in the substituent [179]:

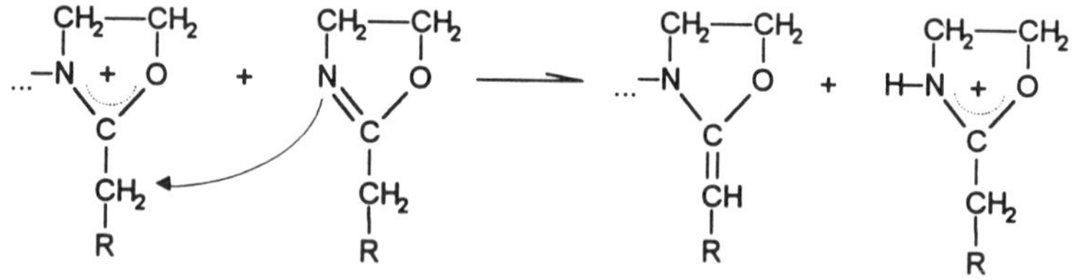

(143)

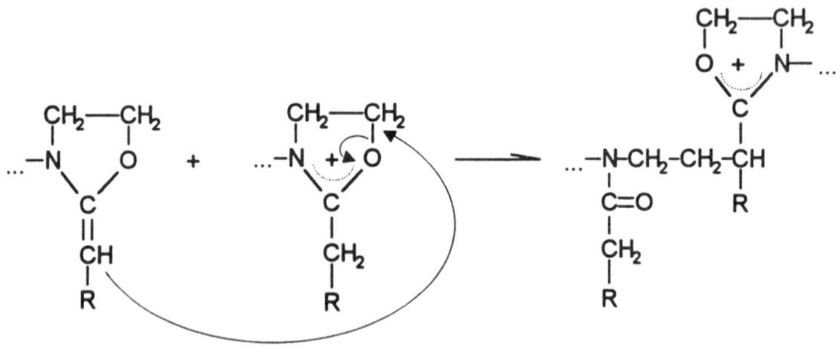

(144)

It should be stressed that growing species with diffused charge distributed along —N—C—O— group are relatively weak electrophiles. Thus, as discussed in Section II.B.6.c, ionic active species may exist when coupled with relatively nucleophilic counterions such as Br^- or I^-. The collapse of ion pairs into covalent alkyl halide end groups does proceed in the system, but these covalent species still react (although slower than ionic species) with highly nucleophilic monomer re-forming ionic species:

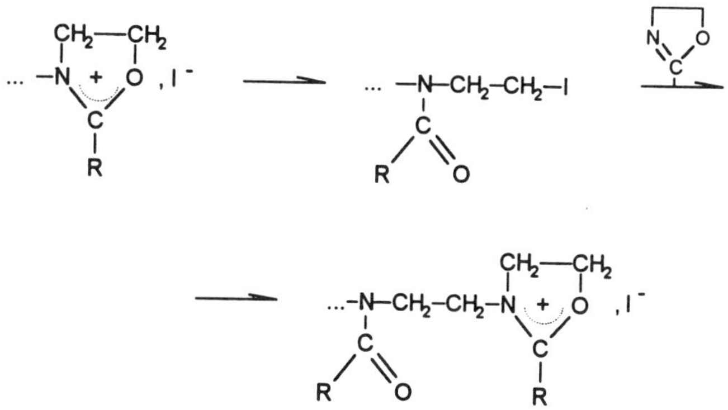

(145)

Both types of species have been directly observed by ^{1}H NMR and the rate constants of propagation has been determined [180].

More recent efforts investigating the kinetics and mechanism of cationic polymerization of oxazolines are aimed at the preparation of various types of functional polymers as described in a recent review [181] (cf., Section IV.D). One of the examples is the development of efficient di- and tetrafunctional initiators of oxazolines polymerization, allylic or benzylic dihalides and allylic tetrahalides [182], e.g.:

(146)

Linear (difunctional) and star-shaped (tetrafunctional) polyoxazolines with M_n up to 8000 and $M_w/M_n < 1.3$ were prepared using these initiators [182].

Recently, the new double isomerization polymerization of oxazolines has been described [183]. Oxazolines substituted at 2-position with cyclic imino group may, depending on the kind of initiator used, undergo conventional polymerization giving structure a, or may undergo double isomerization polymerization. In the latter process, the isomerization of propagating species occurs: the 2-oxazoline ring rearranges to a 5-membered cyclic urea unit and the cyclic imine moiety undergoes ring opening, leading to repeating unit b:

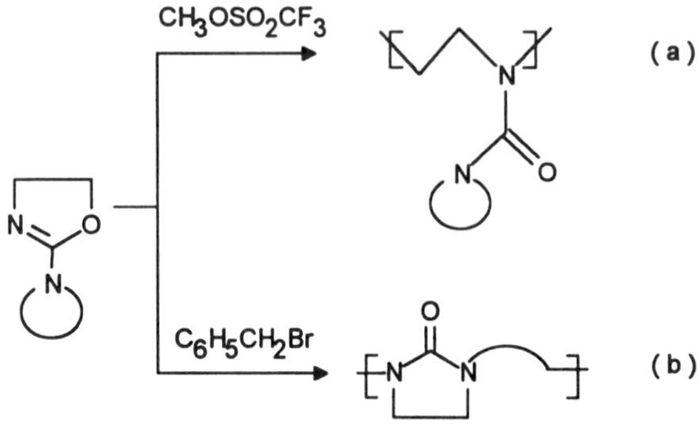

(147)

Polymers with M_n up to 8300 with $M_w/M_n \sim 1.3$ were formed with almost quantitative yields by route b.

G. Cyclic Esters

Cyclic esters contain an ester (or orthoester) group within a ring. Typical examples of the monomers belonging to this group are shown below:

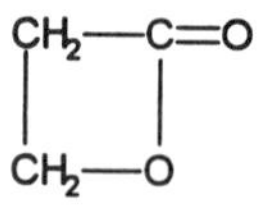

β-propiolactone
(in internal ester of
β-hydroxypropionic acid)

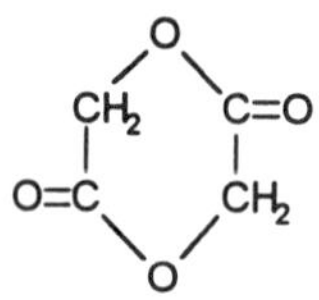

glycolide
1, 4 - dioxane - 2, 5 - dione
(cyclic dimer of α-hydroxyacetic acid)

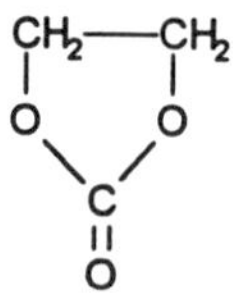

1, 3 - dioxolan - 2 - one
(ethylene carbonate)

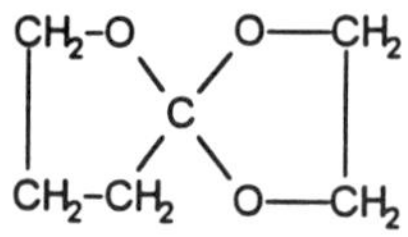

1, 4, 6 - trioxa - spiro[4, 4]nonane
(spiro orthoester)

(148)

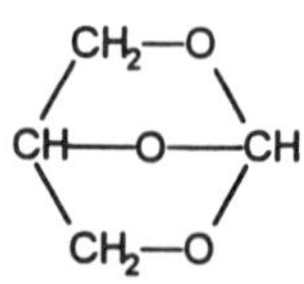

2, 6, 7 - trioxabicyclo [2, 2, 1] heptane
(bicyclic orthoester)

Monomers listed above polymerize by the cationic mechanism. For some groups of monomers (lactones, carbonates) anionic or coordinate mechanism also operates and, from a synthetic point of view, this is the preferred method of converting cyclic esters into linear polyesters. The cationic polymerization of lactones, glycolide and it substituted analog, lactide, as well as spiroorthoesters and bicyclic orthoesters has been studied in some detail.

1. *Lactones*

Both initiation and propagation reaction in lactone polymerization may involve an endo- or exocyclic oxygen atom:

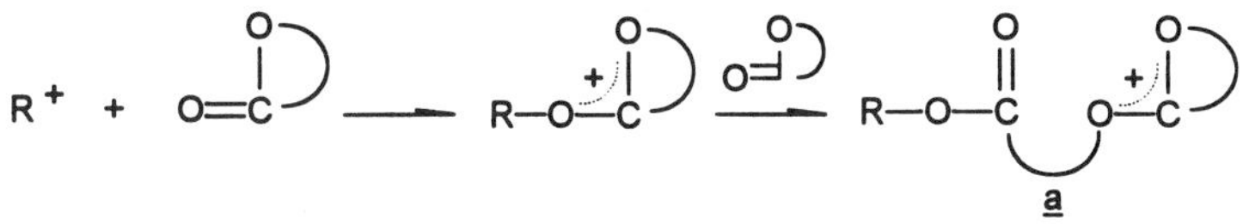
(149)

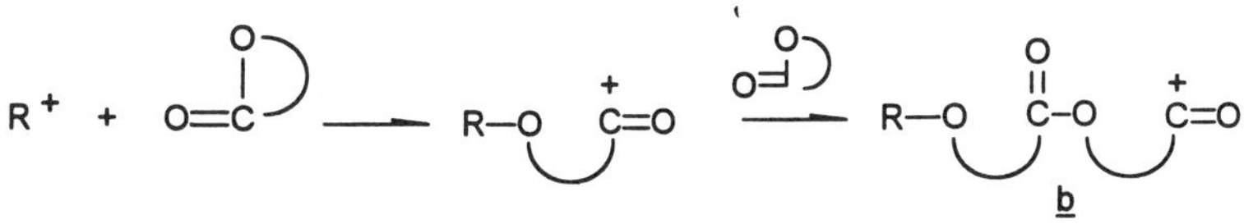
(150)

The detailed study of the mechanism of polymerization initiated with various alkylating agents showed that, in the polymerization of 4-, 6-, and 7-membered lactones, reaction a, i.e., attack on endocyclic oxygen with following opening of acyl-oxygen bond, proceeds exclusively [184–187].

Acylating agents (e.g., oxycarbenium salts) give both types of ring-openings. Active species of a-type re-form species a, and active species of type b (acylating species) form both a and b species; after several further propagation steps, all b species are converted into a species [184].

Cationic polymerization of 4-membered lactone, β-propiolactone, initiated with R—C═O^+, $SbCl_6^-$ gives high molecular weight (M_n up to 300,000) polymers and DP_n is close to this calculated as $[M]_0/[I]_0$ ratio [188]. No cyclic oligomers were detected indicating that chain transfer to polymer is not important in this system. High molecular weight polymers of β-propiolactone were also obtained with CF_3COOH and $(CF_3CO)_2O/AlCl_3$ initiators, whereas Lewis acid initiated polymerization yielded only low molecular weight polymers [189]. β-Propiolactone is cancerogenic, however, and this precludes the broader application of its living cationic polymerization. In the cationic polymerization of higher lactones (δ-valerolactone, ϵ-caprolactone), transesterification and cyclization are already significant [190]. The rate of formation of cyclics is lower, however, than the rate of propagation and this kinetic effect may be used to limit the content of cyclic fraction in polymers. The lower tendency of poly-β-propiolactone chain to participate in chain transfer reaction (especially intramolecular) may be related to higher stiffness of the chain.

As discussed already for cationic polymerization of oxiranes, cyclization can be eliminated if polymerization is performed under the conditions at which the activated monomer mechanism operates. This approach was used for cationic polymerization of ϵ-caprolactone and other higher lactones [191]. Thus, in the polymerization of ϵ-caprolactone in the presence of ethylene glycol (EG) and $(C_2H_5)_3O^+$, PF_6^- catalyst, linear increase of molecular weight with conversion was observed up to $M_n \sim 3000$ and polymers with $DP_n = [M]_0/[EG]_0$ and relatively narrow molecular weight distribution ($M_w/M_n \sim 1.3$) were obtained. No cyclic oligomers were detected in reaction products. Similar results were obtained for polymerization of δ-valerolactone and β-butyrolactone. Kinetic studies of the AM polymerization of lactones have been reported [192].

On the other hand, cationic lactone polymerization may be also used to prepare cyclic oligomers of specific ring size with reasonable yield. Polymerization of bicyclic lactone, 6,8-dioxabicyclo[3.2.1]octan-7-one, generates a repeating unit containing ester and ether bonds:

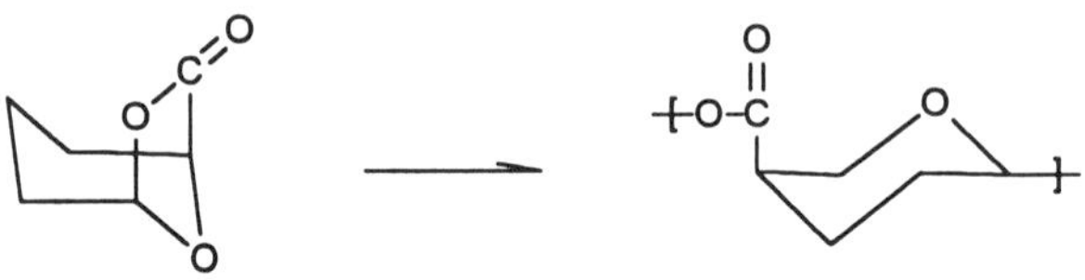

(151)

Polymerization with $BF_3 \cdot Et_2O$ as initiator at $-60°$ C in $CHCl_3$ solution initially gives polyester with $M_n \sim 10{,}000$. At the later stages polymer is degraded to cyclic oligomers. If the polymerization is carried out at slightly higher temperature ($-40°$ C), cyclic oligomers are present from the beginning of the process. Depending on the solvent used, cyclics with specific ring sizes are formed preferentially [193,194]. Thus, cyclic tetramer was obtained with a yield up to 11% in CH_3CN solvent, whereas 27% of cyclic pentamer was formed in $C_3H_7NO_2$ solution. These synthetic macrocyclic oligomers are analogous to some naturally occurring cyclic tetraesters participating in regulation of transport of K^+ ions through cell walls. The synthetic macrocycles show also the specific selectivity for cations [195].

2. *Glycolide, Lactide*

Ring-opening polymerization of cyclic dimers of α-hydroxyacids, namely glycolide and lactide, leads to high molecular weight polyesters:

(152)

Resulting poly(α-hydroxyacids) are important biomaterials used as resorbable sutures and prostheses [196]. The mechanism of polymerization is not well established. Polymerization may be initiated with Lewis acids (SbF_3, $ZnCl_2$, $SnCl_4$); however, other typical cationic initiators (e.g., triethyloxonium or triphenylcarbenium salts) fail to initiate polymerization [197]. Thus, it is not clear whether polymerization proceeds by typical cationic mechanism or rather involves the coordination mechanism. The chain transfer to polymer resulting in transesterification was postulated [198,199] and confirmed later by detailed ^{13}C NMR studies of lactide copolymers [200].

3. *Cyclic Carbonates*

Cationic polymerization of cyclic carbonates, e.g., ethylene carbonate resembles cationic polymerization of lactones. The postulated polymerization scheme is shown by Eq. (153) (for initiation with $CF_3SO_3CH_3$) [201,202]:

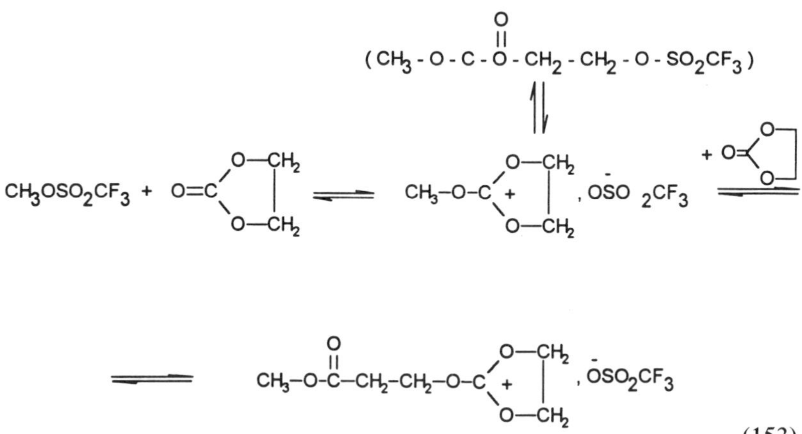

(153)

The isolated polyesters contain various fractions of ether units along the chain, indicating the partial decarboxylation [203]. The exact mechanism of decarboxylation has not yet been elucidated, thus it is not clear whether decarboxylation occurs directly during the propagation step or is the result of the following reactions involving polymer chain units.

It has been found that decarboxylation may be completely eliminated if cationic polymerization of cyclic carbonates is initiated with alkyl iodide or bromide. It is believed that polymerization proceeds with the participation of covalent active species favoring propagation over side reactions leading to CO_2 elimination [204]. It is interesting to note, that BF_3-initiated polymerization of some cyclic carbonates leads to high molecular weight polymers ($M_n > 10^5$) [205].

4. *Cyclic Orthoesters*

Cationic polymerization of spiroorthoesters is of interest because, in contrast to the vast majority of polymerization processes, conversion of spiro monomer into linear polymer is accompanied by expansion in volume. In the first report on the cationic polymerization of 1,4,6-trioxaspiro[4.4]-nonane:

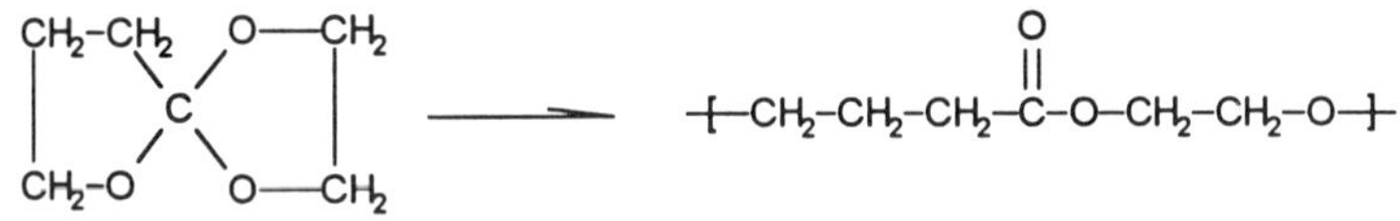

(154)

0.1% expansion in volume was observed [206]. Since then, cationic polymerization of several fused and spiroorthoesters has been studied, but the mechanism of the polymerization and the microstructure of polymers are still not well understood [207]. More detailed analysis of polymer microstructure has revealed that originally postulated regular structure of the chain is an oversimplification and polymers contain not only head-to-head units but also units having two consecutive ester or ether linkages.

The mechanism involving nonconcurrent opening of both rings was proposed to account for this observation [208]. More recently, it was confirmed that at mild conditions (0°, $SnCl_4$ or $BF_3{\cdot}Et_2O$ as initiator) polymers resulting from a single ring opening may be prepared according to the following scheme [209]:

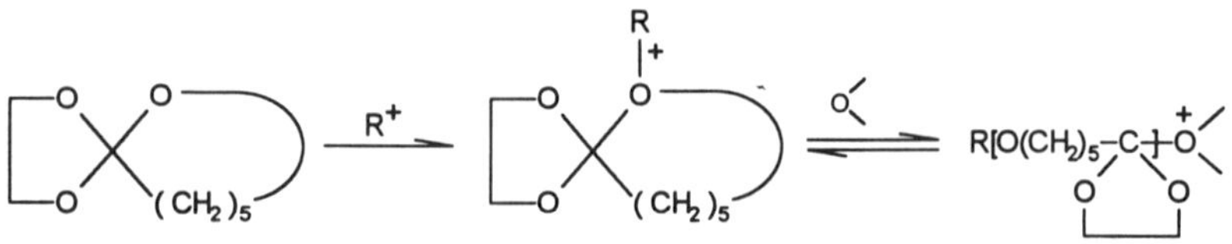

(155)

This is similar to polymerization of bicyclic orthoesters reported earlier where, depending on reaction conditions, polymers resulting from the opening of one or both rings could be prepared [210].

Polymerization of 2,6,7-trioxabicyclo[2.2.1]-heptane at $-78°$ C with $BF_3 \cdot Et_2O$ as initiator proceeds with opening of only one of the two fused rings. Upon heating of the polymer solution in the presence of triflic acid, the opening of the second ring proceeds. Polymerization of bicyclic monomer at 80° C gives directly linear polymer, as shown by the following scheme [210]:

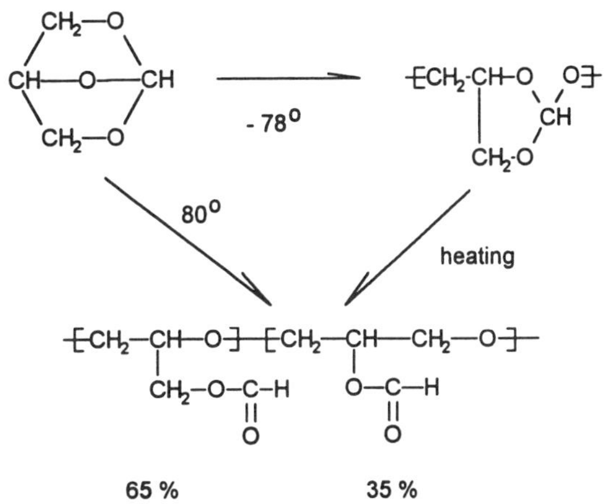

(156)

The same behavior was observed also for polymerization of other bicyclic esters. In all cases, opening of one ring leads to the decrease of volume (shrinkage), whereas the opening of both rings leads to slight increase of volume (expansion) [211].

Polymerization of spiroorthoesters proceeding with single ring opening is a reversible process; the possibility of participation of the chain unit (other than participation of the terminal unit in depropagation) in the process was not considered. Polymerization of spiromonomers has been reviewed [207].

Sulfur analogs of spiroorthocarbonates, namely spirotetrathiocarbonates, also undergo cationic polymerization [212]. A 5-membered monomer gives the mixture of poly(ethylene sulfide) and ethylene trithiocarbonate:

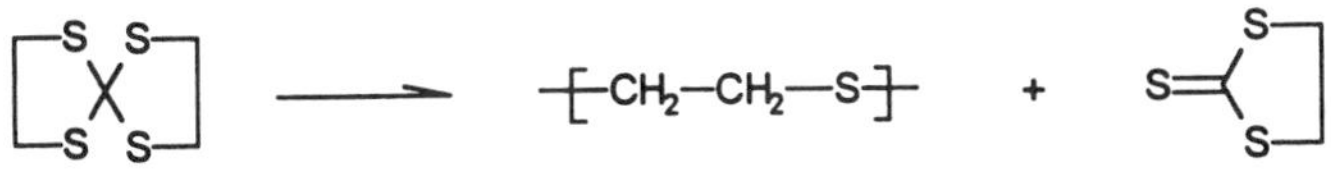

(157)

whereas polymerization of a 7-membered ring leads to polymers containing three different types of repeating units:

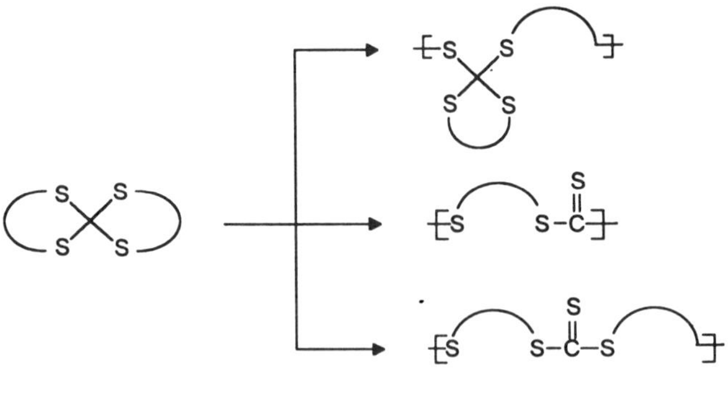

(158)

H. Cyclic Amides

Although several cyclic amides (lactams) can be polymerized by cationic mechanism, this method of polymerization is of little practical importance because the anionic or hydrolytic polymerization provides much more convenient route to corresponding polyamides. Polyamides obtained by cationic polymerization of lactams are less stable and oxidize faster than those obtained by anionic polymerization [213].

Cationic polymerization is the only route, however, for the polymerization of *N*-substituted (*N*-alkyl and *N*-acyl) lactams; the resulting polyamides have much lower melting points than their unsubstituted analogs due to the absence of hydrogen bonding.

1. Lactams

There are essentially two mechanisms of chain growth in the cationic polymerization of lactams. Either the cationically activated monomer reacts with neutral growth center or the neutral monomer molecule reacts with cationic active centers located at the end of the growing chain [214]:

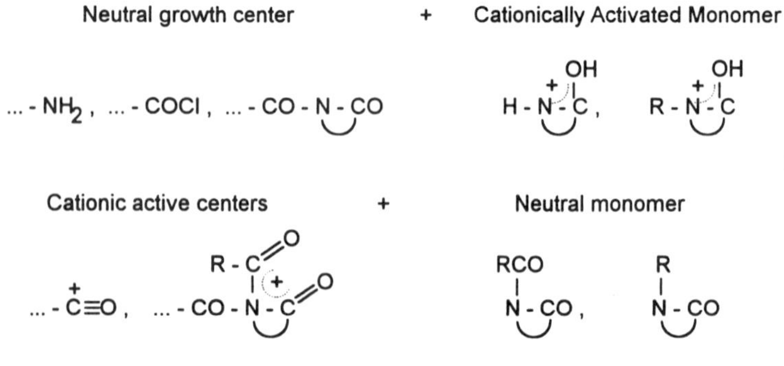

(159)

In the polymerization of unsubstituted lactams, propagation proceeds mainly by the activated monomer mechanism:

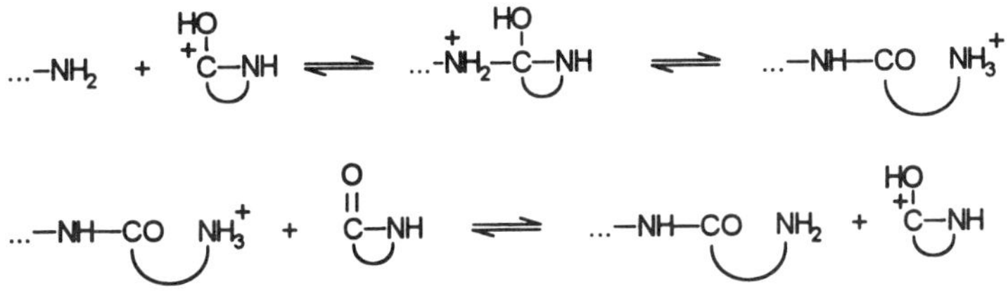

(160)

The amino group is more basic and more nucleophilic than amide groups in the monomer and in the repeating unit of the chain. Thus, activated monomer reacts preferentially with the terminal amino group, but, at the same time, the concentration of activated monomer re-formed at each step is lower than the starting concentration of initiator, because a significant fraction of protons is involved in protonation of the amino group [214]. The Eq. (160) represents only the main reaction leading to formation of polymer. This is accompanied by several side reactions, including those resulting in cyclization and scrambling (transamidation) [215,216].

The main termination reaction is the formation of strongly basic amidine groups, reducing the concentration of acid, required for activation of the monomer [214]:

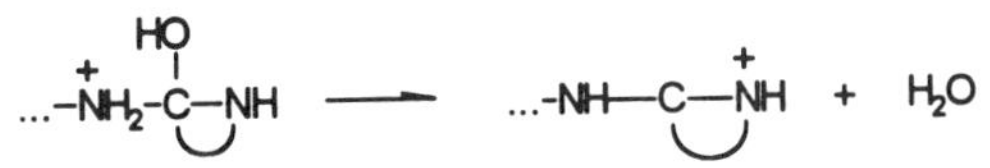

(161)

This undesirable side reaction is avoided in the polymerization of *N*-substituted lactams i.e., *N*-alkyl, *N*-aryl, or *N*-acyl lactams. Polymerization of *N*-substituted lactams proceeds differently, thus in the presence of hydrogen chloride the following sequence of reactions takes place [217].

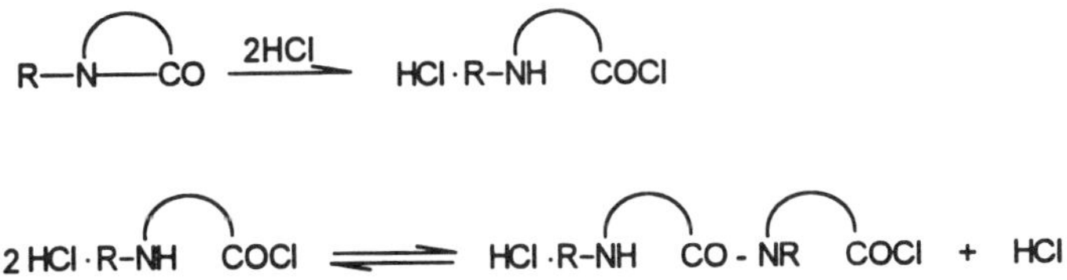

(162)

Thus propagation involves an acylation of nitrogen atom of amido group with acyl chloride end group.

Polyamides prepared by cationic polymerization of lactams contain a significant fraction of cyclic oligomers. Reactions leading to formation of macrocycles are apparently fast as compared with propagation, because equilibrium distribution of cyclic oligomers is attained [218].

Back-biting reaction occurring during cationic polymerization of lactams is detrimental to preparation of block copolymers of two different lactams by sequential polymerization. Block copolymers can be obtained only in those systems in which the rate of polymerization of the second monomer is much higher than the rate of chain transfer to polymer resulting in transamidation [219].

J. Cyclic Phosphorus-Containing Compounds

1. Pentavalent Phosphorus Compounds

Polymerization of cyclic esters of phosphoric acid (cyclic phosphates) is interesting from the synthetic point of view, because the resulting polymers have the sequence of atoms of the chain identical with this, which appears in important biopolymers such as nucleic or teichoic acids. Both 5- and 6-membered cyclic phosphates undergo polymerization.

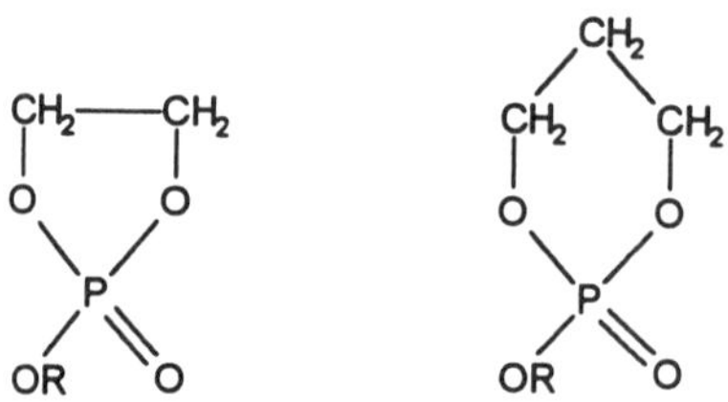

(163)

Polymerization of a 6-membered ring, 2-alkoxy-2-oxo-1,3,2-dioxaphospholane, offers an interesting example of system in which, with increasing the size of the substituent R, polymerization passes from enthalpy-driven (R = CH_3-) to entropy-driven (R = C_2H_5-, C_3H_7-, $(CH_3)_3Si$-) [220,221].

Cyclic phosphates polymerize by both cationic and anionic mechanisms. Anionic polymerization is the method of choice for preparation of high molecular weight polymers because in cationic polymerization the reaction of exocyclic ester group leads to chain transfer:

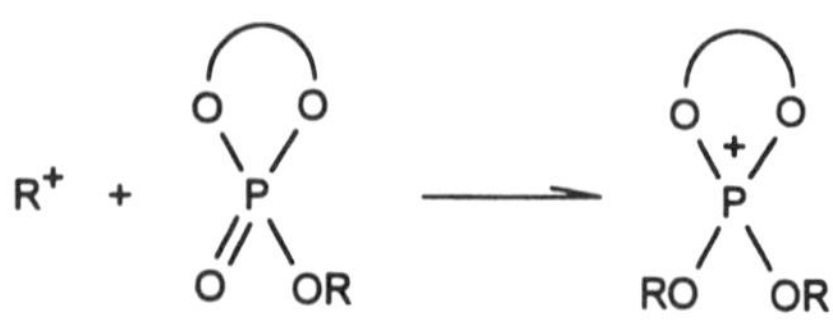

(164)

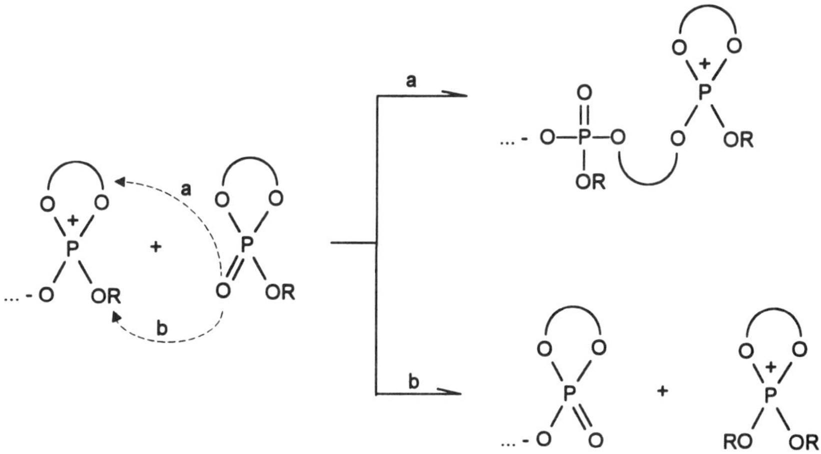

(165)

Thus, the molecular weights of the polymers obtained by cationic polymerization are rather low ($M_n < 2 \cdot 10^3$).

Cationic polymerization of several other cyclic phosphates has been reviewed [222].

2. *Trivalent Phosphorus Compounds*

Cationic polymerization of cyclic compounds containing trivalent phosphorus atom has also been reported. Monomers include cyclic phoshites (a) [223,224], phosphonites (b) [225–227], and deoxophosphones (c) [228]:

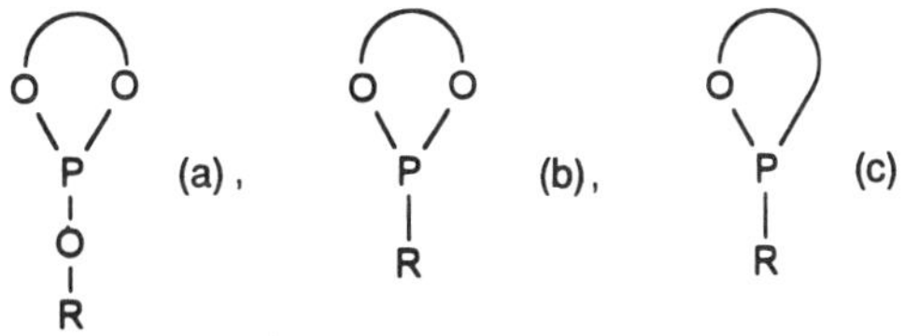

(166)

Polymerization of monomers a and b proceed mostly via Michaelis-Arbuzov type rearrangement involving cyclic phosphonium intermediates to form polyphosphinates or polyphosphonates.

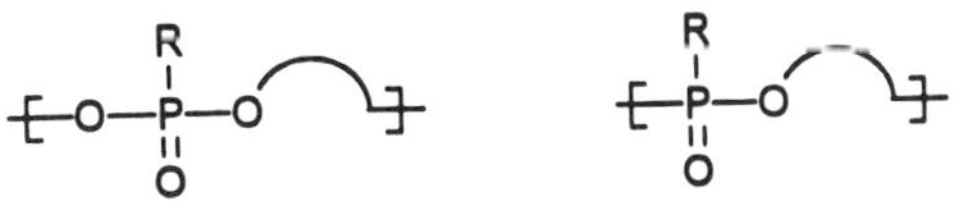

(167)

The cationic polymerization of phoshirane (3-membered cyclic phosphine) leading to polymeric phosphine has been reported [229]:

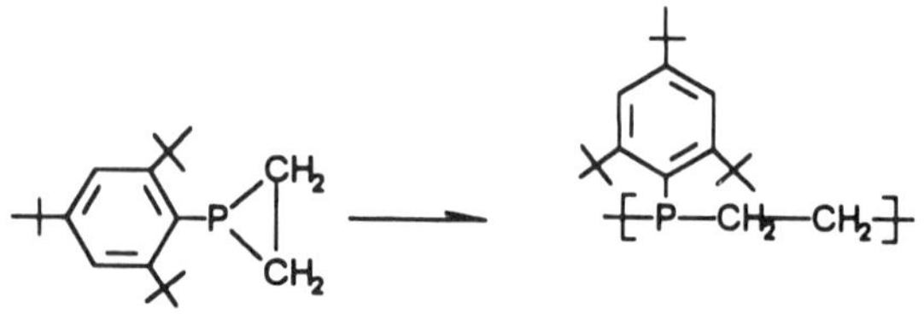

(168)

Polymerization was initiated with alkylating agents such as CH_3I, $C_6H_5CH_2Br$, or $CH_3OSO_2CF_3$. Depending on the nucleophilicity of the counterion, propagation proceeds on phosphonium ion active species (triflate counterion) or on covalent alkyl bromide species.

K. Phosphazenes

Phosphazenes are cyclic compounds built of alternating phosophorus and nitrogen atoms. Polymerization of phosphazene rings has been observed [230,231] and the importance of this inorganic synthetic polymer has been recognized. The early polymers, however, were cross-linked, intractable materials. Only in 1965, the preparation of linear, soluble high molecular weight polyphosphazene by ring-opening polymerization of hexachlorocyclotriphosphazene was reported [232].

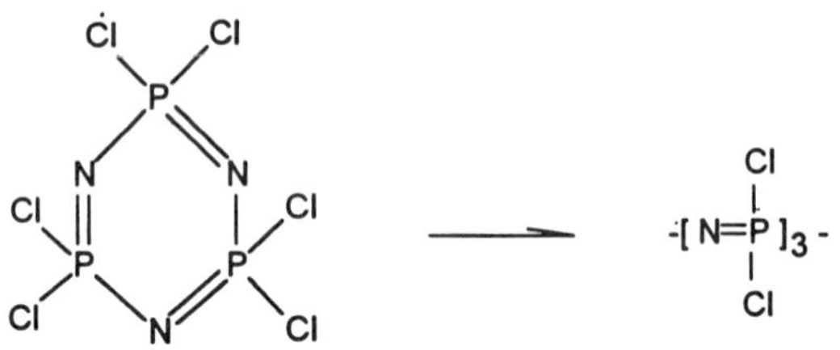

(169)

Research efforts have been concentrated rather on optimization of the polymerization conditions and the properties of the products; thus, in spite of the multitude of reports concerning this system, it is still rather difficult to give a coherent picture of the reaction mechanism.

Polymerization is carried out either in bulk at high temperature (~250° C) or in solution at ambient temperatures. In bulk at 250° C polymerization proceeds even without added catalyst (initiator). In the presence of catalysts, the temperature may be lowered without sacrificing the reaction rate. Very different types of catalysts have been used, but the best results are usually obtained with Lewis acids [233]. Small amounts of water increase the rate of polymerization, but large amounts may suppress the polymerization [234]. Such behavior is typical for number of cationic ring-opening polymerizations, when water acts as coinitiator. These facts, as well as the rise of conductivity at polymerization temperature, indicate that indeed cationic species are involved in the propagation.

The polymerization is believed to proceed according to the scheme:

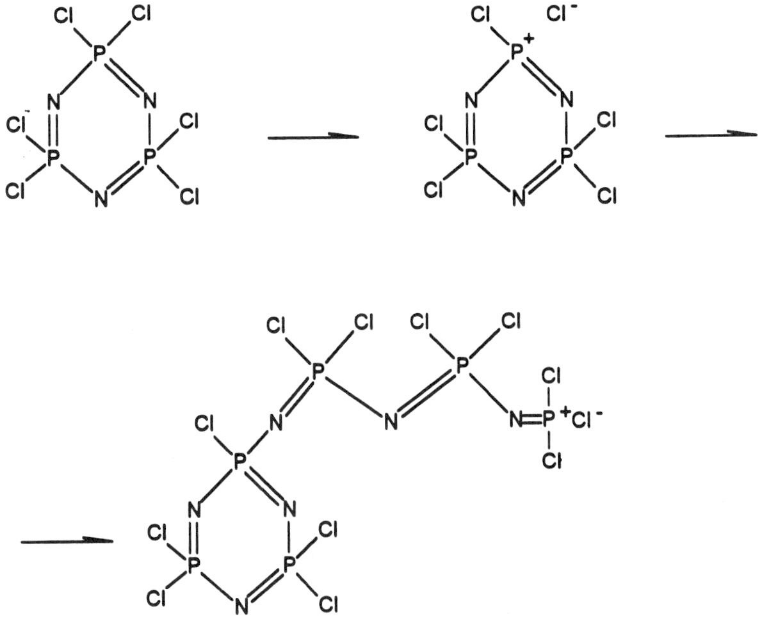

(170)

It is suggested, that gelation, observed at higher conversion during polymerization of phosphazenes is due to the reaction of growing chain end with chain unit [235,236]:

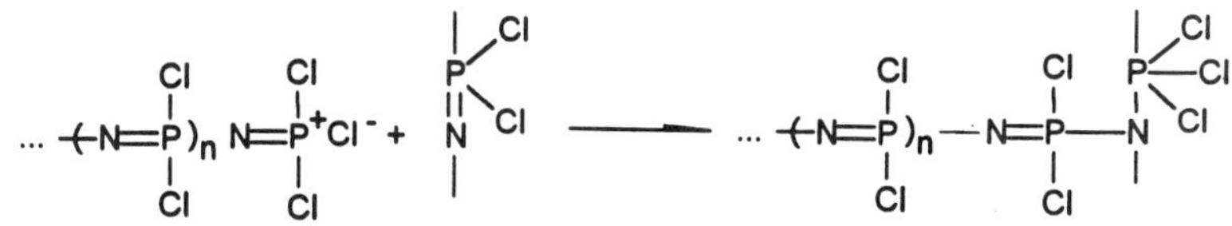

(171)

It has been shown that, in contrast to the previous assumption, the presence of P-halogen bond is not a requirement for polymerization. Thus, phosphazenes containing P—OCH_2CF_3 group were polymerized, apparently by a mechanism involving ionization of P—OCH_2CF_3 bond, facilitated by Lewis acid catalysts [237].

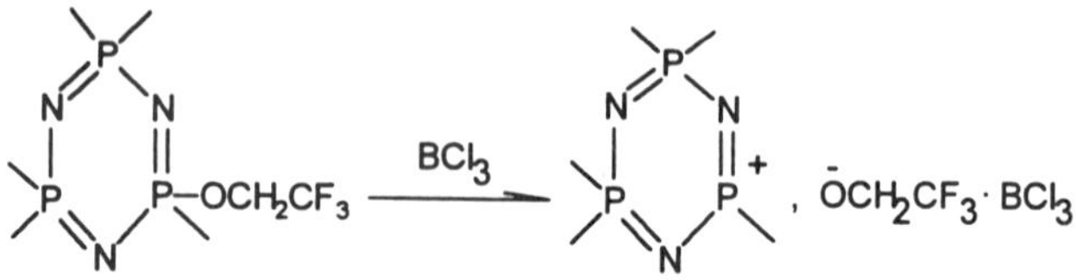

(172)

In the fully organosubstituted bicyclic phosphazenes, the presence of the second ring, as in structure shown below, imposes additional strain of phosphazene ring [238].

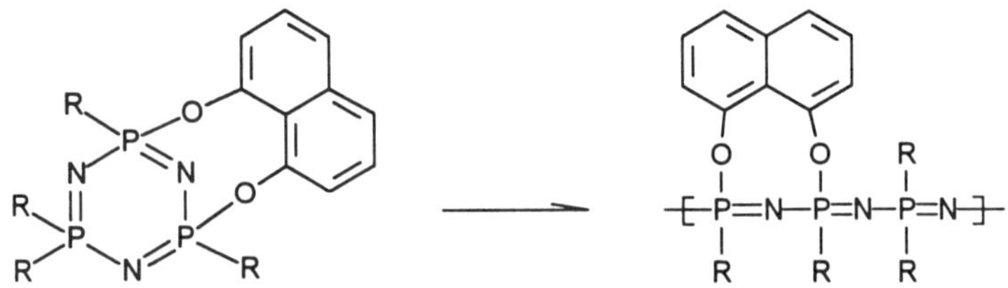

(173)

Thus, although organosubstituted phosphazenes typically would give only cyclic oligomers by ring scission and recombination (end-to-end cyclization), bicyclic phosphazenes, at the conditions of thermal polymerization, give linear polymers in addition to cyclic fraction. The rate of polymerization increases in the presence of catalytic amounts of phosphazene containing P—Cl bond; thus, it is believed that the mechanism is cationic in nature, as shown in Eq. (170). The same behavior was observed for phosphazenes containing the transannular ferrocenyl group [237,239].

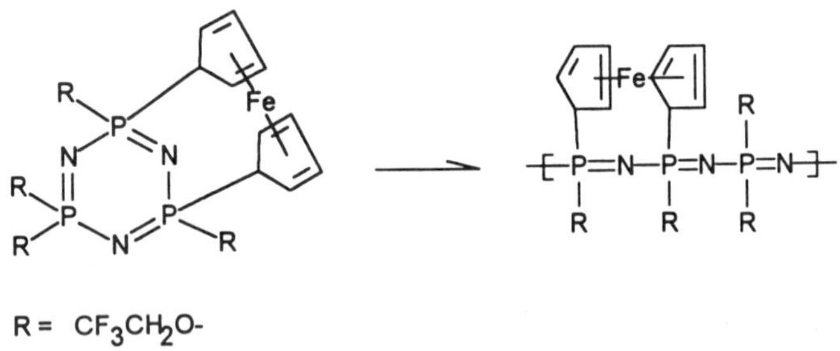

(174)

L. Cyclic Siloxanes

Cyclic siloxanes are composed of several —Si—O— units arranged in a ring. The most commonly studied monomers are methyl-substituted cyclic trimer (D_3) and cyclic tetramer (D_4); however, polymerization of differently substituted cycles (e.g., D_4^H) has also been studied.

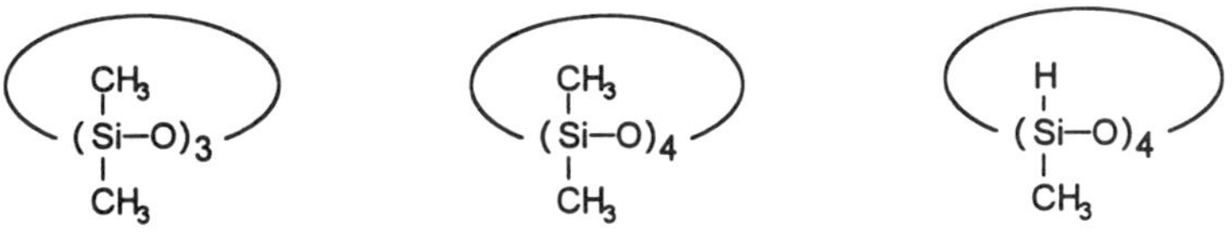

D_3 D_4 D_4^H

(175)

Both protonic and Lewis acids initiate ring-opening polymerization of siloxanes. In spite of the practical importance [240] and the very extensive studies of this system, the mechanism of the polymerization is not completely understood. Even the nature of ionic propagating species is still a matter of controversy. The silicenium [241] or oxonium [242,243] ion structure has been postulated.

Ring-opening polymerization of cyclic siloxanes is invariably accompanied by cyclization and considerable effort has been made to understand and eventually control the cyclization reaction. In the cationic polymerization of hexamethylcyclotrisiloxane (D_3) initiated by CF_3SO_3H, the cyclic fraction is composed of oligomers D_{3n} and the number of dimethylsiloxane units in each oligomer is a multiple of 3 [244]. Thus, a random back-biting typical for many cationic polymerizations has to be rejected, because it should lead to uniform distribution of ring sizes without observed preference for rings containing $3n$ units.

Two possible explanations for observed distributions are: (1) end-to-end closure of growing linear macromolecules which, for polymerization of D_3, give $DP_n = 3n$ and (2) ring expansion polymerization.

There is considerable evidence, that the first possibility is operative [88]. Thus, the head group (e.g., HO- group in the case of protonic acid initiation) reacts with growing species (oxonium or silicenium ions) giving ring with a size equal to the size of linear molecule and re-forming an initiator.

There is thus a certain analogy between the proposed end-to-end ring closure mechanism and the condensation reaction, which has been shown to proceed parallel to addition polymerization as shown in scheme below [245,246]:

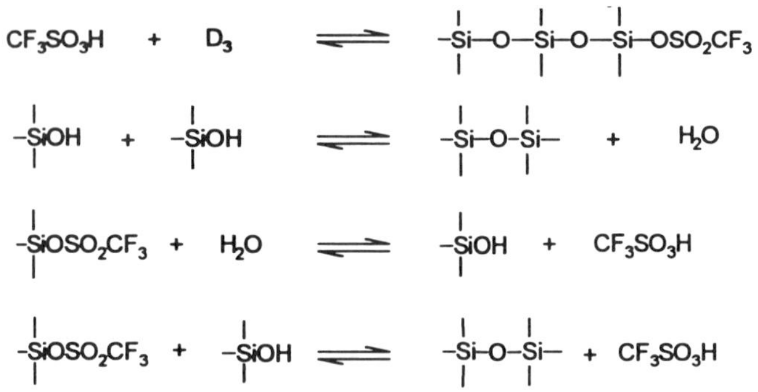

(176)

It is generally agreed that both processes, namely addition polymerization (the nature of active species still of much debate) and acidolysis/condensation reaction, occur simultaneously in the cationic ring-opening polymerization [247,248], although the contribution of both mechanisms is still a matter of discussion. Kinetics of the acid-catalyzed condensation of silanol groups was studied in detail [249,250].

Because the macrocyclic fraction in cyclic siloxane polymerization is formed by end-to-end reaction involving head HO- groups formed upon initiation with protonic acids, it should be possible to affect the content the cyclic fraction (at least under kinetically controlled conditions) by introducing head group other then HO- group. Indeed, polymerization of D_3 monomer can be initiated by $(CH_3)_2SiCl_2{\cdot}SbCl_5$ system, and the content of cyclic D_6 oligomer is considerably lower than in protonic acid-initiated polymerization [251].

The very straightforward results concerning the mechanism of propagation and cyclization in the polymerization of cyclosiloxanes were obtained by studying the radiation-induced cationic polymerization of 6-, 8-, and 10-membered cyclic siloxanes (D_3, D_4, D_5) [252,253].

In contrast to chemically initiated polymerization, all three monomers polymerized with essentially the same rates, giving the same product distribution. It was concluded that propagation and ring formation proceed on the same type of active species:

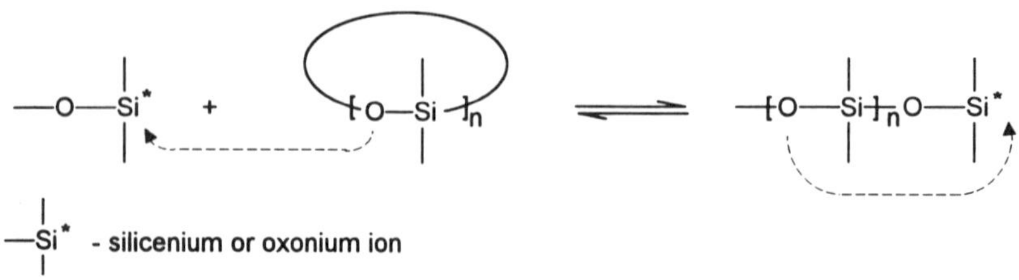

(177)

The simplicity of the system is related to very low concentration of active species and the head groups (of unspecified structure), which makes the end-to-end closure unimportant, and to high purity of the system, allowing only one type of active species to be present. At these conditions, the content of low cycles (up to D_6) in final products was quite large (25–40 mol%).

The formation of large quantities of cyclic oligomers in cationic polymerization of cyclic siloxanes offers interesting possibilities for obtaining topologically distinctive polymer networks [254]. Thus, if relatively large cyclic siloxane oligomers are present in the reaction system where linear polysiloxane chains are linked by suitable multifunctional agents, some of the cyclic species are permanently trapped by chains threading through them [255]. This approach with difunctional coupling agents should lead to a network formed exclusively by cycles threading through cycles. It has been shown experimentally, however, that threading of polysiloxane chain through cyclic siloxane oligomer is possible only when a size of the ring exceeds certain limit, thus no threading was observed for rings composed of 15 —Si—O— units, whereas for rings with 250 units the threading efficiency was ~95%.

Cationic polymerization of cyclic silaethers, octamethyl-1,4-dioxa-2,3,5,6-tetrasilacyclohexane [Eq. (178)] and octamethyl-1-oxa-2,3,4,5-tetrasilacyclopentane [Eq. (179)] was also reported [256]:

(178)

(179)

Cationic polymerization, initiated with strong protonic acids (e.g., triflic acid), occurs purely by opening of the Si—O bond giving high molecular weight regular polysilaethers. For the former monomer, yields up to 80% and M_n up to 40,000 were reported, whereas the latter polymerized with yields up to 90% of polymer and with $M_n \sim 34{,}000$.

IV. SYNTHETIC APPLICATIONS OF CATIONIC RING-OPENING POLYMERIZATION

There are two large-scale industrial processes based on the cationic ring-opening polymerization: production of polyacetals by polymerization of

1,3,5-trioxane and production of poly(tetramethylene oxide)glycols (PTMEG) by polymerization of tetrahydrofuran. There are several other industrial processes using cationically polymerizable monomers described in previous sections, but for some systems (e.g., oxiranes, lactones, lactams, siloxanes), other mechanisms of polymerization (anionic or coordination mechanisms) are preferred.

Still, cationic ring-opening polymerization offers a way to prepare the variety of medium and high molecular weight polymers with diverse structure of the main chain, predictable molecular weights, and required structure of the end groups. This is attained most advantageously, if the polymerization can be conducted as a living process.

The term "living polymerization" has been used quite frequently to describe the systems approaching to various extent the criteria defined originally by Szwarc [97]. The problem of the applicability of this term has been addressed in a few papers [257–259].

Many authors in the field of cationic polymerization (both ring-opening and vinyl) take a rather pragmatic point of view. If, within the certain range of polymerization conditions, molecular weight may be controlled by $[M]_0/[I]_0$ ratio, molecular weight distribution is narrow (this may be not the case for reversible polymerization) and desired terminal groups (or polymer segments as in block-copolymers prepared by sequential polymerization) can be introduced practically quantitatively, then polymerization is called living.

The difference between such "practical" and rigorous definition of living system may be illustrated by the following example. If the ratio of rate constant of propagation k_p to rate constant of transfer or termination (k_t) is 10^2, the system is not living according to the rigorous definition, which requires that chain-breaking reactions are virtually absent. If however this polymerization process is used to prepare oligomers with $DP_n \approx 10$ ($[M]_0/[I]_0 = 10$), the extent of chain-breaking reactions will be still insignificant, and polymers with nearly perfect functionality and $DP_n \approx [M]_0/[I]_0$ can be obtained. Thus, the systems, which by any analytical method appear living under the conditions when only low or medium molecular weight products are formed, may show the detectable deviations from living character at high molecular weight region.

In cationic ring-opening polymerization, there are not too many examples of the systems in which ratios of k_p/k_t are known. In the polymerization of substituted aziridines and substituted thietanes the ratios of rate constants of chain transfer to polymer to the rate constants of propagation have been measured and at least the value obtained for polymerization of *N*-*t*-butylaziridine ($1.2 \cdot 10^4$) [260], indeed indicates the living character of polymerization (providing, that this is the only reaction competing with chain growth, which has been claimed but not unequivocally proven).

The high ratio of k_p/k_t was measured also for cationic polymerization of 3,3-(bis)chloromethyloxetane ($k_p/k_t \sim 10^4$) and indeed polymers with M_n up to $6 \cdot 10^5$ were obtained in this system [124].

The necessary, but not sufficient criterion of the living character of polymerization is the possibility of preparation of high molecular weight polymers ($M_n > 10^5$). This has been achieved in several systems in cationic ring-opening polymerization, e.g., in the polymerization of some cyclic ethers: 3,3-bis(chloromethyl)oxetane, tetrahydrofuran, 1,3-dioxolane, and 1,3,5-trioxane.

In the polymerization of 1,3-dioxolane and tetrahydrofuran it has been shown additionally that concentration of active centers is constant throughout the polymerization (both by direct determination and from analysis of polymerization kinetics). In some other polymerizations, believed to proceed as living processes, only the moderate molecular weights regions ($M_n < 10^5$) were studied; thus, for example, no very high molecular weight polymers were obtained in the polymerization of oxazolines.

As shown earlier in this chapter, there are several systems in cationic ring-opening polymerization, in which quantitative initiation may be achieved and there is essentially no transfer and/or termination. Thus, all the chains have the initiator moiety as the head group and active centers at the growing chain end. These cationic active centers may be deactivated with suitable nucleophiles, leading to specific structure of the terminal end group.

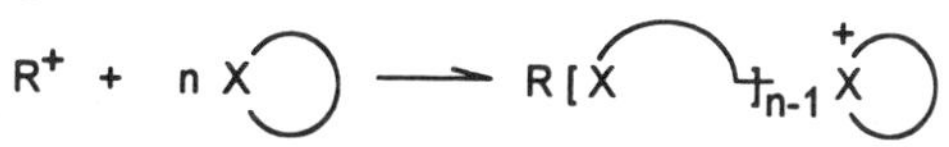

(180)

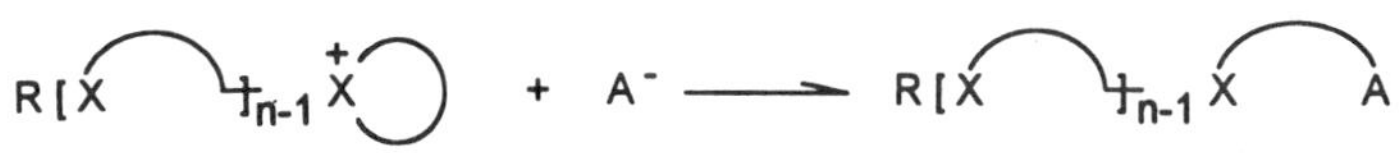

(181)

Equations (180)–(181) outline the general strategy for preparation of polymers containing functional groups. If one functional group is required, it may be introduced either in initiation (head group, R) or in termination process (end group, A). If two identical termini are needed, the nature of R^+ and A^- can be chosen in such a way that they provide the same type of end groups, e.g.:

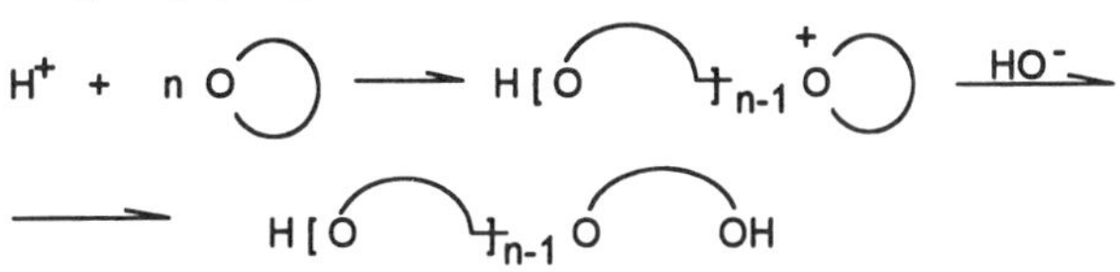

(182)

or bifunctional initiator may be used:

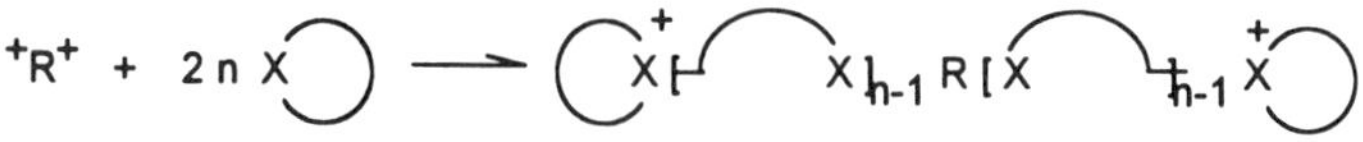

(183)

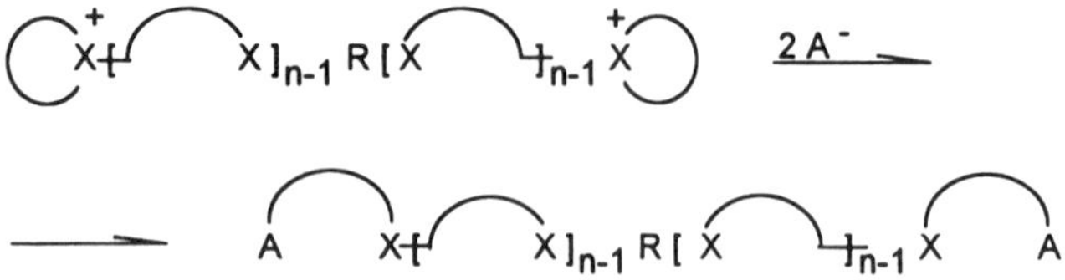

(184)

All the approaches described have been used to prepare functional polymers by cationic ring-opening polymerization. From this point of view, groups of monomers that have been investigated most are cyclic ethers (tetrahydrofuran), cyclic acetals (1,3-dioxolane), cyclic imines (*N*-*t*-butylaziridine), and oxazolines, i.e., these monomers for which the living conditions can be approached.

A. Tetrahydrofuran

The most important telechelic polymer obtained by cationic polymerization is α,ω-diol used in manufacture of polyurethane elastomers. The scheme of the process used on industrial scale is outlined below:

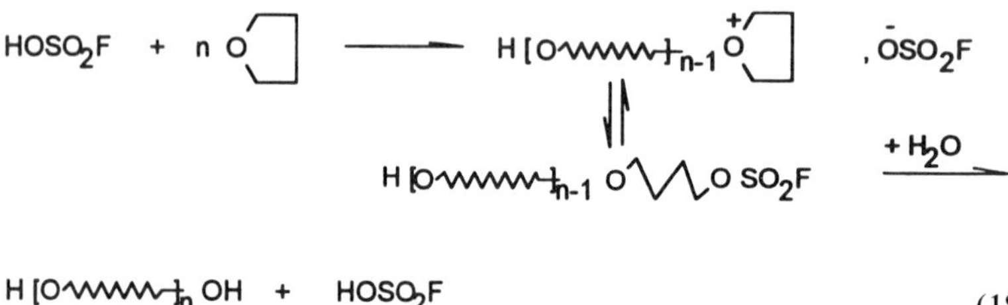

(185)

Other strong acids are also used; the disadvantage of this process is the necessity of using the stoichiometric amounts of acid.

The less straightforward route to dihydroxyl-terminated poly-THF is based on the application of suitable chain transfer agent [261,262], giving polymers terminated with ester groups, which can be easily converted into hydroxyl groups by hydrolysis. Polymerization initiated with catalytic amounts of protic acids (HA) in the presence of acid anhydrides (Ac_2O), gives polymers with $DP_n = ([THF]_0 - [THF]_e)/([HA] + [Ac_2O])$. Originally, high molecular weight poly-THF containing one hydroxyl and one ester group is formed. Eventually, reaction of the hydroxyl group with anhydride and redistribution of molecular weights occurs, leading to products with the degree of polymerization governed by the ratio of concentrations of monomer and anhydride and containing ester groups at both chain ends. When acrylic or methacrylic anhydride is used, this process leads directly to reactive oligomers containing polymerizable groups at both ends [263].

An alternative route to telechelic polymers containing two end groups of the same structure involves the application of difunctional initiators such as:

Bisdioxolenium salts: 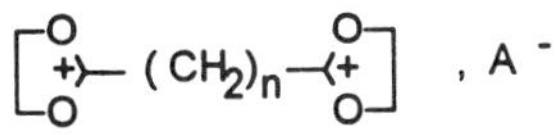 [264]

Difunctional oxocarbenium salts: 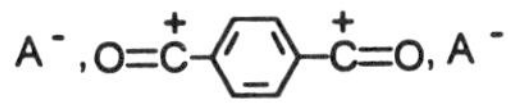 [265]

$$A^-, O{=}\overset{+}{C}(CH_2)_4\overset{+}{C}{=}O, A^- \quad (186)$$

Trifluoromethanosulfonic anhydride: $CF_3SO_2\text{-}O\text{-}SO_2CF_3$ [30]

Resulting polymers containing active species (ionic or covalent) at both ends may be converted into telechelics containing various reactive group by reaction with suitable nucleophiles [266], e.g.:

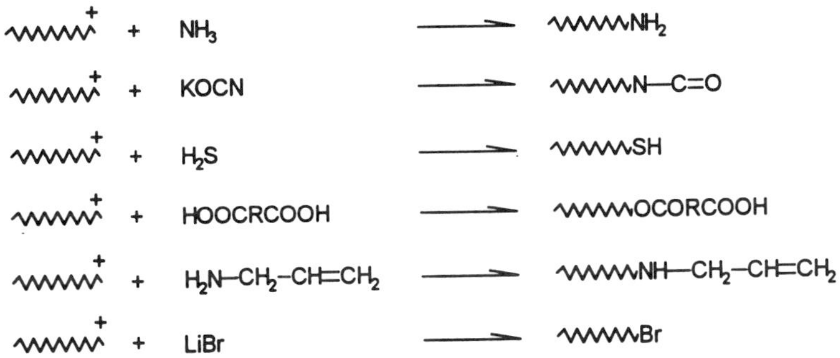

(187)

Stable ionic end groups may be introduced by reaction of difunctional poly-THF with azetidines [267] or tetrahydrotiophene [268]:

 (188)

 (189)

Several monofunctional polymers of tetrahydrofuran-containing polymerizable groups at one chain end (macromonomers) were prepared using the approach outlined in Eqs. (180) and (181). Polymerizable groups were introduced with initiators, e.g.:

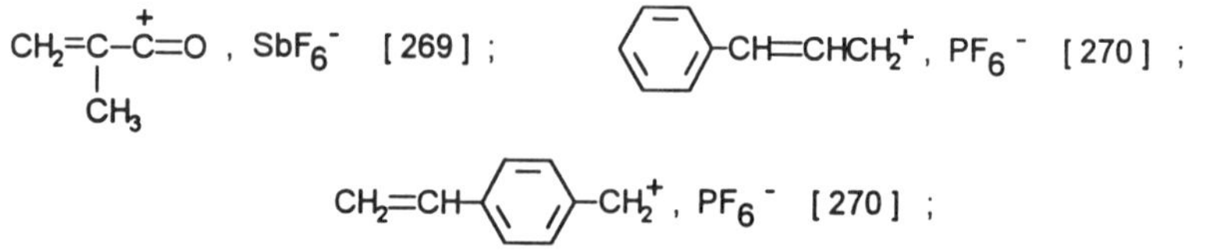

(190)

or with terminating agents, e.g.:

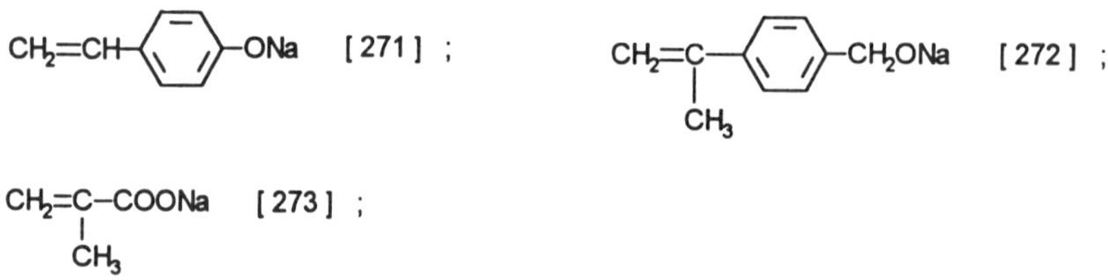

(191)

B. 1,3-Dioxolane

Although several telechelic polymers of 1,3-dioxolane have been prepared by cationic polymerization, their application is limited due to their susceptibility to acid-catalyzed hydrolysis and/or depolymerization. By termination of living mono- and difunctional poly(1,3-dioxolane) with amines or phosphines, polymers containing one or two stable ionic (ammonium, phosphonium) end groups has been prepared [129,274].

Taking advantage of the fast transacetalization (scrambling) accompanying the cationic polymerization of 1,3-dioxolane, the telechelic polymers containing two allyl ether groups were obtained in the polymerization carried out in the presence of bis(allyloxy)methane (the carboxonium active species showed in the scheme for simplicity):

$$\ldots\text{—}OCH_2^+ + CH_2\text{=}CHCH_2OCH_2OCH_2CH\text{=}CH_2 \rightarrow$$
$$\ldots\text{—}OCH_2\text{—}O\text{—}CH_2\text{—}CH\text{=}CH_2 + CH_2\text{=}CH\text{—}CH_2\text{—}O\text{—}CH_2^+ \quad (192)$$

Due to the fast equilibration of the system, the DP_n of the product is equal to the ratio of concentration of monomer and low molecular weight formal. A similar approach was used to prepare poly(1,3-dioxolane) containing two acrylate end groups [275].

$$\ldots\text{—}OCH_2^+ + CH_2{=}CHCOOCH_2CH_2OCH_2OCH_2CH_2OCOCH{=}CH_2 \rightarrow$$
$$\ldots\text{—}OCH_2OCH_2CH_2OCOCH{=}CH_2 + CH_2{=}CHCOOCH_2CH_2OCH_2^+$$

(193)

These products were further used to produce the networks, which could be decross-linked by acid hydrolysis, providing the models for studying the network structure [276].

C. *N-t*-Butylaziridine

Monofunctional polymers of *N-t*-butylaziridine containing polymerizable group (macromonomers) were prepared by initiation of the polymerization with methyltriflate and termination with methacrylic acid [168].

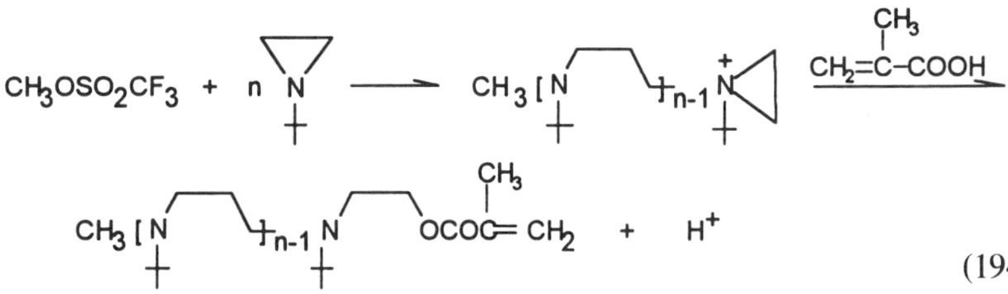

(194)

In a similar way, using bifunctional initiators, polymers terminated at both ends with polymerizable groups were prepared [171].

D. Oxazolines

Although in several papers preparation of relatively high molecular weight polymers has been reported, e.g., from 2-ethyloxazoline ($\acute{\eta}_{sp}/c$ = 3.05 dL/g), 2-*n*-butyloxazoline ($\acute{\eta}_{sp}/c$ = 2.5 dL/g) or 2-phenyloxazoline ($\acute{\eta}_{sp}/c$ = 4.2 dL/g) [277], the cationic polymerization of this group of monomers is employed predominantly for preparation of functional polymers of moderate molecular weights and/or block copolymers.

Differently substituted oxazolines are available through one of the synthetic routes listed below:

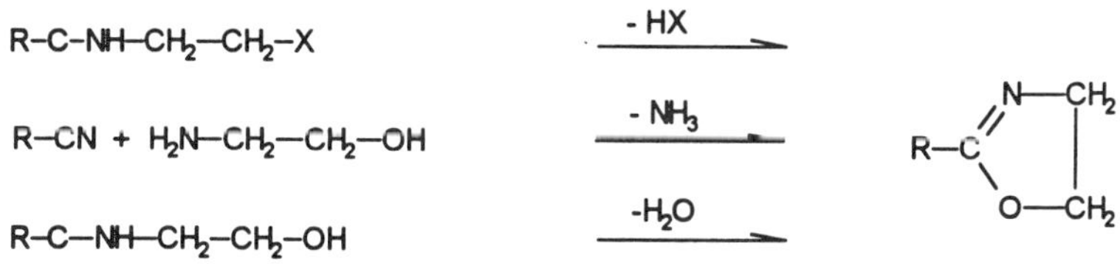

(195)

The cyclization of 2-hydroxyethylamides is used commercially [278]. Various telechelic polymers were prepared using difunctional initiator, bis(oxazolinium salts) [279]:

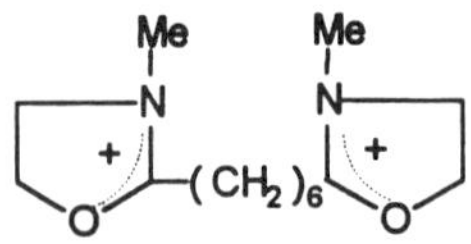

(196)

and terminating the polymerization with water (HO- end groups), ammonia (H_2N- end groups), or n-propylamine (RNH- end groups). Products with M_n in the range from ~1000 to ~4000 were obtained (DP_n values close to $[M]_0/[I]_0$ ratio).

Macromonomers were prepared by polymerizing oxazolines with monofunctional initiators (e.g., methyl p-toluenosulfonate) and terminating the polymerization with salts of acrylic or methacrylic acid. Macromonomers with M_n varying from $M_n \sim 500$ to ~2500 and $M_w/M_n \sim 1.2$–1.4 were obtained; functionalities, however, depended strongly on reaction conditions and the values between 0.99 down to ~0.5 and lower were reported [280].

Alternatively, macromonomers were prepared using vinyl iodoacetate as initiator:

(197)

This system led to macromonomers with M_n from ~600 to ~3000, $M_w/M_n < 1.3$ and functionality between 0.95 and 1.04 [280]. Initiation with β-vinylbenzyliodide or chloride yielded macromonomers containing a styryl group [281].

E. Block Copolymers

The most straightforward method of synthesis of block copolymers is the sequential polymerization, which has been successfully employed in anionic vinyl polymerization. When applied to cationic ring-opening polymerization, this method has several limitations.

Polymerization of 5-, 6-, and 7-membered rings usually proceeds as the reversible process; thus after reaching the ultimate (allowed by thermodynamics) conversion of the monomer A, there is still certain amount of this monomer in equilibrium with growing chains. If monomer A can undergo copolymerization with second monomer added, then the second block will be the copolymer of A and B. This indeed has been observed in the reported synthesis of tetrahydrofuran (THF)-3,3-bis(chloromethyl) oxetane (BCMO) block copolymer. When BCMO was added to living poly-THF (being in equilibrium with its monomer), polymerization proceeded on the poly-THF active species, but monomeric THF was consumed together with BCMO, leading to formation of second block being THF-BCMO copolymer [282].

The other limitation stems from very different structure of heterocyclic monomers and thus very different reactivity of resulting active species. As already discussed, oxonium ions may initiate the polymerization of cyclic amines, but ammonium ions would not initiate the polymerization of cyclic ethers. Thus, the sequential polymerization is possible only when the first monomer is not a stronger nucleophile than the second monomer.

Living difunctional poly-THF was therefore used to initiate the cationic polymerization of *N*-*t*-butylaziridine [170] or oxazoline [283] to give the corresponding A-B-A block copolymers with polyether block B.

Formation of block copolymers in the sequential polymerization may be affected by chain transfer to polymer. As already discussed, in several systems the intramolecular chain transfer to polymer leads to formation of cyclic fraction. Cyclic macromolecules, being neutral, do not participate in further reaction and constitute the homopolymer fraction in resulting copolymer. Intermolecular chain transfer to polymer may lead to disproportionation, i.e., formation of fraction of macromolecules which do not carry active species:

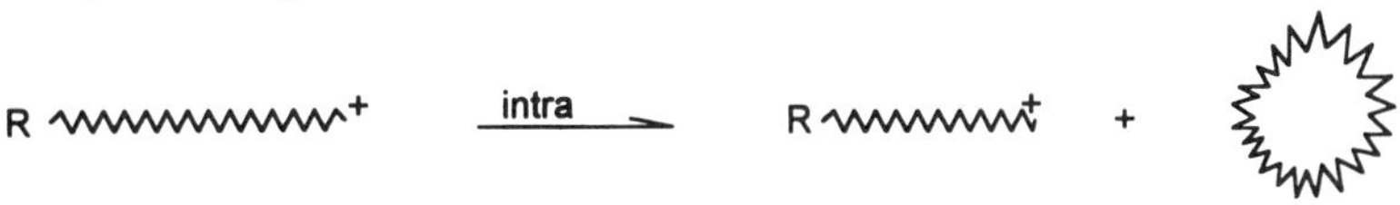

(198)

(199)

Intermolecular chain transfer to polymer leads also to the exchange of segments between macromolecules (scrambling). This may effectively preclude the isolation of block copolymers. This phenomenon is especially pronounced in the polymerization of cyclic acetals.

Thus, in sequential polymerization of two different cyclic acetals, 1,3-dioxolane and 1,3-dioxepane, the sequential polymerization (i.e., polymerization of added second monomer initiated by active species of the first monomer polymerization) may be easily achieved as evidenced by increase of molecular weight [130]. The isolated polymer is not a block copolymer, however, having nearly statistical distribution of both types of units along the chain. In this case, as in the case of 1,3,5-trioxane-1,3-dioxolane system (cf., Section III.B.2), scrambling leads to very fast redistribution proceeding practically parallel to propagation.

All limitations described above do not apply for the systems, which seem to be of considerable practical interest, namely oxazolines. Polymerization of these monomers proceeds practically irreversibly on long-living active species and, as discussed in Section III.F., chain transfer to polymer does not interfere with propagation. On the other hand, due to the possibility of obtaining oxazolines (or 6-membered analogs) with different substituents R:

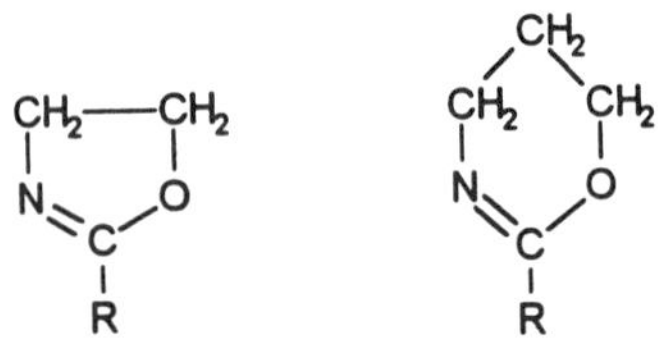

(200)

block copolymers of two oxazolines may contain segments differing considerably in properties. Thus sequential polymerization of *N*-lauryloxazoline and *N*-methyloxazine gives block copolymer composed of poly-*N*-lauroylethyleneimine and poly-*N*-acetyltrimethyleneimine segments. The former one is highly hydrophobic, the later highly hydrophilic, the copolymer thus being the nonionic surfactant [284]. Several block copolymers containing hydrophilic and hydrophobic segments were prepared, including these, containing perfluoroalkyl substituents [285]. It is stated that commercialization of hydrophilic-hydrophobic block copolymers based upon oxazolines is proceeding in Japan [286].

Another possible field of applications involves emulsifiers for polymer blends. The synthetic methods used for preparation of block and graft copolymers for this application illustrates the versatility of the system.

Thus, termination of oxazoline polymerization with potassium hydroxide gave hydroxy-terminated polymers, which in the presence of tin octoate were capable of initiating the anionic polymerization of ϵ-caprolactone [287] giving the corresponding block copolymer, acting as effective compatibiliser.

F. Graft Copolymers

Several graft copolymers were prepared by copolymerization of macromonomers described in the previous section. Another route-to-graft copolymer is based on conversion of suitable groups along the chain of preformed polymer into species capable of initiating the cationic ring-opening polymerization of suitable monomer. Thus, for example, polymerization of tetrahydrofuran was initiated by carbenium ions generated on various chains containing tertiary chlorine atoms by reaction with $AgPF_6$ [288,289]

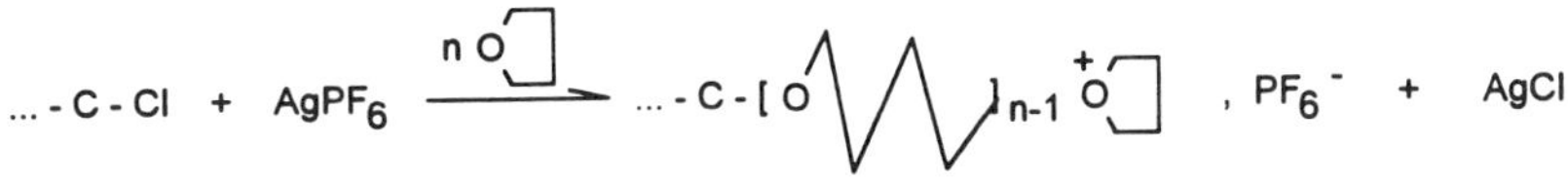

(201)

Due to the high nucleophilicity of oxazoline monomers, their polymerization may be initiated by C—Cl bond even in the absence of silver salt. Thus, chloromethylated polystyrene [290] or polychloroprene [291] were used to prepare corresponding graft copolymers. More recently, polyoxazoline was grafted on poly(vinyl ether)s containing certain proportion of chloroethyl groups [292].

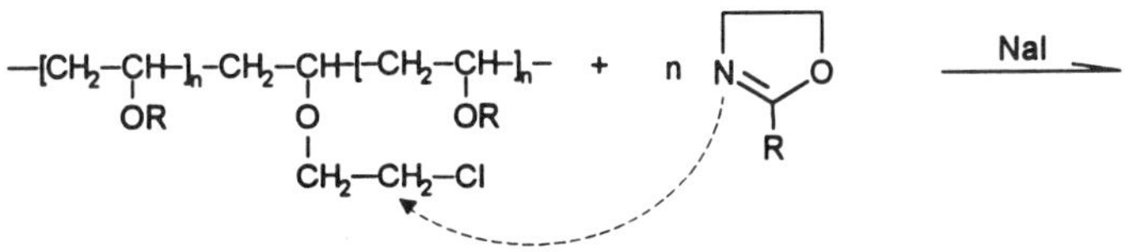

(202)

and on polyethylene containing vinylacetate units [292]:

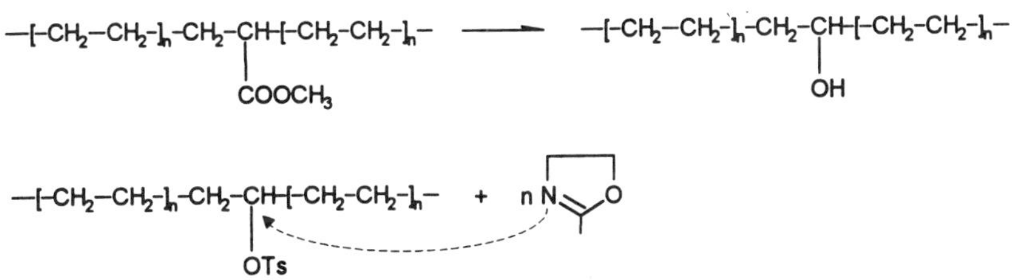

(Ts - tosyl)

(203)

Termination of oxazoline polymerization with 3-aminopropyltriethoxysilane produced polyoxazolines containing $—Si(OR)_3$ end groups. These materials were grafted on the surface of silica gel [293]:

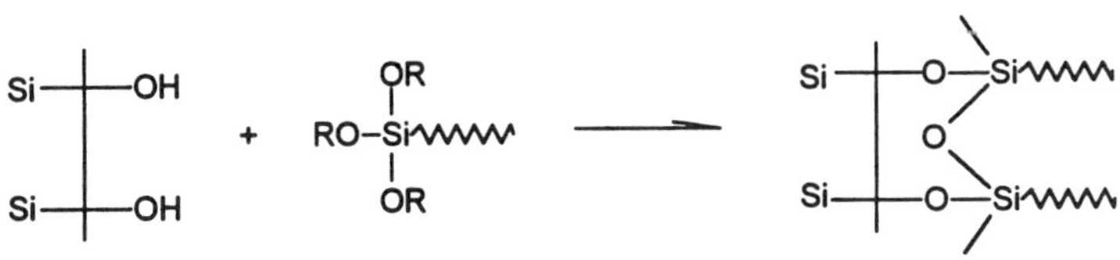

(204)

or subjected to acid-catalyzed cohydrolysis with tetraethoxysilane by sol-gel method to produce glassy, homogenous, transperent modified silica gel.

G. Copolymerization

In the random copolymerization process, both types of active species should be able to participate in the cross-propagation reactions. This imposes certain limitations on the choice of comonomers in the cationic polymerization of heterocyclic monomers. Onium ions, being the active species of these polymerizations, differ considerably in reactivity; thus, as already discussed, oxonium ions initiate the polymerization of cyclic amines, whereas ammonium ions do not initiate the polymerization of cyclic ethers and the corresponding cross-propagation reaction would not proceed:

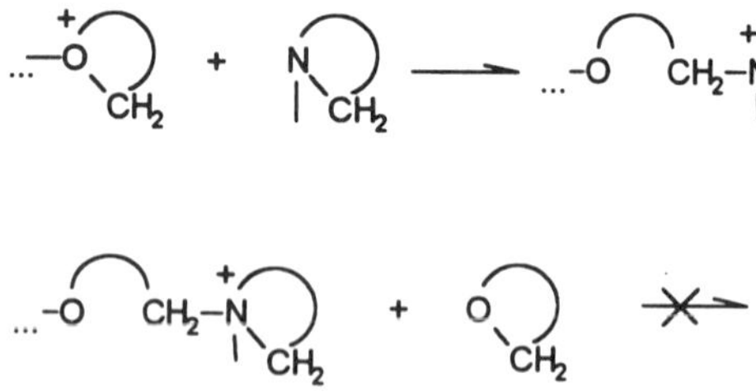

(205)

Thus, random copolymerization of cyclic ethers with cyclic amines is not possible. The other limitation, which will be discussed in more detail in a later part of this section, is the reversibility of homo- and/or cross-propagation steps, when one or both comonomers polymerize reversibly.

The simplest systems involve copolymerization of structurally related pairs of comonomers, polymerizing irreversibly. Copolymerization of different oxetanes [294], thietanes [295], azetidines [296], and oxazolines [297] was studied, the results were interpreted in terms of simple four-parameter copolymerization scheme and the corresponding reactivity ratios for some systems were determined.

In many systems, however, the analysis of the cationic copolymerization of heterocyclic monomers is complicated by two factors: (1) at least some of the homo- and cross-propagation reactions may be reversible; (2) redistribution of the sequences of comonomers within the chain may occur as a result of chain transfer to polymer. Therefore, the conventional treatment involving four irreversible propagation steps is rarely applicable in cationic ring-opening copolymerization. Instead, the diad model should involve four reversible reactions, i.e., eight rate constants

$$
\begin{aligned}
-M_1^* + M_1 &\underset{K_{-11}}{\overset{K_{11}}{\rightleftharpoons}} -M_1 - M_1^* \\
-M_1^* + M_2 &\underset{k_{-12}}{\overset{k_{12}}{\rightleftharpoons}} -M_1 - M_2^* \\
-M_2^* + M_1 &\underset{k_{-21}}{\overset{k_{21}}{\rightleftharpoons}} -M_2 - M_1^* \\
-M_2^* + M_2 &\underset{k_{-22}}{\overset{k_{22}}{\rightleftharpoons}} -M_2 - M_2^*
\end{aligned}
\tag{206}
$$

The copolymer composition equation with reversibility of all propagation steps was derived as a complex function [298,299]:

$$d[M_1]/d[M_2] = f(r_{xy}, K_{xy}, f_{xy}) \tag{207}$$

where r_{xy} = reactivity ratios; K_{xy} = equilibrium constants; f_{xy} = fractions of corresponding chain ends. The number of parameters involved makes it practical application difficult.

The system is considerably simplified if some of the propagation steps may be treated as irreversible reactions [300,301]. This is for example the case of cationic copolymerization of 3,3-bis(chloromethyl)oxetane (BCMO) and tetrahydrofuran (THF) studied experimentally [302,303].

In homopolymerization, the former monomer propagates irreversibly, the homopropagation of THF is highly reversible. It has been pointed out that, for heterocyclic monomers, it is the nature of penultimate unit that governs the reversibility of a given reaction step [304]. Thus, addition of THF to BCMO active center is irreversible, because the backward reaction would require the closure of 4-membered ring. On the other hand, addition of BCMO to THF active centers is reversible:

(208)

(209)

In many instances in cationic ring-opening polymerization, all the reaction steps, however, are reversible. The final composition of copolymer (in equilibrium) is governed then by thermodynamics. Thermodynamic approaches have been developed [305] and recently reviewed [306]. Such thermodynamic approach has been used to analyze the copolymerization of pairs of cyclic acetals (1,3-dioxolane with 1,3-dioxepane and

1,3-dioxolane with 1,3-dioxane) [307], and all involved equilibrium constants were calculated, allowing the prediction of microstructure of copolymers and equilibrium monomer concentrations for any initial composition of the feed.

Some heterocyclic monomers may undergo random copolymerization with vinyl monomers. This is a case of cyclic acetals (e.g., 1,3-dioxolane) which forms the random copolymers with styrene [308,309] or isoprene [310]. Apparently, the oxycarbenium ions, being in equilibrium with tertiary oxonium ions (cf., Section II.B.6.b), are reactive enough to add styrene:

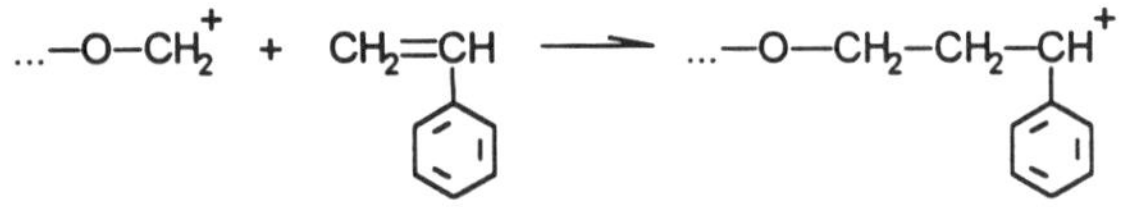

(210)

H. Selected Synthetic Procedures

1. *Preparation of Polytetrahydrofuran/ Poly(1,3,3-trimethylazetidine) Block Copolymer [267]*

Tetrahydrofuran (THF) was refluxed and distilled over sodium wire just before use. Initiator, triflic anhydride (TfAn), was distilled over P_2O_5. To 10 mL (0.123 mol) of THF in a flask fitted with a nitrogen inlet and thermostated at 25° C, 0.0207 mL (0.123 mmol) ot TfAn was added by means of hypodermic syringe with stirring. After 30 min, 0.1 mL of 1,3,3-trimethylazetidine was added. After stirring for 15 min, the reaction mixture was poured into 500 mL of cold water, the precipitated product was filtered, washed with water, and dried under reduced pressure to give bisazetidine-terminated poly-THF in 5.4% yield. M_n was equal to 3900 and M_w/M_n was <1.1.

Of this polymer 0.493 g was dissolved in 5 mL of 1,3,3-trimethylazetidine and the solution was heated in a sealed tube for 18 hr in 80° C. The reaction mixture was then dissolved in benzene and freeze-dried to give 0.721 g of ABA triblock copolymer.

2. *Polymerization of t-Butylaziridine [311]*

t-Butylaziridine (TBA), prepared from ethylene oxide and *t*-butylamine, was dried over CaH_2 and purified by distillation (b.p. = 91° C). Triethyloxonium tetrafluoroborate (TEFB) was prepared according to the Meerwein procedure and purified by several reprecipitations from solution in methylene dichloride into diethyl ether. Solvent, methylene dichloride, was distilled from CaH_2 before use.

Polymerization was carried out in a glass ampoule of 20-mL capacity and provided with a stopcock. The monomer, solvent, and initiator solution were introduced against a current of dry nitrogen. Under the conditions, $[M]_0 = 1.0$ mol/L, $[I]_0 = 2 \cdot 10^{-2}$ mol/L, 99% conversion was reached after 6 hr at 0° C. The polymer precipitated from the reaction mixture. It was filtered off, washed with methylene dichloride, and dried in vacuo. The intrinsic viscosity was equal to 0.08 dL/g. It increases however linearly with increasing $[M]_0/[I]_0$ ratio and at $[M]_0/[I]_0 = 1000$, the intrinsic viscosity is equal to 0.43 dL/g (measured in 1,2-dichlorobenzene solution at 70° C).

The polymer is a white powder with melting point at 142° C. It is soluble in chlorinated hydrocarbons, DMF, and sulfolane above 50° C. At room temperature, it is soluble in aqueous acids, when the concentration of acid is at least equal to the concentration of amino groups.

3. *Polymerization of 2-Phenyl-2-Oxazoline [312]*

a. Preparation of Initiator Complex of *p*-toluenosulfonic acid (TSA) with 2-phenyl-2-oxazoline (PhOx) was prepared by pouring ethanol solution of TSA into diethyl ether solution of PhOx with vigorous stirring. The precipitated mass was filtered and recrystallized from anhydrous diethyl ether. Analysis confirmed 1:1 stoichiometry of the complex.

b. Polymerization Procedure PhOx (2.0 g $= 1.48 \cdot 10^{-2}$ mol) and TSA-PhOx complex (0.0255 g $= 1.48 \cdot 10^{-4}$ mol) were charged into a glass ampoule. After several freeze-evacuate-thaw cycles at liquid nitrogen temperature, the ampoule was placed in an oil bath kept at 150°. The complete conversion was reached after 30 min. The ampoule was cooled, opened, and the reaction mixture was dissoled in small amount of chloroform. The chloroform soltion was poured, with stirring, into a large amount of *n*-hexane. The precipitated product was filtered, washed with diethyl ether, and dried in vacuo at 60° C. The yield was ~100% and M_n (vpo) was equal to 13,500 ($DP_n \sim 100 \sim [M]_0/[I]_0$. The polymer is white, powdery solid with melting point equal to 205° C. It is soluble in chloroform and acetone but insoluble in *n*-hexane or benzene.

4. *Copolymerization of 1,3,5-Trioxane with 1,3-Dioxolane [313]*

A four-necked flask fitted with nitrogen inlet, thermometer, stirrer, and condenser is charged with 3.6 g of poly(ethylene oxide) with $M_n \sim 15{,}000$ (preventing agglomeration of the particles of formed copolymer), 120 ml of cyclohexane ($[H_2O] < 20$ ppm), 144 ml (168 g = 1.87 mol) of molten, freshly distilled 1,3,5-trioxane (TXN) ($[H_2O] < 30$ ppm, $[CH_2O] < 10$ ppm, $[HCOOH] < 10$ ppm) and 7 g (0.094 mol) of 1,3-dioxolane (DXL). The flask is thermostated at 60° C. To initiate the polymerization, 0.1 mol% (with respect to TXN, i.e., 0.14 g) of BF_3 *n*-Bu_2O complex is added, under nitrogen, by means of hypodermic syringe, with constant stirring. Polymerization starts after a short (few minutes) induction period, the temperature raises to 70–75° C and the polymer starts to precipitate. The reaction is terminated after 60 min, by adding the solution of NH_3 in methanol, the polymer is filtered off, washed with methanol, and dried at 70°. A typical yield is 70–80% of copolymer with intrinsic viscosity ~1.5–1.6 dL/g (measured in tetrachloroethane/phenol 3:1 by volume mixture at 90°).

5. *Polymerization of Epichlorohydrin by Activated Monomer Mechanism [114]*

Sixty-two grams of ethylene glycol (EG) was placed in a 5-L three-necked flask fitted with mechanical stirrer, dropping funnel, thermometer, and condenser. 3.0 g of $BF_3 \cdot EG$ complex (0.02 mol of BF_3) was added by means of hypodermic syringe and with constant stirring epichlorohydrin (ECH) was slowly added to the reaction mixture at a rate ~40 g/hr for 48 h. During this time, 2260 g (24.9 mol) of ECH was introduced into the reaction system. The instantaneous molar ratio of [ECH] to [HO-] was <0.5 throughout the reaction (the concentration of unreacted ECH in the reaction mixture was controlled by measuring the refractive index). The external cooling bath was used to keep the temperature at ~20° C. After adding the required amount of ECH, the reaction mixture was kept for additional 24-hr at room temperature. To neutralize the catalyst, 100 g of solid CaO was then added to reaction mixture, the temperature was raised to 60° C, the suspension of CaO was stirred for 2 hr and CaO was filtered off. Polyepichlorohydrin terminated at both ends with hydroxyl groups was obtained in 98% yield. The M_n of the product was equal to 2500 (vpo) ($DP_n = [ECH]_0/[EG]_0$) and M_w/M_n was <1.15. The crude product contains ~0.7% of cyclic oligomers. The content of cyclic oligomers may be reduced to ~0.1% by heating the product to 90° C under vacuum of ~1.5 hPa (~1 mmHg). The purified product is pale yellow, viscous liquid (92,000 cP at 20° C), having the density equal to $d^{20} = 1.369$ g/cm^3 and refractive index $n_D{}^{20} = 1.5125$, which solidifies at ~ −30° C. The chain is composed primarily (>95%) of *h-t* units and the hydroxyl end groups are predominantly (>95%) the secondary ones.

REFERENCES

1. K. Matyjaszewski, *Makromol. Chem. Macromol. Symp. 54/55*: 51 (1992).
2. S. Penczek, P. Kubisa, and K. Matyjaszewski, Cationic ring-opening polymerization—mechanisms, *Adv. Polym. Sci. 37*: 1 (1980).
3. S. Penczek, P. Kubisa, and K. Matyjaszewski, Cationic ring-opening polymerization—synthetic applications, *Adv. Polym. Sci. 68/69*: 1 (1985).
4. K. J. Ivin and T. Saegusa, eds., *Ring-Opening Polymerization*, Vols. I & II, Elsevier, London, 1984.

5. G. Allen and J. C. Bevington, eds., *Comprehensive Polymer Science*, Vol. III, Part 1, Chapters 45–53, Pergamon Press, Oxford, 1989.
6. S. Penczek and P. Kubisa, Cationic-ring opening polymerization, *Ring-Opening Polymerization* (D. J. Brunelle, ed.), Hanser Publishers, Munich, 1993.
7. S. Słomkowski and S. Penczek, *J. Chem. Soc. Perkin Trans. II 1974*: 1718 (1974).
8. P. Kubisa and S. Penczek, *Makromol. Chem. 144*: 169 (1971).
9. S. Penczek and P. Kubisa, *Makromol. Chem. 165*: 121 (1973).
10. A. J. Gordon and R. A. Ford, *The Chemist's Companion, A Handbook of Practical Data, Techniques and References*, Wiley-Interscience, New York, 1972, p. 60.
11. G. A. Olah and P. Von R. Schleyer, eds., *Carbonium ions*, Wiley-Interscience, New York, 1973, Vol. IV, p. 1697.
12. G. A. Olah, G. K. Surya Prakash, and J. Sommer, *Superacids*, Wiley-Interscience, New York, 1985, p. 36.
13. M. Bednarek, K. Brzezińska, J. Stasiński, P. Kubisa, and S. Penczek, *Makromol. Chem. 190*: 929 (1989).
14. G. A. Tsigdinos, *Topic Curr. Chem. 76*: 1 (1978).
15. H. Dannenberg and W. R. Harp, Jr., *Anal. Chem. 28*: 86 (1956).
16. A. J. Bannister and A. W. Luke, *J. Polym. Sci. Polym. Chem. Ed. 30*: 2653 (1992).
17. M. Baaz, V. Gutman, O. Kunze *Monatsch. Chem. 93*: 1142 (1962).
18. G. A. Olah and P. Von R. Schleyer, eds., *Carbonium ions*, Wiley-Interscience, New York, 1973, Vol. IV, p. 1501.
19. G. A. Olah, ed., *Friedel-Crafts and Related Reactions*, Interscience Publishers, New York, 1965, Vol. I, p. 677.
20. C. V. Pittman, Jr., S. P. McManus, and J. V. Larsen, *Chem. Rev. 72*: 357 (1972).
21. G. A. Olah and S. J. Kuhn, *J. Org. Chem. 26*: 237 (1961).
22. R. Szymański, unpublished, mentioned in S. Penczek, P. Kubisa, and K. Matyjaszewski, Cationic Ring-Opening Polymerization—Mechanisms, *Adv. Polym. Sci. 37*: 52 (1980)
23. H. Meerwein, *Houben-Weyl Methoden der Organischen Chemie* (M. Müller, ed.), 4th ed., Vol. VI/3, Georg Thieme Verlag, Stuttgart, 1965, p. 325.
24. G. A. Olah, ed., *Friedel-Crafts and Related Reactions*, Interscience Publishers, New York, 1965, Vol. IV, p. 43.
25. F. Klages and H. Meuresch, *Berichte 85*: 863 (1952).
26. R. Szymański, M. Wieczorek, P. Kubisa, and S. Penczek, *Chem. Commun. 1976*: 33 (1976).
27. M. P. Dreyfuss, J. C. Westphal, and P. Dreyfuss, *Macromolecules 1*: 437 (1968).
28. I. Kuntz and M. T. Melchior, *J. Polym. Sci. A7*: 1959 (1969).
29. F. Ashfar-Taromi, M. Scheer, P. Rempp, and E. Franta, *Makromol. Chem. 179*: 849 (1976).

30. S. Smith and A. J. Hubin, *J. Macromol. Sci. A7*: 1399 (1973).
31. G. A. Olah, O. Farooq, C. X. Li, A. M. Morteza, and J. Aklonis, *J. Appl. Polym. Sci. 45*: 1355 (1992).
32. G. A. Olah, G. K. Surya Prakash, and J. Sommer, *Superacids*, Wiley-Interscience, New York, 1985, p. 27.
33. D. Bethel and V. Gold, *Carbenium Ions*, Academic Press, London, 1967, p. 49.
34. G. A. Olah, *Friedel-Crafts Chemistry*, Wiley-Interscience Publications, New York, 1973, p. 247.
35. G. A. Olah, ed. *Friedel-Crafts and Related Reactions*, Interscience Publishers, New York, Vol. I, 1965, p. 702.
36. G. A. Olah, G. K. Surya Prakash, and J. Sommer, *Superacids*, Wiley-Interscience, New York, 1985, p. 184.
37. G. A. Olah, ed., *Friedel-Crafts and Related Reactions*, Interscience Publishers, New York, 1965, Vol. I, p. 699.
38. G. L. Collins, R. K. Greene, F. M. Berardinelli, and W. V. J. Garruto, *J. Polym. Sci. Polym. Lett. Ed. 17*: 667 (1979).
39. J. Stasiński and G. Dmowska, *Makromol. Chem. Rapid Commun. 8*: 535 (1987).
40. R. Hoene and K. H. W. Reichert, *Makromol. Chem. 177*: 3545 (1976).
41. Y. Zhu and Q. Sheng, *Nanjing Daxue Xuebao 27*: 488 (1991), CA *117*(4): 27190 h (1993).
42. M. A. H. Talukder, *ACS Polym. Prep. 31*(1): 93 (1990).
43. J. S. Hrkach and K. Matyjaszewski, *Macromolecules 25*: 2070 (1992).
44. J. Crivello and M. Fan, *Makromol. Chem. Macromol. Symp. 54/55*: 179 (1992).
45. J. Crivello, *Adv. Polym. Sci. 62*: 1 (1984).
46. F. Abdoul-Rasoul, A. Ledwith, and Y. Yagci, *Polymer 19*: 31 (1970).
47. J. V. Crivello and J. H. W. Lam, *Macromolecules 10*: 1307 (1977).
48. J. V. Crivello, *Ring-Opening Polymerization* (D. J. Brunelle, ed.) Hanser Publishers, Munich, 1993, p. 157.
49. S. Penczek, J. Wieteszka, and P. Kubisa, *Makromol. Chem. 37*: 225 (1966).
50. S. Penczek and H. J. Chapiro, *J. Chem. Phys. 59*: 696 (1962).
51. S. Okamura and H. Hayashi, *Makromol. Chem. 47*: 230 (1961).
52. G. Kiss, K. Kiss, A. J. Kovacs, and J. C. Wittmann, *J. Appl. Polym. Sci. 26*: 2485 (1981).
53. H. Sawada, *Thermodynamic of Polymerization*, Marcel Dekker Inc., New York, 1976, p. 131.
54. S. Penczek, *Comprehensive Polymer Science* (G. Allen and J. C. Bevington, eds.), Pergamon Press, Oxford, 1989, Vol. 3, p. 719.
55. K. J. Ivin and T. Saegusa, *Ring-Opening Polymerization* (K. J. Ivin and T. Saegusa, eds.), Elsevier Applied Science Publishers, London, 1984, p. 1.
56. T. G. Croucher and R. E. Welton, *Polymer 17*: 205 (1976).
57. S. Penczek and K. Matyjaszewski, *J. Polym. Sci. Polym. Symp. 56*: 255 (1976).

58. F. S. Dainton and K. J. Ivin, *Trans. Faraday Soc. 46*: 331 (1950).
59. F. S. Dainton and K. J. Ivin, *Q. Rev. 12*: 81 (1958).
60. A. Greenberg and J. F. Liebman, *Strained Organic Molecules—Organic Chemistry, A Series of Monographs*, Academic Press Inc., New York, 1978, Vol. 38, p. 13.
61. H. R. Allcock, *Heteroatom Ring Systems and Polymers*, Academic Press, New York, 1967, p. 85.
62. J. D. Cox, *Tetrahedron 19*: 1175 (1963).
63. S. Coffey, ed., *Rodd's Chemistry of Carbon Compounds*, Elsevier Scientific Publishing Company, Amsterdam, 1973, Vol. IV, p. 2.
64. H. L. Finke, D. W. Scott, M. E. Gross, and G. Waddington, *J. Am. Chem. Soc. 78*: 5469 (1956).
65. K. Hiraoka and P. Kebarle, *J. Am. Chem. Soc. 99*: 360 (1977).
66. K. Hiraoka and P. Kebarle, *J. Am. Chem. Soc. 99*: 366 (1977).
67. S. Penczek and P. Kubisa, Cationic ring-opening polymerization, *Ring-Opening Polymerization* (D. J. Brunelle, ed.), Hanser Publishers, Munich, 1993, p. 30.
68. T. Saegusa and S. Matsumoto, *J. Polym. Sci. A1 6*: 1599 (1968).
69. K. Brzezińska, W. Chwiałkowska, P. Kubisa, K. Matyjaszewski, and S. Penczek, *Makromol. Chem. 178*: 2491 (1977).
70. K. Matyjaszewski and S. Penczek, *Makromol. Chem. 182*: 1735 (1981).
71. A. Ledwith and D. C. Sherrington, *Adv. Polym. Sci. 19*: 1 (1975).
72. K. Matyjaszewski, S. Słomkowski, and S. Penczek, *J. Polym. Sci. Polym. Chem. Ed. 17*: 2413 (1979).
73. K. Matyjaszewski, S. Słomkowski, and S. Penczek, *J. Polym. Sci. Polym. Chem. Ed. 17*: 69 (1979).
74. K. Brzezińska, K. Matyjaszewski, and S. Penczek, *Makromol. Chem. 179*: 2387 (1978).
75. M. Szwarc, *Carbanions, Living Polymer and Electron Transfer Processes*, Interscience Publishers, New York, 1968, p. 407.
76. S. Penczek and R. Szymański, *Polym. J. 12*: 617 (1980).
77. A. M. Buyle, K. Matyjaszewski, and S. Penczek, *Macromolecules 20*: 269 (1977).
78. K. Matyjaszewski, *J. Macromol. Sci. Rev. Macromol. Chem. Phys. C26*: 1 (1986).
79. S. Kobayashi and T. Saegusa, *Encyclopedia of Polymer Science and Technology* (H. F. Mark, N. M. Bikales, G. G. Overberger, and G. Menges, eds.), Wiley-Interscience, 1986, 2nd ed., Vol. 4, p. 525.
80. T. Baran, K. Brzezińska, K. Matyjaszewski, and S. Penczek, *Makromol. Chem. 184*: 2497 (1983).
81. T. Saegusa, S. Kobayashi, and M. Ischiguro, *Macromolecules 7*: 958 (1974).
82. T. Saegusa, S. Kobayashi, and A. Yamada, *Makromol. Chem. 177*: 2271 (1976).
83. N. Kuhn, *Kolloid Z. 68*: 2 (1934).
84. P. J. Flory, *Statistical Mechanics of Chain Molecules*, Interscience, New York, 1969, Chapter 11.

85. H. Jacobson and W. H. Stockmayer, *J. Chem. Phys. 18*: 160 (1950).
86. J. F. Brown and G. M. J. Slusarczuk, *J. Am. Chem. Soc. 79*: 897 (1957).
87. J. A. Semlyen and P. V. Wright, *Polymer 10*: 643 (1965).
88. J. Chojnowski, P. Ścibiorek, and M. Kowalski, *Makromol. Chem. 178*: 1351 (1977).
89. J. M. Andrews and J. A. Semlyen, *Polymer 13*: 142 (1972).
90. G. Pruckmayr and T. K. Wu, *Macromolecules 10*: 877 (1977).
91. D. V. Worsfold and A. M. Eastham, *J. Am. Chem. Soc. 79*: 900 (1957).
92. S. Kobayashi, K. Morikawa, and T. Saegusa, *Polym. J. 11*: 405 (1979).
93. S. Penczek and S. Słomkowski, *Comprehensive Polymer Science* (G. Allen and J. C. Bevington, eds.), Pergamon Press, Oxford, 1989, Vol. III., Part I, p. 725.
94. C. Rentsch and R. C. Schulz, *Makromol. Chem. 178*: 2535 (1977).
95. Y. Kawakami and Y. Yamashita, *Macromolecules 12*: 399 (1979).
96. R. Szymański, P. Kubisa, and S. Penczek, *Macromolecules 16*: 1000 (1983).
97. M. Szwarc, *Nature* (*Lond.*) *176*: 1168 (1956).
98. F. R. Jones and P. H. Plesch, *Chem. Commun. 1969*: 1230, 1231 (1969).
99. E. A. Arnett, *Progress in Physical Organic Chemistry*, Interscience Publishers, New York, 1963, Vol. I, p. 233.
100. K. Matyjaszewski, *J. Macromol. Sci. Rev. Macromol. Chem. Phys. C26*: 1 (1986).
101. H. D. Hermann, E. Fischer, and K. Weissermell, *Makromol. Chem. 90*: 1 (1966).
102. S. Kobayashi, K. Morikawa, and T. Saegusa, *Polym. J. 11*: 405 (1979).
103. R. J. Kern, *J. Org. Chem. 33*: 388 (1968).
104. K. Ito, N. Usami, and Y. Yamashita, *Polym. J. 11*: 171 (1979).
105. J. Dale and K. Daasvatn, *J. Chem. Soc. Chem. Commun. 1976*: 295 (1976).
106. S. Penczek, P. Kubisa, K. Matyjaszewski, and R. Szymański, *Cationic Polymerization and Related Processes* (E. J. Goethals, ed.), Academic Press, New York, 1984, p. 139.
107. K. Brzezińska, R. Szymański, P. Kubisa, and S. Penczek, *Makromol. Chem. Rapid Commun. 7*: 1 (1986).
108. S. Penczek, P. Kubisa, and R. Szymański, *Makromol. Chem. Macromol. Symp. 3*: 203 (1986).
109. M. Wojtania, P. Kubisa, and S. Penczek, *Makromol. Chem. Macromol. Symp. 6*: 201 (1988).
110. P. Kubisa, *Makromol. Chem. Macromol. Symp. 13/14*: 203 (1988).
111. M. Bednarek, P. Kubisa, and S. Penczek, *Makromol. Chem. Suppl. 15*: 49 (1989).
112. T. Biedroń, M. Bednarek, P. Kubisa, and S. Penczek, *Makromol. Chem. Macromol. Symp. 32*: 155 (1990).
113. M. Bednarek, T. Biedroń, P. Kubisa, and S. Penczek, *Makromol. Chem. Macromol. Symp. 42/43*: 475 (1991).
114. T. Biedroń, P. Kubisa, and S. Penczek, *J. Polym. Sci. Polym. Chem. Ed. 29*: 619 (1991).
115. S. Penczek, *ACS Polymer Prep. 29-2*: 38 (1988).

116. F. Lagarde, L. Reibel, and E. Franta, *Makromol. Chem. 193*: 1087 (1992).
117. J. Grazulevicius and J. Kublickas, *Eur. Polym. J. 27*: 1411 (1991).
118. A. G. Rajendran and H. J. Timpe, *J. Polym. Sci. Polym Chem. Ed. 29*: 1491 (1991).
119. J. B. Rose, *J. Chem. Soc. 1956*: 542 (1956).
120. M. R. Bucquoye and E. J. Goethals, *Makromol. Chem. 179*: 1681 (1978).
121. Y. Arimatsu, *J. Polym. Sci. 4*: 728 (1966).
122. P. E. Black and D. J. Worsfold, *Can. J. Chem. 54*: 3325 (1976).
123. G. P. Aleksiuk, V. V. Shamanin, A. F. Podolsky, L. V. Alferova, and V. A. Kropachev, *Polym. J. 13*: 23 (1981).
124. S. Penczek and P. Kubisa, *Makromol. Chem. 130*: 186 (1969).
125. M. D. Dreyfuss, J. Westphal, and P. Dreyfuss, *Macromolecules 1*: 437 (1968).
126. J. Guzman, L. Garrido, and E. Riande, *Macromolecules 17*: 2005 (1984).
127. P. Dreyfuss, *Polytetrahydrofuran*, Gordon and Breach, New York, 1982.
128. A. Stolarczyk, P. Kubisa, and S. Penczek, *Bull. Acad. Pol. Sci. Ser. Chem. 22*: 431 (1974).
129. P. Kubisa and S. Penczek, *Makromol. Chem. 179*: 445 (1978).
130. W. Chwiałkowska, P. Kubisa, and S. Penczek, *Makromol. Chem. 183*: 753 (1982).
131. S. Penczek, P. Kubisa, and K. Matyjaszewski, Cationic ring-opening polymerization—synthetic applications, *Adv. Polym. Sci. 68/69*: 41 (1985).
132. R. Szymański and S. Penczek, *Makromol. Chem. 183*: 1587 (1982).
133. M. Zieliński, S. Słomkowski, K. Matyjaszewski, P. Kubisa, J. Chojnowski, and S. Penczek, *Makromol. Chem. 181*: 1469 (1980).
134. P. H. Plesch and P. H. Westermann, *J. Polym. Sci. C16*: 3837 (1968).
135. C. Rentsch and R. C. Schulz, *Makromol. Chem. 178*: 2535 (1977).
136. P. Kubisa and S. Penczek, *Makromol. Chem. 180*: 1821 (1979).
137. H. Krüger, H. Pasch, H. Much, V. Gancheva, and R. Velichkova, *Makromol. Chem. 193*: 1975 (1992).
138. E. Franta and L. Reibel, *Makromol. Chem. Macromol. Symp. 47*: 141 (1991).
139. V. Jaacks, *Adv. Chem. Series 91*: 371 (1969).
140. D. Fleischer and R. C. Schulz, *Makromol. Chem. 176*: 677 (1975).
141. M. Kucera and J. Pichler, *Polymer 5*: 371 (1964).
142. N. Lu, G. L. Collins and N. L. Yang, *Makromol. Chem. Macromol. Symp. 42/43*: 425 (1991).
143. I. Ishigaki, Y. Morita, K. Nishimura, and A. Ito, *J. Appl. Polym. Sci. 18*: 1927 (1974).
144. M. Hasegawa, K. Yamamoto, T. Shiwaku, and T. Hashimoto, *Macromolecules 23*: 2629 (1990).
145. V. Jaacks, *Makromol. Chem. 99*: 300 (1966).
146. S. Okamura, K. Hayashi, and Y. Kitanishi, *J. Polym. Sci. 58*: 925 (1962).
147. B. Wunderlich, *Adv. Polym. Sci. 5*: 568 (1968).
148. M. Okada, *Prog. Polym. Sci. 16*: 1027 (1991).
149. T. Uryu, *Makromol. Chem. 178*: 1929 (1977).

150. F. Michael, O. E. Broddle, and K. Reinkins, *Liebig's Ann. 1974*: 124 (1974).
151. H. K. Hall, Jr., F. De Blauwe, L. J. Carr, V. S. Rao, and G. S. Reddy, *J. Polym. Sci. Polym. Symp. 56*: 101 (1976).
152. J. Kops and C. Schuerch, *J. Polym. Sci. C11*: 119 (1965).
153. C. C. Yu, and C. Schuerch, *Polym. Lett. 1*: 163 (1963).
154. C. Schuerch, *Encyclopedia of Polymer Science and Technology* (H. F. Mark and N. M. Bikales, eds.), Interscience Publishers, New York, 1976, Suppl., Vol. I, p. 510.
155. E. Gipstein, E. Wellisch, and O. J. Sweeting, *J. Polym. Sci. 4*: 995 (1966).
156. R. Simmons, E. J. Goethals, and N. Spassky, *Makromol. Chem. 179*: 1851 (1979).
157. D. Van Ooteghem and E. J. Goethals, *Makromol. Chem. 175*: 1513 (1974).
158. R. Simmons and E. J. Goethals, *Makromol. Chem. 179*: 1689 (1978).
159. W. M. Van Craeynest and E. J. Goethals, *Makromol. Chem. 179*: 2613 (1978).
160. E. J. Goethals and W. Drijvers, *Makromol. Chem. 165*: 329 (1973).
161. E. J. Goethals, W. Drijvers, D. Van Ooteghem, and A. M. Buyle, *J. Macromol. Sci. Chem. A7*: 1375 (1973).
162. W. Kern and E. Brenneisen, *J. Prakt. Chem. 159*: 193 (1941).
163. C. R. Dick and G. E. Ham, *J. Macromol. Sci. A4*: 1301 (1970).
164. G. M. Lukovkin, V. S. Pshezhetsky, and A. Murtazaeva, *Eur. Polym. J. 9*: 559 (1973).
165. G. D. Jones, D. C. McWiliams, and N. A. Braxter, *J. Org. Chem. 30*: 1994 (1965).
166. W. A. Ponomarenko, W. A. Tshekulaeva, I. A. Bogatsheva, and N. B. Ignatenko, *Vysokomol. Soed. B. 23*: 230 (1981).
167. E. J. Goethals, E. M. Schacht, P. Bruggeman, and P. Bossaer, *ACS Symp. Ser. 59*: 1 (1977).
168. A. Munir and E. J. Goethals, *J. Polym. Sci. Polym. Lett. Ed. 19*: 1985 (1981).
169. E. J. Goethals and M. Vlegels, *Polym. Bull. 4*: 521 (1981).
170. E. J. Goethals, M. Van de Velde, and A. Munir *Cationic Polymerization and Related Processes* (E. J. Goethals, ed.), Academic Press, London, 1984, p. 387.
171. M. Van de Velde and E. J. Goethals, *Makromol. Chem. Macromol. Symp. 6*: 271 (1986).
172. E. J. Goethals, R. Deveux, and L. Vandenberghe, *Makromol. Chem. Rapid Commun. 3*: 515 (1982).
173. G. E. Ham, *Polymeric Amines and Ammonium Salts* (E. J. Goethals, ed.), Pergamon Press, Oxford, 1980, p. 1.
174. E. H. Schacht and E. J. Goethals, *Makromol. Chem. 175*: 3447 (1974).
175. T. Saegusa, Y. Nagura, and S. Kobayashi, *Macromolecules 6*: 495 (1973).
176. E. M. Schacht, P. Bossaer, and E. J. Goethals, *Polym. J. 9*: 329 (1977).
177. E. J. Goethals and E. M. Schacht, *J. Polym. Sci. Polym. Lett. Ed. 11*: 497 (1973).

178. E. F. Razvodovskii, A. A. Berlin, A. Nekrasow, A. V. Puschayeva, Z. M. Puschova, and N. S. Enikolopian, *Vysokomol. Soed. A15*: 2219 (1973).
179. M. Litt, A. Levy, and J. Herz, *J. Macromol. Sci. Chem. 9*: 703 (1975).
180. T. Saegusa, *Makromol. Chem. 175*: 1199 (1974).
181. S. Kobayashi, *Prog. Polym. Sci. 15*: 751 (1990).
182. S. Kobayashi, H. Uyama, Y. Narita, and I. Ishiyama, *Macromolecules 25*: 3232 (1992).
183. M. Miyamoto, K. Aoi, S. Yamaga, and T. Saegusa, *Macromolecules 25*: 5111 (1992).
184. A. Hofman, R. Szymański, S. Słomkowski, and S. Penczek, *Makromol. Chem. 185*: 655 (1984).
185. H. R. Kricheldorf, J. M. Jonte, and R. Dunsig, *Makromol. Chem. 187*: 771 (1986).
186. A. Hofman, S. Słomkowski, and S. Penczek, *Makromol. Chem. 188*: 2027 (1987).
187. H. R. Kricheldorf, R. Dunsig, and A. Serra, *Macromolecules 20*: 2050 (1987).
188. A. K. Khomyakow and E. B. Lyudvig, *Dokl. Akad. Nauk SSSR 201*: 877 (1971).
189. H. Cherdron, H. Ohse, and F. Korte, *Makromol. Chem. 56*: 179, 187 (1962).
190. K. Ito, M. Tomida, and Y. Yamashita, *Polym. Bull. 1*: 569 (1979).
191. Y. Okamoto, *Makromol. Chem. Macromol. Symp. 42/43*: 117 (1991).
192. B. A. Rosenberg, *Makromol. Chem. Macromol. Symp. 60*: 177 (1992).
193. I. Tajima, M. Okada, and H. Sumitomo, *Macromolecules 14*: 1180 (1981).
194. M. Okada and H. Sumitomo, *Makromol. Chem. Suppl. 14*: 29 (1985).
195. K. Tajima, M. Okada, and H. Sumitomo, *J. Am. Chem. Soc. 103*: 4096 (1981).
196. M. Vert, P. Christel, F. Chabot, and J. Leroy, *Macromolecular Biomaterials* (G. W. Hastings and P. Ducheyne, eds.), CRC Press Inc., Boca Raton, FL, 1984, p. 119.
197. G. S. Sania, E. B. Lyudvig, and E. S. Petrov, *Vysokomol. Soed. B19*: 121 (1977).
198. O. K. Gilding and A. M. Reed, *Polymer 20*: 1459 (1979).
199. V. W. Dittrich and R. C. Schulz, *Angew. Makromol. Chem. 15*: 109 (1971).
200. F. Chabot, M. Vert, S. Chapelle, and P. Gragner, *Polymer 24*: 53 (1983).
201. H. R. Kricheldorf, B. Weegen-Schulz, and J. Jenssen, *Makromol. Chem. Macromol. Symp. 60*: 119 (1992).
202. H. R. Kricheldorf, R. Dunsig, and A. Serra, *Makromol. Chem. 188*: 2453 (1987).
203. H. R. Kricheldorf and J. Jenssen, *J. Macromol. Sci. Chem. 26*: 631 (1989).
204. T. Ariga, T. Takata, and T. Endo, *J. Polymer Sci. A31*: 581 (1993).
205. A.-Ch. Albertsson and M. Sjoeling, *J. Macromol. Sci. Pure Appl. Chem. A29*: 43 (1992).
206. W. J. Bailey and R. L. Sun, *ACS Polym. Prep. 13*: 400 (1972).
207. F. R. Brady Jr., *J. Macromol. Sci. Rev. Macromol. Chem. Phys. C32*: 135 (1992).

208. K. Matyjaszewski, *J. Polym. Sci. 22*: 29 (1984).
209. S. Chikaoka, T. Takata, and T. Endo, *Macromolecules 24*: 6557 (1991).
210. H. K. Hall Jr. and Y. Yokoyama, *Polym. Bull. 2*: 281 (1980).
211. Y. Yokoyama, A. Buyle-Padias, E. A. Bratoeff, and J. H. K. Hall Jr, *Macromolecules 15*: 11 (1982).
212. T. Takata and T. Endo, *Macromolecules 21*: 2314 (1988).
213. H. Sekiguchi and B. Coutin, *Handbook of Polymer Synthesis* (H. R. Kricheldorf, ed.), Marcel Dekker Inc., 1992, Part A, p. 807.
214. J. Sebenda, *Makromol. Chem. Macromol. Symp. 13/14*: 97 (1988).
215. G. Bertalan, J. Rusznak, P. Anna, M. Ivicz-Boros, and G. Maroal, *Polym. Bull. 19*: 539 (1988).
216. R. Puffr and J. Sebenda, *Makromol. Chem. Macromol. Symp. 3*: 249 (1986).
217. B. Masar and J. Sebenda, *Collect. Czech. Chem. Commun. 40*: 93 (1975).
218. P. H. Hermans, D. Heikens, and P. F. Van Velden, *J. Polym. Sci. 30*: 67 (1958).
219. J. Sebenda, *Makromol. Chem. Macromol. Symp. 13/14*: 97 (1988).
220. G. Łapienis and S. Penczek, *Macromolecules 7*: 166 (1974).
221. G. Łapienis and S. Penczek, *Macromolecules 10*: 301 (1977).
222. G. Łapienis and S. Penczek, *Ring-Opening Polymerization* (K. J. Ivin and T. Saegusa, eds.), Elsevier, New York, 1984 Vol. 2, p. 919.
223. T. Shimadzu, T. Hakozaki, T. Kagiya, and F. Fukui, *J. Polym. Sci. B3*: 871 (1965).
224. T. Kawakami, K. Miyata, and Y. Yamashita, *Polym. J. 11*: 175 (1979).
225. A. N. Pudovik and T. V. Evstafeev, *Yysokomol. Soed. 6*: 2139 (1964).
226. H. J. Harwood and N. K. Palel, *Macromolecules 1*: 233 (1968).
227. S. Kobayashi, M. Y. Huang, and T. Saegusa, *Polym. Bull. 4*: 185 (1981).
228. S. Kobayashi, M. Suzuki, and T. Saegusa, *Polym. Bull. 4*: 315 (1981).
229. J. Kadokawa and S. Kobayashi, *Polym. Prep. Jpn.* (*Engl. Ed.*) *41*: 889.
230. J. Liebig, *Justus Liebig Ann. Chem. 11*: 139 (1834).
231. H. M. Stokes, *J. Am. Chem. Soc. 18*: 629 (1896).
232. H. R. Allcock and R. L. Kugel, *J. Am. Chem. Soc. 87*: 4216 (1965).
233. H. G. Horn and F. Kolmann, *Makromol. Chem. 183*: 1833 (1982).
234. H. R. Allcock, J. E. Gardner, and K. M. Smeltz, *Macromolecules 8*: 36 (1975).
235. H. R. Allcock, *Angew. Chem. Int. Ed. Engl. 16*: 147 (1977).
236. R. E. Singler, N. S. Schneider, and G. L. Hagnauer, *Polym. Eng. Sci. 15*: 321 (1975).
237. I. Manners, G. H. Riding, J. A. Dodge, and H. R. Allcock, *J. Am. Chem. Soc. 111*: 3067 (1989).
238. H. R. Allcock and M. L. Turner, *Macromolecules 26*: 3 (1993).
239. H. R. Allcock, J. A. Dodge, I. Manners, and G. H. Riding, *J. Am. Chem. Soc. 113*: 9596 (1991).
240. W. Patnode and D. F. Wilcock, *J. Am. Chem. Soc. 68*: 358 (1946).
241. E. V. Kogan, A. G. Ivanova, V. O. Reikksfeld, N. I. Smirnov, and V. N. Gruber, *Polym. Sci. USSR, Engl. Transl. 5*: 249 (1963).
242. T. G. Kendrick, *J. Chem. Soc. 1965*: 2027 (1965).

243. K. A. Andrianow and S. E. Yakushkina, *Vysokomol. Soed. 1*: 613 (1959).
244. J. Chojnowski, M. Mazurek, M. Ścibiorek, and L. Wilczek, *Makromol. Chem. 175*: 3299 (1974).
245. G. Sauvet, J. J. Lebrun, and P. Sigwalt, *Cationic Polymerization and Related Processes* (E. J. Goethals, ed.), Academic Press, London, 1984, p. 237.
246. L. Wilczek and J. Chojnowski, *Macromolecules 14*: 9 (1981).
247. L. Wilczek and J. Chojnowski, *Makromol. Chem. 184*: 77 (1983).
248. D. Souverein, A. Laborgne, G. Sauvet, and P. Sigwalt, *Eur. Polym. J. 16*: 861 (1980).
249. S. Rubinsztajn, M. Cypryk, and J. Chojnowski, *Macromolecules 26*: 5389 (1993).
250. M. Cypryk, S. Rubinsztajn, and J. Chojnowski, *J. Organomet. Chem. 91*: 446 (1993).
251. P. Sigwalt and V. Stannett, *Makromol. Chem. Macromol. Symp. 32*: 217 (1990).
252. D. M. Naylor, V. Stannett, A. Deffieux, and P. Sigwalt, *Polymer 31*: 954 (1990).
253. D. M. Naylor, V. Stannett, A. Deffieux, and P. Sigwalt, *Polymer 32*: 1084 (1991).
254. S. J. Clarson, J. E. Mark, and J. A. Semlyen, *Polym. Commun 28*: 151 (1987).
255. L. C. De Bolt and J. E. Mark, *Macromolecules 20*: 2369 (1987).
256. L. Chojnowski, J. Kurjata, and S. Rubinsztajn, *Makromol. Chem. Rapid Commun. 9*: 469 (1988).
257. P. Rempp, *Makromol. Chem. Macromol. Symp. 60*: 209 (1992).
258. R. I. Quirk, *Polym. Int. 27*: 359 (1992).
259. S. Penczek, P. Kubisa, and R. Szymański, *Makromol. Chem. Rapid Commun. 12*: 77 (1991).
260. E. J. Goethals, *Makromol. Chem. Macromol. Symp. 42/42*: 51 (1991).
261. B. A. Rosenberg, T. I. Ponomarieva, L. D. Narkevitch, and N. S. Enikolopyan, *Dokl. Akad. Nauk. SSSR 175*: 365 (1967)
262. W. Stix and W. Heitz, *Makromol. Chem. 180*: 1367 (1979).
263. H. J. Kress, and W. Heitz, *Makromol. Chem. Rapid Commun. 2*: 1367 (1979).
264. Y. Yamashita and K. Chiba, *Polym. J. 4*: 200 (1973).
265. E. Franta, L. Reibel, J. Lehmann, and S. Penczek, *J. Polym. Sci. Polym. Symp. Ser. 56*: 139 (1976).
266. S. Smith and A. J. Hubin, *U.S. Patent 3 824 198* (1974).
267. Y. Tezuka and E. J. Goethals, *Eur. Polym. J. 18*: 991 (1982).
268. E. J. Goethals and F. D'Haese, *Br. Polym. J. 20*: 103 (1988).
269. J. Sierra-Vargas, J. G. Zilliox, P. Rempp, and E. Franta, *Polym. Bull. 3*: 83 (1980).
270. E. J. Burgess, A. V. Cunliffe, D. H. Richards, and E. Thompson, *Polymer 19*: 334 (1978).

271. R. Asami, M. Takakaki, K. Kyuda, and E. Asakura *Polym. J. 15*: 139 (1983).
272. J. Sierra-Vargas, P. Mason, G. Beinert, P. Rempp, and E. Franta, *Polym. Bull. 7*: 277 (1982).
273. M. Takaki, R. Asami, and T. Kuwabara, *Polym. Bull. 7*: 521 (1982).
274. E. J. Goethals, R. R. De Clercq, H. C. De Clercq, and P. J. Hartmann, *Makromol. Chem., Macromol. Symp. 47*: 151 (1991).
275. R. R. De Clercq and E. J. Goethals, *Macromolecules 25*: 1109 (1992).
276. E. J. Goethals, *Makromol. Chem. Macromol. Symp. 48/49*: 427 (1991).
277. T. G. Bassiri, A. Levy, and M. Litt, *J. Polym. Sci. Polym. Lett. 5*: 871 (1967).
278. M. Kaiser, *U.S. Patent 4 203 900* (1980).
279. S. Kobayashi, E. Masuda, D. Shoda, and Y. Shimano, *Macromolecules 22*: 2878 (1989).
280. H. Uyama and S. Kobayashi, *Macromolecules 24*: 614 (1991).
281. R. C. Schulz and S. Schwartzenbach, *Makromol. Chem. Macromol. Symp. 13/14*: 495 (1988).
282. T. Saegusa, S. Matsumoto, and Y. Hashimoto, *Macromolecules 3*: 377 (1970).
283. S. Kobayashi, H. Uyama, E. Ihara, and T. Saegusa, *Macromolecules 23*: 1586 (1990).
284. M. Litt, J. Herz, and E. Turi, *Block Copolymers* (S. L. Aggarwal, ed.), Plenum Press, New York, 1970, p. 313.
285. M. Miyamoto, K. Aoi, and T. Saegusa, *Macromolecules 22*: 3540 (1989).
286. J. E. McGrath, *Makromol. Chem. Macromol. Symp. 42/43*: 69 (1991).
287. G. Sinai-Zingde, A. Verma, Q. Lin, A. Brink, J. M. Bronk, H. Marand, and J. E. McGrath, *Makromol. Chem. Macromol. Symp. 42/43*: 329 (1991).
288. P. Dreyfuss and J. P. Kennedy, *J. Polym. Sci. Polym. Symp. 56*: 129 (1976).
289. Y. Ecstein and P. Dreyfuss, *J. Polym. Sci. Polym. Chem. Ed. 18*: 1799 (1980).
290. T. Saegusa, S. Kobayashi, and A. Yamada, *Macromolecules 8*: 390 (1975).
291. T. Saegusa, A. Yamada, and S. Kobayashi, *Polym. J. 11*: 53 (1979).
292. S. Kobayashi, T. Shinano, and T. Saegusa, *Polym. J. 23*: 1307 (1991).
293. Y. Chujo, E. Ihara, S. Kure, K. Suzuki, and T. Saegusa, *Makromol. Chem. Macromol. Symp. 42/43*: 303 (1991).
294. M. Bucquoye and E. J. Goethals, *Eur. Polym. J. 14*: 323 (1978).
295. E. J. Goethals, *J. Polym. Sci. Polym. Symp. 56*: 271 (1976).
296. E. J. Goethals, E. H. Schacht, Y. E. Bogaert, S. I. Ali, and Y. Tezuka, *Polym. J. 12*: 571 (1980).
297. T. Kagiya, T. Matsuda, M. Nakato, and R. Hirata, *J. Macromol. Sci. A6*: 1631 (1972).
298. P. Wittmer, *Multicomponent Polymer System* (N. A. J. Platzer, ed.), American Chemical Society, Washington, DC, 1970, p. 140.
299. A. A. Durgaryan, *Vysokomol. Soed. 8*: 790 (1966).
300. G. Lowry, *J. Polym. Sci. 42*: 463 (1960).
301. K. J. Ivin, *Pure Appl. Chem. 4*: 271 (1962).

302. Y. Yamashita, H. Kasahara, K. Suyama, and M. Okada, *Makromol. Chem. 117*: 242 (1968).
303. P. Kubisa and S. Penczek, *J. Macromol. Sci. Chem. 7*: 1569 (1973).
304. S. Penczek, *Bull. Acad. Pol. Sci. Ser. Sci. Chem. 20*: 437 (1972).
305. J. A. Howell, M. Izu, and K. F. O'Driscoll, *J. Polym. Sci. A1 8*: 699 (1970).
306. R. Szymański, *Prog. Polym. Sci. 17*: 917 (1992).
307. R. Szymański, *Makromol. Chem. 192*: 2943 (1991).
308. M. Okada, Y. Yamashita, and Y. Ischii, *Makromol. Chem. 94*: 181 (1966).
309. Y. Kawakami, J. Suzuki, and Y. Yamashita, *Polym. J. 9*: 519 (1977).
310. M. Okada, S. Ikai, K. Chiba, M. Hirata, and Y. Yamashita, *Polym. J. 1*: 1 (1970).
311. P. K. Bossaer and E. J. Goethals, *Makromol. Chem. 178*: 2983 (1977).
312. T. Kagiya and T. Matsuda, *J. Macromol. Sci.-Chem. A5*: 1265 (1971).
313. S. Penczek, *Makromol. Chem. 172*: 243 (1973).

7

Step-Growth Electrophilic Oligomerization and Polymerization Reactions

VIRGIL PERCEC and DALE H. HILL Case Western Reserve University, Cleveland, Ohio

I. INTRODUCTION

Step-growth electrophilic oligomerization and polymerization reactions are investigated much less than chain-growth electrophilic polymerizations. Nevertheless, they can be used for the generation of many polymers and oligomers that can not be obtained by alternative synthetic methods. This chapter discusses these polymerization reactions and is organized according to the nature and reactivity of the electrophilic propagating species. The presentation of each type of propagating species and the resulting polymer is intended to illustrate the most important recent developments rather than be comprehensive.

Benzylic carbenium ions are the first type of propagating species considered. The types of polymers derived from benzylic carbenium ions includes the following: linear poly(vinylbenzene), polyindanes, polybenzyls, phenol-formaldehyde resins, soluble anthracene polymers, and soluble ladder-type poly(phenylene)s. In addition, three important cyclic oligomers, cyclotriveratrylene, cyclotetraveratrylene, and calixarenes, are obtained from related reactions involving benzylic species. The synthesis of hyperbranched polymers containing cyclotetraveratrylene cores is included within this discussion. Benzylic carbenium ions are also the propagating species in the synthesis of poly(phthalicylidenearylene)s and in a transalkylation reaction which forms a trimer. The next type of propagating species considered is that with a positively charged sulfur atom. This includes sulfonium cations and sulfonylium cations that are used in the synthesis of polysulfides and polysulfones, respectively. A discussion of the participation of acyl cations in the synthesis of poly(aryl ether ketone)s follows. Phenoxenium ions are considered as propagating species in one of the mechanisms suggested for the oxidative polymerization of 2,6-dimethylphenol to yield poly(2,6-dimethylphenylene oxide)s. Cation radical species have been postulated as intermediates in the synthesis of polysulfides, polyaryl ethers, polypyrroles, polythiophenes, polyanilines, poly(*p*-phenylene vinylene), and polyphenylenes. In one case dual propagation pathways involving cation radicals and neutral cations are operative. The electrophilic pathway led to transalkylated polymers. Finally zwitterionic polymerization reactions are considered.

II. BENZYLIC CARBENIUM IONS AS PROPAGATING SPECIES

A. Linear Poly(vinylbenzene)

Benzylic carbenium ions have been implicated as the propagating species in the cation polymerization of linear divinylbenzene derivatives. The synthesis of linear polymers from divinylbenzene (DVB) and derivatives of DVB is significant because bifunctional olefins (and acetylenes, heterocycles etc.) usually give insoluble cross-linked polymers in chain polymerizations. The remarkable exception is provided by cyclopolymerization of diolcfins [1]. A notable exception is the linear polymer with pendant isopropenyl groups obtained via the anionic polymerization of 1,4-diisopropenylbenzene when monomer conversion is kept below 50% [2]. There are two different types of linear polymers obtainable from the cationic polymerization of divinylbenzene derivatives. The first type contains unsaturated units in the main chain [3–7]. The second type contains indane units in the main chain [8–24]. The initial steps leading to both polymers are the same, i.e., protonation of the vinyl group to give a carbenium ion followed by dimerization with the vinyl group of a second molecule. The dimer can then undergo a β-hydrogen elimination reaction giving the α,β-unsaturated group. Alternately, the dimer can undergo an internal Friedel-Crafts alkylation reaction, followed by deprotonation to give the indane structure. The exact product that is formed is determined by which reaction (i.e., β-hydrogen elimination or tne internal Friedel-Crafts alkylation) is faster. This is determined by the substrate structure (including steric and electronic properties), the reaction initiator, and the exact reaction conditions (e.g., temperature, concentration). For example, in the case of the dimerization of styrene with CF_3SO_3H the β-hydrogen elimination (Fig. 1, path A) is much faster than the internal Friedel-Crafts alkylation (path B) [4,5,25].

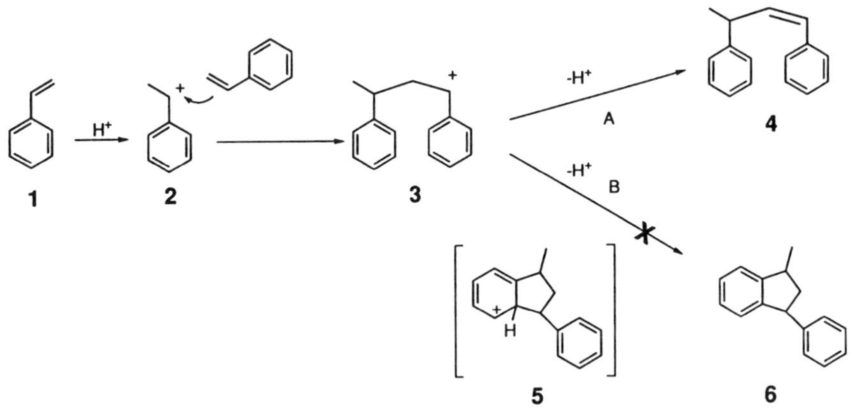

Figure 1 CF_3SO_3H catalyzed dimerization of styrene.

The CF_3SO_3H-initiated polymerization of DVB proceeds in a similar manner with the stepwise dimerization resulting in a product with α,ω-vinyl groups that can participate in further reactions [Eq. (1)] [3].

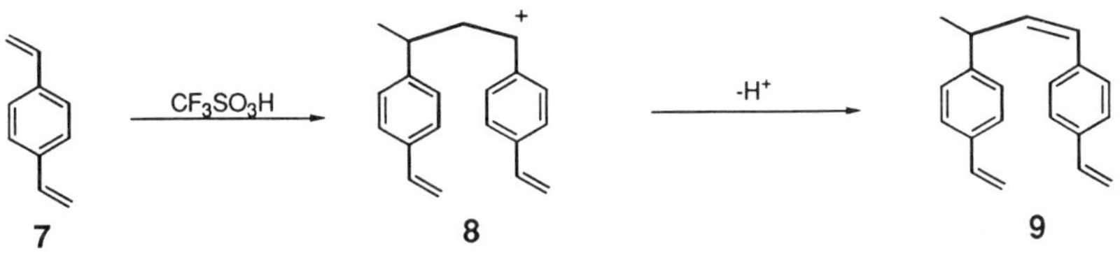

(1)

The structure of the product formed is determined by the reaction conditions and by the structure of the divinylbenzene substrate. Where the DVB has one α-alkyl substituent present, the benzylic carbenium ion that is formed after dimerization is further stabilized by two alkyl groups. As a result of this added stabilization the elimination reaction is slowed and the carbenium ion is long lived enough to participate in an internal Friedel-Crafts alkylation reaction. Thus the indane structure is favored to a greater extent [8].

Linear polymers of DVB, with unsaturated groups in the main chain, have been obtained by the CF_3SO_3H- or $AcClO_4$-initiated polymerization in benzene or CH_2Cl_2 [Eq. (2)] [3].

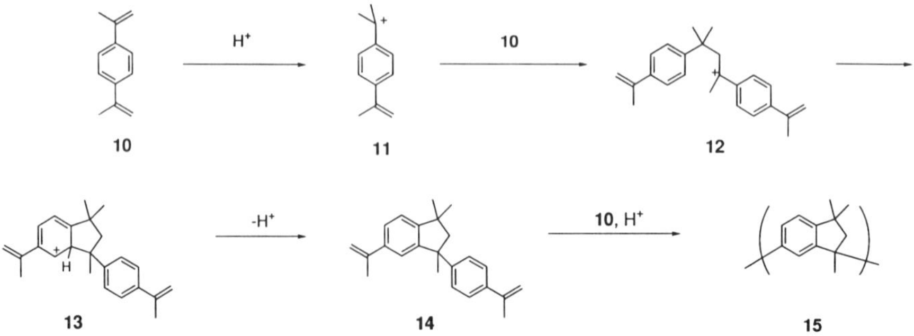

Figure 2 Polymerization of 1,4-diisopropenylbenzene to give poly[1,7(1,3,3-trimethyl)indanylene]. (From Ref. 8.)

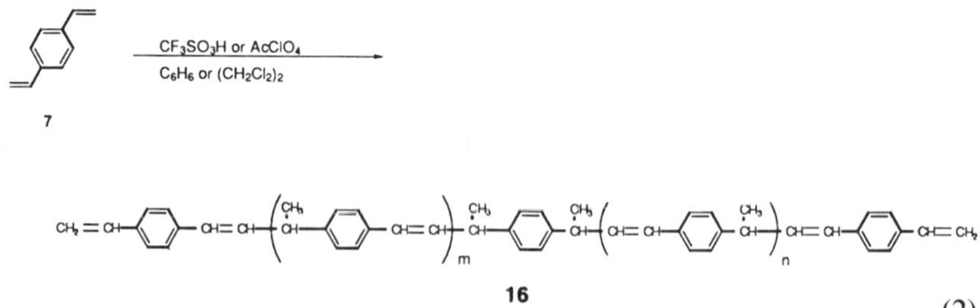

(2)

The synthesis of these linear polymers was first accomplished by using a reaction initially reported for the linear dimerization of styrene with acylium salts (such as $AcClO_4$) or oxoacids (such as CF_3SO_3H) [4,5]. The M_n increased linearly from 0 to 80% conversion. Above 80% conversion a sharp increase of M_n occurred. This dependence is consistent with a step-growth reaction. Polymerizations were performed at low concentrations of monomer (0.10 *M*). This was necessary because the products of polymerization at higher concentrations contained pendant vinyl groups which often participated in cross-linking reactions. The highest molecular weights were obtained by sequentially adding monomer (up to seven times) after 70–80% DVB consumption. Only "dimerization" of vinyl groups occurred due to frequent chain transfer and facile β-proton elimination from the propagating cation. Consequently, polymers with linear unsaturated backbones with terminal vinyl groups were obtained. Linear telechelic poly(DVB)s with substituted styrene end caps were synthesized by the polymerization of DVB in the presence of substituted styrenes [Eq. (3)] [6].

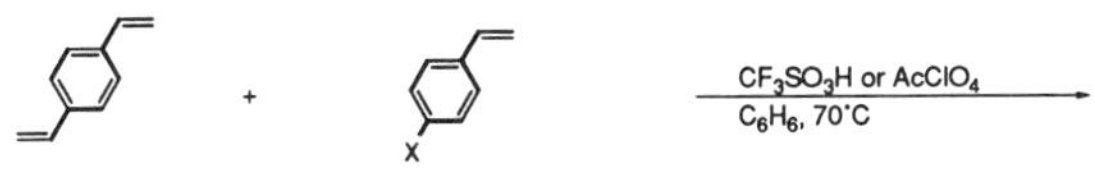

7

17	X
17a	CH_3
17b	$OCOCH_3$
17c	OH
17d	CH_2Cl

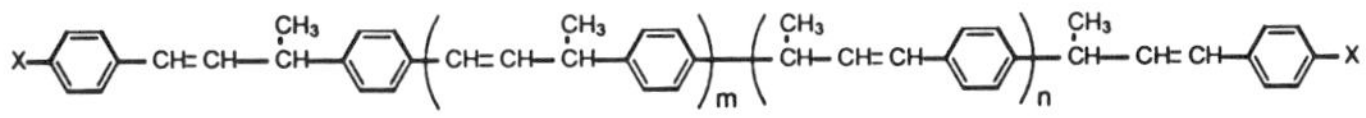

18	X
18a	CH_3
18b	$OCOCH_3$
18c	OH
18d	CH_2Cl

(3)

The linear dimerization of styrene and linear polymerization DVB has been accomplished at lower temperatures (<70° C) and higher monomer concentrations (>0.10 *M*) without side reactions using Pd^{2+} derivatives as cationic initiators [Eq. (4)] [7]. The specific initiators used for this reaction were $Pd(PPh_3)_2(BF_4)_2$, $Pd(Ph_2PCH_2CH_2PPh_2)(BF_4)_2$, and $Pd(2,6\text{-}t\mathrm{Bu}_2C_5H_3N)_2(MeNO_2)_2(BF_4)_2$. All of these derivatives were air-stable solids. The dimerization of styrene could be performed with or without solvent (acetonitrile or acetonitrile-chloroform) (Eq. 4).

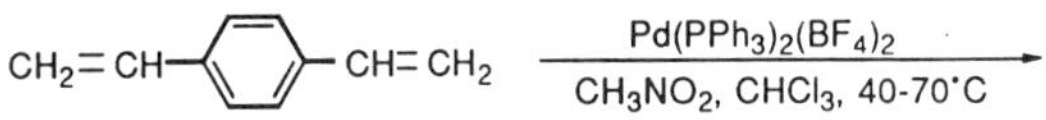

7

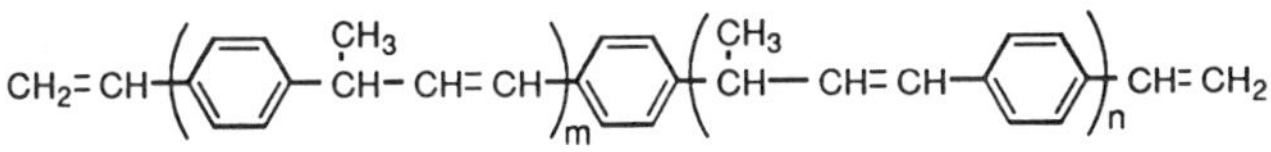

19

(4)

These Pd complexes are unique in several respects. No cross-linked polymer was detected in the reaction with *p*-divinylbenzene. A completely soluble linear polymer was obtained. No indane units were formed from the reaction of styrene with $Pd(PPh_3)_2(BF_4)_2$. α-Methylstyrene is usually much more reactive than styrene in reactions involving carbenium ion intermediates. However, the treatment of a mixture of styrene and α-methylstyrene with $Pd(PPh_3)_2(BF_4)_2$ resulted in the selective dimerization of styrene.

The products of this reaction can be rationalized by considering the reaction mechanism. A possible mechanistic scheme is shown in Fig. 3. The electrophilicity of the carbenium ion formed upon initial dimerization

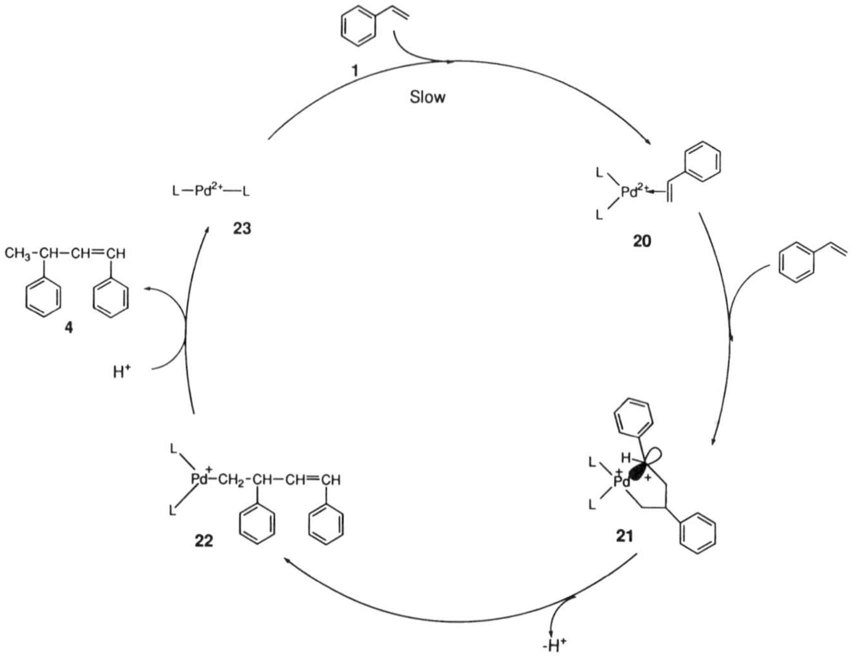

Figure 3 Pd^{2+}-catalyzed linear dimerization of styrene. (From Ref. 7.)

of styrene (**21**) is reduced by the ability of Pd to donate electron density through a d-orbital. Therefore this carbenium ion does not participate in the electrophilic reactions which would give oligomeric or indane products. Instead, elimination of the β-hydrogen results in the formation of an unsaturated group **22**. Reaction of **22** with a proton gives the linear dimer of styrene and regenerates the original catalyst. It is important to note that although the electrophilicity of carbenium ion **21** is reduced, electrophilic attack can still occur if a sufficiently electron-rich olefin is present. For example, polymer was obtained when *p*-methoxystyrene was reacted with the Pd^{2+} catalysts [7].

The linear polymerization of DVB was accomplished with $Pd(PPh_3)_2(BF_4)_2$ in acetonitrile-chloroform solvent at temperatures ranging from 40 to 70° C. The polymer contained structural units derived from both head-tail and head-head coupling. The polymerization proceeded by a step-growth mechanism. The molecular weight increased exponentially as a function of time. This reaction was also used for a one-pot synthesis of a telechelic polymer. This polymer was synthesized by reacting the catalyst with DVB until the desired molecular weight poly(DVB) formed. A telechelic polymer was obtained upon the addition of *p*-(chloromethyl)styrene to endcap the growing polymer chains [Eq. (5)] [7].

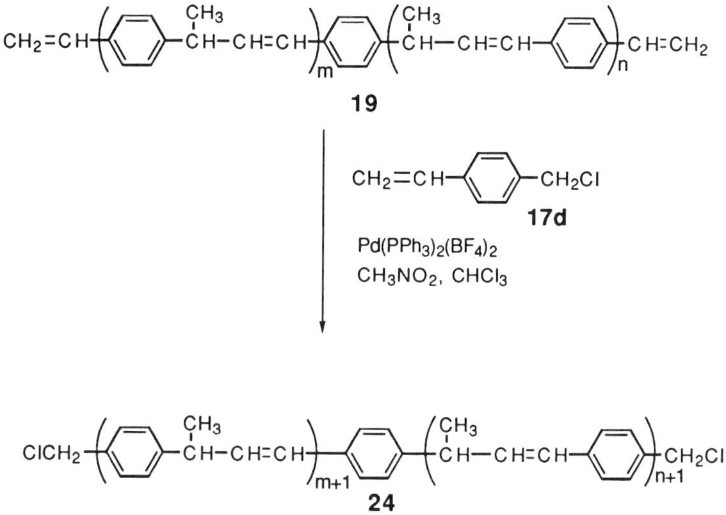

(5)

B. Polyindanes

Polyindanes have been obtained from a variety of monomers which are capable of forming benzyl carbenium ions stabilized with dialkyl groups upon the appropriate initiation. Linear polymers containing indane units in the main chain have been obtained by the BCl_3-initiated polymerization of 1,4-diisopropenylbenzene [Eq. (6)] [8]. Polyindanes with structural unit **15** have also been obtained when difunctional monomers capable of forming α,α-dimethylbenzylic carbenium ions upon treatment with an initiator [8–12] (Fig. 4).

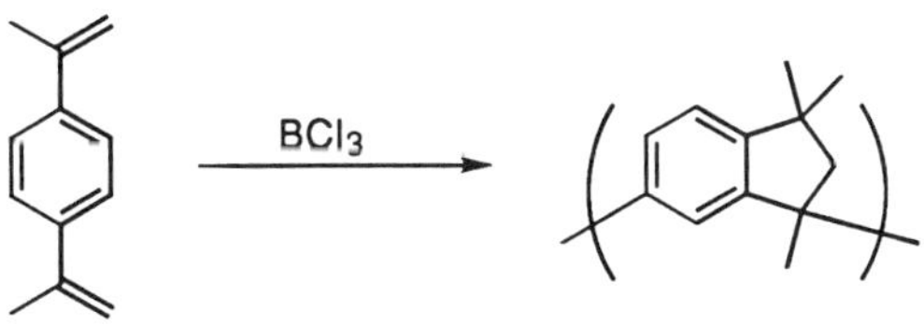

10 **15**

(6)

Common initiators are Bronsted acids (such as H_2SO_4, CF_3CO_2H) and Lewis acids (such as BCl_3 and $AlCl_3$) [8,11]. The formation of linear unsat-

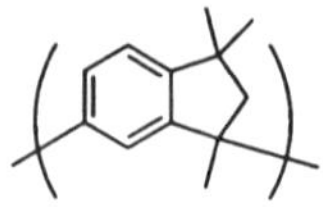

15

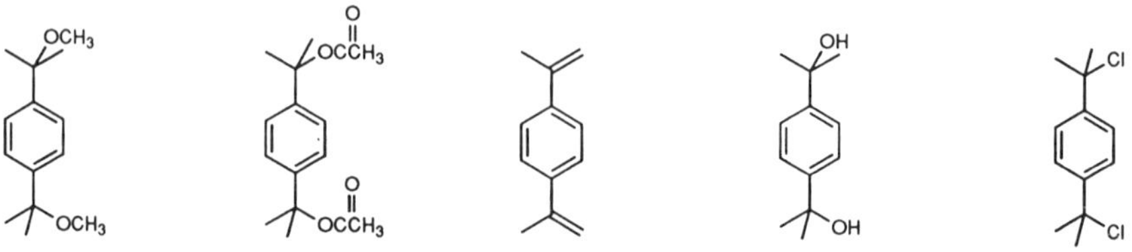

25 **26** **10** **27** **28**

Figure 4 Monomers which can be polymerized to give poly[1,7(1,3,3-trimethyl)-indanyl], **15**. (From Ref. 8.)

urated units along the backbones is avoided by doing the polymerization reactions above the ceiling temperature [8]. Cross-linking reactions are avoided by keeping low monomer concentrations [8]. Thus, the formation of the indane unit is favored.

Cross-linking reactions require an intermolecular Friedel-Crafts reaction between tertiary cations and positions of the phenyl group made less reactive by the presence of bulky groups on the indane structure. Electrophilic attack of benzylic cation **29** (Fig. 5) upon unsubstituted positions of the phenyl group of **15** is sterically inhibited.

The uniformity of the polymer structure obtained is determined by the exact reaction conditions used, with the initiator having a large influence [11]. Polymers with main chain structural uniformities greater than 99% units **15** and less than 1% units **31** (Fig. 6) have been obtained using $AlCl_3$, H_2SO_4, or CF_3CO_2H as initiators [11].

The molecular weight is limited by the formation of unreactive end groups. End groups with structures **33**, **34**, and **36** have been proposed [Eqs. (7) and (8)]. These end groups are formed by the acid (Bronsted or Lewis)-catalyzed cleavage of a linkage between an indanyl group and an adjacent indanyl or phenyl group [11].

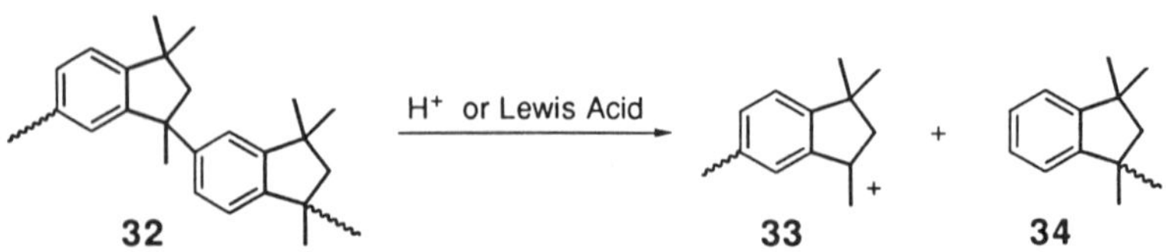

32 **33** **34** (7)

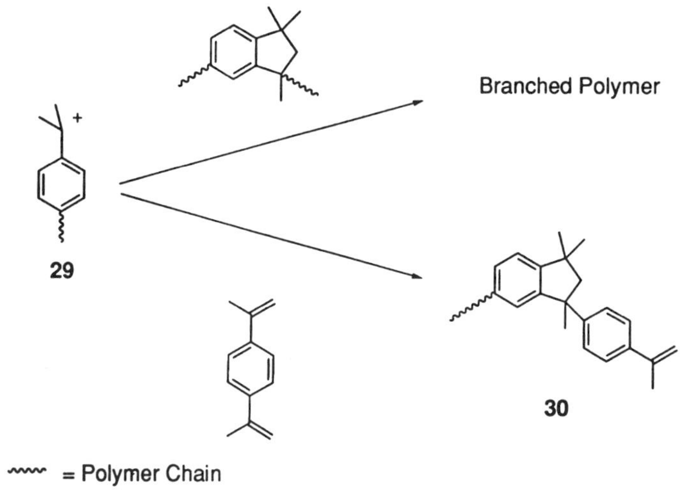

Figure 5 Electrophilic attack of benzylic carbenium ion upon monomer or unsubstituted positions of the phenyl group of the indane polymer unit.

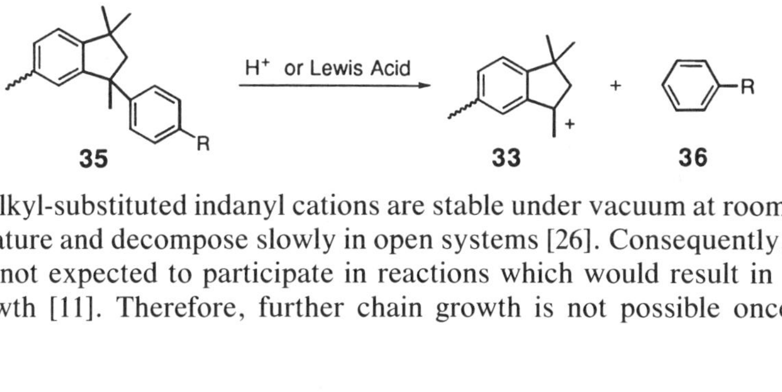

(8)

p-Alkyl-substituted indanyl cations are stable under vacuum at room temperature and decompose slowly in open systems [26]. Consequently, they are not expected to participate in reactions which would result in chain growth [11]. Therefore, further chain growth is not possible once end

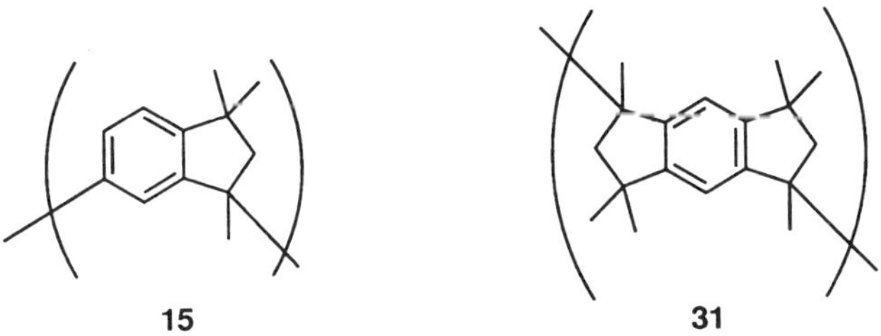

Figure 6 Polyindane structural units obtained in polymerization of 1,4-diisopropylbenzene. (From Ref. 11.)

groups **33**, **34**, or **36** form. Thus it was difficult to obtain polymers with M_w greater than 20,000 [11].

Two general methods have been used in the synthesis of telechelic polymers. One method involved the addition of an endcapping agent to the polymerization reaction. The endcapping agent contained one alkene group that participated in the indane structure-forming reaction and a group which did not participate in the polymerization. This group could be transformed later into a reactive group [Eq. (9)] [13].

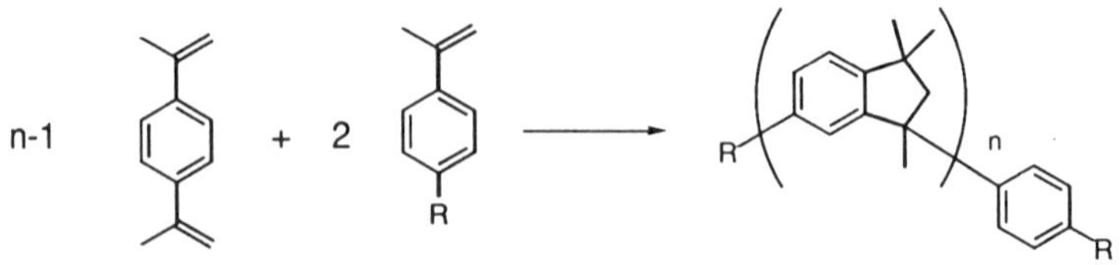

(9)

10

37	R
37a	CH_3
37b	NO_2
37c	CO_2H

38	R
38a	CH_3
38b	NO_2
38c	CO_2H

The R group had a large influence on the structure of the resulting polymer. Both indane and unsaturated units were present (Fig. 7).

The electronic nature of R determined the ease of protonation of **40** and **41**. The reprotonation of the vinyl group of **40** and **41** can allow the thermodynamically favored indane to form. However, when R is an electron-withdrawing group such as NO_2 or CO_2H protonation was inhibited and the polymer had a large proportion of unsaturated units [13].

The other method for synthesizing telechelics was through the step-growth polymerization of two indane monomers containing acid chloride and amine groups [Eq. (10)] [13].

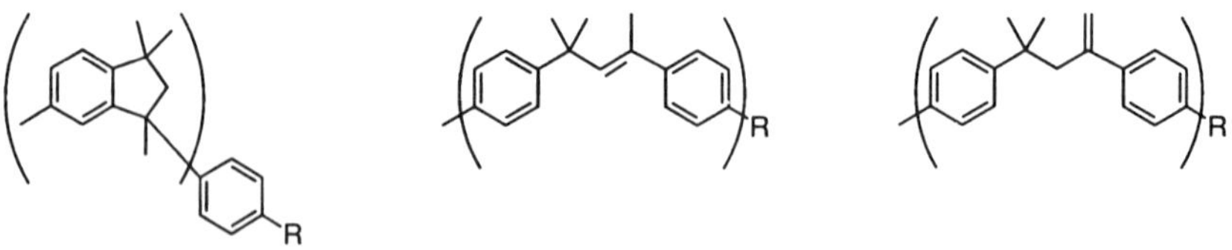

39	R
39a	CH_3
39b	NO_2
39c	CO_2H

40	R
40a	CH_3
40b	NO_2
40c	CO_2H

41	R
41a	CH_3
41b	NO_2
41c	CO_2H

Figure 7 Structural units obtained in the polymerization of **10** with encapping agent **37a–c**. (From Ref. 13.)

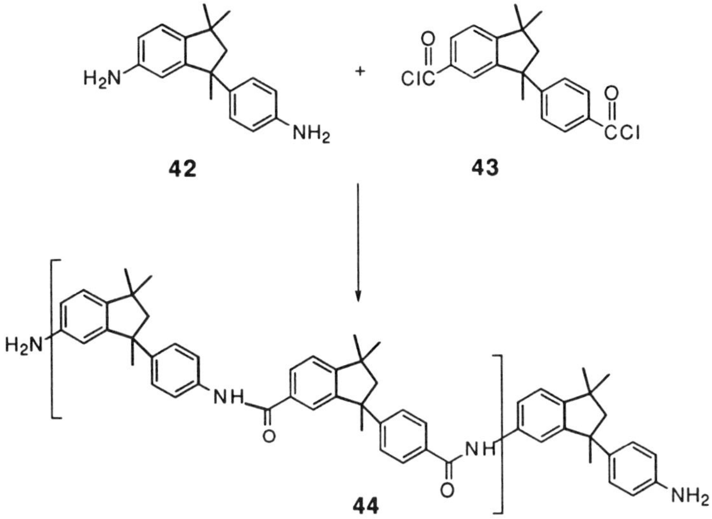

42 + 43 → 44 (10)

Polyindane **15** is very brittle. To obtain less brittle materials, polymers with lower *Tg* values were synthesized [8]. A range of glass transition temperatures (293° C to −10° C) can be obtained by varying the type of alkyl group attached to the indane [16]. For example, the glass transition temperature Tg is lowered from 246° C to 26° C by using a hexyl rather than a methyl substituent on the indane [15]. The thermal stability was affected less dramatically. Two percent weight loss was lowered from 420° C to 340° C when the methyl group has changed to a hexyl group [15]. A series of polymers was made by varying the length of the alkyl group attached to monomer **45** (Fig. 8) [15]. Once again the formation of unreac-

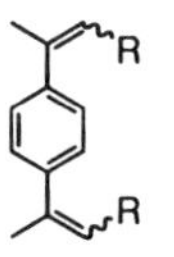

45	R	45	R
45a	H	**45e**	n-C_4H_9
45b	CH_3	**45 f**	n-C_5H_{11}
45c	n-C_2H_5	**45g**	n-C_7H_{15}
45d	n-C_3H_7		

Figure 8 Substituted monomers used for lowering the Tg of the polyindane. (From Refs. 15 and 16.)

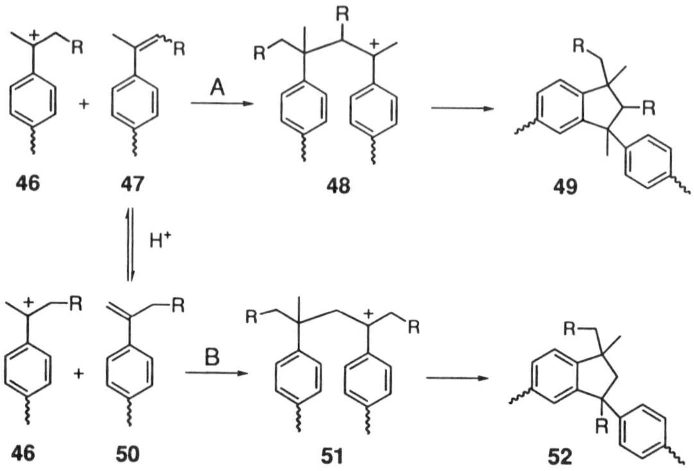

Figure 9 Isomeric polyindane structural units formed as result of the acid-catalyzed isomerization of monomers. (From Ref. 15.)

tive end groups limited the molecular weights (M_n = 2800–4300 g/mol, DP = 10–27) [15].

These polymers contained indane structural units **49** and **52** (Fig. 9). This is because of the facile isomerization of **47** and **50** under the acidic reaction conditions. The relative proportion of **47** to **50** is determined by the equilibrium constant. Thus the two isomeric structural units are formed as a result of the electrophilic addition of **46** to **47** or to **50**.

Cyclohexyl-substituted monomer **53** containing the cyclohexyl group exists in three isomeric forms in the presence of catalytic amounts of acid [Eq. (11)]. The polymer obtained from the cyclohexyl-containing monomer contains **52** exclusively as the structural unit [16]. In this case there is a high degree of steric hindrance in the transition state that would lead to the formation of **49**.

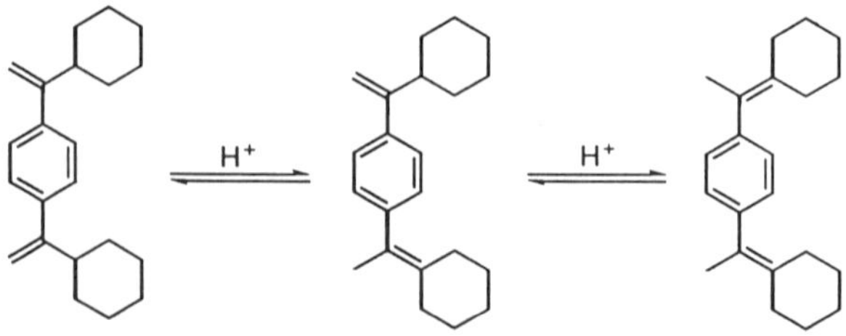

(11)

53 **54** **55**

Polyketones containing indane structural units have been prepared by Friedel-Crafts acylation [Eq. (12)] [17]. However, oligomers and/or insoluble polymers were obtained.

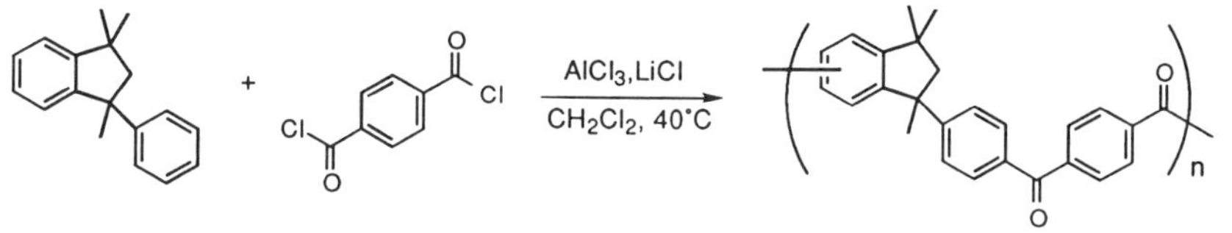

56 **57** **58**

(12)

High molecular weight soluble polymers were not obtained due to cross-linking reactions. These cross-linking reactions involved partial cleavage of the indane ring and multiple acylation reactions [17].

The polymerization of bis(1,1-diphenyl vinyl) monomers **59a–d** gave polymers with indane units **61** and **62** and unsaturated unit **60** in the main chain (Fig. 10) [20]. Indane unit **61** or unsaturated unit **60** could be selectively obtained under the appropriate reaction conditions [20].

Structural unit **60** was favored by acids such as $SnCl_4$, BF_3OEt_2, FSO_3H, $ClSO_3H$, and 70% $HClO_4$. Indane unit **61** was favored by $TiCl_4$,

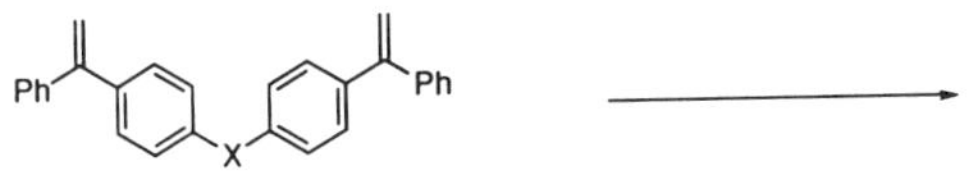

59	X
59a	CH_2
59b	CH_2CH_2
59c	O
59d	S

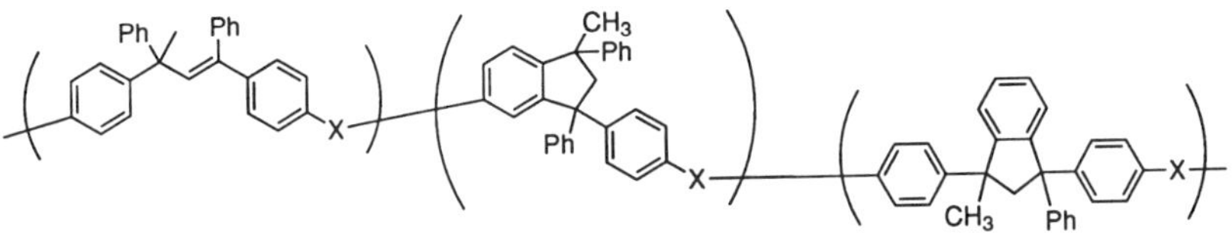

60	X
60a	CH_2
60b	CH_2CH_2
60c	O
60d	S

61	X
61a	CH_2
61b	CH_2CH_2
61c	O
61d	S

62	X
62a	CH_2
62b	CH_2CH_2
62c	O
62d	S

Figure 10 Indane structural units obtained by the polymerization of 1,1-diphenyl vinyl monomers **59a–d**. (From Ref. 20.)

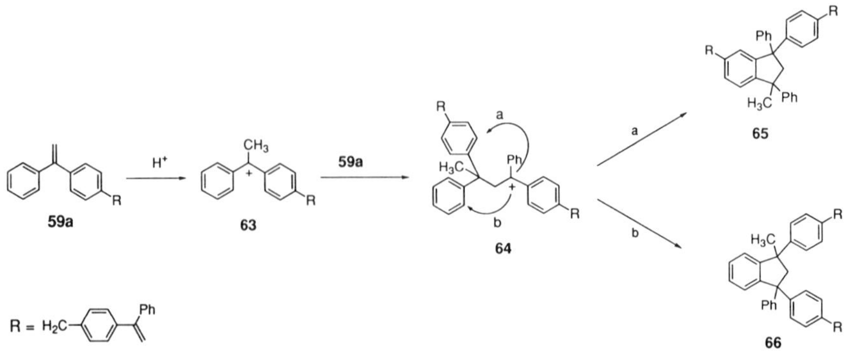

Figure 11 Reaction pathways leading to formation of two types of indane units. (From Ref. 20.)

$SbCl_5$, $FeCl_3$, and $AlCl_3$. The reaction conditions had a large influence on structure when using CF_3SO_3H or H_2SO_4. The unsaturated groups of polymers with no indane units could be converted to the indane structure by treatment with CF_3SO_3H (30° C, 25 hr in CH_2Cl_2) [20]. The reaction pathways leading to the formation of the two types (**61** and **62**) of indane unit are illustrated in Fig. 11.

The thermal stability of polyindanes obtained from **59a–d** varied with the nature of the group *X*. The order of stability for the four *X* groups was $O > CH_2 > CH_2CH_2 > S$. The polymer obtained from the polymerization of **67** had stability equivalent to that obtained from **59c**.

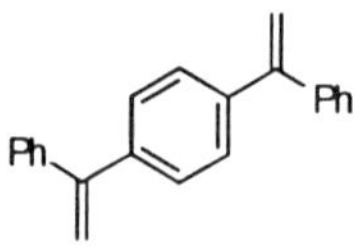

67

Polymers with indane units in the main chain have also been obtained by the photoinitiated cationic polymerization of bis(4-isopropenylphenoxy)alkanes [Eq. (13)] [23,24].

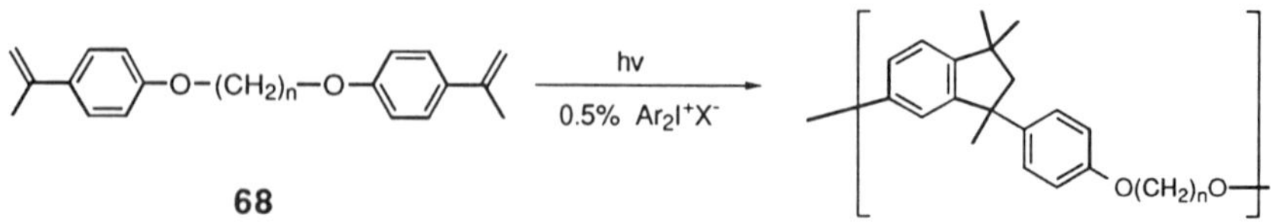

68 **69**

(13)

To solubilize the initiator, an alkoxy substituent was attached to one or both rings of the $Ar_2I^+ X^-$ initiator. For example, (4-octyloxyphenyl)phenyl-iodonium hexafluoroantimonate, **70**, was a good initiator for the reaction [23].

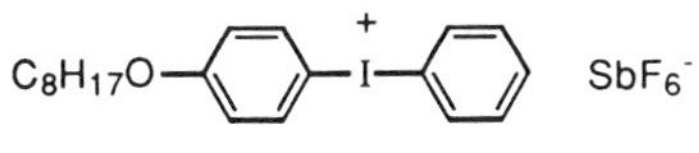

70

The proposed mechanism was identical with that in acid-catalyzed reactions except for the initiation step. Photolysis of the iodonium salt yields cations and cation radicals that react with traces of water or the monomer to form HX [23]. The Bronsted acid HX then functions similarly to other Bronsted acids in the polymerization reactions. 1,3-Diisopropenylbenzene has also been polymerized in a photoinitiated cationic reaction using **70** as the initiator [Eq. (14)] [9].

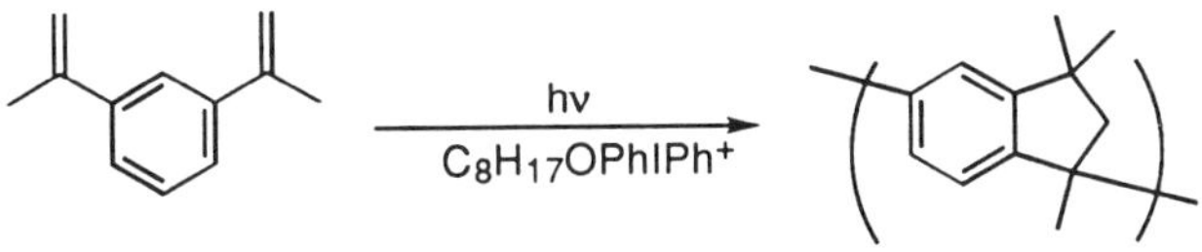

71 **72** (14)

C. Polybenzyls

The synthesis of polybenzyls by carbenium ion intermediates has been summarized in several reviews [27,28]. Polybenzyls obtained from the polymerization of benzyl chloride are amorphous, highly branched polymers [29] (Fig. 12).

One exception to the amorphous structure has been reported [30]. Crystalline polybenzyl was obtained from the low temperature (− 125° C) polymerization of benzyl chloride. However, the reaction was difficult to reproduce [31,32]. Consequently this procedure is not an effective method for the synthesis of linear polybenzyls. The usual amorphous, highly branched structure is formed as a result of a lack of positional selectivity and multiple substitution of the arene rings. Similar polymeric structures are obtained upon the polymerization of other nonsubstituted benzyl halides and benzyl alcohol [29]. The highly branched structure is a consequence of the involvement of benzyl carbenium ions in the Friedel-Crafts reaction. Benzyl substituents activate the monosubstituted phenyl groups toward further benzylation reaction. However, monomers containing alkyl substituents that sterically hinder substitution at the ortho position have been polymerized to linear polybenzyls. For example, the following

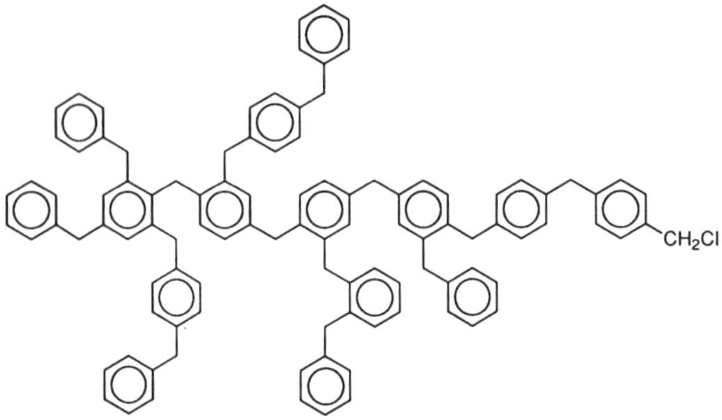

73

Figure 12 Example of typical polybenzyl structure obtained by the polymerization of benzyl chloride. (From Ref. 28.)

monomers have been polymerized to give crystalline linear substituted polybenzyls: α-chloroethylbenzene [30,33], α-ethylbenzyl chloride [34], α,α-dimethylbenzyl chloride [35,36], 2,5-dimethylbenzyl chloride [30,31,32,37], and 2,3,5,6-tetramethylbenzyl chloride [31] (Fig. 13). Higher yields and higher molecular weights of poly(α-methylbenzene) was obtained by moderating the activity of $AlCl_3$ with nitroethane [33].

Linear substituted polybenzyls can also be obtained by copolymerization reactions where *o*-substitution is hindered. For example the linear polybenzyl **78** was obtained by copolymerization of 1,2,4,5-tetramethylbenzene and α,α'-dichloroxylene [38] (Fig. 13).

D. Phenol-Formaldehyde Polymers

Benzylic cations have also been implicated in the formation of phenol-formaldehyde polymers. The synthesis of phenol-formaldehyde polymers, including the involvement of carbocations, has been reviewed [39]. The involvement of benzylic cations in this reaction sequence is shown in Fig. 14. The first step of the acid-catalyzed reaction is the formation of a hydroxymethyl carbocation. Electrophilic substitution of phenol produces an *o*- or *p*-hydroxymethylphenol **91**. Rapid protonation of the hydroxymethyl, followed by loss of H_2O, results in the formation of benzylic

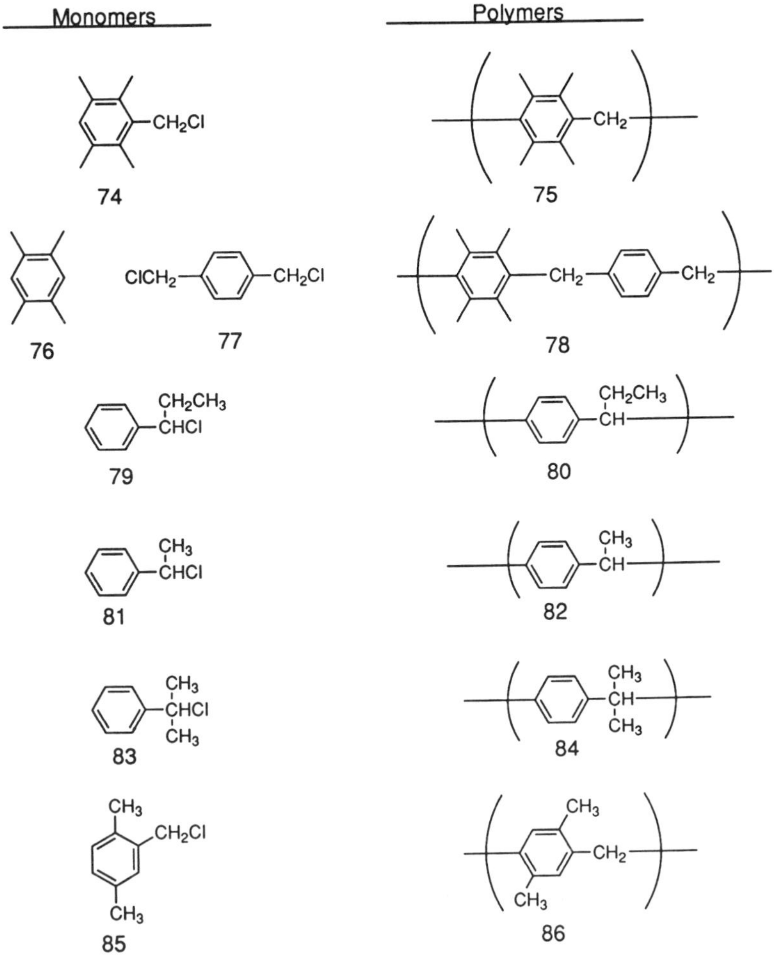

Figure 13 Synthesis of linear polybenzyls.

carbenium ion **93**. Reaction with phenol results in the formation of dihydroxydiphenylmethane **95a**. Three isomeric dihydroxydiphenylmethanes are formed as a result of *o*- and *p*-substitution in the successive electrophilic substitution reactions. The concentration of **95** increases because these substituted phenols are less reactive than phenol. Participation of **95** in further reactions results in the formation of chains containing 2–13 phenolic units joined by methylene groups.

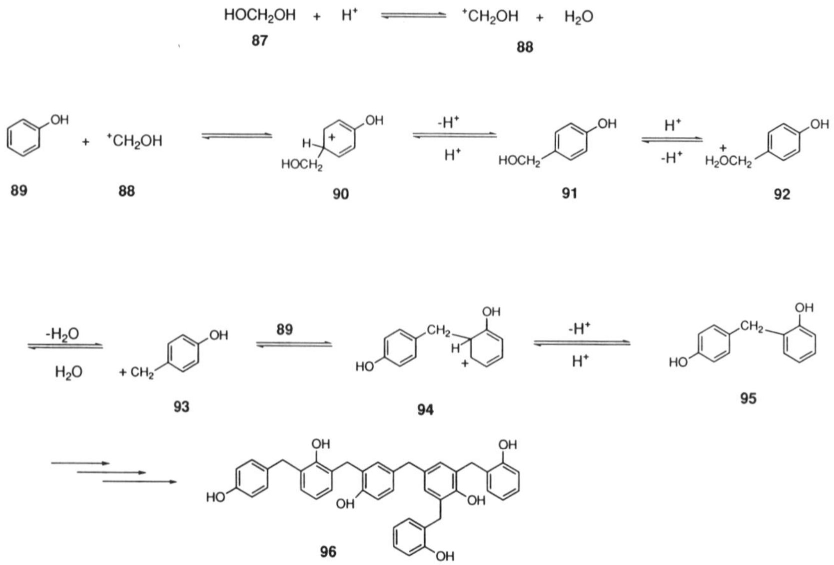

Figure 14 Acid-catalyzed formation of a phenol-formaldehyde resin.

E. Anthracene-Containing Polymers

The first class of soluble polymers containing anthracene units in the main chain was synthesized by a method involving cationic propagating species [40]. This polymer contains 2,6(7)-dihydroxy[1,3,5(6),7(8)-tetramethylanthracene] units [40] and has shown promise as a photoreactive polymer [41]. Previous reports of the synthesis of poly(9,10-dimethyleneanthracene) resulted in intractable and insoluble polymers [42–45]. The synthesis of poly(9,10-trimethylene anthracene) from the soluble precursor poly(9,10-trimethylene dihydroanthracene) resulted in an insoluble polymer with a low degree of polymerization [46]. The only example of the synthesis of high molecular weight polymers containing anthracene units in the main chain used 9,10-dihydro-9,10-ethanoanthracene monomers [47,48]. The resulting polyamides or polyesters were soluble and were converted to the corresponding insoluble anthracene polymers by the thermally induced elimination of ethylene [47,48]. The synthesis of soluble anthracene polymers was accomplished by the copolymerization of 4,4′-bis(2,6-dimethylphenoxy-4-phenyl)sulfone, **97**, with methylene chloride in an electrophilic substitution reaction catalyzed by $AlCl_3$ in CH_2Cl_2 (Fig. 15) [40]. Thus in this reaction CH_2Cl_2 plays both the role of monomer and solvent.

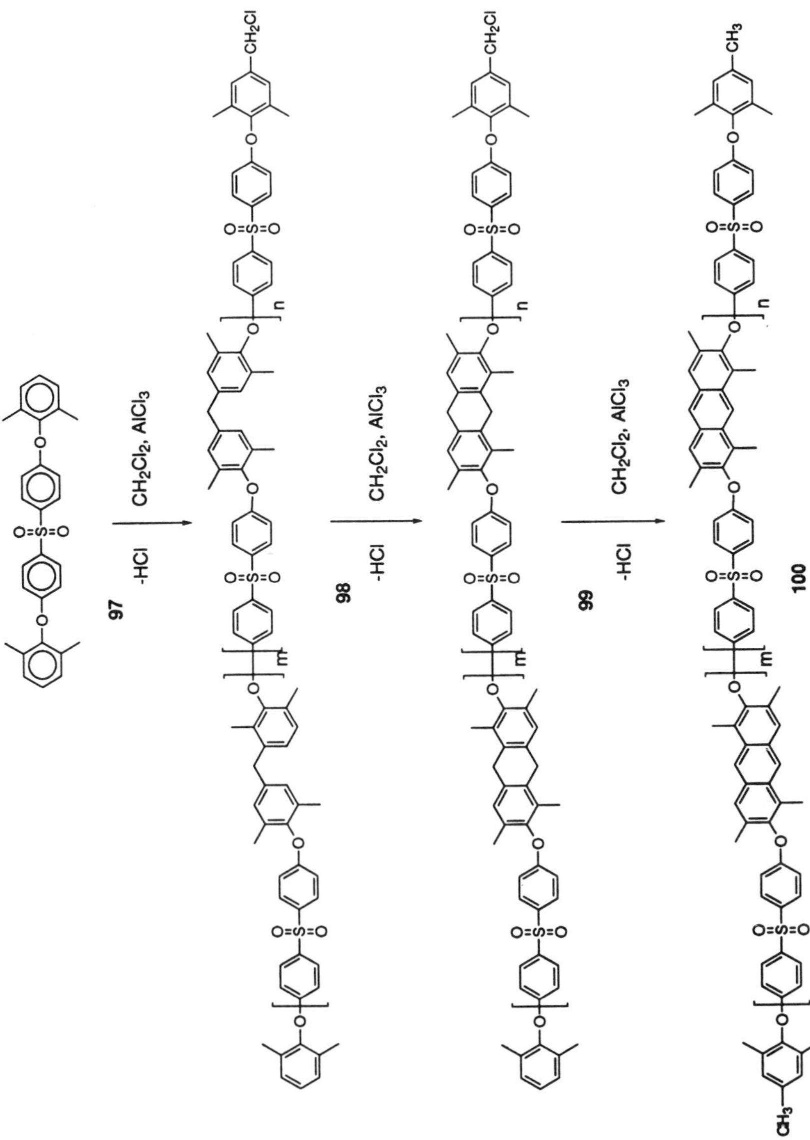

Figure 15 Synthesis of soluble anthracene polymer by $AlCl_3$-mediated copolymerization of 4,4′-bis(2,6-dimethylphenoxy-4-phenyl)sulfone with methylene chloride. (From Ref. 40.)

4,4′-Bis(2,6-dimethylphenoxy-4-phenyl)sulfone (DMPPS) has two characteristics that facilitate this reaction. This monomer is very nucleophilic and readily undergoes two electrophilic substitution reactions. Branching and cross-linking reactions are sterically hindered by the methyl groups. The first electrophilic substitution reaction of DMPPS with methylene chloride produces a substituted benzyl chloride derivative that is more electrophilic than methylene chloride. Therefore, this benzyl chloride derivative reacts rapidly with another DMPPS molecule (or growing polymer chain end) to form a substituted diphenyl methane unit. A large excess of methylene chloride favors the further reaction of the diphenyl methane unit. A sequence of two electrophilic substitution reactions on this unit results in the formation of a dihydroanthrylene structural unit. The dehydrogenation of this unit results in the formation of the corresponding anthrylene unit. The dehydrogenation reaction is catalyzed by $AlCl_3$. Hydride transfer to benzylic carbocations can also participate in dehydrogenation reactions to produce terminal or pendant methyl groups. This is a termination reaction that limits the molecular weight of the resultant polymer. A large excess of $AlCl_3$ was used to ensure the completion of the aromatization reaction [40].

The resulting aromatic polyether sulfone contains two types of tetramethyl-substituted anthracene units. One isomer was derived from a diphenyl methane unit in which the methylene linkage was formed as the result of meta-substitution of a DMPPS molecule and the para-substitution of the other DMPPS. The other isomer was formed as result of para-substitution of both DMPPS molecules [40]. The polymer was crystalline (T_m = 280° C) [41].

The cationic polymerization of 9-vinyl anthracene (**101**) was originally reported to proceed by 1,2-chain growth to produce a substituted vinyl-type polymer [49]. However it was later shown that the actual structure obtained was poly-9,10-dimethyleneanthracene (**105**) [45]. This polymer was obtained by the sequence of reactions shown in Fig. 16 [45,50].

F. Phenylene-Type Ladder Polymers

An electrophilic substitution reaction has been used for the key ladder-forming step in the synthesis of soluble ladder-type poly(phenylene)s [51–53]. These aromatic polymers have a ribbon-like rigid, planar structure. They are of interest because of their optical and electronic properties [51,54,55]. The preparation of these polymers was accomplished by two basic steps. The first step was the construction of a substituted poly(*p*-phenylene) backbone. The ladder structure was obtained by a subsequent intramolecular electrophilic ring closure reaction. For example, the syn-

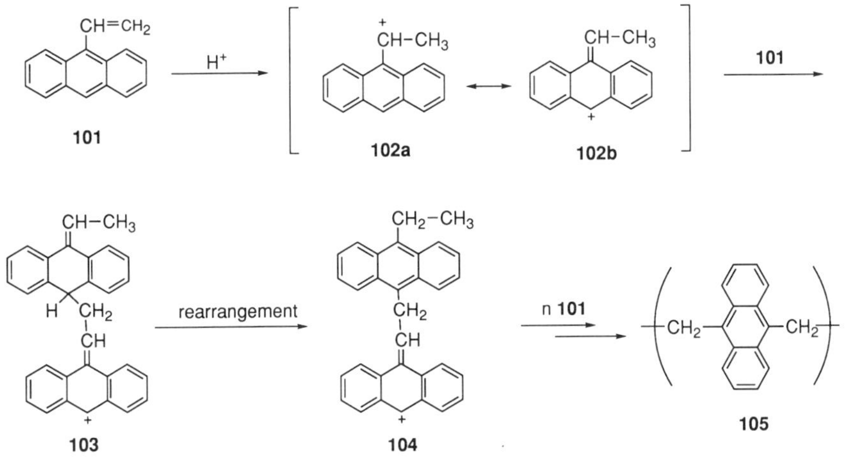

Figure 16 Formation of poly-9,10-dimethyleneanthracene by the cationic polymerization of 9-vinyl anthracene. (From Ref. 45.)

thesis of soluble phenylene-type ladder polymer **110** is shown in Fig. 17 [51]. The substituted polyphenylene backbone was prepared by the Pd(0)-catalyzed coupling reaction of 2,5-dihexyl-1,4-phenylenediboronic acid (**106**) with 1,4-bis(4-decylbenzoyl)-2,5-dibromobenzene (**107**) in the presence of potassium carbonate. The presence of alkyl substituents on both coupling partners was necessary to obtain a soluble product. The electron-withdrawing benzoyl groups activated the dibromobenzene compound in the coupling reaction. This facilitated high conversions and the formation of polymers with number average molecular weights of up to 9200. The carbonyl groups were then reduced to hydroxyl groups with $LiAlH_4$. The addition of $BF_3 \cdot Et_2O$ to a CH_2Cl_2 solution of **109** effected the quantitative intramolecular ring-closure reaction to give the ladder polymer **110**. Analysis of polymers **108–110** by ^1H and ^{13}C NMR spectroscopy indicated well-defined regular structures. No unreacted carbonyl or hydroxyl groups were detected in **110**. Several factors were crucial in enabling the synthesis of a soluble ladder polymer by this method. First, the open chain poly(*p*-phenylene) precursor was solubilized by alkyl groups on both coupling partners. Second, the reduction of the carbonyl groups took place quantitatively with no chain cleavage. Finally, the ladder-forming intramolecular electrophilic substitution reaction occurred quantitatively with no chain cleavage [51].

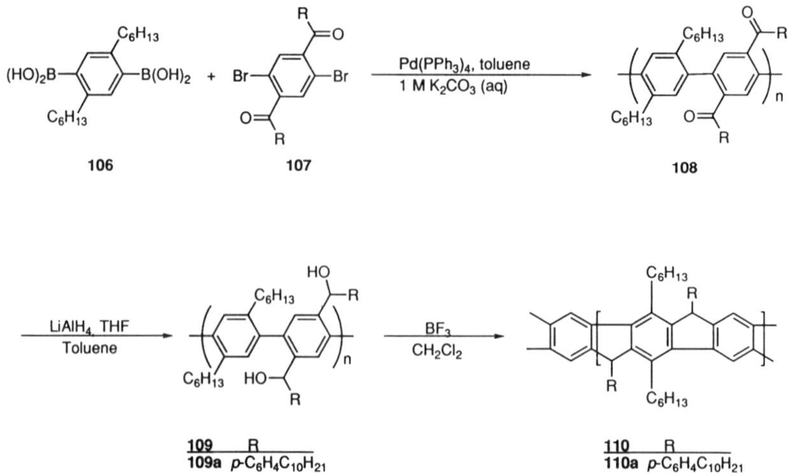

Figure 17 Synthesis of soluble ladder-type poly(phenylene)s by a two-step process: Construction of the Poly(p-phenylene) backbone and intramolecular ring closure. (From Ref. 51.)

Desirable electronic properties were obtained as a result of the structure of the planar ladder-like polymer backbone. However, the presence of alkyl side chains decreased the relative amount of the electronically active chromophore in the polymer. Therefore an alternative route was developed for the synthesis of **110** which had a reduced number of alkyl side chains [52]. In the previous synthetic scheme (Fig. 17) the least soluble polymer was **108**. Therefore the lower limit of alkyl group content of **110** was determined by the solubility of **108**. An alternative route was devised in which the open-chain precursor polymer was synthesized by the coupling of **106** and 2,5-dibromoterephthalic dialdehyde (**111**) instead of **107** (Fig. 18). The resulting poly(2,5-diformyl-1,4-phenylene-2′,5′-dihexyl-1′,4′-phenylene) (**112**) was completely soluble. The highest molecular weight fraction (M_n = 5100) was obtained in 48% yield after two reprecipitations. Reaction of **112** with a Grignard reagent converted the aldehyde group into a secondary alcohol. Once again, the ladder structure was obtained by an intramolecular electrophilic reaction. In the case of **110b** (R = C_6H_5) a regular structure was obtained with no cleavage of the polymer chain. However, the amount of ring closure is reduced 10–15% by side reactions in the reaction of polymers **109c** (R = $C_{10}H_{21}$) and **109d** (R = 1,4—C_6H_4—$N(CH_3)_2$) with BF_3 [52].

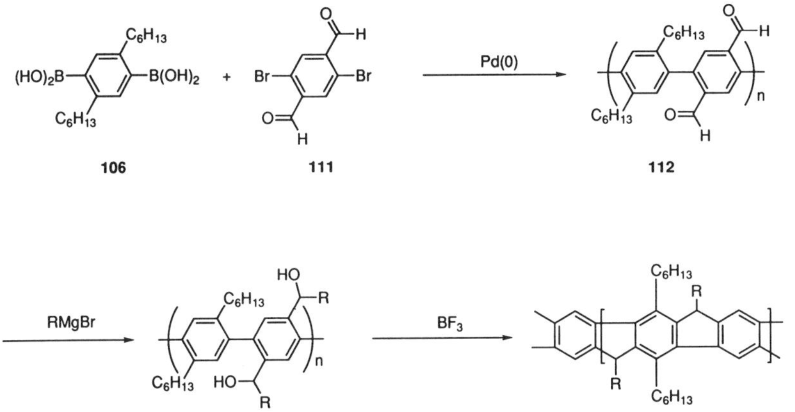

109	R
109b	C_6H_5
109c	$C_{10}H_{21}$
109d	p-$C_6H_4N(CH_3)_2$

110	R
110b	C_6H_5
110c	$C_{10}H_{21}$
110d	p-$C_6H_4N(CH_3)_2$

Figure 18 Synthesis of soluble ladder-type poly(phenylene)s with a reduced amount of alkyl side chains. (From Ref. 52.)

The synthesis of fully unsaturated ribbon-type molecules was also accomplished [53]. The dehydrogenation of the methylenic bridge hydrogens of **110** resulted in the formation of an unstable polymer [56]. This was because the degenerate ground states of **110** destabilized this species. Therefore, a polymer with more stable ground state (**117**) was synthesized [53] (Fig. 19). This polymer was derived the dehydrogenation of a polymer (**116**) having alternating para- and meta-phenylene units instead of the para-phenylene subunits of **110**. The synthetic scheme was similar to that used for the synthesis of **110a** (Fig. 18) with the dehydrogenation accomplished by treatment of **116** with DDQ. The resulting polymer (**117**) was stable in an inert atmosphere. Films of the polymer were relatively stable in air, with a slow decomposition taking place over several weeks [53].

G. Cyclotriveratrylene, Cyclotetraveratrylene, and Cyclotetraveratrylene-Containing Polymers

The acid-catalyzed reaction of veratrole and formaldehyde yields a trimer with a formula $(C_9H_{10}O_2)_3$ named cyclotriveratrylene (CTV) [57–62]. CTV exists in a "crown" conformation. CTV is formed as the result of the condensation reaction of the veratryl cation. The veratryl cation is

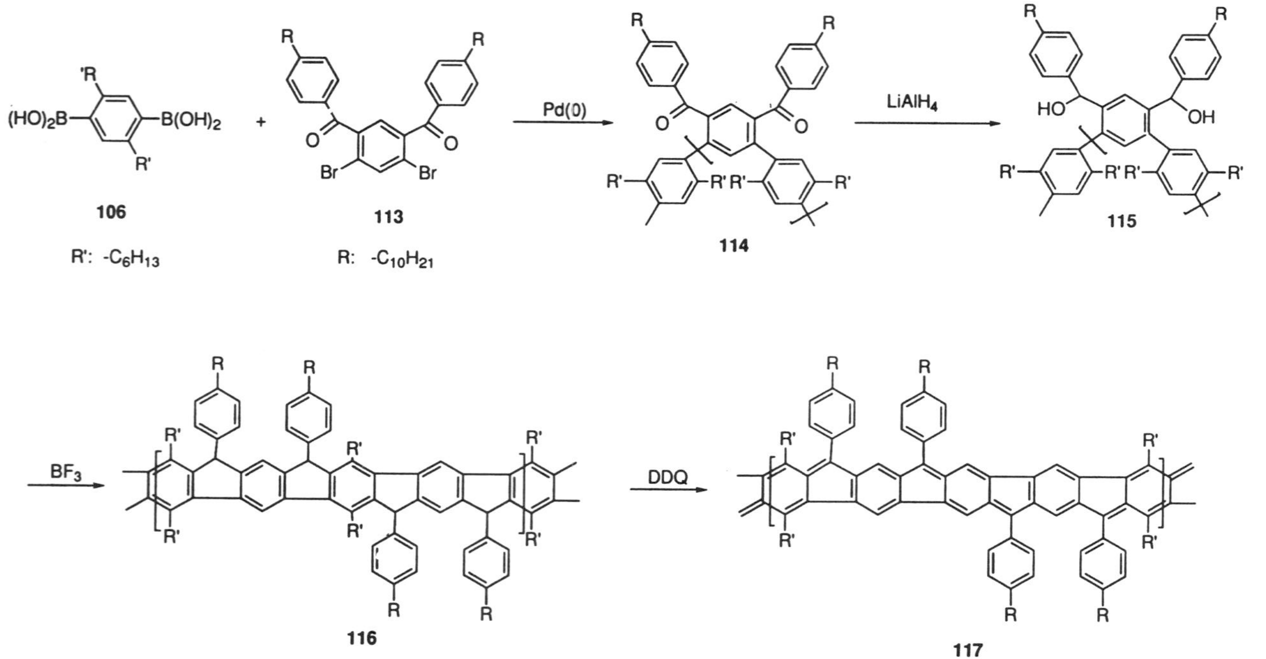

Figure 19 Synthesis of fully unsaturated ribbon-type polymeric backbone. (From Ref. 53.)

usually generated in the presence of strong Bronsted or Lewis acids. Various precursors to veratryl cations can be used, with common precursors being veratrole and formaldehyde [57–59], a veratryl alcohol [63,64], or an *N*-tosylate of veratrylamine [65,66]. The condensation reactions to prepare CTV usually also produce the cyclic tetramer cyclotetraveratrylene (CTTV) as a by product. The acid-catalyzed synthesis of CTV from veratryl alcohol begins with protonation of veratryl alcohol **118** followed by loss of H_2O to form the veratryl carbenium ion **119** (Fig. 20). The participation of **119** and **118** in an electrophilic substitution reaction gives a veratryl dimer **120**. Protonation of the benzylic hydroxyl group followed by elimination of H_2O generates benzylic carbenium ion **121**. This carbenium ion reacts with **118** to give the veratryl trimer **122**. The carbenium ion generated from **122** can participate in a cyclization reaction to form CTV, **126** or can further react with **118** to form the veratryl tetramer **124**. Cyclization of **125** produces CTTV, **127**. Other pathways for the formation of CTTV, including the reaction of **121** with **120**, are possible.

This electrophilic cyclooligomerization reaction has been developed into a method for the synthesis of hyperbranched liquid-crystalline polymers containing discotic mesogens in the branching point. Both CTV [67–71] and CTTV [72–74] alkyloxy and alkanoyloxy derivatives have attracted interest because of their ability to form discotic or columnar liquid crystalline phases. Mesogens based on CTV form pyramidic [68] mesophases because of the rigid crown conformation of CTV. This rigidity causes these mesophases to order very slowly. Consequently their formation is inhibited by impurities. In contrast, pyramidic mesophases are not obtained from CTTV-based mesogens because of the conformational flexibility of CTTV. Mesophases of flexible octakis(*n*-alkyloxy)tetrabenzocyclododecatrene (CTTV-*n*) derivatives form very quickly. These phases form in the presence of small amounts of impurities. Therefore, these cyclotetramerization reactions were utilized for the in situ mesogen-forming polymerization reactions.

A systematic study using a wide variety of reaction conditions was completed to determine the conditions which favor the selective formation of CTV or CTTV [75]. CTV was determined to be the kinetic cyclization product and CTTV was determined to be the thermodynamic product. Therefore, it was possible to determine reaction conditions which favored predominant formation of the desired product. CTV was formed preferentially by two methods. The treatment of 3,4-bis(methyloxy)benzyl chloride with stoichiometric amounts of $AgBF_4$ in methylene chloride resulted in the formation of CTV in a 91:9 ratio to other products. In this case 9% of the product was CTTV and higher molecular weight products. The cyclotrimerization of 3,4-bis(methyloxy)benzyl alcohol with superacids

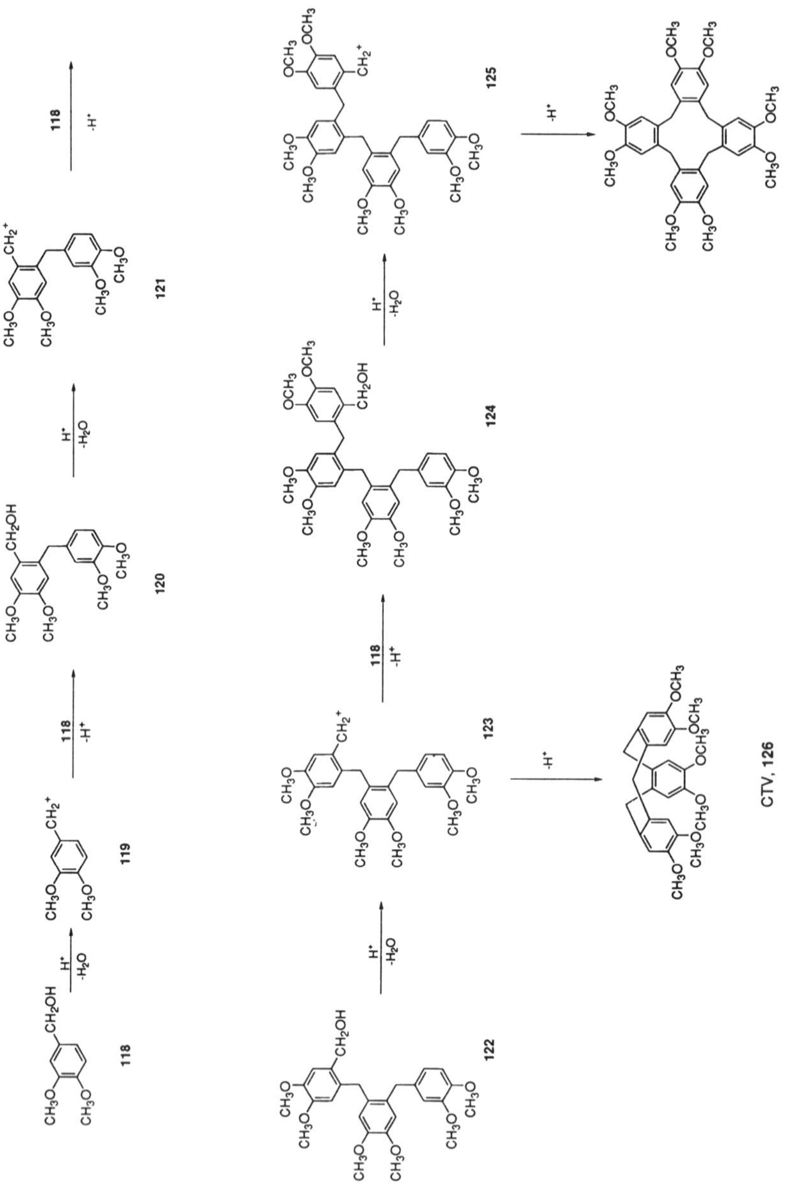

Figure 20 Acid-catalyzed synthesis of cyclotriveratrylene (CTV) and cyclotetraveratrylene (CTTV).

$HClO_4$ and CF_3SO_3H in a nonsolvent also resulted in the high-yield formation of CTV. Reaction conditions that favored the selective formation of the CTTV were also determined. The highest yield of CTTV was obtained by the reaction of 3,4-bis(methyloxy)benzyl alcohol in a dilute CH_2Cl_2 solution using a large excess of the weaker acid CF_3CO_2H.

The cyclooligomerization of 3,4-bis(methyloxy)benzyl alcohol was performed in a series of acids (listed in order of increasing acidity): CF_3CO_2H, CH_3SO_3H, *p*-CH_3PhSO_3H, H_2SO_4, $HClO_4$, CF_3SO_3H. The rate of the cyclooligomerization reaction increased with increasing acid concentration and strength. As the acid concentration and/or strength was increased the concentration of electrophilic species increased (Table 1).

The reaction did not proceed with a catalytic amount (13%) of CF_3CO_2H in CH_2Cl_2. However, the reaction proceeded in high yield (98–100%) in the presence of excess CF_3CO_2H. CTTV was formed preferentially under high-dilution conditions when a large excess of CF_3CO_2H is used. In this case 55% isolated yield of CTTV was obtained. The stronger acids were only required in catalytic amounts for the cyclooligomerization.

Precipitation of the product from the reaction medium occurred when the solvent was the acid itself, water, CH_3SO_3H, H_2SO_4, $HClO_4$, or CF_3SO_3H in CH_2Cl_2. A slight turbidity of the reaction solution was detected when using *p*-toluene sulfonic acid. These reaction conditions favored the formation of the kinetic product (CTV) as the main product. The use of solvent in the cyclization of *N*-veratrylethanolamine-*N*-tosylate resulted in increased formation of the minor product CTTV. Thus a kinetic product distribution was indicated. Therefore the effect of increasing the solubility of the reaction mixtures by the addition of dimethyl sulfoxide (DMSO) was investigated. The addition of DMSO to reactions involving CH_3SO_3H decreased the rate of reaction and decreased the reaction yield. However, the ratio of CTV to CTTV was essentially unaffected. A decrease in yield also occurred with added DMSO when the strong acid $HClO_4$ was used. In this case the addition of a larger quantity of DMSO was necessary to affect the reaction. The relative amount of CTTV was slightly increased when the strong acids $HClO_4$ or CF_3SO_3H were used to catalyze the reaction. In summary, the weak base DMSO decreases the rate of reaction and only affects the product distribution when the reaction is catalyzed by strong acids that have higher concentrations of ion pairs. The product distribution is affected very little by changes in reaction times, temperature, and by the addition of small amounts of water or methanol. However, the strength of the acid had a marked influence on the product distribution. The weakest acid investigated (CF_3CO_2H) favored CTTV formation. The synthetic procedure for the predominant

Table 1 Product distribution of the cyclooligomerizations of 3,4-bis(methoxy)benzyl alcohol[a]

No.	Acid	Mol of acid/ mol of OH	Solvent	mL of Solvent/ mmol of OH	Mol of DMSO/ mol of OH	Temp.	Time (h)	Yield (%)	Product distribution CTV:CTTV:other
1	CF_3COOH	0.13	CH_2Cl_2	1.68	0	reflux	6	0	
2	CF_3COOH	2.18	CH_2Cl_2	1.51	0	ambient	2	98	27:49:24
3	CF_3COOH	2.18	CH_2Cl_2	1.51	0	reflux	2	99	24:48:28
4	CF_3COOH	2.18	H_2O	1.51	0	reflux	2	99	37:13:50
5[b]	CF_3COOH	10.9	CH_2Cl_2	16.0	0	ambient	4	90	10:65:25
6	CF_3COOH	10.9	H_2O	0.84	0	reflux	2	100	22:25:53
7	CF_3COOH	10.9	H_2O	0.84	0	reflux	6	100	22:26:52
8	CH_3SO_3H	0.13	CH_2Cl_2	1.68	0	reflux	6	100	40:33:27
9	CH_3SO_3H	0.13	CH_2Cl_2	1.68	0.21	reflux	6	76	35:35:30[c]
10	CH_3SO_3H	0.13	CH_2Cl_2	1.68	0.84	reflux	6	73	25:21:54[c]
11	TsOH-H_2O	0.13	CH_2Cl_2	1.68	0	reflux	2	99	48:30:22
12	TsOH-H_2O	0.62	CH_2Cl_2	1.68	0	reflux	2	98	53:29:18
13	H_2SO_4	0.13	CH_2Cl_2	1.68	0.21	reflux	6	76	40:22:38[c]
14	H_2SO_4	0.13	CH_2Cl_2	1.68	0.84	reflux	6	80	43:22:35[c]
15	70% $HClO_4$	0.13	CH_2Cl_2	1.68	0	reflux	2	96	49:24:27
16	70% $HClO_4$	0.13	CH_2Cl_2	1.68	0[d]	reflux	2	97	46:31:23
17	70% $HClO_4$	0.13	CH_2Cl_2	1.68	0[e]	reflux	2	99	45:33:21
18	70% $HClO_4$	0.13	CH_2Cl_2	1.68	0.21	reflux	2	97	47:35:18
19	70% $HClO_4$	0.13	CH_2Cl_2	1.68	0.20[e]	reflux	2	90	32:24:44[c]
20	70% $HClO_4$	0.13	CH_2Cl_2	1.60	1.1	reflux	2	0	
21	70% $HClO_4$	0.13	CH_2Cl_2	1.18	4.1	reflux	2	0	
22	70% $HClO_4$	0.13	CH_2Cl_2	1.68	0	reflux	6	99	44:31:25
23	70% $HClO_4$	0.13	CH_2Cl_2	1.68	0.21	reflux	6	93	41:37:22
24[b]	70% $HClO_4$	175	H_2O	1.68	0	ambient	2	83	66:10:24
25	CF_3SO_3H	0.12	CH_2Cl_2	1.67	0	ambient	2	97	53:31:16
26	CF_3SO_3H	0.12	CH_2Cl_2	1.68	0	reflux	6	95	48:28:24
27	CF_3SO_3H	0.12	CH_2Cl_2	1.68	0.20	reflux	6	99	41:34:25
28	CF_3SO_3H + CF_3COOH	0.12 + 13.0	CH_2Cl_2	1.51	0	ambient	2	98	25:48:27
29[b]	CF_3SO_3H + CF_3COOH	0.30 + 10.9	CH_2Cl_2	16.0	0	ambient	2	99	31:48:21

[a] 6.0 mmol $ArCH_2OH$.
[b] 0.60 mmol $ArCH_2OH$.
[c] Other products include some unreacted starting material.
[d] Also contains 0.79 mmol of MeOH/mmol of $ArCH_2OH$.
[e] Also contains 1.5 mmol of H_2O/mmol of $ArCH_2OH$.
Source: Ref. 75.

formation of cyclic tetramers from 3,4-dialkyloxybenzyl alcohol was applied to the synthesis of a complete series of octaalkyloxy tetrabenzocyclododecatetraene (CTTV-*n*) derivatives containing alkyloxy substituents ($n = 4$–15). Two methods were used for the synthesis of these compounds: the alkylation of 2,3,6,7,10,11,14,15-octahydroxytetrabenzo[a,d,g,j]cyclododecatetraene (CTTV-OH) and the direct cyclotetramerization of 3,4-(dialkyloxy)benzyl alcohol using excess CF_3CO_2H in CH_2Cl_2 (Fig. 21).

In addition to crystalline and isotropic phases, all of these compounds except CTTV-4 displayed an enantiotropic columnar mesophase. The phase-transition temperatures of CTTV-*n* compounds prepared by cyclotetramerization were lower than those of the compounds prepared by the etherification of CTTV-OH. The melting and isotropization temperatures decreased smoothly as the length of the alkyl group increased. The CTTV-forming reaction was also applied as a novel polymerization reaction that constructed CTTV-*n* molecules during the polymerization process. This resulted in the formation of hyperbranched liquid-crystalline polyethers containing disk-like mesogens.

The discotic ordered hexagonal (D_{ho}) phase of a main-chain liquid-crystalline polymer containing disk-like mesogenic groups is shown in Fig. 22 [76]. This phase consists of columns of the disk-like groups arranged in a hexagonal arrangement.

These seven columns are statistically interconnected by the main chain of the polymer. The disks can be connected within the same plane or in different planes. The connections can also be within the same column or between adjacent columns.

The syntheses of poly(CTTV) was accomplished by the polymerization of a 3,4-bis(*n*-alkyloxy)benzyl alcohol (**130**) with a α,ω-bis{[2-(alkyloxy)-5-(hydroxymethyl)phenyl]oxy}alkane (**135**) [76].

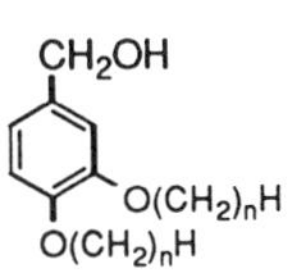

130

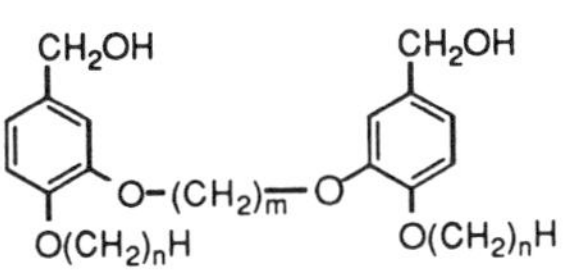

135

As an example, the synthesis of poly(CTTV)7/7,16(x,y) is shown in Fig. 23. Both interdisk and intradisk links formed. Symmetric monomeric CTTV-*n* (i.e., **136**) was removed by reprecipitation. The high ratio of 1(n) to 2(n,m) results in a polymer containing some branching due to linkages between some CTTV units to three other CTTV units.

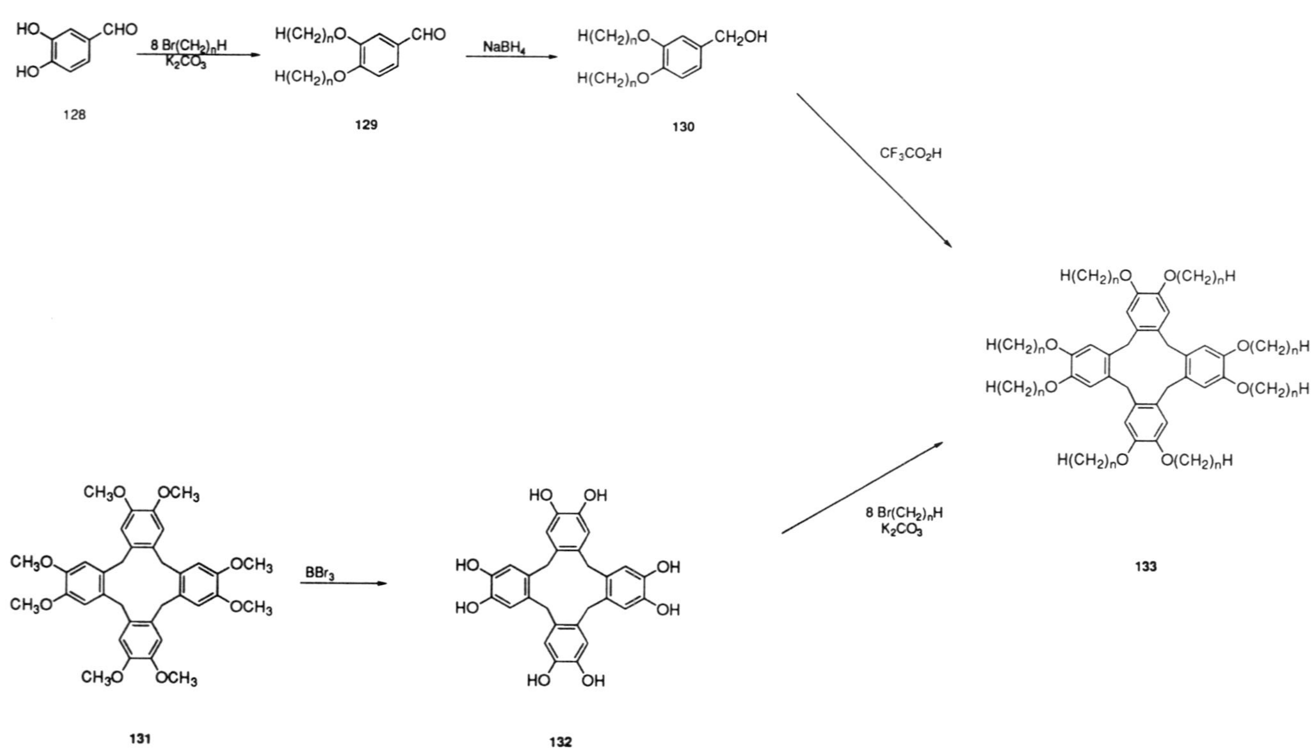

Figure 21 Two routes for the synthesis of CTTV-*n*: Cyclotetramerization of 3,4-(dialkyloxy)benzyl alcohols or etherification of CTTV-OH. (From Ref. 74.)

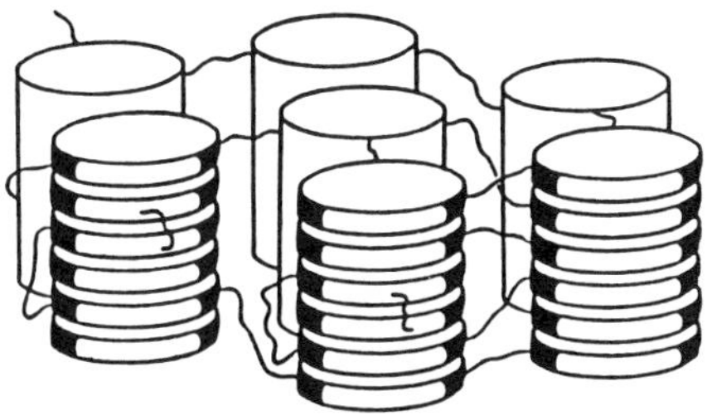

a

134

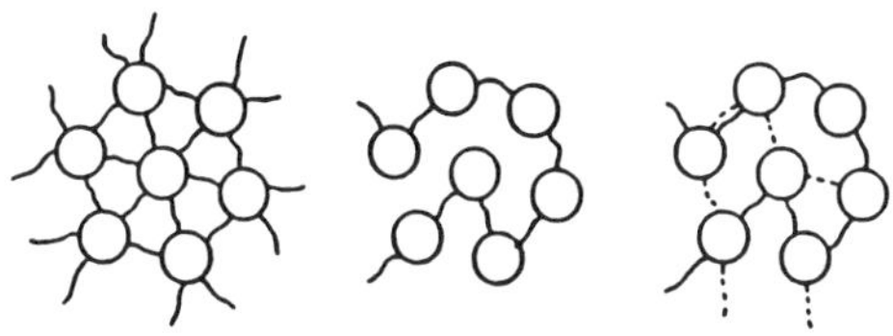

b c d

Figure 22 The discotic ordered hexagonal (Dho) phase of a main-chain liquid-crystalline polymer containing disk-like mesogenic groups: (a) viewed from the side of the columns; (b) viewed from the top of the columns; (c) top disk layer of the columns; (d) top two disk layers of the columns. (From Ref. 76.)

There are 14 CTTV isomeric units possible from the reaction of 3,4-bis(alkyloxy)benzyl alcohols **130** and **135**. This is illustrated for the reaction of **138** and **139** in Fig. 24.

H. Calixarenes

Calixarenes [77] are defined as [1.*n*]metacyclophanes with its basic structural unit consisting of phenolic groups linked by ortho-methylene groups. Two examples are shown in Fig. 25. Several reviews are available on this subject [77–79]. These compounds can be synthesized by the acid- or base-catalyzed condensation reaction of a substituted phenol with formaldehyde or an aldehyde (Fig. 26). Calixarenes have also been synthesized by a stepwise reaction that sequentially add phenolic groups followed by a cyclization step. More efficient convergent synthesis have also been developed [80–82]. The bowl-like structural conformation generally

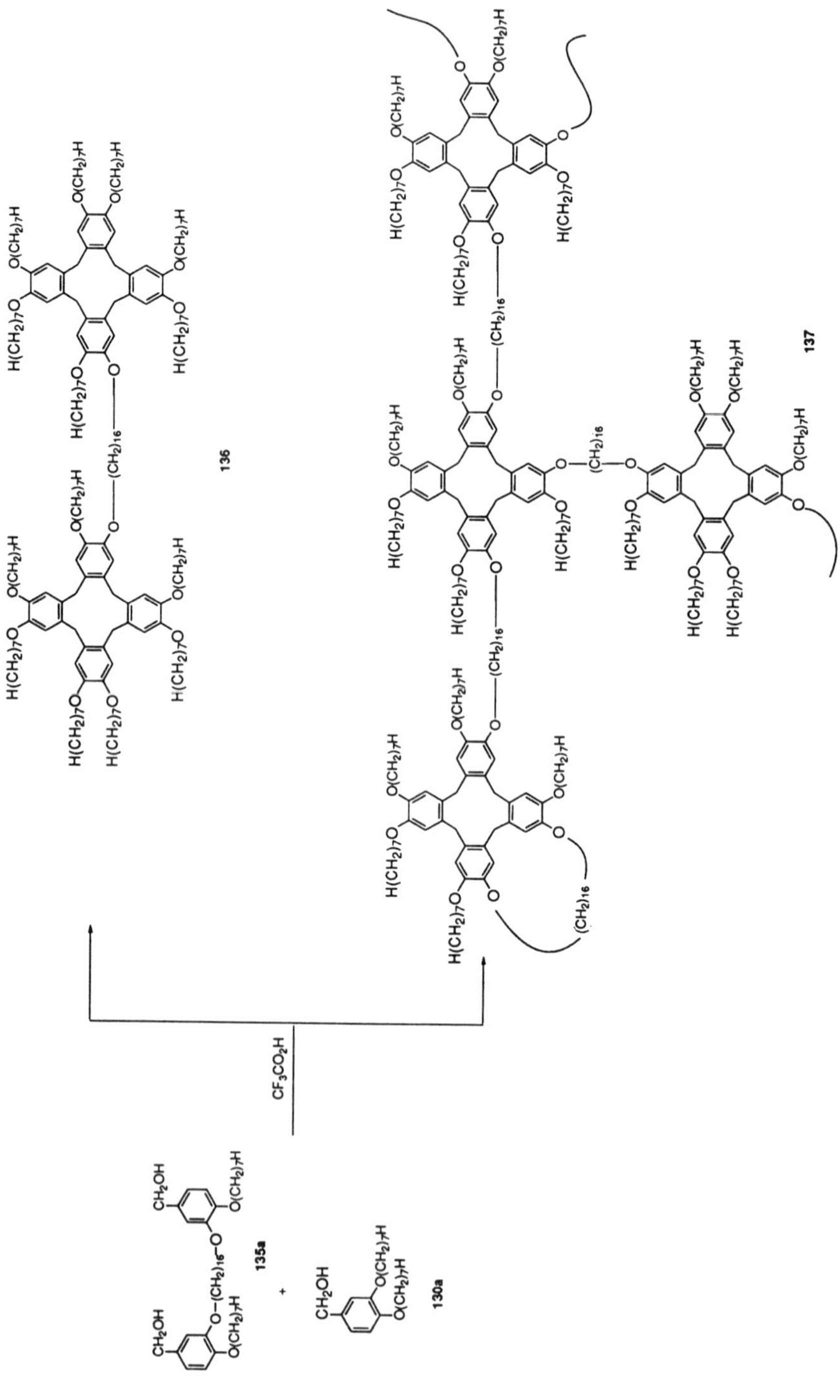

Figure 23 Synthesis of twin discotic dimer and branched oligomers from CTTV-7 mesogens connected with a hexadecane spacer. (From Ref. 76.)

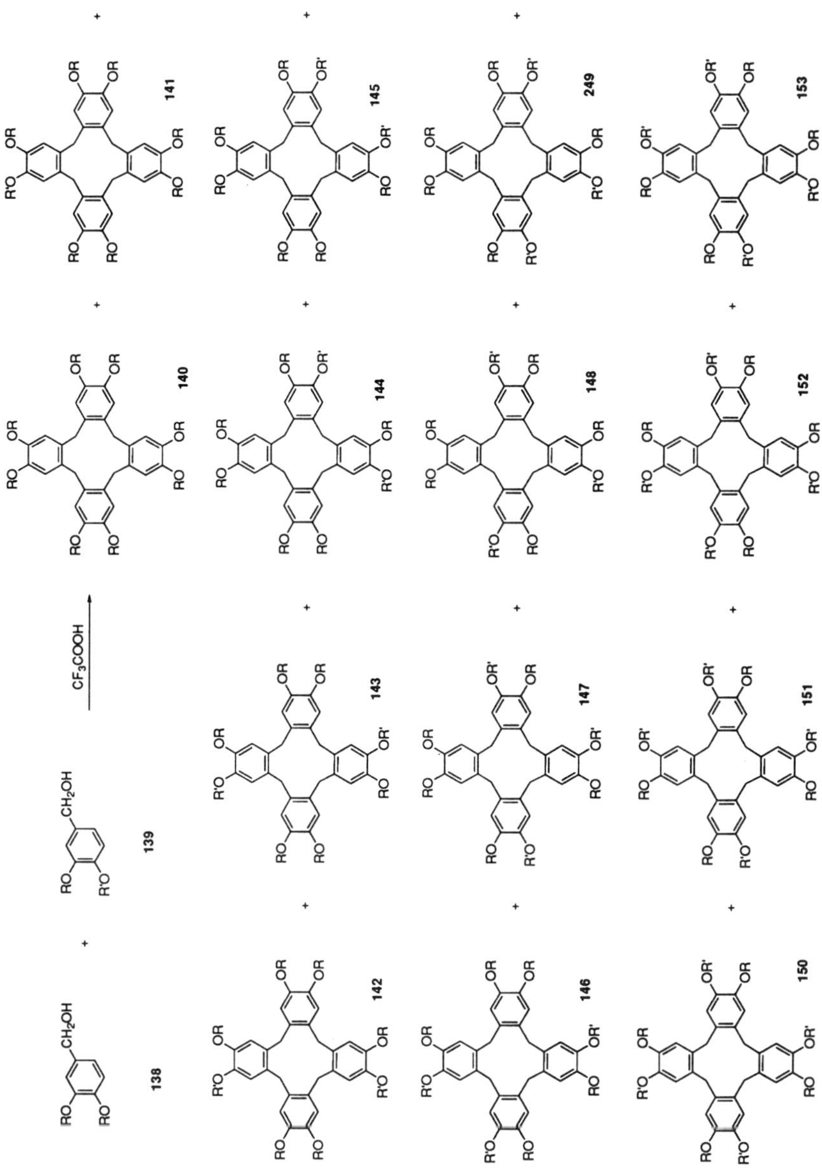

Figure 24 Isomeric CTTV units possible from cyclotetramerization of 3,4-(dialkyloxy)benzyl alcohols **133** and **134**. (From Ref. 76.)

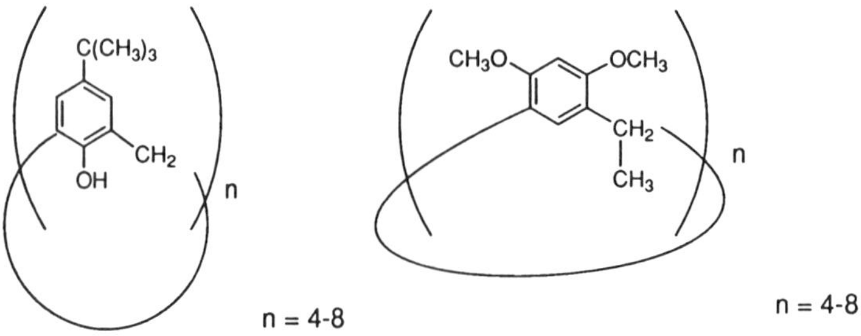

154 **155**

Figure 25 Examples of calixarene structure

adopted by calixarenes makes them important compounds in the area of molecular recognition chemistry and in the preparation of liquid crystals [83–90].

I. Poly(phthalicylidenearylene)s

Poly(phthalicylidenearylene)s are aromatic polymers with phthalide and phenylene groups in the polymer backbone. These polymers are of interest because of their high chemical and thermal stabilities as well as their electrical properties [91,92]. Poly(phthalidylidenearylene)s have been obtained from the homopolycondensation of 3-aryl-3-chlorophthalides via an electrophilic substitution reaction. For example, 3-(4-biphenyl)-3-chlorophthalide can be polymerized to give poly(3,3′-phthalidylidene-4,4′-biphenylylene) [Eq. (15)] [93]. The best results (M_w = 46,000–60,000) were obtained using $SbCl_5$ or $InCl_3$ as catalysts at 100–110° C in nitrobenzene.

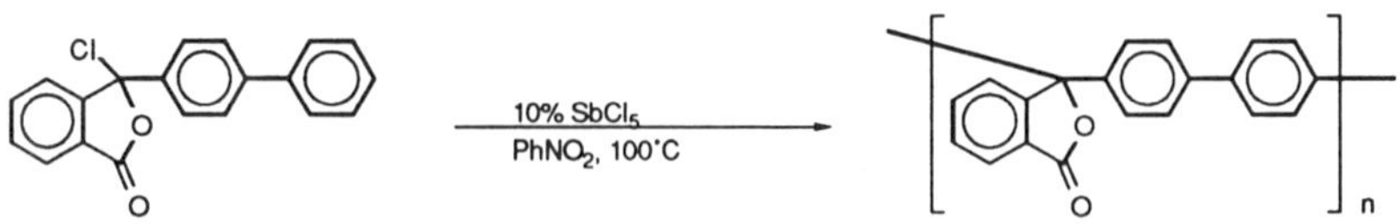

164 **165**

(15)

Alternatively, they have been synthesized by the polycondensation of 3-chlorophthalides or phthaloyl dichloride with aromatic hydrocarbons [Eq. (16)] [94]. However, polymers with irregular structural units were produced. The irregular structure is the result of para- and meta-substitution of the oxydiphenylene unit.

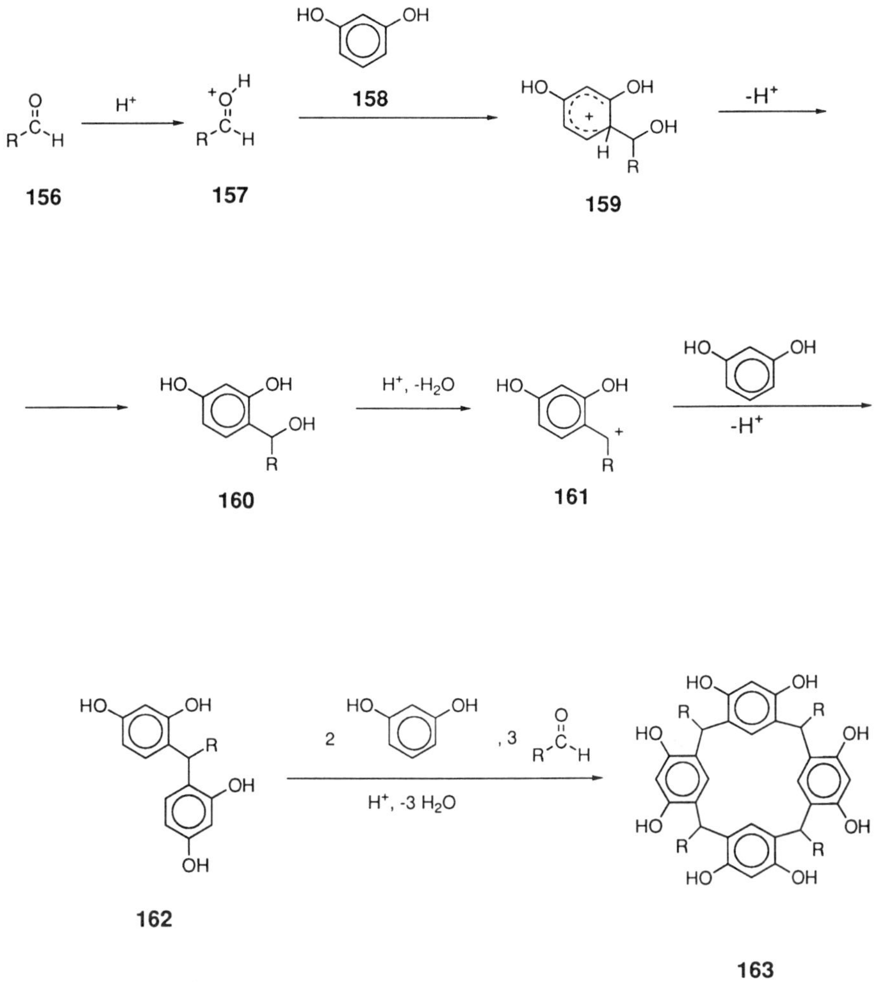

Figure 26 Acid-catalyzed synthesis of calixarenes

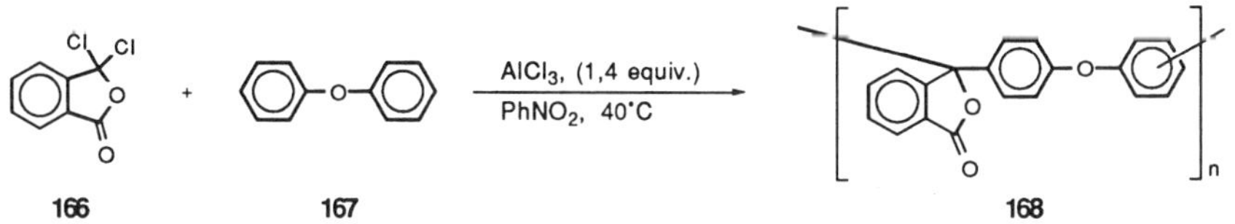

(16)

The polymerization of monomers with substituents at the 4-position of the phthalide group resulted in the formation of polymers with two types of repeat units. One unit contained the substituent at the original 4-position. In the other repeat unit the substituent was isomerized to the 7-position [Eq. (17)] [95].

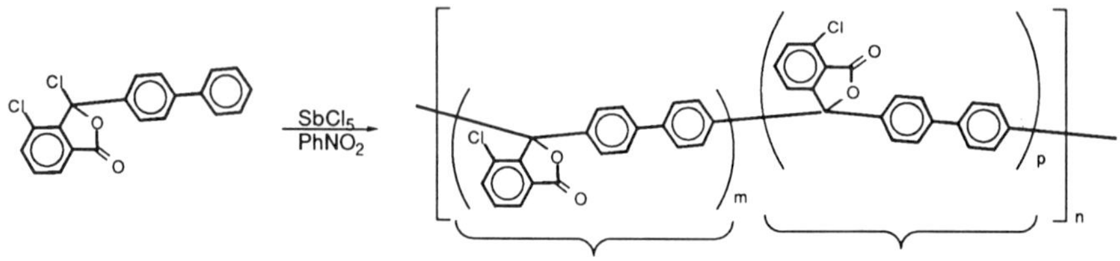

169 170

(17)

The formation of the isomerized product was rationalized by the sequence of reactions shown in Fig. 27. A carbocation with a chlorosubstituent at the 4-position is formed by the reaction of the monomer with the Friedel-Crafts catalyst. Structural unit A is formed by the direct reaction of this carbocation with the aromatic ring of another monomer molecule (or polymer end group). Alternatively, an isomerization reaction can occur to produce a carbocation with the chlorosubstituent at the 7-position. Reaction of this carbocation with substrate leads to the formation of repeat unit B.

Poly(triarylcarbinol)s have been synthesized by a similar method (Fig. 28) [96].

J. Transalkylation Reactions

Transalkylation reactions under acidic conditions have been reported for 2,2-bis(4-hydroxyphenyl)propane with 2,6-dialkylphenol [97], 2,6-diphenylphenol [98], diphenyl ether [99], and with 4-bromophenyl phenyl ether [99]. For example, the reaction of 2,2-bis(4-hydroxyphenyl)propane and a 2,6-disubstituted phenol is shown in Fig. 29.

This reaction apparently proceeds through the intermediacy of a benzylic carbenium ion. The possibility of using this reaction as a polymerization reaction was investigated [99]. An efficient transalkylation reaction between **185** and diphenyl ether (**167**) would produce polymer structure **189** and phenol [Eq. (18)].

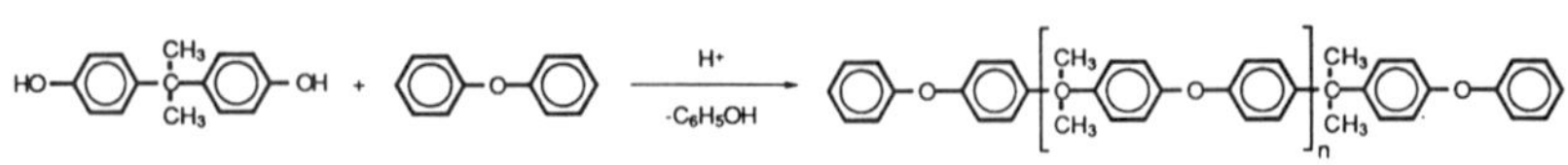

185 167 189

(18)

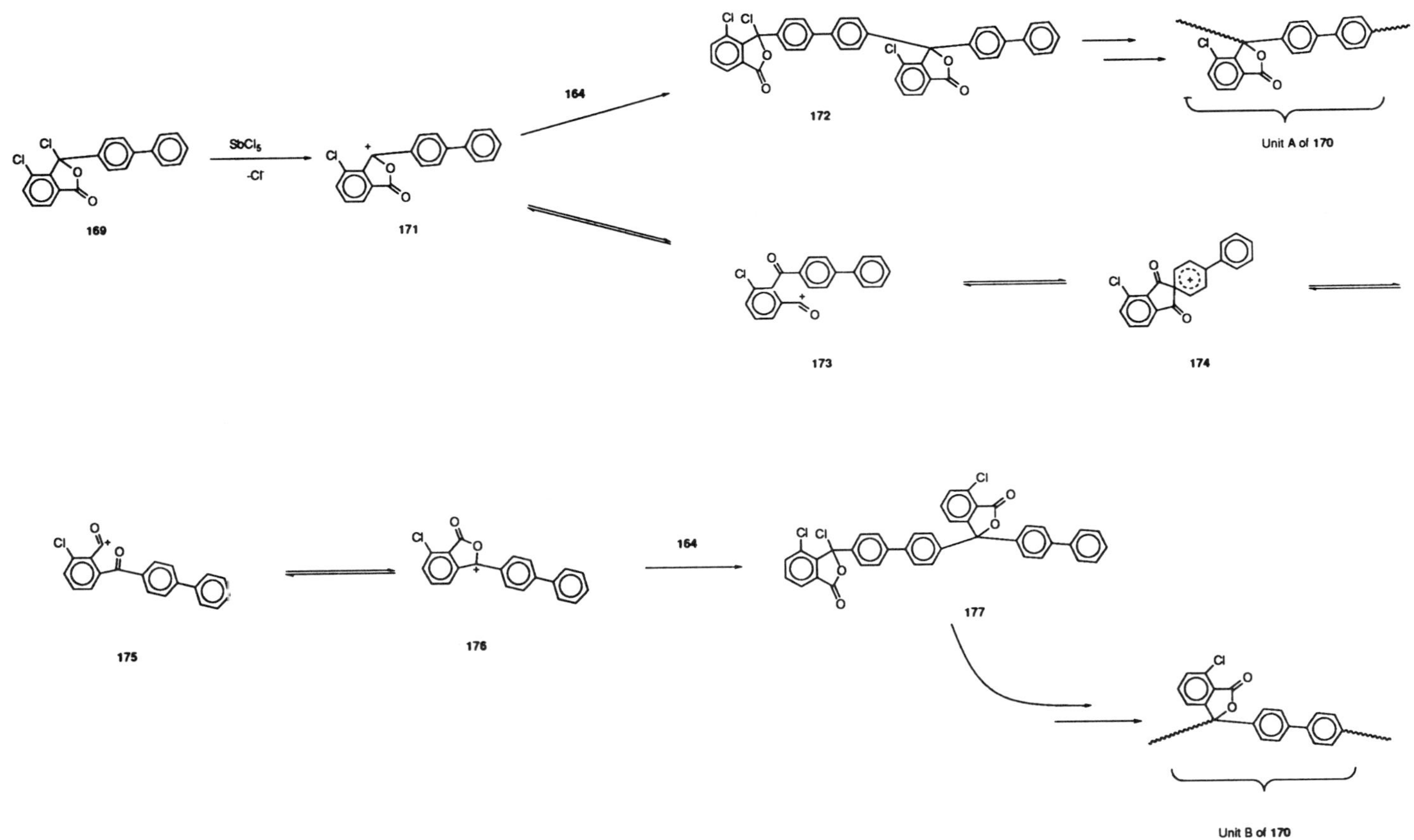

Figure 27 Mechanism of the isomerization of phthalide group in the synthesis of poly(3,3′-phthalidylene-4,4′-biphenylylene)s. (From Ref. 95.)

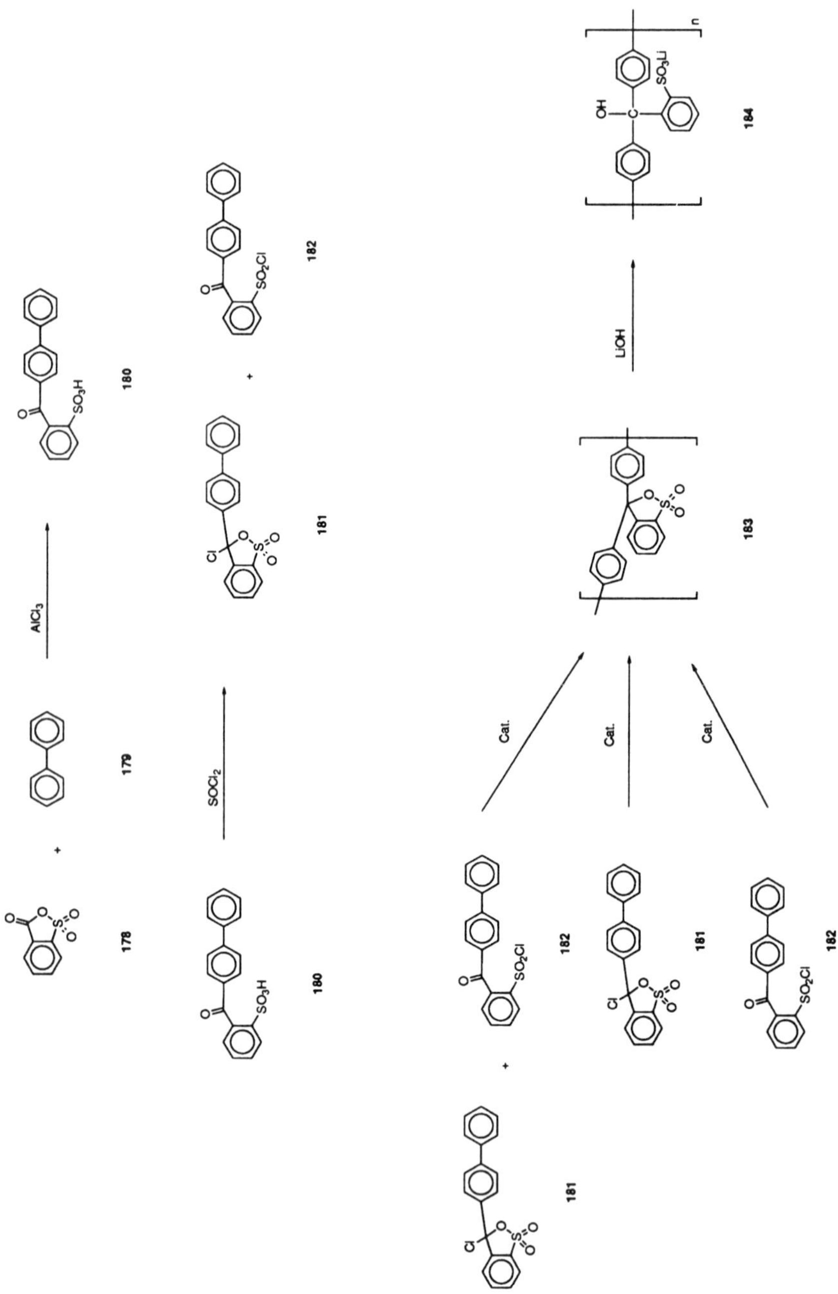

Figure 28 Synthesis of poly(triarylcarbinol)s. (From Ref. 96.)

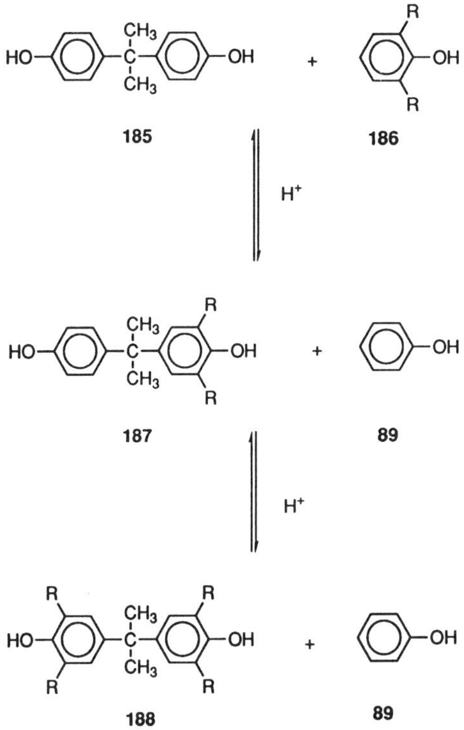

Figure 29 Transalkylation reactions of **185** and **186**. (From Ref. 99.)

The effectiveness of several different protic acid catalysts was evaluated under various reaction conditions. The treatment of **185** and excess **167** with Nafion at 100° C resulted in the formation of 6,6′-dihydroxy-3,3,3′,3′-tetramethyl-1,1′-spirobiindan (**190**) [Eq. (19)]. This product was also obtained when CH_3SO_3H was used as the acid catalyst at 65° C [Eq. (20)].

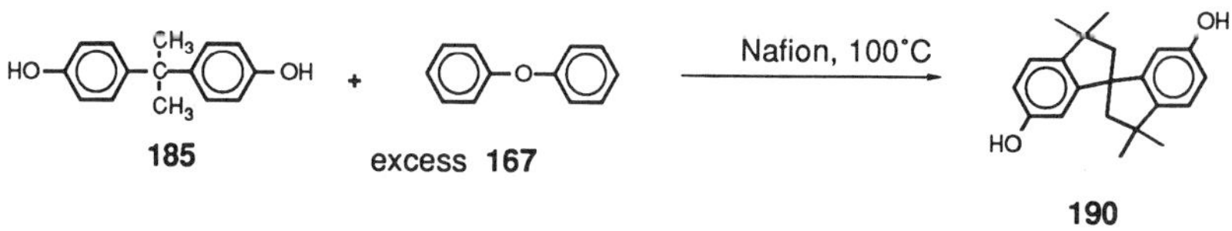

(19)

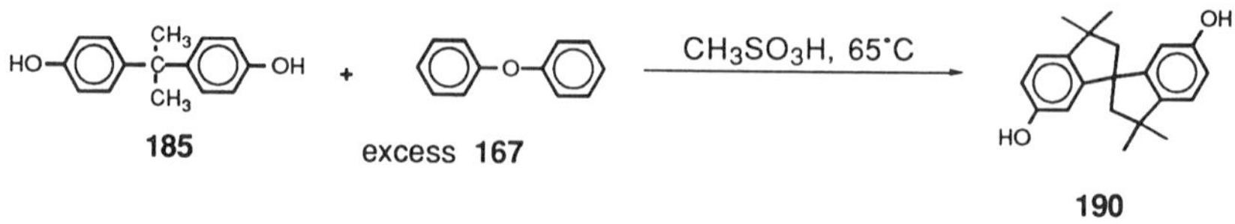

(20)

However, the use of CH_3SO_3H as catalyst at room temperature resulted in the formation of monotransalkylated and ditransalkylated products [Eq. (21)]. The reaction was slow, with reaction times ranging from 2 to 3 days. The reaction could be made specific for either product by adjusting the acid concentration. Higher concentrations favored the formation of the ditransalkylated product.

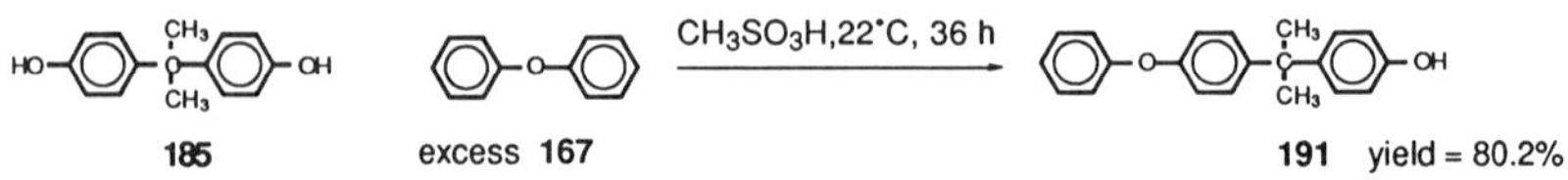

(21)

When CH_3SO_3H was used as initiator in 3,5-dimethylphenol at 65° C ditransalkylated product was the major product (71.1%). In addition a small amount of oligomer (n = 1, 9.3%) and monotransalkylated (6.7%) products were formed [Eq. (22)].

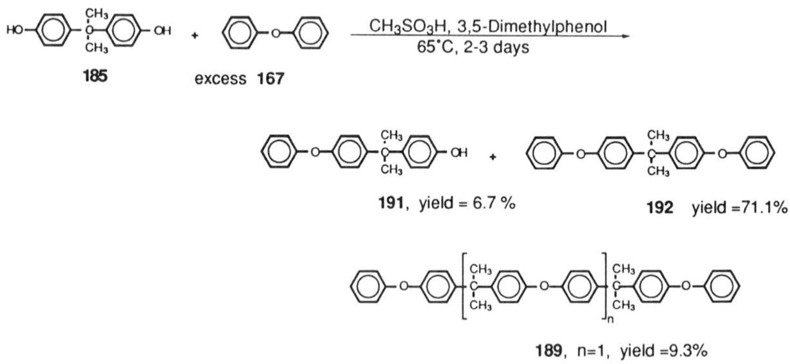

(22)

III. SULFUR-BASED CATIONS AS PROPAGATING SPECIES

A. Sulfonium Ions in the Synthesis of Poly(phenylene Sulfide)s

Cationic propagating species that contain positively charged sulfur atoms include sulfonium and sulfonylium cations. Sulfonium cations have been suggested to be involved in the synthesis of poly(phenylene sulfide) (PPS).

PPS has chemical, thermal, electronic, and mechanical properties that make it of interest as a high-performance engineering plastic [100]. However, the high crystallinity of PPS causes solubility problems in the synthesis of high molecular weight PPS. The original method for obtaining PPS was the reaction of *p*-halothiophenoxy alkali metal salts at high temperatures [101]. A commercial process for the synthesis of PPS involves the high temperature, high-pressure polymerization reaction of *p*-dichlorobenzene and sodium sulfide in NMP [Eq. (23)] [102–104]. The mechanism of this reaction has been controversial [100,105]. One suggestion was that the propagation steps of this reaction involved sulfenyl radicals, aromatic radicals, and cation radicals (see below) [106,107]. A later study accounted for all of the experimental data with an S_NAr mechanism [108].

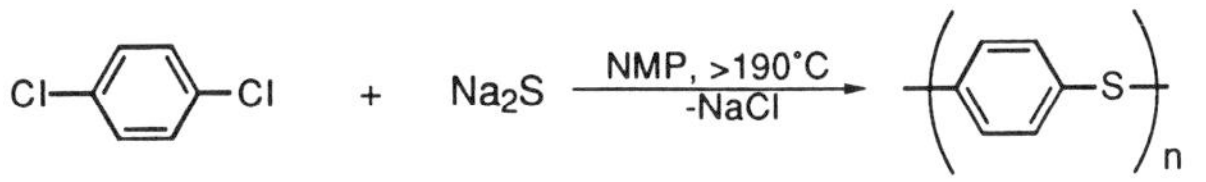

(23)

193 194

Cationic-oxidative polymerization reactions have been developed to synthesize PPS under less severe conditions. For example, PPS has been obtained by a cationic-oxidative polymerization of thiophenol or diphenyl disulfide at room temperature. However, the resulting PPS has a low molecular weight due to premature precipitation. The cationic-oxidative polymerization reaction has been performed electrolytically [109] [Eq. (24)], with Lewis acids [110] [Eq. (25)], with 2,3-dichloro-5,6-dicyano-*p*-benzoquinone (DDQ) [111] [Eq. (26)], and with O_2 in the presence of a catalytic amount of $VO(acac)_2$ [112] [Eq. (27)].

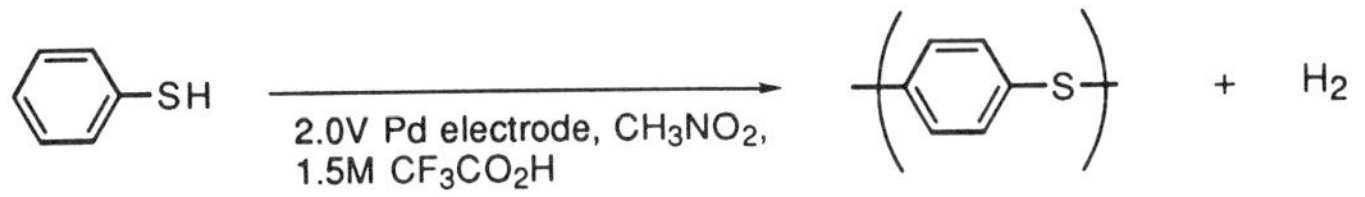

(24)

195 194

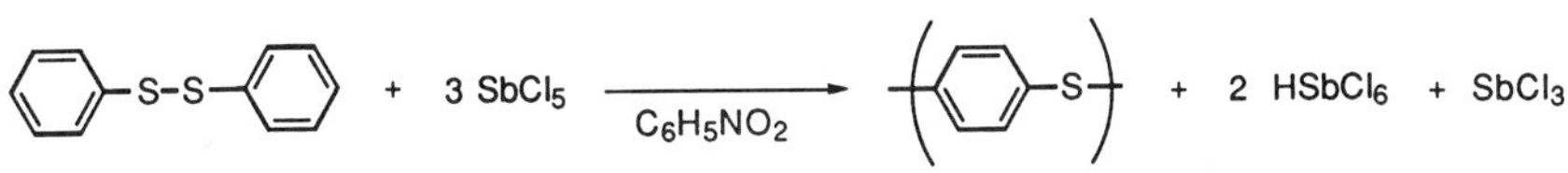

196 194

(25)

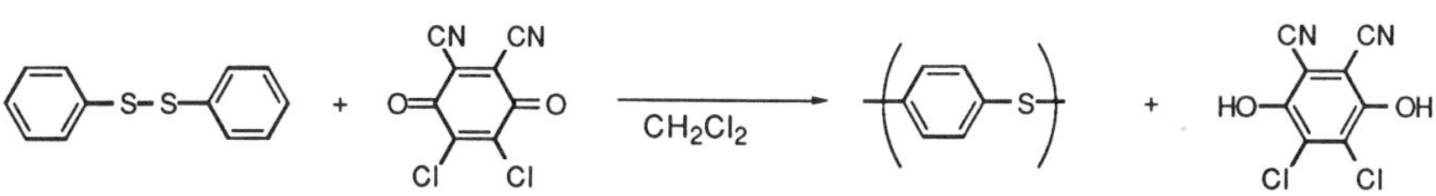

196 DDQ 194 DDH

(26)

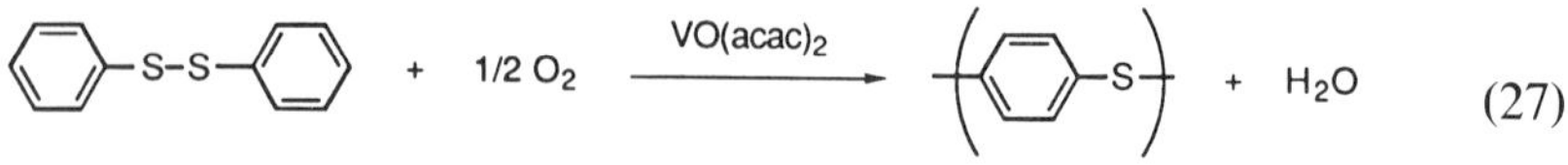

196 194

These cationic-oxidative polymerization reactions produced linear polymers. The anodic oxidation of thiophenol gives diphenyl sulfide. However, when 1.5 *M* CF_3CO_2H was added to the reaction cell and the oxidation was performed with a Pd electrode at 2.0 V in CH_3NO_2, no salts were detected by atomic absorption analysis (10^{-3}% sensitivity limit) [109]. The product's melting point (T_m) was between 180° C and 190° C. The melting point was used to estimate that the molecular weight was greater than 10^3. The reaction could also be performed with H_2SO_4 or stannic chloride as the acid, although molecular weight and reaction yields were affected [109]. The reaction was inhibited by basic solvents such as CH_3OH, DMF, and H_2O [109].

Linear PPS (MW > 10^3) has also been obtained in 88% yield by the reaction of equimolar amounts of diphenyl sulfide and $SbCl_5$ in nitrobenzene at room temperature for 24 hr [110]. Other Lewis acids ($AlCl_3$, $TiCl_4$, $MoCl_5$) were also used. Substituted PPS, including 2,6-disubstituted PPS, were also obtained by this method [Eq. (28)] [110].

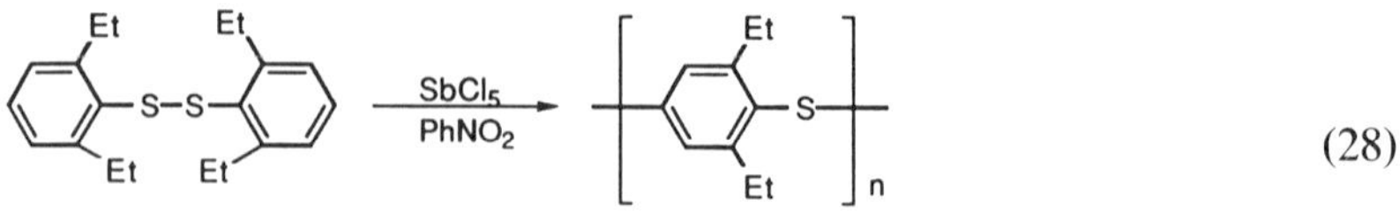

197 198

The effect of alkyl group substituents upon the molecular weight of PPS was investigated [113]. A 50% yield and low molecular weight (M_n = 450) were obtained with no alkyl substituents (R = R′ = H) when the polymerization was performed with $AlCl_3$ in CH_3NO_2 [Eq. (29)].

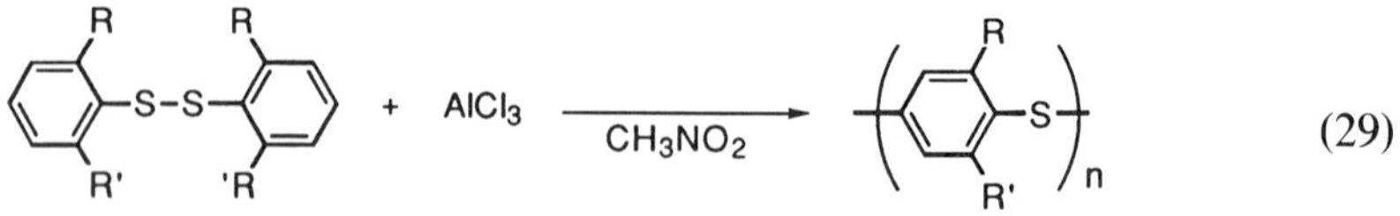

199 200

The yield of PPS decreased when the reaction temperature was increased. This was partially due to the formation of the by-product thianthrene. Improved yields (86%) and molecular weights (M_n = 10,000) were obtained when two ethyl substituents (R = R′ = Et) were present. The increase in yield was attributed to both an increase in polymer solubility and a decrease in the formation of thianthrene by-products [113]. The presence of 2,6-dialkyl substituents increases the solubility of the polymer and suppresses the thianthrene formation. Examples of polymers synthesized by this method are shown in Fig. 30 [113].

The most effective Lewis acids were $FeCl_3$, $SbCl_5$, and $MoCl_5$. They were effective when used in a 2-fold ratio to monomer. Some other Lewis acids were effective, although a 10 equimolar ratio was required for $AlCl_3$, $TiCl_4$, and $BF_3{\cdot}OEt_2$ [113].

The function of $SbCl_5$ in the polymerization of diphenyl disulfide is as an oxidizing agent, rather than that of an initiator or catalyst (Fig. 31). The quantitative oxidation of diphenyl disulfide with $SbCl_5$ gives phenyl bis(phenylthio)sulfonium ion **203** [114]. This cation electrophilically attacks the phenyl ring of **196** to form the intermediate complex **204**. Elimination of **196** and H^+ gives **205**. The polymerization then proceeds by repeated oxidation of the S—S bond of oligomers followed by electrophilic substitution reactions to give PPS with structures **208** and **209** [115]. Thus, disulfide linkages are present in the main chain of PPS synthesized by this method [115].

The intermediate **203** was proposed partly on the basis of the reactivity of the nonpolymerizable dimethyl disulfide with $SbCl_5$ [Eq. (30)]. The methyl bis(methylthio)sulfonium ion, **211**, was isolated and characterized [115].

$$3\ CH_3{-}S{-}S{-}CH_3 + 3\ SbCl_5 \longrightarrow 2\ CH_3{-}\overset{\oplus}{S}(SCH_3){-}S{-}CH_3 + SbCl_6^- + SbCl_3 \qquad (30)$$

210 **211**

The phenyl bis(phenylthio) sulfonium cation, **203**, was detected by ^{13}C NMR analysis of a reaction mixture of diphenyl disulfide and $SbCl_5$ in CH_2Cl_2 at $-40°$ C. No signals due to radical species were detected by ESR. The generation of **203** in the presence of excess diphenyl disulfide leads only to the formation of trimer **205** (Fig. 32). The equimolar addition

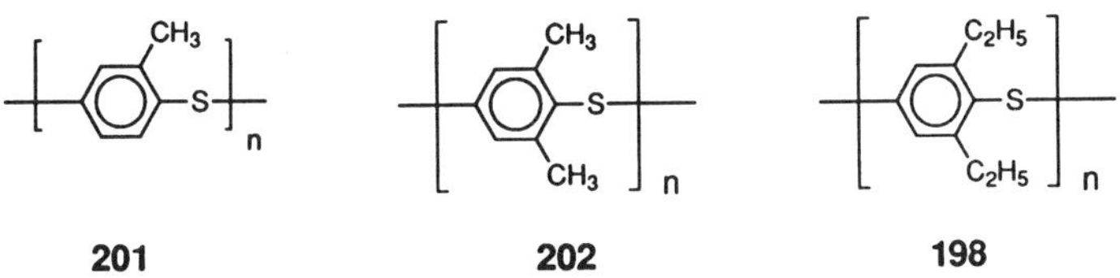

201 **202** **198**

Figure 30 PPS with alkyl substituents and enhanced solubility. (From Ref. 113.)

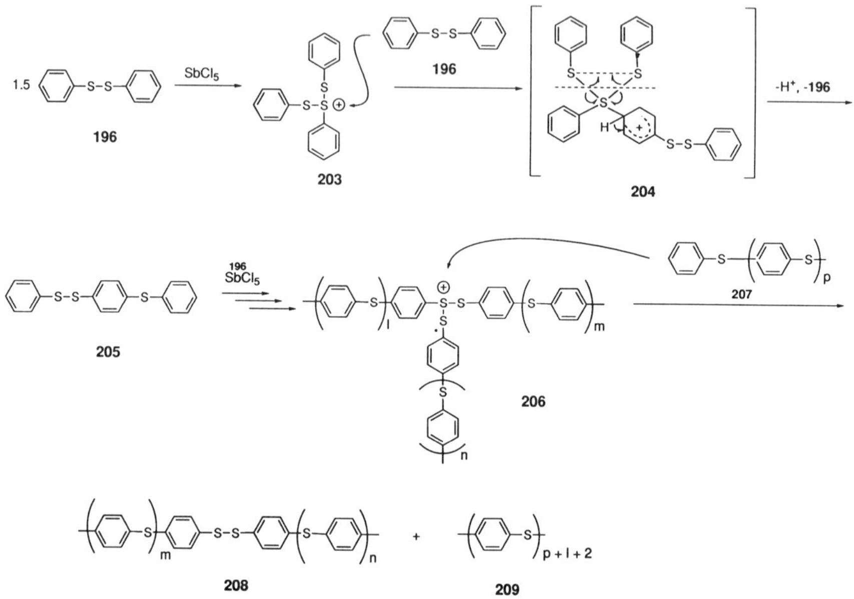

Figure 31 Mechanism of the $SbCl_5$ Mediated Synthesis of Polysulfides (Ref. 115).

of $SbCl_5$ to this solution yields PPS. Thus a stepwise polymerization mechanism is indicated rather than a chain reaction. Also consistent with a cationic mechanism is the suppression of the polymerization reaction in basic solvents.

Quinone-oxidizing agents were also effective oxidizing agents. High yields (95%) of ultrapure PPS were obtained via the oxidative polymeriza-

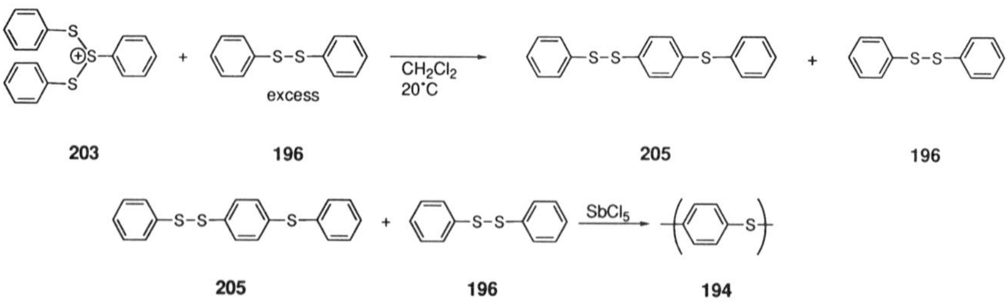

Figure 32 Evidence of a stepwise polymerization mechanism by the formation of trimer **205** upon treatment of the phenyl bis(phenylthio) sulfonium cation with excess diphenyl disulfide. (From Ref. 115.)

tion of diphenyl disulfides with 2,3-dichloro-5,6-dicyano-*p*-benzoquinone (DDQ) in CH_2Cl_2 at room temperature [111]. A catalytic amount of CF_3SO_3H was necessary for the oxidation reaction. Molecular weights were low ($M_n > 10^3$) due to the polymers low solubility in CH_2Cl_2. The mechanism proceeds similarly to the previous reactions with the phenyl-bis(phenylthio)sulfonium cation being formed through the intermediacy of a charge transfer complex (Fig. 33).

The oxidative polymerization of diphenyl disulfide with O_2 in the presence of a catalytic amount of $VO(acac)_2$ also produced high yields (95%) of PPS [112,116]. Reactions were performed in CF_3SO_3H and $(CF_3CO)_2O$ in $ClCH_2CH_2Cl$ at room temperature. Number-average molecular weights (M_n) of approximately 10^3 were obtained. Once again, the low molecular weights were obtained because of the insolubility of PPS in the reaction medium at room temperature [117]. There is a 1.0 V potential gap between O_2 and the disulfide. This potential gap was bridged through the use of $VO(acac)_2$ which has readily accessible oxidation states of III, IV, and V (Fig. 34).

The synthesis of high molecular weight PPS via a cationic oxidative polymerization route was inhibited by premature precipitation of PPS. A method for the synthesis of high molecular weight PPS from a soluble precursor under mild conditions was developed. Poly(sulfonium cation) was used as a precursor to PPS [Eq. (31)] [118].

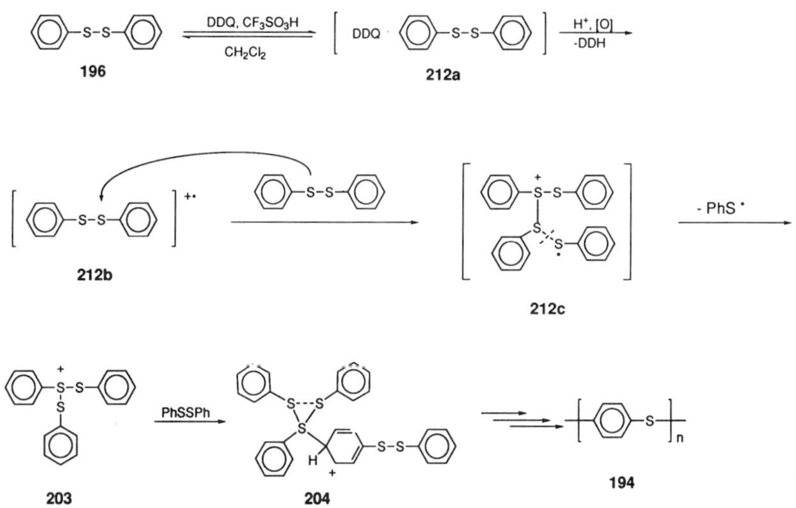

Figure 33 Polymerization of diphenyl sulfide mediated by DDQ. (From Refs. 111 and 114.)

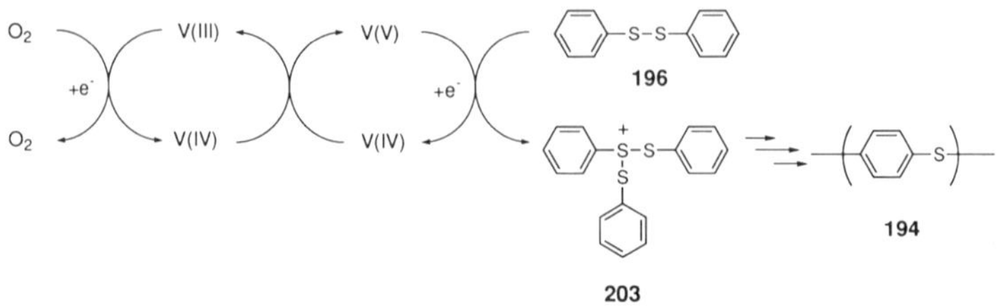

Figure 34 VO(acac)$_2$-catalyzed oxidative polymerization synthesis of PPS. (From Ref. 116.)

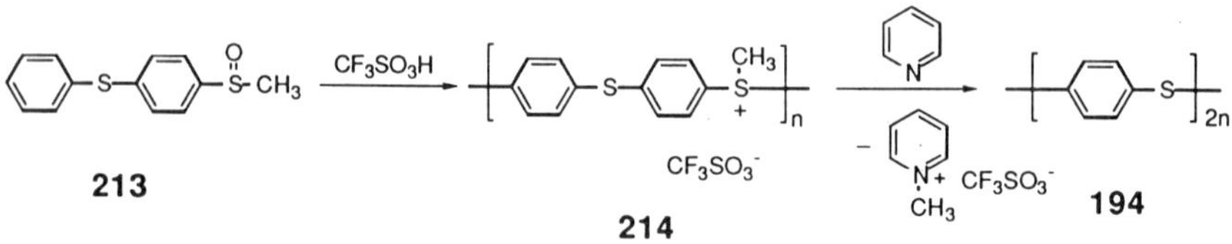

(31)

Poly[methyl (4-phenylthio)phenylsulfonium trifluoromethanesulfonate] (**214**) was prepared by stirring methyl(*p*-thiophenoxy)phenyl sulfoxide in CF_3SO_3H at room temperature for 24 hr followed by precipitation in water. The polymerization was influenced by the acidity of the mixture. The polymerization was accelerated by dehydration using P_2O_5. The structure was confirmed to contain 1,4-phenylene units by spectroscopic (IR, NMR) methods. Linear PPS was obtained by demethylating **214** in pyridine at 115° C for 5 hr. The weight-average molecular weight determined by GPC was 2×10^5 [118]. The preparation of **214** with a ClO_4^- counterion has been accomplished by the oxidative polymerization of **213** with DDQ (or electrochemically) followed by precipitation from perchloric acid [119].

The addition of P_2O_5 to the reaction mixture resulted in higher molecular weights ($M_w = 2.4 \times 10^5$) [120]. The P_2O_5 reacted with the water that was formed during the course of the polymerization. However, M_w/M_n was 5.6 due to the high viscosity of the reaction mixture. Lower melting points ($T_m = 260°$ C vs. $T_m = 280°$ C for commercial Ryton V-1) were obtained. This was presumably due to the existence of a small amount of sulfonium cations. It was also possible that ortho substituents in the main chain could cause a decrease in T_m. However, no evidence of their presence was detected by NMR. CH_3SO_3H was not acidic enough for the polymerization reaction to occur effectively.

A polymer with a sulfone group on every other unit of the chain was obtained by the oxidation of **214** with H_2O_2 [Eq. (32)]. Thus, poly(sulfonyl-1,4-phenylenethio-1,4-phenylene) (PSPT) was synthesized in the presence of H_2O_2 and CF_3CO_2H in $CHCl_3$ via a poly(sulfonium cation) [121].

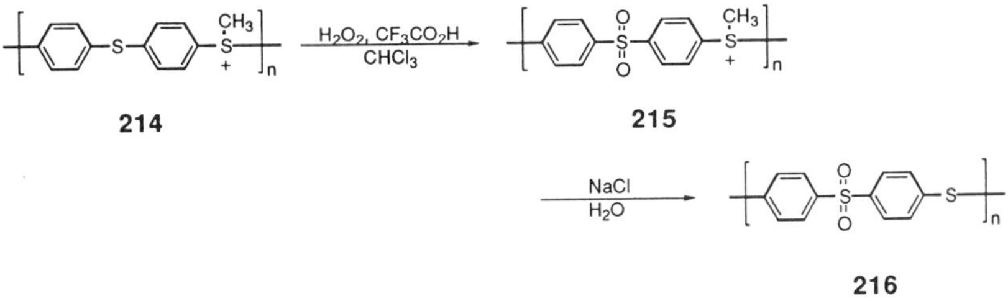

(32)

Molecular weights of $M_w = 23{,}500$ ($M_w/M_n = 2.4$) were obtained. Higher molecular weights ($M_w = 1.9 \times 10^5$) were obtained using high molecular weight polycation under the same conditions [121].

Arylated and alkylated poly(*p*-phenylene sulfide) derivatives have been synthesized by the method shown in Eq. (33). Sulfonium ion polymers, such as **217**, have potential applications as electronic components and lithographic materials [122].

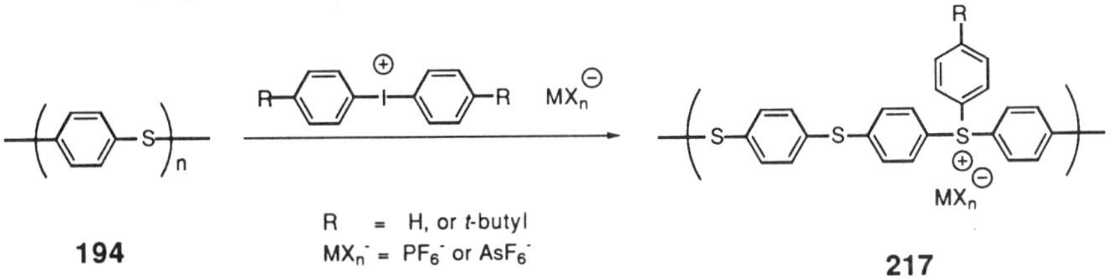

(33)

The solubility problems have also been averted by polymerizing methyl phenyl sulfide to give a soluble polymer which can be dealkylated to give PPS [Eq. (34)]. In this case, the Ce(III)-catalyzed oxidative polymerization used O_2 as the oxidant [123].

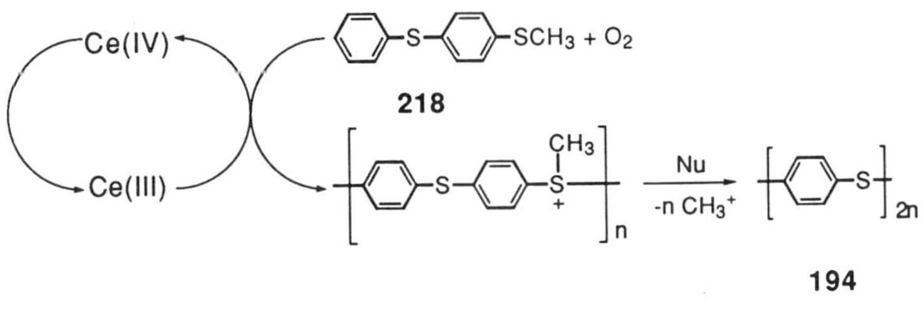

(34)

The structural properties of PPS have been improved by blending it with other polymers. The addition of a perfluoroalkane polymer with PPS has been reported to increase the toughness and sliding properties of the PPS blend [124]. The improvement of these properties in PPS alloys has been demonstrated to be a function of the degree of dispersibility. Therefore, a PPS block copolymer containing perfluoroalkane blocks is expected to possess improved properties. This type of block copolymer was obtained by the reaction of monomers containing central perfluoroalkane segments connected to terminal arylthio groups with S_2Cl_2 in the presence of DDQ in CH_2Cl_2 [125]. The oxidative polymerization of α,ω-bis(phenyleneoxy-1,4-phenylenethio)perfluorohexane (**220**) gave the corresponding sulfide-linked block copolymer (**221**) [Eq. (35)]. The reaction proceeded quantitatively with M_w = 57,500.

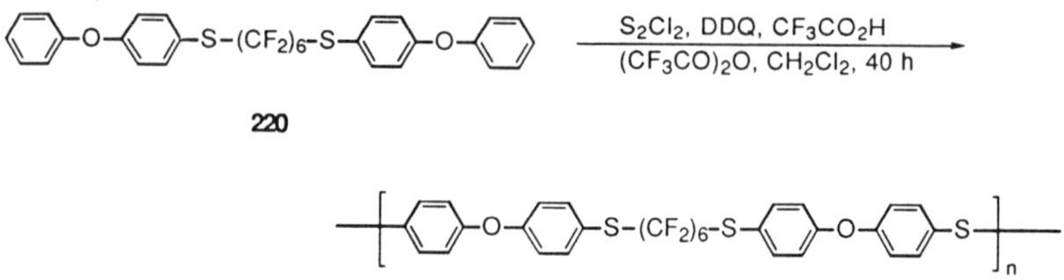

220

221

yield = 100%; M_w = 57,500; M_w/M_n = 2.6

(35)

A much lower molecular weight (M_w = 1700) was obtained in the oxidative polymerization of α,ω-bis(phenylenethio)perfluorohexane, **222** [Eq. (36)]. In this case the strong electron-withdrawing effect of the perfluorohexyl group inhibits the reaction by lowering the electron density at the *p*-position of the phenyl group.

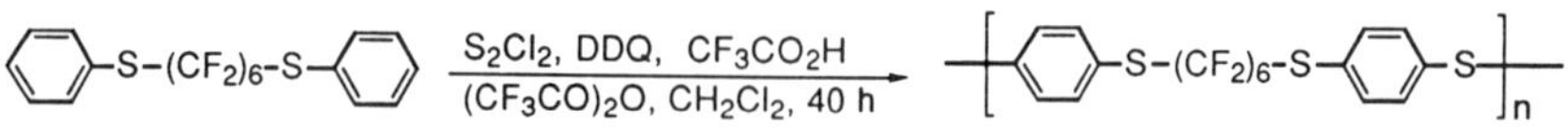

222 **223**

yield = 65%; M_w = 1700; M_w/M_n = 1.6

(36)

The oxidation of the sulfide groups of copolymer **221** to sulfone groups was accomplished by reaction with hydrogen peroxide in trifluoroacetic acid and $CHCl_3$. A lowering of the molecular weight occurred during the oxidation of **221** to **224**. This was due to the cleavage of disulfide bonds in the polymer chain [Eq. (37)]

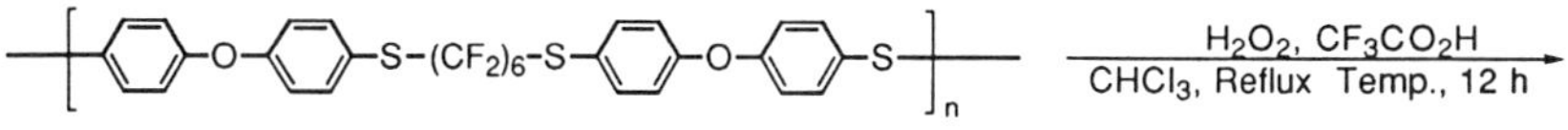

221

$M_w = 57{,}500$, $M_w/M_n = 2.6$, $T_g = 34°C$

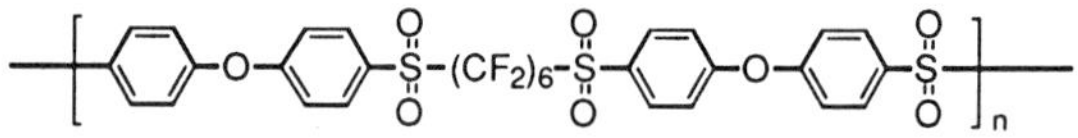

224

yield = 92%, $M_w = 48{,}500$, $M_w/M_n = 2.1$, $T_g = 147°C$ (37)

A triblock copolymer consisting of one central perfluorinated block and terminal polysulfide blocks was synthesized by the procedure shown in Fig. 35.

Oligo(*p*-phenylene sulfide)s with reactive end groups have recently been prepared by an oxidative polymerization method [126]. Polymers with chloro, bromo, iodo, and carboxyl end groups were obtained.

B. Sulfonylium Cations in the Formation of Polysulfones

Aromatic polysulfones are a commercially important class of thermoplastic polymers [127]. They have highly desirable qualities such as chemical inertness, thermal stability, and flame retardency [128,129]. Although a number of methods are available for the synthesis of polysulfones [130,131,132], step polymerization methods are the most widely used industrially [127]. Polysulfones have been synthesized with the involvement of sulfonylium cations as propagating species.

Polysulfones can be prepared by Friedel-Crafts sulfonylation reactions [133–150]. The carbenium ion intermediates can be generated from the Lewis acid-catalyzed reaction of involving arene sulfonyl chloride [Eq. (38)], or by the acid-catalyzed reaction involving arene sulfonic acid [Eq. (39)].

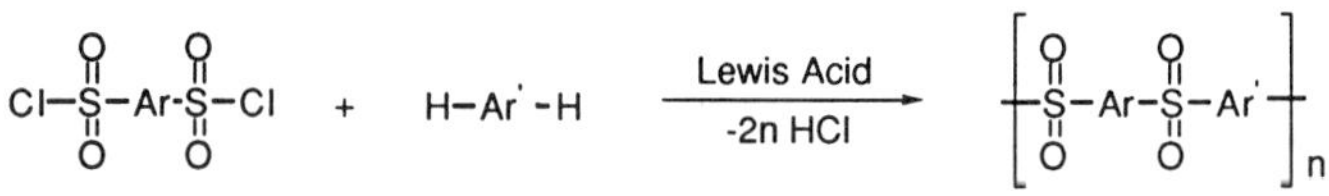

228 **229** **230** (38)

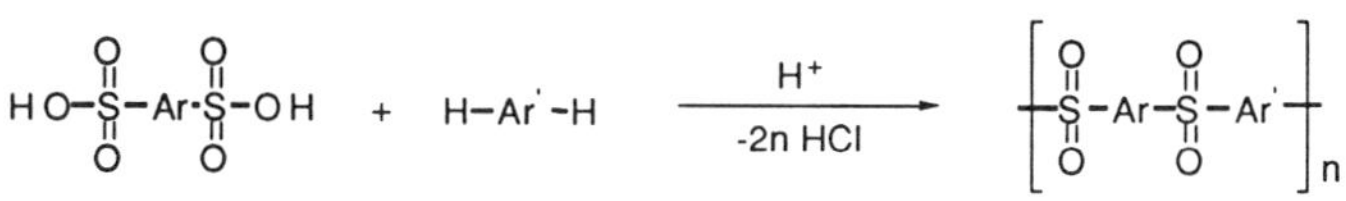

231 **229** **230** (39)

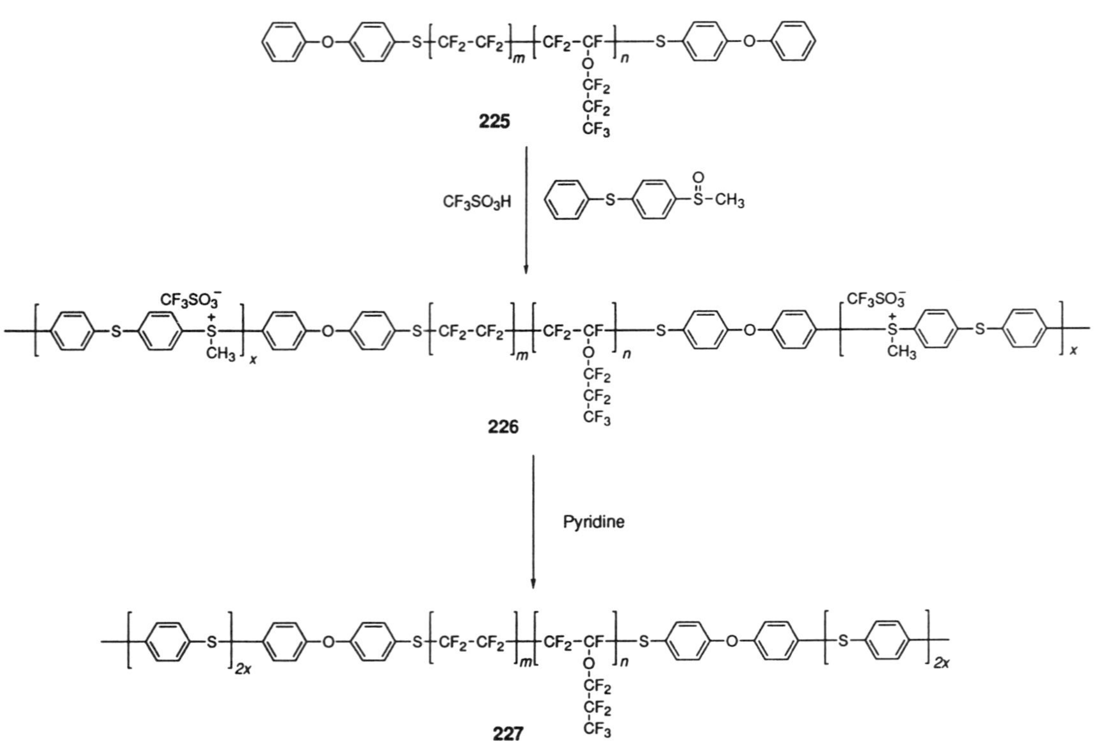

Figure 35 Synthesis of block copolymer with central perfluorinated block and terminal PPS blocks. (From Ref. 125.)

A typical Friedel-Crafts sulfonylation reaction is illustrated in Eq. (40). This reaction involves the reaction of the sulfonylium cation $PhSO_2^+$ with benzene to form an intermediate carbocation (**233**) [Eq. (40)]. Deprotonation of **233** gives the sulfone (**234**). Electron-donating substituents activate the ortho and para groups of the phenyl ring by resonance effects. Conversely electron-withdrawing groups deactivate the phenyl ring and are meta-directors. Reaction at the para-position is generally favored in the polysulfonylation reaction for steric reasons.

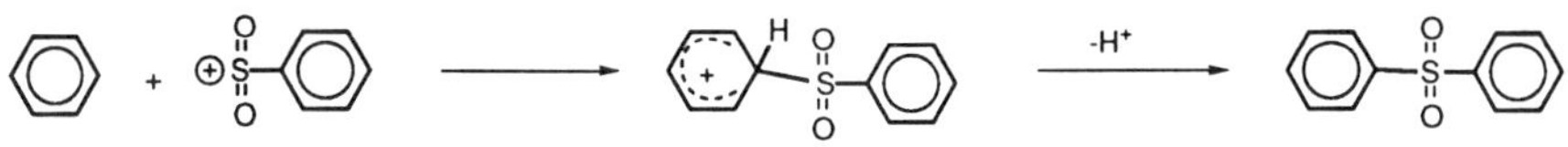

232 **233** **234**

(40)

Lewis acids such as $AlCl_3$, $AlBr_3$, $FeCl_3$, and $SbCl_5$ have been used [151]. Friedel-Craft sulfonylation reactions typically require more than one mole equivalent of Lewis acid per sulfonyl halide. This is apparently due to the coordination of one equivalent of the Lewis acid with the sulfonyl group.

The aromatic substrate which undergoes reaction is deactivated by the sulfonyl group toward further electrophilic substitution reactions. Polymers are generally synthesized from monomers having aromatic groups which are separated by some type of spacer that prevents conjugation by resonance between the two rings.

For example, H-Ar′-H in Eq. (38) could be diphenyl ether [Eq. (41)] [138]. However, it is important to note that in this case the unsubstituted phenyl group of the product of the first electrophilic substitution reaction is now less reactive. This product will give a different ratio of *o*- and *p*-substitution. More regular polymers can be obtained from the polymerization reactions shown in Eqs. (42) and (43). High selectivity (>99% *p*-substitution) was obtained in the polymerization of **239** [Eq. (43)] [138,139].

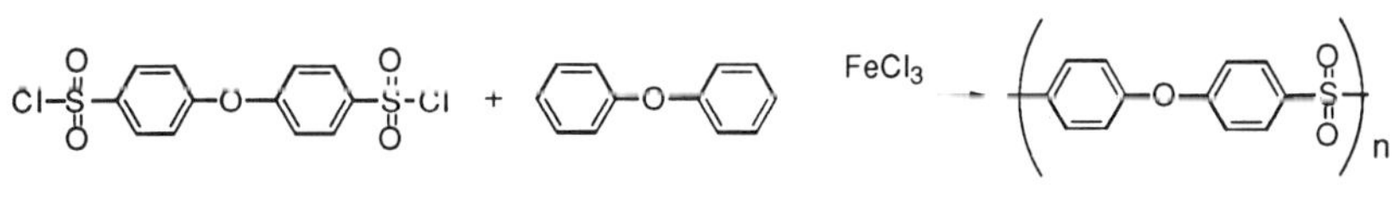

235 **167** **236**

(41)

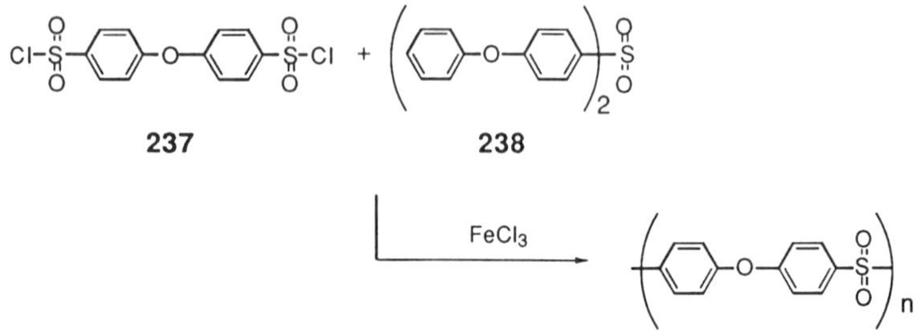

237 238 236

(42)

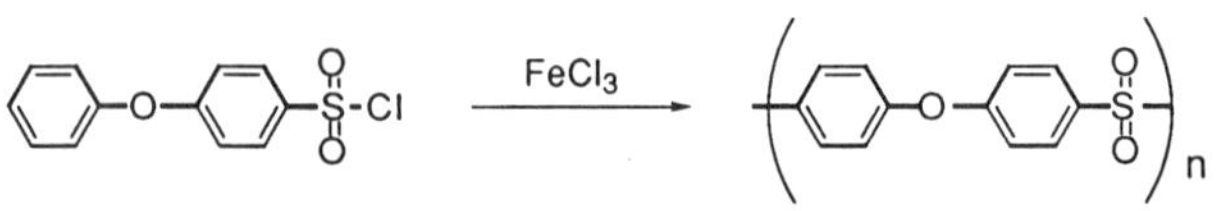

239 236

(43)

Examples of other AB monomers have been polymerized to yield polysulfones are the following: α-naphthalenesulfonyl chloride, p-thiophenoxybenzenesulfonyl chloride, p-phenylbenzenesulfonyl chloride, and 2-dibenzofuransulfonyl chloride [131].

The Lewis acid catalyst has been reported to undergo a substitution reaction at the carbon ortho to the oxygen of the ether linkage [Eq. (44)]. This reaction is more significant above 120° C. The substitution at two adjacent o-carbons results in the formation of a robust six-membered cyclic structure, **242** [152]. Facile hydrolysis of **241** during work-up gave **240**. Alternatively, **241** can participate in cross-linking reactions.

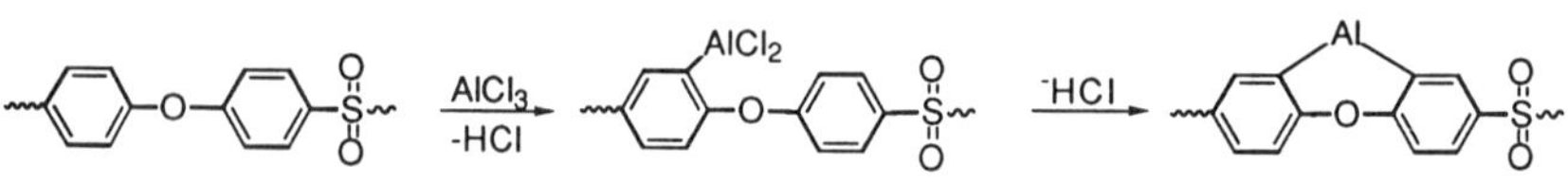

240 241 242

(44)

The use of high catalyst concentrations and/or high temperatures to obtain high molecular weights can lead to undesired products that are the result of increased ortho (or even meta) substitution, cross-linking reactions, and the formation of undesired organometallic species. Low catalysts concentrations (0.1–5 mol%) have been used to obtain high molecular weight sulfones with more regular structures. These reactions typically involve $FeCl_3$ and can be performed in typical Friedel-Crafts solvents or in bulk [127].

Polysulfonates can also be obtained using the condensation reaction of aryl sulfonic acids with arenes [Eq. (45)] [134].

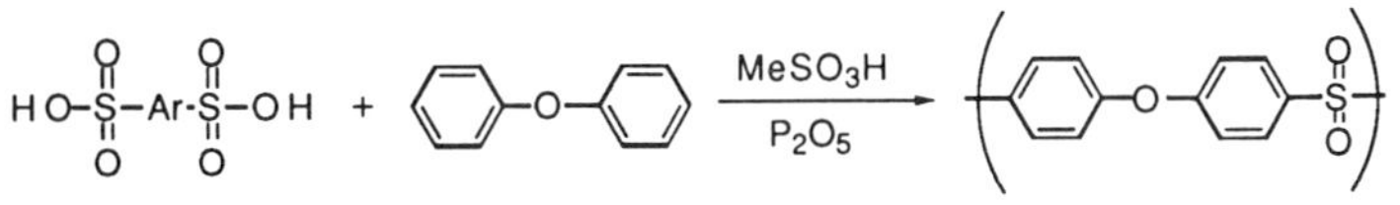

231 167 236

(45)

IV. ACYLIUM CATIONS IN THE SYNTHESIS OF POLY(ARYLETHER KETONE)S

Acyl cations are involved as propagating species in the synthesis of poly(ether ketone)s. Poly(ether ketone)s are a class of thermoplastic crystalline polymers that have many desirable properties that make them useful as high-performance engineering materials [153,154]. The poly(ether ketone)s with the most useful properties are actually para-linked poly(arylether ketone)s (PAEKs). They have excellent chemical resistance to oxidation and hydrolysis, high thermal stability, and many useful mechanical properties. Unlike some other materials with similar properties they are readily melt processable using conventional equipment. In addition, their mechanical properties are not affected deleteriously by most solvents. These polymers are usually crystalline. PAEKs contain arene groups joined by ether and carbonyl linkages. For example, two commercial poly(ether ketone)s are PEK and PEEK (Fig. 36).

Two general methods, electrophilic and nucleophilic, predominate the synthesis PAEKs [153,154]. The nucleophilic route to PAEKs was used in the polymerization of various combinations of bis-electrophiles

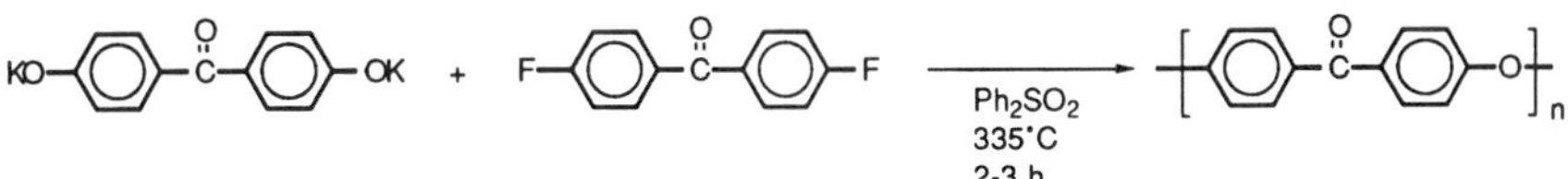

245 246 243

(46)

such as bis-4-fluorophenyl ketone and bisphenolates such as the potassium salt of bis-4-hydroxyphenylketone [155] [Eq. (46)] or by homopolymerization of various AB monomers such as **247** [Eq. (47)]. This subject has been reviewed extensively [153,154].

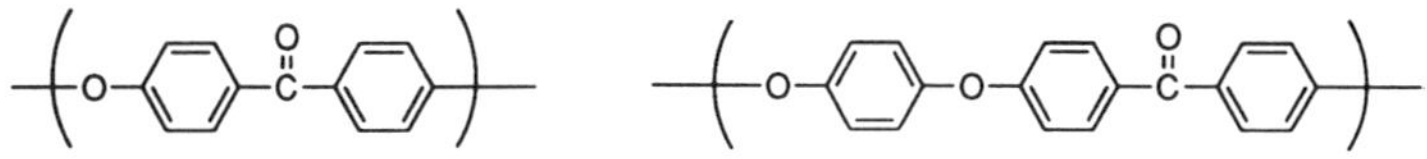

PEK **243** PEEK **244**

Figure 36 Structures of PEK and PEEK.

247 243 (47)

One of the problems in PAEK synthesis is maintaining solubility of the polymer long enough to obtain high molecular weights. A number of methods have been used to increase the polymer's solubility during the polymerization process. Two general methods that have been successfully used under electrophilic reaction conditions are (a) the protonation of carbonyl groups and (b) the presence of excess $AlCl_3$.

The first report of a totally aromatic PAEK [156] [Eq. (48)] and the first report of PEK [157] [Eq. (49)] involved electrophilic reactions. However, low molecular weight polymers were obtained due to their crystallization from the reaction mixture.

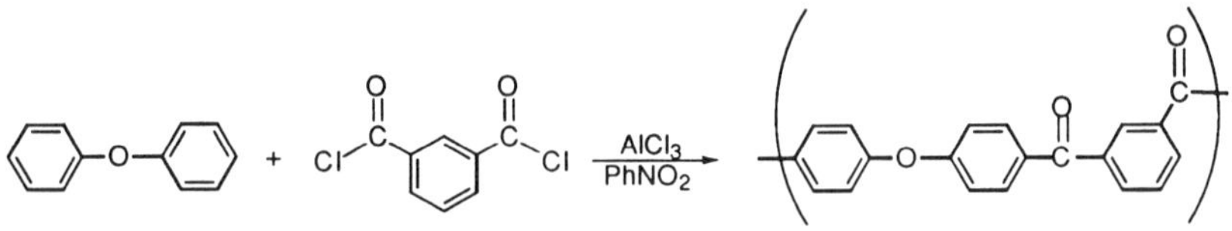

167 248 249

(48)

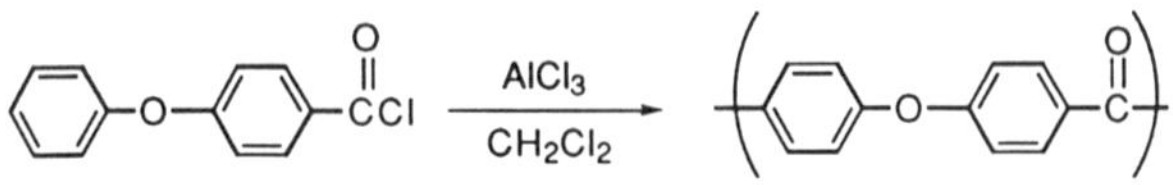

250 243 (49)

PEK was also prepared in polyphosphoric acid [Eq. (50)] [158]. The protonation of the carbonyl groups helped to maintain solubility.

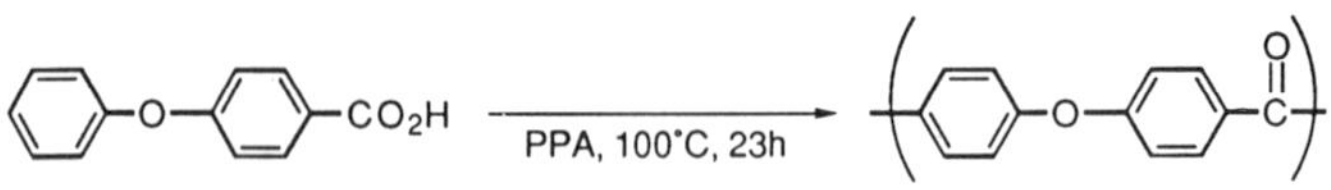

251 243 IV = 0.53 (50)

The solubility was greatly enhanced by the use of liquid HF. HF was found to be an excellent solvent for PAEKs and was used as a polymerization solvent in conjunction with $BF_3{\cdot}HF$ as the catalyst [Eq. (51)] [159].

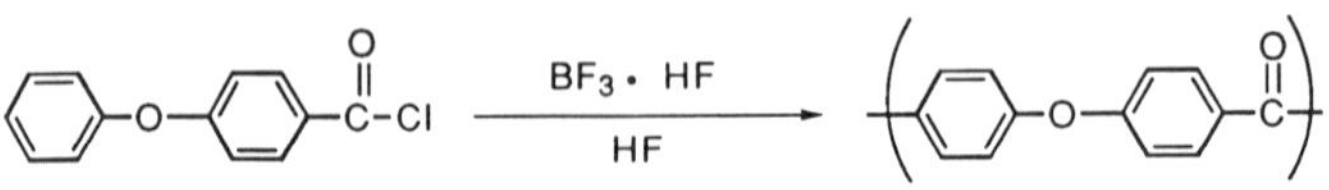

250 243 IV = 2.3-1 (51)

Another approach to preventing premature precipitation of PAEK is to modify its structure. For example, the crystallinity of the resulting PAEKs were decreased by using 1,1′-dialkyl-substituted diphenyl ethers [Eq. (52)] [160].

252 248 253

(52)

The addition of excess $AlCl_3$ to the reaction mixture results in a significant increase in the solubility of the growing polymer chain. However, $AlCl_3$ is also involved in a number of undesired reactions. Complexation of $AlCl_3$ with the terminal aryloxy group inhibits chain growth. Side reactions such as ortho-substitution and cross-linking decrease the desired physical properties of the polymer. Alkylation of the para-aryloxy group effectively endcaps the polymer and limits molecular weight. Meta-alkylation provides sites for cross-linking reactions to occur. These side reactions increase in frequency as the reaction temperature and time are increased. The significance of these undesired reactions can be reduced by the addition of a Lewis base to the reaction mixture [161].

The Lewis base acts as a controlling agent and functions to suppress side reactions which would lead to structures that would degrade or cross-link during the elevated temperatures of melt processing. Typical reaction conditions involve the use of one equivalent of Lewis acid per equivalent carbonyl group, plus one equivalent per equivalent of Lewis base, plus a catalytic amount (0.05–0.03 equivalent Lewis acid/acid halide) and a nonprotic solvent. The exact role of the Lewis base is not clearly understood. The Lewis acid/Lewis base complex functions as a solvent for the polymer-Lewis acid complex. Thus it facilitates keeping the polymer in solution or in a reactive gel state. The Lewis base also reduces the alkylation of the growing polymer with solvents such as dichloroethane in the terminal ortho- and para-positions. It does so by competing with these solvents for the Lewis acid. In the absence of Lewis base the amount of solvent can be used to control the polymerization. In any case the Lewis acid/Lewis base complex apparently has a large affect on the catalytic activity of the Lewis acid. High molecular weight means polymers with inherent viscosity greater than 0.6. To obtain useful mechanical properties the viscosity must be greater than 0.6, and for melt processability the viscosity should be no higher than 2.0. The molecular weight can also be controlled by end-capping agents [161].

The acid CF_3SO_3H has been used as a solvent and catalyst to promote the rapid polycondensation of carboxylic acids and aromatic ethers at room temperature to give PAEK. Premature polymer precipitation did not occur in CF_3SO_3H. The substitution reaction takes place at the para-position to the ether linkage (Fig. 37) [162].

The reaction apparently proceeds by the electrophilic attack of an acylium ion or protonated mixed anhydride $[ArCO(H)OSO_2CF_3]^+$, upon the para-position of an aromatic ether (Fig. 37). Loss of a proton results in the formation of **256**. The nonsubstituted aryl group of the diphenyl ether was found to be much less reactive toward electrophilic substitution. This group is deactivated by protonation of the keto group in the strongly acidic environment. Therefore, monomers must be designed so that this type of resonance effect does not inhibit substitution at the second site of substitution [Eq. (53)] [162].

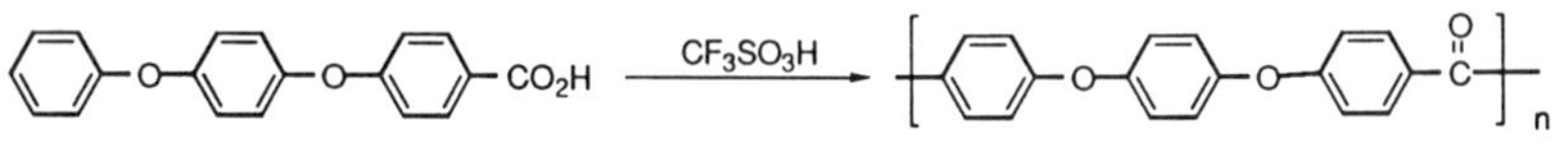

258 259

(53)

Polyaryl(ether-ketone-carbaborane)s have also been synthesized by using CF_3SO_3H as both solvent and catalyst to give linear amorphous polymers [Eq. (54)] [163]. These polymers have low molecular weights (IV = 0.24–0.59) and low thermolytic weight loss upon pyrolysis (10–15% at 1000° C)

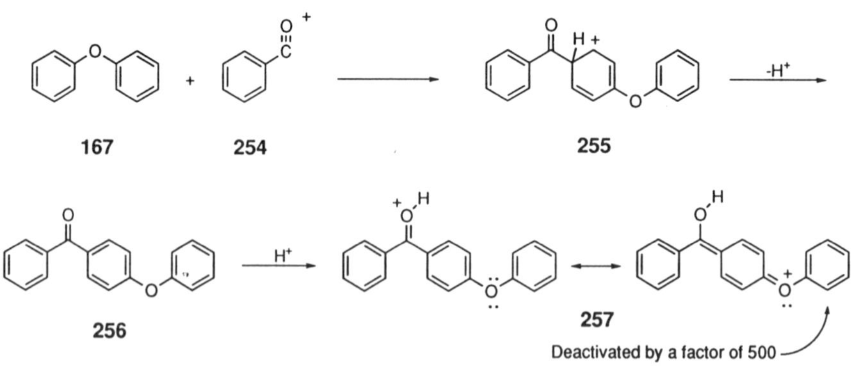

Figure 37 Inhibition of substitution reactions on both phenyl groups of diphenyl ether. (From Ref. 162.)

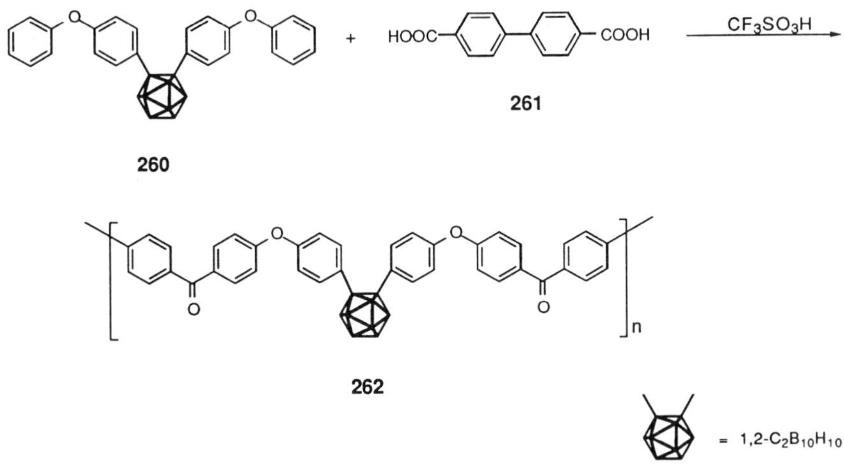

260 261 262

(54)

High molecular weight PAEKs incorporating other groups in the polymer backbone have been synthesized using the $AlCl_3$-Lewis base catalyst system (Fig. 38). For example, PAEKs containing the following main chain groups have been synthesized: imide [164,165], amide [164,166,167], sulfone [164], ester [166], azo [164,166,167], quinoxaline [164,166,167], benzimidazole [166], aliphatic [164], fluoroaliphatic [164], and fluoroaromatic [164].

PAEKs with high molecular weights can be obtained by the polycondensation of dicarboxylic acids containing phenyl ether structures with diphenoxybenzene in P_2O_5/CH_3SO_3H (PPMA) [Eq. (55)]. The self-con-

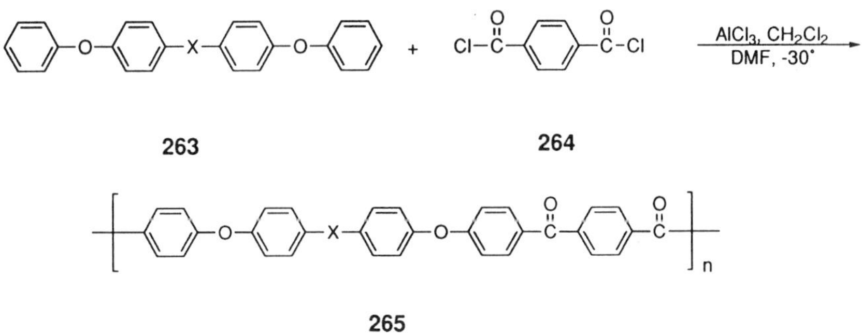

263 264 265

Figure 38 Synthesis of PAEKs incorporating various types of functional groups. Types of *X* groups incorporated: imide, amide, sulfone, ester, azo, quinoxaline, aliphatic, fluoroaliphatic, fluoroaromatic. (From Ref. 164.)

densation of 3-phenoxybenzoic acid in PPMA also results in polymer formation. Terephthalic and isophthalic acids could not be used for this reaction [168]. An related reaction was performed in PPMA to synthesize polyketones [169] containing various groups including dibenzo-18-crown-6 [170], liquid crystalline mesogens [171], and totally aromatic groups [172].

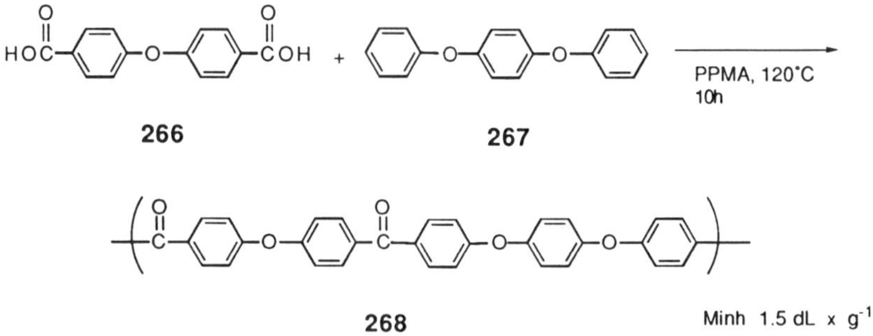

266 267 268

(55)

V. PHENOXENIUM IONS IN THE SYNTHESIS OF POLY(2,6-DIMETHYLPHENYLENE OXIDE)S

Aromatic polyethers can be obtained by the oxidative coupling polymerization of 2,6-disubstituted phenols [173]. The reaction is initiated by passing oxygen through a solution containing the phenol, an amine, and a catalytic amount of a copper (I) salt in an organic solvent. High molecular weight linear poly(phenylene oxide)s (PPO) are obtained in the polymerization of 2,6-dimethylphenol [173]. Usually small amounts of diphenoquinone (DPQ) by-products form [173]. The amount of this by-product formed increases with the increasing size of the substituent groups. DPQ is the only product when 2,6-di-*tert*-butylphenol is reacted [Eq. (56)].

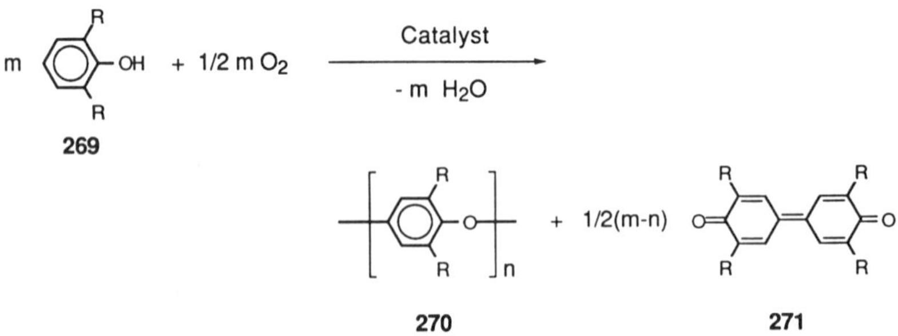

269 270 271

(56)

The telechelic α,ω-bis(2,6-dimethylphenol)-poly(2,6-dimethylphenylene oxide) (PPO-2OH) [174–182] is of interest as a precursor in the synthesis of block copolymers [175] and thermally reactive oligomers [179]. The synthesis has been accomplished by five methods. The first synthetic method was the reaction of a low molecular weight PPO with one phenol chain end with 3,3′,5,5′-tetramethyl-1,4-diphenoquinone. This reaction occurred by a radical mechanism [174]. The second method was the electrophilic condensation of the phenyl chain ends of two PPO-OH molecules with formaldehyde [177,178]. The third method consists of the oxidative copolymerization of 2,6-dimethylphenol with 2,2′-di(4-hydroxy-3,5-dimethylphenyl)propane [176–178]. This reaction proceeds by a radical mechanism. A fourth method was the phase transfer-catalyzed polymerization of 4-bromo-2,6-dimethylphenol in the presence of 2,2-di(4-hydroxy-3,5-dimethylphenyl)propane [181]. This reaction proceeded by a radical-anion mechanism. The fifth method developed was the oxidative coupling polymerization of 2,6-dimethylphenol (DMP) in the presence of tetramethyl bisphenol-A (TMBPA) [Eq. (57)] [182].

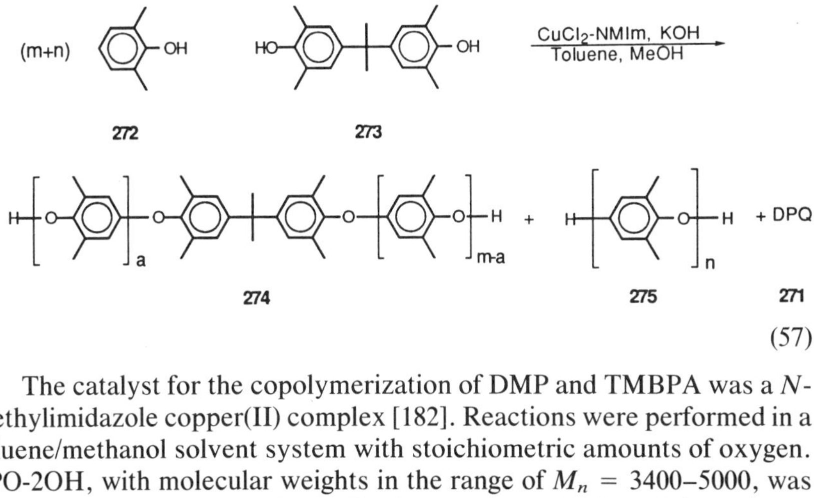

272 273

274 275 271

(57)

The catalyst for the copolymerization of DMP and TMBPA was a *N*-methylimidazole copper(II) complex [182]. Reactions were performed in a toluene/methanol solvent system with stoichiometric amounts of oxygen. PPO-2OH, with molecular weights in the range of M_n = 3400–5000, was obtained as the major product. The homopolymer PPO and the quinone dimer DPQ were identified as side-reaction products. The formation of these side products was reduced by slowly adding the DMP during the course of the polymerization reaction. This lowered the initial concentration of DMP. Thus PPO-2OH was obtained which contained only 0.5–0.6% DPQ and with no PPO detectable by ^{1}H NMR [182].

Imidazole copper(II) complexes had previously functioned as catalysts in the oxidative coupling polymerization of DMP [183–186]. The mechanism which was proposed for this reaction involved the participa-

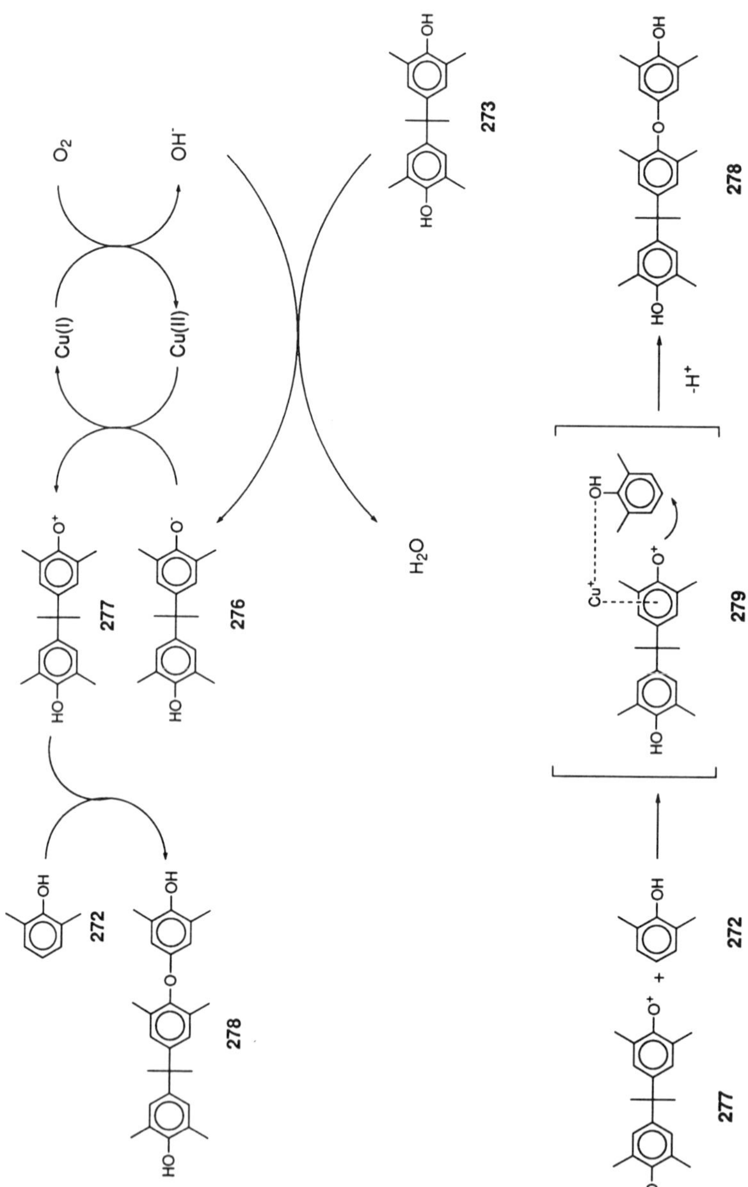

Figure 39 Mechanism of PPO formation when reaction is initiated by the addition of base to TMBPA. (From Ref. 182.)

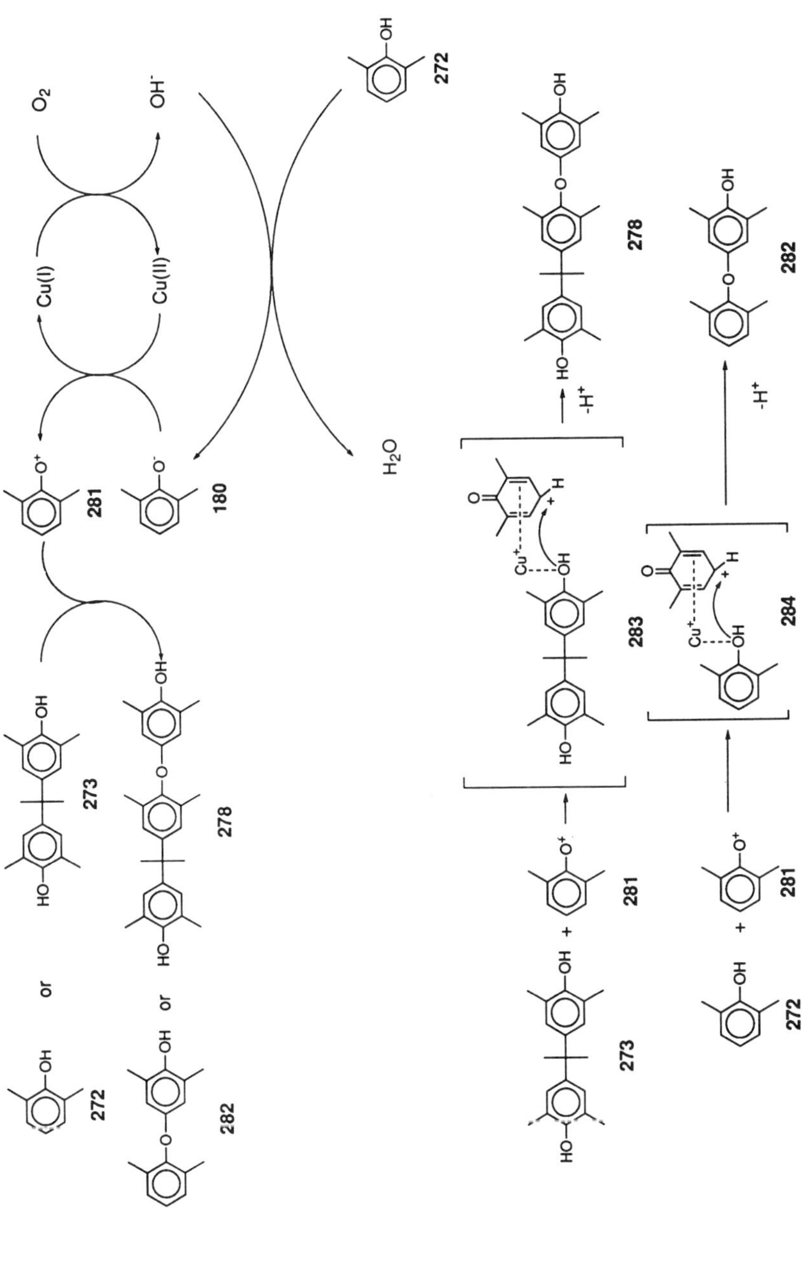

Figure 40 Mechanism of PPO formation when reaction is initiated by the addition of base to DMP. (From Ref. 182.)

tion of intermediate phenoxenium species [187]. A similar mechanism involving a phenoxenium ion intermediate was proposed for the formation of PPO-2OH [182]. Two reaction pathways are apparently operative (Figs. 37 and 38). One pathway involves the initial formation of the phenolate anion of TMBPA (**276**) by deprotonation of **273** with a base (Fig. 39). This anion readily coordinates to Cu(II) and is oxidized to the phenoxenium ion **277**. The Cu(I)-coordinated phenoxenium ion (**279**) then electrophilically attacks the para-position of a DMP molecule to give **278**. The product then participates in further reaction cycles to give the PPO-2OH product. This reaction scheme does not account for the formation of the PPO byproduct. Therefore the competing pathway as shown in Fig. 40 was used to account for PPO formation. This pathway involves the initial formation of the DMP anion (**280**). Once this anion is oxidized to the phenoxenium ion (**281**) it can electrophilically react with a DMP or TMBPA molecule. Thus both PPO and PPO-2OH can be formed by this reaction pathway.

VI. CATION RADICALS

A. Polysulfides

The reaction of *p*-dichlorobenzene and disodium sulfide forms poly(thio-1,4-phenylene) [Eq. (23)]. One of the various mechanisms suggested for this polymerization is a single-electron transfer (SET) process involving the participation of cation radicals (Fig. 41) [106].

B. Aromatic Polyethers and Polyarylenes by the Scholl Reaction

1. *Introduction to Scholl Reaction*

Aromatic polyethers, including poly(ether sulfone)s and poly(ether ketone)s, have been synthesized by the Scholl reaction. In the Scholl reaction a Friedel-Crafts catalysts is used to effectuate the coupling of two aromatic groups to form an aryl-aryl bond, accompanied by the elimination of two aromatic hydrogens [Eq. (58)] [188–190]. This reaction proceeds under oxidative reaction conditions by a cation-radical mechanism [191,192].

$$\underset{\mathbf{297}}{\mathrm{Ar{-}H}} \xrightarrow[-2H^+]{-e^-} \underset{\mathbf{298}}{\mathrm{Ar{-}Ar}} \qquad (58)$$

The first step of the reaction is a single-electron oxidation of the aromatic substrate to form a cation-radical [193,194]. Oxidizing agents for this reaction include Bronsted acids, oxidative Lewis acids, halogens (i.e., Br_2), metal salts (e.g., $Ti(CF_3CO_2)_3$), electron-donor-acceptor complexes, irra-

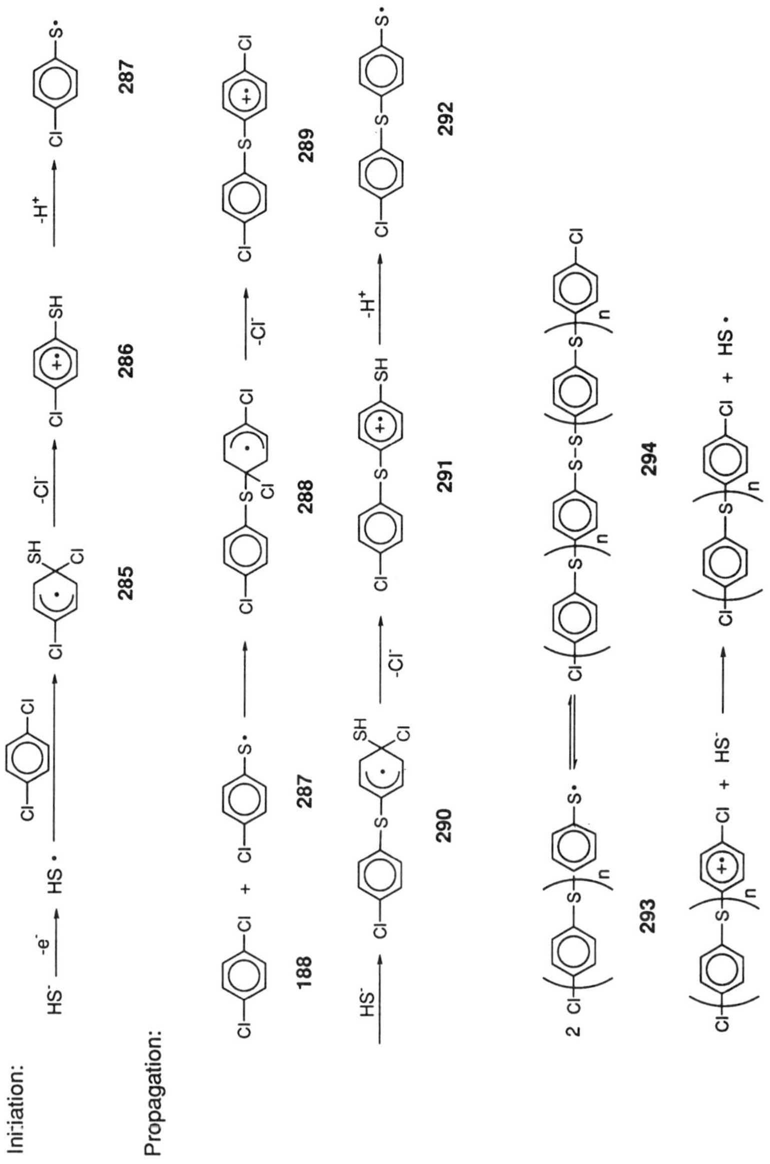

Figure 41 Proposed cation-radical mechanism of PPS formation. (From Ref. 106.)

diations (e.g., x-ray), zeolite surfaces, and anodic electrochemical oxidation [193–197].

Electrochemical methods were used to obtain kinetic information concerning the cation-radical dimerization of anisole (and related compounds). Two mechanisms were consistent with data: A radical-radical coupling (RRC) mechanism and a radical-substrate coupling (RSC) mechanism (Fig. 42) [198].

The first step is the single-electron oxidation of anisole to give cation radical **300**. The two mechanisms are distinguished from each other by the subsequent steps. In the RRC mechanism, two cation radicals dimerize to form a dimeric dication **301**. Elimination of two protons completes the reaction. In the RSC mechanism, the initial cation radical undergoes electrophilic attack of anisole to form a dimeric cation radical. Single-electron transfer from **303** to **300** generates dimeric dication **304** and anisole. Elimination of H^+ from the dication gives the coupled product. A later study indicated that the RSC mechanism was the actual reaction pathway. At high substrate concentrations the reaction of the cation radical with substrate was the rate-determining step. At low substrate concentrations the oxidation of the dimeric cation radical was the rate-determining step [199,200].

The first type of polymer obtained by the Scholl reaction was poly(*p*-phenylene)s [201,202]. Branched soluble poly(phenylene)s [203] and low molecular weight poly(binaphthylene oxide) [192] have been prepared. The first report of the synthesis of high molecular weight polymers by the Scholl reaction was the polymerization of a series of α,ω-bis(1-naphthoxy-)alkanes [204]. Subsequently, the utility of the Scholl reaction as a general method for the synthesis of a wide variety of high molecular weight aromatic polyethers has been established. Polymers have been obtained from monomers containing terminal bis(1-naphthyloxy), bis(2-naphthyloxy), bis(phenoxy), and bis(phenylthio) groups. The central units of these polymers contained nonnucleophilic aromatic groups, *n*-alkane, and diethylene oxide groups.

2. *Bis(1-naphthyloxy) Monomers*

a. Types of Monomers A wide variety of bis(1-naphthyloxy) monomers are capable of undergoing the Scholl reaction. High molecular weight polymers have been obtained by the cation-radical polymerization of bis(1-naphthyloxy) monomers containing alkane, diethylene oxide, and aromatic groups [Eqs. (59)–(61) and Fig. 43].

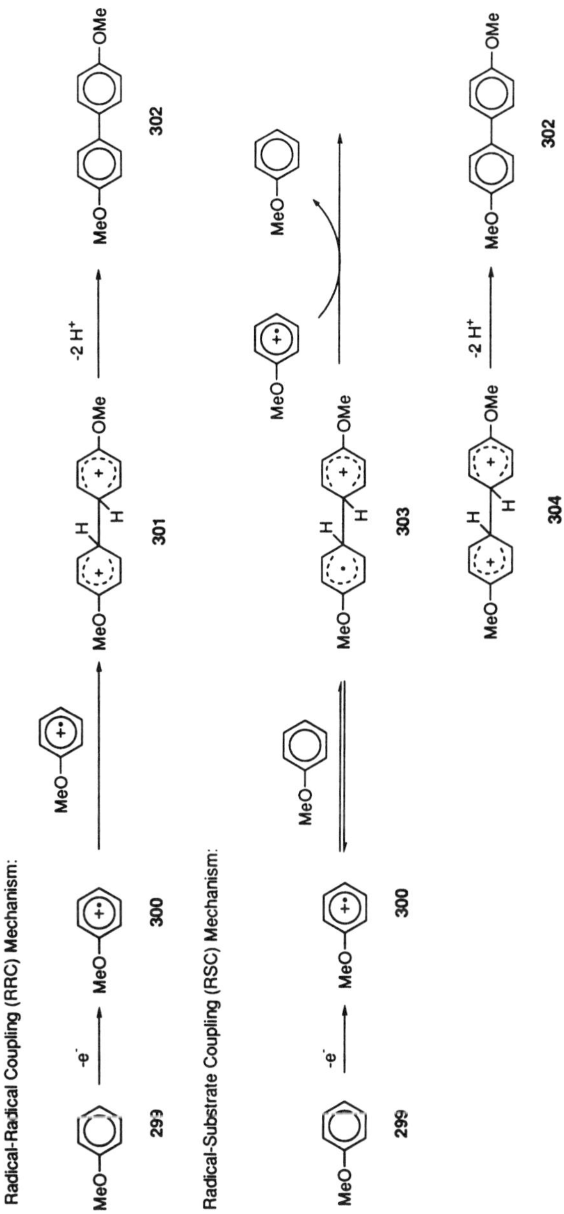

Figure 42 The radical-radical coupling mechanism and the radical-substrate mechanisms. (From Ref. 198.)

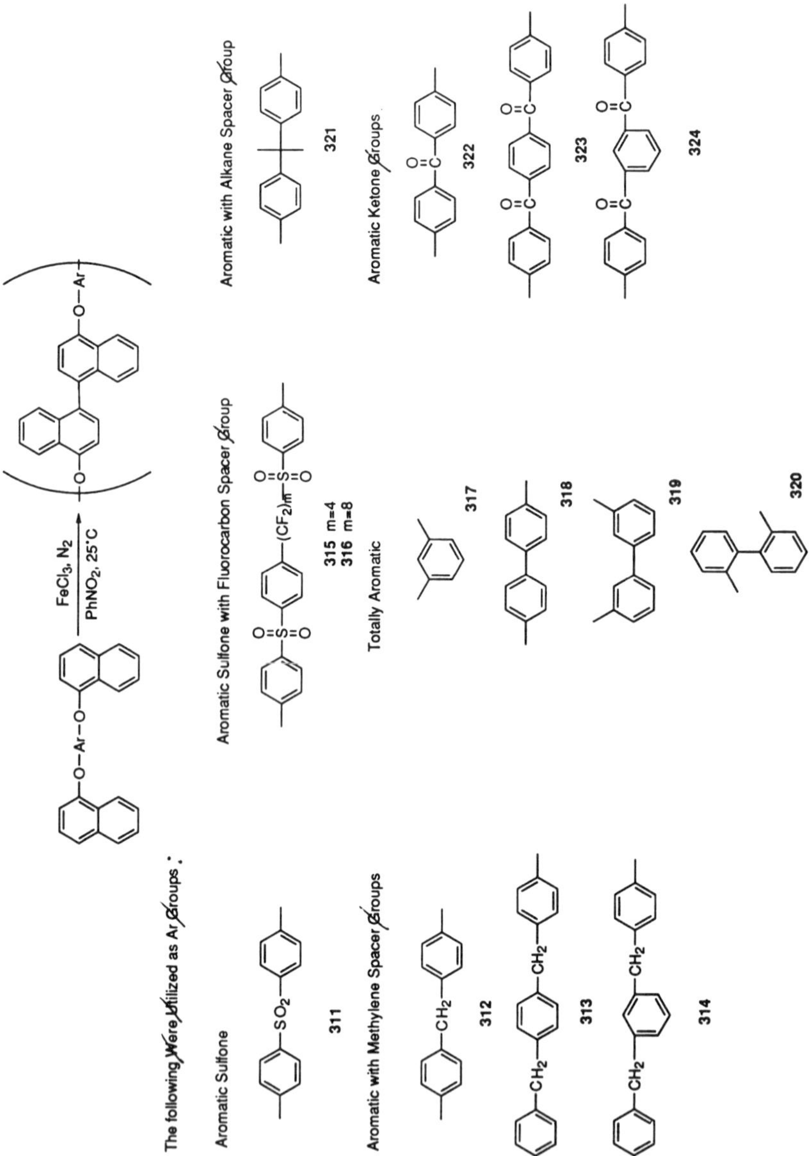

Figure 43 Types of groups incorporated into bis(1,1′-binaphthoxy) monomers polymerized by the Scholl reaction.

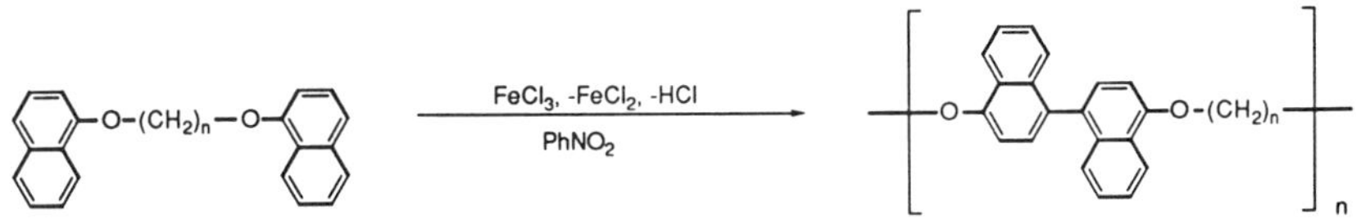

305 306

(59)

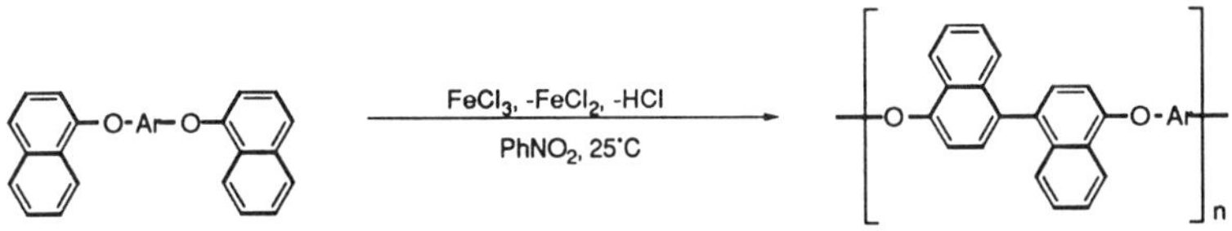

307 308

(60)

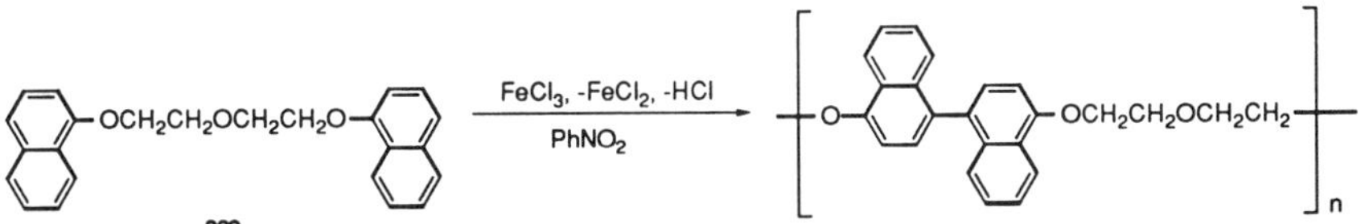

309 310

(61)

A variety of functional groups have been incorporated into the central aromatic group including ketones, sulfones, sulfones with fluorocarbons, and methylene groups (Fig. 43). In addition, high molecular weight polymers have been obtained from bis(1-naphthyloxy) monomers with totally aromatic central units. Lower molecular weight polymers were obtained from the polymerization of a bis(1-naphthyloxy) with a central 2,2-propyl group. The monomers polymerized include the following: bis[4-(1-naphthyloxy)phenyl]sulfone [205,206], 4,4′-bis(1-naphthyloxy)benzophenone [205,207], α,ω-bis[4-(1-naphthyloxyphenyl)sulphonyl]perfluoroalkanes [208], bis[4-(1-naphthyloxyphenyl)methane [207], 1,3- and 1,4-bis[4-(1-naphthyloxyphenyl)methyl]benzene [207] and 2,2′- and 3,3′-bis(1-naphthyloxy)biphenyl [209], 1,3-bis(1-naphthyloxy)benzene [209], and α,ω-bis(1-naphthyloxy)alkanes [204,206,210]. In summary, the polymerization reaction occurs in the presence of sulfone, ketone, alkane, arene, and fluorocarbon functional groups.

b. Reaction Conditions and Oxidizing Agents Polymerizations proceeded in dry nitrobenzene at 25° C with $FeCl_3$ as the oxidant. Nucleophilic solvents (e.g., tetrahydrofuran, pyridine) inhibited the reaction. A stoichiometric amount of $FeCl_3$ was used for the polymerization reactions.

Monomers containing SO_2 or CO groups required twice the stoichiometric amount due to coordination of these groups with the Lewis acid. Other oxidizing Lewis acids also facilitated the reaction. For example, the following oxidants were used: $AlCl_3/CuCl_2$, $AlCl_3/CuCl/O_2$, $AlCl_3/FeCl_3$, $FeCl_3$:tris(3,6-dioxaheptyl)amine. In addition to the use of Lewis acids as oxidizing agents, several other oxidants initiated the polymerization reaction. The use of stoichiometric amounts of DDQ, as an electron acceptor, also resulted in polymer formation [Eq. (62)] [211].

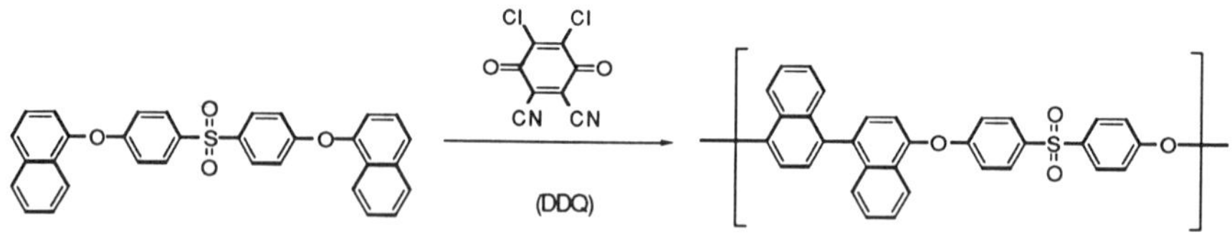

311 325

M_n = 5000 g/mol
M_w/M_n = 1.6
Yield: 85%

(62)

In addition, the polymerization could be performed catalytically in the presence of air. For example, 4,4′-bis(naphthoxy)diphenyl sulfone was polymerized using atmospheric oxygen as the oxidant, 1 mol% vanadylacetylacetonate and 10% triflic acid in $CHCl_3$ [Eq. (63)] [211].

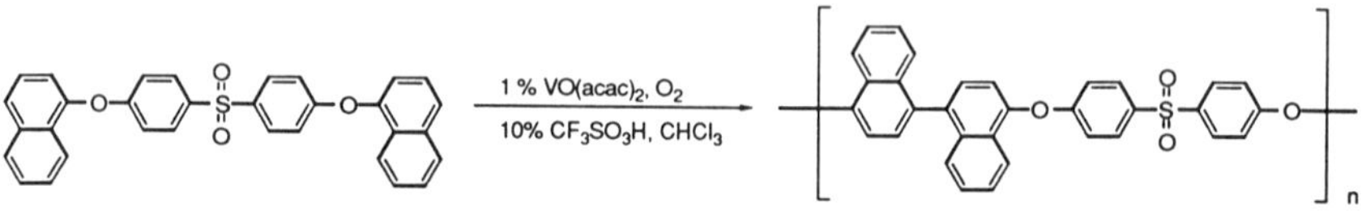

311 325

32% Yield based on Monomer
3200% Yield Based on $VO(acac)_2$
M_n = 4300 g/mol

(63)

The role of VO(acac) is presumably identical with that of the oxidative polymerization of diphenyl disulfide. The V(IV) catalysts disproportionate to give a V(II) and V(III) species. The V(V) species oxidizes monomer **311** producing cation-radical **328** and V(IV). V(III) is readily oxidized in triflic acid by oxygen to regenerate the V(IV) catalyst [211].

The polymerization takes place with coupling at the C_4 position of the 1-naphthoxy group. The restricted rotation about the C_4—C_4 binaphthyl bond of the poly(ether sulfone)s and poly(ether ketone)s resulted in higher glass transition temperatures than the analogous biphenyl polymers. The bulk of the binaphthyl group reduces the crystallinity of the polymer thus inhibiting premature precipitation during the polymerization. Consequently, high molecular weight aromatic polyether sulfones and aromatic polyether ketones were obtained. These polymers were soluble in organic solvents such as chloroform, methylene chloride, and THF.

c. Mechanism The reaction is initiated by the formation of an electron donor-acceptor (EDA) complex **327** of monomer **326** and $FeCl_3$ (Fig. 44). A single-electron transfer (SET) reaction results in the formation of cation radical **328**. The reaction propagates by dimerization of **328** and **326** to form cation-radical dimer **331**. There are two possible pathways from cation radical **331** which lead to dimer **335**. One pathway involves a single-electron transfer reaction to give dimeric dication **330**. Two species are present that can function as oxidants, **328** and $FeCl_3$. The elimination of two protons gives the neutral dimer **335**. Alternatively, **331** can lose a proton to form dimeric radical **333**. Oxidation of **333** gives dimer cation **334** which eliminates a proton to give **335**. The termination of the polymerization may occur by abstraction by the cation radical **328** of Cl^- from its counter-ion $FeCl_4^-$ to give radical **329**. Oxidation of **329** followed by loss of a proton gives **330** as a chain end.

3. *Polyarylenes*

The Scholl reaction has also been applied to the synthesis of soluble unsubstituted polyarylenes containing alternating binaphthylene and biphenylene units [212]. These polymers were obtained from the polymerization of bis(1-naphthyl)biphenyls in nitrobenzene using $FeCl_3$ (typically 2.4–3.6 equivalents) as the oxidant (Fig. 45). They were soluble in $CHCl_3$. Their solubility was enhanced by the noncrystallizability of the bulky 1,1′-binaphthyl group and in some cases by the nonlinearity of the polymer backbone [212].

The polymerizations proceeded at 25° C except 4,4′-bis(1-naphthyl)biphenyl. In this case the monomer had insufficient solubility at room temperature. The polymerization of this monomer at 80° C resulted in the formation of significant amounts (26–78%) of $CHCl_3$-insoluble fractions. The highest molecular weight was obtained with **337** (M_n = 4000; M_w/M_n = 2.5; DP = 10). Thus this polymer has 40 aromatic rings in each polymer chain.

Polymers **340–343** have a deep red or brown color probably related to the presence of conjugated polynuclear units. A similar color was noted in the polymers derived from cation-radical polymerization of 1,1′-binaphthyl and *o*-terphenyl. In these cases the color was attributed to the presence of cation-radicals of perylene and triphenylene structural units [193,201,203]. Perylene units are possible in polymers **340–343** (Fig. 46).

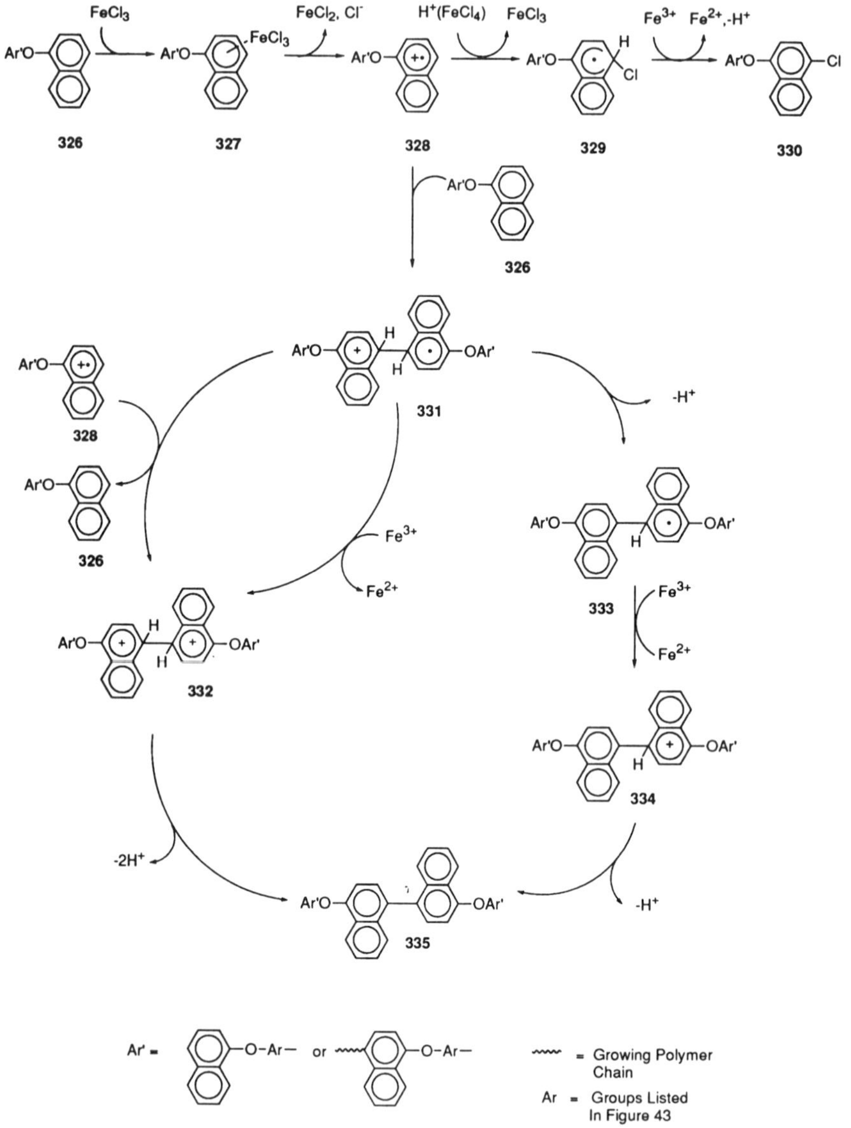

Figure 44 Mechanism of the Scholl polymerization reaction of bis(1,1′-binaphthoxy) monomers.

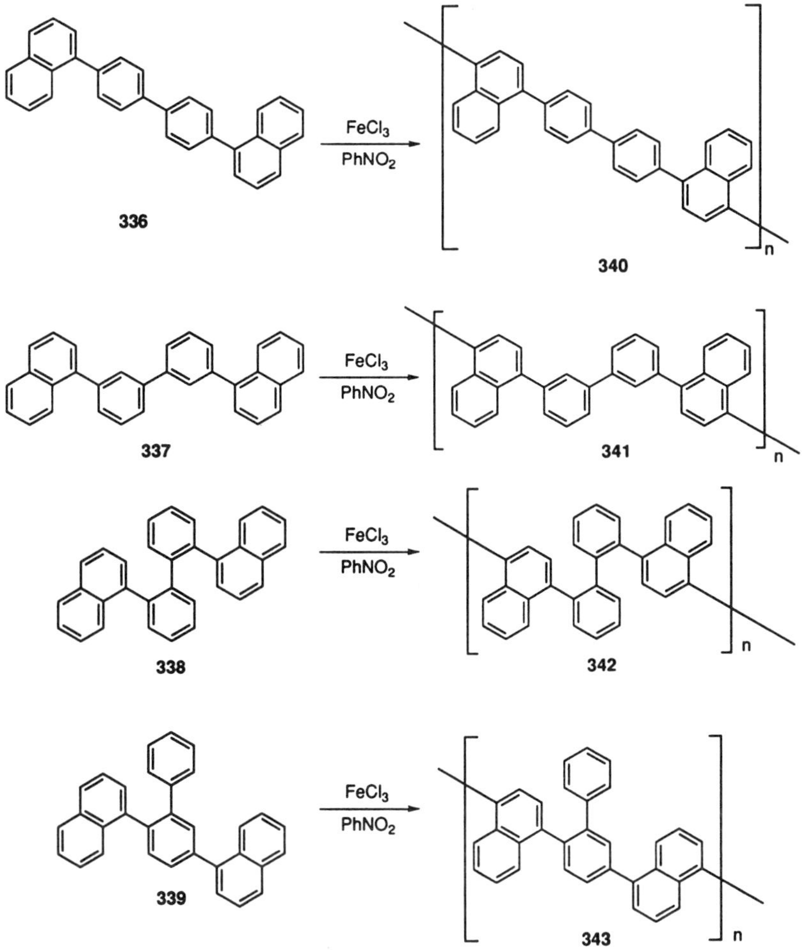

Figure 45 Synthesis of polymers with alternating binaphthylene and biphenylene units. (From Ref. 212.)

Triphenylene units can also form in polymers **342** and **343** because they contain *o*-terphenyl groups (Fig. 47).

4. *Bis(2-naphthyloxy) Monomers*

A bis(2-naphthyloxy) monomer (i.e., 4,4′-bis(2-naphthyloxy)diphenyl sulfone) has also been polymerized by the Scholl reaction [Eq. (64)] [213].

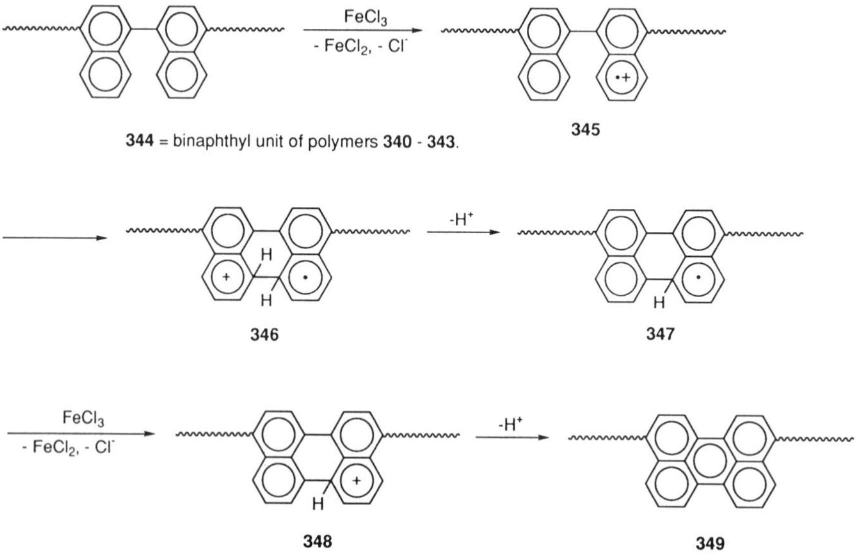

Figure 46 Mechanism of perylene unit formation for polymers **340–343**. (From Ref. 212.)

The structure of the polymer repeat unit derived from coupling at the most reactive position (**356**) is shown in Eq. (64).

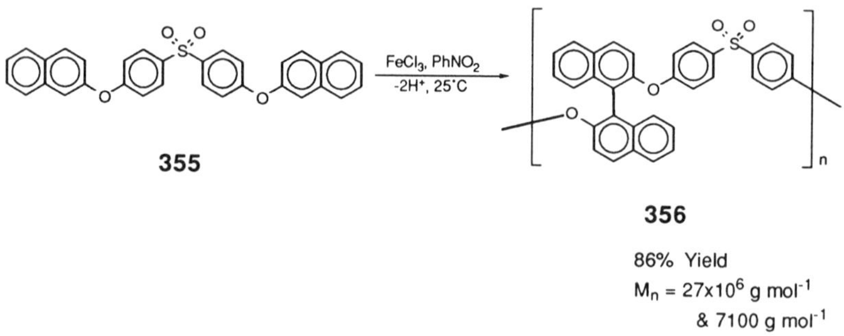

(64)

The mechanism suggested for the polymerization of **355** leading this structure is shown in Fig. 48. The oxidation of **355** by single-electron transfer from $FeCl_3$ gives cation-radical **357**. Coupling of **357** with monomer **354** results in the formation of the dimeric cation radical **358**. Then

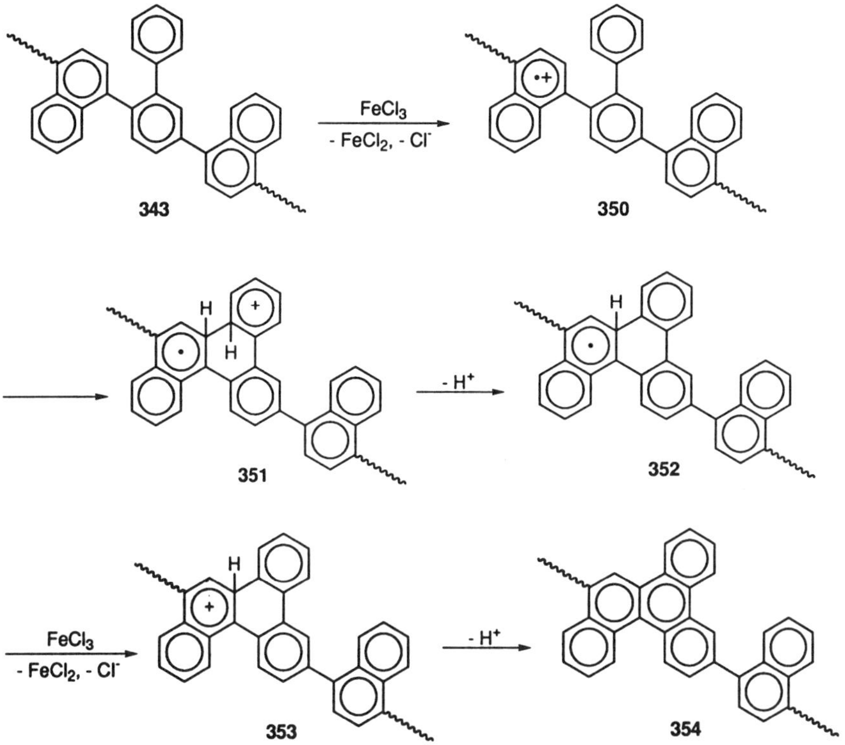

Figure 47 Mechanism of triphenylene formation in polymer **343**. (From Ref. 212.)

358 can undergo a SET oxidation by either **357** or Fe^{3+} to form a dimeric dication, **359**, which upon the elimination of two protons results in the formation of the neutral dimer **362**. Alternatively, **358** loses one proton, resulting in the formation of radical dimer **360**, which is then oxidized to cation dimer **361**. Deprotonation of **361** yields **362** [213].

Reaction at the C_1 position predominates during early stages of polymerization. However, at number-average molecular weights of approximately 1000, intermolecular reactions at less reactive positions (C_5, C_8, and C_4) become significant. This is due to the decreased concentration of unreacted C_1 positions. Reaction at these positions yields a branched polymer fraction of very high molecular weight and a small amount of an insoluble three-dimensional polymer network. This is in addition to a fraction of intermediate molecular weight. Therefore, a multimodal weight

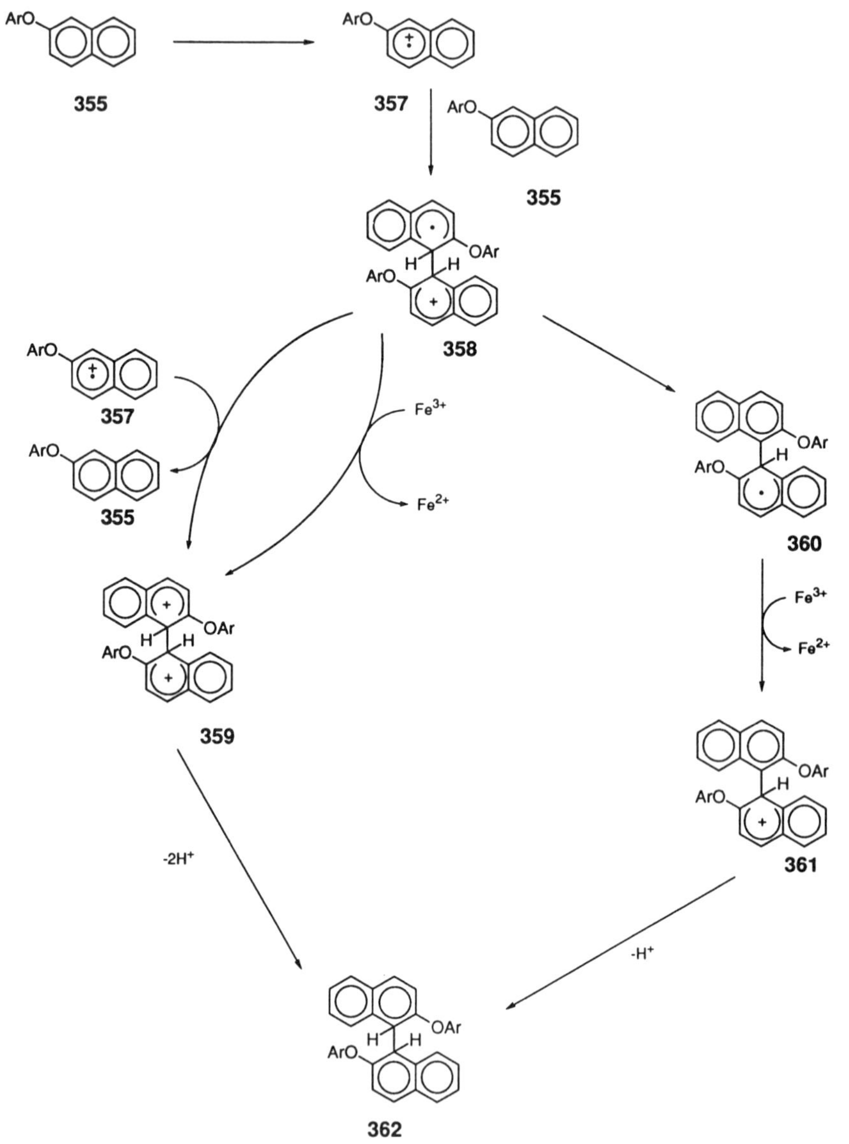

Figure 48 Radical-cation polymerization of aromatic polyethers. (From Ref. 213.)

distribution was obtained. Number-average molecular weights (M_n) of 10^6 g mol^{-1} and <10,000 were obtained for the highest and lowest soluble molecular weight fractions. For example, the branched polymer structure obtained by the coupling reaction of two polymer chains at the C_5 position of both naphthyl groups is shown in Fig. 49.

5. *Bis(phenoxy) and Bis(phenylthio) Sulfone Monomers*

Soluble aromatic polyethers were obtained by the oxidative polymerization of 4,4′-bis(phenoxy)diphenyl sulfone, of its substituted derivatives and 4,4′-bis(phenylthio)diphenyl sulfone [Eq. (65)] [210]. The solubility and yield of the polymers was increased by the presence of methyl, *tert*-butyl, and methoxy group substituents on the phenoxy group of **365** (Fig. 50).

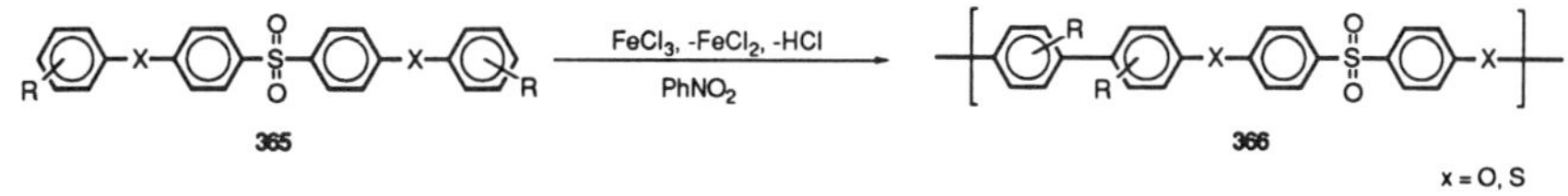

(65)

Higher yields (48% vs. 13%) were obtained with phenylthio derivative **365b** than with phenoxy derivative **365a**. This is rationalized by considering the reaction mechanism (Fig. 51) [210].

The polymerization begins with the oxidation of monomer by $FeCl_3$ to generate cation radical **367**. Monomer **365b** has a lower oxidation potential than **365a**, because the sulfur atom of **365b** is more electron donating than the oxygen. The second step for the polymerization is the nucleophilic attack of monomer on the electrophilic cation-radical intermediate to form 1,1′-dihydrobiphenyl intermediate **368**. The stronger electron-do-

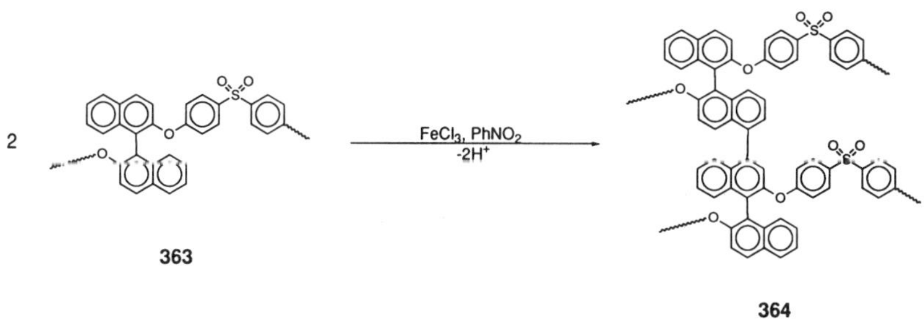

Figure 49 The formation of branched polymer by the coupling reaction of two polymers at the C_5 position of both binaphthyl groups of **363**. (From Ref. 213.)

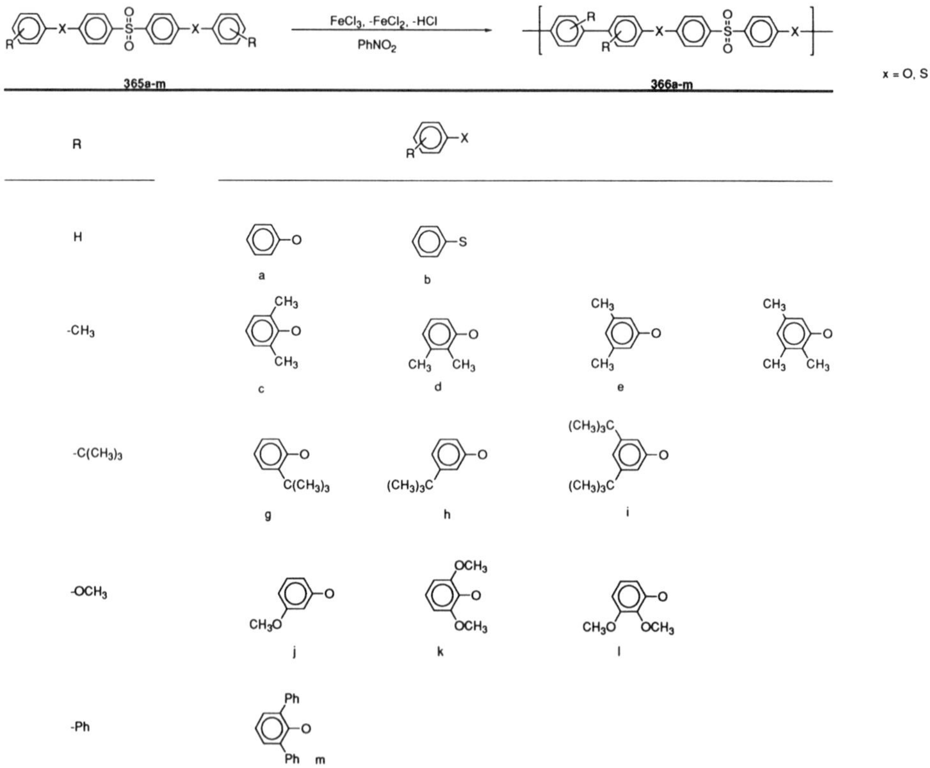

Figure 50 Structure of substituted derivatives of 4,4′-bis(phenoxy)diphenyl sulfone. (From Ref. 210.)

nating capability of sulfur (versus oxygen) makes **365b** more nucleophilic than **365a** in this reaction. Therefore, the rate constants of the first two steps are higher with **365b**. However, the cation radical **367b** is more stable, less reactive than **367a** due to the superior electron-donating capabilities of sulfur versus oxygen. The greater reactivity of monomer **365b**, as evidenced by reaction yields, indicated that the differences in oxidation potentials and nucleophilicities are larger than the differences in stability between the two propagating species. The polymerizability of 4,4′-bis(phenoxy)diphenyl sulfone increased when electron-donating substituents were present. These groups lowered the oxidation potential and increased the nucleophilicity of the monomers. These substituents also stabilized the corresponding cation-radical dimer [210].

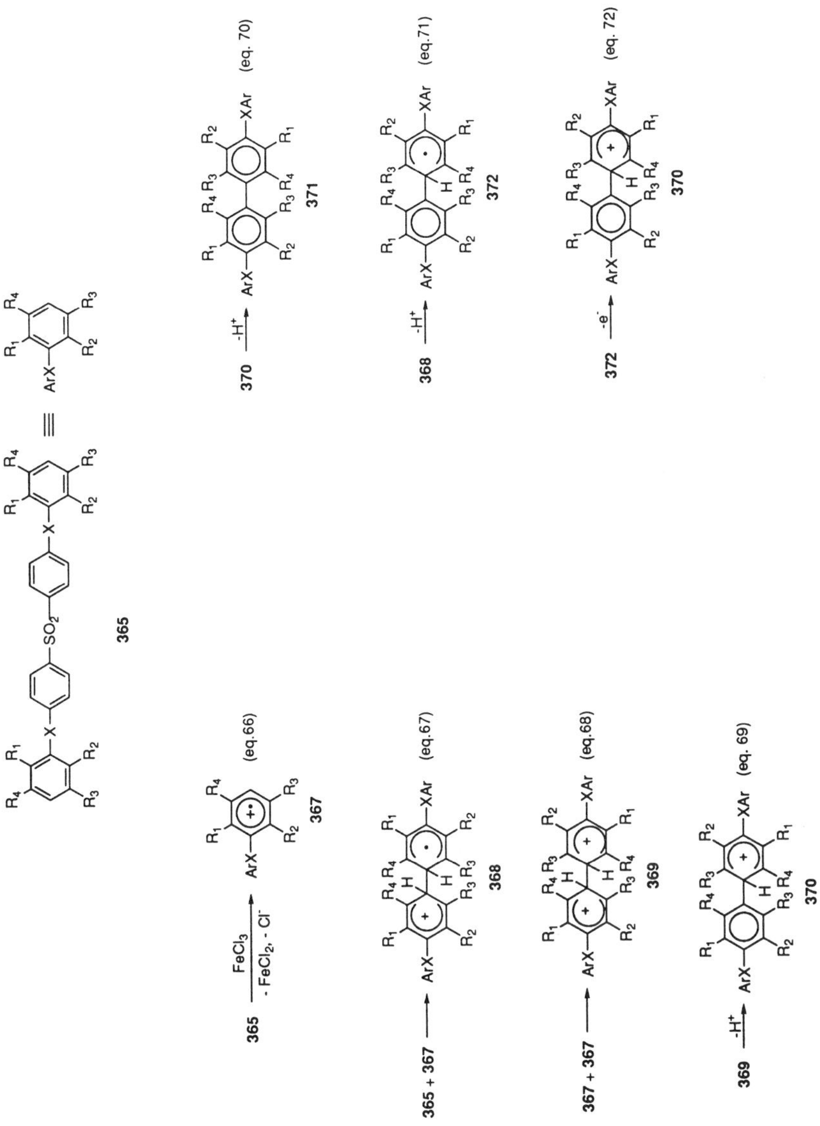

Figure 51 Mechanism of oxidative polymerization of derivatives of 4,4′-bis(phenoxy)diphenyl sulfone and 4,4′-bis(phenylthio)diphenyl sulfone. (From Ref. 210.)

In summary, the polymerization of 4,4′-bis(phenoxy)diphenyl sulfone and its derivatives substituted with various electron-donating groups, and 4,4′-bis(phenylthio)diphenyl sulfone produced polymers of low molecular weight. The trends in the polymerizability of these monomers indicated that the nucleophilicity and oxidizability of these monomers are the most important factors [210].

6. *Bis(phenoxy) and Bis(phenylthio) Alkane Monomers*

The polymerization of 1,5-bis(phenoxy)pentane, 1,5-bis(phenoxy)pentane substituted with electron-donating groups, and 1,5-bis(phenylthio)pentane was investigated to gain insights on the effect of an alkanedioxy central group versus a diphenyl sulfone group on the polymerization of bisphenoxy and bisphenylthio monomers in the Scholl reaction [Eq. (73)] [214].

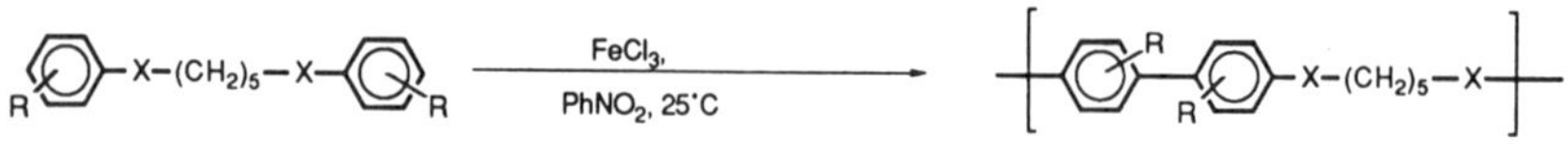

373 374

x = O, S

(73)

A series of polymerizations were performed to determine the effect of changing the diphenyl sulfone central group of the bis(phenoxy) and bis(phenylthio) monomers to a pentanedioxy group. The polymers obtained from methyl-substituted 1,5-bis(phenoxy)pentane contained structural units derived from proton transfer reactions [214]. Thus diphenyl methane, 1,2-diphenyl ethane, and benzyl chloride units were detected (Fig. 52).

The polymerizability of monomers having a 1,5-pentanedioxy central group were lower than monomers containing a diphenyl sulfone group. This was due to a difference in the reactivity and concentration of the cation-radical propagating species in the polymerization [214].

This difference can be understood by considering the stability and electronic nature of the single-electron oxidation products **375** [Eq. (82)] and **367a** [Eq. (83)].

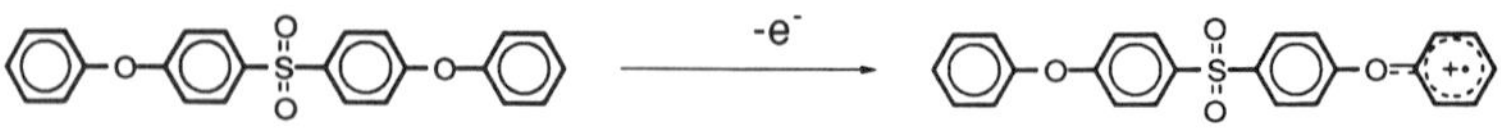

365a 367a

(82)

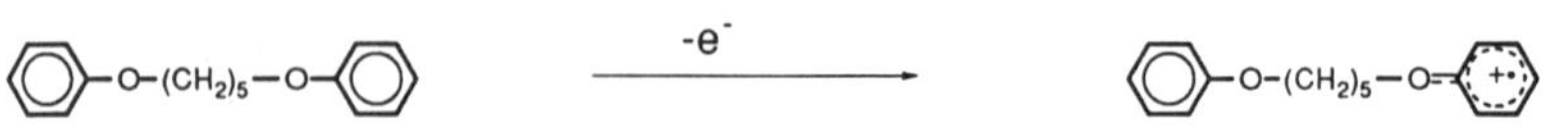

373 375

(83)

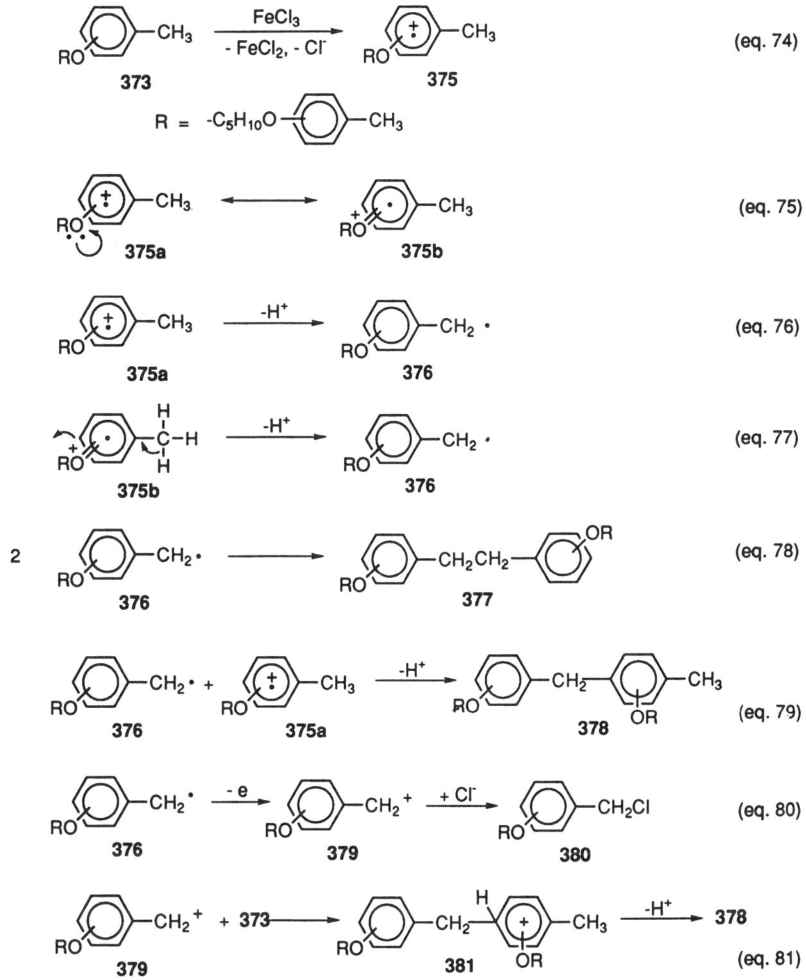

Figure 52 Mechanism of the proton transfer reactions during the cation-radical polymerization of methyl-substituted 1,5-bis(phenoxy)pentane monomers. (From Ref. 214.)

A consideration of the resonance structures of cation radicals **375** and **367a** explains the differences in reactivity (Fig. 53). Cation-radical **375** containing a 1,5-pentanedioxy group is strongly stabilized by the oxonium contributing resonance structure **B**. The corresponding oxonium ion **B′** with a central sulfone group **367a** is less stabilizing. This is because the

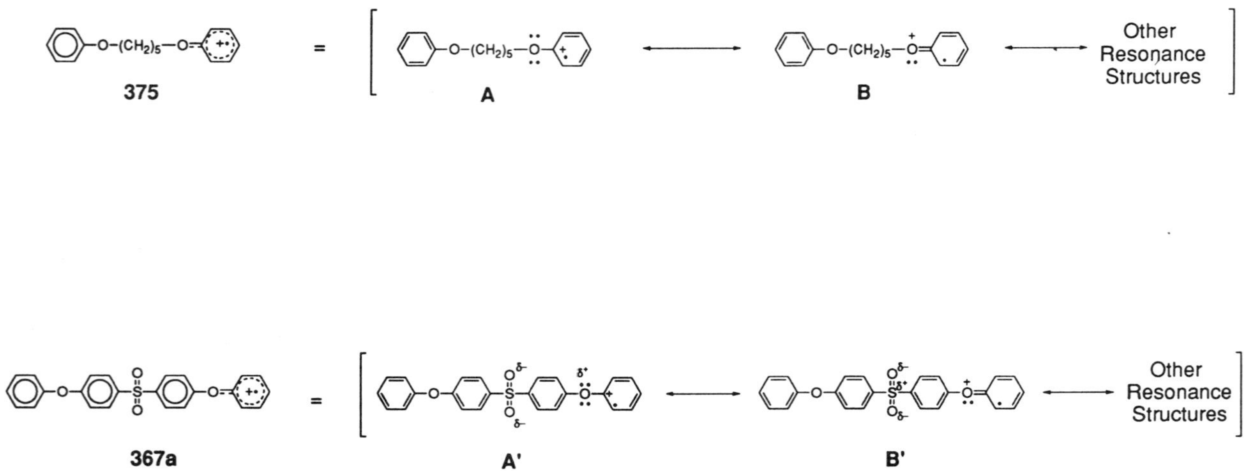

Figure 53 Resonance structures of radical cations **375** and **367a**. (From Ref. 214.)

effect of the partial positive charge on sulfur is transmitted by resonance to the oxonium ion. Therefore, the sulfone acts as an electron-withdrawing group. As a result, resonance structure **B** is a stronger contributor to the structure of **375** than **B′** is to the structure of **367a**. The increased relative importance of **B** relative to **B′** means that **375** is less reactive than **367a**. Oxonium ion structures **B** and **B′** provide stabilization to the cation radical because every atom has a full octet of electrons. As the oxonium ion structure becomes a more important contributing resonance structure, the reactivity of the cation radical decreases.

C. Dual Propagation Pathways: Cation Radical and Electrophilic

The $FeCl_3$-initiated oxidative polymerizations of both 2,2-bis[4-(1-naphthoxy)]phenyl propane (**321**) and 2,2-bis[4-(1-naphthyl)phenyl] propane (**382**) involve two different propagation pathways [215]. If the oxidative polymerization of 2,2-bis[4-(1-naphthoxy)]phenyl propane proceeded like the other bis(1-naphthoxy) monomers (see Figs. 43 and 44) the polymer repeat unit would contain an isopropenylidene group linking two phenyl groups (Fig. 54).

However, in addition to this unit (A), two other types of isopropenyl units (B and C) were detected [215]. These units contained an isopropenylidene group between a phenyl and naphthyl group (unit B) and an isopropenylidene group between two naphthyl groups (unit C). Therefore the structure of the polymer indicates that two distinct propagation reactions are involved. The first pathway is the cation-radical dimerization of the naphthyl groups to give a dinaphthyl structure containing unit A. This propagation reaction generates $H^+[FeCl_4]^-$. The $H^+[FeCl_4]^-$ initiates a second propagation reaction. This transalkylation reaction generates structural units containing isopropylidenic groups inserted both between phenyl and naphthyl (B) and between two naphthyl units (C).

The first propagation reaction proceeds by the mechanism outlined in Fig. 55. Each coupling reaction produces two protons. Although the reaction vessel is continuously purged with nitrogen not all of the HCl is removed. The facile reaction of HCl with ferric chloride produces the stronger acid $H^+[FeCl_4]^-$. This acid catalyzes the second propagation reaction which produces transalkylated product (Figs. 55 and 56). Monotransalkylation (Fig. 55) is initiated by protonation of a phenyl ring ipso to the isopropylidenic unit. Cleavage of the phenyl isopropylidene bond of the cationic σ-complex results in the formation of a segment with a terminal phenyl group **387** and an isopropylidene carbenium ion **388**. A terminal isopropyliene group (**389**) can form upon elimination of H^+. How-

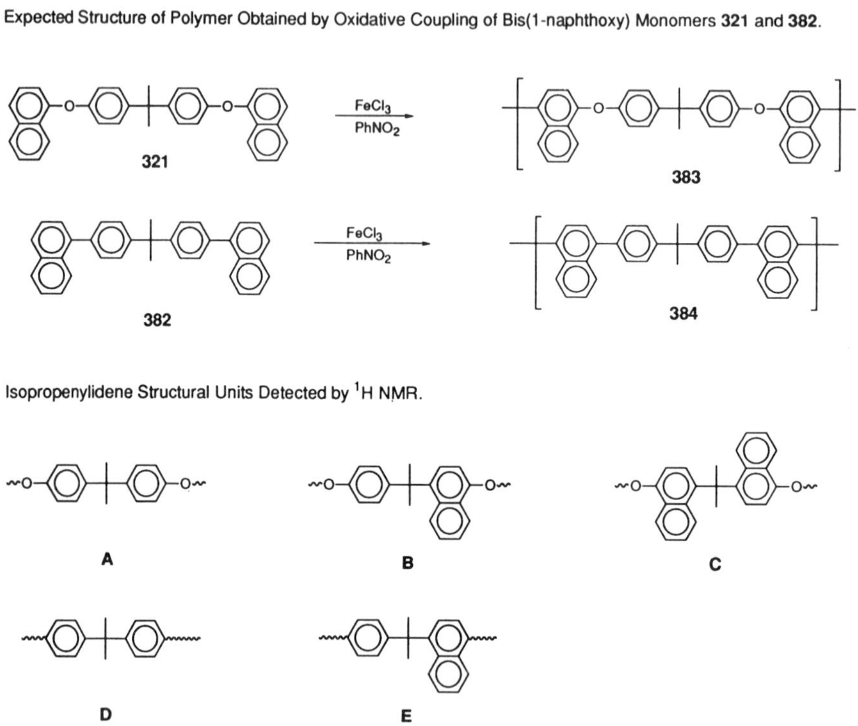

Figure 54 Isopropenyl structural units expected and obtained from the oxidative polymerization of 2,2-bis[4-(1-naphthoxy)phenyl]propane and 2,2-bis[4-(1-naphthyl)phenyl]propane. (From Ref. 215.)

ever, in the acidic reaction medium the equilibrium constant strongly favors **388**. Electrophilic attack by **388** upon a nucleophilic naphthoxy group of the monomer or the polymeric termini to form σ-complex **391** is favored. Elimination of H^+ produces the 2-phenyl-2-naphthylpropane unit (i.e., unit B of **392**). The monotransalkylated structure can participate in a reaction leading to the formation of a ditransalkylated structure. Protonation of the ipso position of the phenyl group of unit B generates a σ-complex. Cleavage of the phenylisopropylidene bond produces a phenoxy-terminated chain end **394** and the carbenium ion **395**. Electrophilic attack of this carbenium ion upon a naphthyloxy group followed by elimination of H^+ gives the 2,2-dinaphthylpropane unit (i.e., unit C of **397**). Transalkylated isopropylidene units accounted for 22% of all isopropyl-

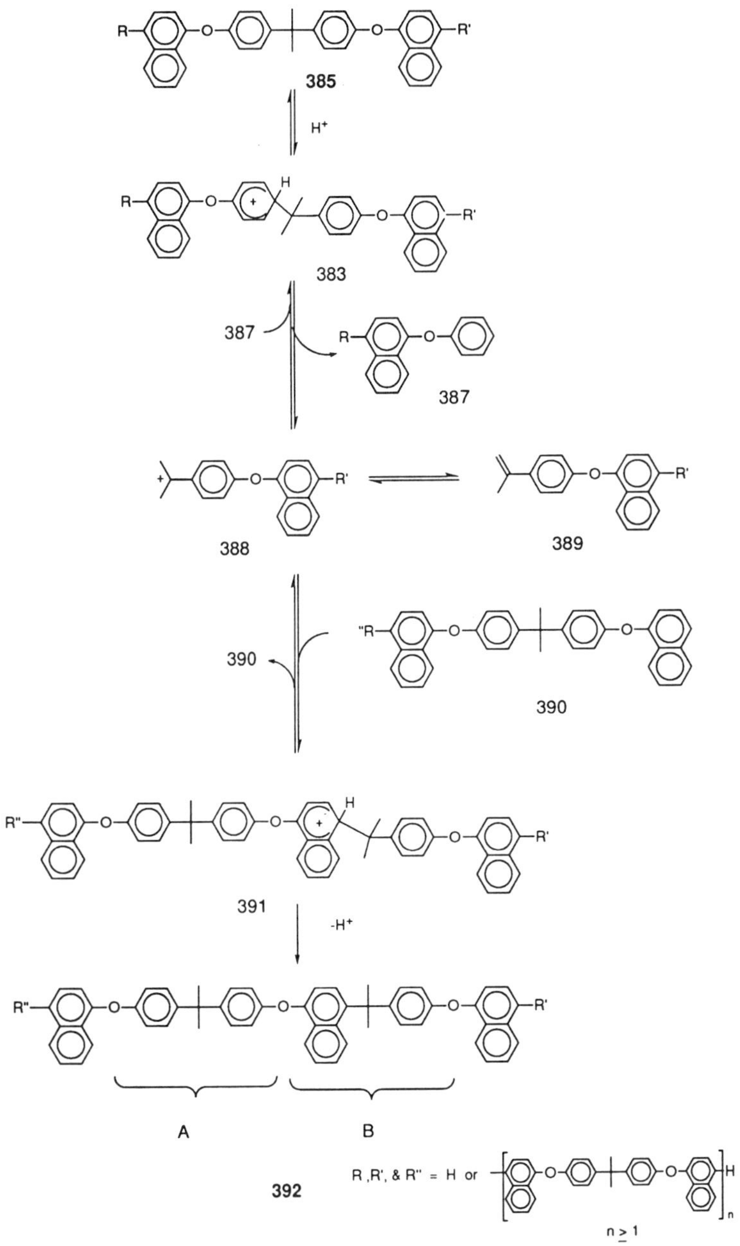

Figure 55 Mechanism of transalkylation reaction to give structural unit B. (From Ref. 215.)

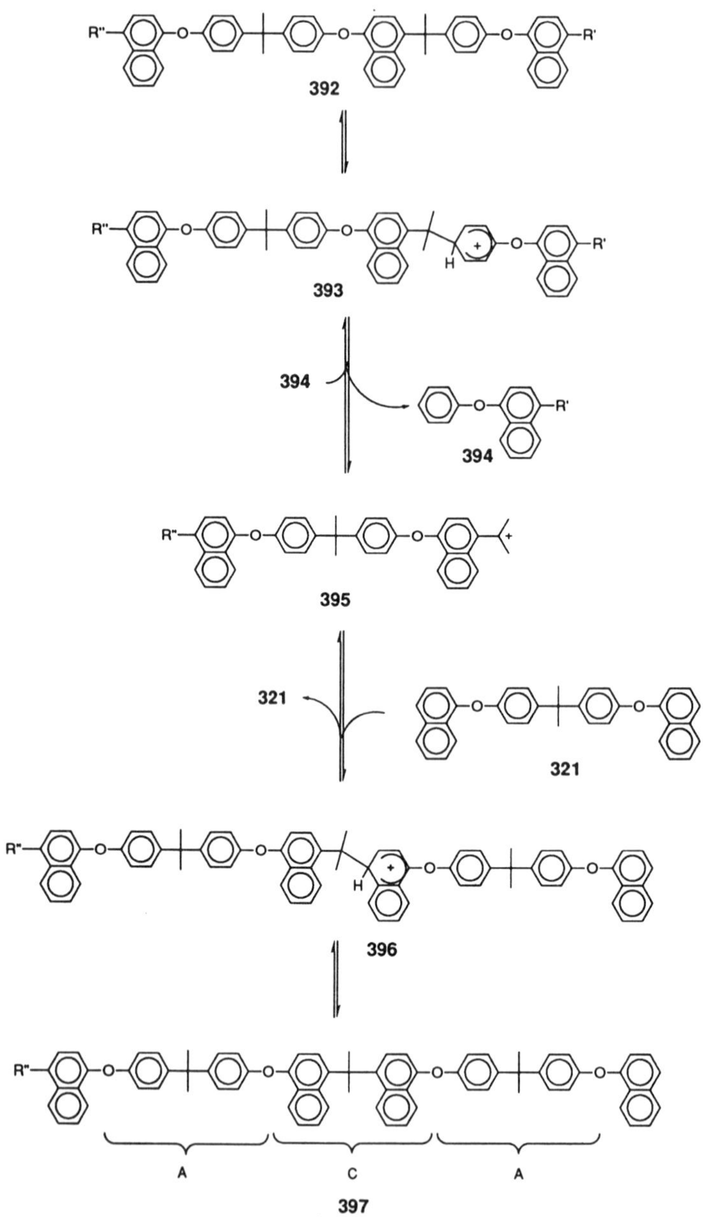

Figure 56 Mechanism of transalkylation reaction to give structural unit C. (From Ref. 215.)

idene units. The monotransalkylated unit comprised 18% of the structure with ditransalkylated units accounting for the remaining 4%. Because the naphthoxy group is more nucleophilic than the phenoxy group, protonation of unit B of **392** occurs predominately at the ipso position of the napthalene ring. Reaction at this position simply generates **391**. The formation of phenoxy chain ends occurs in both the mono- and ditransalkylation reactions. These groups are much less reactive. The oxidative cationic polymerization of similar bis(phenoxy)monomers resulted in oligomer formation. Thus the formation of these groups results in the formation of polymers with low molecular weights. The polymerization of 2,2-bis[4-(1-naphthyl)phenyl]propane also resulted in low molecular weight polymers. In this case 14% of the isopropylidene units were monotransalkylated. No ditransalkylated units were detected. The higher susceptibility of **321** as compared with **382** to the transalkylation reaction can be rationalized because of the higher nucleophilicity of 1-naphthoxy than 1-naphthyl and the phenyl groups [215].

D. Polypyrroles

The discovery that doped forms of polypyrroles conduct electrical current has spurred a great deal of synthetic activity related to polypyrroles [216–218]. Reviews are available on various aspects of the synthesis and properties of polypyrroles [219,220]. In addition, summaries of important aspects of polypyrroles are included in several reviews on electrically conducting polymers [221–226]. Polypyrrole has been synthesized by chemical polymerization in solution [227–231], chemical vapor deposition (CVD) [232,233], and electrochemical polymerization [234–240]. The polymer structure consists primarily of units derived from the coupling of the pyrrole monomer at the 2,5-positions [Eq. (84)]. However, up to a third of the pyrrole rings in electrochemically prepared polypyrrole are not coupled in this manner [241].

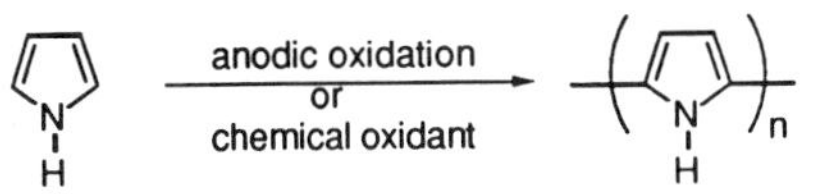

398 **399** (84)

Although some mechanistic details are still controversial, it has been established that the oxidative polymerization (chemically or electrochemically) of pyrrole and pyrrole derivatives proceeds via an $E(CE)_n$ mechanism which involves cation-radical propagating species. The most commonly accepted mechanism of polypyrrole formation is illustrated in Fig. 57 [237,242]. The polymerization begins with the one-electron oxidation of pyrrole to produce cation radical **399**. This cation radical has been

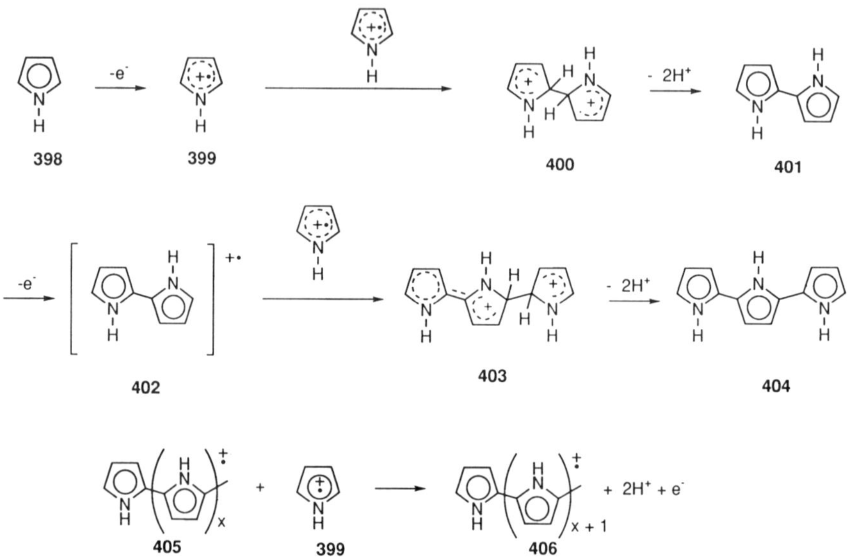

Figure 57 Mechanism of polypyrrole formation involving the coupling reaction of two cation radicals. (From Refs. 237 and 242.)

detected by fast-scan cyclic voltammetry under the polymerization conditions [243]. Two cation radicals couple to form dication dimer **400**. The subsequent loss of two protons yields neutral aromatic dimer **401**. This dimer is then readily oxidized to a cation-radical dimer (**402**) because of its proximity to the electrode surface (in the electrochemical polymerization) and its lower oxidation potential compared with the monomer. Cation radical dimer **402** couples with another cation radical of monomer, dimer, or oligomer. It is important to note that because the electrochemical polymerization takes place at the anode surface the local concentration of monomeric and oligomeric cation-radical species is high compared with the corresponding neutral species. Thus coupling reactions of cation-radical species are favored. As the oligomers grow, their oxidation potential decreases. Consequently, the resulting cation radicals become more stable and less reactive. At this stage oligomer growth probably occurs by the coupling of monomeric cation radicals with the oligomeric cation radicals. In summary, chain propagation occurs by a series of repeated steps of (a) an oxidation to form a cation radical, (b) the coupling of two cation radicals to form a dication followed by (c) the elimination of two protons to form an aromatic structure.

In principle, the cation-radical propagating species can participate in the chain-growth reaction by coupling with another cation radical or with a neutral substrate. Chain growth in the previously considered mechanism occurs by the radical-coupling reaction of two cation radicals. The other possibility that the coupling step involves the reaction of a cation radical with an neutral pyrrole substrate has been proposed (Fig. 58) [240,244]. The reaction is initiated by the oxidation of pyrrole to a cation radical (**399**). This species attacks the neutral pyrrole monomer to produce a cation-radical dimer (**407**). The neutral aromatic dimer (**401**) is formed upon the loss of two protons and an electron. Chain growth continues by repeated coupling reactions between the cation radical oligomer and neutral monomer.

An electrochemical study of the reaction kinetics of several substituted pyrroles has indicated that the carbon-carbon bond formation step proceeds by the coupling reaction of two cation radicals rather than the coupling of a cation radical with a neutral substrate molecule [245]. This study also indicated neutral radicals, formed by the deprotonation of the cation radicals before the carbon-carbon bond-forming step, were not involved in the coupling step.

Although considerable progress has been made, many unresolved questions still exist concerning the reaction mechanism. Questions concerning where and how these processes occur in relationship to the electrode surface are also under investigation [246].

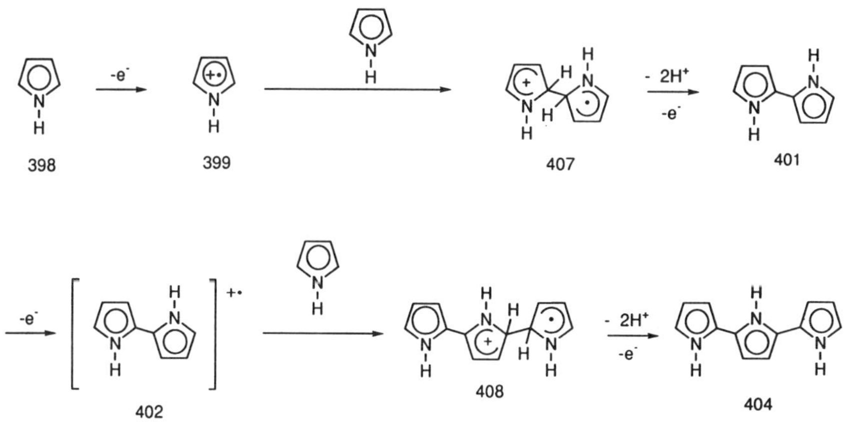

Figure 58 Mechanism of polypyrrole formation via the coupling reaction of a cation radical with neutral pyrrole monomer. (From Refs. 240 and 244.)

E. Polythiophenes

Interest in the synthesis of polythiophenes has burgeoned since the early 1980s because of their usefulness as electrically conducting polymers. The stability of polythiophene toward oxygen and water vapor in both the oxidized (conducting) and reduced state imparts special significance to this polymer. High molecular weight polythiophene and its substituted derivatives are most commonly synthesized by one of three methods: electrochemical [Eq. (85)] [202,236,247,249], chemical oxidation [250–253], or an organometallic coupling reaction [254–258].

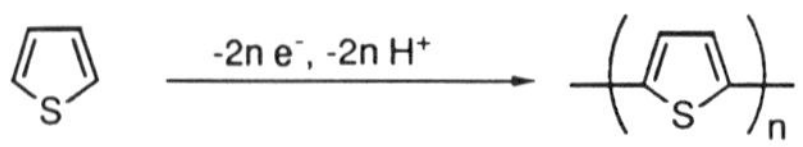

409 410 (85)

Summaries on the synthesis, properties, and uses of polythiophenes are included in two general reviews on polythiophenes [259,260]. A synopsis of important aspects of polythiophenes are also included in several reviews on various aspects of conducting polymers [221–226]. Cation radicals are the propagating species in both electrochemical and chemical oxidative polymerizations of thiophene and its derivatives. The polymer obtained by this method is linked primarily by α,α-linkages. However, other types of linkages (α,β and β,β) are present in varying amounts (Fig. 59). Substituted thiophene derivatives can couple in a ''head-to-tail'' or ''head-to-head'' manner.

Several mechanisms have been proposed for both electrochemical and chemical oxidative synthesis of polythiophenes. The proposed mechanisms are similar to those proposed for polypyrrole formation. The first step of the polymerization is the oxidation of the thiophene monomer to a cation radical. The subsequent steps are controversial. There are several possibilities. The cation radical can couple with another cation radical or with a neutral species. Alternatively, the cation radical can deprotonate to form a neutral radical. This radical can then couple with another radical or with a neutral species. Several of these possibilities are discussed below.

A mechanism in which the chain-growth step occurs by the radical coupling of two of these cation radicals to form a dication dimer is illustrated in Fig. 60. The reaction begins with oxidation of the monomer to form cation radical **411**. The radical coupling of two cation radicals produces the dication dimer **412**. Two protons are eliminated to form aromatic dimer **413**. This dimer is readily oxidized to cation radical **414** due to its close proximity to the electrode and its ease of oxidation. Cation-radical dimer **414** then couples with a cation radical monomer **411** to form a dicat-

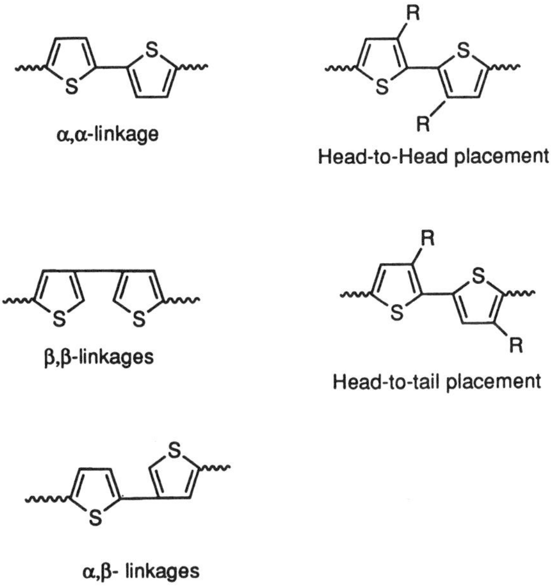

Figure 59 Various linkages between thiophene units.

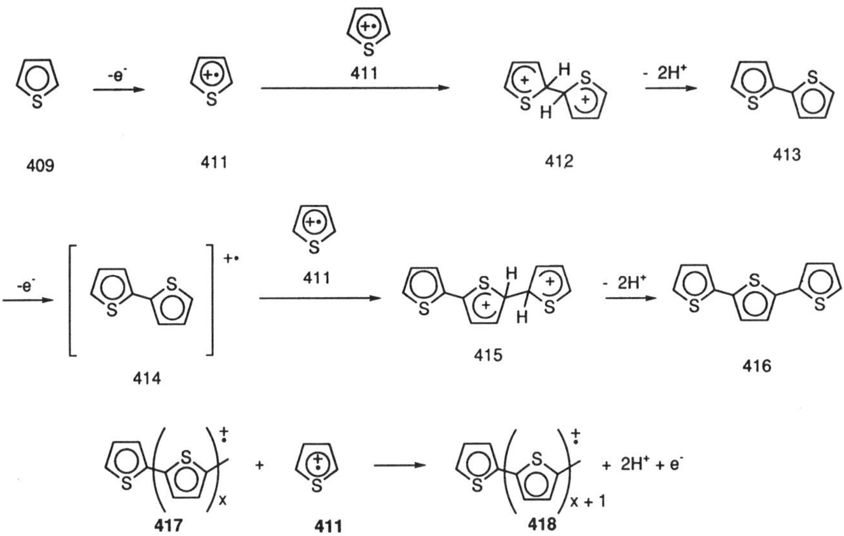

Figure 60 Mechanism of polythiophene formation via coupling reaction of two cation radicals. (From Refs. 222 and 250.)

ion trimer **415**. The aromatic trimer **416** is obtained upon elimination of two protons. The polymerization continues by a repeated sequence of (a) an electrochemical oxidation to generate **417** and **411** followed by (b) the coupling of **417** and **411** and (c) the loss of two protons from the dication to form an aromatic structure. This sequence continues until precipitation of the polythiophene occurs [202,222,249–251,261].

Another mechanistic possibility is the attack of the thiophene cation radical (**420**) upon a neutral thiophene monomer (**419**) to form a cation-radical dimer (**421**) [247]. The oxidation and loss of two protons leads to formation of the neutral dimer (**422**). Once again, rapid oxidation of the dimer occurs upon its formation due to its close proximity to the electrode surface and its lower oxidation potential. The cation-radical dimer (**423**) which is formed then reacts with another monomer molecule in a similar series of steps to produce the trimer **425**. A kinetic study of the electrochemical polymerization of thiophene and 3-alkylthiophenes led to the proposal of this mechanism (Fig. 61) [247]. The rate-determining step in this series of reactions is the oxidation of the thiophene monomer. The reaction is first order in monomer concentration. The addition of small amounts of 2,2′-bithiophene or 2,2′:5′,2″-terthiophene to the reaction resulted in a significant increase in the rate of polymerization and in a lowering of the applied potential necessary for the polymerization reaction. In this case the reaction was 0.5 order in the concentration of the additive.

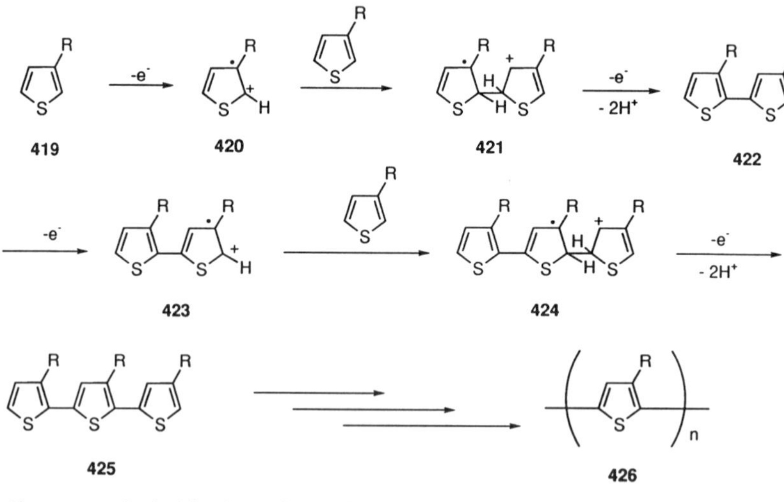

Figure 61 Polythiophene formation via mechanism involving electrophilic attack of cation radical upon neutral thiophene monomer. (From Ref. 247.)

The coupling reaction of a radical species with a neutral monomer substrate [262] has been proposed for the polymerization of 3-alkyl thiophenes using $FeCl_3$ as an oxidizing agent (Fig. 62) [263]. According to this mechanistic scenario, coupling is initiated by the oxidation of an 3-alkyl thiophene with $FeCl_3$ to cation radical **420**. This species deprotonates to form radical species **427** (or **428**) which then couples with monomer to form dimeric radical **429**. The neutral dimer **422** is obtained by the oxidation and deprotonation of this species. The reaction then proceeds in a similar manner with chain growth occurring by the reaction of a neutral radical with monomer. For oxidation to occur, it was essential that the $FeCl_3$ was present in the solid state. The radical mechanism was proposed on the basis of the structure of the resulting polymer, the crystal structure of $FeCl_3$, and quantum chemical computations of possible thiophene intermediates [263].

A procedure has recently been developed for the regioselective polymerization of an 3-(4-octylphenyl)thiophene with $FeCl_3$ [253]. The slow addition of a slurry of $FeCl_3$ in $CHCl_3$ to a solution of the monomer resulted in the formation of a polymer with 94% head-to-tail linkages. The slow addition of $FeCl_3$ keeps the ratio of Fe^{3+} to Fe^{2+} low during the

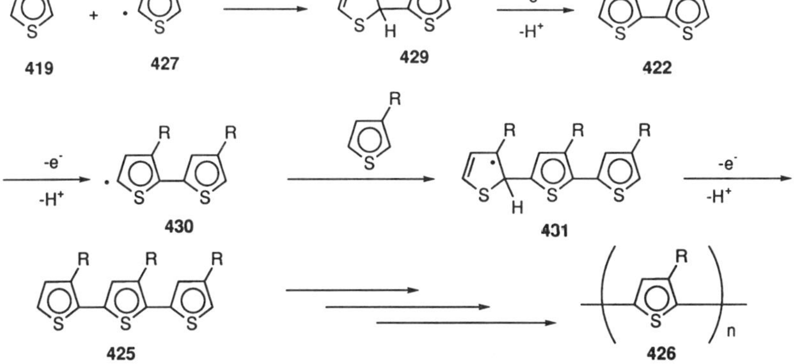

Figure 62 Polythiophene formation via mechanism involving radical attack upon neutral thiophene monomer. (From Ref. 263.)

polymerization process. The lower oxidation potential of the reaction medium results in a more selective polymerization. A mechanism was proposed that involves initiation by a thiophene radical cation. Once oligomers formed (**432**) the propagation steps were proposed to involve the coupling of oligomeric dications (**433**) with neutral monomer substrates (**419**) (Fig. 63). This mechanism was proposed rather than the radical-monomer coupling mechanism in part because the high selectivity of the coupling process indicated the involvement of cationic species. Also the growing polymer chain was not expected to be neutral in the highly oxidizing medium [253].

F. Polyanilines

Polyaniline (PANI) and its derivatives can also be synthesized by electrochemical or chemical methods that generate cation radicals as propagating species. An interest in electrically conducting polymers has been responsible for a recent surge in interest in polyaniline [264–266]. Its advantages as a conducting polymer include the low cost of aniline and the stability of polyaniline toward moisture, oxygen, and elevated temperatures. Several reviews on polyanilines are available [267–269] and general aspects of polyanilines are included in several reviews on conducting polymers [221–226] [Eq. (86)]

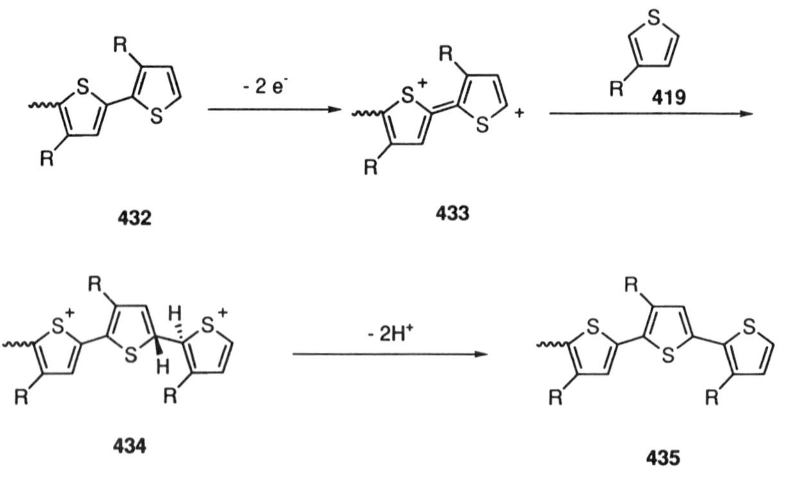

Figure 63 Mechanism for the propagation of POPT by $FeCl_3$ via carbocationic mechanism. (From Ref. 253.)

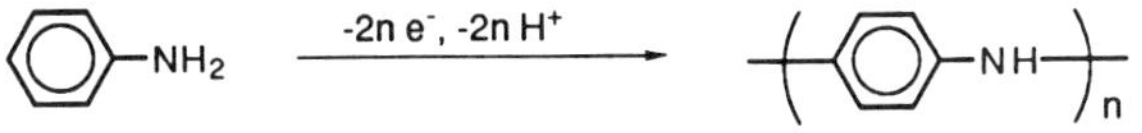 (86)

436 437

Polyaniline exists in various forms which differ in the oxidation state and degree of protonation of the main chain. The average oxidation state of the polymer main chain can be varied chemically or electrochemically. The structures of several common oxidation states are shown in Fig. 64. The degree of protonation can be varied by treatment with acid or base. In addition to the structure obtained from head-to-tail coupling at the para-position, other structural units can be obtained. These units can be obtained by substitutions at other positions of the phenyl group as well as head-to-head and "tail-to-tail" coupling (Fig. 65). The polymerization conditions determine the structural characteristics, oxidation state, and

General Structure of Polyaniline

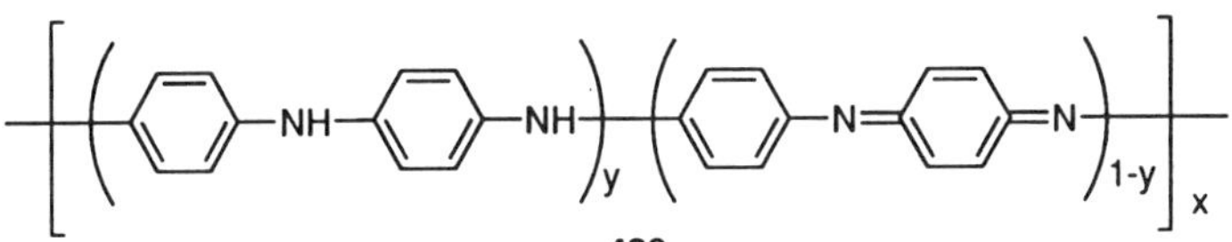

438

y=1 Completely Reduced (leucoemeraldine)

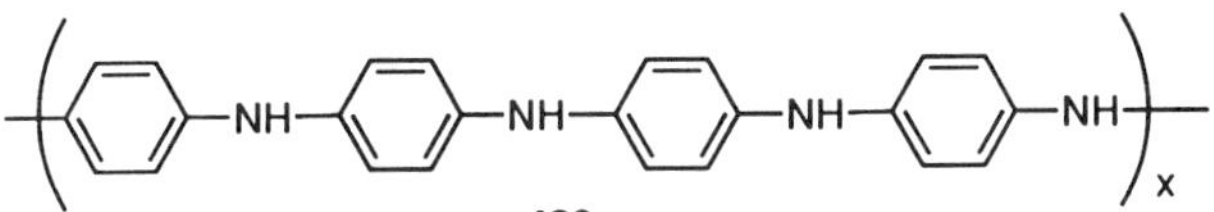

439

y=0.5 Half-oxidized (emeraldine)

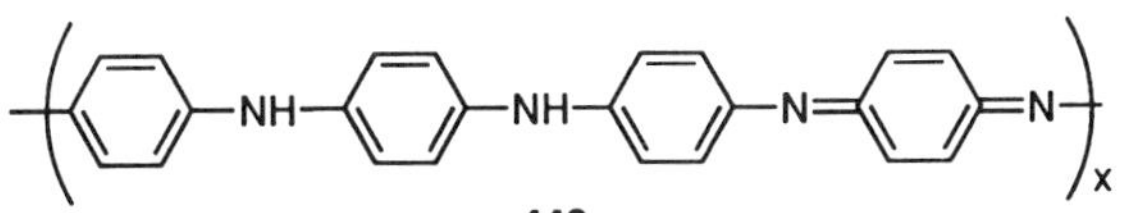

440

y=0 Completely oxidized (perigraniline)

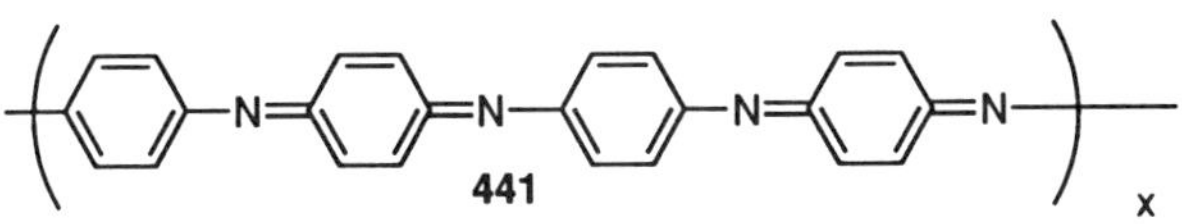

441

Figure 64 Common oxidation states of polyaniline (base forms are shown). (From Ref. 265.)

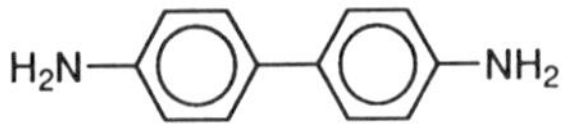

442

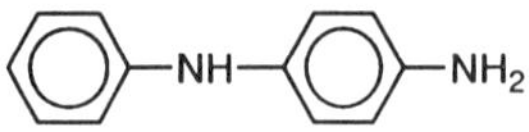

443

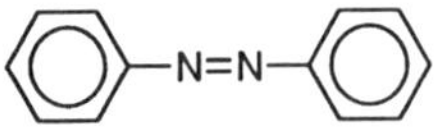

444

Figure 65 Structures of dimers derived from "tail-to-tail", "head-to-tail," and "head-to-head" coupling of aniline units. (From Refs. 273 and 275.)

degree of protonation. The oxidative polymerization of aniline is usually accomplished electrochemically [267,270–276] or with chemical oxidizing agents [266,270,271,277–279]. Polyaniline has also been synthesized by a gas-phase plasma method [280], layered in $V_2O_5nH_2O$ xerogels [281], a vapor deposition method [282], and at an air-water interface by the Langmuir-Blodgett (LB) technique [283,284].

The mechanism of polyaniline formation is an area of active research and controversy. The wide range of reaction conditions used in polyaniline synthesis and the resulting differences in the structure and characteristics of the polymers has probably contributed to the proposal of many different mechanisms. The majority of the proposed mechanisms begin with the oxidation of aniline to a cation radical (**445**). Two of these cation radicals couple to form *N*-phenyl-*p*-phenylenediamine (**443**). The oxidation of the aniline monomers to form dimeric species is the slow step in the polymerization [271,285,286]. The subsequent steps of polymer growth are under discussion.

One proposed mechanism for the electrochemical polymerization of aniline is shown in Fig. 66 [287]. Aniline is oxidized to cation radical **445** which dimerizes to form dication dimer **446**. Deprotonation ($-2H^+$) of **446** gives **443**. The oxidation of dimer **443** gives cation radical **447** which is further oxidized to diiminium dication **448**. The coupling of **448** with **445** followed by the loss of two protons gives **449**. The addition of aniline units to the polymer chain continues in a similar manner by the coupling of the terminal diiminium dication group with monomer cation radical **445**.

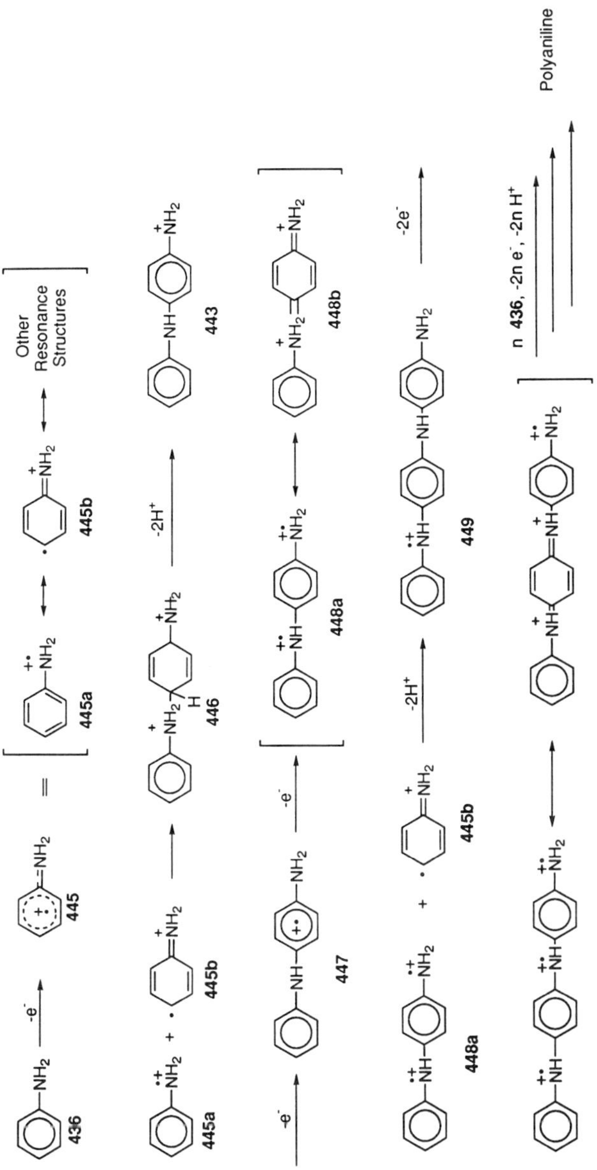

Figure 66 Mechanism for polymerization of aniline involving the coupling of cation radicals. (From Ref. 287.)

Another possibility is that the polymer growth step involves the reaction of a terminal dication or cation group of the growing polymer with neutral monomer. A comprehensive mechanism involving this step is outlined in Fig. 67 [288]. The reaction begins with the oxidation of aniline to cation radical **445**. Dimerization of this species can occur in a head-to-tail, head-to-head, or tail-to-tail manner to produce *N*-phenyl-*p*-phenylenediamine (**443**), *N,N'*-diphenylhydrazine (**452**), or benzidine (**442**), respectively. The transformation of **452** to **442** via the benzidine rearrangement occurs in the acidic reaction solution. The dimers are rapidly oxidized because of their lower oxidation potential (versus aniline) and their proximity to the electrode. The two-electron oxidation of **443** produces quinoidal diiminium ion **448**. There are two reaction pathways available to **448** that involve reactions with neutral monomer to produce the aniline trimer **454**. The electrophilic attack of **448** upon aniline followed by loss of two protons results in the formation of **454**. Alternatively, the deprotonation of **448** produces a nitrenium ion **453** which can also electrophilically react with aniline to give **454**. Stable nitrenium cations have been detected in other reaction mediums [289]. In a similar manner, quinoidal diiminium ion **455** can react directly with aniline to form aniline trimer **457**. The deprotonation of **455** can occur first to produce **456** which then reacts with aniline to form **457**.

This mechanism was proposed based on the effect of small amounts of additives upon the polymerization reaction of aniline and alkyl ring-substituted anilines. In the presence of additives such as *p*-aminodiphenylamine, benzidine, and *p*-phenoxyaniline the polymerization rate was more rapid at oxidation potentials below that of aniline. Therefore, the polymerization reaction could be performed at lower applied potentials in electrochemical polymerizations or with weaker oxidants in chemical oxidative polymerizations. These additives have lower oxidation potentials than aniline and can be oxidized to quinoidal diiminium ions. They also have at least one sterically accessible amino group. Thus, the polymerizations were initiated by the oxidation of these additives. The neutral aniline monomer then reacted with the oxidized additive. Polymer growth continued by a repeated cycle of (a) oxidation of the growing polymer, (b) coupling with aniline, and (c) deprotonation. At higher switch potentials (i.e., above the oxidation potential of aniline) the amount of dimeric species generated from the monomer would be expected to be greater than the amount of additive. Therefore, if the switch potential was gradually increased to a sufficiently high level, the difference in the two rate constants (with and without additive) would be expected to decrease to zero. This type of decrease was detected and the rate constants were identical at a switch potential above 0.9 V vs. SCE. In summary, the results are consis-

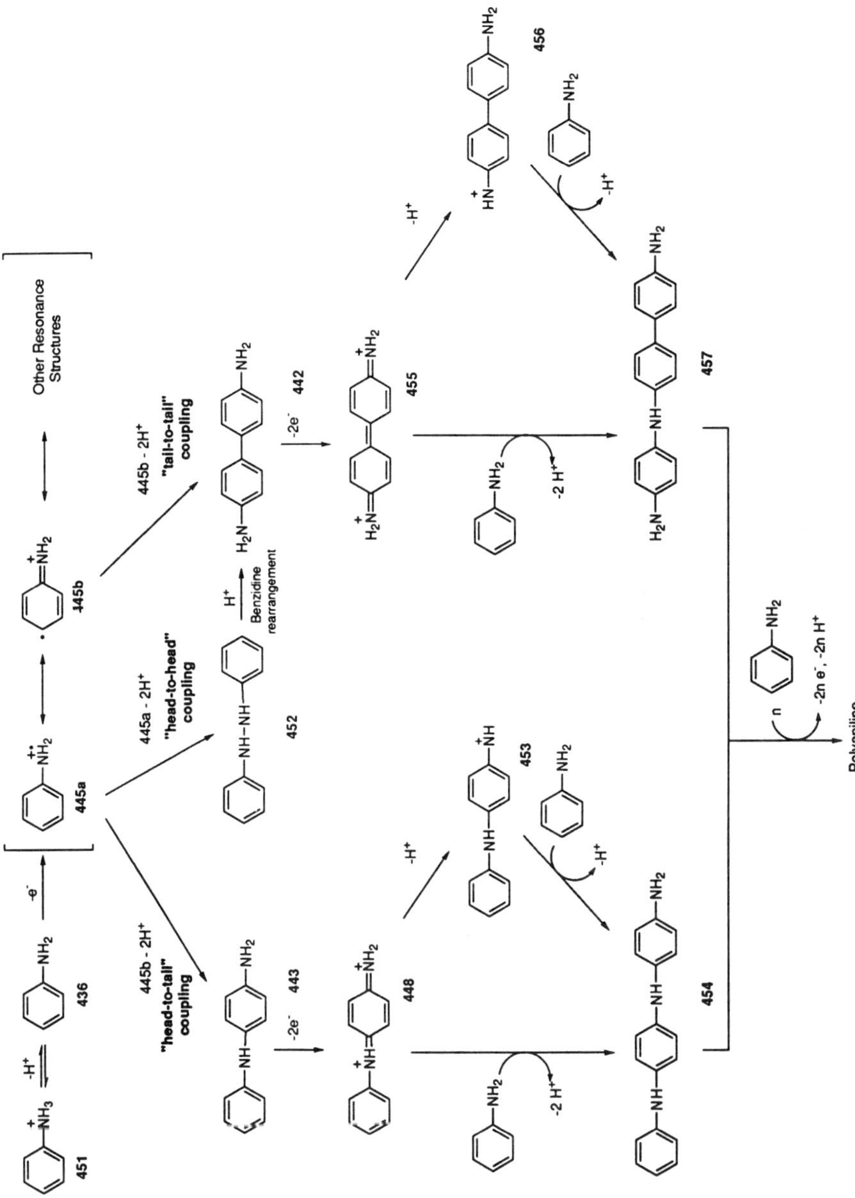

Figure 67 Mechanism of polyaniline formation involving the coupling of a dication or cation with neutral monomer. (From Ref. 288.)

tent with the proposed mechanism in which polymer growth occurs by the coupling reaction of neutral monomer with a growing polymer chain containing a terminal iminium or nitrenium group. Part of the rationale for proposing the participation of a neutral monomer in this mechanistic scheme was that the polymerization proceeds at applied potentials which are insufficient to oxidize the monomer. In addition to the mechanistic insights provided by this investigation, it also represents an improvement in the methods available for the synthesis of polyanilines. This is because fewer undesired side reactions are expected when the polymerization is performed at lower oxidation potentials or with weaker chemical oxidants [290].

The mechanism of the coupling reaction of aniline with the oxidized polymer remains an area of active investigation. As proposed in the previous mechanism (Fig. 67), aniline may couple directly with the oxidized polymer [288]. Another possibility is that aniline may be oxidized by the polymer before coupling [291]. This alternative was supported by an investigation of the kinetics and thermodynamics of the polymerization of 2-pentadecylaniline monolayers at fluid surfaces [292]. A large negative activation area and an increase in reaction rate with applied surface pressure were determined. These factors indicate that (a) the activated complex is bimolecular and (b) that like charges approach during activated complex formation. This intensification of charge during activated complex formation is expected when two cation radicals combine. The coupling reaction of **443** with aniline is shown in Fig. 68. The oxidation of **443** followed by deprotonation and further oxidation yields **453**. The acid-catalyzed oxidation of aniline by **453** yields **447** and **445**. The coupling of these two species followed by deprotonation gives **454**.

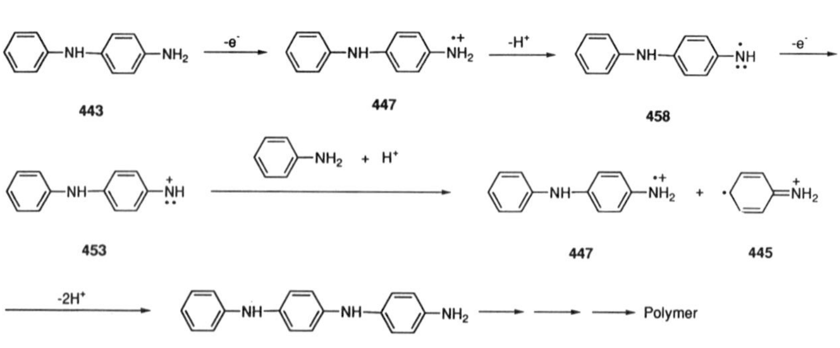

Figure 68 Mechanism of polyaniline formation involving oxidation of aniline by polymer before coupling. (From Ref. 292.)

G. Poly(*p*-phenylene Vinylene)

There has been a great deal of interest in poly(*p*-phenylene vinylene) (PPV) due to its ability of conduct electricity when in its doped form [293]. The early synthetic routes used for PPV synthesis resulted in the formation of oligomers. A significant factor that limited the molecular weight was the premature precipitation of the oligomers from solution. This problem was overcome by the development of a method to synthesize high molecular weight PPV from a soluble precursor polymer. This method involved the polymerization of bis-sulfonium salts to give high molecular weight, water-soluble polyelectrolytes [294]. These polyelectrolytes could be converted to PPV by treatment with excess base or thermolysis [295]. This conversion step involved the elimination of dialkyl sulfide and HCl. High molecular weight PPV films have also been fabricated by this method (Fig. 69) [296–300].

Two proposed mechanisms for the polymerization of α,α'-bis(dialkylsulfonio)-*p*-xylene dihalides (**463**) are shown in Fig. 70 [302]. Both mechanisms begin with the abstraction of an α-proton to produce ylid **464**. The 1,6-elimination of dialkyl sulfide produces the *p*-xylylene pseudo-diradical **465**. One mechanism involves the formation of polymer from this species via the dimerization of **465** to give the dication diradical **466**. This species was proposed to grow rapidly to high molecular weight polymer (**467**) by head-to-tail additions to both ends [301]. An alternative mechanism involving polymer formation from **465** via a putative anionic mechanism has been proposed [302]. There were two main factors in the proposal of

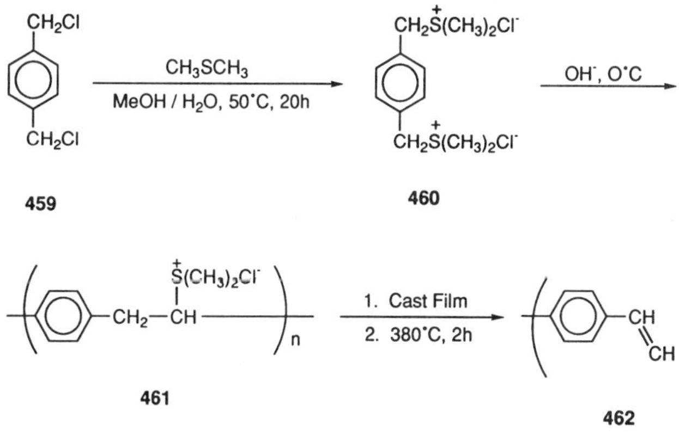

459 460 461 462

Figure 69 Poly(*p*-phenylene vinylene) synthesis from a soluble precursor polymer. (From Ref. 298.)

this mechanism. First, there was no direct evidence that radical species were actually present. There was no decrease in the ESR signal intensity of the radical trap TEMPO when the polymerization was performed in its presence. Conversely, no ESR signal was detected when the reaction was performed in the presence of 0.1 to 1.0 equivalents of the spin-trap phenyl-*N*-(*tert*-butyl)nitrone. Transient radicals would be expected to be converted to stable nitrosyl radicals in the presence of this spin trap. Furthermore, the polymerization was unaffected when it was performed with styrene present. No polystyrene was detected under these conditions. Because there was no experimental evidence of the participation of radicals, a polymerization mechanism that was more consistent with the mechanism of other *p*-xylylenes [303,304] was proposed for the conversion of **465** to polymer [302]. The highly polar reaction conditions and the polarity of the sulfur carbon bond of **463** would favor the nucleophilic attack of **465** upon **463** in a putative anionic chain polymerization (Fig. 70).

H. Polyphenylenes

Poly(*p*-phenylene) (PPP) is the simplest aromatic rigid rod-like polymer. The synthesis of high molecular weight PPP is of interest both for the purpose of providing a simple model for rigid rod-like polymers as well as for the synthesis of useful new materials. PPP has been the subject of several reviews [194,202,305,306]. Poly(phenylene)s can be prepared via cation-radical intermediates. These intermediates have been implicated as the key propagating species in the oxidative coupling of benzene and substituted benzenes in the presence of a Lewis acid catalysis and an oxidizing agent [307]. One method for the synthesis of PPP is the reaction of benzene with $AlCl_3$ and $CuCl_2$ [Eq. (87)] [194,308–310].

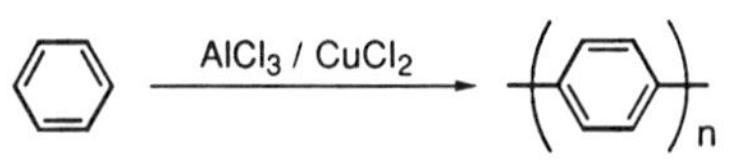

(87)

469 **470**

The reaction begins with the one-electron reduction of benzene to form a benzene cation radical **471** (Fig. 71) [307]. Other benzene molecules have been proposed to associate in a coordinative manner with this cation radical [307,311]. As each additional benzene molecule associates the cation radical becomes further delocalized. Eventually the chain of associated benzene molecules becomes so long that the terminal benzenes have too little cation-radical character to sustain further propagation. At this point the upper limit of chain length for the original cation radical is reached. When viewed from the side, the chain of coordinatively associated benzene molecules has an appearance similar to stairs. The formation

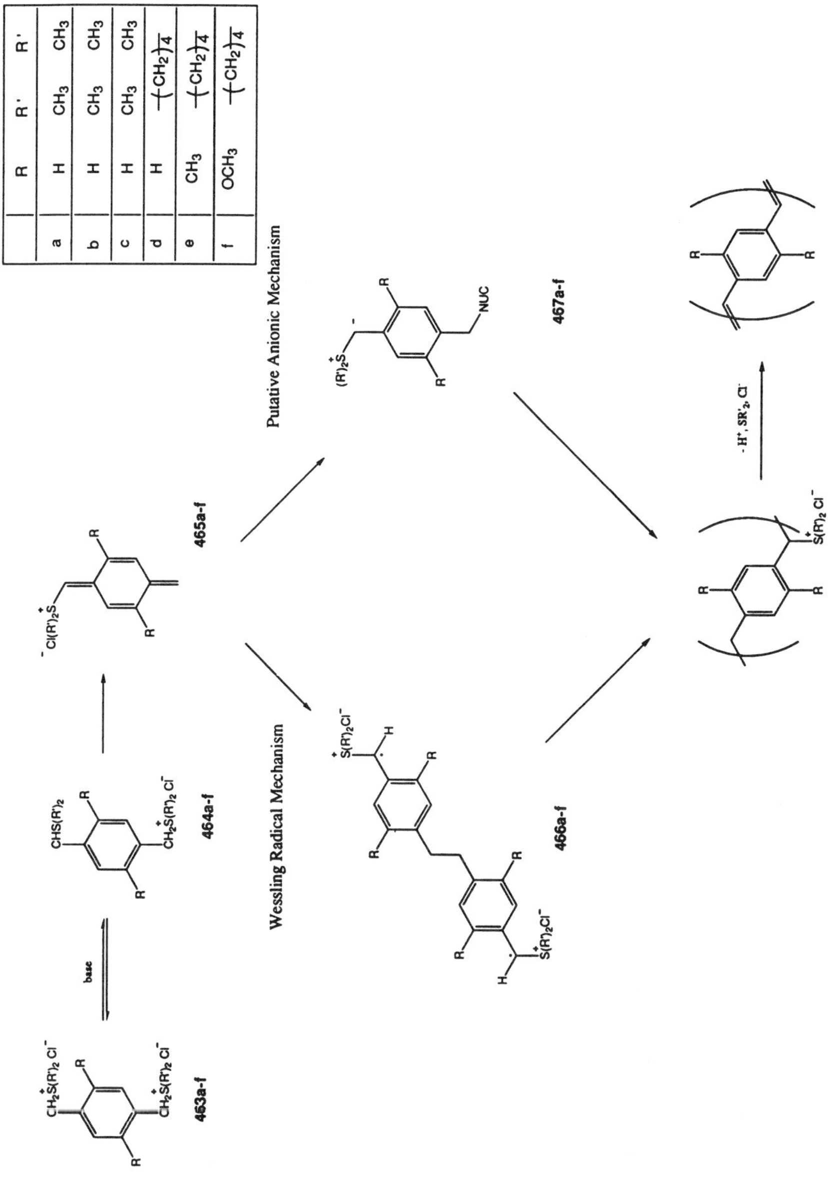

	R	R'	R'
a	H	CH_3	CH_3
b	H	CH_3	CH_3
c	H	CH_3	CH_3
d	H	$-(CH_2)_4-$	
e	CH_3	$-(CH_2)_4-$	
f	OCH_3	$-(CH_2)_4-$	

Figure 70 Two proposed mechanisms for the synthesis of PPV. (From Ref. 302.)

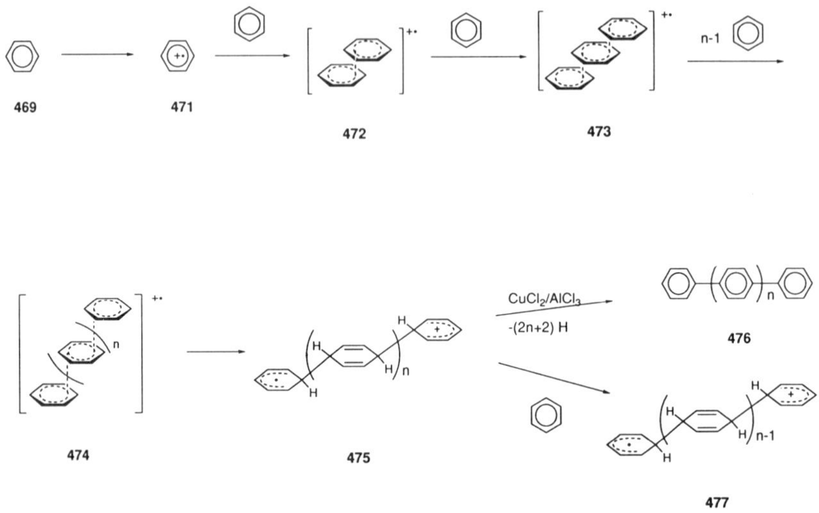

Figure 71 Radical cation stair-step mechanism of polyphenylene synthesis. (From Refs. 307 and 311.)

of covalent σ-bonds linking the benzene molecules leads to an oligomeric (1,4-dihydrobenzene) cation-radical structure with a hydrobenzene radical at one terminus and a hydrobenzene cation at the other terminus. A one-electron oxidation and the loss of two protons is necessary to aromatize the terminal phenyl groups [307,311].

The 1,4-dihydrobenzene groups may rearrange giving the conjugated oligomeric 1,3-dihydrobenzene structure before aromatization. In any case, oxidation of the dihydrobenzene groups followed by loss of two protons results in formation PPP. This reaction is fundamentally different from most polymerization reactions in that the electronic nature of the propagating species is actually delocalized throughout the entire oligomer chain during the chain-growth period, rather than being concentrated upon the chain termini. Depolymerization (i.e., **475** to **477**) can also occur before the aromatization step. The phenylene chains are typically 13–15 units long [307]. Irregular structures are obtained due to substitution at positions other than the para-position.

Another mechanism has been proposed for the oxidative polymerization of benzene in HF/SbF$_5$ in the presence of an applied potential [312]. In this proposed mechanism, the coupling step occurs after protonation

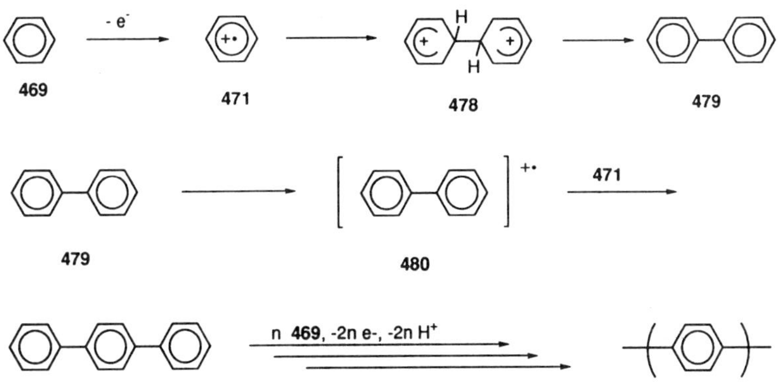

Figure 72 Proposed mechanism for electropolymerization of benzene involving the stepwise coupling of cation-radical species. (From Ref. 313.)

of benzene by the electrophilic attack of the protonated benzene cation upon another benzene molecule to form a cationic dimer.

A mechanism involving the coupling of cation radicals has also been considered for the electropolymerization of benzene compounds [306,313]. This mechanism occurs by a sequence of events similar to those proposed for the electropolymerization of pyrroles. The first step is the oxidation of benzene to a cation radical (**471**). Two of these cation radicals combine to form a dication dimer (**478**). The neutral aromatic dimer (**479**) is formed upon loss of two protons. This dimer is then reoxidized to a cation radical (**480**). Chain growth is accomplished by the coupling reaction of this cation radical with other cation radicals followed by deprotonation to form aromatic structures. Polymer growth continues by this sequence of steps until precipitation from solution occurs (Fig. 72).

VII. ZWITTERIONIC POLYMERIZATIONS

Zwitterionic propagating species have been implicated in both homopolymerization and copolymerization reactions. An early example of the involvement of zwitterions in an alternating copolymerization was provided by the copolymerization of *p*-benzoquinone diazides with tetrahydrofuran [Eq. (88)] [314,315]. This reaction could be thermally or photochemically initiated. A similar 1:1 alternating copolymer was obtained from the thermally or photochemical decomposition of 3,5-dimethylbenzene-1,4-diazooxide in THF [316].

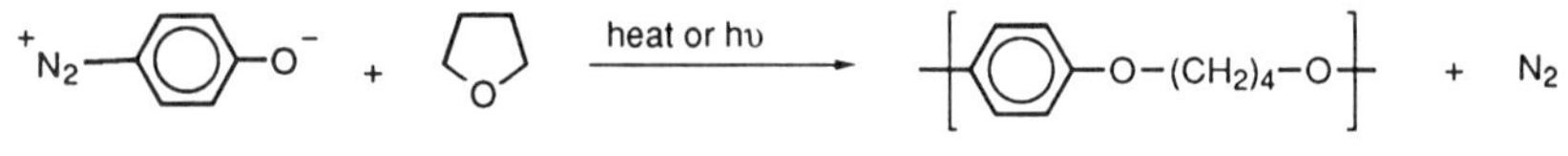

482 483 484

(88)

Several types of zwitterionic polymerization reactions have been identified. One type is spontaneously initiated upon mixing a nucleophilic monomer with an electrophilic monomer. Another type involves the thermal generation of a reactive zwitterion. Finally, zwitterions that are sufficiently stable to be isolated have been polymerized.

A. Spontaneous Initiated Alternating Copolymerization of Nucleophilic-Electrophilic Monomer Pairs

Zwitterions are the propagating species in a number of spontaneous polymerizations that occur upon mixing electron donor and acceptor monomers. Several reviews of spontaneous polymerizations via zwitterion intermediates are available [317–323]. In this novel type of polymerization reaction there is no need to add an initiator or catalyst. Alternating copolymers have been obtained in some cases when at least one of the monomers contains a heterocycle. In these polymerizations a nucleophilic monomer (M_N) and an electrophilic monomer (M_E) react to form a zwitterion [Eq. (89)]. This zwitterion is referred to as a "genetic zwitterion."

$$\underset{485}{M_N} + \underset{486}{M_E} \longrightarrow \underset{487}{{}^{+}M_N\text{-}M_E^{-}} \tag{89}$$

$$\underset{487}{{}^{+}M_NM_E^{-}} \xrightarrow{{}^{+}M_NM_E^{-}} \underset{488}{{}^{+}M_NM_E\text{-}M_NM_E^{-}} \xrightarrow{n\,{}^{+}M_NM_E^{-}} \underset{489}{{}^{+}M_N\text{-}(M_EM_N)_{n+1}\text{-}M_E^{-}} \tag{90}$$

$$\underset{489}{{}^{+}M_N\text{-}(M_EM_N)_{n+1}\text{-}M_E^{-}} \xrightarrow{{}^{+}M_N\text{-}(M_EM_N)_m\text{-}M_E^{-}} \underset{490}{{}^{+}M_N\text{-}(M_EM_N)_{n+m+1}\text{-}M_E^{-}} \tag{91}$$

Polymer chain growth is initiated by the dimerization reaction of two molecules of zwitterion **487** to form oligomeric zwitterion **488** [Eq. (89)]. Polymer chain growth continues by the addition of other zwitterions. These zwitterions can be the genetic zwitterion **487**, oligomeric zwitterions, or macrozwitterions. The reaction of two macrozwitterions results in a large increase in the molecular weight. In some polymerizations, free monomer can also react with the propagating species.

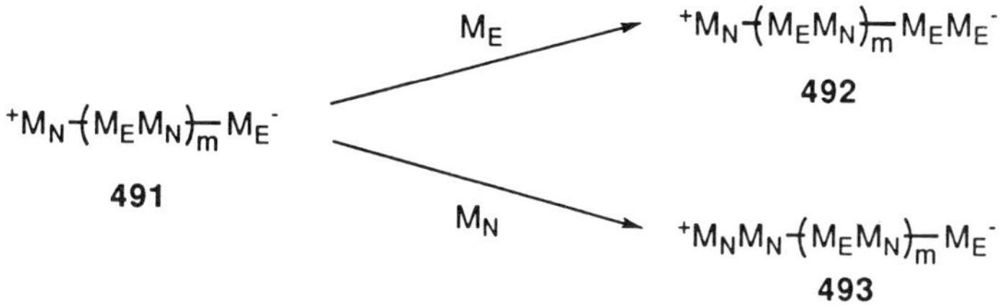

(92)

Therefore, an alternating copolymer is produced when the propagating species react exclusively with other zwitterions [Eqs. (90)–(91)].

The first example of this type of polymerization reaction was provided by the reaction of 2-oxazoline and β-propiolactone to give an alternating copolymer [324] [Eq. (93)].

(93)

494 495 496

A variety of monomer pairs have been used in spontaneous zwitterion polymerizations. Examples of nucleophilic and electrophilic monomer pairs, genetic zwitterions, and polymer structures are shown in Table 2.

A comprehensive mechanistic study of the reaction of 2-methyl-2-oxazoline and acrylic acid [338] confirmed the initiation and propagation steps of the originally suggested mechanism [339] (Fig. 73). In addition two termination reactions responsible for the low molecular weight were identified. The reaction of **534** and **507** results in the formation of an unstable zwitterion **544**. The genetic zwitterion **545** is obtained after a proton transfer reaction. The dimerization of **545** occurs via a ring-opening nucleophilic attack of the carboxylate anion upon the C_5 position of the oxazoline ring. Polymer growth occurs by the coupling reaction of various zwitterions. For example, the reaction of the growing polymer chain **547** and the genetic zwitterion (**545**) is shown in Fig. 73. Two termination reactions were identified. The reaction of zwitterion **547** with acrylic acid resulted in the formation of a polymer (or oligomer) with unreactive chain ends (**549**). A polymer (or oligomer) with unreactive chain ends (**551**) was also formed by the reaction of **547** with a salt formed by the reaction of 2-methyl-2-oxazoline and acrylic acid. The importance of these termination reactions is enhanced by the high concentration of **543** and **507** relative to propagating species.

Table 2 Examples of nucleophilic and electrophilic monomers that participate in spontaneous zwitterionic polymerization

Entry	M_N	M_E	Genetic Zwitterion	Polymer Structure	Reference
1	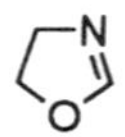 **494**	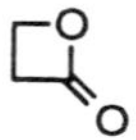 **495**	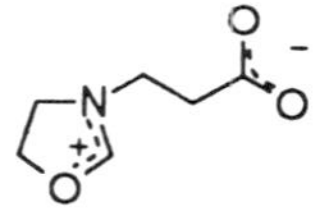 **497**	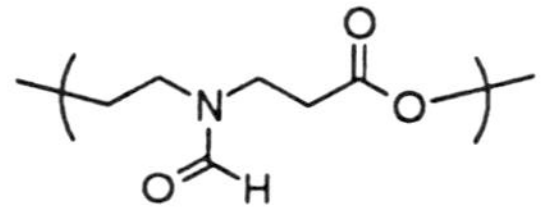 **496**	[325]
2	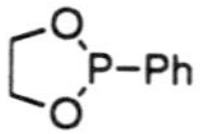 **498**	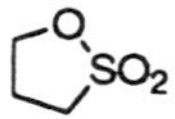 **499**	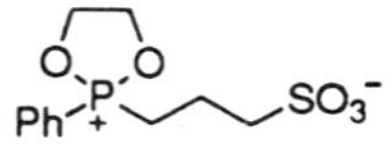 **500**	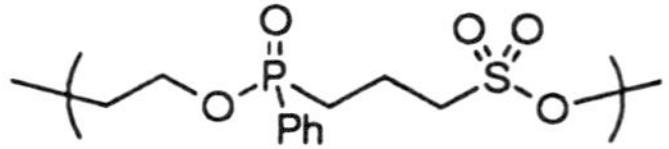 **501**	[326]
3	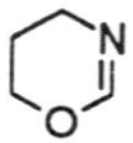 **502**	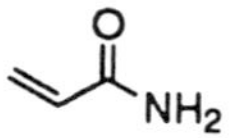 **503**	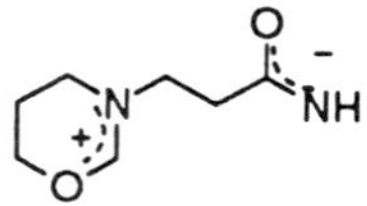 **504**	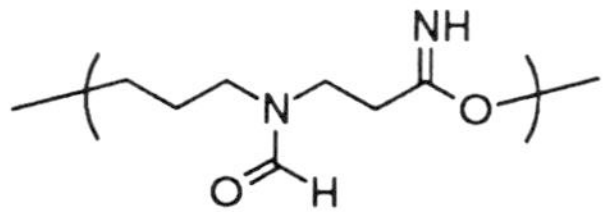 **505**	[327]
4	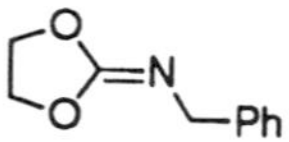 **506**	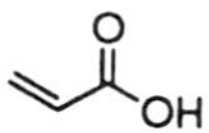 **507**	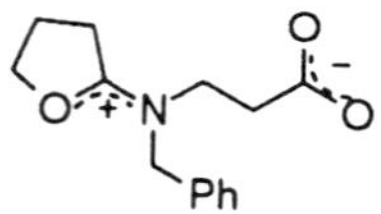 **508**	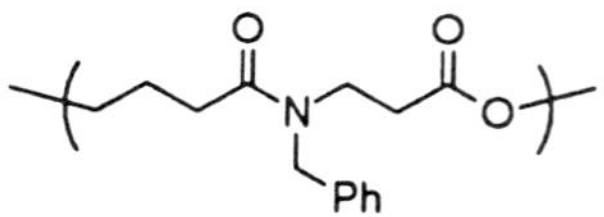 **509**	[328]

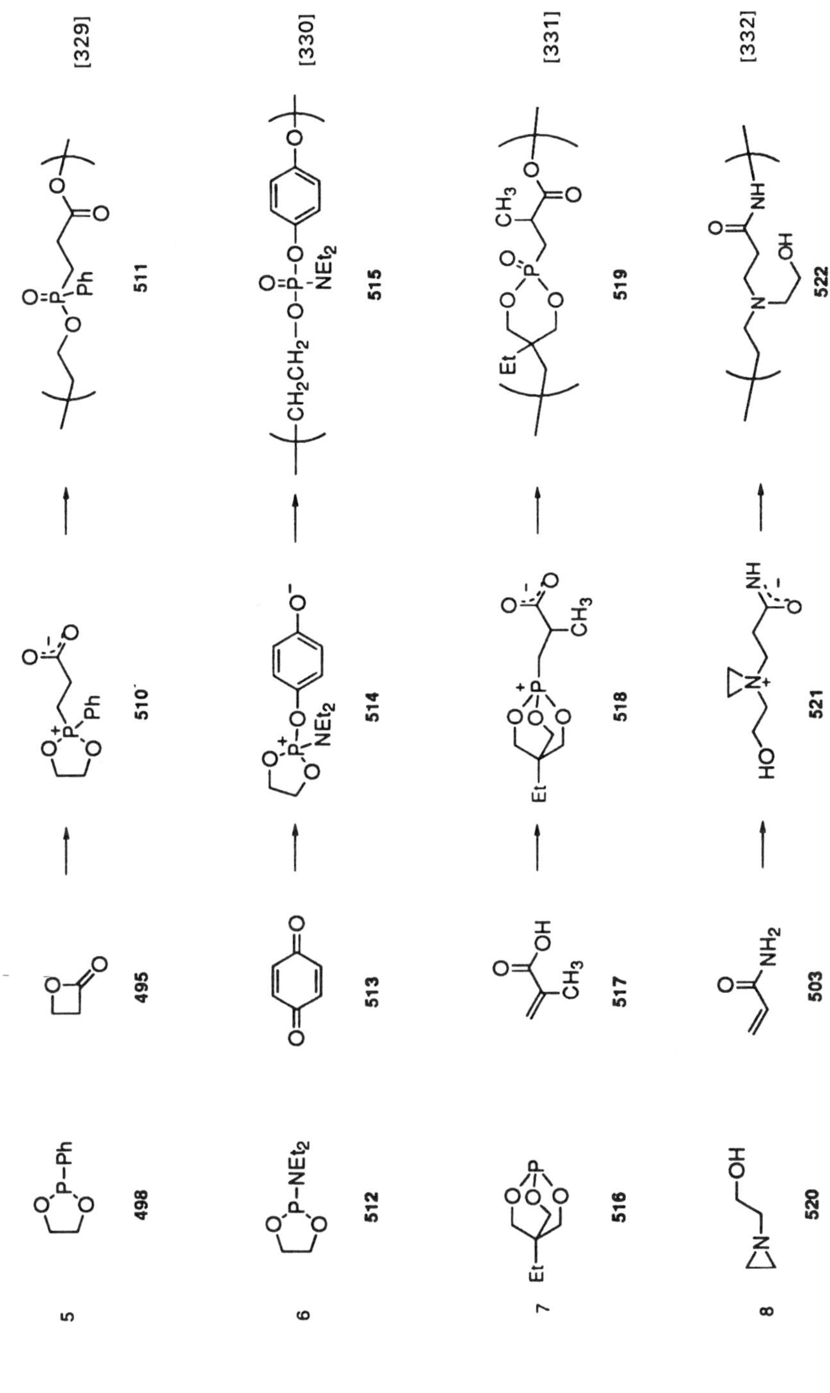
5
498
495
510
511
[329]
6
512
513
514
515
[330]
7
516
517
518
519
[331]
8
520
503
521
522
[332]

Table 2 (Continued)

Entry	M_N	M_E		Genetic Zwitterion		Polymer Structure	Reference
9	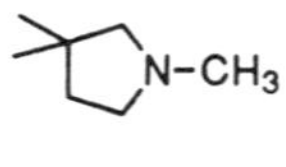 **523**	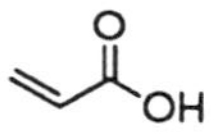 **507**	→	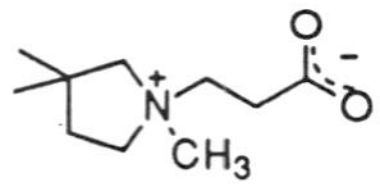 **524**	→	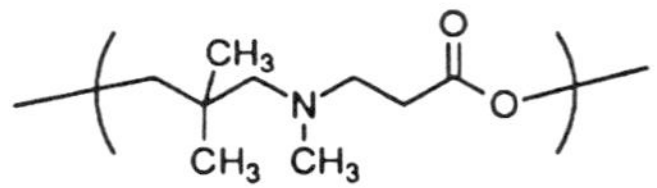 **525**	[333]
10	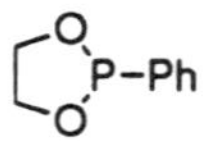 **498**	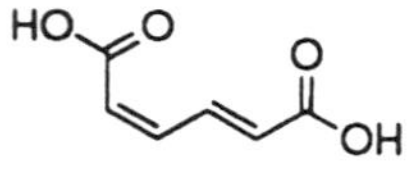 **526**	→	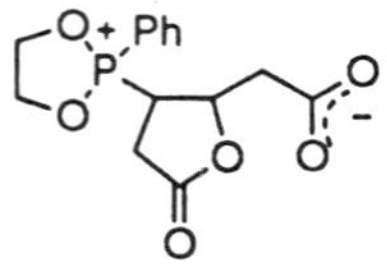 **527**	→	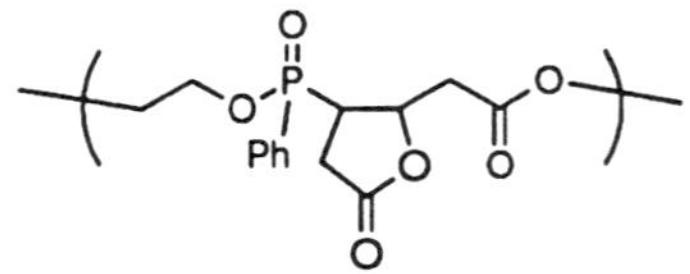 **528**	[334]
11	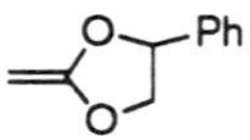 **529**	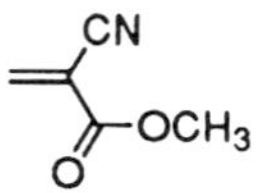 **530**	→	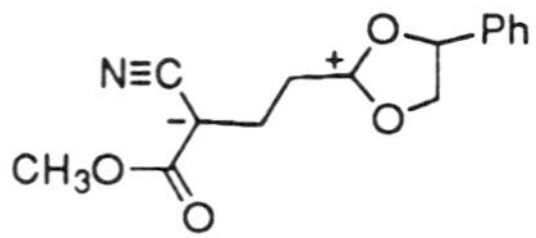 **531**	→	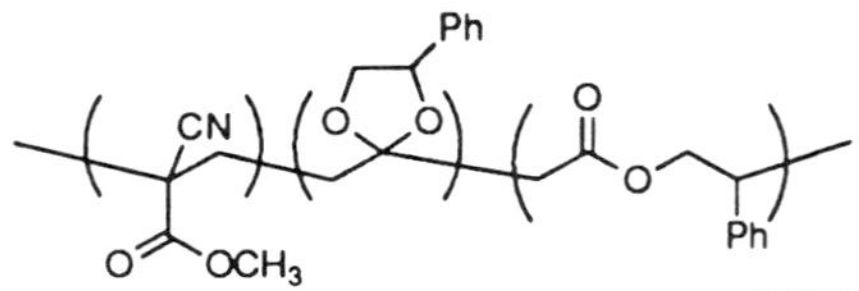 **532**	[335]
12	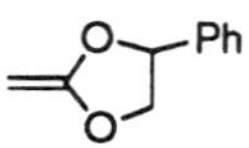 **529**	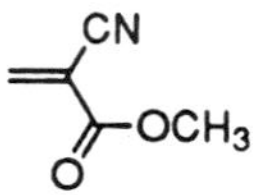 **530**	→	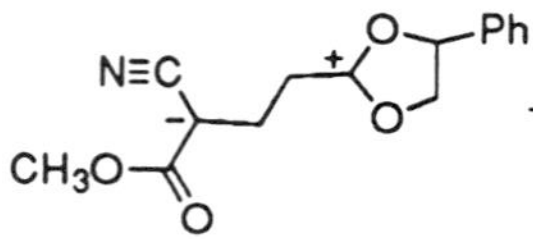 **531**	Bu_4NBr →	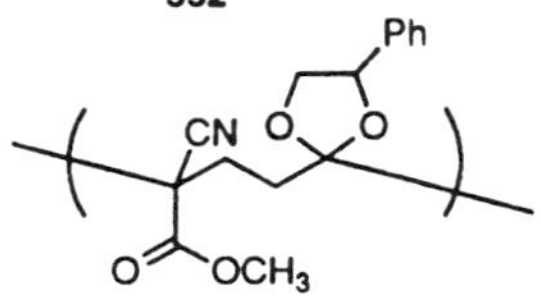 **533**	[335]

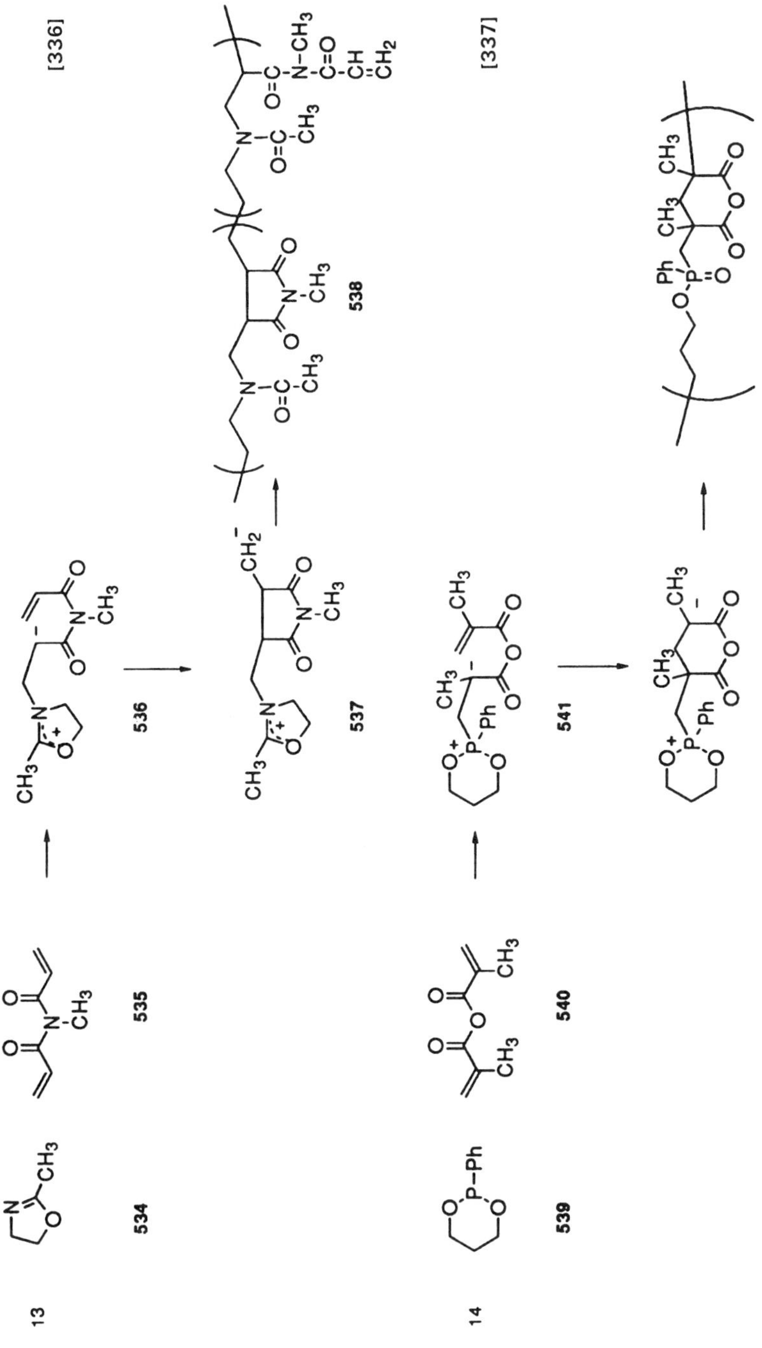

13 — 534, 535, 536, 537, 538 [336]

14 — 539, 540, 541, 542, 543 [337]

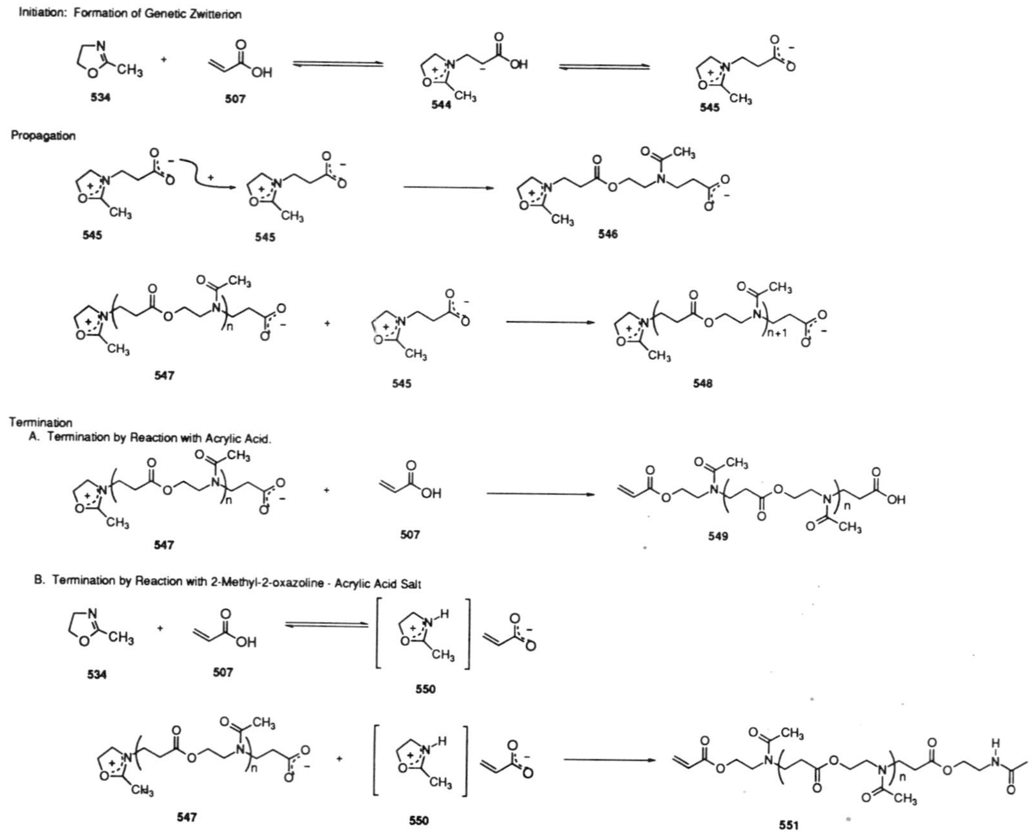

Figure 73 Mechanism of the reaction of 2-methyl-2-oxazoline and acrylic acid. (From Ref. 338.)

These particular termination reactions are not possible in the zwitterionic polymerization of 2-{[2-(2-oxazolin-2-yl)propyl]thio}ethanethiol (**552**) [Eq. (94)] and 2-(2-oxazolin-2-yl)propanethiol. Higher molecular weights were obtained (M_n from 4,100 to 42,900). Polymerizations were performed in bulk and in DMF at temperatures ranging from 100° C to 200° C. Oxazoline and thiol end groups (**554a**) were detected in the polymerization of **552**. Olefinic end groups (**554b**) were also identified in the polymers prepared from **552** at higher temperatures (140° C and higher in solution, 120° C and higher in bulk). The predominate end groups formed in bulk polymerizations of **552** at 170° C and higher were oxazoline and olefin end groups.

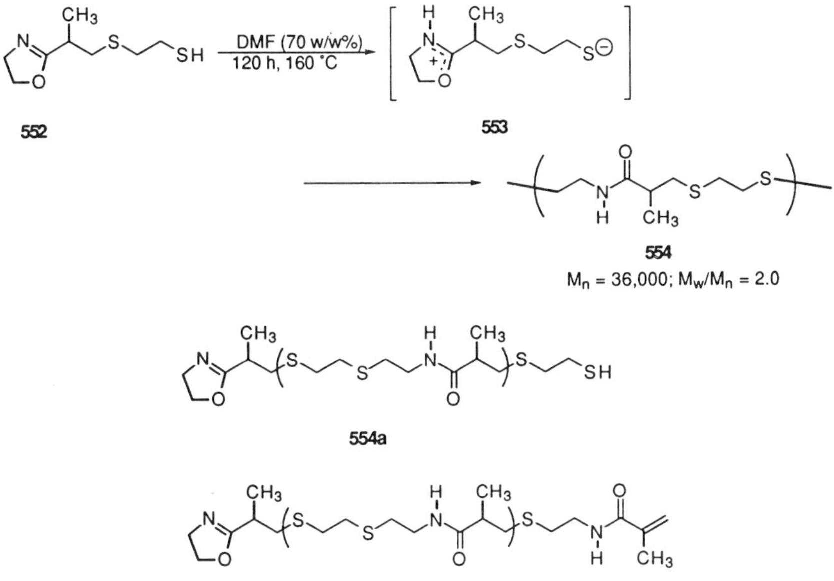

(94)

B. Polymerization of Preformed Zwitterionic Monomers

The polymerization of zwitterions with sufficient stability to be isolated and purified has been demonstrated. These reactions of stable zwitterions provide well-defined examples of the involvement of zwitterionic propagating species. The first example of the polymerization of a stable zwitterion used a tetrahydrothiophenium arene oxide zwitterion **555** [Eq. (95)] [340–342].

555 556 (95)

Zwitterion 1-[(4-carboxyphenyl)methyl]tetrahydrothiophenium hydroxide inner salt (**557**) was polymerized at temperatures ranging from 70° C to 200° C [343]. The copolymer contained oxycarbonyl-1,4-phenylenemethylene (A) and oxycarbonyl-1,4-phenylenemethylenethiotetramethylene (B) repeat units [Eq. (96)]. These units were distributed randomly in the copolymer.

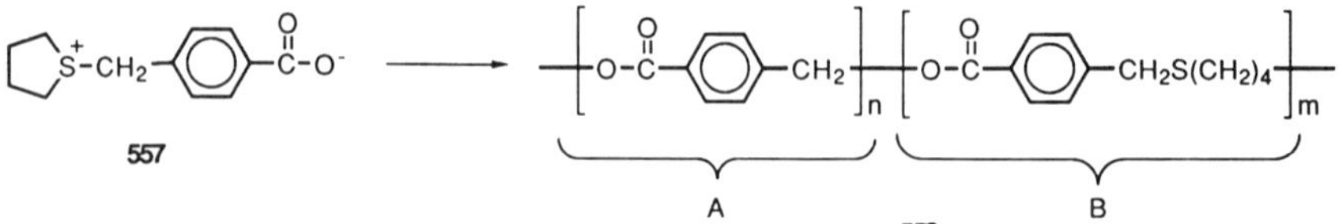

557 558

(96)

The number-average molecular weight varied with the reaction temperature. The highest molecular weight for the bulk polymerization of CH_3OH-free zwitterion was obtained at 100° C (M_n = 32,700, M_w/M_n = 3.1, 92% yield). The ratio of repeat units A/B was 3.0 at this temperature. The ratio of A/B was increased to 99 when the polymerization was done at 200° C. However, the molecular weight was low (M_n = 5000) at this temperature. These bulk polymerizations were performed in sealed tubes. The molecular weight was increased by performing the reaction under dynamic vacuum (M_n = 27,200 at 100° C and M_n = 41,000 at 130° C). Polymers obtained by solution polymerization had significantly lower molecular weights and yields (M_n = 7400, 68% yield at 100° C).

The polymerization was initiated by attack of the carboxylate anion on either the benzylic carbon or on the α-carbon of the tetrahydrothiophenium ring (Fig. 74). Reaction at the benzylic carbon was favored. The product of the initiation step contained reactive groups at both ends. Thus propagation of the polymerization proceeded stepwise by reactions at both ends of the growing polymer chain. Before termination reactions, the growing polymer chain has the structure **562** (Fig. 74). Because the polymer contains reactive groups at both termini, polymer growth can continue after deactivation of one of the groups.

Polymerization could be terminated by several mechanisms. Cyclization of the growing oligomer terminates chain growth. Reaction of the reactive chain ends with water resulted in the formation of hydroxyl (tetrahydrothiophenium ring termini) and carboxylic acid end groups.

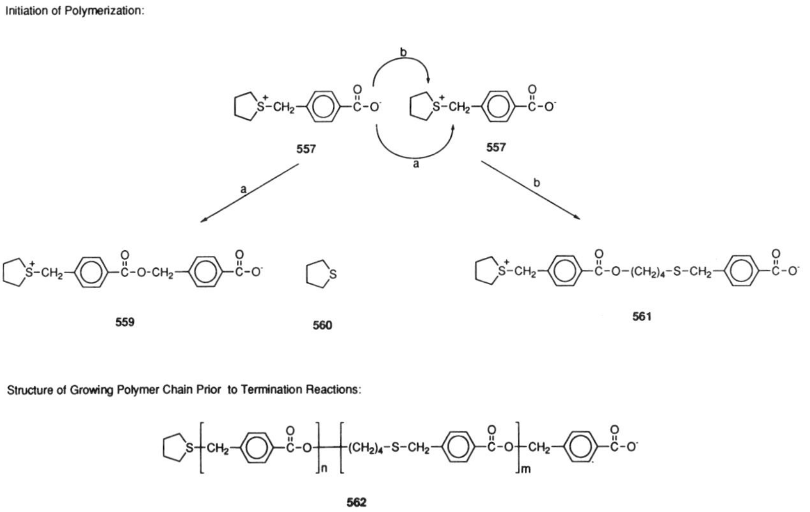

Figure 74 Initiation of polymerization of the zwitterion 1-[4-carboxyphenyl)-methyl] tetrahydrothiophenium hydroxide inner salt. (From Ref. 343.)

Some of the carboxylic acid groups were esterified to methyl ester end groups when methanol was present.

The polymerization of tetrahydro-1-[4-hydroxy-3-(2-hydroxyethoxy)-phenyl]thiophenium hydroxide inner salt was carried out in bulk and in solution. Higher yields (88.3–90.6%) and number-average molecular weights (M_n = 51,300–67,800) were obtained in bulk. Lower polymer yields (maximum 19%) and lower molecular weights (maximum M_n = 6930) were obtained in solution. The major products in solution were cyclic oligomers. Increased polymer yields and molecular weights were obtained by the addition of trifluoroacetic acid salts to the solution.

Three termination reactions were identified: intramolecular cyclization to form cyclic oligomers, reaction with water to form alcohol and phenol end groups, and β-elimination resulting in vinyl and phenol end groups. The majority of end groups formed in the bulk polymerization were vinyl and phenol groups. In the absence of additives, cyclization reactions were the most important termination reaction in solution. The major end groups for the linear polymer formed in solution were alcohol and phenol groups [344] (Fig. 75).

Stable zwitterions containing ammonium and carboxylate ionic centers have also been polymerized. The bulk zwitterionic polymerization of

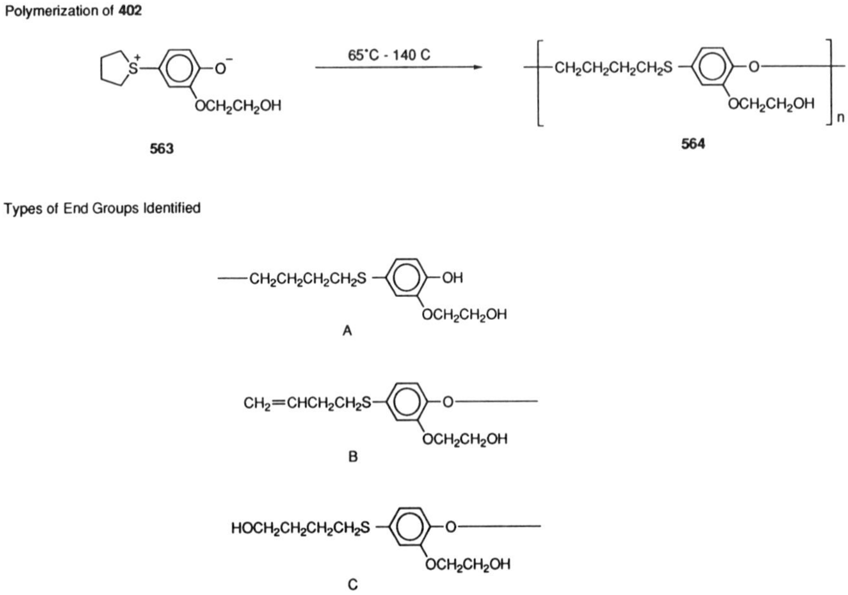

Figure 75 Polymerization of 1-[4-hydroxy-3-(2-hydroxyethoxy)phenyl]thiophenium hydroxide inner salt, **563**. (From Ref. 344.)

1-[(4-carboxyphenyl)methyl]quinuclidinium hydroxide inner salt (**565**) was performed at temperatures ranging from 170° C to 260° C [Eq. (97)] [345]. The random copolymers had number-average molecular weights ranging from 4000 to 17,000. The highest M_n was obtained at 180° C while continuously evacuating the reaction vessel at a pressure of 0.1 mm Hg.

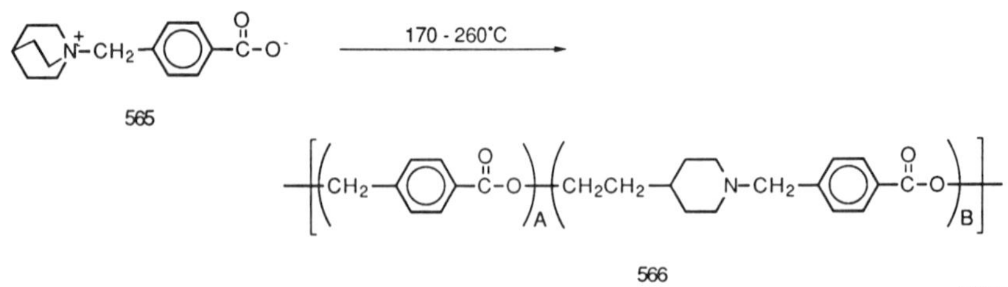

(97)

ACKNOWLEDGMENT

Financial support by the National Science Foundation (DMR-92-067181) is gratefully acknowledged.

REFERENCES

1. G. B. Butler, *Cyclopolymerization and Cyclocopolymerization*, Marcel Dekker, New York, 1992.
2. P. Lutz, G. Beinert, and P. Rempp, *Makromol. Chem. 183*: 2787–2797 (1982).
3. H. Hasegawa and T. Higashimura, *Macromolecules 13*: 1350–1354 (1980).
4. M. Sawamoto, T. Masuda, H. Nishii, and T. Higashimura, *J. Polym. Sci. Part C. Polym. Lett. Ed. 13*: 279–282 (1975).
5. T. Higashimura and H. Nishii, *J. Polym. Sci. Part A. Polym. Chem. Ed. 15*: 329–339 (1977).
6. T. Higashimura, S. Aoshima, and H. Hasegawa, *Macromolecules 15*: 1221–1225 (1982).
7. Z. Jiang and A. Sen, *J. Am. Chem. Soc. 112*: 9655–9657 (1990).
8. O. Nuyken, G. Maier, D. Yang, and M. B. Leitner, *Makromol. Chem. Macromol. Symp. 60*: 57–63 (1992).
9. D.-H. Suh and J. V. Crivello, *Chem. Mater. 5*: 210–213 (1993).
10. J. V. Crivello and A. Ramdas, *Am. Chem. Soc. Div. Polym. Chem. Polym. Preprints 32*(3): 175–176 (1991).
11. T. Dittmer, F. Gruber, and O. Nuyken, *Makromol. Chem. 190*: 1771–1790 (1989).
12. T. Dittmer, F. Gruber, and O. Nuyken, *Makromol. Chem. 190*: 1755–1770 (1989).
13. O. Nuyken, D. Yang, F. Gruber, and G. Maier, *Makromol. Chem. 192*: 1969–1979 (1991).
14. O. Nuyken, M. B. Leitner, and G. Maier, *Makromol. Chem. 192*: 3071–3078 (1991).
15. O. Nuyken, M. B. Leitner, and G. Maier, *Makromol. Chem. 193*: 487–500 (1992).
16. G. Maier, M. B. Leitner, and O. Nuyken, *Makromol. Chem. 193*: 3083–3096 (1992).
17. G. Maier, D. Yang, and O. Nuyken, *Polym. Bull. 28*: 135–141 (1992).
18. J. Nishimura, N. Tanaka, and S. Yamashita, *J. Polym. Sci. Part A. Polym. Chem. Ed. 18*: 1203–1211 (1980).
19. R. G. Neville, *U.S. Patent 3,725,346* (1973); *Chem Abstr. 79*: 19380b (1973).
20. J. Nishimura, N. Tanaka, N. Hayashi, and S. Yamashita, *J. Polym. Sci. Part A. Polym. Chem. Ed. 18*: 515–526 (1980).
21. J. V. Crivello and A. M. Carter, *Am. Chem. Soc. Div. Polym. Chem. Polym. Preprints 32(3)*: 177–178 (1991).
22. J. V. Crivello, D.-H. Suh, and A. M. Carter, *Am. Chem. Soc. Div. Polym. Mat. Eng. 67*: 258–259 (1992).
23. J. V. Crivello and A. Ramdas, *J. Macromol. Sci. Pure Appl. Chem. A29*: 753–774 (1992).
24. J. Crivello, *J. Polym. Sci. Part A. Polym. Chem. Ed. 31*: 1847–1857 (1993).
25. T. Higashimura, M. Hiza, and H. Hasegawa, *Macromolecules 12*: 217–222 (1979).

26. K. Matyjaszewski and P. Sigwalt, *Macromolecules 20*: 2679–2689 (1987).
27. C. P. Tsonis, *Comprehensive Polymer Science* (G. Allen and J. C. Bevington, eds.), Pergamon, Oxford, 1989, Vol. 5, Chap. 26, pp. 455–464.
28. R. W. Lenz, *Carbocation Polymerization* (J. P. Kennedy and E. Marechal, eds.), Wiley, New York, 1989, pp. 395–407.
29. R. L. Shriner and A. Berger, *J. Org. Chem. 6*: 305–318 (1941).
30. J. P. Kennedy and R. B. Isaacson, *J. Macromol. Chem. 1*: 541–552 (1966).
31. G. Montaudo, F. Bottino, S. Caccamese, P. Finocchiaro, and G. Bruno, *J. Polym. Sci. Part A-1 8*: 2453–2464 (1970).
32. J. Kuo and R. W. Lenz, *J. Polym. Sci. Part A. Polym. Chem. Ed. 15*: 119–130 (1977).
33. J. E. Chandler, B. H. Johnson, and R. W. Lenz, *Macromolecules 13*: 377–380 (1980).
34. J. Skura and R. W. Lenz, *Polym. Bull. 2*: 31–36 (1980).
35. A. Fritz, *Chem. Technol. 2*: 687–689 (1972).
36. A. Fritz and R. W. Rees, *J. Polym. Sci. Part A-1 10*: 2365–2378 (1972).
37. N. A. Peppas and G. N. Valkanas, *J. Polym. Sci. Polym. Chem. Ed. 12*: 2567–2579 (1974).
38. W. C. Overhults and A. D. Ketley, *Makromol. Chem. 95*: 143 (1966).
39. A. Knop, V. Böhmer, and L. A. Pilato, *Comprehensive Polymer Science* (G. Allen and J. C. Bevington, eds.), Pergamon, Oxford, 1989, Vol. 5, Chap. 35, pp. 611–665.
40. V. Percec and J. M. Tingerthal, *J. Polym. Sci. Part A. Polym. Chem. Ed. 25*: 2577–2583 (1987).
41. J. M. (Tingerthal) Bruss, M. R. V. Sahyun, E. Schmidt, and D. K. Sharma, *J. Polym. Sci. Part A. Polym. Chem. Ed. 31*: 987–994 (1993).
42. J. H. Golden, *J. Chem. Soc.* 3741–3748 (1961).
43. R. Z. Greenly, Ph.D. Dissertation, (C.S. Marvel), University of Illinois (1961).
44. J. Coudanne, M. Brigodiot, and E. Marechal, *Eur. Polym. J. 19*: 627–631 (1983).
45. R. H. Michel, *J. Polym. Sci. Part A 2*: 2533–2545 (1964).
46. D. Bender, H. Unterberg, and K. Müllen, *Angew. Chem. Int. Ed. Engl. 25*: 444–446 (1986).
47. A. H. Frazer, B. C. Anderson, and T. Fukunaga, *J. Polym. Sci. Part A. Polym. Chem. Ed. 23*: 2791–2802 (1985).
48. W. F. Hale, A. G. Farnham, R. N. Johnson, and R. A. Clendinning, *J. Polym. Sci. Part A-1 5*: 2399–2414 (1967).
49. E. D. Bergmann and D. Katz, *J. Chem. Soc.* 3216–3217 (1958).
50. C. Bunel, S. Cohen, J. P. Laguerre, and E. Marechal, *Polym. J. 7*: 320–325 (1975).
51. U. Scherf and K. Müllen, *Makromol. Chem. Rapid Commun. 12*: 489–497 (1991).
52. U. Scherf and K. Müllen, *Macromolecules 25*: 3546–3548 (1992).
53. U. Scherf and K. Müllen, *Polymer 33*: 2443–2446 (1992).

54. U. Scherf, A. Bohnen, and K. Müllen, *Makromol. Chem. 193*: 1127–1133 (1992).
55. K. Müllen and U. Scherf, *Makromol. Chem. Macromol. Symp. 69*: 23–33 (1993).
56. U. Scherf and K. Müllen, *Synthesis* 23–38 (1992).
57. G. M. Robinson, *J. Chem. Soc.* 267–276 (1915).
58. A. S. Lindsey, *Chem. Ind.* 823–824 (1963).
59. A. S. Lindsey, *J. Chem. Soc.* 1685–1692 (1965).
60. H. Erdtman, F. Haglid, and R. Ryhage, *Acta Chem. Scand. 18*: 1249–1254 (1964).
61. A. Goldup, A. B. Morrison, and G. W. Smith, *J. Chem. Soc.* 3864–3865 (1965).
62. A. Colett, *Inclusion Compounds* (J. L. Atwood, J. E. D. Davies, and D. D. MacNicol, eds.), Academic Press, London., 1984, Vol. 2, pp. 96–112.
63. J. D. White and B. D. Gesner, *Tetrahedron Lett* 1591–1594 (1968).
64. J. D. White and B. D. Gesner, *Tetrahedron 30*: 2273–2277 (1974).
65. B. Umezawa, O. Hoshino, H. Hara, and J. Sakakibara, *Chem. Pharm. Bull. Jpn. 16*: 177–178 (1968).
66. B. Umezawa, O. Hoshino, H. Hara, K. Ohyama, S. Mitsubayashi, and J. Sakakibara, *Chem. Pharm. Bull. Jpn. 17*: 2240–2244 (1969).
67. J. Malthête and A. Collet, *Nouv. J. Chim. 9*: 151–153 (1985).
68. H. Zimmermann, R. Poupko, Z. Luz, and J. Billard, *Z. Naturforsch. Teil A 40*: 149–160 (1985).
69. A. M. Levelut, J. Malthête, and A. Collet, *J. Phys. 47*: 351–357 (1986).
70. J. Malthête and A. Collet, *J. Am. Chem. Soc. 109*: 7544–7545 (1987).
71. J. Malthête, A. Collet, and A. M. Levelut, *Liq. Cryst. 5*: 123–131 (1989).
72. H. Zimmermann, R. Poupko, Z. Luz, and J. Billard, J. *Liq. Cryst. 3*: 759 (1988).
73. W. Kranig, H. W. Spiess, and H. Zimmermann, *Liq. Cryst. 7*: 123–129 (1990).
74. V. Percec, C. G. Cho, and C. Pugh, *J. Mater. Chem. 1*: 217–222 (1991).
75. V. Percec, C. G. Cho, and C. Pugh, *Macromolecules 24*: 3227–3234 (1991).
76. V. Percec, C. G. Cho, C. Pugh, and D. Tomazos *Macromolecules 25*: 1164–1176 (1992).
77. C. D. Gutsche, *Calixarenes* (J. F. Stoddart, Series ed.), Monographs in Supramolecular Chemistry, Royal Soc. Chem., Cambridge, 1989, Vol. 1.
78. J. Vicens and J. Böhmer, eds. *Calixarenes. A Versatile Class of Macrocyclic Compounds* Academic Publishers, Dordrecht, 1990.
79. C. D. Gutsche, *Synthesis of Macrocycles*; *The Design of Selective Complexing Agents* (R. M. Izatt and J. J. Christensen, eds.), John Wiley & Sons, New York, 1987, pp. 93–165.
80. K. H. No and C. D. Gutsche, *J. Org. Chem. 47*: 2713–2719 (1982).
81. H. Kämmerer, G. Happel, and B. Mathiasch, *Makromol. Chem. 182*: 1685–1694 (1981).
82. V. Böhmer, P. Chhim, and H. Kämmerer, *Makromol. Chem. 180*: 2503–2506 (1979).

83. D. J. Cram, S. Karbach, H. E. Kim, C. B. Knobler, E. F. Maverick, J. L. Ericson, and R. C. Helgeson, *J. Am. Chem. Soc. 110*: 2229–2237 (1988).
84. A. G. S. Högberg, *J. Org. Chem. 45*: 4498–4500 (1980).
85. A. G. S. Högberg, *J. Am. Chem. Soc. 102*: 6046–6050 (1980).
86. W. J. Burke, W. E. Craven, A. Rosenthal, S. H. Ruetman, C. W. Stephens, and C. Weatherbee, *J. Polym. Sci. 20*: 75–88 (1956).
87. J. L. Ballard, W. B. Kay, and E. L. Kropa, *J. Paint Technol. 38*: 251–262 (1966).
88. F. Bottino, G. Montaudo, and P. Maravigna, *Ann. Chim. 57*: 972–991 (1967).
89. T.-T. Wu and J. R. Speas, *J. Org. Chem. 52*: 2330–2332 (1987).
90. T. Asao, S. Ito, and N. Morita, *Tetrahedron Lett. 29*: 2839–2842 (1988).
91. I. V. Novoselov, M. G. Zolotukhin, G. I. Nikiforova, A. P. Kapina, N. M. Shishlov, A. N. Latchinov, and Y. A. Sangalov, *Makromol. Chem. 194*: 973–982 (1993).
92. A. N. Lachinov, Y. Zherebov, and M. G. Zolotukhin, *Synth. Met. 59*: 377–386 (1993).
93. M. G. Zolotukhin, V. A. Kovardakov, S. N. Salazkin, and S. R. Rafikov, *Vysokomol. Soedin. Ser. A 26*: 1212–17 (1984); *Chem. Abstr. 101*: 152430 (1984).
94. M. G. Zolotukhin, A. A. Panasenko, V. S. Sultanova, E. A. Sedova, L. V. Spirikhin, L. M. Khalilov, S. N. Salazkin, and S. R. Rafikov, *Makromol. Chem. 186*: 1747–1753 (1985).
95. M. G. Zolotukhin, E. A. Sedova, Y. L. Sorokina, S. N. Salazkin, Y. A. Sangalov, V. S. Sultanova, A. A. Panasenko, L. M. Khalilov, and R. M. Muslukhov, *Makromol. Chem. 191*: 1477–1485 (1990).
96. M. G. Zolotukhin, S. S. Akhmetzyanov, N. M. Shishlov, A. N. Latchinov, V. S. Sultanova, L. V. Spirikhin, and L. M. Khalilov, *Makromol. Chem. 193*: 975–981 (1992).
97. V. Mark and C. V. Hedges, *U.S. Patent 4,560,808* (1985); *Chem. Abstr. 104*: 225358p (1986).
98. Z. Y. Wang and A. S. Hay, *Synthesis* 471–472 (1989).
99. Z. Y. Wang, N. Berard, I. Wisniewska, and A. S. Hay, *J. Org. Chem. 55*: 4966–4969 (1990).
100. J. F. Geibel and R. W. Campbell, *Comprehensive Polymer Science* (G. Allen and J. C. Bevington, eds.), Pergamon, Oxford, 1989, Vol. 5, Chap. 32, pp. 543–560.
101. R. W. Lenz and W. K. Carrington, *J. Polym. Sci. 41*: 333–358 (1959).
102. J. T. Edmonds and H. W. Hill, Jr. *U.S. Patent 3,354,129* (1967); *Chem. Abstr. 68*: 13598 (1968).
103. R. W. Campbell and J. T. Edmonds, *U.S. Patent 4,038,259*; *Chem. Abstr. 87*: 854v (1977).
104. H. W. Hill, Jr., *Ind. Eng. Chem. Prod. Res. Dev. 18*: 252–253 (1979).
105. A. Fradet, *Bull. Soc. Chim. Belg. 98*: 635–641 (1989).
106. W. Koch and W. Heitz, *Makromol. Chem. 184*: 779–792 (1983).

107. W. Koch, W. Risse, and W. Heitz, *Makromol. Chem. Suppl. 12*: 105–123 (1985).
108. D. R. Fahey and C. E. Ash, *Macromolecules 24*: 4242–4249 (1991).
109. E. Tsuchida, H. Nishide, K. Yamamoto, and S. Yoshida, *Macromolecules 20*: 2315–2316 (1987).
110. E. Tsuchida, K. Yamamoto, H. Nishide, and S. Yoshida, *Macromolecules 20*: 2030–2031 (1987).
111. E. Tsuchida, K. Yamamoto, M. Jikei, and H. Nishide, *Macromolecules 23*: 930–934 (1990).
112. E. Tsuchida, K. Yamamoto, M. Jikei, and H. Nishide, *Macromolecules 22*: 4138–4140 (1989).
113. K. Yamamoto, S. Yoshida, H. Nishide, and E. Tsuchida, *Bull. Chem. Soc. Jpn. 62*: 3655–3660 (1989).
114. E. Tsuchida, K. Yamamoto, and H. Nishide, *Makromol. Chem. Macromol. Symp. 59*: 1–16 (1992).
115. E. Tsuchida, K. Yamamoto, H. Nishide, S. Yoshida, and M. Jikei, *Macromolecules 23*: 2101–2106 (1990).
116. K. Yamamoto, E. Tsuchida, H. Nishide, M. Jikei, and K. Oyaizu, *Macromolecules 26*: 3432–3437 (1993).
117. E. Tsuchida, K. Yamamoto, M. Jikei, E. Shouji, and H. Nishide, *J. Macromol. Sci. Chem. 11*: 1285–1298 (1991).
118. K. Yamamoto, E. Shouji, H. Nishide, and E. Tsuchida, *J. Am. Chem. Soc. 115*: 5819–5820 (1993).
119. K. Yamamoto, E. Shouji, F. Suzuki, S. Kobayashi, and E. Tsuchida, *J. Org. Chem. 60*: 452–453 (1995).
120. E. Tsuchida, E. Shouji, and K. Yamamoto, *Macromolecules 26*: 7144–7148 (1993).
121. E. Tsuchida, K. Yamamoto, and E. Shouji, *Macromolecules 26*: 7389–7390 (1993).
122. B. M. Novak, E. Hagen, S. Hoff, and A. Viswanathan, *Macromolecules 27*: 1985–1986 (1994).
123. E. Tsuchida, F. Suzuki, E. Shouji, and K. Yamamoto, *Macromolecules 27*: 1057–1060 (1994).
124. K. Kawashima and T. Adachi, *Jpn. Patent 62,232,457* (1987); *Chem. Abstr. 108*: 22927h (1988).
125. E. Tsuchida, K. Kamamoto, E. Shouji, J. Katoh, M. Matsuo, and H. Kamiya, *Macromolecules 27*: 7504–7510 (1994).
126. E. Tsuchida, K. Yamamoto, K. Oyaizu, F. Suzuki, A. S. Hay, and Z. Y. Wang, *Macromolecules 28*. 409–416 (1995).
127. F. Parodi, *Comprehensive Polymer Science* (G. Allen and J. C. Bevington, eds.), Pergamon, Oxford, 1989, Vol. 5, Chap. 33, pp. 561–591.
128. O. B. Searle and R. H. Pfeiffer, *Polym. Eng. Sci. 25*: 474–476 (1985).
129. H. Buchert and G. Blinne, *Synthetic 16*: 41 (1985).
130. N. Tokura, *Encyclopedia of Polymer Science & Technology* (N. M. Bikales, H. F. Mark, and N. G. Gaylord, eds.), Wiley, New York, 1960, Vol. 9, pp. 460–485.

131. H. R. Kricheldorf, *Handbook of Polymer Synthesis, Part A* (H. R. Kricheldorf, ed.), Marcel Dekker, New York, 1992, pp. 545–615.
132. J. B. Rose, *Polymer 15*: 456–465 (1974).
133. H. A. Vogel, *British Patent 1,163,975* (1969); *Chem. Abstr. 71*: 113498 (1969).
134. S. M. Cohen, and R. H. Young, *J. Polym. Sci. Part A. Polym. Chem. Ed. 4*: 722–727 (1966).
135. M. E. A. Cudby, R. G. Feasey, S. Gaskin, M. E. B. Jones, and J. B. Rose, *J. Polym. Sci. Part C 22*: 747–760 (1967).
136. D. M. White and G. D. Cooper, *Kirk-Othmer Encyclopedia of Chemical Technology*, 3rd ed. (M. Grayson and D. Eckroth, eds.), Wiley, New York, 1982, Vol. 18, p. 605.
137. N. J. Ballintyn, *Kirk-Othmer Encyclopedia of Chemical Technology*, 3rd ed. (M. Grayson and D. Eckroth, eds.), Wiley, New York, 1982, Vol. 18, p. 832.
138. M. E. A. Cudby, R. G. Feasey, B. E. Jennings, M. E. B. Jones, and J. B. Rose, *Polymer 6*: 589–601 (1965).
139. M. E. A. Cudby, R. G. Feasey, S. Gaskin, V. Kendall, and J. B. Rose, *Polymer 9*: 265–281 (1968).
140. ICI Ltd., *Belg. Patent 639,634* (1964); *Chem. Abstr. 63*: 700f (1965).
141. H. A. Vogel, *Ger. Patent 1,745,085* (1973); *Chem. Abstr. 80*: 121, 522 (1974).
142. M. E. B. Jones, *U.S. Patent 4,008,203* (1977); *Chem. Abstr. 86*: 172, 380 (1977).
143. E. J. Bourne, M. Stacey, J. C. Tatlow, and J. M. Tedder, *J. Chem. Soc.* 718–720 (1951).
144. B. M. Graybill, *J. Org. Chem. 32*: 2931–2933 (1967).
145. H. J. Sipe, Jr., D. W. Clary, and S. B. White, *Synthesis* 283–284 (1984).
146. L. Field, *J. Am. Chem. Soc. 74*: 394–398 (1952).
147. E. J. Bourne, M. Stacey, J. C. Tatlow, and J. M. Tedder, *J. Chem. Soc.* 2976–2979 (1949).
148. T. E. Tyobeka, R. A. Hancock, and H. Weigel, *J. Chem. Soc. Chem. Commun.* 114–115 (1980).
149. M. Ueda, *Jpn. Patent 85 228 541* (1985); *Chem Abstr. 104*: 187100 (1986).
150. J. B. Rose, *Eur. Patent Appl. 49 070* (1982); *Chem Abstr. 97*: 6999 (1982).
151. W. E. Truce and C. W. Vriesen, *J. Am. Chem. Soc. 75*: 5032–5036 (1953).
152. J. J. Eisch and W. C. Kaska, *J. Am. Chem. Soc. 84*: 1501–1502 (1962).
153. P. A. Staniland, *Comprehensive Polymer Science* (G. Allen and J. C. Bevington, eds.), Pergamon, Oxford, 1989, *Vol. 5*, pp. 483–497.
154. J. B. Rose, *Recent Advances in Mechanistic and Synthetic Aspects of Polymerization* (M. Fontanille and A. Guyot, eds.), NATO ASI Series D. Reidel Publ., Boston, 1987, pp. 207–221.
155. T. E. Attwood, P. C. Dawson, J. L. Freeman, L. R. J. Hoy, J. B. Rose, and P. A. Staniland, *Polymer 22*: 1096–1103 (1981).
156. W. H. Bonner, *U.S. Patent 3,065,205* (1962); *Chem. Abstr. 58*: 5,806f (1963).
157. I. Goodman, J. E. McIntyre, and W. Russel, *Br. Patent Appl. 971,227* (1964); *Chem. Abstr. 61*: 14,805a (1964).

158. Y. Iwakura, K. Uno, and T. Takiguchi, *J. Polym. Sci. Part A-1 6*: 3345–3355 (1968).
159. B. M. Marks (Dupont Ltd.), *U.S. Patent 3,441,538* (1969); *Chem. Abstr. 67*: 44371 (1969).
160. J. Lee and C. S. Marvel, *J. Polym. Sci. Part A. Polym. Chem. Ed. 21*: 2189–2195 (1983).
161. V. Jansons and H. C. Gors (Raychem Corp.), *PCT Int. Appl. WO 84/03,892* (1984); *Chem. Abstr. 102*: 204,469u (1985).
162. H. M. Colquhoun, D. F. Lewis, *Polymer 29*: 1902–1908 (1988).
163. H. M. Colquhoun, J. A. Daniels, I. R. Stephenson, and K. Wade, *Polym. Commun. 32*: 272–274 (1991).
164. P. J. Horner and R. H. Whiteley, *J. Mater. Chem. 1*: 271–280 (1991).
165. C. J. Borrill and R. H. Whiteley, *J. Mater. Chem. 1*: 655–661 (1991).
166. K. J. Dahl, P. J. Horner, H. C. Gors, V. Jansons, and R. H. Whiteley, *PCT Int. Appl. WO 86/02,368* (1986); *Chem. Abstr. 106*: 33661b (1987).
167. K. J. Dahl, P. J. Horner, H. C. Gors, V. Jansons, and R. H. Whiteley, *Eur. Patent Appl. 178,185* (1986). Cited in Ref. 164.
168. M. Ueda and M. Sato, *Macromolecules 20*: 2675–2678 (1987).
169. M. Ueda and T. Kana, *Makromol. Chem. Rapid Commun. 5*: 833–836 (1984).
170. M. Ueda, T. Kano, T. Waragai, and H. Sugita, *Makromol. Chem. Rapid Commun. 6*: 847–850 (1985).
171. T. D. Shaffer and V. Percec, *Polym. Bull. 14*: 367–374 (1985).
172. N. Yonezawa, S. Miyata, T. Nakamura, S. Mori, Y. Ueha, and R. Katakai, *Macromolecules 26*: 5262–5263 (1993).
173. A. S. Hay, H. S. Blanchard, G. F. Endres, and J. W. Eustance, *J. Am. Chem. Soc. 81*: 6335–6336 (1959).
174. D. M. White, *J. Polym. Sci. Part A. Polym. Chem. Ed. 19*: 1367–1383 (1981).
175. D. M. White and G. R. Loucks, *Am. Chem. Soc. Div. Polym. Chem. Polym. Prep. 25(1)*: 129–130 (1984).
176. W. Heitz, W. Stix, H. J. Kress, W. Koch, and W. Risse, *Am. Chem. Soc. Div. Polym. Chem. Polym. Prep. 25(1)*: 136–137 (1984).
177. W. Risse, W. Heitz, D. Freitag, and L. Bottenbruch, *Makromol. Chem. 186*: 1835–1853 (1985).
178. W. Koch, W. Risse, and W. Heitz, *Makromol. Chem. Suppl. 12*: 105–123 (1985).
179. H. Nava and V. Percec, *J. Polym. Sci. Part A. Polym. Chem. Ed. 24*: 965–990 (1986).
180. W. Heitz, *Angew. Makromol. Chem. 145/146*: 37–68 (1986).
181. V. Percec and J. H. Wang, *Polym. Bull. 24*: 493–500 (1990).
182. W. Chen, G. Challa, and J. Reedijk, *Polym. Commun. 32*: 518–523 (1991).
183. W. Chen and G. Challa, *Eur. Polym. J. 26*: 1211–1216 (1990).
184. W. Chen and G. Challa, *Polymer 31*: 2171–2177 (1990).
185. W. Chen and G. Challa, *React. Polym. 14*: 63–73 (1991).
186. W. Chen, G. Boven, and G. Challa, *Macromolecules 24*: 3982–3987 (1991).

187. W. Chen and G. Challa, *Am. Chem. Soc. Div. Polym. Chem. Polym. Preprints 32*: 361–362 (1991).
188. R. Scholl and C. Seer, *Annalen 394*: 11 (1912).
189. R. Scholl and C. Seer, *Ber. 55*: 330 (1922).
190. M. Tisler, *Org. Prep. Proc. Int. 18*: 17 (1986).
191. A. T. Balaban and C. D. Nenitzescu, *Friedel-Crafts and Related Reactions* (G. A. Olah, ed.), Interscience, New York, 1964, Part 2, p. 979.
192. G. A. Clowes, *J. Chem. Soc. C* 2519–2526 (1968).
193. P. Kovacic and M. B. Jones, *Chem. Rev. 87*: 357–379 (1987).
194. A. J. Bard, A. Ledwith, and H. J. Shine, *Adv. Phys. Org. Chem. 13*: 155–278 (1976).
195. R. C. Larock, *Comprehensive Organic Transformations*, VCH, New York, 1989, p. 45.
196. V. Ramamurthy, J. V. Caspar, and D. R. Corbin, *J. Am. Chem. Soc. 113*: 594–600 (1991).
197. J. V. Caspar, V. Ramamurthy, and D. R. Corbin, *J. Am. Chem. Soc. 113*: 600–610 (1991).
198. K. Bechgaard, O. Hammerich, N. S. Moe, A. Ronlan, U. Svanholm, and V. D. Parker, *Tetrahedron Lett.* 2271–2274 (1972).
199. V. D. Parker, *Adv. Phys. Org. Chem. 19*: 131–189 (1983).
200. O. Hammerich and V. D. Parker, *Adv. Phys. Org. Chem. 20*: 55–189 (1983).
201. J. G. Speight, P. Kovacic, and F. W. Koch, *J. Macromol. Sci. Rev. Macromol. Chem. C5*: 295–386 (1971).
202. J. Bargon, S. Mohmand, and R. J. Waltman, *IBM J. Res. Dev. 27*: 330–341 (1983).
203. N. Billow and L. J. Miller, *J. Macromol. Sci. A1*: 183–197 (1967).
204. R. G. Feasey, A. Turner-Jones, P. C. Daffurn, and J. L. Freeman, *Polymer 14*: 241–249 (1973).
205. V. Percec and H. Nava, *J. Polym. Sci. Part A. Polym. Chem. Ed. 26*: 783–805 (1988).
206. V. Percec and J. H. Wang, *Polym. Bull. 25*: 9–16 (1991).
207. V. Percec, J. H. Wang, and S. Okita, *J. Polym. Sci. Part A. Polym. Chem. Ed. 29*: 1789–1800 (1991).
208. V. Percec, J. H. Wang, Y. Oishi, and A. E. Feiring, *J. Polym. Sci. Part A. Polym. Chem. Ed. 29*: 965–976 (1991).
209. V. Percec, S. Okita, and J. H. Wang, *Macromolecules 25*: 64–74 (1992).
210. V. Percec, J. H. Wang, and Y. Oishi, *J. Polym. Sci. Part A. Polym. Chem. Ed. 29*: 949–964 (1991).
211. V. Percec and J. H. Wang, *Makromol. Chem. Macromol. Symp. 54/55*: 337–356 (1992).
212. V. Percec and S. Okita, *J. Polym. Sci. Part A. Polym. Chem. Ed. 30*: 1037–1049 (1992).
213. V. Percec and J. H. Wang, *J. Mater. Chem. 1*: 1051–1056 (1991).
214. V. Percec, J. H. Wang, and Y. Oishi, *J. Polym. Sci. Part A. Polym. Chem. Ed. 30*: 439–448 (1992).

215. V. Percec, J. H. Wang, and S. Okita, *J. Polym. Sci. Part A. Polym. Chem. Ed. 30*: 429–438 (1992).
216. A. F. Diaz, K. K. Kanazawa, and G. P. Gardini, *J. Chem. Soc. Chem. Commun.* 635–636 (1979).
217. A. F. Diaz, J. I. Castillo, J. A. Logan, and W.-Y. Lee, *J. Electroanal. Chem. 129*: 115–132 (1989).
218. K. K. Kanazawa, A. F. Diaz, R. H. Geiss, W. D. Gill, J. F. Kwak, J. A. Logan, J. F. Rabolt, and G. B. Street, *J. Chem. Soc. Chem. Commun.* 854–855 (1979).
219. A. Deronzier and J.-C. Moutet, *Acc. Chem. Res. 22*: 249–255 (1989).
220. G. B. Street, *Handbook of Conducting Polymers* (T. A. Skotheim, ed.) Marcel Dekker, New York, 1986, Vol. 1, pp. 265–291.
221. R. J. Waltman and J. Bargon, *Can. J. Chem. 64*: 76–95 (1986).
222. A. F. Diaz and J. C. Lacroix, *New J. Chem. 12*: 171–180 (1988).
223. D. Bloor, *Comprehensive Polymer Science* (G. Allen and J. C. Bevington, eds.), Pergamon, Oxford, 1989, Vol. 2, pp. 687–705.
224. J. R. Reynolds, *Chemtech. 18*: 440–447 (1988).
225. H. Naarmann and P. Strohriegel, *Handbook of Polymer Synthesis*, Part B (H. R. Kricheldorf, ed.) Marcel Dekker, New York, 1992, pp. 1353–1435.
226. A. O. Patil, A. J. Heeger, and F. Wudl, *Chem. Rev. 88*: 183–200 (1988).
227. R. E. Myers, *J. Electron. Mat. 15*: 61–69 (1986).
228. T. Yamamoto, K. Sanechika, and A. Yamamoto, *J. Polym. Sci. Part C. Polym. Lett. Ed. 18*: 9–12 (1980).
229. J. A. Walker, L. F. Warren, and E. F. Witucki, *J. Polym. Sci. Part A. Polym. Chem. Ed. 26*: 1285–1294 (1988).
230. Y. E. Wang and S. Miyata, *Conjugated Polymers and Related Materials* (W. R. Salaneck, I. Lundström, and B. Rånby, eds.) Oxford Sci. Pub., New York, 1993, pp. 149–159.
231. S. Machida, S. Miyata, and A. Techagumpuch, *Synth. Met. 31*: 311–318 (1989).
232. K. Yoshino, S. Hayashi, and R. Sugimoto, *Jpn. J. Appl. Phys. 23*: L899–L900 (1984).
233. T. Ojio and S. Miyata, *Polym. J. 18*: 95–98 (1986).
234. K. K. Kanazawa, A. F. Diaz, W. D. Gill, P. M. Grant, G. B. Street, G. P. Gardini, and J. F. Kwak, *Synth. Met. 1*: 329–336 (1980).
235. A. Diaz, *Chem. Scr. 17*: 145–148 (1981).
236. G. Tourillon and F. Garnier, *J. Electroanal. Chem. 135*: 173–178 (1982).
237. E. M. Geniès, G. Bidan, and A. F. Diaz, *J. Electroanal. Chem. 149*: 101–113 (1983).
238. A. F. Diaz and J. Bargon, *Handbook of Conducting Polymers* (T. A. Skotheim, ed.), Marcel Dekker, New York, 1986, Vol. 1, pp. 81–115.
239. A. F. Diaz, K. K. Kanazawa, J. I. Castillo, and J. A. Logan, *Conductive Polymers* (R. B. Seymour, ed.), Plenum Press, New York, 1981, pp. 149–153.
240. T. Inoue and T. Yamase, *Bull. Chem. Soc. Jpn. 56*: 985–990 (1983).

241. P. Pfluger and G. B. Street, *J. Chem. Phys. 80*: 544–553 (1984).
242. E. M. Geniès, G. Bidan, and A. F. Diaz, *J. Electroanal. Chem. 149*: 101–113 (1983).
243. C. P. Andrieux, P. Audebert, P. Hapiot, and J.-M. Savéant, *J. Am. Chem. Soc. 112*: 2439–2440 (1990).
244. S. Asavapiriyanont, G. K. Chandler, G. A. Gunawardena, and D. Pletcher, *J. Electroanal. Chem. 177*: 229–244 (1984).
245. C. P. Andrieux, P. Audebert, P. Hapiot, and J.-M. Savéant, *J. Phys. Chem. 95*: 10158–10164 (1991).
246. R. John and G. C. Wallace, *J. Electroanal. Chem. 306*: 157–167 (1991).
247. T. Wei, C.-C. Chan, J. Tian, G.-W. Jang, and K. F. Hsueh, *Chem. Mater. 3*: 888–897 (1991).
248. R. J. Waltman, J. Bargon, and A. F. Diaz, *J. Phys. Chem. 87*: 1459–1463 (1983).
249. A. F. Diaz, J. Crowley, J. Bargon, G. P. Gardini, and J. B. Torrance, *J. Electroanal. Chem. 121*: 355–361 (1981).
250. G. Kossmehl, *Makromol. Chem. Macromol. Symp. 4*: 45–64 (1986).
251. M. Mermilliod-Thevenin and G. Bidan, *Mol. Cryst. Liq. Cryst. 118*: 227–233 (1985).
252. M. Leclerc, F. M. Diaz, and G. Wegner, *Makromol. Chem. 190*: 3105–3116 (1989).
253. M. R. Andersson, D. Selse, M. Berggren, H. Järvinen, T. Hjertberg, O. Inganäs, O. Wennerström, and J.-E. Österholm, *Macromolecules 27*: 6503–6506 (1994).
254. T. Yamamoto, K. Sanechika, and A. Yamamoto, *Chem. Lett.* 1079–1082 (1981).
255. K. Sanechika, T. Yamamoto, and A. Yamamoto, *J. Polym. Sci. Part C. Polym. Lett. Ed. 20*: 365–371 (1982).
256. C. Z. Holtz, P. Kovacic, and I. A. Khoury, *J. Polym. Sci. Part A. Polym. Chem. Ed. 21*: 2617–2628 (1983).
257. M. Kobayashi, J. Chen, T. C. Chung, F. Moraes, A. J. Heeger, and F. Wudl, *Synth. Met. 9*: 77–86 (1984).
258. T. Yamamoto, *Prog. Polym. Sci. 17*: 1153–1205 (1992).
259. J. Roncali, *Chem. Rev. 92*: 711–738 (1992).
260. G. Tourillon, *Handbook of Conducting Polymers* (T. A. Skotheim, ed.), Marcel Dekker, New York, 1986, pp. 293–350.
261. J.-C. LaCroix and A. F. Diaz, *J. Electrochem. Soc. 135*: 1457–1463 (1988).
262. P. Lang, F. Chao, M. Costa, and F. Garnier, *Polymer 28*: 668–674 (1987).
263. V. M. Niemi, P. Knuuttila, J.-E. Österholm, and J. Korvola, *Polymer 33*: 1559–1562 (1992).
264. A. G. MacDiarmid, J.-C. Chiang, M. Halpern, W.-S. Huang, S.-L. Mu, N. L. D. Somasiri, W. Wu, and S. I. Yaniger, *Mol. Cryst. Liq. Cryst. 121*: 173–180 (1985).
265. J.-C. Chiang and A. G. MacDiarmid, *Synth. Met. 13*: 193–205 (1986).
266. A. G. MacDiarmid, J.-C. Chiang, A. F. Richter, N. L. D. Somasiri, and

A. J. Epstein, *Conducting Polymers* (L. Alcacer, ed.), Reidel Publishing Co., Dordrecht, Holland, 1987, p. 105.
267. E. M. Geniès, A. Boyle, M. Lapkowski, and C. Tsintavis, *Synth. Met. 36*: 139–182 (1990).
268. E. M. Geniès, M. Lapkowski, and C. Tsintavis, *New J. Chem. 12*: 181–196 (1988).
269. A. G. MacDiarmid, *Conjugated Polymers and Related Materials* (W. R. Salaneck, I. Lundström, and B. Rånby, eds.) Oxford University Press, Oxford, 1993, pp. 73–98.
270. M. Leclerc, J. Guay, and L. H. Dao, *Macromolecules 22*: 649–653 (1989).
271. D. M. Mohilner, R. N. Adams, and W. J. Argersinger Jr., *J. Am. Chem. Soc. 84*: 3618-3622 (1962).
272. R. L. Hand and R. F. Nelson, *J. Am. Chem. Soc. 96*: 850–860 (1974).
273. A. F. Diaz and J. A. Logan, *J. Electroanal. Chem. 111*: 111–114 (1980).
274. J. Bacon and R. N. Adams, *J. Am. Chem. Soc. 90*: 6596–6599 (1968).
275. Y. Wei, W. W. Focke, G. E. Wnek, A. Ray, and A. G. MacDiarmid, *J. Phys. Chem. 93*: 495–499 (1989).
276. Y.-B. Shim, M.-S. Won, and S.-M. Park, *J. Electrochem. Soc. 137*: 538–544 (1990).
277. D.-P. Kang and M.-S. Yun, *Synth. Met. 29*: E343–E348 (1989).
278. M. Inoue, R. E. Navarro, and M. B. Inoue, *Synth. Met. 30*: 199–207 (1989).
279. A. Pron, F. Genoud, C. Menardo, and M. Nechtschein, *Synth. Met. 24*: 193–201 (1988).
280. R. Hernandez, A. F. Diaz. R. Waltman, and J. Bargon, *J. Phys. Chem. 88*: 3333–3337 (1984).
281. K. Uvdal, M. Löglund, P. Dannetun, L. Bertilsson, S. Stafström, W. R. Salaneck, A. G. MacDiarmid, A. Ray, E. M. Scherr, T. Hjertberg, and A. J. Epstein, *Synth. Met. 29*: E451–E456 (1989).
282. M.G. Kanatzidis, C.-G. Wu, H. O. Marcy, and C. R. Kannewurf, *J. Am. Chem. Soc. 111*: 4139–4141 (1989).
283. H. Zhou, R. Stern, C. Batich, and R. S. Duran, *Makromol. Chem. Rapid Commun. 11*: 409–414 (1990).
284. R. S. Duran and H. C. Zhou, *Polymer 33*: 4019–4023 (1992).
285. Y. Wei, Y. Sun, and X. Tang, *J. Phys. Chem. 93*: 4878–4881 (1989).
286. D. E. Stilwell and S.-M. Park, *J. Electrochem. Soc. 135*: 2254–2262 (1988).
287. E. M. Geniès and C. Tsintavis, *J. Electroanal. Chem. 195*: 109–128 (1985).
288. Y. Wei, G.-W. Jang, C.-C. Chan, K. F. Hsueh, R. Hariharan, S. A. Patel, and C. K. Whitecar, *J. Phys. Chem. 94*: 7716–7721 (1990).
289. E. M. Genies and M. Lapkowski, *J. Electroanal. Chem. 236*: 199–208 (1987).
290. T. Kobayashi, H. Yoneyama, and H. Tamura, *J. Electroanal. Chem. 177*: 293–297 (1984).
291. K. Tzou and R. V. Gregory, *Synth. Met. 47*: 267–277 (1992).
292. R. R. Bodalia and R. S. Duran, *J. Am. Chem. Soc. 115*: 11467–11474 (1993).

293. G. E. Wnek, J. C. W. Chien, F. E. Karasz, and C. P. Lillya, *Polymer 20*: 1441–1443 (1979).
294. R. A. Wessling and R. G. Zimmerman, *U.S. Patent 3,401,152* (1968); *Chem. Abstr 69*: 87,735q (1968).
295. R. A. Wessling and R. G. Zimmerman, *U.S. Patent 3,706,677* (1972); *Chem. Abstr. 78*: 85306n (1973).
296. S. Antoun, D. R. Gagnon, F. E. Karasz, and R. W. Lenz, *J. Polym. Sci. Part C. Polym. Lett. Ed. 24*: 503–509 (1986).
297. T. Granier, E. L. Thomas, D. R. Gagnon, F. E. Karasz, and R. W. Lenz, *J. Polym. Sci. Part B. Polym. Phys. Ed. 24*: 2793–2804 (1986).
298. D. R. Gagnon, J. D. Capistran, F. E. Karasz, R. W. Lenz, and S. Antoun, *Polymer 28*: 567–573 (1987).
299. I. Murase, T. Ohnishi, T. Noguchi, and M. Hirooka, *Polymer Commun. 25*: 327–329 (1984).
300. I. Murase, T. Ohnishi, T. Noguchi, M. Hirooka, and S. Murakami, *Mol. Cryst. Liq. Cryst. 118*: 333–336 (1985).
301. R. A. Wessling, *J. Polym. Sci. Part D. Polym. Symp. 72*: 55–66 (1985).
302. P. M. Lahti, D. A. Modarelli, F. R. Denton III, R. W. Lenz, and F. E. Karasz, *J. Am. Chem. Soc. 110*: 7258–7259 (1988).
303. S. Iwatsuki, T. Itoh, T. Sato, and T. Higuchi, *Macromolecules 20*: 2651–2654 (1987).
304. M. W. Brandt, J. E. Mulvaney, and H. K. Hall Jr., *Macromolecules 21*: 1553–1556 (1988).
305. M. B. Jones and P. Kovacic, *Comprehensive Polymer Science* (G. Allen and J. C. Bevington, eds.), Pergamon, Oxford, 1989, Vol. 5, Chap. 27, pp. 465–472.
306. L. M. Goldenberg and P. C. Lacaze, *Synth. Met. 58*: 271–293 (1993).
307. S. A. Milosevich, K. Saichek, L. Hinchey, W. B. England, and P. Kovacic, *J. Am. Chem. Soc. 105*: 1088–1090 (1983).
308. M. B. Jones and P. Kovacic, *Encyclopedia of Polymer Science and Engineering* (J. I. Kroschwitz, ed.), Wiley, New York, 1987, Vol. 10, pp. 670–683.
309. H. Naarmann, M. Beaujean, R. Merenyi, and H. G. Viehe, *Polym. Bull. 2*: 683–689 (1980).
310. P. Kovacic and A. Kyriakis, *J. Am. Chem. Soc. 85*: 454–458 (1963).
311. P. Kovacic and W. B. England, *J. Polym. Sci. Part C. Polym. Lett. Ed. 19*: 359–362 (1981).
312. G. Brilmyer and R. Jasinski, *J. Electrochem. Soc. 129*: 1950–1954 (1982).
313. M.-C. Pham, S. Aeiyach, J. Moslih, P. Soubiran, and P.-C. Lacaze, *J. Electroanal. Chem. 277*: 327–336 (1990).
314. J. K. Stille and P. Cassidy, *J. Polym. Sci. Part C. Polym. Lett. 1*: 563–565 (1963).
315. J. K. Stille, P. Cassidy, and L. Plummer, *J. Am. Chem. Soc. 85*: 1318–1321 (1963).
316. T. Kunitake and C. C. Price, *J. Am. Chem. Soc. 85*: 761–764 (1963).

317. T. Saegusa, *Chem. Tech. 5*: 295–299 (1975).
318. T. Saegusa, S. Kobayashi, and Y. Kimura, *Pure Appl. Chem. 48*: 307–315 (1976).
319. T. Saegusa, *Angew. Chem. Int. Ed. Eng. 16*: 826–835 (1977).
320. T. Saegusa, S. Kobayashi, Y. Kimura, and T. Yokoyama, *Ring Opening Polymerization* (T. Saegusa and E. Goethals, eds.), Am. Chem. Soc., Washington D.C., ACS Symposium Series, Vol. 59, 1977, Chap. 23, pp. 332–343.
321. S. Kobayashi and T. Saegusa, *Alternating Copolymers* (J. M. G. Cowie, ed.), Plenum Publishing Co., New York, 1985, pp. 189–238.
322. H. K. Hall, Jr. and A. B. Padias, *Acc. Chem. Res. 23*: 3–9 (1990).
323. H. K. Hall, Jr., *Angew. Chem. Int. Ed. Eng. 22*: 440–455 (1983).
324. T. Saegusa, H. Ikeda, and H. Fujii, *Macromolecules 5*: 354–358 (1972).
325. T. Saegusa, S. Kobayashi, and Y. Kimura, *Macromolecules 7*: 1–4 (1974).
326. T. Saegusa, S. Kobayashi, and J. Furakawa, *Macromolecules 10*: 73–75 (1977).
327. T. Saegusa, S. Kobayashi, and Y. Kimura, *Macromolecules 8*: 374–376 (1975).
328. T. Saegusa, Y. Kimura, K. Sano, and S. Kobayashi, *Macromolecules 7*: 546–549 (1974).
329. T. Saegusa, Y. Kimura, N. Ishikawa, and S. Kobayashi, *Macromolecules 9*: 724–727 (1976).
330. H. Kobayashi, H. Ohama, and Y. Kodaira, *Jpn. Patent 71-02352* (1971); *Chem. Abstr. 74*: 14269 (1971).
331. J. S. Clovis and F. R. Sullivan, *U.S. Patent 3,755,270* (1973); *Chem. Abstr. 80*: 27846 (1974).
332. B. L. Rivas and I. Concha, *Polym. Bull. 30*: 191–197 (1993).
333. T. Saegusa, Y. Kimura, S. Sawada, and S. Kobayashi, *Macromolecules 7*: 956–958 (1974).
334. S. Kobayashi, J. Kadokawa, H. Uyama, S. Shoda, and S. Lundmark, *Macromolecules 25*: 5861–5866 (1992).
335. T. Yokozawa, K. D. Ando, and T. Endo, *Macromolecules 27*: 1978–1980 (1994).
336. J. Kadokawa, Y. Matsumsura, and S. Kobayashi, *Macromol. Chem. Phys. 195*: 3689–3698 (1994).
337. S. Lundmark, J. Kadokawa, and S. Kobayashi, *Macromolecules 25*: 5873–5877 (1992).
338. G. Odian and P. A. Gunatillake, *Macromolecules 17*: 1297–1307 (1984).
339. T. Saegusa, Y. Kimura, and S. Kobayashi, *Macromolecules 10*: 236–239 (1977).
340. M. J. Hatch, M. Yoshimine, D. L. Schmidt, and H. B. Smith, *J. Am. Chem. Soc. 93*: 4617–4618 (1971).
341. D. L. Schmidt, H. B. Smith, M. Yoshimine, and M. J. Hatch, *J. Polym. Sci. A-1 10*: 2951–2966 (1972).
342. D. L. Schmidt, *Ring-Opening Polymerizations* (T. Saegusa and E. Goethals, eds.) Am. Chem. Soc., Washington, D.C., ACS Symposium Series Vol. 59, 1977, Chap. 22, pp. 318–331.

343. P. Gunatillake, G. Odian, and D. L. Schmidt, *Macromolecules 22*: 1522–1529 (1989).
344. G. Odian, M. P. O'Callaghan, C.-K. Chien, P. Gunatillake, M. Periyasamy, and D. L. Schmidt *Macromolecules 23*: 918–925 (1990).
345. C. L. Lu, P. Gunatillake, and G. Odian, *Macromolecules 25*: 4464–4468 (1992).

8

Industrial Cationic Polymerizations: An Overview

JEAN-PIERRE VAIRON and NICOLAS SPASSKY Laboratoire de Chimie Macromoléculaire, URA 24, Université Pierre et Marie Curie, Paris, France

I. INTRODUCTION

Industrial polymers obtained by cationic polymerization are based either on alkenes or on heterocyclic monomers. More than three dozen of commercial polymers and copolymers, with different grades and compositions, are shared between these two main families, and their current production is close to 3 million metric tons per year, which represents roughly 3% of the overall amount of synthetic polymers. It is in fact difficult to get a more precise picture of the market because, besides the well-known major products (polybutenes, butyl type elastomers, polyacetal resins) representing two-thirds of the production, a wide variety of specialty polymers are synthesized each of them in a much more limited amount (10^4–10^5 tons per year) and the companies are particularly sparing of information. For example, it is striking that the overall consumption of the so-called "hydrocarbon resins" involving alkenes, dienes, vinyl aromatic monomers, and terpenes is comparable to that of butyl rubbers. Moreover it must be noticed that some of the considered polymers and copolymers are also, and sometimes mainly, obtained through other types of polymerization, and the part related to cationic processes is indistinguishable. This is the case for silicones which are not considered in the above-mentioned world production estimate.

As a consequence of this broad variety of polymers obtained by cationic processes, often in limited production scale, the literature offers very few overall surveys of industrial cationic polymerization. There is no compilation considering both alkenes and ring-opening polymerizations, and the last but complete and still pertinent review on alkene carbocationic industrial polymerization is that of Kennedy and Marechal [1] published in the early 80s. Recent and often excellent monographs are available for specific products, but due to the lack of disclosable information they are in most cases limited to a general description of the old conventional processes. Obviously a lot of jealously protected improvements were brought to these empirical processes, particularly concerning polymerization control and polymer recovery, but obtaining details supposes an exhausting circuitous route inside the patent literature and commercial documentation. The situation is understandably similar for new processes and products.

In the present overview, we tried to gather most of the industrial cationic polymerizations for alkenes, as well as for carbonyl and heterocyclic monomers. Our main objective was not to insist on the conventional processes which are already detailed in the specific reviews quoted all along the text, but to outline the up-to-date improvements, new products, and market trends every time the corresponding information had been made available.

II. COMMERCIAL POLYMERS FROM ALKENES (VINYLIC MONOMERS)

A. Isobutene-Based Polymers and Copolymers

The major industrial production of polymers obtained by cationic polymerization of alkenes is related to the 2-methylpropene (isobutene or isobutylene) homo- and copolymers which can be classified into three families:

1. The *Polybutenes*, random isobutene-butene copolymers predominantly composed of isobutene units, manufactured directly from the butanes-butenes C_4 refinery stream, with low molar masses (MM) ranging from light to highly viscous liquids.
2. The *Polyisobutenes*, homopolymers of purified isobutene and available on a very wide range of molar masses from liquid short oligomers to high or ultra-high MM rubbery materials.
3. The *butyl rubber* (IIR, as abbreviated in the ISO nomenclature for isobutene-isoprene rubber) random copolymer of isobutene with 0.5–2.5 wt% isoprene, and its *haloderivatives* (XIIR).

Conceptually the chemistries of all these polymerizations are similar because they are related to isobutene, which is by far the main component, and the compositions, structures, molar masses, and physical states of the corresponding products are controlled by a variation of solvent, temperature, initiator, and feed composition. Isobutene is a very reactive cationic monomer, and its first polymerization to oligomers, initiated by strong acids mixture, had already been reported in 1873 [2]. Although some early studies were published [3–6], industrial interest really started after the IG Farbenindustrie Cy showed in 1931, with a system initiated by BF_3, that lowering the reaction temperature could produce high molar mass polymers [7]. This led to fundamental studies of the reaction [8], and at the end of the thirties, the Standard Oil Development Company (now Exxon Chemical) discovered the "Butyl," introducing into the polyisobutene chain the unsaturations allowing vulcanization [9,10]. After the recognition of the suitability of this rubber for tire inner tubes, a rapid industrial development occurred, accelerated during World War II by the shortage of natural rubber.

Over the past four decades numerous complete and pertinent reviews have been devoted to isobutene-based polymers [1,11–19] and this presentation will be limited to a survey of the main current processes, products, applications, and economic aspects.

1. Polybutenes

This generic name is commonly, but improperly, used to designate a family of copolymers obtained from unsaturated hydrocarbons of petroleum refinery C_4 fractions, and the products obviously must not be confused with the crystalline isotactic 1-polybutene obtained through Ziegler-Natta catalysis. The commercial polybutenes are clear, water-white, odorless, oily to highly viscous liquids, with number-average molecular masses ranging from about 300 to 2500.

a. Manufacture The butane-butene stream ("B-B") is, after butadiene removal, composed of about: 15–45% isobutene; 15–25% 1-butene; 10–20% Z/E 2-butenes; 3–40% isobutane; 10–15% *n*-butane depending on whether it is originated from catalytic or stream cracking [20].

Cationic copolymerization of isobutene and butenes in solution of their saturated homologs is performed directly on the C_4 fraction, generally using initiation by $AlCl_3$ "activated" by a protogen, supposing that a cocatalysis is really needed for this system. Few detailed informations are disclosed concerning the actual processes but examination of early [21–24] and more recent literature [25] indicates that the reaction can be realized over a wide range of temperature (−45 to +80° C). Industrial processes probably operate between −10 to +30° C, under a pressure of 0.04–0.6 MPa (0.4–6 bars) depending on temperature. Conventional polybutenes seem to be most generally prepared using HCl (or H_2O)-$AlCl_3$ initiator-coinitiator system. The Lewis acid is introduced into the reactor at a concentration varying from 0.1 to 1 wt% with respect to the charge either as an hydrocarbon slurry of solid aluminum chloride, or as a dispersed (or soluble?) hydrocarbon/HCl/$AlCl_3$ complex [24]. Recently developed reactive polybutenes appear to be based on homogeneous initiation by BF_3 or, preferably, its complexes like BF_3-EtOH [25]. Average molar masses of the polymer are monitored by varying the reaction temperature, protogen concentration, catalyst concentration (i.e., relative rate of catalyst addition), and composition of the incoming C_4 stream which can be enriched first in isobutene.

Descriptions of the continuous processes have been reported in some recent reviews [1,17,18]. After elimination of butadiene, the C_4 stream (raffinate-1) is desulfurized with aqueous caustic soda, washed, and thoroughly dried over alumina or silica gel, and then cooled before entering a reactor which is generally of the agitated type. The "activator" (protogen or initiator in the Kennedy's terminology [1]) is added with the hydrocarbon feed at a concentration that is a few percent of the Lewis acid concentration. When water is used, the process does not necessarily need a specific addition, because the control of drying process can afford the required trace amounts of water for a catalytic initiation.

The Lewis acid (coinitiator) is introduced directly in the reactor at a concentration in the 0.1–1 wt% range with respect to isobutene, depending on the desired average molar masses, either as a solution or as a slurry in preformed light polymer fraction. The residence time, depending on the activity of the initiating system, seems to be about 0.5–1 h or more for $AlCl_3$-based processes and shorter (15–30 min) for the BF_3-based ones. Recycling of the polymerization mixture (and catalyst slurry [26]) is reported at least for $AlCl_3$ systems. Conversion of isobutene varies from 80% to completion, whereas only a minor part of butenes, almost exclusively 1-butene, is incorporated. Analysis of monomers stream before and after reaction shows that the chains are composed of 85–95% of isobutene units, and incorporation of butenes, which are transfer agents and/or poisons [1], is more important in low molar mass products. The outcoming gaseous fraction is fed back to refinery. The reaction is exothermic ($\Delta H_{ss} = -54$ kJ mol^{-1} [27]) and temperature of the polymerizing medium is regulated by an external heat exchanger. The polymer solution in the saturated hydrocarbon fraction is then washed with aqueous base to remove the catalyst, "dewatered," and sent to successive devolatilizers. The butanes and remaining olefins (butenes, isobutene) are flashed off from polymer at about 200° C, first under 3–4 bars, and then at atmospheric pressure. Eventually the short oligomers, dimers to pentamers, with higher boiling temperatures and low flash points, are separated from the main polymer fraction under vacuum.

b. Structure and Basic Properties As indicated before, polybutenes are clear, colorless, and virtually odorless light fluid to highly viscous fluids with kinematic viscosities at 50° C ranging from 20 centistokes (cSt) to $7 \cdot 10^4$ cSt, which corresponds roughly to number-average molar masses from 340 to 2200 ($DP_n \approx 6$–40). Molar masses distributions (Fig. 1) are close to the most probable value ($M_w/M_n \approx 2$) for polymers obtained from $AlCl_3$ initiation), and appear slightly lower, in the case of BF_3 catalysis ($M_w/M_n \approx 1.5$–1.8) [28,34]. Specific densities (0.830–0.920 $g \cdot cm^{-3}$ at 15° C) and flash points (120–250° C, open cup, ASTM D92) depend on molar mass.

Due to the large differences in reactivities of the comonomers the chains are mostly composed of isobutene units with minor amounts of 1-butene and traces of the even less reactive Z-2-butene. They are linear and present several types of unsaturations. Spontaneous termination and transfer involving proton abstraction lead to the expected and largely predominant exo/endo terminal double bonds (A, B) but some other tri- and tetrasubstituted olefinic structures (C, D) together with internal vinylidenes were also detected by ^1H and ^{13}C NMR spectroscopy [30–34].

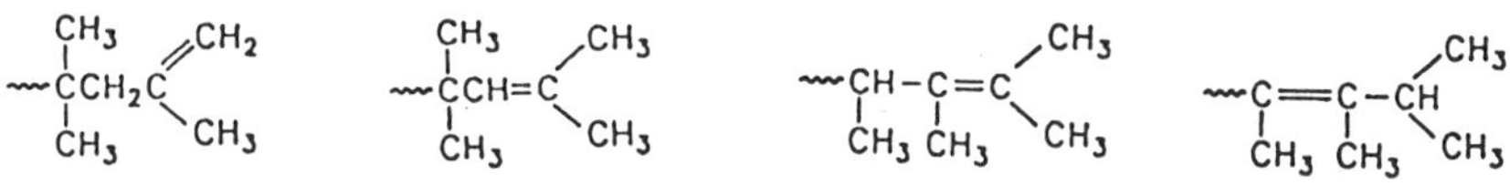

A **B** **C** **D**

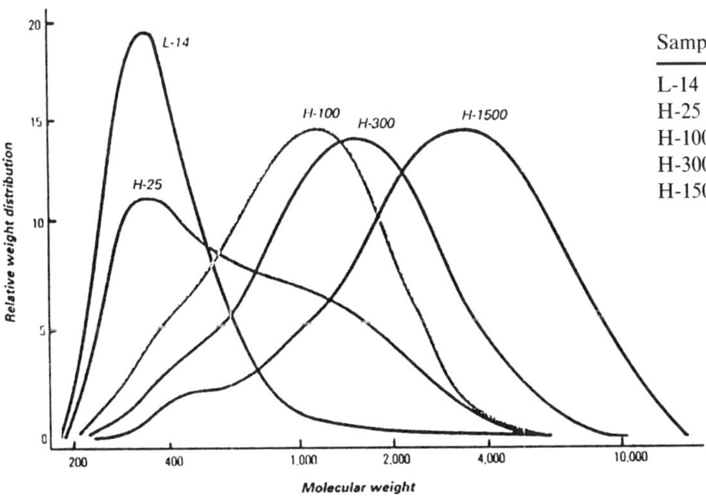

Sample	$\overline{Mn}$
L-14	320
H-25	610
H-100	920
H-300	1290
H-1500	2060

Figure 1 Molar mass distribution of some commercial polybutenes. (From Ref. 28, courtesy of Amoco Chemical Company.)

The nature of the Lewis acid and the residence time in the reactor drastically affect the ratio exo/endo of these terminal unsaturations. $AlCl_3$-based initiation leads to about 10% of vinylidenes with respect to total unsaturations, whereas the use of BF_3-based catalysts, allowing a shorter contact time between the acid and the polymer and thus limiting isomerization, increases this exo structure to 70% or more [25,29]. The vinylidene double bond is much more reactive than the endo one, and favors considerably the further terminal functionalization of the oligomers; for example, maleination, epoxidation, ozonolysis, reaction with phosphorus pentasulfide, etc. (Fig. 2).

These structural considerations are valid for polybutenes as well as for the polyisobutenes obtained from pure monomer (next section), and new products with enhanced reactivity appeared recently on the market (Ultravis-BP Chem., Glissopal-BASF). Both are based on BF_3 initiation, present 70–90% of vinylidene unsaturations, and, with respect to conventional $AlCl_3$-based polymers, lead to faster conversion, higher yields, and increased selectivity in chemical modification. As an example, the polybutenylsuccinic anhydride (PIBSA, Fig. 2) can be obtained either through an "ene" addition of maleic anhydride onto polybutene, or through the reaction of maleic anhydride with chlorinated polybutene [35]. The direct condensation is realized thermally with good rates and yields when the terminal unsaturation is essentially the reactive vinylidene, whereas the chlorine route has to be chosen in the other case.

Polybutenes and low molar mass polyisobutenes are chemically stable, nondrying (which do not cross-link on the exposure to air), and their

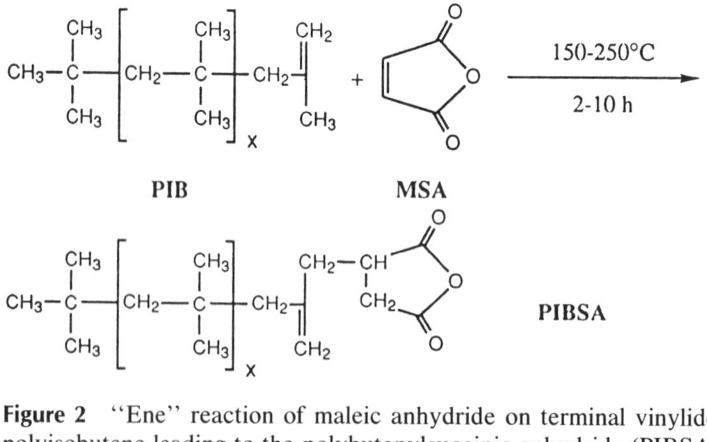

Figure 2 "Ene" reaction of maleic anhydride on terminal vinylidene group of polyisobutene leading to the polybutenylsuccinic anhydride (PIBSA).

resistance to oxidation is excellent at light and moderate heat. They do not leave carbon residue when they are volatilized or thermally depolymerized ($T \geq 270°$ C), and have no corroding action on common metals. They are soluble in hydrocarbons, chlorinated solvents, and diethyl ether; insoluble in strong polar solvents such as alcohols, ketones, acetonitrile, and dimethylformamide; and are obviously hydrophobic. Due to the nonpolar structure of the chain, polybutenes present excellent dielectrical properties. They are nonstaining, stable to high shear, and show excellent lubricating ability. They are nontoxic and meet the requirements for food contact applications.

c. Uses, Producers, and Economic Aspects The versatile properties of polybutenes and low molar mass polyisobutenes allow a wide variety of applications, ranging from additives in lubricants, fuel, caulks and sealants, pressure-sensitive and hot-melt adhesives, coatings, cling films for food wrapping, personal care products, and also as concrete sealers, rubber modifiers, resin plasticizers, asphalt and bitumen modifiers, cable insulators, impregnants for dielectrics, etc.

The demand for these products is increasing, but even if new applications are regularly introduced, the most important still concerns lubrication. Owing to their nonstaining, low-carbon residue, high-shear characteristics, and their availability over a wide range of viscosities, polybutenes are used as thickeners to adjust motor oil viscosities, waterborne lubricating emulsions in metal-working operations, special lubricants for compressors, and partial or substantial components in low-smoke, two-cycle engine oils. The possibility of chain end-capping by polar group opened to these polymers the field of lubricating oil and fuel active additives. The polybutenylsuccinic anhydride (PIBSA) is used as corrosion inhibitor, and its further reaction with oligopolyalkylenimines leads to polybutenylsuccinimides (Fig. 3), widely used as ash-free dispersants to avoid or limit the formation of sludges in motor oils. The BF_3-based polybutenes (or polyisobutenes) with high content of terminal vinylidene allow the synthesis of PIBSA without resorting to the chlorine process. The resulting oil and fuel additives are thus chlorine free and respect the current trend toward more environmentally friendly lubricants.

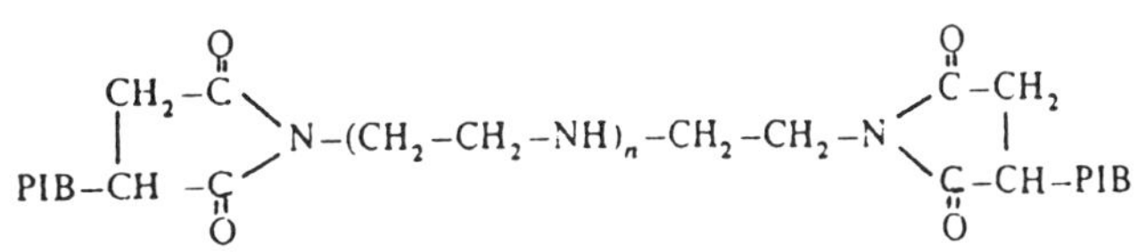

$1 \leq n \leq 4$

Figure 3 General formula for polybutenylsuccinimides

Polybutene derivative formed in the reaction with phosphorus pentasulfide (Fig. 4) is used as detergent and wear-resistance improver additive. The 1995 world capacity for production of polybutenes, including low molar mass polyisobutenes, is estimated at about $750 \cdot 10^3$ metric tons. Half of the market is shared between Exxon Chemical ("Parapol") and Amoco Chemical ("Indopol"). BP Chemicals ("Hyvis," "Napvis," "Ultravis"), the first European producer, ranks in the third world position, followed by Lubrizol (additives manufacturer), BASF (liquid PIB "Oppanol" B1, B3, "Glissopal"), and some other companies in Japan, Korea, Canada, Argentina, etc. Distribution of the overall market corresponds approximately to 45% in the Americas, 40% in Europe, and 15% in Far East Asia. Nothing is reported in literature concerning polybutene synthesis in Central Europe and Community of Independent States (CIS). Prices of polybutenes are not easily available because they depend strongly on the specific use, the correlated performances, and the tonnage of the product. Nevertheless an average of 1 U.S. dollar per kg seems to be a current basic estimate for these low-cost materials.

2. *Polyisobutenes*

Polyisobutenes are prepared from pure monomer (>99.5%). Depending on polymerization temperature and solvent, polymers with a wide variety of molar masses can be obtained as the feed is cleared from *n*-butenes which leads to transfer and/or termination. Polyisobutenes are typically classified in three categories:

1. Liquid polymers with low molar mass up to 2300.
2. Low to medium molar mass semiliquid polymers ($M_n = 10^4–8 \cdot 10^4$).
3. High to ultra-high molar mass elastomeric polymers ($M_n = 10^5$ to several millions).

a. Monomer Production and Purification Most important part of pure isobutene is obtained from the petroleum C_4 stream after catalytic cracking (or naphta steam cracking) and butadiene removing (raffinate-I). In this case, extraction is based on the higher reactivity of the isobutene unsaturation than that of *n*-butenes [17,20]. A second source of pure isobu-

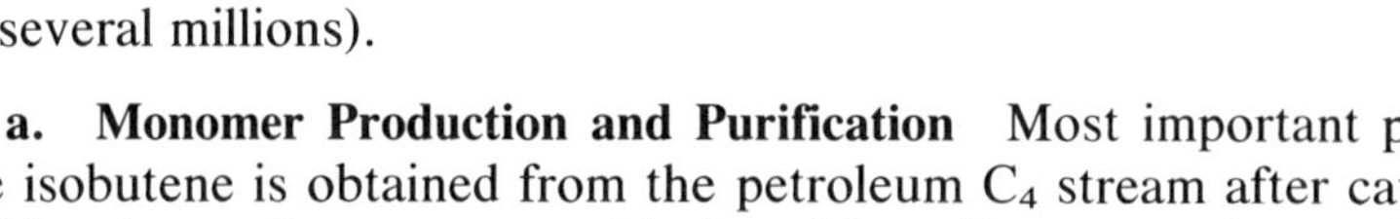

Figure 4 Derivative of polybutene and phosphorus pentasulfide

tene is the *t*-butanol which is obtained as a by-product in the industrial synthesis of propylene oxide and then dehydrated over fixed-bed alumina catalyst at about 400° C [36,37].

Separation of isobutene from C_4 stream consists of selectively preparing a chemical intermediate which can be isolated from the *n*-butenes and which can easily regenerate the monomer. The first and still widely used method is the acidic hydration of isobutene into *t*-butanol by sulfuric acid (≈50–60 wt%) at low temperature (20–40° C) [38,39]. The butenes are removed by stripping, and high-temperature steam treatment of the main phase containing the *t*-butanol regenerates high purity (>99%) isobutene. A variant of this process consists of low-temperature separation of *t*-butanol which is distilled and then dehydrated over alumina to give polymerization grade isobutene (>99.7%) [40]. In a more recent process, hydration of isobutene is performed under less aggressive conditions using acid ion-exchange resins and is followed by medium temperature catalytic dehydration [20,41,42].

A newly developed isobutene recovery is based on its selective and quantitative etherification into methyl *t*-butyl ether (MTBE) by treatment of the C_4 stream with methanol in the presence of a cation-exchange resin (e.g., Amberlyst 15), followed by catalytic splitting of the isolated ether [43–45]. MTBE was, and still is, primarily produced as a high-octane blending component for unleaded gazoline.

b. Manufacture and Basic Characteristics of Polyisobutenes Processes for production of low molar mass polyisobutenes are similar to those used for polybutenes [12,17,18]. To limit and control the molar mass of the product, the reaction is performed in nonpolar medium at rather high temperature. In the BASF batch technique, the polymerization is initiated by BF_3/CH_3OH and takes place at the refluxing temperature of isobutane used as solvent (−10° C). The Exxon continuous process operates at similar temperature in hexane solution, using for initiation a slurry of $AlCl_3$ in hexane [46]. Purity level of the isobutene feed, and particularly the remaining amount of butenes, is an essential parameter governing the molar mass of the polymer. As already discussed in the previous section, the liquid PIB have M_n in the range 300–2500 (BASF Oppanol B1-B3 [47], Exxon Parapol 950-1300-2225 [48]). Structural features of the polymer, and particularly nature of unsaturations, are similar to those of polybutenes.

Medium molar mass polyisobutenes are obtained by using higher purity isobutene, by lowering the temperature to ≈ −25 to −40° C, and eventually by operating in the presence of a chain-terminating agent (diisobutenes) [12]. The semiliquid or soft-resin polymers present typical M_v in the range $40–85 \cdot 10^3$ ($M_n = 8–13 \cdot 10^3$).

High molar mass polyisobutenes are produced in noticeably different continuous processes. Temperature is decreased to about −100° C and higher grade monomer is needed. In the BASF so-called "belt" process (Fig. 5), a cooled 1:1 liquid mixture of ethylene and isobutene is sprayed, together with a solution of BF_3 in liquid ethylene, on a 16- to 18-m-long and 35-cm-wide closed loop steel belt moving on drive rollers inside a gas-tight tubular housing [12,18]. A rapid and quantitative polymerization (>98%) takes place at the boiling temperature of ethylene (−103.7° C). Heat of polymerization is removed by ethylene vaporization. The polymer is continuously scraped off from the steel belt, and the gases remaining inside (ethylene, unreacted isobutene, and BF_3) are removed in a twin-screw extruder. Flash-off ethylene is recycled after purification.

The Exxon slurry process is similar to that used for butyl rubber synthesis (see next section). Polymerization proceeds in polar medium (methyl chloride) with $AlCl_3$ as catalyst, at temperature as low as −95° C. This allows the attainment of high molar masses but the purity of isobutene is still a key factor particularly to obtain ultra-high molar masses. These high masses lie in the 10^5–$6 \cdot 10^5$ range for M_n with a rapidly increasing polydispersity (M_v/M_n from 2 to 4). The ultra-high masses correspond to $M_n \geq 8 \cdot 10^5$ with $M_v/M_n \approx 7$–8, as for Oppanol B246 [18,47]. Medium to high molar masses polyisobutenes are clear, tacky polymers ranging from semiliquid to slow-flowing soft rubbery materials with a T_g about −70° C [49]. The chain is regular and, providing high enough molar mass, is able to crystallize upon stretching [50,51] with a helical conformation and 8 monomer units in the repeating distance along the chain. The refractive index is $n_D^{20} = 1.5077$. As for butyl rubber (next section), gas permeability of polyisobutene is low and about 20 times less than observed for

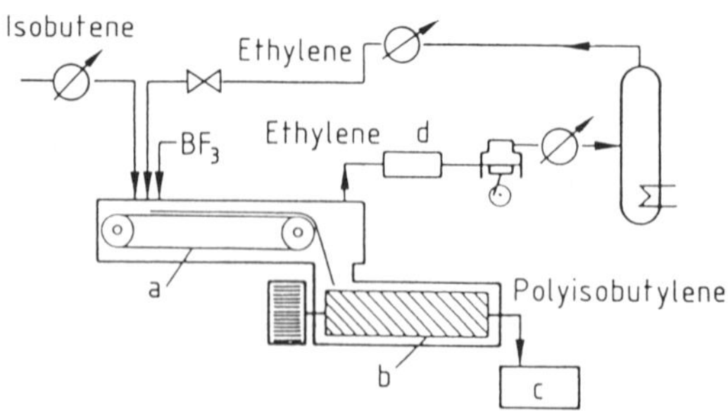

Figure 5 BASF belt process for high molar mass polyisobutene. (a) moving steel belt; (b) twin-screw extruder; (c) finishing unit; (d) purification of flash-off ethylene. (From Ref. 18, courtesy of VCH Publishers, Weinheim.)

natural rubber [17]. Thermal and chemical resistances are similar to those of polybutenes.

c. Uses, Producers, and Economic Aspects The most important production concerns the low molar mass polyisobutenes. Taking into account the new BASF plant in Antwerp (onstream in 1994), the world capacity can be estimated about 10^5 metric tons per year, which represents 15% of total polybutenes capacity. Properties and uses are indistinguishable from those of polybutenes and have already been discussed in the previous section.

Medium and high molar mass polyisobutenes are used essentially as basic materials for sealants and adhesives, as flexibility improvers for waxes and bitumen and as impact additives for thermoplastics. The low toxicity allows their use in chewing gum formulations. Detailed production is not disclosed but it is probably limited to a few tenths of kilotons. Main producers are BASF (Oppanol B1-3, B10-50, B80-246) and Exxon Chemical (Parapol, Vistanex LM and MM).

3. *Butyl and Halobutyl Rubbers*

As previously noticed, butyl rubber (IIR), poly(methylpropene-*co*-2-methyl-1,3-butadiene), is a random copolymer of isobutene and 0.7–2.2 mol% of isoprene. The industrial slurry process used all over the world consists in a low-temperature copolymerization initiated by $AlCl_3$ in methylchloride. In contrast to 1,3-butadiene, isoprene copolymerizes readily with the more reactive isobutene. Reactivity ratios of the pair isobutene-isoprene, $r_1 = 2.5 \pm 0.5$ and $r_2 = 0.4 \pm 0.1$, measured at the conditions of industrial process [10], show that the copolymerization behaves ideally ($r_1 \cdot r_2 = 1$), and, at the used low concentration of isoprene, isolated units of this latter comonomer are randomly distributed along the chain with 90% 1,4-*trans*-enchainment [52,53]:

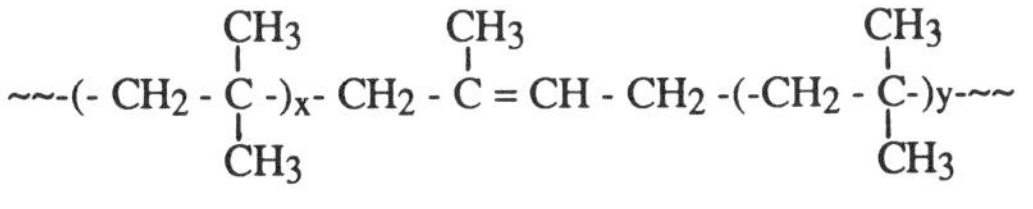

Grades of butyl differ by the level of unsaturation, molar masses characterized by Mooney viscosity ML 1 + 8 (100 or 125° C), and the characteristics of the eventually added stabilizer (staining or nonstaining). Butyl rubber, which ranks third in total synthetic elastomers consumed, has unique properties and applications, due to its low gas permeability, to its high hysteresis, and to its low level of unsaturation, sufficient for vulcanization but still providing excellent resistance to oxygen and ozone.

Halogenation of the isoprenyl unit provides enhanced cure reactivity to the rubber molecule, increases the covulcanizability and adhesion to unsaturated tire elastomers, and opens the field of tubeless tire innerliners.

Large-scale manufacture of butyl rubber started during World War II, in the scope of the U.S. Government rubber-procurement program, and the actual process is essentially similar to the historical one [9]. Brominated butyl (BIIR) was introduced in the 1950s by Goodrich Chemical Co. [54–56] but substantial commercial development occurred only in 1971 when the Polysar Ltd continuous and economic manufacturing process based on elemental bromine came onstream [57]. Production of chlorinated butyl (CIIR) was introduced on a commercial scale by Exxon Chemical in 1961.

a. Manufacture and Basic Properties of Butyl Rubber The butyl *slurry process*, considered as the typical example of industrial cationic polymerization, has been reviewed extensively [1,12–19,58,59].

A schematic flow chart of this process is shown in Fig. 6. The feed is composed of 25–35 wt% purified isobutene and the required amount of isoprene (0.4–1.25 wt%), dissolved in cooled liquid methyl chloride. Purity of supplied isobutene is variable and preliminary drying and fractionation in a two-tower system remove water, *n*-alkenes, *t*-butanol, and diisobutenes. The polymerization grade (>99.5%) fresh isobutene is then mixed in a feed blend drum with isoprene (purity >98%) and a recycle stream of diluent and unreacted monomers. A concentrated catalyst solu-

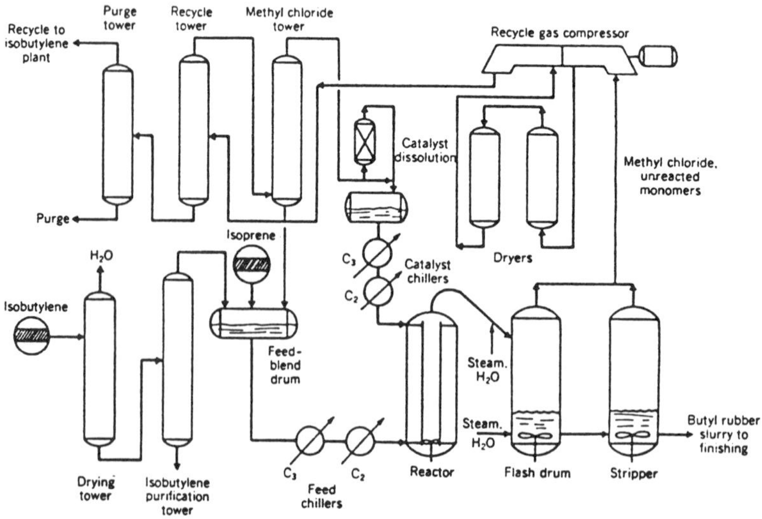

Figure 6 Schematic flow chart of slurry butyl process. (From Ref. 17, courtesy of Wiley Interscience, New York.)

tion is prepared by passing liquid methyl chloride over solid $AlCl_3$ at 30–45° C and is then diluted at the desired level (0.2–0.3 wt%). Water is generally added in low concentration (0.02–0.2 wt%) to control the activity of initiating system. Both catalyst and feed streams are chilled at about −95° C and continuously introduced into the reactor. Cooling is generally obtained with a two-stage equipment, liquid ethylene for lowest temperature, and propylene for intermediate. Reactants are mixed vigorously by an axial flow pump located at the bottom of the vessel, and heat of polymerization is removed by boiling ethylene circulating in concentric rows of tubular exchangers. The reaction, performed at −100° C under a few bars pressure, is almost instantaneous and the overall rate is governed by the rate of introduction of the catalyst solution. The copolymer demixed as a fine slurry which continuously overflows at the top of reactor. Polymer particles are far below the T_g and rigid, preventing their agglomeration, and the suspension is stable in the higher density medium. Nevertheless, adding block copolymers as stabilizing agents in the liquid phase has been considered [60]. Conversion is 70–95% with respect to isobutene and somewhat lower for isoprene, depending on the required grade of copolymer. Residence time is about 30 min. Molar mass and isoprene content are controlled by the amount of remaining transfer or terminating agents in the feed and by monomer concentrations in the liquid phase. Several reactors are used alternatively, as after 1 or 2 days of continuous operation fouling occurs and rubber agglomerates on the surface of exchangers, limiting the circulation of slurry and thus the temperature control of the polymerization.

Slurry is transferred from reactor to a stirred flash drum where it is mixed with steam and hot water (≈60° C, 1.5 bar). Methyl chloride and unreacted monomers are flashed off overhead and recycled, whereas particles agglomerate as coarse crumbs, the size of which is controlled by addition of zinc stearate. The suspension is then stripped to eliminate traces of volatiles, and the rubber is separated by filtration, dewatered by extrusion, dried and sent to a finishing unit for baling, packaging, and weighing.

A continuous *solution process* [61] has been developed and is operated in the USSR (now CIS). Copolymerization is performed in 10–12 wt% comonomers solution in isopentane or other C_5–C_7 hydrocarbons, at temperatures between −90 and −50° C, initiated by an aluminum alkyl halide/water catalyst. The catalyst is eliminated by adding a low amount of methanol to the reactor effluent stream, and the polymer is extracted by contacting the solution with hot water and steam. Further operations are similar to those of slurry process.

Vulcanization of regular butyl rubber can be performed either by the classical sulfur-ZnO-accelerator method, by phenolic resin cure with stannous chloride activation, or by *p*-quinone dioxime cure [17,59], but due to the low level of unsaturation it is noticeably slower than for conventional highly unsaturated polydienes. The second method is used when a good thermal aging uncompatible with sulfur cross-links is needed; as, for example, for tire-curing bladders. It is supposed to involve the reaction of an allylic hydrogen of the isoprene unit with the methylol groups of the resin. The last method, which supposes an oxidation step of the dioxime (Fig. 7) forming the cross-linking *p*-dinitrosobenzene [62], is used when low-temperature cure is needed as for caulks or electrical cables insulation.

About 40 different grades of regular butyl rubbers are available [63], depending on molar mass ($M_v \approx 3\text{–}6 \cdot 10^5$, polydispersity $M_w/M_n \approx 3\text{–}5$) and on unsaturation level. For 0.7 mol% of isoprene the molar mass of subchain in the cross-linked rubber is ≈8000, and ≈2500 for the highest level (2.2%), which is much more than in polydienes or natural rubber networks. They are shared in two main families characterized by their range of Mooney viscosity, typically ML(1 + 8)(100° C) 41–57 and 60–80.

Viscosity average molar mass, as measured in CCl_4 at 25° C, is related to intrinsic viscosity by the following relationship [64]:

$$[\eta]\ 100\ \text{cm}^3 \cdot \text{g}^{-1} = 1.07\ 10^{-4}\ M_v^{0.78}$$

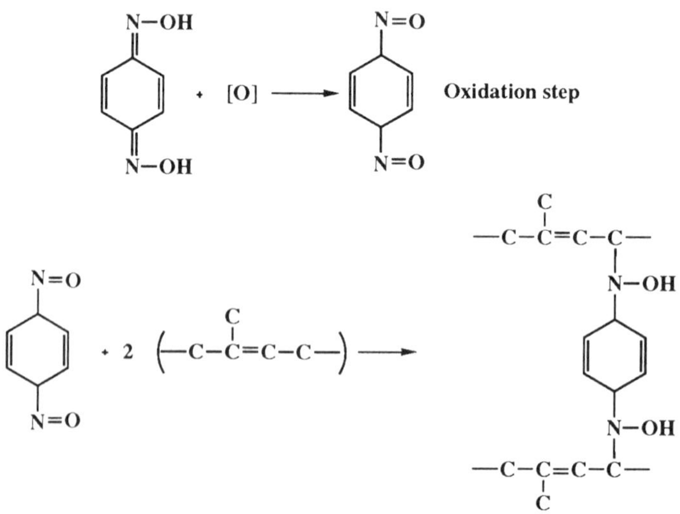

Figure 7 *p*-Benzoquinone dioxime cross-linking of butyl rubber. (From Ref. 59.)

The copolymer is amorphous with a T_g varying from $-67°$ C for the lowest unsaturation content to $-75°$ C for the highest. Due to the distribution of isoprene units along the chain, it does not crystallize upon extension. Density (0.917 g·cm^{-3}) and refractive index (n_D^{20} = 1.508) are close to those of polyisobutene. Elongation at break is in the range 600–750% and tensile strength is about 17–20 MPa.

Ozone and oxygen resistance, low gas permeability, and high damping characteristics are the most outstanding properties of butyl rubber with respect to those of polydienes. The first one and the related thermal stability result from the predominantly saturated nature of the chains. The two others originate in the restricted rotational ability of the chain units due to the presence of the side methyl groups, which can account for the low diffusivity of gases and for the substantial increase of the viscous term in the complex compliance. Diffusivities of O_2 (0.081 10^{-8} cm^2·sec^{-1} at 25° C) and of N_2 (0.045 10^{-8} cm^2·sec^{-1}) in butyl are about 20 times lower than those observed in natural rubber [17].

Slightly cross-linked grades (XL) of butyl are commercially available. They are obtained by terpolymerization of isobutene, isoprene, and divinylbenzene (Fig. 8). Main characteristics with respect to regular butyl are high resistance to sag and cold flow, high green strength, and good recovery after deformation in the uncured state.

Because cross-linking is moderate the terpolymer is still partially soluble (typically 50%) and contains variable amounts of unreacted vinylbenzene side groups which allow further curing by peroxides. Star-branched butyl polymers (SB grades) were developed recently [65]. The copolymerization of isobutene and isoprene is performed in presence of a styrene-butadiene–styrene block copolymer which acts as a branching agent (grafting onto). The resulting copolymer is claimed to be a blend of linear butyl chains with 10–15 wt% of star-branched macromolecules bearing butyl arms and it can be halogenated as regular butyl. Molar mass distribu-

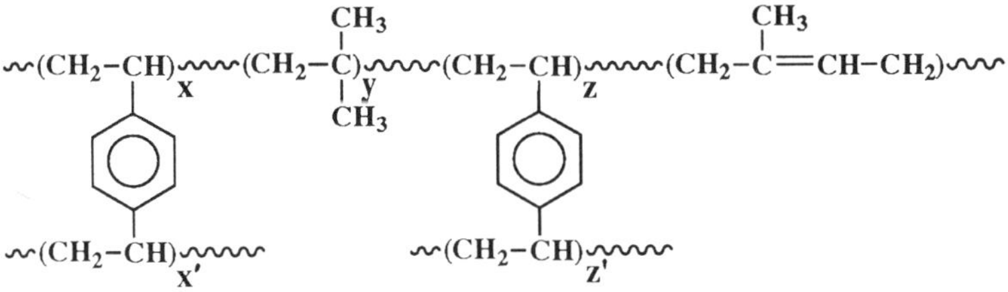

Figure 8 General formula for cross-linked terpolymer of isobutene, isoprene, and divinylbenzene.

tion is bimodal with a main peak corresponding to the major linear fraction and a high mass peak associated with star fraction (Fig. 9).

In order to keep the same Mooney viscosity as in regular butyl rubber, the average mass of linear chains, which leads to faster polymer stress relaxation, is reduced accordingly. High molar mass star-branched fraction is represented as ≈150-nm size clusters overlapping through their butyl arms shells, which could explain the observed improved green strength of this new material with respect to regular butyl. On the other hand, lowering molar mass of main linear fraction results in enhanced processability. Cross-link density after curing is also reported to be higher in SB relatively to regular butyl grades at equivalent level of unsaturation.

b. Halogenation of Butyl Rubber and Basic Characteristics of Bromo (BIIR) and Chloro (CIIR) Derivatives The pioneering work of R. F. Morrissey in the early 1950s [54,55] showed that halogenation drastically increases the cure reactivity of butyl, affording a rubber covulcanizable with conventional polydienes used in tire industry while keeping the low permeability and stability characteristics of the former. This led the Goodrich Co to commercialize a brominated butyl (Hycar 2202) obtained by milling the regular rubber with *N*-bromosuccinimide and which was available until 1969 [19,58,66]. In 1961, the Exxon Chemical Co introduced commercially the chlorinated butyl obtained with elemental chlorine, showing that the reaction takes place via substitution [67–70]. Bromination via elemental bromine was developed in 1971 by Polysar Ltd [57]. Both companies have been producing the two halobutyl rubbers since 1980.

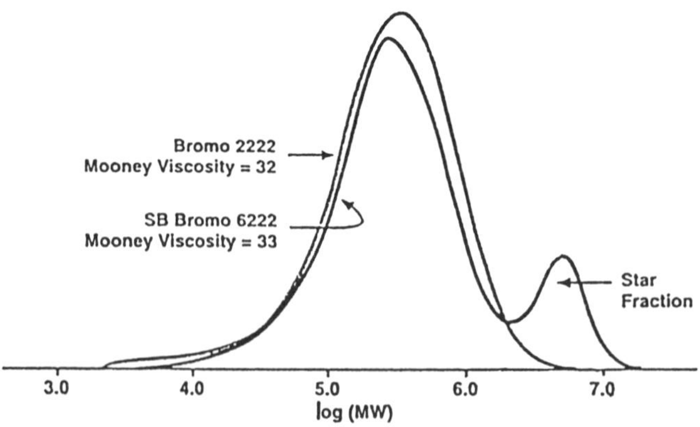

Figure 9 Compared molar mass distributions of regular and star-branched bromobutyls. (From Ref. 65.)

Even if direct bulk-phase halogenation in extruder has been recently reported [71], this reaction is performed more generally in nonpolar hydrocarbon solution using a two-stage process [16,17,19,59]. The butyl rubber is dissolved in hexane starting either on line from the reactor cold slurry or from dry particles obtained separately by chopping and grinding the finished rubber crumbs or bales. In the first case the dissolution of the fine particles is rapid, but implies solvent replacement in a stirred drum with overhead flash-off of methyl chloride and unreacted monomers. Further stripping and flash concentration are necessary to obtain a 20–25 wt% rubber solution free from traces of the halogenated diluent. The second route is technologically simpler but supposes longer dissolution periods (hours) and leads to lower concentration of rubber. Halogenation is obtained by contacting the thoroughly stirred polymer solution with the elemental halogen at 40–65° C, in a 1:1 stoichiometry relative to unsaturated units (Fig. 10). The reaction with chlorine is rapid (a few seconds) and is controlled by using gaseous chlorine or its dilute solution. Reaction with bromine is about 5 times slower, and bromine can be introduced safely in pure liquid form. Evolved HCl or HBr is then neutralized by contacting the medium with stirred dilute aqueous caustic solution, and the organic phase is separated by settling. Stabilizers, which prevent dehydrohalogenation during further treatment, and antioxidants are introduced, and the solution is transferred to a flash drum in which hexane is vaporized by steam and hot water under 2–3 bar pressure. A water slurry

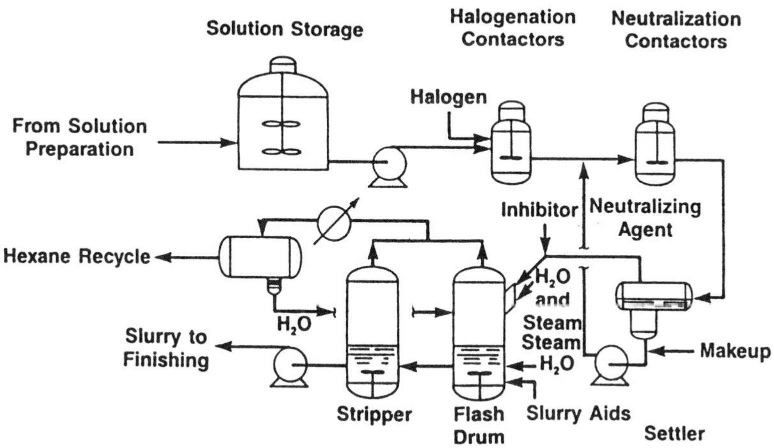

Figure 10 Schematic flow chart of halogenated butyl process. (From Ref. 59, courtesy of International Thomson Pub.–Van Nostrand Reinhold Co.)

of crumbs is obtained which is further stripped to reduce residual heptane to acceptable level. Final recovery of the halogenated polymer is similar to that of regular butyl process. Halogenation is not quantitative and leads to ≈0.9 halogen atom per isoprenyl unit for CIIR and ≈0.75 for BIIR. More than 30 commercial grades with different Mooney viscosity are available and typically contain 1.2–1.3 wt% of chlorine for CIIR and 2–2.1 wt% of bromine for BIIR [72].

The halogenation reaction proceeds in darkness and is reasonably considered as ionic. It has been shown that if chlorine gives primarily free radical addition on mono- and di-substituted alkenes, it gives ionic substitution products with tri- and tetra-substituted alkenes [73]. A model compound study, together with NMR analysis of commercial chloro and bromobutyl samples, confirmed that the reaction on isoprenyl unit leads predominantly to the exomethylene-substituted structure A, and this is explained by steric hindrance due to the tetra-substituted carbon in β-position which favors proton elimination rather than the nucleophilic attack of halide counter ion in the second phase of addition (Fig. 11, Table 1) [74,75].

The thermodynamically favored endo structure C (*cis* + *trans*) is about 25% in reaction products from model small molecule and less than 10% from polymers, whereas few addition product (D) is obtained. Allylic rearrangement of A appears easier for the brominated molecule and gives the endo structure B.

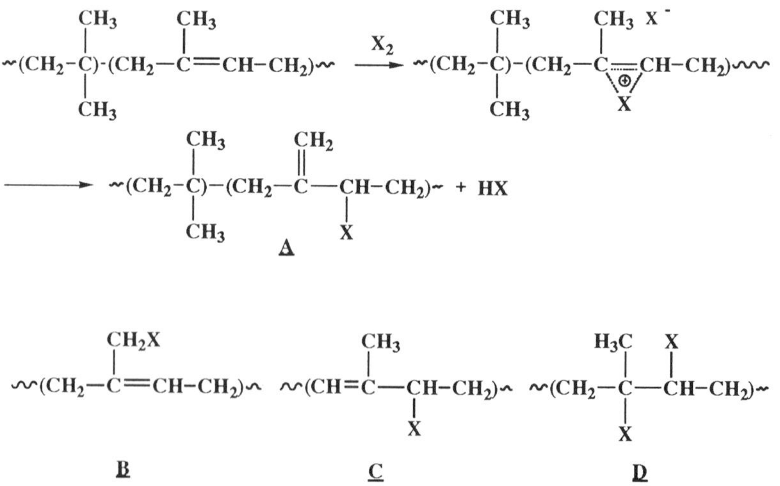

Figure 11 General scheme for butyl rubber halogenation and observed structures

Table 1 Approximate structural composition (%) of halogenated isoprenyl units in halobutyl rubbers

Structure	BIIR	CIIR
A	71 (26)[a]	90
B	14 (59)	–
C	5	9
D	10	1

[a] Values in parentheses indicate the range of possible allylic rearrangements.
Source: Refs. 74 and 75.

Most of physical and chemical properties of regular butyl and halobutyl are similar. Nevertheless, dehydrohalogenation, more sensitive for brominated polymer, is observed at high temperature and can occur during processing. The reaction is autocatalyzed by the evolved acid and stabilization consists of trapping this acid by calcium stearate and epoxidized soybean oil, which are added in the last phase of production process.

The presence of both unsaturation and allylic halogen makes the curing much faster for halogenated than for regular butyl. The reaction is even faster for bromo than for chloro derivative, due to the more reactive C—Br bonds and to the amount of rearranged form B. Thus vulcanization is still efficient in the presence of highly unsaturated polydienes like natural rubber, polybutadiene, or styrene-butadiene rubber, which is not the case for regular butyl. Fast to moderate curing can be achieved with a great variety of systems including those used for butyl and specific recipes for CIIR (ZnO/tetramethyl thiuram disulfide, diamines) and for BIIR (dicumyl peroxide/bis-maleimide) [59].

c. Other Butyl-Type Copolymers of Isobutene A new type of butyl elastomer referenced as Exxpro-XP 50, where isoprene is replaced by *p*-methylstyrene (PMS), has been developed recently and commercially introduced by Exxon Chemical Co [76,77]. Copolymers are still prepared in methyl chloride at $-90°$ C with initiation by an aluminum-based Lewis acid. For PMS contents up to 20 wt%, the slurries are butyl-like and polymerization can be achieved in the same reactor as for butyl synthesis (Fig. 12). Owing to the close reactivity ratios observed ($r_{IB} = 0.99$, $r_{PMS} = 1.43$), random copolymers can be prepared over the entire composition range, from elastomeric products ($T_g \leq -45°$ C for PMS $\leq$ 20 wt%) to hard glassy thermoplastics with T_g values above 100° C for high PMS contents. The above-considered XP 50 elastomer typically ranges from 2.5 to 11 wt% of *p*-methylstyrene.

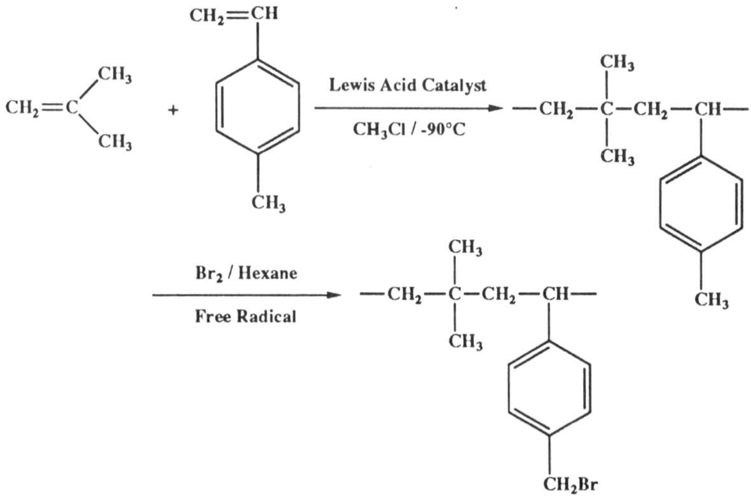

Figure 12 General scheme for preparation of brominated *p*-methylstyrene-isobutene elastomer. (From Ref. 76.)

Selective radical bromination of the *p*-methyl group by elemental bromine is performed in solution either thermally, photolytically, or in the presence of radical initiators. The reaction does not lead to any change in molar mass or distribution, and the only potential side reaction, which has to be controlled by adjusting the reaction conditions, is debromination between two *p*-bromobenzyl moieties. Under similar conditions radical chlorination leads to substitution on the benzylic site as well as on the methylene and methyl groups of isobutene units, with changes in molar mass.

Vulcanization by conventional halobutyl-curing systems appears more efficient in generating cross-links from benzylic bromine than observed with allylic halogen; this is explained by the absence of partial dehydrohalogenation as observed for the allylic halogen functionality. In addition, nucleophilic substitution can easily transform the benzylic halide in a variety of other functionalities like polar groups, radiation-sensitive moieties, grafting sites, etc., opening a broad field of applications for this new butyl-type rubber.

Unique physical properties of halobutyl are retained in brominated *p*-methylstyrene-isobutene copolymer, but, due to absence of backbone unsaturation, the degradation under irradiation is decreased markedly and the resistance to ozone is outstanding. These improvements, together with thermally stable C-C cross-links, lead to significantly enhanced heat aging and fatigue performances.

d. Uses, Producers, and Economic Aspects The tire industry consumes about 85% of butyl and halobutyl rubber production. The most important applications are linked to low-gas and moisture permeability, heat, ozone, chemical-aging resistances, and high damping properties. Regular grades are used for manufacture of inner tubes, tire-curing bladders, and automotive suspension bumpers, together with a variety of secondary applications.

Halobutyls, which can easily covulcanize with high-unsaturation rubbers, are used in innerliners for tubeless tires as component for sidewalls of passenger-car tires and for heat-resistant truck inner tubes. Other applications involve a variety of sealant tapes, caulks, and pharmaceutical closures.

The overall production capacity is estimated about 760,000 metric tons in 1994 [78], which corresponds to an increase of $\approx$10% during the last decade [19]. In Europe and North America most of passenger cars are equipped with tubeless tires and the demand for halobutyl is much higher than for regular grades. A reverse situation is observed in Asia and former Eastern countries where the road network is less adapted for tubeless tires, and the world productions for IIR and XIIR are presently roughly balanced with an increasing demand for halogenated grades. Respective productions of bromo and chlorobutyls are also similar. Approximate capacities, together with producers and plant locations, are given in Table 2. Exxon Chemical Co and Polysar-Bayer are by far the most important manufacturers. A new Exxon line with a capacity of 10^4 tons per year came onstream in 1994 in Baytown to produce the poly(*p*-methylstyrene-co-isobutene) Exxpro rubber. Current prices are about 3 U.S. dollars per kg for butyl and halobutyl rubbers, and 3.7 dollars for the new Exxpro rubber.

B. Hydrocarbon Resins

The generic name "hydrocarbon resins" designates several families of low molar mass polymers (M_n from 600 to 10^4) obtained by polymerization of petroleum, coal tar, and turpentine distillates [80–82]. In most cases, these products are obtained by cationic polymerization of mixtures either of aliphatic and/or aromatic mono and diolefins present in the more or less enriched C_5 and C_9 feedstreams, or of pure aromatic monomers generally of the styrene type. They are complex mixtures of polymers ranging from viscous liquids and tacky fluids to hard, brittle thermoplastics, and are used as additives in adhesives, printing inks, rubbers, coatings, etc. [80–82]. They are obviously amorphous and are characterized by their softening point (0 to $\approx$150° C), determined by standardized methods (i.e.,

Table 2 Estimated Production Capacities (1995) for Butyl Rubbers [79,19] in Metric Tons Per Year

<table>
<tr><th>Country</th><th>Producer</th><th>Plant Location</th><th>Reg. Butyl IIR</th><th>Halobutyls XIIR</th></tr>
<tr><td>USA</td><td>Exxon Chem. Co</td><td>Baytown (TX)</td><td>131,000</td><td>–</td></tr>
<tr><td></td><td></td><td>Baton Rouge (LA)</td><td></td><td>109,000</td></tr>
<tr><td>CAN</td><td>Bayer-Polysar</td><td>Sarnia (ONT)</td><td colspan="2">120,000</td></tr>
<tr><td>BELG</td><td>Bayer-Polysar</td><td>Antwerp</td><td colspan="2">95,000</td></tr>
<tr><td>FR</td><td>Exxon Chem. Fr.</td><td>N.Dame de Gravenchon</td><td>53,000</td><td></td></tr>
<tr><td>UK</td><td>Exxon Chem. Ltd</td><td>Fawley</td><td></td><td>62,000</td></tr>
<tr><td>JAP</td><td>Japan Butyl</td><td>Kawasaki</td><td>75,000</td><td>–</td></tr>
<tr><td></td><td></td><td>Kashima</td><td></td><td>25,000</td></tr>
<tr><td>CIS</td><td>Trade Company: Raznoimport</td><td>Nizhnekamsk</td><td>50,000</td><td></td></tr>
<tr><td></td><td></td><td>Togliatti (solution process)</td><td>35,000</td><td>–</td></tr>
<tr><td>ROM</td><td></td><td>Pitesti</td><td>5000</td><td>–</td></tr>
</table>

"ring-ball" RB, ASTM E 28-67; Mettler, ASTM D 3461-76). Depending on the purity of the feed, the resins can be either water-white (from pure vinylaromatic monomers), amber-like pale yellow, or brown (from crude distillates or concentrates). Color is evaluated from the standard scales, as for example Gardner (ASTM D-1544-80) for dark color and Hazen (ASTM D-1209-62) for light color. When unmodified, these resins are non-polar, hydrophobic, and soluble in most common aromatic and aliphatic hydrocarbon solvents. Their compatibility for blending with wax, EVA, rubbers, or other resins is determined by the cloud point method.

Historically the very first prepared were the coal-tar "indene-coumarone" resins which came on the market in the early 1900s and are still produced even if they are now largely surpassed by other types. They were followed by the terpene resins obtained from turpentine distillates, and then in the 1940s by the petroleum resins prepared from C_5 and C_8-C_{10} streams. The so-called "pure monomer" resins were developed only recently.

1. Indene-Coumarone Resins

These resins are prepared from indene-coumarone fraction of coal-tar heavy naphta, which, as a typical composition, includes indene ($\approx$50 wt%), coumarone (benzofuran in IUPAC nomenclature, ≈7%), their 2-methyl-substituted derivatives (≈3%), styrenic monomers (≈7%), dicyclopentadiene (≈5%), and nonpolymerizable alkylbenzenes (≈28%) [80–84]. The IC resins are in fact complex copolymers in which indene is the major component.

indene 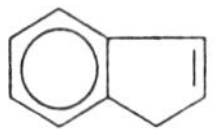

benzofuran (coumarone) 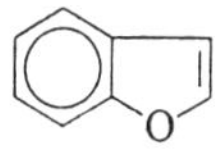

Polymerization is performed by contacting the phenol-free crude, diluted or not in aromatic naphta, with $AlCl_3$ or BF_3. At the end of the reaction, the initiator is deactivated by alkaline washes or lime. Solvent and unreacted feed are separated from the resin by distillation. The composition, molar mass, properties of the copolymer depend on feedstock content, type of initiator, and temperature of operation (−10 to +50° C). The colored products (Gardner 4–18) range from liquids at room temperature to solids with softening point as high as 150° C, and their density is close to 1.1 g·cm^{-3}. Indene-coumarone resins are essentially composed of aromatic units and are comparable to aromatic C_8-C_{10} petroleum resins (next section), but they are darker even if the actually used Lewis acid initiation leads to lighter colored products with higher softening points than those obtained from sulfuric acid as in the early processes.

2. *Petroleum Resins*

The petroleum resins are based on aliphatic olefins, dienes, and aromatic monomers, respectively, present in the C_5-C_6 and C_8-C_{10} streams derived from petroleum steam cracking [80–82,85]. Diolefins, like cyclopentadiene, its thermal dimer and their methyl derivatives, are also present in the C_5 fraction. Dicyclopentadiene (DCPD) concentrates are obtained by heat soaking, separated by distillation from lighter C_5 components, and then thermally polymerized at ≈250° C for several hours. The DCPD resins are colored, rigid oligomers resulting from the Diels-Alder bicyclic enchainment of cyclopentadienyl moieties, with softening points ranging from 90 to 180° C. We shall focus here on the aliphatic and aromatic petroleum resins obtained by cationic polymerization.

a. **Aliphatic Petroleum Resins** The olefinic C_5 fraction from petroleum steam cracking has a variable composition depending on the source and further separations. After reduction to <0.2% of the cyclodienes level because they are gel precursors, piperylene (1,3-pentadiene) and isoprene (2-methyl 1,3-butadiene) are the major dienic components of the stream which also contains monoolefinic and saturated analogs like pentene, cyclopentene, methylbutenes, pentane, etc. Further distillations allow the variation of piperylene and isoprene contents and the control the diolefin-monoolefin ratio as well as the nature of the monoolefins.

Polymerization is performed either in batch or continuously, contacting the feedstream with 0.4–1.5 wt% of Lewis acid at controlled temperature variable from 0 to 100° C (more typically about 20–50° C). Polymerization or residence time is about 1 hr and the resin yield varies from 10 to 40 wt%. Residual water or possibly added HCl [86,87] act as promoters, as well as tertiary halides [88]. The efficiency of the Lewis acid has been reported to be in the order [87],

$$AlCl_3 \approx AlBr_3 > TiCl_4 > BF_3\text{-}Et_2O > SnCl_4$$

Initiation by $AlCl_3$ leads to somewhat higher resin yields and lower oligomer content (1–2%), which explains that this Lewis acid is generally preferred. Nevertheless, BF_3 complexes (e.g., with Et_2O, phenol, etc.) are used when lower molar mass and tackifying properties are requested [89]. Isoprene and monoolefins like 2-methylbutenes decrease the softening point of the resin and improve their tackifying ability, as it has been shown when they are used as comonomers in piperylene polymerization [90,91].

Rate of polymerization, yield, molar masses, and distribution are depending on the feed composition, type and concentration of initiator, residence time, and reaction temperature. As for the IC resins, the catalyst is deactivated by alkaline washes and the resin is separated by evaporation of volatiles. Aliphatic hydrocarbon resins (density = 0.97–0.98 $g \cdot cm^{-3}$) range from viscous oils to hard, brittle solids. Softening points varies from ≈0 to 190° C with typical values around 70–100° C. The basic C_5 products are colored from light to dark brown (Gardner 4–8), but lighter-colored resins (Gardner 2) are obtained from piperylene-enriched feeds. They are highly soluble in hydrocarbon solvents and present a wide range of compatibility with other resins and waxes. Their degree of unsaturation is lower, and thus their light and heat stabilities are higher than for IC resins. Their ranges of solubility and compatibility are also broader.

b. Aromatic Petroleum Resins The C_8–C_{10} fraction (b.p. ≈140–200° C at atmospheric pressure) from petroleum cracking, includes essentially vinyl aromatic monomers like vinyltoluenes (*o*-, *m*-, *p*-methylstyrenes), styrene, α-methylstyrene, indene, methylindenes, and dicyclopentadiene together with nonpolymerizable alkylbenzenes. Composition depends on the source of petroleum feed, on the type of cracking, and on further purification and enrichment, but styrenic monomers (≈25 wt%) and indene (≈20 wt%) are the major components. Polymerization is performed similarly as for aliphatic resins, with $AlCl_3$ or BF_3 initiation. The brown resins (Gardner 7–11) are soft to hard solids with a density about 1.1 $g \cdot cm^{-3}$. Their properties and uses are comparable to those of aromatic IC resins, but their degree of unsaturation is lower, which improves their light and heat stabilities. Their ranges of solubility and compatibility are also broader.

c. Resins from Pure Monomers More recent developments concern water-white thermoplastic resins based on pure vinyl aromatic monomers (styrene, α-methylstyrene) and their co- or terpolymers with vinyltoluenes in a broad variety of compositions [80,82]. Polymerization is performed in toluene or xylenes solution ($\geq$1:1 v/v) within a 15–40° C temperature range. It is generally, but not exclusively, initiated by gaseous BF_3 or BF_3 complexes ($\leq$1wt%) with controlled residual water (50–100 ppm) as promoter. The reaction time is 3–5 hr and resin yield is 90–98%. Number-average molar mass ranges from 700 to 4000, depending on feed composition and polymerization conditions. Softening points vary from 40 to 170° C.

3. *Polyterpene Resins*

Polyterpene resins are related to the oldest reported polymerization, as they were first observed in 1789 by Bishop Watson by treatment of turpentine with sulfuric acid [92]. Commercial polyterpene resins are synthesized by cationic polymerization of β- and α-pinenes extracted from turpentine, of *d,l*-limonene (dipentene) derived from kraft-paper manufacture, and of *d*-limonene extracted from citrus peels as a by-product of juice industry [1,80,82,93]. The batch or continuous processes are similar for the three monomers. The solution polymerization is generally performed in mixed xylenes or high boiling aromatic solvent, at 30–55° C, with $AlCl_3$-adventitious water initiation. The purified feedstream (72–95% purity, depending on monomer) is mixed in the reactor with solvent and powdered $AlCl_3$ (2–4 wt% with respect to monomer), and then stirred for 30–60 min. After completion of the reaction, the catalyst is deactivated by hydrolysis, and evolved HCl is eliminated by alkaline aqueous washes. The organic solution is then dried, and the solvent is separated from the resin by distillation.

Mechanisms of polymerization and resulting chain structures are complex for the three monomers. A detailed presentation is irrelevant here, and for more information the reader is referred to the pertinent review by Ruckel and Arlt [93].

In the case of β-pinene (Fig. 13), it is proposed that the protonic initiation creates a cyclic tertiary carbenium ion which rearranges into a more stable pendant species, and the successive addition-isomerization processes lead to a chain structure with alternating isobutyl and cyclohexenyl subunits. The presence of one unsaturation per repeating unit has been confirmed by polymer analysis.

Observed molar mass of poly(β-pinene) are low (M_n from 600 to 2000), due to important chain transfer and termination processes, either spontaneous or involving alkylation of the aromatic solvent. Transfer might also involve a sterically hindered nonpropagating camphenium ion

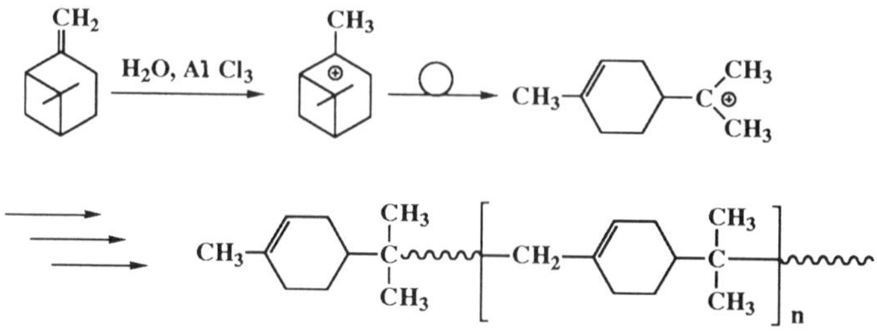

Figure 13 Mechanism of β-pinene cationic polymerization. (From Ref. 93.)

resulting from a different rearrangement of the β-pinene tertiary carbenium ion, leading thus to camphenic chain end groups (Fig. 14) [94].

In the case of *d*- or *d,l*-limonene polymerization, the chain bears, on the average, one unsaturation per two monomer units and this indicates that both endo- and exocyclic double bonds of the monomer are involved. The tertiary cyclic carbocation could propagate via cycloaddition and lead to bicyclic saturated units in the chain (Fig. 15).

The α-pinene molecule possesses an endocyclic double bond and its reactivity is much lower in the cationic polymerization process. The same endo- and exocyclic carbenium ions as for β-pinene are formed but propagation on the endocyclic unsaturation is sterically hindered and a large amount of transfer resulting dimer is formed. Addition of catalyst adjuvants such as antimony trichloride improves considerably the resin yield. They are supposed to stabilize the cyclic carbenium, increasing its lifetime and favoring propagation with respect to transfer [93]. Analysis of polymer

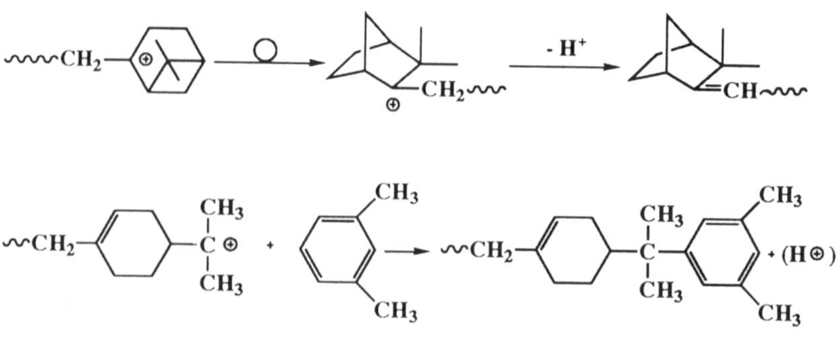

Figure 14 Transfer and/or termination processes in β-pinene cationic polymerization. (From Refs. 93 and 94.)

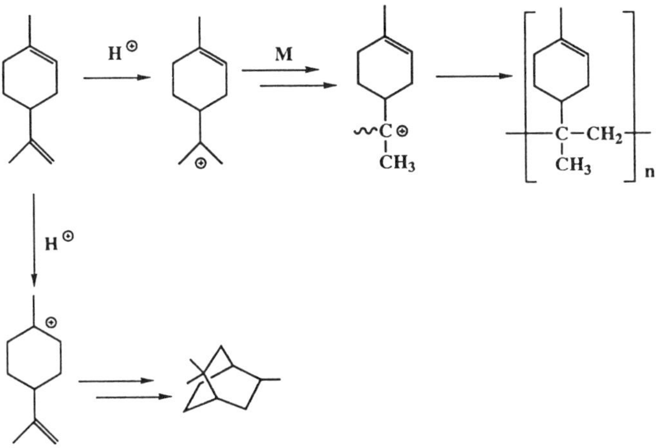

Figure 15 Possible propagation routes for *d*- or *d,l*-limonene cationic polymerization. (From Ref. 93.)

shows that approximately two-thirds of the chain units bear an unsaturation, the remaining part corresponding to saturated (2:2:1) bicyclic structures (Fig. 16).

Average molar masses are lower for polylimonene and poly(α-pinene) ($M_n \approx$ 500–900, tetramer to hexamer) than for poly(β-pinene), but roughly similar softening points are observed which agrees with a higher chain rigidity for the former oligomers. Commercial polyterpene resins are light colored (Gardner 1–5), soluble in common solvents, and compatible with

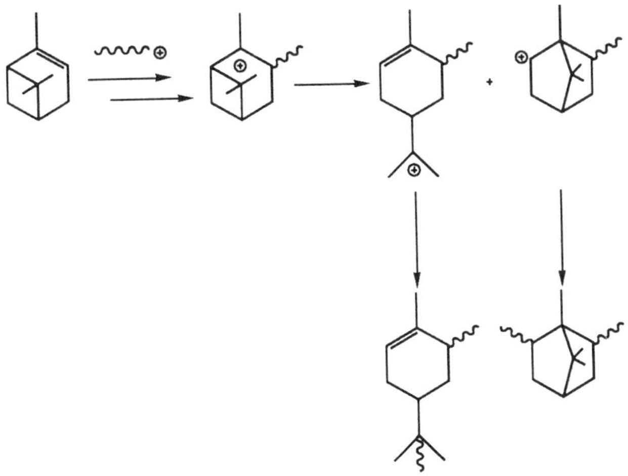

Figure 16 Chain structures in α-pinene cationic polymers. (From Ref. 93.)

a wide variety of polymers and resins. Their softening point range from 18 to 135° C for the three major types of resins. Most of their uses are based on their outstanding tackifying properties.

4. *Uses, Producers, and Economic Aspects of Hydrocarbon Resins*

Hydrocarbon resins are available on various forms, from solid (powder, flakes, beads, etc.) to solutions and dispersions. They are most generally used as additives in adhesives (pressure-sensitive, hot-melt, solvent), in caulks and sealants, in printing inks (letterpress, lithographic, gravure), rubber processing, and in protective coatings. Depending on their nature and softening point they are used as binders, tackifiers, processing aids, water repellents, film-forming agents, etc. Aliphatic/aromatic petroleum resins and still produced coumarone-indene resins are used extensively in hot-melt adhesives and coatings, inks, rubber compounding, sealants, and paints. Water-white pure monomer resins have the same applications but are chosen whenever higher color specifications are requested. Dicyclopentadiene resins are used in printing inks, paints, and adhesives. Terpene resins are used almost exclusively as light color tackifiers of rubber or EVA copolymers in pressure-sensitive and hot-melt adhesive formulations.

Solubility, compatibility, and properties of all hydrocarbon resins can be improved by chemical modification and introduction in the chains of the needed polar or functional moieties, either during the synthesis of the resin or on the final product. The most important modified resins result from:

1. Synthesis in presence of phenol and alkylphenols, which leads to Friedel-Crafts alkylation by the cationic propagating chain ends (Aromatic C9, IC, pure monomer resins, polyterpenes).
2. Reaction with maleic anhydride (DCPD resins).
3. Hydrogenation (DCPD resins).

Mixing of aliphatic and aromatic resins, or petroleum and terpene resins, allows a variety of formulations with modulated properties.

There are more than 35 producers all over the world; an exhaustive list was provided in a recent review [82]. World consumption of hydrocarbon resins has been estimated ≈750 kt in 1994 which corresponds to an increase of 7.8% with respect to 1993. About 65% of the resins were used in adhesive formulations. Production was 350 kt for aromatic (C9 + IC) resins, 274 kt for aliphatic (C5 + DCPD), and 128 kt for the more rapidly developing water-white pure monomer resins, which corresponds on average to about 80% of world plant capacities [95]. Production of polyterpene resins appears limited to 25–30 kt per year.

The market of hydrocarbon resins is roughly shared between America (48%), Europe (27%), and Asia (25%). The current basic price is about 3 U.S. dollars per kg.

C. Polyvinylethers

Polymerization of vinyl ethers (VE) has been the subject of a considerable amount of theoretical studies. These monomers can be polymerized through radical initiation but the reaction is very slow and leads only to oligomers. Cationic polymerization initiated by a wide variety of Lewis acids is much more efficient and definitely preferred for homopolymer synthesis. Detailed theoretical aspects, and particularly recent developments concerning the controlled/living cationic polymerization of these monomers, have been discussed as well in previous exhaustive review [1,13,98,99] as in the present book (Chapters 4 and 5), and they will no longer be considered here.

Monomer syntheses and their industrial polymerization and copolymerization were initiated in the thirties by IG Farbenindustrie, now BASF, and this company is still the main producer. Other companies, as GAF Corporation and Union Carbide in the United States or ICI in the United Kingdom came on the market at a later time. Actually BASF is the only producer of homopolymers, whereas copolymers are manufactured by BASF and GAF Co. We shall not consider here the synthesis of the industrial copolymers of methylvinyl ether and maleic anhydride (Gantrez-GAF, Sokalan-BASF), of isobutylvinyl ether and vinyl chloride (Laroflex MP 15-60, BASF), and of isobutylvinyl ether and acrylic monomers (Acronal, BASF) as they are produced by radical copolymerization.

The homopolymers obtained by cationic route and which are of some commercial importance are the poly methyl (PVM), ethyl (PVE), isobutyl (PVIB) and octadecyl (PVOD) vinylethers.

$$\sim\sim(-CH_2-\underset{\displaystyle O\text{-}R}{\underset{|}{CH}}-)_n\sim\sim \qquad \text{with } R = \ -CH_3,\ \ -C_2H_5,\ \ -\overset{\displaystyle CH_3}{\overset{|}{\underset{\displaystyle CH_3}{\underset{|}{C}}}}H_2\text{-}CH_3,\ \ -C_{18}H_{37}$$

1. *Manufacture*

If the theoretical aspects of the cationic polymerization of vinyl ethers are well documented, the literature remains particularly reserved concerning industrial syntheses of the commercial homopolymers.

Even if new processes of synthesis were developed from alcohols, like catalytic vinylation with ethylene or vinyl exchange with vinyl acetate, the major commercial route for VE monomers seams to be still the Reppe method based on reaction, in basic conditions, of acetylene with the corresponding alcohols [96,97,100]:

$$HC{\equiv}CH + R{-}CH_2OH \xrightarrow[120°C - 180°C]{HO^- \text{ or } RO^-} RO{-}CH{=}CH_2$$

Industrial homopolymerization can be performed either in bulk or in solution, with conventional initiators like BF_3 and its complexes, $AlCl_3$, or $SnCl_4$, on a wide range of temperature. Molar mass of the polymer critically depends on the purity and dryness of the medium and on reaction temperature, the higher the temperature the lower the molar mass. Increasing the polarity of the solvent could also increase the molar mass of the polymer, but the only disclosed industrial solution process was the BASF-conveyor belt system (see Section II.A.2) operating in liquid nonpolar propan for polyisobutylvinyl ether synthesis. This process is no longer in use. Bulk processes were reviewed recently [100]. They operate batchwise or continuously, and a recipe for a soft-resin of poly(methylvinyl ether), molar mass $M_n \approx 20{,}000$, indicates that the reaction is performed at 15° C, with initiation by BF_3-$2H_2O$ at a concentration of $\approx 3 \cdot 10^{-4}$ mol·L^{-1}. In a first step, a part (5%) of the monomer (b.p. = 6° C) and the required volume of a 3 wt% catalyst solution in dioxane is introduced in a gas-tight stirred-tank reactor, equipped with jacket exchanger and reflux condenser, and cooled at 12° C. The strongly exothermic polymerization ($\Delta H_{ss} = -67$ kJ·mol^{-1}) takes place rapidly, and then the major parts of the reactants are simultaneously and continuously pumped for 10–16 hr into the reactor. The temperature is thus controlled and maintained at ca. 15° C. The polymerization is almost quantitative (yield ≈97%). At the end of the reaction, the temperature is increased to 60 and 90° C for several hours, the residual monomer is flashed off, and the polymer is recovered as a hot melt or converted into water, ethanol or toluene solutions. Same procedure may be applied for the other vinyl ethers. A continuous bulk process is also described briefly for poly ethyl and isobutylvinyl ethers. The polymerization, performed with a $\approx 10^{-3}$ mol·L^{-1} initiator concentration and at 98° C, leads to more or less highly viscous oils depending on monomer (M_n = 1400 for PVE and 17,000 for PVIB). In both processes, varying the polymerization temperature and initiator concentration should obviously control the molar masses and the physical state.

2. *Basic Properties, Uses and Economic Aspects*

Commercial poly(vinyl ethers) range from viscous oils to soft tacky resins and elastomeric solids (specific viscosity η_{sp} = 0.1–6). Due to the polymerization conditions (initiator, low-polarity medium), they are amorphous and atactic with a majority of isotactic triads [96,97]. Glass-transition temperature is −34° C for PVM, −42° C for PVE, and −19° C for

Table 3 Solubility characteristics of commercial poly(vinyl ethers)

Solvent	PVM	PVE	PVIB
Water	+[a]	–[b]	–
Methanol, ethanol	+	+	–
Isopropanol, *n*-butanol	+	+	+
Aromatic hydrocarbons	+	+	+
Aliphatic hydrocarbons	–	+	+
Chlorinated solvents	+	+	+
Diethyl ether	–	–	+
Ketones, esters	+	+	+

[a] + = soluble.
[b] – = insoluble.

PVIB. Low molar mass ($M_n = 2\text{–}3 \cdot 10^3$) but long side chains poly(octadecylvinyl ether) has a wax consistency with a melting point of 50° C. Poly(methylvinyl ether) is soluble in water below 28° C (cloud point), which is explained by strong water solvation of the chains. At higher temperature hydrogen bonds are destroyed and the polymer precipitates. Increasing the length of the side group suppresses the solubility in water and more generally decreases the solubility in polar solvents (Table 3). Poly(octadecylvinyl ether) is compatible with most of commercial waxes.

Poly(vinyl ethers) have extensive uses as nonmigrating tackifiers in adhesives and particularly pressure-sensitive tapes and labels, as viscosity-index improvers in lubricants and as plasticizers in coatings, films, and elastomers. They are available and used in dry state, solution or as blends with a wide variety of compatible polymers and copolymers.

As stated above, the two current producers are BASF and GAF Co. The former produces both homopolymers (2/3) and copolymers (1/3), whereas the latter produces only copolymers. The overall annual production (1994) is about 20,000–25,000 tons, divided into 60% of copolymers and 40% of homopolymers. The current average price is about 5–6 U.S. dollars per kg for homo- and copolymers.

III. COMMERCIAL POLYMERS WITH HETEROATOMS IN THE MAIN CHAIN

A. Polyethers

1. *Polyoxiranes*

a. Polyepichlorohydrin Elastomers Historically, poly(ethylene oxide) reported by Staudinger in 1933 was the first high molecular weight polyepoxide to be prepared [101]. Studies on substituted polyethers (—OCH(R) CH_2—) in which the stereochemistry of the macromolecular chain is influencing the physical properties of the products started deeply in the 50s. An important development occurred when Pruitt and Baggett (Dow Chem. Co.) reported in 1955 a new ferric catalyst obtained by reaction of propylene oxide with ferric chloride [102]. This catalyst polymerized propylene oxide to a high molecular weight partly crystalline polymer. Price and Osgan [103] studied the stereochemistry of polymerization of racemic and optically active propylene oxide with this initiator and basic initiators.

In polyethers the presence of the oxygen atoms contributes greatly to the chain flexibility, and thus elastomeric behavior may be expected [104]. High molecular weight products are required to be processed by industrial conventional ways. Vandenberg discovered in 1957 a new family of catalysts which was particularly effective for high molecular weight polyoxiranes. These new catalysts were derived from the reaction of organometallic compounds such as aluminum trialkyls, e.g., $AlEt_3$, with water. A particularly versatile catalyst was obtained by combining $AlEt_3$, water, and acetyl acetone [105–107]. Vandenberg noted that this new catalyst polymerized epichlorohydrin to a rubbery, high molecular weight polymer containing more than 97% of head-to-tail sequences. Amorphous homopolymers and copolymers derived from epichlorohydrin appeared to be especially interesting oil-resistent rubbers. This combination of properties make them particularly interesting for commercialization. They are available from Hercules, Inc. under trademark Herclor and from B.F. Goodrich Chem. Co., a license of Hercules, Inc., under the trademark Hydrin.

Mechanism of polymerization. The $AlEt_3/H_2O$/acetylacetone catalyst called also "chelate Vandenberg" catalyst is a stable and soluble product having the structure presented in Fig. 17.

The mechanism proposed by Vandenberg for polymerization of epoxides [106] involves two metal (aluminum) atoms to make it possible to obtain a backward attack at the primary carbon atom of the epoxide. In this mechanism, the epoxide is coordinated to one aluminum atom before its attack by the growing chain. The latter is coordinated to an adjacent aluminum atom (Fig. 18).

Although in the case of ECH the chelate catalyst $AlEt_3/H_2O$/acac = 1:0.5:1 leads to almost the same type of polymers as the cationic $AlEt_3/H_2O$ = 1:0.5 one, it is not always so [106,107] and a distinction must be made between the first catalyst involving a coordination mechanism and the second one which is cationic.

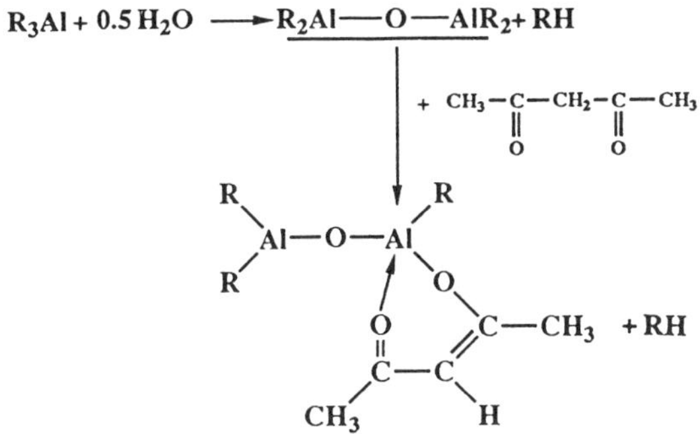

Figure 17 "Chelate Vandenberg" catalyst.

Structure of epichlorohydrin elastomers. The commercial ECH-based available elastomers have the following structure:

1. Epichlorohydrin (ECH) homopolymer or CO (ASTM)

   $(\text{-OCH(CH}_2\text{Cl)-CH}_2\text{-})_n$

2. Epichlorohydrin (ECH)-ethylene oxide (EO) copolymer or ECO (ASTM)

   $(\text{-OCH(CH}_2\text{Cl)-CH}_2\text{-})_n\ (\text{-OCH}_2\text{CH}_2\text{-})_n$

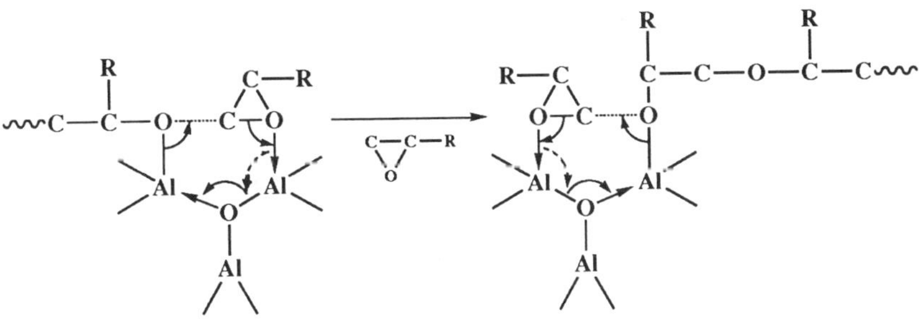

Figure 18 Vandenberg mechanism for polymerization of epoxides.

3. Epichlorohydrin (ECH)—allylglycidylether (AGE) copolymer or GCO (ASTM)

$$(\text{-O}\underset{\text{CH}_2\text{Cl}}{\underset{|}{\text{CH}}}\text{-CH}_2\text{-})_n\ (\text{-O}\underset{\text{OCH}_2\text{CH=CH}_2}{\underset{|}{\text{CH}}}\text{-CH}_2\text{-})_n$$

4. Epichlorohydrin (ECH)-ethylene oxide (EO), allylglycidyl ether (AGE) terpolymer or GECO (ASTM)

$$(\text{-O}\underset{\text{CH}_2\text{Cl}}{\underset{|}{\text{CH}}}\text{-CH}_2\text{-})_n\text{-}(\text{-OCH}_2\text{CH}_2\text{-})_n\text{-}(\text{-O}\underset{\text{OCH}_2\text{CH=CH}_2}{\underset{|}{\text{CH}}}\text{-CH}_2\text{-})_{0.025n}$$

ECH elastomers have more than 97% of head-to-tail sequences, they are amorphous (no cristallinity detected by DSC or x-ray analysis) and with a stereorandom distribution of enantiomeric units. These polymers are thus atactic. In ECH/EO copolymer the mole ratio of both monomers was kept to approximately 1. In the terpolymer the amount of AGE is small (around 2.5%).

The molecular weights estimated are 4.5×10^5 for ECH polymer and 1.4×10^6 for ECH/EO copolymers. Homopolymer ECH (1) and copolymer ECH/EO (2) are solvent-resistant rubbers. Copolymers with AGE are S-vulcanizable rubbers.

Nomenclature. Trivial name = epichlorohydrin (ECH); ASTM name = chloromethyloxirane (CO); IUPAC name = oxy(chloromethyl) ethylene.

Commercial manufacturers and trademarks. The commercial producers and the corresponding trademarks of different ECH elastomers were available a few years ago and described in a specialized reviews [108–110]. In recent years, the situation was changed since Hercules, Inc. is no more producing elastomers and those of BF Goodrich Co. are now a part of Nippon Zeon Co. The estimated production is around 10,000 tons per year.

In United States Zeon Chem. is producing 8,400 tons per years of Herclor H (Hydrin 100 of Goodrich or CO), Herclor C (Hydrin 200 or ECO), and Herclor T (Hydrin 400 or GECO), whereas in Japan Nippon Zeon produces 2000 tons per year of Gechron 1000 (CO), Gechron 1100 (GCO), Gechron 2000 (ECO), and Gechron 3100 (GECO).

Basic properties. The basic properties of ECH elastomers can also be found in detail in the above-mentioned reviews [108–110]. The homo- and copolymers have a high specific gravity (1.4–1.5). The Mooney viscosities are of the same range as for other commercial elastomers. Ozone, heat, fuel and oil resistance are good. An excellent resistance to vapor permeation by hydrocarbons, fluorocarbons, and air was observed.

The 38% of chlorine content found in the homopolymer creates the fuel resistance and promotes flame retardancy. The ECH/EO copolymer has improved low-temperature properties, e.g., flexibility. The very polar chloromethyl group responsible of the oil resistance properties is also the cross-linking site in these elastomers.

Technology, processing, curing. The polymerization is usually carried out in solution (benzene, toluene) at 40–130° C using Vandenberg catalysts ($AlEt_3/H_2O$ and $AlEt_3/H_2O$/acetylacetone). The uniform product composition is obtained by adjustment in the catalyst feed rate. The products are isolated by aqueous coagulation. Before coagulation stabilizers of a hindered phenol type are added to the polymer solution.

Standard rubber fabrication techniques, i.e., extension, calendering, frictioning, injection, etc., may be used with all ECH elastomers. The cross-linking techniques proceed by nucleophilic displacement of the chlorine atom from the chloromethyl group. Ethylenethiorea, mercaptothiadiazole, trithiocyanuric acid, and many others are used as curing agents. Acid acceptors such as metal oxides (lead oxide, magnesium oxide, etc.), lead salts (carbonate, phthalate, etc.), metal stearates (calcium, barium, etc.), and others are used as scavengers for HCl by-product. In the case of terpolymer with allylic unsaturation more conventional cross-linking agents such as sulfur or peroxides are used.

Reinforcing fillers such as carbon black, alumina, and silica are used as well as not reinforcing fillers (calcium carbonate, talc, etc.).

Plasticizers are used for low-temperature improvement. Typically, they are diesters and ethers; DOP (di-2-ethylhexyl phthalate) appears to give the best results.

Blends of epichlorohydrin elastomers have not met commercial success.

Epichlorohydrin elastomers may be chemically modified by nucleophilic substitution reactions at the side-chain chloromethyl group. Numerous substituted products have found applications as flame retardants, flocculating agents, selectively permeable membranes, photosensitive materials, etc.

Applications. Epichlorohydrin elastomers have good low-temperature flexibility, a broad range of temperature utilization, outstanding resistance to ozone, air and heat aging, high resistance to a variety of fuels, and good chemical resistance. Therefore, they have found applications particularly in the automotive industry and in many oil-field specialities.

b. Poly(Propylene Oxide) Elastomers Poly(propylene oxide) is one the simplest substituted polyethers which are known to be potentially excellent elastomers due to their chain flexibility coming from the presence of the oxygen atom in the main chain.

The usual methods of synthesis of poly(propylene oxide) lead to low molecular weight compounds which can be hydroxyl-ended [111]. Poly(-oxypropylene) polyols are used in the synthesis of rubbery polyurethanes [112,113].

Polyether elastomers require high molecular weight products. The latter can be obtained using coordination-type catalysts. Especially effective catalysts are those reported by Vandenberg based on the reaction of organoaluminum compounds with water and subsequent combination with acetyl acetone [104]. These catalysts were already described in the case of epichlorohydrin elastomers. They allow the preparation of a wide range of polyethers [104]. Amorphous propylene oxide unsaturated epoxide copolymers were prepared in an early work by Vandenberg and Robinson (Hercules, Inc.) and recognized as potential elastomers. Gruber et al. (General Tire) at the same time [114] published on preparation and properties of the same type of copolymers. Several applications of patents were deposited but finally Vandenberg was awarded priority. The new elastomer which is a sulfur-curable copolymer of propylene oxide (PO) and allylglycidyl ether (AGE) was commercialized in 1972 under the trademark of Parel and has a current abbreviation of GPO. Properties and applications of Parel elastomers were discussed in detail in some specialized papers [109,115,116].

Structure of GPO. The basic chemical structure of GPO is represented as follows.

$$(\text{-OCH - }\underset{\displaystyle CH_3}{\underset{|}{C}}H_2\text{-})_{0.94}\;(\text{-O - }\underset{\displaystyle CH_2\text{-}OCH_2\text{-}CH{=}CH_2}{\underset{|}{C}}H\text{ - }CH_2\text{-})_{0.06}$$

The commercial product has a random structure with head-to-tail enchainments. Only small amounts of crystallinity are present. The allyl glycidyl units provide sites for sulfur-cross-linking and also reduce the stereoregularity of the polymer.

The mechanism of polymerization is the same one as already described for epichlorohydrin elastomers.

Manufacture. Propylene oxide is copolymerized with allyl glycidyl ether in an aliphatic, aromatic, or chlorinated hydrocarbon solution using Vandenberg-type catalysts. A complete conversion and a uniform copolymer is obtained containing about 6% of AGE.

After inactivation treatment, the catalyst is removed, and phenolic antioxydants and other stabilizers are added. Cross-linking curing is realized on unsaturated pendant groups. Peroxides are avoided because they cause chain scission and therefore systems with sulfur as cross-linker and zinc oxide, 2-mercaptobenzothiazole and tetramethylthiuram monosulfide as accelerators are used.

Vulcanizate copolymers have almost equivalent modulus, elongation, and hardness properties as natural rubber and chloroprene, but tensile properties are lower. Low temperature flexibility is outstanding and aging performance good. GPO elastomers can accomodate high levels of carbon black reinforcement. Although it is not ordinarly necessary they can also accomodate ester plasticizers like dioctylphthalate, but not unsaturated ones, because the latter can react during the curing process and therefore decreases very much some properties.

Properties of GPO. GPO has excellent low-temperature properties, excellent dynamic properties ressembling that of natural rubber, a good ozone resistance and good heat aging resistance [116].

GPO has a low-glass transition temperature (−75° C) and a low density (1.25). The determination of molecular weights by the usual ways and some solution properties data have been reported [117,118]. However, due to high molecular weight of the industrial polymers, the Mooney viscosity is difficult to determine and was replaced by viscosity obtained by measurements using an oscillating disc rheometer (ODR) which gives apparently meaningful numbers at 100° C [115].

Very useful properties of GPO include outstanding room temperature hysteresis and good dynamic properties over a wide temperature range. For example in measurements of dynamic modulus the flatness of the curve is observed between −40° C and 140° C and this property is maintained even after aging the polymer 7 days at 150° C, which is not the case with natural rubber.

Uses of GPO. Because of its good heat resistance which combines with good rubber properties GPO had important applications in motor mounts. It was added to other rubbers to improve resistance to heat and atmosphere aging.

c. Curing of Epoxy Resins Epoxy resins are characterized by the presence of epoxy groups on aliphatic, cycloaliphatic, or aromatic backbones. Epoxy rings are able to react with various curing agents leading to cross-linked insoluble and intractable materials. These materials include other constituents such as fillers, solvents, plasticizers, and accelerators which are necessary for improvement of processing and modification of some properties of cured resins. Epoxy resins were first recognized in 1936 in Switzerland by De Trey Frères and in United States by De Voe and Reynolds, but they were commercialized after the World War II on the base of high molecular weight resins prepared from bisphenol A and epichlorohydrin. CIBA-AG, Shell Chemical Co. (which was the only supplier of epichlorohydrin), Union Carbide (later Bakelite Co.), then the Dow Chemical Co. and Reichold Chemical Inc. were marketing epoxy resins. In the 1960s glycidated novolacs resins with various formulations were manufactured and marketed by these companies.

Epoxy resins have great and wide commercial applications and are used in many industries. About 45% of production is used in protective coatings and the remainder in various applications such as composites, laminates, casting, molding, tooling, construction, adhesives, photoresists, and others.

A considerable number of detailed descriptions on synthesis, production, and applications of epoxy resins exists. Because the aim of this chapter is the application of cationic initiators, and more particularly photoinitiators, to the polymerization of epoxies leading to cross-linked products (curing reaction), only litterature dealing with these aspects will be cited. For the general aspects of epoxy resins the scientific and patent literature may be found in detailed reviews [119,120] and classical books [121–123].

Photocationic initiators. Epoxy resins can be cross-linked by compounds containing active hydrogen, e.g., carboxylic acids, anhydrides, amines, phenols etc., or by the ionic polymerization process. Lewis acids such as BF_3 and usually a crystalline complex of BF_3 with amines, e.g., $BF_3 \cdot NH_2C_2H_5$, can be used for curing reaction at 80–100° C [124].

The curing by photoinitiated cross-linking and particularly by cationic photoinitiation is of great interest for many technological applications. Cationic photoinitiators have been developed in the past 20 years particularly in the group of Crivello and Lam [125]. Many excellent reviews and books cover the field of cationic photoinitiators. In the present paper limited literature is given to facilitate finding other necessary references [126–129].

The main advantages of cationic photoinitiators is that they have high reaction rates and require a low energy. They can operate at a low temperature, they are not inhibited by oxygen, they do not promote the polymerization of epoxy groups in the dark, and they are often stable at elevated temperatures. Some disadvantages exist: that is, inhibition by bases, chain-transfer reaction by water, and the presence of acids in cured products.

A large number of cationic photoinitiators are known. The most significant from the commercial point of view are aryldiazonium, diaryliodonium, triarylsulfonium, and ferrocenium salts. These salts possess anions of very low nucleophilicity which do not terminate the polymerization process. Nonionic cationic photoinitiators such as organosilanes, latent sulfonic acids, and some other miscellaneous compounds are also used.

Most of the known cationic photoinitiators produce acid species in an irreversible reaction, and once formed these species continue to promote the polymerization reaction even after the end of irradiation. This behavior is of living type and is in contrast to the radical photoinitiated process which rapidly ends in the dark. Cationic photoinitiators producing acid species by a reversible reaction do not show living type behavior, stop reaction after the dark, but may be interesting in some resist applications. It must be added that photoinitiators can form directly cations on irradiation or, in certain cases, the cation is produced by interaction between the photoinitiator and a sensitizer involving the excitation of the former.

Formulation: applications. The epoxy resins that are used most in the formulation of cationic epoxy coatings include glycidylether resins derived from bisphenol A and higher oligomers, cycloaliphatic epoxy resins (e.g., those containing a cyclohexene oxide moiety), hexahydrophthalic diglycidyl esters, multifunctional novolacs, and epoxidized polysiloxanes. Cyclohexene oxide derivatives and epoxidized alkenes are the most reactive to acids, whereas glycidyl ethers are much less reactive. For this reason, to accelerate the cure of bisphenol A diglycidyl ether and also to improve the film hardness, amounts up to 15% of cycloaliphatic diepoxide are added. Considerable cure acceleration can also be obtained by addition of polyols (e.g., addition of 4% ethylene glycol accelerates the cure rate by a factor of 2). The addition of reactive diluents such as vinyl ethers is another way to enhance reactivity. Vinyl and propenyl ethers are very reactive monomers with cure rates comparable with those observed in free radical polymerization of acrylates.

In practice the liquid formulation containing epoxy monomer, photoinitiators and other components or additives (fillers, pigments, etc.) are coated onto the substrate which may be glass, plastic, paper, leather, wood, etc. The coating is then irradiated with light of an appropriate wavelength. Acid begins to be produced immediately or in a few seconds. Sometimes heating is required.

For some applications epoxy resins containing two functional groups were used. These systems contain light-sensitive groups capable of being cross-linked by light and other groups which can further be cross-linked by heating. Epoxy resins containing chalcone groups are examples of such difunctional monomers which can find applications as photoresults especially as solder masks.

Although the commercial importance of light-induced cationic polymerization is much less than that of the corresponding free radical one, there are specific advantages in using cationic photoinitiators, particularly the lower volume shrinkage and the lack of inhibition by oxygen. The most important area of applications is surface coatings, but the use in photoresists and in printing plates for silk screen printing inks, etc., is well established. The worldwide capacity of production of epoxy resins is estimated at over 5×10^5 tons per year. CIBA-GEIGY, Dow, and Shell account for 70% of the overall capacity.

2. Polyoxetanes

a. Poly[3,3-bis(chloromethyl)oxetane] (PBCMO) The ring strain associated with the four-membered ether ring allows the easy and practically irreversible polymerization of many substituted oxetanes [130–132]. 3,3-Bis(chloromethyl) oxetane (BCMO) was the first to be polymerized [132] and in the 1950s Hercules, Inc. started production of polymers under the trade name of Penton [133]. This polymer was marketed on a large scale for about 15 years and then was withdrawn in the 1970s. PBCMO has also been produced in the USSR under the name of Pentaplast.

Polymerization-structure-properties. BCMO is polymerized by the cationic way in different inert solvents. Alkylaluminum compounds yield polymers of high molecular weight. Commercial polymers may reach molecular weights of 250,000–300,000.

The structural unit is:

$$[\text{-O-CH}_2\text{-}\underset{\displaystyle CH_2Cl}{\overset{\displaystyle CH_2Cl}{\overset{|}{\underset{|}{C}}}}\text{-CH}_2\text{-}]$$

The chlorine content is 45.5% per weight. Due to neopentyl-type structure the chlorine atoms are very resistant to dehydrochlorination. This, combined with high crystallinity of these materials, gives them self-extinguishing properties and high chemical resistance. PBCMO is a thermoplastic with a $T_m = 181°$ C and $T_g = 5°$ C. It has outstanding thermal properties and mechanical properties comparable to those of nylon 6 [134]. The chemical resistance is excellent; it resists the attack of alkalis and many inorganic acids, e.g., concentrated H_2SO_4 in which it is stable up to 120° C. Only a few polar solvents dissolve it at elevated temperatures. A particular property of PBCMO is the low water-absorption value and for this reason PBCMO is useful for articles that require sterilization.

Manufacture. Due to its low melting viscosity, PBCMO can be easily produced by conventional injection molded procedures. Coatings can be applied by the usual techniques as a solution, dispersion, or dry powder. Heavy coatings to complex metal objects are applied by the fluidized bed method.

A continuous polymerization process has been developed by Hercules Co. using triethylaluminum catalyst at elevated temperatures [135].

Applications. PBCMO is used especially as material resistant in corrosive atmospheres at moderately elevated temperatures. Useful applications are found as adhesives, coatings, sheeting, lining pipes, tanks, etc. It must also be noted that BCMO and other oxetanes are also used to prepare polyglycols which are used in polyurethanes, polyesters, and polyamide-type elastomers.

3. *Polyoxolanes*

a. Polytetrahydrofuran Tetrahydrofuran (THF) is one of the most studied cyclic ethers and it is an exemplary system for the kinetics and mechanism of polymerization. The first report on polymerization of THF was published in Germany in the late 1930s by Meerwein [136]. Interest for polytetrahydrofuran (PTHF) started mainly after World War II with many groups of investigators and hundreds of publications. Detailed reviews have been published on the results of these investigations [137–146]. Information can also be found in books and reviews treating cationic ring-opening polymerization.

High molecular weight PTHF has excellent elastomeric properties but its price is four to five times higher than that of usual rubbers. However, low molecular weight glycols, which are easily prepared from THF and are very useful for the preparation of polyurethanes and polyester thermoplastic elastomers, has been commercially developed inspite of these high costs.

The commercially produced poly(tetramethylene ether) glycols (PTMEG) usually have molecular weights of 1000 and 2000. Few products with lower (650) and higher (2,900) molecular weights were prepared on a smaller scale.

In 1995, the production of PTMEG is about 150,000 tons per year. The producers are: Du Pont (65,000 tons per year, expected 85,000 tons per year in 1997), BASF (46,000 tons per year), QO Chemicals (9000 tons per year), Gus (4000 tons per year), Mitsubishi (4000 tons per year), Hodoyaga (3000 tons per year), Sanyo (1000 tons per year), and Asahi (1000 tons per year).

Structure, properties. PTHF is a linear elastomer with the following repeat unit: [—$OCH_2CH_2CH_2CH_2$—]. Poly(tetrahydrofuran) is also named poly(tetramethylene oxide) (PTMO) and the official name is poly(-oxy-1,4-butane diyl). For PTHF diols the usual names are poly(tetramethylene ether) glycols (PTMEG).

PTHF elastomers have a zigzag planar conformation. Some of typical physical properties are $T_m = 43°$ C, $T_g = -86°$ C, density approximately 1. Other properties are comparable to those of usual rubbers. PTHF elastomers are soluble in many solvents (THF, aromatic and chlorinated hydrocarbons, esters, ketones, liquid sulfur dioxide, etc.). Aliphatic hydrocarbons in general are nonsolvents.

PTHF diols have molecular weights between 600 and 3000 with lower melting points (20–40° C) and density below 1. In addition to previously mentioned solvents, they are also soluble in ethyl ether, alcohols, acetone, and slightly in water.

PTHF and PTMEG like other monomeric ethers are sensitive to oxidation, forming peroxides. For this reason common antioxidants, e.g., pyrocatechol, are added to the polymer to inhibit these reactions.

Manufacture. At the present time only PTMGE diols are produced commercially on a large scale. In the United States Du Pont PTHF diols are marketed under the name of Terethane, Quaker Oats glycols are marketed under the name of Polymeg, and in Europe for BASF the trade name is poly THF.

In the traditional method of preparation of PTMEG diols strong protonic acids like H_2SO_4 or HSO_3F containing SO_3 or $HClO_4$ are used as initiators. It is known that polymerization has a living character and thus its termination can be obtained by adding, for example, water. This also assumes hydroxyl end groups. This termination is necessary for stabilization of the polymer, because PTHF may depolymerize when heated during the drying stage. However, the control of molecular weight is not very efficient in this traditional procedure because hydroxyl groups can interact during the process with growing oxonium ions and the result is formation of high molecular weight products which are undesirable for the purposes. Moreover, the acids used as initiators are not recoverable and they must be eliminated, because they degrade the polyglycols formed. Thus, this process is costly and has several disadvantages.

It is possible to overcome these problems by taking advantage of the transfer reactions. The latter occur with added small molecules and this allows the control of molecular weight and of the nature of end groups. A combination of a strong acid like nafion SO_3H together with acetic anhydride produces the formation of PTHF with diacetate end groups. The latter are hydrolyzed in the termination step and allow the formation of well-controlled low molecular weight PTMEG. Such a procedure is used by Du Pont Co.

It has already been mentioned that PTMEG is subject to oxidative and thermal degradation and thus antioxidants should be added and the storage surveyed. Overheating and formation of THF monomer should be avoided. In addition, PTMEG is hygroscopic which is a great inconvenience for the later applications, e.g., of polyurethanes. For this reason, PTMEG must be stored in closed tanks under nitrogen atmosphere. These tanks in mild or stainless steel are equipped with internal or external heating to maintain the temperature at approximately 50° C. The pressure must also be controlled and therefore different conservation and emergency vents are provided. PTMEG solidifies at room temperature and consequently the storage in a heated room is necessary before transfer in tank trucks or containers to be shipped. Special installations of unloading, storage, and transfer are used.

Applications. The main application for PTMEG is their use as "soft segment" in elastomeric materials such as polyurethanes, polyesters, or polyamides [143]. Commercially, the most important are polyurethanes and represent 80% of the market [144]. They are prepared by a one-step process with methylene diphenylene-4,4′-diisocyanate (MDI) and the corresponding diol. The thermoplastic polymer obtained has excellent elastomeric properties, good hydrolytic stability, high abrasion resistance, strength, toughness, and high water vapor permeability.

Very often, PTHF with isocyanate end groups are commercially synthesized first. Their trade names are Adiprenes for Du Pont products and Vibrathanes for Uniroyal products. Further, the synthesis of polyurethanes is completed by addition of polyols, diamines, or moisture. Spandex fibers are the largest use areas (almost 40%), followed by thermoplastics and castable urethanes. The industrial applications include packaging, tubing, rollers, wheels, automotive parts, cable coating and guides, encapsulating and potting compounds. The elastomers prepared with PTMEG have excellent fungus and microbial resistance, which in addition to their biocompatibility and hydrolytic stability, make them very useful for medical applications such as encapsulants for pacemakers, catheter tubing, and components for artificial heart devices.

b. Block Copolymers

Poly(ether-b-ester). Polyester block copolymers are produced by melt polymerization procedure. For example, Du Pont commercialized Hytrel which contains 55% of soft segment containing PTMEG and 45% of hard segment coming from 1,4-butanediol and dimethyl terephthalate. The properties vary depending on the overall balance of different reagents. Hard terephthalic ester segments can be partially replaced by esters of isophthalic or sebacic acid. AKZO NV also marketed a series of thermoplastic copolyesters based on butanediol, terephthalic acid, and polytetrahydrofuran under the trade name of Arnitel. These polymers have good low- and high-temperature performances, good chemical and weathering resistance, etc. They are applied in tubings, automotive components, sporting goods, sealings, etc. Syntheses and applications of such block copolymers has been reviewed [145].

Different interesting types of random, block, graft, or star copolymers have been prepared in past years but none of them have achieved commercial importance [137,138,141].

Poly(ether-b-amide). In these particular thermoplastic elastomers the hard blocks consist of aliphatic polyamides, whereas the soft segments are formed of aliphatic polyethers.

The general formula of polyether block amide (PEBA) is shown below:

$$HO(-\underset{\underset{O}{\|}}{C}-PA-\underset{\underset{O}{\|}}{C}-O-PE-O-)_nH$$

where PA is a polyamide hard segment and PE is a polyether soft segment. The latter is usually derived from THF. Their synthesis was described [146]. PEBAs were introduced on the market in the early 1980s by ATO Chemie (now Elf Atochem). Different types of polyamides can be used: 6, 6/6, 11, 12, 6/11, and 6/12. Poly- and copolyether glycols are derived from THF and from propylene or ethylene oxide. These PEBA received the trade name of PEBAX. ATO Chemie developed a two-step process in the first step of which a dicarboxylic oligoamide is prepared, which is reacted in a second step with the polyether diol [147]. Chemisches Werke Hüls produce a thermoplastic polyamide elastomer under the trade name of Vestamid which is synthesized in a single process by charging at once laurolactam, dodecanedioic acid, and poly THF in a reactor and carrying the polycondensation at high temperature. Copolymers with statistical distribution of hard and soft segments are obtained [148]. EMS also produces similar products based on laurolactam under the trade names Grilamid ELY and Grillon ELX. In the 1980s, Ube patented several processes differing somewhat from the previous ones, starting from polyether with amino end groups and leading to products with the trade name of Ube-PAE.

PEBA exhibit a two-phase (crystalline and amorphous) structure and can be classified as a flexible nylon. Physical, chemical, and thermal properties can be modified by appropriate combination of different amounts of polyamide and polyether blocks [149]. Hydrophilic PEBAs can be prepared which can have specific applications in medical devices. Similarly to other thermoplastic elastomers, the polyamide-based ones find applications in automotive components, sporting goods conveyor belting, adhesives, and coatings [150]. In recent years the world consumption was approximately 6400 tons per year with about 80% in Western Europe and the rest equally split between the United States and Japan [143].

The elastomeric polyamides are relatively expensive (for Pebax about \$10.00/kg for low-quality grades and about \$17.00 for high-quality grades; for Grilamid ELY the prices are almost double), but because in finished articles the density is lower, the wall thickness much lower, and the molding cycles shortened, the final cost may be lower than that of corresponding goods in rubber.

B. Polyacetals

Acetal resin is a general term used for high molecular weight polymers and copolymers derived from formaldehyde. The name of polyoxymethylene conforms better with the structure of the repeat unit $(—OCH_2—)_n$. In copolymers, this structure is occasionally interrupted by a comonomer unit, e.g., oxyethylene unit.

The first observation of formaldehyde polymers was done almost 150 years ago [151]. The first systematic studies on linear formaldehyde polymers was performed by Staudinger and Kern in the 1920s [152] but the products obtained were insufficiently thermally stable. An intensive work realized by Du Pont starting from the 1940s resulted in the production at the end of the 1950s of a high molecular weight, thermally stable polyoxymethylene which was thermoplastically processable [153,154]. Thus, the commercialization of formaldehyde polymers occurred some 100 years after their first observation. The decisive step was the blocking of unstable hemiacetal terminal groups by acetate end capping. The Du Pont polymer was introduced on the market under the trade name Delrin and was produced with capacity of 7000 tons per year in 1959. Shortly after Celanese Co. [155] (presently Hoechst Celanese) in the United States and Hoechst [156] in Europe developed a resin based on the copolymerization of trioxane with small amounts of cyclic ethers, such as ethylene oxide, or cyclic acetals. The respective tradenames of Celanese copolymer was Celcon and that of Hoechst Hostaform. It must be remembered that the formation of a white polymer powder from the sublimation of trioxane was observed as early as 1922 [157].

After the first production of homopolymer and copolymers a rapid expansion of production of acetal resins by different companies occurred worldwide starting in the 1960s. Acetal copolymer was produced by Polyplastics (joined company of Hoechst and Daicel) in Japan, by BASF and Degussa in Europe, by Asahi, Mitsubishi in Japan in the 1970s. Plants were built in Poland and the Soviet Union; recently in the 1990s new plants were constructed in Asia, Korea, and Taïwan. The names and the capacity of production of these acetal resins is given in the following text.

The polyoxymethylenes are presently widely used in different areas. Approximatively one-third of the market is represented by homopolymers and two-thirds by copolymers. Homopolymers are produced by anionic polymerization of formaldehyde using amines, alkoxides, and other types of anionic initiators. The details of these polymerizations will not be discussed in this paper, although some of their properties will be compared to those of copolymers which are obtained by cationic copolymerization of trioxane with cyclic ethers or cyclic esters. Comprehensive reviews on general aspects of synthesis and properties of acetal resins are available [158–162].

1. Copolymers Production Process

The production process starts with the synthesis of monomer. Trioxane monomer is produced from aqueous formaldehyde in the presence of strong acid catalyst. Formaldehyde is usually prepared by oxidation of methanol, although in the Asahi process it is prepared from methylal. Trioxane is a white, chemically stable crystalline solid, but it depolymerizes to formaldehyde in the presence of acids at elevated temperatures. The purification of trioxane is carried out by crystallization or distillation.

The copolymerization between trioxane and suitable comonomers (ethylene oxide, 1,3-dioxolane, diethylene glycol formal, 1,4-butane diol formal in amounts of 2–5% by weight) is performed using cationic initiators. The cationic initiators could be Lewis acids, such as BF_3 or its etherate $BF_3 \cdot Bu_2O$ which was used, for example by Celanese (the mechanism of this reaction was studied in detail [163,164]) or protic acids such as perchloric acid, perfluoroalkane sulfonic acids and their esters and anhydrides. Heteropoly acids were used and also a series of carbenium, oxocarbenium salts, onium compounds, and metal chelates. To regulate the molecular weight chain-transfer agents, such as methylal and butylal, are added.

Usually, the copolymerization is carried out in bulk and the reaction is very rapid, completed in few minutes. Although the comonomer, e.g., ethylene oxide, is preferentially consumed at the beginning, the copolymer obtained has a statistical distribution of removal units, which is due to the occurring transacetalization reaction.

The next step in the process consists of removing the unstable terminal hemiacetal and formal groups responsible of depolymerization reactions which may easily occur by unzippering mechanism: $(—CH_2CH_2O—) - (—CH_2—O—)_nH$.

In the case of copolymer resins several techniques are available to reach a stable carbon-carbon link. One of them consists of heating the raw copolymer. At melting unzippering occurs with release of formaldehyde until reaching the stable terminal —C—C— bond. Another method is based on degradation of hemiacetal units by working in alkaline media. To avoid other types of degradation, e.g., by oxygen and peroxy radicals, classical antioxydants (hindered phenols, amines) are added as well as scavengers for release formaldehyde (nitrogen-containing compounds) and also to prevent acidic degradation (polyamides, triazines, melamine, alkaline-earth salts).

2. *Producers, Trade Names*

As already stated the two-thirds of acetal resins commercialized on the market are represented by copolymers prepared by the cationic process and one-third by homopolymers obtained anionically. The capacity of production can be estimated to be about 450,000 tons per year. The homopolymer is mostly produced by Du Pont under the trade name of Delrin

(100×10^3 tons per year) and by Asahi Chem. (Tenac) (20×10^3 tons per year).

For copolymers the commercial situation of the 1990s is given in Table 4.

3. *Polymer Properties*

The physical and chemical properties of homopolymers obtained anionically and of copolymers are not very different. In fact homopolymers have higher crystallinity and higher tensile strength. Acetal resins are crystalline materials (m.p. of copolymers = 165° C) with crystallinity almost 60% for copolymers and a high density 1.41–1.43. The molecular weights are in the range of 20,000–90,000. Polyoxymethylenes have high toughness, hardness, rigidity, and stiffness. They also have good lubricity, low coefficient of friction, and good electrical and dielectric properties. They are stable to various chemicals, solvents, stable to the alkaline environment, but not resistant to acid or oxidizing agents. They have a good resistance to fuels, oils, greases and brake fluids.

Polyoxymethylenes have a low water absorption (0.6% at 20° C). Copolymers are very stable to long exposure in hot water and their properties remain almost unchanged after 1 year of immersion at 82° C. In order to be protected against UV radiation, suitable additives such as benzophenone or benzotriazole may be included.

Table 4 Production Capacities and Trade Names for Acetal Copolymers

Producer	Trade name	Capacity (10^3 tons/yr)
Hoechst Celanese (USA)	Celcon[a]	68
Hoechst (Europe)	Hostaform	50
BASF (USA)	Ultraform	16
BASF (Europe)	Ultraform	30
Polyplastics (Japan)	Duracon	75
Mitsubishi (Japan)	Iupital	35
Asahi (Japan)	Tenac-C	15
Korea Eng. Plast. (Korea)	Kepital	20
Lucky-Goldstar Int. (Korea)	Lucel	10
Russia, Poland, China	–	~10

[a] Kematal (Europe).
Source: Ref. 160.

4. Uses

Acetal resins are competitive for replacing metals and their alloys (zinc, aluminum, magnesium, etc) in several uses. Their main areas of application are the automotive sector (fuel pumps and gauges, caps, etc.), domestic and sanitory equipment, mechanical engineering (valves, pistons, etc.), machinery (conveyors, gears, etc.), and the electrical and electronics industry. In addition to classical materials, impact–resistance-modified polyoxymethylenes were introduced on the market with improved properties of high-energy absorption under impact. They are mostly based on blends of polyoxymethylene and thermoplastic polyurethane elastomers [160]. Acetal block copolymers and super drawn polyoxymethylene fibers were developed in Japan [161]. It is expected that acetal applications and production will be extended in forthcoming years.

It is interesting to notice also that the internal recycling of polyoxymethylene waste is possible. Hoechst has developed a process in which the waste is decomposed into the monomer, which further may be purified and repolymerized [160] and the cycle can be repeated several times without affecting the quality.

C. Polyalkylene Imines

Polyalkylene imines is the usual name for polymers obtained by polymerization of cyclic imine monomers containing secondary or tertiary amine groups. Iminoesters like oxazolines also lead to linear polyalkylene imines. Polyalkylene imines are water-soluble polyamines with a weak base character and offer a variety of physicochemical properties of industrial interest. The simplest monomer ethyleneimine or aziridine is a volatile liquid, flammable, and toxic. It was first produced commercialy in late 1930s by IG Farben Industry by a cyclization process from 2-chloroethylamine with base. The polyethyleneimines were used in paper products, water-proofing textiles, and some coatings. After World War II, the production continued in Germany with Badische Anilin and Soda Fabrik and in the United States with Chemirad, then, in the 1960s, with Dow Chemical Co., which was an important producer of ethyleneimine monomer (EI) and polymer (PEI), but ceased their production in 1978. Presently, the producers are BASF in Germany, Cordova Chemical Co. in the United States, and Nippon Soda in Japan. 1-Substituted aziridines which were previously made by Dow Chemical Co. and Union Carbide are now mainly produced by Cordova Chemical Co. As concerning 2-substituted 2-oxazolines and their polymers, from which linear polyethyleneimines can be prepared via acidic hydrolysis, they are supplied by Dow Chemical Co.

and Chemisches Werkes Huls. The four-membered ring imines (azetidines) and their polymers have not been commercialized.

The mechanism of polymerization of cyclic imines and the properties of different types polymers obtained have been studied extensively and number of reviews are available [165–169].

1. *Aziridine and Substituted Aziridine Polymers*

There are three general methods of synthesis of aziridines:

Gabriel method based on the cyclization of 2-haloalkyl amines with a base. This method was used for the first commercial production of ethyleneimine.

Wenker method in which vicinal amino alcohols are first converted into 2-aminoethyl hydrogen sulfate esters; the latter are then treated with base leading to corresponding aziridine derivative.

Dow process which consisted of an amination of ethylene dichloride with ammonia in the presence of calcium oxide. Several *N*-substituted aziridines with different alkyl-substituted groups are presently available from Cordova Chemical Co.

a. Branched Polyethyleneimines The polymerization of aziridine proceeds through cationic mechanism using different types of initiators (protonic acids, Lewis acids, onium salts, etc.). All of them lead to the same initiating species through proton transfer from the monomer. As a consequence of the proton transfer from the initial active species, branched structures due to tertiary amine termination are found. Thus, by this way commercial *branched PEI* are produced. Cordova Chemical Co. commercializes PEI with different molecular weights (MW) which are marketed under the trade names of Corcat P-3 (MW 300), Corcat P-6 (MW 600), Corcat P-12 (MW 1200), and the corresponding Corcat P-18, P-150, P-600, etc. The degree of branching usually reaches one of every three nitrogen atoms with tertiary amino groups; as shown in the scheme, the ratio between primary, secondary, and tertiary groups is approximatively 1:2:1.

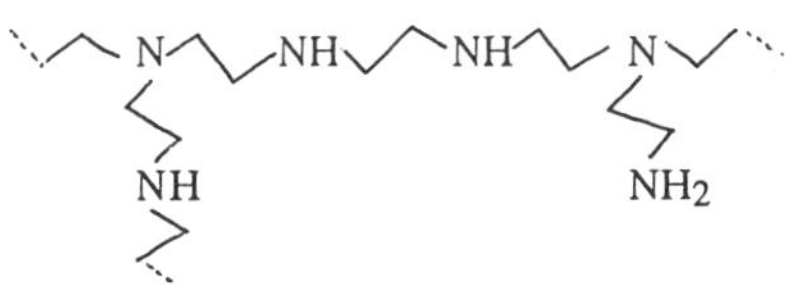

Figure 19 Branched structure of polyethyleneimine.

The polyethyleneimines are weak bases and they exhibit a cationic character which depends on the degree of protonation. Usually 80% of the nitrogen atoms are available for protonation, but in the presence of polyanions, stoichiometric polyelectrolyte complexes are formed with increasing of protonation up to 90%.

PEI is stable, not toxic, and can be handled without danger. It has a good thermal stability with no substantial decomposition until 300° C. PEI has unique colloid-chemical properties due to its high affinity for anionic dissolved material, e.g., polyanions or negatively charged suspended solids. Therefore PEI is used for flocculation of anionic colloids suspended in aqueous solution. The most important field of commercial application for PEI concerns the manufacture of paper and board. This polymer enhances the dewatering rate of the pulp and improves the drying velocity. This provides economic advantage by saving the energy in the drying section and increasing the speed of production. In addition, it improves the retention of paper finer, fillers, dyes, pigments, etc. and therefore reduces the amounts of introduced additives and fillers. To obtain all these effects only very small amounts (between 0.01 and 1% of the weight of dry pulp) of added PEI are necessary. Other applications include uses in textile manufacture, in coatings, and in water purification. The ability of PEI to form complexes with heavy metals finds application in the field of textile dyeing and in anion exchanges.

Biological and catalytic activities are among other fields of numerous investigations and applications.

b. Linear Polyethyleneimines Linear PEI have physical properties very different from branched PEI. The latter are soluble in water, the former precipitate from water in the form of insoluble crystallohydrates. Anhydrous linear PEI exists as double-stranded helical chains which are stabilized by interchain hydrogen bonds.

There are three different ways to prepare linear PEI [170]. One is based on polymerization of oxazoline and the subsequent hydrolysis of the side-chain amide group. This method will be discussed in the next chapter on oxazolines. In a second method aziridine is polymerized at low temperature leading to a mixture of branched PEI, which remains in solution, and a linear PEI, which precipitates [171]. A two-stage process consisting of formation in a first stage of oligomers, which are further polycondensed, was also described. A third method uses the polymerization of EI in which the proton is protected by tetrapyranyl group. After polymerization, the protective group is removed by acid hydrolysis and the free base obtained after neutralization [172].

The properties and applications of linear PEI are described in the polyoxazoline chapter.

c. *N*-Substituted Polyaziridines *N*-Substituted aziridines are also polymerized by cationic initiators. The mechanism of polymerization is simplified by the absence of proton transfer. However, a termination reaction may occur by nucleophilic attack of the nitrogen of the monomer on the polymer molecule forming branched or cyclic quaternary ions which are unreactive toward propagation and thus correspond to a termination reaction. The nature of the nitrogen substituent determines the rate of termination reaction. The ratio k_p/k_t expresses the "living character" of the reaction [173]. Some *N*-substituted monomers with substituents C_2H_5, $(—CH_2—)_2C_6H_5$, $(—CH_2—)_2CN$, $(—CH_2—)_2OH$, $(—CH_2—)_2OCO$ $C(CH_3)\ C = CH_2$, et al., were produced by Dow Chem., Union Carbide, and presently some of them are available from Cordova. Low molecular weight polymers were reported. Their different uses are described in numerous patents. (See Ref. 167 for some information.) Depending on the nature of substituent, these products are water soluble or not. Due to the living character of polymerization of *N*-*t*-butylaziridine, block copolymers can be produced with poly THF or poly(dimethylsiloxane) sequences. Only fundamental research was performed on these products.

2. *Oxazoline Polymers*

Oxazolines are iminoesters which can be simply obtained by catalytic or heat isomerization from *N*-acylaziridines.

Commercially oxazolines are produced by cyclodehydration of 2-hydroxyethyl amides on alumina [174] on other catalysts [175].

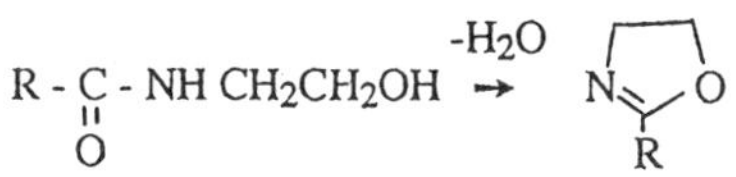

Oxazolines are much less toxic than aziridines and therefore these monomers are preferred for the fabrication, for example, of linear polyethylene imines.

Oxazoline monomers have been commercialized since the mid-1970s and mainly produced by Dow Chemical Co. and Chemisches Werkes Hüls. The principal monomer supplied is 2-ethyl-2-oxazoline.

Unsubstituted and substituted oxazolines are easily polymerized with cationic initiators. The polymerization reaction is a ring-opening reaction with isomerization and it has been studied and reviewed extensively [176,177].

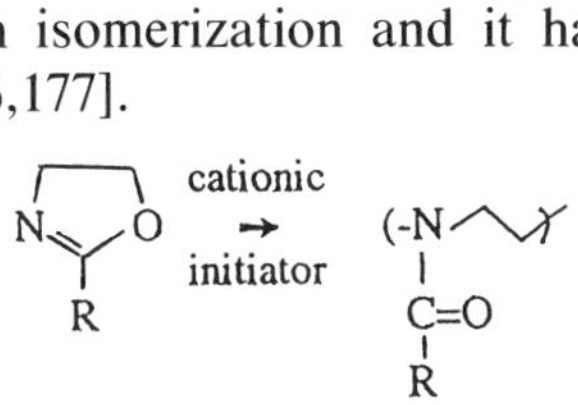

N-Substituted polyethyleneimines of various molecular weights (from a few thousand to millions) can be obtained. The nature of substituent strongly affects the physical properties of the polymers, for example, their crystallinity and solubility. The commercially available 2-ethyl-2-oxazoline is produced as a polymer by Dow Chemical Co. in different molecular weights (50,000, 200,000, 500,000). The polymers are mostly soluble in water and other solvents. They have good thermal stability, thermoplasticity, and low toxicity. Various applications in textile, paper, plastics, and adhesive industries were patented [167].

Oxazoline polymerization has a living type character and therefore a great variety of block and graft polymers can also be produced.

A major interest, early foreseen [178] of poly(N-acylethyleneimine)s obtained by the cationic polymerization of 2-substituted oxazolines, was their ability to lead to linear polyethyleneimines by hydrolysis reaction. An almost complete basic hydrolysis was performed on polymers prepared using BF_3-Et_2O catalyst [179].

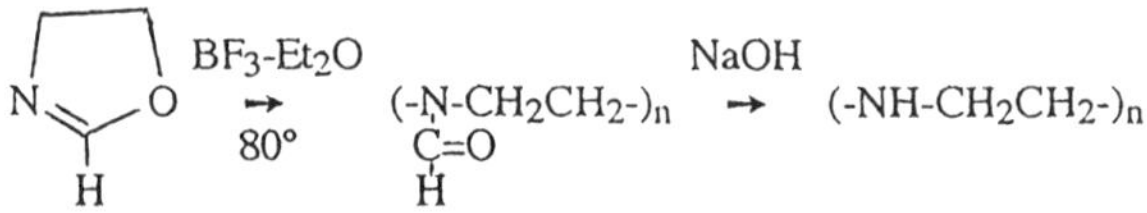

Linear PEI have properties very different from those of branched PEI. They precipitate from water solution forming insoluble crystallohydrates. The structure of the latter determined by x-rays reveals a system of double-stranded helices stabilized by H bonds [180]. Three hydrated forms have been identified: hemihydrate, sesquihydrate, and dihydrate.

Poly(N-acylethyleneimine)s are also a source for production of totally unsubstituted or partially N-acylated linear poly(ethyleneimine)s. In that case acid hydrolysis is preferred. The physical properties of obtained polymers depend on the degree of hydrolysis. Below 70% of deacylation the polymers are amorphous; otherwise they are crystalline. Partially hydrolyzed polymers have applications in paper manufacture, adhesives, coatings, ion-exchange resins, textile manufacture, and water treatment.

D. Polyphosphazenes

In recent years great interest was focused on the area of inorganic and organometallic polymers. Polyphosphazenes are polymers with inorganic backbone of alternating phosphorus and nitrogen atoms and pending substituents of various natures (chlorine, organic groups, etc.).

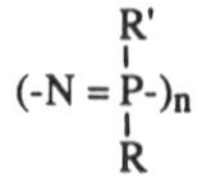

Presently more than 300 different stable polyphosphazenes with different R and R′ groups have been synthesized. The combination of inorganic backbone providing particular properties not found in organic polymers and other physical, chemical, and biological properties coming from substituents make these materials very attractive for various applications.

Historically, the first report on a high polymer of phosphazene dates from 1897 when Stokes [181] described the formation of a rubbery material on heating of hexachlorocyclotriphosphazene. This "inorganic rubber" was a cross-linked polydichlorophosphazene. Almost 70 years after, Allcock and Kugel [182,183] reported the preparation of a soluble poly(dichlorophosphazene) from ring-opening polymerization of the cyclic trimer. This polymer is a key macromolecular intermediate for the preparation of a great variety of other polyphosphazenes by nucleophilic substitution of chlorine groups.

The elastomers with two mixed substituents were prepared in 1968 [184] and the technological development of the chloropolymer began at the Firestone Tire and Rubber Laboratories [185].

Until recent time different polyphosphazenes were produced by Ethyl Co. [186] in the United States, Atochem in France, and other groups in Japan, Italy, and Russia. Very recently Ethyl and Atochem have stopped their production.

1. *Production Processes of Polydichlorophosphazene*

The basic product of all various commercial polyphosphazenes is polydichlorophosphazene. Two companies, Ethyl Corporation and Elf Aquitaine (Atochem) had pilot plants to produce the chloropolymer.

Ethyl Corporation uses the Firestone-licensed process which involves the thermal polymerization of hexachlorocyclotriphosphazene. The mechanism of the thermal polymerization of the trimer has been discussed [187,188]. The cationic nature seems to be substantiated by the effectiveness of Lewis acid catalysts, such as BCl_3 [189]. The process seems to ressemble the "living" type because addition of fresh monomer increases the molecular weight. The cross-linking reactions have been minimized [189]. However, hexachlorocyclotriphosphazene is expensive and in commercial processes one avoids isolating this monomer during the running of the reaction between phosphorus pentachloride and ammonium chloride in chlorobenzene at high temperature (130° C) during few days. An overall yield of 60% on the basis of PCl_5 introduced is obtained.

The Elf Aquitaine process is quite different because it is based on the polycondensation reaction of dichlorophosphoryltrichlorophosphazene $Cl_3P{=}N{-}P(Cl_2){=}O$, the latter being prepared from ammonium sulfate phosphorus trichloride and chlorine [190,191]. This condensation reaction produces with an almost quantitative yield a purely linear polymer (no cross-linking, no residual monomer) with a good control of the molecular weight.

2. *Polymer Modification*

Polydichlorophosphazene is a highly reactive polymer. The chlorine atoms can be replaced by a great variety of side groups leading to different polyorganophosphazenes aryloxy, fluoroalkoxide, etc. symmetrically or unsymmetrically substituted. The physical properties strongly depend on the nature of various substituents.

The substitution reaction is carried out in solution in organic media (THF or benzene) at reflux. With nonbulky nucleophiles such as alkoxides RO Na (R = $+CH_3+$, C_2H_5, CF_3CH_2+, C_6H_5+), amines $R'NH_2$ (R = C_2H_5, n-C_4H_9, C_6H_5) the substitution is almost quantitative. With bulky substituents more vigorous conditions and longer times are necessary [192]. With organometallic reagents (RMgX, RLi) secondary reactions occur together with chain degradation. Other ways are preferred to obtain alkyl- or aryl-substituted polyphosphazenes. By simultaneous reaction of two different nucleophiles mixed alkoxy and aryloxy polyphosphazenes can be obtained.

3. *Properties*

The physical properties of phosphazene polymers greatly depend on the nature of substituents. They, however, have in common the inherent flexibility of the main chain and a high dipole moment.

For halogen or alkoxy-substituted polymers T_g values are low (between -60 and $-100°$ C). Alkoxy polyphosphazene with $C_1 - C_3$ substituents are elastomers. Higher alkoxy- or aryloxy-substituted polyphosphazenes are thermoplastics. Fluoroalkoxypolyphosphazenes exhibit a good stability toward diluted acids and bases. Some of them have outstanding thermal stability and good flame-retardant properties.

4. *Production, Application*

Recently only two firms were producing polyphosphazenes: Ethyl Corp. (trade names Eypel F and Eypel A) and Atochem (trade name Orgaflex AMF). Both of them are no longer producing. The production was approximately 25–50 tons per year for each of them. Eypel F, also called FZ elastomer, is a mixed fluoroalkoxypolyphosphazene elastomer with a wide service temperature range (-65 to $+175°$ C). Polyphosphazenes have a low-temperature flexibility and high-temperature stability and also fuel, oil, and chemical resistance. Eypel A is a mixed aryloxy polyphosphazene with R = —OC_6H_5 and R′ = —OC_6H_4—p C_2H_5. This polymer has good fire resistance and has applications in foams for aerospace, military, and commercial uses.

Among new applications [192,193] attention has been focused on the biocompatible, bioactive, and biodegradable properties. Dopamine and several enzymes, e.g., trypsine, have been covalently bound to polyphosphazene chain. Also anestisics, steroids, and antibacterial agents may be linked to polyphosphazene with promising pharmaceutical applications.

In another approach crystallizable side groups like TCNQ and phthalocyanine were attached to the polymer chain, as well as mesogenic groups leading to liquid crystalline properties. Also side groups with transition metals have been linked in view of macromolecular catalysts, electronically conducting properties, etc. All these functional polymers are at an exploratory stage in laboratories.

E. Silicones

The first observation of the formation of a silicone oil was reported more than 120 years ago [195], but it was just after the World War II that the first industrial applications started at General Electric and Dow Corning, mainly after the discovery of direct synthesis of organochlorosilanes from halogenosilanes by Rochow [196]. The structure of silicones was first recognized by Robinson and Kipping [197].

The term silicones defines compounds with repeating unit —Si—O— in which silicon atoms are bearing one or several various organic groups, the most usual ones being methyl or phenyl. The most important industrial products are methylsilicones. Silicones may include four different structural groups designated by mono (M), di (D), tri (T), and Q (tetra). The various combinations of these groups defines the main families of silicone polymers and their subsequent applications. A description of these structures is given in Table 5.

The linear silicone fluids consists of D units. Silicone elastomers and rubbers are composed of D units bearing as side-chain or end-chain crosslinkable functional groups (vinyl, silanol). Highly branched silicone resins combine T units and D units and in other cases Q and M units. In this way a very great variety of products with a broad application range is obtained.

1. *General Properties and Applications of Silicones*

Silicones exhibit various outstanding properties which lead to many applications.

Si—O bond is very stable at high temperatures and in aging; for this reason it has automotive and aerospatial applications. The high chain mo-

Table 5

Functionality Symbol	Abbreviation	Structural Unit	Application
(Mono) M	$R_3SiO_{0.5}$	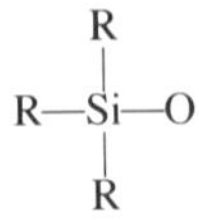	• Chain end-protecting groups • Silicone fluids
(Di) D	R_2SiO	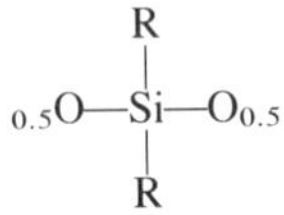	• Linear polysiloxanes • Silicone fluids, rubbers, elastomers
(Tri) T	$RSiO_{0.5}$	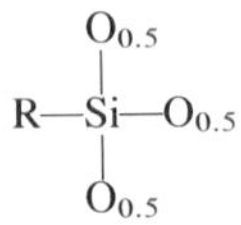	• Branched polysiloxanes • Silicone resins for paints, impregnating materials, etc.
(Tetra) Q	SiO_2	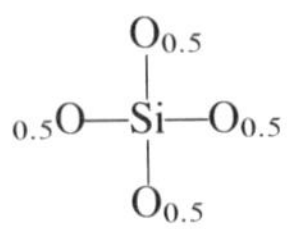	• Branched polysiloxanes • Silicone resins

bility of Si—O bond leads to products with very low T_g ($\sim -120°$ C) and make silicones suitable for low-temperature rubbers. Silicones are chemically inert and not toxic and thus are used in the biomedical field. Their tensioactive and surface properties find application in various surfactants and coatings. Silicone elastomers are highly permeable and dissolve many gases and thus are used as gas-separative membranes, in contact lenses, etc. Their excellent electrical properties make them interesting in the field of electrical insulation.

Presently thousands of different industrial products have been developed. They include fluids, resins, and rubbers of different formulations which are used in various industries and in our daily life. It is not the object of this article to enter into the detail of various aspects of synthesis, production, grades, properties, applications of all silicone derivatives because many of them are not directly linked to cationic synthesis processes. Most of the information may be found in excellent fundamental books and reviews covering silicones.

Description of industrial applications and technological advances may be found in the following reviews that appeared in the past 10 years [198–203]. Some of these reviews [198–202] also include some fundamental aspects of synthesis of polysiloxanes. More detailed descriptions of the latter are given in specialized papers [204–207].

2. *Synthesis by Cationic Polymerization, Applications, Production*

Linear and cyclic polysiloxanes are generally produced by reacting organodichlorosiloxanes with water. The resulting mixture of oligomeric siloxanes may be directly polymerized into linear polysiloxanes or converted into cyclic siloxanes which are then polymerized.

The most usual starting cyclic material is octamethylcyclotetrasiloxane (D_4). The ring-opening polymerization of cyclosiloxanes is an equilibrating reaction and can be promoted by anionic or cationic initiators. The main part of industrial processes is using anionic polymerization, but cationic polymerization has a small but significant part (10–15%). The description of cationic polymerization of cyclic siloxanes may be found Chapter 6 and in general reviews [204–207]. The first high molecular weight polydimethylsiloxanes (PDMS) were obtained by polymerization of D_4 with H_2SO_4 [208]. Other Brönsted and Lewis acids were used and some of them received industrial applications in preparation of oils and gums. Trifluoromethanesulfonic acid was shown to be an efficient initiator for the polymerization of hexamethylcyclotrisiloxane (D_3) and studied by the group of Chojnowski in Lødz [209]. The polymerization of D_4 with the same initiator was examined by the group of Sigwalt in Paris [210]. The current activity in cationic polymerization of cyclosiloxanes was reviewed by Sigwalt [211]. The mechanism of cationic polymerization is complex and includes addition polymerization and acidolysis-condensation process. The formation of macrocyclics is observed. Acidolysis-condensation process seems to play a minor role in the case of D_3 polymerization ($\sim$10%) and a major role in the polymerization of D_4.

Commercial processes are using strong acids, but also inhomogeneous catalysts such as clays activated with H_2SO_4, for example. Functional siloxane copolymers are obtained by polymerizing D_4 with appropriately substituted siloxanes bearing methyl, phenyl, vinyl fluoroalkyl, and hydrogen groups. Silicone fluids have the already mentioned properties of good performances at high and low temperatures, hydrophobicity, lubricity, surface activity, physiological inertness, dielectric behavior, solubility of gases, etc. and are used as refrigerants, heat-transfer media, waterproofing agents, protective coatings, antifoams, antiflocculents, lubricants, paint additives, dielectric coolants. Other applications of some polysiloxanes obtained by the cationic process may be found as additives in formulation of elastomers.

It is rather difficult to distinguish the part of products obtained by cationic process implied in the general production of silicone products including fluids, rubbers, and resins. The global production of silicones can be estimated at 600,000 tons per year and the participation of products obtained by cationic way is approximately 10–15% of the total. For the same reason, it is not possible to give any trade name. The main producers of silicone products are Dow Corning, General Electric in the United States, Bayer (Germany), Hüls (Germany), Rhône-Poulenc (France), Wacker (Germany) and again Dow Corning (United Kingdom) in Europe, Shinetsu, Toshiba, Toray (Dow Corning) in Japan. Other companies such as Goldschmidt or OSI are not themselves preparing the raw materials but processing silicone products. The United States has the largest market (~40%) in silicone products followed by Europe and Japan. Some other countries, e.g., Brazil, China, India, Italy, Mexico, Poland, Russia, and Spain, are also producing silicones but in much smaller amounts.

IV. CONCLUSION

It is clear from the present survey that a significant number of commercially available polymers and copolymers are produced by alkene and ring-opening cationic processes. Thus even if cationic polymerization has the reputation of an unextinguishable source of brain-storming parties for academic scientists, it is the only route for elaboration of some major structural or technical polymeric materials and there is no doubt concerning its present and future commercial importance.

Several conclusions have to be drawn. The first is related to the obvious gap between the empiricism and even archaism of most of industrial cationic polymerization processes and the level of fundamental science devoted for decades to these reactions. Previous chapters in this volume clearly illustrate the situation. This feature was pointed out in the early book of Kennedy and Marechal [1], and the explanation based on the very favorable price/performances characteristics of the products is still realistic. Nevertheless it is noteworthy that recent improvements or new processes based on more scientific approaches led to a better control of the polymerization, of polymer structure, and to high-performance commercial products which will increasingly occupy the market. This is the case for the recently marketed "reactive" BF_3-based polybutenes with high content of exomethylenic chain ends, for the strongly developing "pure monomer" hydrocarbon resins (+8% in 1994), or for the new benzyl halide-based halobutyl rubber, and it is revealing that these products represent the three families of cationically prepared industrial polymers which are by far the most important for the volume of production and thus for economics.

The second conclusion concerns the future. From the view point of the "classical" approach of cationic polymerization, technological transfer of the tremendous amount of acquired fundamental knowledge will progressively lead to the reconsideration of most old processes. In every family, new products in the sense of a better definition of structure, molar mass range, distribution, and reactivity, should emerge from a more reasoned game with nature and concentration of initiator, with monomer purity, with polarity of the medium, with reaction temperature, with chain/end functionalization, etc. Corresponding applied research and plant investments do not necessarily imply a crippling increase in costs, as demonstrated by the current prices of the three new products quoted above. From the viewpoint of prospective technologies, there is no doubt that controlled cationic polymerization will lead, in a more or less immediate future, to a variety of macromolecular architectures like well-defined block, graft, or even dendritic structures. Here "controlled" is understood as mastering of elementary events of the polymerization, including the living-like behavior of the system already discussed in previous chapters. The general concepts of cationic macromolecular engineering were presented, for vinylic or alkene-type polymers, in the recent and well-documented book of J. P. Kennedy and B. Ivan [212] in which the first author, who is undoubtedly the pioneer in this field and further in that of "living" cationic polymerization, rationalizes the approach he proposed to the community more than two decades ago and the tremendous amount of elegant work he and his school realized during the corresponding period. His obstinate and enthusiastic vision certainly originated the currently exploding fundamental and applied research in the domain of controlled cationic polymerization and gave to this old reaction a new and promising youngness. Should we consider that, confronted with the obsolescence of too many of its processes, the industry of cationic polymerization will overcome its excess of prudence? Certainly, yes.

REFERENCES

1. J. P. Kennedy and E. Marechal, *Carbocationic Polymerization*, Wiley-Interscience New York, 1982, Chap. 10.
2. A. M. Buttlerow and V. Gorjainow, *Chem. Ber. 6*: 561 (1873).
3. S. W. Lebedev and E. P. Filonenko, *Ber. Dtsch. Chem. Ges. 58*: 163–168 (1925).
4. S. W. Lebedev and G. G. Koblianski, *Ber. Dtsch. Chem. Ges. 63*: 103–112 (1930).

5. E. Ricard, *U.S. Patent 1,745,028* (1930).
6. H. D. Elkington, *Brit. Patent 358,068* (1930) (to De Bataafse).
7. M. Otto and M. Müller-Cunradi, *Ger. Patent 641,284* (1931) (to IGFarbenindustrie).
8. R. M. Thomas, W. J. Sparks, P. K. Frölich, M. Otto, and M. Müller-Cunradi, *J. Am. Chem. Soc. 62*: 276 (1940).
9. R. M. Thomas and O. C. Slotterbeck, *U.S. Patent 2,243,658* (1941) (to Standard Oil Development Co).
10. R. M. Thomas and W. J. Sparks, *U.S. Patent 2,356,128* (1944) (to JASCO).
11. R. M. Thomas and W. J. Sparks, Butyl rubber, *Synthetic Rubber* (G. S. Whitby, ed.), Wiley, New York, 1954, Chap. 24, pp. 838–891.
12. H. Güterbock, *Polyisobutene*, Springer Verlag, Berlin, 1959.
13. P. H. Plesch, *The Chemistry of Cationic Polymerization*, Pergamon Press, Oxford, 1963.
14. J. P. Kennedy and E. G. M. Tornqvist, *Polymer Chemistry of Synthetic Elastomers* Wiley, New York, 1968.
15. J. P. Kennedy, *Cationic Polymerization of Olefins: A Critical Inventory*, Wiley, New York, 1975.
16. F. P. Baldwin and R. H. Schatz, Butyl rubber, *Kirk-Othmer Encyclopedia of Chemical Technology*, 3rd ed., John Wiley & Sons, New York, 1984, Vol. 8, pp. 470–484.
17. E. N. Kresge, R. H. Schatz, and H-C Wang, Isobutene polymers, *Encyclopedia of Polymer Science and Engineering* (*EPSE*), 2nd ed., Wiley-Interscience, New-York, 1987, Vol. 8, pp. 423–448.
18. K. S. Whiteley, T. G. Heggs, H. Koch, R. L. Mawer, and W. Immel, Polyisobutene, *Ullman's Encyclopedia of Industrial Chemistry*, 5th ed., VCH Publishers, Weinheim, 1992, Vol. A21, pp. 555–561.
19. J. Duffy and G. J. Wilson, Synthesis of butyl rubber by cationic polymerization, Vol. A23, pp. 288–294; Halobutyl rubber, Vol. A23, 314–318, *Ullmann's Encyclopedia of Industrial Chemistry*, 5th Ed., VCH Publishers, Weinheim, 1993.
20. F. Obenaus, W. Droste, and J. Neumeister, Butenes, *Ullmann's Encyclopedia of Industrial Chemistry*, 5th ed., VCH Publishers, Weinheim, 1985, Vol. A4, p. 487.
21. D. Mark and A. R. Orr, *Pet. Refiner 35*(12): 185 (1956).
22. R. A. Labine, *Chem. Eng. 67*: 98 (1960).
23. I. F. Levine and L. T. Folsom, *U.S. Patent 2,484,384* (1949) (to California Research Co.).
24. B. L. Evering, E. L. d'Ouville, and D. R. Carmody, *U.S. Patent 2,407,873* (1946) (to Standard Oil of Indiana).
25. J. N. R. Samson, *WO No. 85/01942* (1984) (to BP Chemicals Ltd).
26. W. K. Jackson, *U.S. Patent 2,957,930* (1960) (to Cosden Petroleum Co.).
27. R. H. Biddulph, P. H. Plesch, and P. P. Rutherford, *Polymer 1*: 521 (1960).
28. Amoco Chemical Co., *Technical Bulletin PB12-N D0394* (1994).
29. BASF A. G., Technical Document ESB/MA G100, and Glissopal Technical Information Ti/ES 1186f (1995).

30. I. Puskas, E. M. Banas, and A. G. Nerheim, *J. Polymer Sci. Symp. 56*: 191 (1976).
31. M. Tessier and E. Marechal, *Polym. Bull. 10*: 152 (1983).
32. M. K. Mishra, E. R. Santee Jr., and J. P. Kennedy, *Polym. Bull. 15*: 469 (1986).
33. S. Nemes, J. Si, and J. P. Kennedy, *Polym. Bull. 23*: 597 (1990).
34. J. Spevacek, L. Toman, and P. Vlcek, *Polym. Bull. 34*: 461 (1995).
35. J. Weill and B. Sillion, *Rev. Inst. Fr. Pet. 40*(1): 77 (1985).
36. *Eur. Chem. News 37*(1004): 23 (1981).
37. H. R. Grane and I. E. Katz, *U.S. Patent 3,665,048* (1970) (to Atlantic Richfield Co.).
38. G. T. Baumann and M. R. Smith, *Hydrocarbon Process. Pet. Refiner. 33*(5): 156 (1954).
39. A. M. Valet, J. P. Michaux, J. J. Bellec, and H. R. Sheely, *Hydrocarbon Process. Pet. Refiner. 41*(5): 119 (1962).
40. H. Kroper, K. Schlomer, and H. M. Weitz, *Hydrocarbon Process. Pet. Refiner. 48*(9): 195 (1969).
41. *Chem. Eng. N.Y. 90*(25): 60 (1983).
42. F. Obenaus, B. Greving, H. Balke, and B. Scholtz, *Eur. Patent A0082937* (1983) (to Hüls AG).
43. R. Tesei, V. Fattore, and F. Buonomo, *German Patent 2,534,544* (1976) (to SNAM Progetti).
44. V. Fattore, M. M. Mauri, G. Oriani, and G. Paret, *Hydrocarbon Process. 60*(8): 101 (1981).
45. *Oil Gas J. 77*(1): 76 (1979).
46. H. G. Goering and V. F. Minstretta, *U.S. Patent 2,739,143* (1956) (to Esso Research & Engineering).
47. BASF AG, Technical Document TI/ED 1280 e (1987).
48. Exxon Chemical, Technical Document AS069 (1992).
49. L. A. Wood, *Rubber Chem. Technol. 49*: 189 (1976).
50. C. S. Fuller, C. J. Frosch, and H. R. Pape, *J. Am. Chem. Soc. 62*: 1950 (1940).
51. A. M. Liquori, *Acta Crystallogr. 8*: 345 (1955).
52. C. Y. Chu and R. Vukov, *Macromolecules 18*: 1423 (1985).
53. D. M. Cheng, I. J. Gardner, H. C. Wang, C. B. Frederick, and A. H. Dekmezian, *Rubber Chem. Technol. 63*: 265 (1990).
54. R. A. Crawford and R. T. Morrissey, *U.S. Patent 2,631,984* (1953) and *U.S. Patent 2,681,899* (1954) (to B. F. Goodrich Chemical Co.).
55. R. T. Morrissey, *Ind. Eng. Chem. 47*: 1562 (1955).
56. R. T. Morrissey, *U.S. Patent 2,816,098* (1957) (to Goodrich Chemical Co.).
57. J. Walker, R. H. Jones, and G. Feniak, Philadelphia Rubber Group Technical Meeting, Oct. 12, 1972.
58. W. D. Gunter, Butyl and halogenated butyl rubbers, *Developments in Rubber Technology-2* (A. Whelan and K. S. Lee, eds.), Applied Science Publishers, London 1981, Chap. 6, p. 155.
59. J. V. Fusco and P. Hous, Butyl and halobutyl rubbers, *Rubber Technology*

(M. Morton, ed.), 3rd ed., Van Nostrand Reinhold Company, Inc., 1987, p. 284.
60. K. W. Powers et al., *U.S. Patent 4,252,710* (1981), *U.S. Patent 4,474,924* (1982), *U.S. Patent 4,358,560* (1982) (to Exxon Research & Engineering Co.).
61. N. V. Scherbakova et al., *Can. Pat. 1,019,095* (1977), and *East German Patent 104,985* (1974).
62. A. B. Sullivan, *J. Org. Chem. 31*: 2811 (1966).
63. *The Synthetic Rubber Manual*, 12th ed., International Institute of Synthetic Rubber Producers, Inc., Houston, TX, 1992, Table 11.
64. J. Brandrup and E. H. Immergut, eds., *Polymer Handbook*, 2nd ed., Wiley, New York, 1975.
65. Exxon SB, Butyl and Halobutyl Technical Document 202-0992-1003-A (1992).
66. D. L. Hertz Jr., *Handbook of Elastomers* (A. K. Bhowmick and H. L. Stephens, eds.), Dekker, New York, 1988, Chap. 13, p. 445.
67. F. B. Baldwin, D. J. Buckley, I. Kuntz, and S. B. Robison, *Rubber Plastics Age 42*: 500 (1961).
68. F. P. Baldwin and R. M. Thomas, *U.S. Patent 2,964,489* (1960) (to Esso Research & Engineering Co.).
69. B. R. Tegge, F. P. Baldwin, and G. E. Serniuk, *U.S. Patent 3,023,191* (1962) (to Esso Research & Engineering Co.).
70. J. A. Johnson and E. D. Luallin, *U.S. Patent 4,484,072* (1966) (to Esso Research & Engineering Co.).
71. R. C. Kowalski, *Chem. Eng. Prog. 85(5)*: 67 (1989).
72. *The Synthetic Rubber Manual*, 12th ed., International Institute of Synthetic Rubber Producers, Inc., Houston, TX, 1992, Table 12.
73. M. L. Poutsma, *J. Am. Chem. Soc. 87*: 2172, 4285 (1965).
74. A. Tongerloo and R. Vukov, Proceedings of the International Rubber Conference, Venice, 1979, pp. 70–79.
75. R. Vukov, *Rubber Chem. Technol. 57*: 275 (1984).
76. H. C. Wang and K. W. Powers, Proceedings of the ACS Rubber Division Meeting, Toronto, 1991, Paper 80.
77. D. F. Kruze and J. V. Fusco, Proceedings of the ACS Rubber Division Meeting, Nashville, 1992, Paper 76.
78. *Rubber Statistical Bulletin*, Vol. 45–49, Dec./Jan (1990–1994), International Rubber Study Group, York House, Empire Way, Wembley, HA9 0PA, U.K.
79. *Worldwide Rubber Statistics 1992*, International Institute of Synthetic Rubber Producers, Inc., 2077 S. Gessner Road, Suite 133, Houston, TX 77063-1123.
80. W. Vredenburgh, K. F. Foley, and A. N. Scarlatti, Hydrocarbon resins, *Encyclopedia of Polymer Science and Engineering (EPSE)*, 2nd ed., Wiley-Interscience, New York, 1987, Vol. 7, pp. 758–782.
81. E. W. Flick, *Industrial Synthetic Resins Handbook*, Noyes Publishers, Park Ridge, NJ 1989, pp. 214–384.

82. G. Collin, R. Mildenberg, and M. Zander, Resins, Synthetic, *Ullmann's Encyclopedia of Industrial Chemistry*, 5th ed., VCH Publisher, Weinheim, 1993, Vol. A23, pp. 89–96.
83. P. O. Powers, Coumarone-indene resins, *Encyclopedia of Polymer Science and Technology (EPST)*, Wiley-Interscience, New York, 1966, Vol. 4, pp. 272–294.
84. P. O. Powers, Coumarone-indene resins, *Kirk-Othmer Encyclopedia of Chemical Technology (EPST)*, 2nd ed., Wiley-Interscience, New York, 1966, Vol. 11, pp. 242–262.
85. J. Findley, Petroleum resins, *Encyclopedia of Polymer Science and Technology* (*EPST*), Wiley-Interscience, New York, 1966, Vol. 9, pp. 863–860.
86. F. M. Benitez and M. F. English, *U.S. Patent 4,419,503* (1983) (to Exxon Research & Engineering Co.).
87. A. Lepert, *U.S. Patent 4,078,132* (1978) (to Exxon Research & Engineering Co.).
88. A. Lepert, *U.S. Patent 4,068,062* (1978) (to Exxon Research & Engineering Co.).
89. R. A. Osborn and H. L. Bullard, *U.S. Patent 4,098,983* (1978) (to Goodyear Tire and Rubber Co.).
90. H. A. Pace and V. J. Anhorn, *U.S. Patent 3,577,398* (1971) (to Goodyear Tire and Rubber Co.).
91. D. R. St Cyr, *U.S. Patent 3,813,357* (1974) (to Goodyear Tire and Rubber Co.).
92. B. R. Watson, *Chem. Assays* (*Lond.*) *3*: 5 (1789).
93. E. R. Ruckel and H. G. Arlt, Jr., Polyterpenes resins, *Naval Stores, Production, Chemistry Utilization* (D. F. Zinkel and J. Russell, eds.), Pulp Chemicals Association, 60 East 42nd Street, New York, NY 10165, 1989, Chap. 13, pp. 510–530.
94. A. Takata, T. Otsu, and M. Imoto, *Kogyo Kagaku Zasshi 69*: 715 (1966).
95. *Chemical Marketing Reporter*, Jan. 9, 1995, p. 37.
96. D. H. Lorenz and N. Bikales, Vinyl ether monomers and polymers, *Encyclopedia of Polymer Science and Technology (EPST)*, Wiley-Interscience, New York, 1971, Vol. 14, pp. 504–521.
97. M. Biswas, A. Mazumdar, and P. Mitra, Vinyl ether polymers, *Encyclopedia of Polymer Science and Engineering (EPSE)*, 2nd ed., Wiley-Interscience, New-York, 1989, Vol. 17, pp. 447–468.
98. E. V. Hort and R. C. Gasman, Vinyl ether monomers and polymers, *Kirk-Othmer Encyclopedia of Chemical Technology*, 3rd ed., John Wiley & Sons, New York, 1983, Vol. 23, pp. 937–960.
99. T. Higashimura and M. Sawamoto, Carbocationic polymerization: vinyl ethers, *Comprehensive Polymer Science* (Sir G. Allen, ed.), 1989, Vol. 3, pp. 673–696.
100. G. Schröder, Poly(vinyl ethers), in *Ullmann's Encyclopedia of Industrial Chemistry*, 5th ed., VCH Publisher, Weinheim, 1993, Vol. A22, pp. 11–15.
101. H. Staudinger and H. Lohmann, *Am. Chem. 505*: 41 (1933).

102. M. E. Pruitt and J. M. Baggett (Dow Chem. Co.), *U.S. Patent 2,706,181* (April 12, 1955).
103. C. C. Price and M. Osgan, *J. Am. Chem. Soc. 78*: 4787 (1956).
104. C. C. Price, *Chemist 38*: 131 (1961).
105. E. J. Vandenberg (Hercules, Inc.), *U.S. Patent 3,135,705* (May 11, 1959).
106. E. J. Vandenberg, *J. Polym. Sci. Part A-1 7*: 525 (1969).
107. E. J. Vandenberg, *Coordination Polymerization* (C. C. Price and E. J. Vandenberg, eds.), Plenum Press, New York, 1983, p. 11.
108. D. L. Hertz, Jr. *Handbook of Elastomers* (A. K. Bhowmick and H. L. Stephens eds.), M. Dekker, New York, 1988, Chap. 13, p. 459.
109. R. W. Body and V. L. Kyllinstad, *EPSE*, 2nd ed., J. Wiley-Interscience, New York, 1986, Vol. 6, p. 307.
110. J. Dunn, *Ulmann's Encyclopedia of Industrial Chemistry*, VCH Publisher, Inc., Weinheim, 1993, Vol. A23, Chap. Synthetic Rubber, p. 297.
111. F. E. Bailey, Jr. and J. V. Koleske, eds., *Alkylene Oxides and Their Polymers*, Marcel Dekker, Inc., New York, 1991.
112. C. C. Price (University of Notre-Dame), *U.S. Patent 2,866,774* (December 30, 1958).
113. L. C. Pizzini and J. T. Patton, Jr., *EPST*, J. Wiley-Interscience, New York, 1967, Vol. 6, p. 145.
114. E. E. Gruber, D. A. Meyer, G. H. Swart, and K. V. Weinstock, *Ind. Chem. Product Resin Div. 3*: 194 (1964).
115. D. A. Berta and E. J. Vandenberg, *Handbook of Elastomers*, Dekker, Inc., New York, 1988, Chap. 19, p. 643.
116. C. R. Boss, *Polyethers*, ACS Symposium Series 6, (E. J. Vandenberg, ed.), American Chemical Society, Washington, 1975, Chap. 8, p. 120.
117. G. Allen, C. Booth, and M. N. Jones, *Polymer 5*: 195 (1964).
118. G. Allen, C. Booth, and C. C. Price, *Polymer 8*: 397 (1967).
119. L. V. Adams and J. A. Gannon, Epoxy resins, *EPSE*, 2nd ed., J. Wiley-Interscience Inc., New York, 1986, Vol. 6, p. 322.
120. J. W. Muscopf and S. B. Mc Collister, *Ulmann's Encyclopedia of Industrial Chemistry*, VCH Publisher Inc., Weinheim, 1993, Vol. A9, Chap. Epoxy Resins, p. 547.
121. C. A. May and Y. Tanaka, eds., *Epoxy Resins Chemistry and Technology*, Marcel Dekker Inc., New York, 1973.
122. H. Lee and K. Neville, eds., *Handbook of Epoxy Resins*, McGraw-Hill, Inc., New York, 1967 (reprinted 1982).
123. B. Ellis, ed., *Chemistry and Technology of Epoxy Resins*, Chapman and Hall, New York, 1992.
124. J. J. Harris and S. C. Temin, *J. Appl. Polym. Sci. 10*: 523 (1966).
125. J. V. Crivello and J. H. W. Lam, *Macromolecules 10*: 1307 (1977); *ACS Symp. Ser. 114*: 1 (1979).
126. F. Lohse and H. Zweigel, Photocrosslinking of epoxy resins, *Adv. Polym. Sci. 78*: 62 (1986).
127. J. V. Crivello, J. L. Lee, and D. A. Conion, *Makromol. Chem. Macromol. Symp. 13/14*: 145 (1988).

128. J. V. Crivello and K. K. Dietliker, *Chemistry and Technology of UV and EB Formulation for Coatings, Inks and Paints* (P. K. T. Oldring, ed.), SITA Technology Ltd, London, 1991, Vol. III, Chap. III, p. 327.
129. G. E. Green, B. P. Stark, and S. A. Zahin, *J. Macro. Sci. Rev. Macro. Chem. C21(2)*: 187 (1981–82).
130. S. Okamura, H. Miake, and N. Shimazaki, *EPST*, J. Wiley-Interscience, New York, 1968, Vol. 9, p. 668.
131. M. P. Dreyfuss and P. Dreyfuss, *EPSE*, 2nd ed., J. Wiley-Interscience, New York, 1987, Vol. 10, p. 653.
132. A. C. Farthing and R. J. W. Reynolds, *J. Polym. Sci. 12*: 503 (1954).
133. G. E. Hulse (Hercules, Inc), *U.S. Patent 2,722,493* (Nov. 1, 1995); W. H. Schilling (Hercules, Inc.), *Brit. Patent 769,116* (Feb. 27, 1957).
134. H. Boardman, *EPST*, J. Wiley-Interscience, New York, 1968, Vol. 9, p. 702.
135. Hercules Powder Co., *Brit. Patent 858,789* (Jan. 18, 1961); *Chem. Abst. 55*: 11920 (1961).
136. H. Meerwein *Ger. Patent 741,478* (June 21, 1939).
137. P. Dreyfuss and M. P. Dreyfuss, *Adv. Polym. Sci. 4*: 528 (1967).
138. P. Dreyfuss and M. P. Dreyfuss, *EPST*, J. Wiley-Interscience, New York, 1970, Vol. 13, p. 670.
139. T. Saegusa, S. Kobayashi, *Prog. Polym. Sci. (Japan) 6*, 107 (1973).
140. S. Penczek, P. Kubisa, and K. Matyjaszewski, *Adv. Polym. Sci. 37*: 1 (1980); *68/69*: 1 (1985).
141. P. Dreyfuss, *Poly(tetrahydrofuran)*, Gordon & Breach, New York, 1982.
142. P. Dreyfuss, M. P. Dreyfuss, and G. Pruckmayr, *EPSE*, J. Wiley-Interscience, New York, 1989, 2nd ed., Vol. 16, p. 649.
143. G. H. Bohlmann and J. Wakin, *Thermoplastic Elastomers*, SRI International-Economics Program 104A (1993).
144. J. K. Backus et al., *EPSE*, 2nd ed., John Wiley-Interscience, New York 1988, Vol. 13, p. 243.
145. G. Holden, *EPSE*, 2nd ed., John Wiley-Interscience, New York, 1986, Vol. 5, p. 417.
146. G. Deelens, P. Foy, and E. Maréchal, *Eur. Polym. J. 13*: 337, 343, 353 (1977).
147. G. Deelens, P. Foy, and C. Jungblut (Ato-Chimie), *Fr. 2 273021* (1974).
148. S. Muncu, K. Burzin, R. Feldmann, and R. Feinauer, *Angew. Makromol. Chem. 74*: 49 (1978).
149. L. Fischer, *Medical Device Technology*, Sept. 1994, p. 32.
150. J. R. Flescher, Jr., *Modern Plastics*, Sept. 1987, p. 100.
151. A. M. Butlerov, *Ann. 3*: 242 (1852).
152. W. Kern, *Die hoch molekularen organischen Verbindungen Kautchuk und Cellulose* (H. Staudinger, ed.), Springer Verlag, Berlin, 1932.
153. R. N. MacDonald (E. I. Du Pont de Nemours & Co., Inc.), *U.S. Patent 2,768,994* (Oct. 30, 1956).
154. C. E. Schweitzer, R. N. MacDonald, and J. O. Punderson, *J. Appl. Polym. Sci. 1*: 158 (1959).

155. C. T. Walling, F. Brown, and K. W. Bartz (Celanese Corp.), *U.S. Patent 3,027,352* (March 27, 1962).
156. K. Küllmar, E. Fisher, and K. Weissermel (Hoechst), *DE 1124/703* (1959).
157. D. L. Hammick and A. R. Boeree, *J. Chem. Soc. 121*: 2738 (1922).
158. J. C. Bevington, *EPST*, J. Wiley-Interscience, New York, 1964, Vol. 1, p. 609.
159. T. J. Dolce and J. A. Grates, *EPSE*, 2nd ed., J. Wiley & Sons, New York, 1985, Vol. 1, p. 42.
160. G. Sextro, *Ulmann's Encyclopedia of Industrial Chemistry*, VCH Publisher, Inc., Weinheim, 1992, Vol. A21, p. 591.
161. J. Masamoto, *Prog. Polym. Sci. 18*: 1 (1993).
162. J. Masamoto, T. Iwaisoko, K. Yoshida, K. Matsuzaki, K. Kagawa, and H. Nagashara, *Makromol. Chem. Macromol. Symp. 42/43*: 409 (1991).
163. G. L. Collins, R. K. Green, F. M. Berardinelli, and W. H. Ray, *J. Polym. Sci. Polym. Chem. Ed. 19*: 1597 (1981).
164. Na Lu, G. L. Collins, and N. L. Yang, *Makromol. Chem. 42/43*: 425 (1991).
165. E. J. Goethals, *Ring-Opening Polymerization* (K. J. Ivin and T. Saegusa; eds.), Elsevier, 1984, Vol. 2, Chap. 10, p. 715.
166. A. Goldstein, *EPST*, J. Wiley & Sons, New York, 1964, vol. 1 p. 734.
167. D. A. Tomalia and G. R. Killat, *EPSE*, 2nd ed., J. Wiley & Sons, New York, 1985, Vol. 1, p. 680.
168. D. A. Tomalia, *Handbook of Polymer Synthesis* (H. R. Kricheldorf, ed.), M. Dekker, Inc., New York, 1992, Part A, Vol. 13, p. 748.
169. G. E. Ham, *Polymeric Amines and Ammonium Salts* (E. J. Goethals, ed.), Pergamon Press, Oxford, 1980, p. 1.
170. E. J. Goethals, *Comprehensive Polymer Science*, Pergamon Press, Oxford, 1988, Vol. 3, Chap. 52, p. 837.
171. P. A. Gembitski, A. I. Chmarin, N. A. Kheshleva, and D. S. Zhuk, *Vysokomol. Soed. Ser. A 20*: 1505 (1978).
172. K. Weyts and E. J. Goethals, *Polym. Bull. (Berl.) 19*: 13 (1988).
173. E. J. Goethals, E. H. Schacht, P. Bruggeman, and P. Bossaert, *ACS Symp. Ser 59*: 1 (1977).
174. W. Seeliger, E. Aufderbaar, W. Diepers, R. Feinauer, R. Nehring, W. Thier, and H. Hellmann, *Angew. Chem. Int. Ed. Engl. 5*: 875 (1966).
175. M. Kaiser (Dow Chem. Co.), *U.S. Patent 4,203,900* (May 20, 1980).
176. S. Kobayashi, T. Saegusa, *Ring-Opening Polymerization* (K. J. Ivin and T. Saegusa, eds.), Elsevier, 1984, Vol. 2, Chap. 11, p. 761.
177. T. Saegusa and S. Kobayashi, *Polymeric Amines and Ammonium Salts* (E. J. Goethals, ed.), Pergamon Press, Oxford, 1980, p. 55.
178. D. A. Tomalia and D. P. Shutz, *J. Polym. Sci. Part A1 4*: 2253 (1966).
179. T. Saegusa, M. Ikeda and H. Fujii, *Macromolecules 5*: 108 (1972).
180. Y. Chatani, T. Kobataka, H. Tadokoro, and R. Tanaka, *Macromolecules 15*: 170 (1982).
181. H. N. Stokes, *Am. Chem. J. 19*: 782 (1897).
182. H. R. Allcock and R. L. Kugel, *J. Am. Chem. Soc. 87*: 4216 (1965).

183. H. R. Allcock, R. L. Kugel, and K. J. Valan, *Inorg. Chem. 5*: 1709 (1965).
184. S. H. Rose, *J. Polym. Sci. Part B 6*: 837 (1968).
185. D. P. Tate, *J. Polym. Sci., Polym. Symp. 48*: 33 (1974).
186. H. Penton, *Eur. Rubber J. 168*: 20 (1986).
187. H. R. Allcock and R. J. Best, *Can. J. Chem. 42*: 447 (1964).
188. G. L. Haugnauer, *J. Macromol. Sci. Chem. A16*: 385 (1981).
189. M. S. Senett, G. L. Haugnauer, R. L. Singler, and G. Davies, *Macromolecules 19*: 959 (1986); 22: 4235 (1989).
190. M. Meloni, R. De Jaeger, E. Puskaric, and J. Hendel, *Makromol. Chem. 183*: 1137 (1982).
191. G. D'Halluin, R. De Jaeger, J.-P. Chambrette, and Ph. Potin, *Macromolecules 25*: 1254 (1992).
192. H. R. Allcock, M. N. Mang, A. A. Dember, and K. J. Wynne, *Macromolecules 22*: 4179 (1989).
193. H. R. Allcock, *EPSE*, 2nd ed., J. Wiley & Sons, New York, 1988, Vol. 13, p. 31.
194. H. R. Allcock, *Inorganic and Organometallic Polymers*, ACS Symp. Series 360 (M. Zeldin, K. J. Wynne, and H. R. Allcock, eds.), American Chemical Society, Washington, DC, 1988, Chap. 129, p. 250.
195. A. Ladenburg, *Ann. Chem. 164*: 300 (1872).
196. E. G. Rochow, *An Introduction to the Chemistry of Silicones*, J. Wiley and Sons, New York, 1951.
197. R. Robison and F. S. Kipping, *J. Chem. Soc. 106*: 40 (1914).
198. W. Noll, *Chemistry and Technology of Silicones*, Academic Press, New York, 1984.
199. P. Sigwalt, *L'Actualité Chim. 3*: 45 (1986).
200. K. E. Polmanteer, *Handbook of Elastomers* (A. K. Bhowmick and H. Stephens, eds.) M. Dekker, New York, 1988, Chap. 17, p. 551.
201. *Les Silicones—Production et Applications*, Techno Nathan, 1988.
202. H. H. Moretto, M. Schulze, and G. Wagner, *Ulmann's Encyclopedia of Industrial Chemistry*, VCH Publisher Inc., Weinheim, 1993, Vol. A24, p. 57.
203. J. Ackermann, *Ulmann's Encyclopedia of Industrial Chemistry*, VCH Publisher, Inc., Weinheim, 1993, Vol. A23, p. 306.
204. P. W. Wright, *Ring-Opening Polymerization* (K. J. Ivin and T. Saegusa, eds.), Elsevier, New York, 1984, Vol. 2, p. 324.
205. B. Hardman and A. Torkelson, *EPSE*, 2nd ed., J. Wiley-Interscience, New York, 1989, Vol. 15, p. 204.
206. T. C. Kendrick, B. M. Parbhoo, and J. W. White, *Comprehensive Polymer Science* (G. C. Eastmond, A. Ledwith, S. Russo, and P. Sigwalt, eds.), Pergamon, New York, 1989, Vol. 4, chap. 25, p. 459.
207. J. C. Saam, *Silicon based Polymer Science* (J. M. Ziegler and F. M. Gordon Fearon, eds.), Advances in Chemistry Series, 224, American Chemical Society, Washington DC, 1990, Chap. 3.
208. W. Patnode and D. F. Wilcock, *J. Am. Chem. Soc. 68*: 358 (1946).

209. J. Chojnowski and L. Wilczek, *Makromol. Chem. 180*: 117 (1979); and other papers by these authors.
210. J.-J. Lebrun, G. Sauvet, and P. Sigwalt, *Makromol. Chem. Rapid Commun. 3*: 757 (1982); and other papers by Sigwalt et al.
211. P. Sigwalt, *Polym. J. 19*: 567 (1987).
212. J. P. Kennedy and B. Ivan, *Designed Polymers by Carbocationic Macromolecular Engineering—Theory and Practice*, Hanser Publishers, Munich, 1992.

Index